XXIII International Workshop on High Energy Physics and Quantum Field Theory (QFTHEP 2017)

EPJ Web of Conferences Volume 158 (2017)

Yaroslavl, Russia
26 June - 3 July 2017

Editors:

Dmitri Melikhov
Igor Volobuev

ISBN: 978-1-5108-5053-8

Printed from e-media with permission by:

Curran Associates, Inc.
57 Morehouse Lane
Red Hook, NY 12571

Some format issues inherent in the e-media version may also appear in this print version.

Printed by Curran Associates, Inc. (2017)

For additional information, please contact EDP Sciences – Web of Conferences
at the address below.

EDP Sciences – Web of Conferences
17, Avenue du Hoggar
Parc d'Activité de Courtabœuf
BP 112
F-91944 Les Ulis Cedex A
France

Phone: +33 (0) 1 69 18 75 75
Fax: +33 (0) 1 69 28 84 91

contact-edps@webofconferences.org

Additional copies of this publication are available from:

Curran Associates, Inc.
57 Morehouse Lane
Red Hook, NY 12571 USA
Phone: 845-758-0400
Fax: 845-758-2633
Email: curran@proceedings.com
Web: www.proceedings.com

TABLE OF CONTENTS

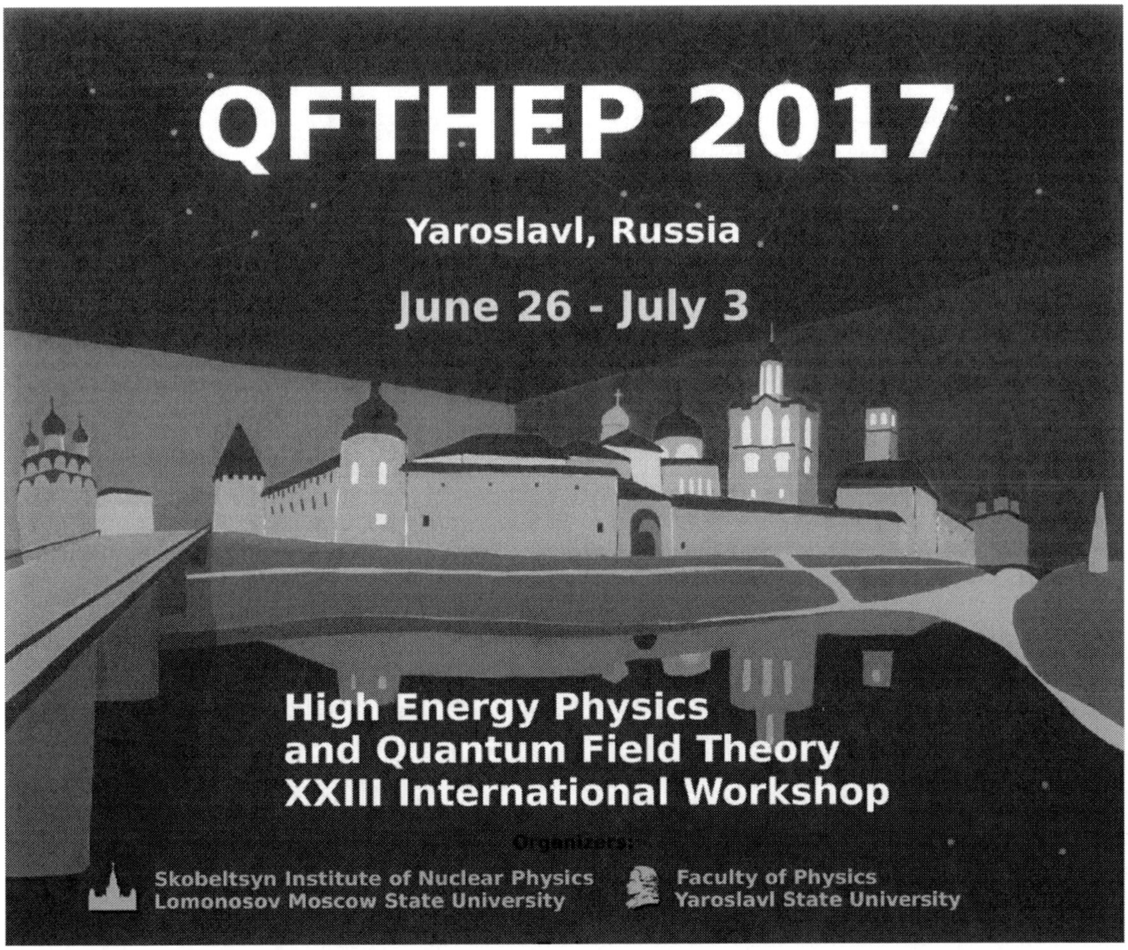

The XXIII International Workshop

High Energy Physics and Quantum Field Theory

June 26– July 3, 2017

Yaroslavl, Russia

Conference webpage: http://qfthep.sinp.msu.ru/

Editors: Dmitri Melikhov and Igor Volobuev

International Advisory Committee

- A. Andrianov (SPbSU, St. Petersburg)
- A. Bondar (BINP, Novosibirsk)
- F. Boudjema (LAPPTH, Annecy)
- D. Denisov (Fermilab, Batavia)
- A. Djouadi (Un.PSUD, Paris)
- J. Ellis (CERN, Geneva)
- K. Ellis (University of Durham)
- J. Fujimoto (KEK, Tsukuba)
- F. Gianotti (CERN, Geneva)
- I. Ginzburg (NMI, Novosibirsk)
- R. Heuer (CERN, Geneva)
- C. Hill (Fermilab, Batavia)
- W. Hollik (MPI, Munich)
- P. Jenni (CERN, Geneva)
- L. Lipatov (PNPI, Gatchina)
- V. Matveev (JINR, Dubna)
- J. Mnich (DESY, Hamburg)
- V. Novikov (ITEP, Moscow)
- P. Osland (Univ. of Bergen)
- M. Panasyuk (SINP MSU, Moscow)
- V. Rubakov (INR RAN, Moscow)
- R. Tenchini (CERN, Geneva)
- N. Tyurin (IHEP, Protvino)
- G. Weiglein (DESY, Hamburg)

Organizing Committee

D.V.Skobeltsyn Institute of Nuclear Physics M.V.Lomonosov Moscow State University:
E. Boos (co-chairman), V. Savrin (co-chairman), A. Baskakov, A. Berezhnoy, V. Bunichev, G. Chernoluzhskiy, L. Dudko, M. Dubinin, L. Gladilin, A. Gribushin, A.Kozachuk, I. Lokhtin, O. Lukina, D. Melikhov, M. Merkin, N. Nikitin, M. Perfilov, E. Rakhmetov, M. Smolyakov, A. Snigirev, S. Vernov, I. Volobuev

P.G.Demidov Yaroslavl State University:
S. Kaschenko, A. Parkhomenko, A. Smirnov, A. Kuznetsov, A. Gvozdev, M. Martynov, A. Egorova

Scientific Program

- Higgs boson studies, searches for new physics and other experimental results from the LHC
- Physics prospects at linear and circular e+e- colliders and super B-factories
- Extensions of the Standard Model and their phenomenological consequences at the LHC and future colliders
- Higher order corrections and resummations for collider phenomenology
- Automatic calculations and Monte Carlo simulations in high energy physics
- Collider physics connections to astroparticle/cosmology
- Modern nuclear physics and relativistic nucleus-nucleus collisions
- Detectors for future experiments in high energy physics

Participants

Alexeyev	Stanislav	Sternberg Astronomical Institute
Altsybeev	Igor	SPbSU
Alvarez Piqueras	Damian	IFIC - Universitat de València, Spain
Andrianov	Aleksandr	SPbSU
Andrianov	Vladimir	SPbSU
Arbuzov	Boris	SINP MSU
Baranova	Natalia	SINP MSU
Baranov	Sergey	P.N.Lebedev Physics Institute
Barranco Navarro	Laura	IFIC (CSIC-UV) Valencia, Spain
Buarque Franzosi	Diogo	II. Physikalisches Institut - Universitaet Goettingen, Germany
Baskakov	Alexey	SINP MSU
Belyaev	Vasily	YSU
Berezhnoy	Alexandr	SINP MSU
Berta	Peter	Institut für Physik Johannes Gutenberg-Universität, Germany
Biryukov	Alexander	Samara University
Boldyrev	Alexey	SINP MSU
Boos	Eduard	SINP MSU
Bunichev	Viacheslav	SINP MSU
Cheng	Huajie	Institute of High Energy Physics, CAS, China
Chernoded	Andrey	MSU
Chernolyzhskii	Georgii	SINP MSU
Cirkovic	Predrag	Institute of Physics Belgrade, University of Belgrade
Danilina	Anna	MSU
Demidov	Sergei	INS RAS
Di Micco	Biagio	Universita' Roma Tre, Italy
DiPetrillo	Karri	Harvard University, USA
Dobrynina	Alexandra	YSU
Dolgopolov	Mikhail	Samara University
Dudko	Lev	SINP MSU
Elliot	Alison	University of Victoria, Canada
Egorov	Vadim	MSU
Eskin	Alexey	Samara University
Feofilov	Grigory	SPbSU
Ganai	Prince	National Institute of Technology Srinagar, India
Godunov	Sergey	ITEP
Gorbunov	Dmitry	INR RAS
Gurskaya	Albina	Samara University
Gvozdev	Alexander	YSU
Gladilin	Leonid	SINP MSU
Issadykov	Aidos	JINR
Ivanov	Aleksandr	SPbSU
Ivanov	Andrew	Kansas State University, USA
Karabanov	Ilya	YSU
Karpishkov	Anton	Samara National Research University
Keyzerov	Sergey	SINP MSU
Kharlamov	Petr	SINP MSU
Kharuk	Natalia	SPbSU
Korosteleva	Elena	SINP MSU
Korotkova	Natalia	SINP MSU
Kovalenko	Vladimir	SPbSU

Kozachuk	Anastasiia	SINP MSU
Krutov	Andrei	Samara University
Kuzmichev	Leonid	SINP MSU
Kuznetsov	Aleksandr	YSU
Kuznetsova	Anna	YSU
Kuznetsov	Mikhail	INR RAS
Kryukov	Alexander	SINP MSU
Lafage	Vincent	IPN, IN2P3, CNRS, France
Lepenik	Christopher	University of Vienna, Austria
Lipatov	Artem	SINP MSU
Liu	Yanwen	University of Science and Technology of China
Lokhtin	Igor	SINP MSU
Lukyashin	Anton	ITEP, MEPhI
Lukina	Olga	SINP MSU
Maevskiy	Artem	SINP MSU
Malyshev	Maxim	SINP MSU
Mandrik	Petr	NRC "Kurchatov Institute" Institute for High Energy Physics
Marakulin	Arthur	INR RAS
Martynov	Mikhail	YSU
McClymont	Laurie	University College London
Melikhov	Dmitri	SINP MSU
Meloni	Davide	Dipartimento di Matematica e Fisica, Roma Tre, Italy
Merkin	Mikhail	SINP MSU
Mosichkin	Anatoly	YSU
Myagkov	Igor	SINP MSU
Nefedov	Maxim	Samara National Research University
Nguyen	Minh Duc	SINP MSU
Nikitin	Nikolai	MSU
Novokhatskii	Alexey	MSU
Ognev	Igor	YSU)
Okorokov	Vitalii	National Research Nuclear University MEPhI
Okrugin	Alexander	YSU
Parkhomenko	Alexander	YSU
Perfilov	Maxim	SINP MSU
Petrova	Elena	SINP MSU
Pismak	Yury	SPbSU
Petrukhin	Anatoly	MEPhI
Polezaev	Roman	Samara national research University
Povarov	Alexander	YSU
Pozdeeva	Ekaterina	SINP MSU
Rakhmetov	Edward	SINP MSU
Ricken	Oliver	Physikalisches Institut, Universität Bonn, Germany
Rumyantsev	Dmitry	YSU
Rusov	Alexander	YSU
Rusov	Aleksey	YSU and University of Siegen
Rykova	Elza	Samara University
Ryzhov	Andrey	IHEP, Protvino
Satunin	Petr	INR RAS
Savin	Vasiliy	YSU
Savrin	Victor	SINP MSU
Shaikhiev	Artur	INR RAS
Shitova	Anastasia	YSU

Shkarovskiy	Sergey	JINR
Shlenev	Denis	YSU
Skugoreva	Maria	Kazan (Volga region) Federal University
Skurikhin	Aleksander	SINP MSU
Smirnov	Alexander	YSU
Smolyakov	Mikhail	SINP MSU
Snigirev	Alexander	SINP MSU
Sokolov	Anatoly	IHEP, Protvino
Solovyanov	Oleg	IHEP (Protvino), NRC KI
Solovieva	Elena	MIPT, LPI
Studenikin	Alexander	MSU & JINR
Taranenko	Arkadiy	National Nuclear Research University MEPhI
Tenchini	Roberto	INFN Pisa, Italy
Tretyakova	Tatiana	SINP MSU
Turchikhin	Semen	JINR
Uglov	Timofey	Lebedev Physics Institute
Vernov	Sergey	SINP MSU
Volkov	Petr	SINP MSU
Volobuev	Igor	SINP MSU
Vorotnikov	Georgii	SINP MSU
Voskresenskaya	Olga	JINR
Williams	Timothy	University of Birmingham, UK
Yazkov	Valeriy	SINP MSU

sciences

EPJ Web of Conferences

Statement of Peer review

In submitting conference proceedings to EPJ Web of Conferences, I certify to the Publisher that I adhere to the **Policy on Publishing Integrity** of the journal in order to safeguard good scientific practice in publishing.

1. All articles have been subjected to peer review administered by the proceedings editors.
2. Reviews have been conducted by expert referees, who have been requested to provide unbiased and constructive comments aimed, whenever possible, at improving the work.
3. Proceedings editors have taken all reasonable steps to ensure the quality of the materials they publish and their decision to accept or reject a paper for publication has been based only on the merits of the work and the relevance to the journal.

Title, date and place of the conference

XXIII International Workshop "High-Energy Physics and Quantum Field Theory" (QFTHEP2017), June 26- July 3, 2017, Yaroslavl, Russia

Proceedings editor(s): Dmitri Melikhov, Igor Volobuev

Date and editor's signature 6.10.2017

DOI: 10.1051/epjconf/201715800001

Preface

The XXIII International Workshop "High Energy Physics and Quantum Field Theory" (QFTHEP2017) was held in Yaroslavl, Russia, from June 26 to July 3, 2017. The 2017 edition of the workshop was organized jointly by the D.V.Skobeltsyn Institute of Nuclear Physics of the M.V.Lomonosov State University (SINP MSU) and the Faculty of Physics of the P.G. Demidov Yaroslavl State University

QFTHEP2017 continues the series of workshops started by SINP MSU in 1985 and aimed at putting together theoreticians and experimentalists for discussing topics of current interest and providing a stimulating environment for exchanging ideas on new developments in high energy physics and research programs for future colliders.

QFTHEP2017 covered a wide range of topics in the modern theoretical and experimental high-energy physics, with the main emphasis on

- Higgs boson studies, searches for new physics and other experimental results from the LHC
- Physics prospects at linear and circular e+e- colliders and super B-factories
- Extensions of the Standard Model and their phenomenological consequences at the LHC and future colliders
- Higher order corrections and resummations for collider phenomenology
- Automatic calculations and Monte Carlo simulations in high energy physics
- Collider physics connections to astroparticle/cosmology
- Modern nuclear physics and relativistic nucleus-nucleus collisions
- Detectors for future experiments in high energy physics

About 100 researchers (with a traditionally large fraction of PhD students and young scientists) participated in the workshop.

On behalf of the Organizing Committee, we have pleasure to thank all participants for their lively scientific discussions and for creaing the exciting atmosphere of this event. The contributions presented in this volume reflect the scientific content of QFTHEP 2017.

We look forward to seeing all participants at the next edition of QFTHEP in 2019!

Dmitri Melikhov and Igor Volobuev (eds)

Higgs Boson Physics at CMS

Predrag Cirkovic[1,*] (on behalf of the CMS Collaboration)

[1] *Institute of Physics Belgrade, Pregrevica 118, P.O. Box 68, 11080 Belgrade, Serbia*

Abstract. An overview of the standard model Higgs boson analyses and beyond standard model Higgs boson searches will be given. A data sample of proton-proton collisions at $\sqrt{s} = 13\,TeV$ collected with the CMS detector at the LHC is used. The measurements of properties of the Higgs boson in the 4ℓ and $\gamma\gamma$ decay channels will be presented. The future prospects will be covered.

1 Introduction

In this report, the subjects will be discussed in the three main sections. After a brief experimental and theoretical overview, a presentation of the latest results of the Standard Model (SM) Higgs boson measurements in the main analysis channels together with the measurements of the Higgs boson properties thereafter, will be discussed in section 2. The latest results of the searches for beyond the Standard Model (BSM) Higgs boson in terms of the charged, MSSM, doubly-charged, invisible and di-Higgs boson production, will be covered in section 3. The future prospects for the SM Higgs boson production in the 4ℓ and $\gamma\gamma$ decay channels will be described in section 4. The summary will be described at the end. The focus will be on the most recent results, such as the Higgs boson mass measurement in the $H \to 4\ell$, fiducial cross section measurements in the 4ℓ and di-photon channels, the latest results on $t\bar{t}H$, and the search for resonant and non-resonant di-Higgs boson production. Many of the searches at CMS [1] presented here have been updated with the full 2016 dataset corresponding to an integrated luminosity of $35.9\,fb^{-1}$.

2 Standard Model (SM) Higgs Boson Searches

Starting with the measurements of the Higgs boson production and decay rates with the combined 7 and 8 TeV CMS dataset [2] and the combined ATLAS and CMS analysis of the LHC pp collision data at $\sqrt{s} = 7$ and $8\,TeV$ [3], the measured signal strenghts, both the individual ones measured by CMS and ATLAS alone and the combined one are compatible with 1.0. The largest uncertainties for the production modes appear in the channel with the lowest cross sections, such as $t\bar{t}H$. Concerning decays, the smallest uncertainties are present in the channels with the cleanest signatures, e.g. $H \to \gamma\gamma$ and $H \to WW$, and the largest uncertainties in $H \to b\bar{b}$ are caused mostly by the presence of large QCD background. Individual signal strengths (μ) have been also measured in the plane of their fermionic and bosonic parts, with the combined CMS [2] and the combined CMS and ATLAS [3] data from the

*e-mail: predrag.cirkovic@cern.ch

LHC pp collision at $\sqrt{s} = 7$ and $8\,TeV$. An overall agreement of the 2D maximum likelihood best fits with the SM expectations has been achieved for all channels within 1σ region.

The same datasets used in Refs. [2] and [3] have been used to measure also the coupling between the Higgs boson and the other SM particles. The measurements of the coupling with muon, τ-lepton, b-quark, W and Z bosons and top quark have been compared with their prediction, which are proportional to the particle mass. The coupling values range between very small values, featuring muons, up to the values close to unity, which characterize the top quark. An overall agreement between the measurements and the theoretical prediction has been noticed here as well. The Run I measurements already put constraints on the couplings and for the measurements with the Run II datasets it is expected to get more stringent constraints. The fits in the 2D plane of the coupling modifiers for vector bosons (κ_V) and fermions (κ_F) show that the results are compatible with their SM expectations (1, 1) for all five Higgs boson decay channels and their combination within one standard deviation (1σ).

2.1 Higgs boson decaying to a pair of photons

$H \to \gamma\gamma$ [4] represents one of the Higgs boson discovery channels, characterized by high resolution with a very small uncertainty on di-photon mass, a clean final state, and a small branching ratio. Boosted decision tree is used for the photon identification, selection of di-photon vertices, and selection of di-photon events. Boosted Decision Tree (BDT) distribution has been flattened according to the total $\gamma\gamma$ signal [4], which is composed of the four Higgs boson production mode components. In this report, the Fig. 5 (left) of [4] has been presented. The distribution of the transformed score of the di-photon multivariate (BDT) classifier for events with two photons satisfying the preselection requirements in data has been compared to the distribution of the simulated signal and simulated background stacked together [4]. The correspondence between the data distribution and the distribution of the total signal plus background stack has been achieved. The score of the BDT classifier is used to select and divide the events into four "untagged" categories according to the di-photon mass resolution and predicted signal over background ratio. The category of the events with the lowest BDT score has been discarded from the analysis. The signal is selected according to the highest transverse energy photons and the background is composed of the irreducible $\gamma\gamma$ component, and the reducible fakes that originate from γ+jet and di-jet events. In Fig. 1 (left), a di-photon mass distribution is presented, for the total signal and background (upper row in Fig. 1, left) and for the background component subtracted (bottom row in Fig. 1, left). The best fit mass is found at $m_H = 125.4\,GeV$ with statistical uncertainty of approximately $0.15\,GeV$ and the systematic uncertainties preliminarily estimated to be between $0.2\,GeV$ and $0.3\,GeV$ (still under study) [4]. In Fig. 1 (right), the results of the signal strength measurements in the $H \to \gamma\gamma$ are presented. The signal strength measurement for the four categories combined provide the value of $1.16^{+0.15}_{-0.14}$, compatible with the SM prediction within $\approx 1\sigma$. The gluon fusion signal strength is the most compatible one, while the $t\bar{t}H$ providing the signal strength of 2.2, corresponds to a $3.3\,\sigma$ excess with respect to the absence of $t\bar{t}H$ production, and agrees with the SM prediction within $1.6\,\sigma$.

The signal strength measured in bosonic and fermionic parts [4] is also consistent with the SM predictions, especially in its fermionic part, which is mostly driven by contribution from gluon fusion process. The 2D likelihood scans of the Higgs boson coupling modifiers have been also performed in $H \to \gamma\gamma$, but here for the couplings to photons and gluons, and couplings to vector bosons and fermions [4]. Both fits are compatible with the SM predictions, within the range of 1σ.

The Fig. 2 summarizes the most recent results on the fiducial cross section measurements performed in $H \to \gamma\gamma$ channel [5]: the differential fiducial cross section measurements for di-photon p_T, in Fig. 2 (left), and the number of additional hadronic jets, in Fig. 2 (right) [5]. For this measurement, CMS requires two high-p_T isolated photons within the fiducial volume defined as: $p_{T1}/m_{\gamma\gamma} < 1/3$,

DOI: 10.1051/epjconf/201715801001

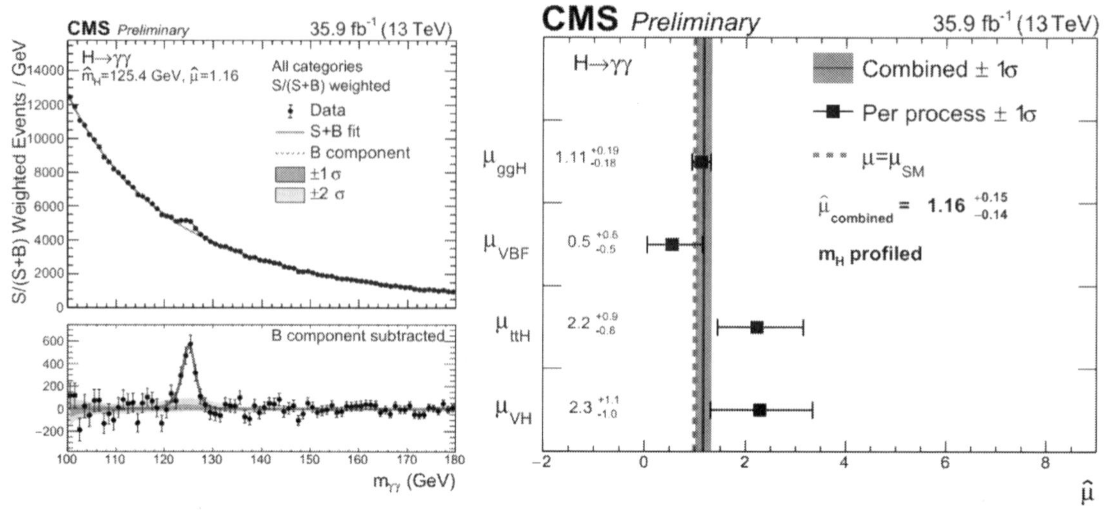

Figure 1. The $m_{\gamma\gamma}$ distribution (left) as weighted sum of all categories [4]. S and B represent the number of signal and background events, respectively. Signal strength modifiers (right) measured for each process (black points) for profiled m_H, compared to the overall signal strength (green band) and to the SM expectation (dashed red line) [4].

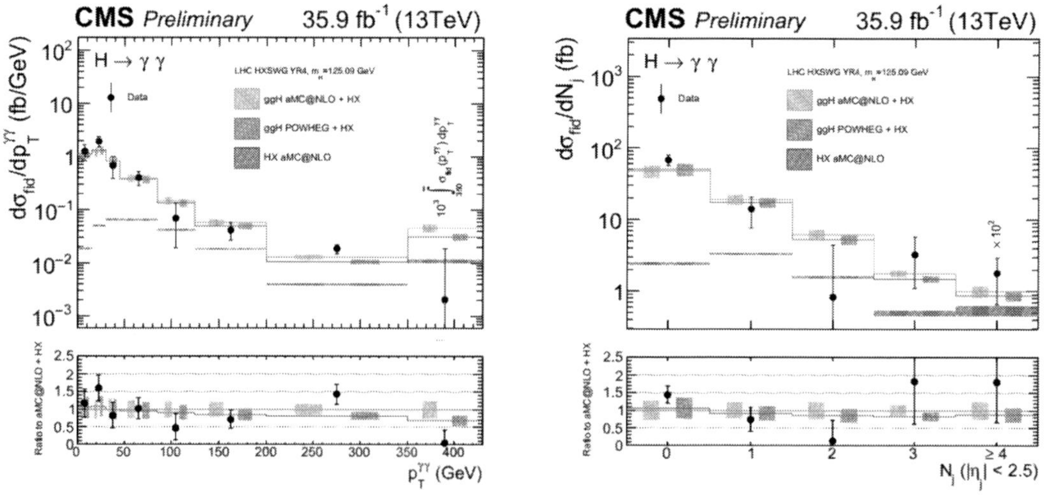

Figure 2. Measured $H \to \gamma\gamma$ differential cross-section (black points) for $p_T{}^{\gamma\gamma}$ (left) and N_{jets} (right) [5]. The measurements are compared to the theoretical predictions, combining the Higgs boson cross sections and branching fraction as in the LHC Higgs Cross Section Working Group [6].

$p_{T2}/m_{\gamma\gamma} < 1/4$, $|\eta_{1,2}| < 2.5$, $Iso_{gen1,2} < 10\,GeV\,(\Delta R = 0.3)$. No classification has been performed according to production mode, which implies the focus on gluon fusion. Classification of events based on mass resolution estimator improved sensitivity by 10%. The signal production cross section is extracted through simultaneous maximum likelihood fit to the di-photon invariant mass in all the

 DOI: 10.1051/epjconf/201715801001

production modes. The unfolding to particle-level cross sections is achieved by extracting the vector of fiducial cross sections directly from the likelihood fit. The result of the total fiducial cross section with the 13 TeV data, measured to be $\hat{\sigma}_{\text{fiducial}} = 84 \pm 11(\text{stat}) \pm 7(\text{syst}) = 84^{+13}_{-12}$ [5] (and compared to the result obtained with the 8 TeV data), shows no significant deviations from the SM predictions and represents the most precise fiducial measurement to date.

2.2 Higgs boson decaying to four leptons

The next important discovery channel is $H \rightarrow ZZ \rightarrow 4\ell$ [7], with the signature consisting of four isolated leptons (electrons or muons), with excellent momentum resolution. The leptons are grouped into pairs where each of them contains the leptons of the same flavour and the mutually opposite sign of the charge. This channel is characterized by a fully reconstructed 4ℓ mass peak, shown in Fig. 3 (left), and a large signal over background ratio ($S/B > 2 : 1$). The main background, estimated from Monte Carlo, is irreducible SM production of ZZ, and the reducible background, estimated from data, is $Z + X$, which is dominated by fakes from Z+jets. A kinematic discriminants (D^{kin}) based on matrix element calculations is used for the signal extraction. The events are split into seven categories according to Higgs boson production modes and based on the number of leptons, number of (b-)jets, transverse missing energy and the selection on the kinematic discriminant. Signal strength in this channel is measured for each of the categories and each of the production modes providing consistent combined results [7], as well as in the plane of its fermionic and bosonic parts, in Fig. 3 (right). All the results show agreement between the measurements and the SM predictions within the present uncertainties.

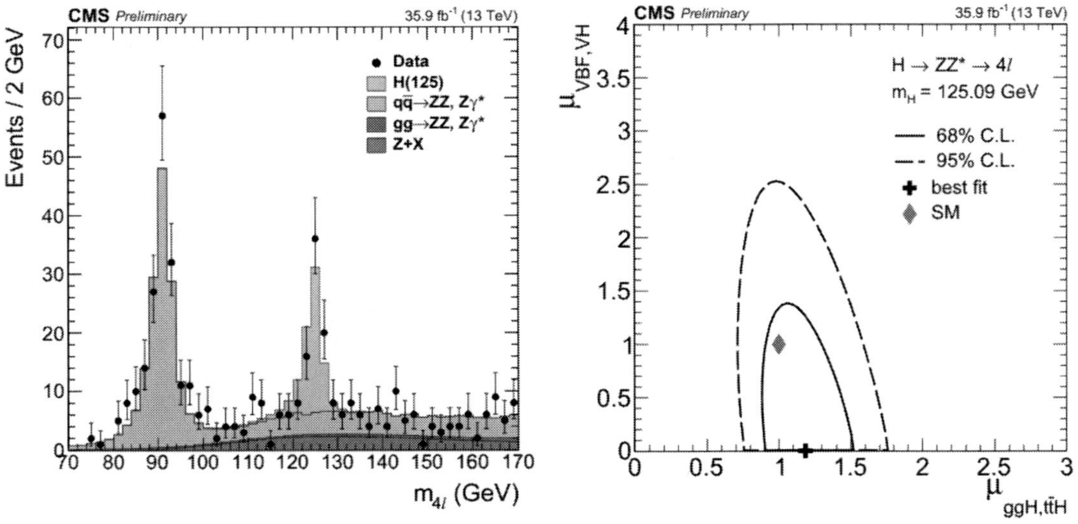

Figure 3. Distribution of the reconstructed four-lepton invariant mass $m_{4\ell}$ in the low-mass range (left) [7]. Points with error bars represent the data and stacked histograms represent expected signal and background distributions. Result of the 2D likelihood scan for the $\mu_{ggF,t\bar{t}H}$ and $\mu_{VBF,VH}$ signal strength modifiers (right) [7], where ggF (or ggH) denotes gluon fusion production mode, $t\bar{t}H$ denotes the production of the Higgs boson associated top quarks, VBF denotes vector boson fusion production, and VH denotes production of the Higgs boson associated to a (weak) vector boson. The solid and dashed contours show the 68% and 95% confidence level (CL) regions, respectively.

DOI: 10.1051/epjconf/201715801001

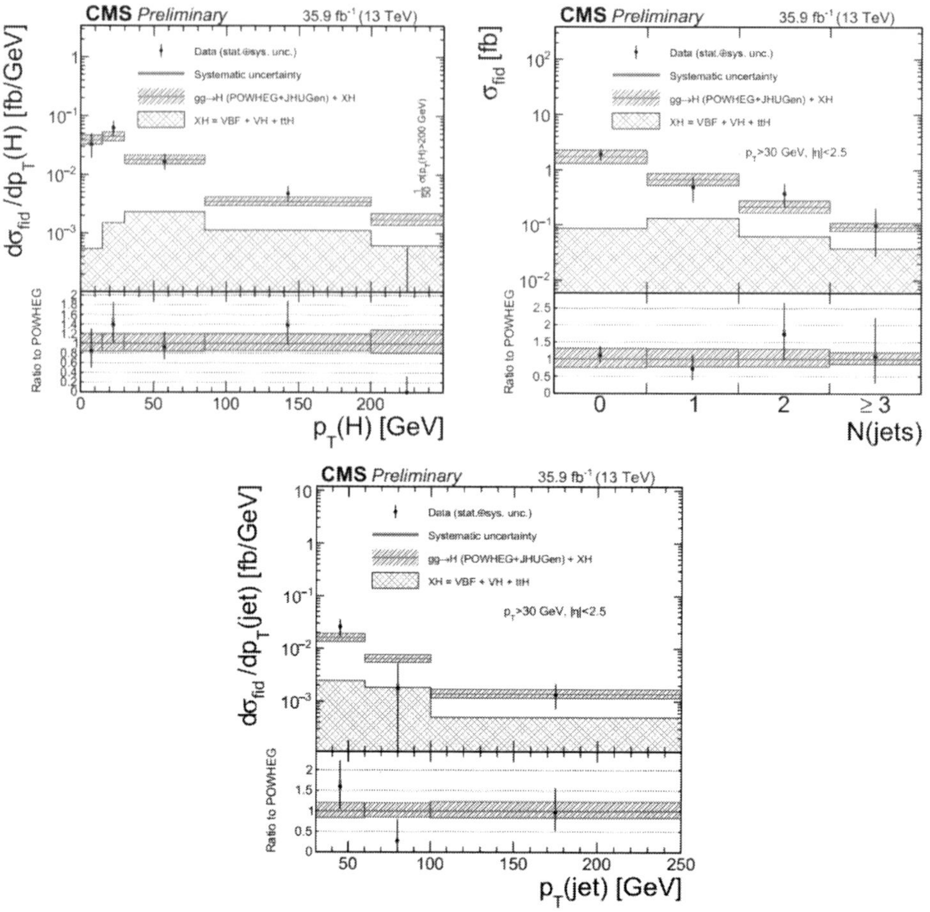

Figure 4. The results of the differential cross section measurement for $p_T(H)$ (top-left), $N(jets)$ (top-right) and $p_T(jet)$ (bottom). The acceptance and theoretical uncertainties in the differential bins are calculated using POWHEG. The sub-dominant component of the signal $(VBF + VH + t\bar{t}H)$ is denoted as XH [7].

Fiducial cross section measurement has been performed in the $H \to ZZ \to 4\ell$ channel. This measurement is characterized by minimal dependence on the assumptions of the relative fraction or kinematic distributions of the separate production modes. The details about the fiducial volume definition in form of the lepton kinematics and isolation requirements and the event topology requirements, are summarized in the Tab. 4 of [7]. This definition is very similar to the one used for the fiducial cross section measurement performed at 7 and 8 TeV [8], however the new measurement provides better agreement with the SM expectations and lower (statistics-dominated) uncertainties. The differential cross sections has been measured in $H \to ZZ \to 4\ell$ versus p_T of the Higgs boson, jet multiplicity and p_T of the jet. The dominant sources of systematic uncertainty are the experimental uncertainties of the lepton identification efficiencies and luminosity measurement, and the theoretical uncertainties are found to be sub-dominant. In order to asses the model dependence of the measurement, the unfolding procedure is repeated using different response matrices created by varying the relative fraction of each SM production mode within its experimental constraints. The uncertainty is determined to be negligible with respect to the experimental systematic uncertainties. The results of the differential cross section measurement in Fig. 4 show a relatively good agreement between data and gluon fusion dominated predictions from POWHEG [9–11] simulation.

2.3 Higgs boson mass and width measurements

Mass measurement performed in 4ℓ channel [7] is based on one-, two- and three-dimensional fit on the following variables: invariant mass of the Higgs boson ($m_{4\ell}$), expected uncertainty on mass ($\mathcal{D}_{mass}$) and the kinematic discriminant ($\mathcal{D}^{kin}_{bkg}$). In Fig. 5 (left), there is a comparison of the 1D likelihood scan as a function of mass for the 1D, 2D, and 3D measurement. The likelihood to be maximized is constructed using the reconstructed transverse momenta of the two leptons forming the Z candidate with the invariant mass closest to the nominal Z boson mass (Z1), the per lepton resolutions, the refitted transverse momentum and the invariant mass calculated from the refitted four momenta. For each event, the likelihood is maximized and the refitted transverse momenta are used to recalculate the four-lepton mass and mass uncertainty, which are denoted as m'_{4l} and D'_{mass}. These distributions are then used in the definition of the likelihood to extract the Higgs boson mass [7]. Compared to the mass measurements performed with the CMS and ATLAS Run I data [12], which combination was already quite precise (less than 0.2%), the new result is even more precise (around 0.17%). The latest measurement presented in Fig. 5 (right), performed in the mass range between 105 and 140 GeV, using the on-shell Higgs boson production and the floating m_H parameter, constrained the Higgs boson width to be less than 1.1 GeV at 95% C.L [7].

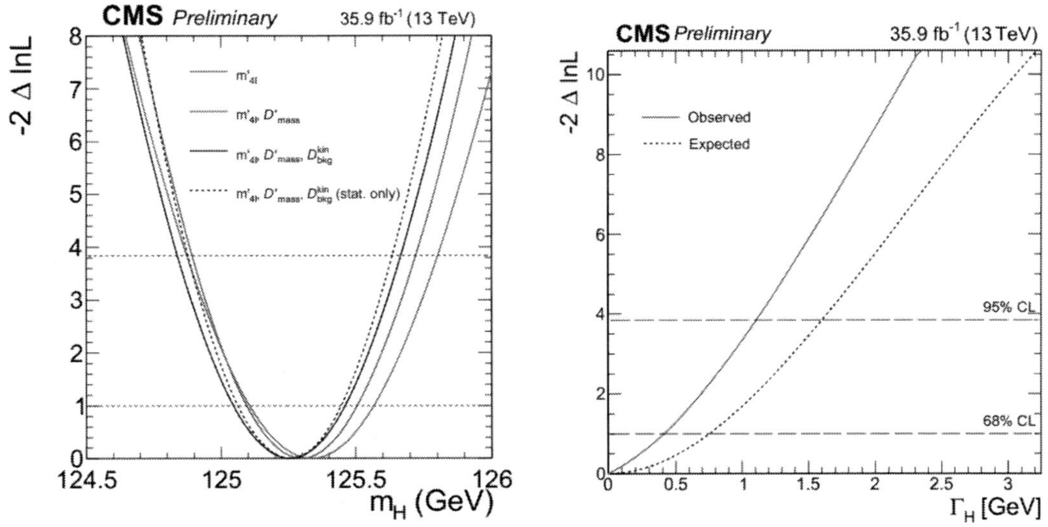

Figure 5. 1D likelihood scan as a function of mass for the 1D, 2D, and 3D measurement (left). The likelihood scans are shown for the mass measurement using the refitted mass distribution with $m(Z1)$ constraint. Solid lines represents the scan with full uncertainties included, dashed lines statistical uncertainty only [7]. Observed and expected likelihood scan of Γ_H using the signal range $105 < m_{4\ell} < 140 GeV$, with m_H floated [7].

2.4 Higgs boson decaying to a pair of tau leptons

Higgs boson decaying to $\tau\tau$ [13] is another discovery channel, because of the large event rate expected in the SM with respect to the other leptonic decay modes, and the smaller contribution from background events with respect to the $H \rightarrow b\bar{b}$ channel. With the four final states ($\mu\tau_h$, $e\tau_h$, $\tau_h\tau_h$, and $e\mu$), split into three categories (0-jet, VBF and boosted), the total number of categories reaches twelve. In each category the two variables are chosen to build two-dimensional (2D) distributions. The search for an excess of the signal events over the expected background involves a 2D global maximum likelihood fit based in all channels, together with control regions for the $t\bar{t}$, QCD multijet, and W+jets backgrounds. Combined observed and predicted $m_{\tau\tau}$ distributions that include the VBF category of the $e\mu$, $e\tau_h$ and $\mu\tau_h$ channels are shown in Fig. 6 (left). The signal is normalized to its best-fit signal strength, while the normalization of the predicted background distributions ($Z \rightarrow \tau\tau$, W+jets, QCD multijet and others) corresponds to the result of the global fit. The mass distributions for a constant range of the second dimension of the signal distributions are weighted according to $S/(S+B)$.The "others" background contribution includes events from diboson, $t\bar{t}$, and single-top quark production, as well as scalar boson decays to a pair of W bosons and Z bosons decaying to a pair of light leptons. The background uncertainty band includes all systematic and statistical sources of background uncertainties, after the global fit. The inset in the top-right corner in Fig. 6 (left) shows the corresponding difference between the observed data and the sum of the expected signal and background distributions.

The final results are presented in the form of the plots of the signal strength per category and the signal strength per channel, with the combined best-fit on the signal strength of 1.06 ± 0.25 [13]. A graph of the p-value versus mass of the Higgs boson, in Fig. 6 (right), presents the observation of decays of the SM scalar boson to pairs of τ leptons, with an observed significance of 4.9 and an expected significance of 4.7 standard deviations for a mass of 125 GeV.

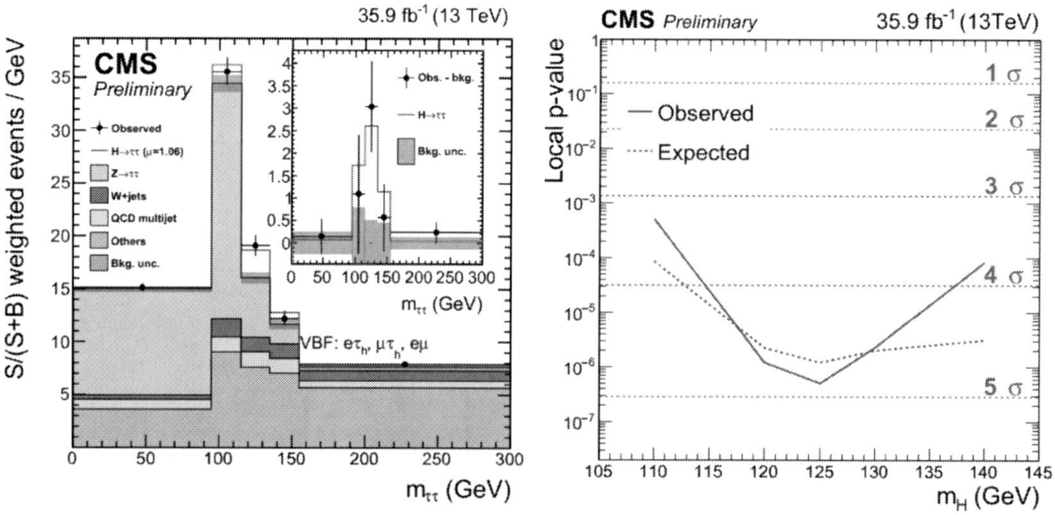

Figure 6. Combined observed and predicted $m_{\tau\tau}$ distributions, together with the inset showing the difference between the observed data and the sum of the expected signal and background distributions (left) [13]. Local p-value and significance as a function of the SM scalar boson mass hypothesis (right). The observation (red) is compared to the expectation (blue) for a scalar boson with a mass $m_H = 125\,GeV$ [13].

2.5 Production of the Higgs boson associated to top quarks

The latest results of the analyses of the production of the Higgs boson associated to a pair of top quarks have been also presented. Concerning its large number of possible final states, the several channels have been considered: $t\bar{t}H$ with $H \to b\bar{b}$ [14], $t\bar{t}H$ with $H \to \gamma\gamma$ [4], $t\bar{t}H$ with a τ lepton [15], 4ℓ [7] and multilepton [16] in the final state. All except the $t\bar{t}H$ with $H \to b\bar{b}$ have been updated with the full 2016 dataset. In $H \to \gamma\gamma$, and the $t\bar{t}H$ to multilepton, the significance exceeds 3σ. The plot in Fig. 7 summarizes the results of the signal strength measurements performed with Run II data from all channels. The comparison with the combined result of the analyses obtained with the Run I CMS and ATLAS data is presented [3].

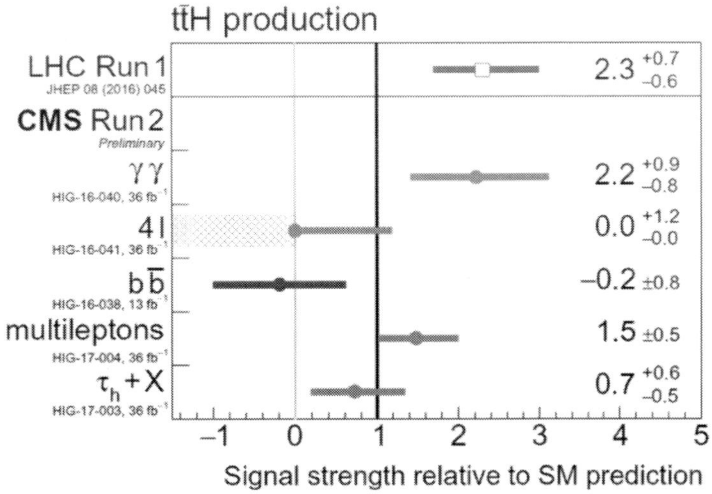

Figure 7. The $t\bar{t}H$ analyses summary plot for the signal strengths relative to the SM prediction.

The $t\bar{t}H$ analysis with $H \to b\bar{b}$ [14] considers the two following final state categories: di-lepton and lepton plus jets. Dilepton channel is characterized by minimal non-$t\bar{t}$ background and minimal jet combinatorics, while the lepton plus jets is characterized by high statistics. In Fig. 8 the results of the signals strength measurements and the 95% CL upper limits on μ are shown.

The $t\bar{t}H$ to multilepton [16] is performed with the Higgs boson decays to WW, ZZ and τ leptons characterized into the three final state categories: two leptons same sign ($2\ell ss$), three leptons (3ℓ) and four leptons (4ℓ). Signal is extracted using the unrolled 2D BDT distributions, where the BDT discriminators have been trained separately against $t\bar{t}$ and $t\bar{t}V$. This channel has provided an evidence for the $t\bar{t}H$ production, giving the result on the signal strength of about 1.5 +/- 0.5, which could be noticed from the excess of events in the di-lepton channel, in Fig. 9 (left), spread across the bins.

Higgs boson decaying to at least one τ [15] lepton in the final state is one of the $t\bar{t}H$ channels complementary to the $t\bar{t}H$ to multilepton channels. It considers the $1\ell + 2\tau_h$, $2\ell ss + 1\tau_h$ and $3\ell + 1\tau_h$ event categories, with the $t\bar{t}Z/W$ and electroweak background, as well as the fakes coming from the non-prompt leptons, jets to τ_h and the charge-flips as the main background components. The results have been observed with the significance of $1.4\,\sigma$, while the expected significance is $1.8\,\sigma$. The upper limit on the signal rate of 2.0 times the SM $t\bar{t}H$ production rate at 95% CL is set [15].

The search for the Higgs boson is also performed in the channel with a single top quark associated to the Higgs boson. This production is characterized by a relatively low cross section with respect to the total $t\bar{t}H$ cross section. In SM, this is the consequence of a destructive interference of the production with the coupling to the W boson and the production with the coupling to the top quark. The analysis of tH production is performed to expose the relative sign of t-H (and W-H) couplings via this interference. The signal with $H \to b\bar{b}$ in the final state is defined as a lepton (electron or muon) plus 3 or 4 b-tagged jets and one non-tagged jet. BDT is used to find jet assignment for the tHq signal

DOI: 10.1051/epjconf/201715801001

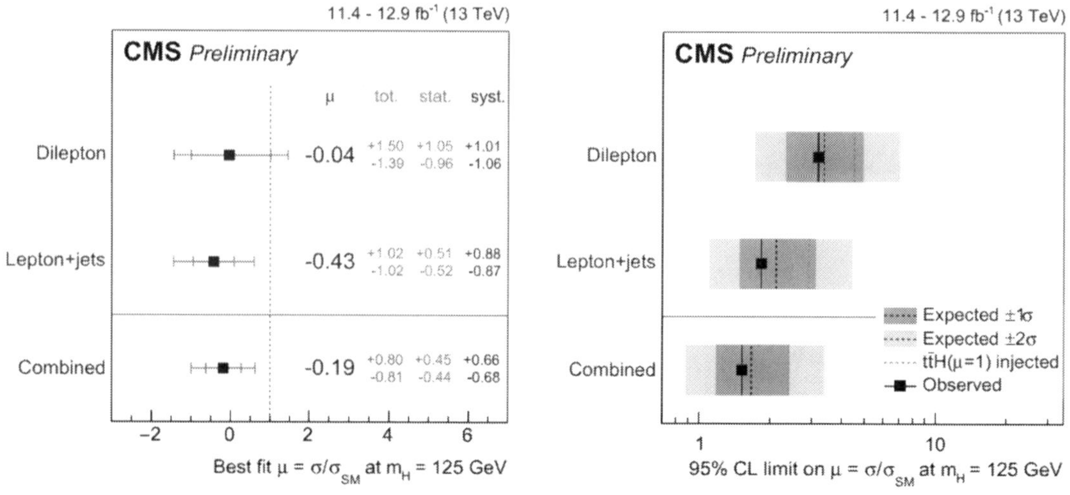

Figure 8. Best-fit values of the signal strength modifiers μ with their $\pm 1\sigma$ confidence intervals, also split into their statistical and systematic components (left), and median expected and observed 95% CL upper limits on μ (right). The expected limits are displayed together with $\pm 1\sigma$ and $\pm 2\sigma$ confidence intervals. Also shown are the limits in case of an injected signal of $\mu = 1$ [14].

and the $t\bar{t}$, which represents the main background process. The final results obtained for SM scenario provide 113.7 (98.6) and for the inverted top coupling scenario: 6.0 (6.4), for the observed (and the expected) 95% CL upper limit on the cross section for the SM tH production, respectively [17]. In the multilepton final state [18], the strategy is to analyse the events with a single-top quark associated Higgs boson, which decays to WW, ZZ or $\tau\tau$, split into the two categories: two lepton same sign and three lepton. The main irreducible backgrounds are $t\bar{t}$+X (where X can take W, Z, H, or γ^*), photon conversion, diboson, tri-bosons, and rare SM processes. Reducible backgrounds include fakes from non-prompt leptons and misidentified objects passing lepton selection, as well as the charge flips from charge mis-IDs and the opposite-sign processes (such as $t\bar{t}/Z$+jets). The tHq process can benefit from a greatly enhanced production cross section in case of anomalous top-Higgs couplings, and the results published in [17] and [18] are used to constrain these couplings.

2.6 VBF production of the Higgs boson

VBF production mode with $H \rightarrow b\bar{b}$ is characterized by significantly larger cross section with respect to VH or $t\bar{t}H$, and a very large QCD multijet background. BDT is used for identification of the VBF-like events with four jets out of which one or two are b-tagged. The analysis, performed with the $2.3\,fb^{-1}$ of data collected at the energy of $13\,TeV$, provided the signal strength of $\mu = -3.7$, and the other measurement, performed with the $13\,TeV$ data combined with the $8\,TeV$ data of roughly $19\,fb^{-1}$, provided the signal of $\mu = 1.3$ [19]. The observed (and expected) upper limits in the absence of a signal are 3.0(5.0) times the SM prediction for the case of the $13\,TeV$ dataset only, and 3.4(2.2) times the SM prediction have been obtained with the combined datasets [19].

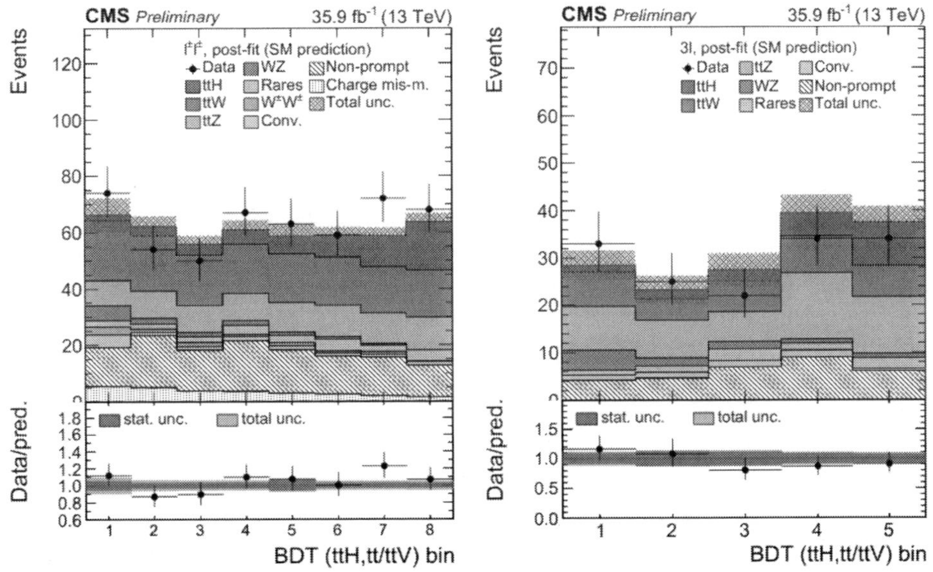

Figure 9. Combination of the BDT classifier outputs in the bins used for signal extraction, for the same-sign di-lepton (left) and three-lepton (right) channels. The distributions are shown after the fit to the data, with all processes constrained to the SM expectation [16].

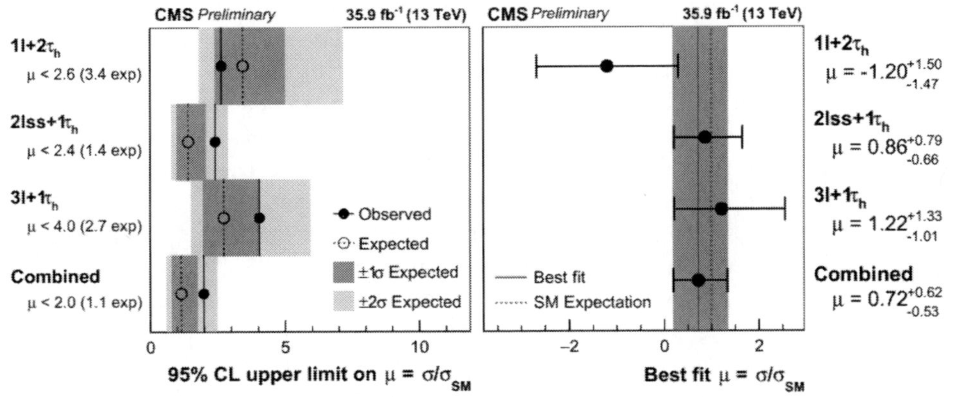

Figure 10. 95% CL upper limits on the $t\bar{t}H$ signal rate (left), obtained in each of the categories $2\ell ss + 1\tau_h$, $3\ell + 1\tau_h$, and $1\ell + 2\tau_h$ individually and for the combination of all three event categories. The expected limits are computed for the background-only ($\mu = 0$) hypothesis. Signal rates μ (right), in units of the SM $t\bar{t}H$ production rate, measured in each of the categories $2\ell ss + 1\tau_h$, $3\ell + 1\tau_h$, and $1\ell + 2\tau_h$ individually and for the combination of all three categories [15].

3 Beyond the Standard Model (BSM) Higgs Boson Searches

Discovery of another scalar boson, neutral or charged, would represent unambiguous evidence for an extended Higgs boson sector. Neutral and charged Higgs bosons appear in many extensions of the SM, of which the simplest are the two Higgs boson doublet models (2HDMs). In a 2HDM, the following five Higgs bosons are predicted: two charged Higgs bosons, $H^{\pm}$, two neutral scalar (CP-even) Higgs

 DOI: 10.1051/epjconf/201715801001

bosons, h and H, and one neutral pseudoscalar (CP-odd) Higgs boson, A. The 2HDM models can be classified into types, depending on the Higgs boson coupling to fermions. In the Type-I 2HDM only one of the Higgs boson doublets couples to fermions, while in a Type-II 2HDM one doublet couples to up-type fermions and the other one to down-type fermions. As a consequence, in Type-I 2HDM the coupling of the charged Higgs bosons to charged leptons and neutrinos is proportional to $\cot\beta$, and in Type-II 2HDM , it is proportional to $\tan\beta$, where $\tan\beta$ represents the ratio of the vacuum expectation values of the two Higgs boson doublets.

The minimal supersymmetric standard model (MSSM), as a special case of a Type-II 2HDM, incorporates a symmetry between bosons and fermions, introducing additional heavier partners for the existing particles, and allowing for possible cancellation of the quadratically divergent self-energy corrections to the Higgs boson mass at high energy, so-called supersymmetry (SUSY). Two free parameters are used to express the MSSM at tree level: the $\tan\beta$ and m_A (the mass of the pseudoscalar Higgs boson). Concerning the values of the $\tan\beta$ parameter, while small and medium values represent the models where the gluon fusion is the dominant production mode, a large value means that couplings to down-type fermions are enhanced. This leads to enhanced rate of decays into τ leptons, and a second dominant production mode, b associated production. Radiative corrections lead to large modifications of the behaviour beyond tree level, which can depend on a large number of different SUSY parameters. These parameters are fixed to choices that result in different benchmark scenarios, which must produce MSSM models consistent with the discovered $125\,GeV$ Higgs boson. The two scenarios considered in [20] are the m_h^{mod+} and $hMSSM$. In the case of the m_h^{mod+} scenario, choices are made for the SUSY parameters which allow the majority of the m_A-$\tan\beta$ plane to be consistent with a $125\,GeV$ scalar Higgs boson. In the case of the $hMSSM$ scenario, the condition of $m_H = 125\,GeV$ is fixed across the whole plane and the radiative corrections adjusted accordingly.

Another direction would lead to the search for a doubly-charged Higgs boson. The experimental observation of a non-zero neutrino masses [21] contradicts the standard model (SM) assumption of a massless neutrino. There are several models that attempt to expand the SM and accommodate neutrino masses. The minimal Type II seesaw mechanism [22] extends the SM particle spectrum with a scalar triplet with $SU(2)_L \times U(1)_Y$ quantum numbers Φ. The triplet has a doubly-charged component $\Phi^{\pm\pm}$, a singly-charged component $\Phi^{\pm}$, and a neutral component Φ^0. The analysis documented in this summary is specific to the left-handed doubly-charged Higgs boson.

A number of models for the physics beyond the SM allow for invisible decay modes of the Higgs boson, such as decays to neutralinos in supersymmetric models [23], graviscalars in models with extra spatial dimensions [24]. More generally, invisible Higgs boson decays can be realized through interactions between the Higgs boson and dark matter (DM) [25]. In Higgs-portal models [26], the Higgs boson acts as a mediator between SM and DM particles allowing for direct production of DM at the LHC. Furthermore, cosmological models proposing that the Higgs boson played a central role in the evolution of the early universe motivate the study of the relationship between the Higgs boson and DM [27].

The BSM physics effects can appear either via anomalous couplings of the Higgs boson or via new particles that can be directly produced or enter in the quantum loop at production. The experimental signature would be an enhancement of the Higgs boson pair (di-Higgs boson) production cross section at a specific invariant mass value (resonant production) or over the whole invariant mass spectrum (non resonant production). The resonant production is predicted by many extensions of the SM such as the Singlet model [28], the Two Higgs boson Doublet Model (2HDM) [29], the minimal supersymmetric standard model (MSSM) [30], and the Warped Extra Dimensions (WED) [31] model. Although the physics motivations and the phenomenology of these theories are very different, the signal is represented as a CP-even scalar particle (S) decaying into a Higgs boson pair, with an intrinsic width that is often negligible with respect to the detector resolution. In the non resonant case, the BSM physics is modeled with an effective Lagrangian by extending the SM Lagrangian with dimension-6 operators [32].

A brief overview of the recent results of the BSM Higgs boson searches, essential for the hunt for new physics, follows in the next sections.

3.1 Search for the charged Higgs boson

This section summarizes the results of the search for a charged Higgs boson in decays to a τ lepton and a neutrino based on $12.9\,fb^{-1}$ recorded in 2016 at $\sqrt{s} = 13\,TeV$ [33]. The analysis considers the production mode $pp \to tbH^{\pm}$ ($pp \to H^{\pm}W^{\mp}bb$) for masses larger (smaller) than the top quark ($m_{H^{\pm}} > mt - mb$) and focuses on the fully hadronic final state. In this analysis, transverse mass of the Higgs boson has been used for signal extraction. The model independent 95% CL limits and MSSM m_h^{mod+} exclusion limits in $m_{H^+} - \tan\beta$ plane have been presented, where no excess has been observed in the mass range from $80\,GeV - 3\,TeV$ and the possibility of the existence of the charge Higgs boson has been excluded in most of the region of interest.

3.2 MSSM Higgs boson with subsequent decays to a pair of tau leptons

In the search for the neutral MSSM Higgs boson with the decays into $\tau\tau$ with $12.9\,fb^{-1}$ of data at $\sqrt{s} = 13\,TeV$ [20] the following final states: $\tau_h\tau_h$, $\mu\tau_h$, $e\tau_h$ and $e\mu$, and the two event categories based on the number of b-tagged jets have been considered. Main backgrounds are $Z/\gamma^*\ \tau\tau$, QCD and $t\bar{t}$, and the total transverse mass is used for the signal extraction. The results have been presented in form of the model independent 95% CL limits and the exclusion limits in the $m_A - \tan\beta$ plane, in the context of the m_h^{mod+} and $hMSSM$ scenario. No excess has been found in the mass range $100\,GeV - 3\,TeV$, and with the complete 2016 dataset, a larger part of the $m_A - \tan\beta$ plane has been excluded.

3.3 Doubly-charged Higgs boson searches

A search for doubly-charged Higgs boson production in three and four lepton final states at $\sqrt{s} = 13\,TeV$ is performed for a narrow di-lepton resonance in pair production and associated production modes: $q\bar{q} \to Z^0/\gamma \to \Phi^{++}\Phi^{--} \to \ell_i^+\ell_j^+\ell_k^-\ell^-l$, and $qq' \to W^{\pm} \to \Phi^{\pm\pm}\Phi^{\mp} \to \ell_i^{\pm}\ell_j^{\pm}\ell_k^{\mp}\nu_l$. Measuring the branching fractions of the doubly-charged Higgs boson would give access to neutrino parameters such as the masses (m_1, m_2, m_3), the Majorana Phases (α_1, α_2, or $\Delta\alpha = |\alpha_1 - \alpha_2|$), the CP-violating angle δ, and the Higgs boson triplet vacuum expectation value. Since the branching fractions are not fixed by the model, model independent limits are set assuming 100% Φ^{++} decays to a single pair of leptons: ee, $e\mu$, $\mu\mu$, $e\tau$, $\mu\tau$, and $\tau\tau$. Limits have been also set for four benchmark points that target different neutrino mass hierarchies. As the final result of this search, no significant excess has been observed. Lower bounds on the $\Phi^{\pm\pm}$ mass are extended to between $800\,GeV$ and $820\,GeV$ for final states with 100% decays to ee, $e\mu$, and $\mu\mu$, and to $714\,GeV$ and $643\,GeV$ for final states with 100% decays to $e\tau$ and $\mu\tau$, respectively. Final states with 100% decays to $\tau\tau$ are excluded up to $535\,GeV$. In addition, benchmark points of the type II seesaw model targeting four possible neutrino mass hypotheses are also probed, with lower bounds between $716\,GeV$ and $761\,GeV$ [34].

3.4 Invisible Higgs boson

Analyses of the invisible Higgs boson decays target different Higgs boson production modes: VBF production with two forward jets, VH production with $Z \to ee/\mu\mu$ [35], $Z \to b\bar{b}$, $W/Z \to q\bar{q}$, or gluon

Table 1. Comparison table of the 13 TeV non-resonant observed (expected) σ/σ_{SM} 95% CL limits for $b\bar{b}\tau\tau$ [37], $b\bar{b}\ell\nu\ell\nu$ [38], $b\bar{b}\gamma\gamma$ [39] and $b\bar{b}b\bar{b}$ [40] channels.

channel	limit	integrated luminosity
$b\bar{b}\tau\tau$	28 (25)	35.9 fb^{-1}
$b\bar{b}\ell\nu\ell\nu$	79 (89)	35.9 fb^{-1}
$b\bar{b}\gamma\gamma$	91 (90)	2.7 fb^{-1}
$b\bar{b}b\bar{b}$	342 (308)	2.3 – 2.7 fb^{-1}

fusion in association with the initial state radiation have been presented. The observed/expected upper limits at 95% CL on $B(H \to invisible)$ of 0.24/0.23 for the combination of 7, 8 and 13 TeV with the 2015 dataset measurements, then 0.40/0.42 for the $Z(ll) + MET$ analysis and 0.53/0.40 for the monojet and $V(jj)$ analyses updated with full 2016 dataset [36] have been shown in this report.

3.5 Di-Higgs boson

In CMS, both di-Higgs boson productions, resonant and non-resonant, have been investigated. Many BSM models predict resonant X→ HH production, while non-resonant SM di-Higgs boson production provides a crucial test of electroweak symmetry breaking. The results of the following five searches for the Higgs boson pair production in proton-proton collisions at $\sqrt{s} = 13\,TeV$ have been reported. The search for pair production of Higgs bosons in the two τ leptons and two bottom quarks final state [37], the search for resonant and non-resonant Higgs boson pair production in the $b\bar{b}\ell\nu\ell\nu$ final state [38] and the search for $H(b\bar{b})H(\gamma\gamma)$ decays [39], besides a pair of leptons or photons in the final state, require a decay of $H \to b\bar{b}$ to keep branching fraction higher. The highest branching fraction is obtained in the search for resonant pair production of Higgs bosons decaying to two bottom quark-antiquark pairs [41] and the search for heavy resonances decaying to a pair of Higgs bosons in the four b quark final state [42], where the first one targets low mass resonances between 260 and 1200 GeV, and the other one targets the masses between 800 and 3000 GeV.

The current status of the results on 13 TeV non-resonant observed (expected) σ/σ_{SM} 95% CL limits (where [37] and [38] have been updated with the full 2016 datasets) is shown in the Tab. 1. A comparison between the summary of the Run-II $\sigma(gg \to X) \times B(X \to HH)$ 95% CL upper limits from the five resonant di-Higgs boson searches [43] and the results of the corresponding Run I analyses has been presented in this report.

4 Higgs Boson Searches Prospects

This section is related to the updates on projections of physics reach with the upgraded CMS detector for HL-LHC [44]. Selected measurements in Higgs boson physics, highlighting the performance of the planned upgrades of the CMS detector, have been presented. Public results are extrapolated to larger data sets 300 and 3000 fb^{-1}. In order to summarize the future physics potential of the CMS detector at the HL-LHC, extrapolations have been made under different uncertainty scenarios: S1, S1+, S2 and S2+, where S1 denotes the scenario with all systematic uncertainties kept constant with integrated luminosity, and S2 denotes theoretical uncertainties scaled down by a factor 1/2 and experimental systematic uncertainties scaled down by the square root of the integrated luminosity (until they reach a defined lower limit based on estimates of the achievable accuracy with the upgraded detector). The "+" denotes taking into account the effects of higher pileup conditions and detector upgrades on the future CMS performance, instead of leaving the performance of the CMS detector unchanged with respect to the reference analysis. In the following text the future prospects for the SM Higgs boson production in the $H \to \gamma\gamma$ and $H \to ZZ \to 4\ell$ analysis channels have been described.

According to the measurements of Higgs boson properties using the $H \to \gamma\gamma$ decay, with 12.9 fb^{-1} of data collected in 2016 [45], the following projections from the analysis have been made: the signal strength (inclusively and per production mode, presented in this report), Higgs boson coupling strength modifiers and the fiducial cross section measurement [44]. Projected symmetrized uncertainties for the $H \to \gamma\gamma$ signal strength relative to the standard model ($\mu_{\gamma\gamma}$), inclusively and per production mode (VBF, $t\bar{t}H$ and ggH) are shown. No projections for the VH signal strength is given because the reference analysis [45] was not sensitive to it. For the 3000 fb^{-1} case the effect of the high pileup conditions of the HL-LHC has been taken into account as degradations to the photon identification efficiency and vertex identification efficiency. In each case the uncertainty for the inclusive signal strength is shown split into statistical uncertainties, experimental systematic uncertainties and theoretical systematic uncertainties component. In S2 and S2+, the experimental uncertainty on the integrated luminosity is reduced to 1.5%.

According to the measurements of Higgs boson properties using $H \to ZZ \to 4\ell$ with 12.9 fb^{-1} of data collected in 2016 [46], the following projections from the analysis have been made: the signal strength per production mode (presented in this report), differential cross section measurement for $p_T(H)$ and the constraints on anomalous $H \to ZZ$ couplings [44]. The projected 68% CL uncertainties on the Higgs boson signal strength for different production modes at 300 and 3000 fb^{-1}, with S1 and S2 have been presented. In S2, the experimental uncertainties on the integrated luminosity and the lepton identification efficiency are reduced to 1.5% and 1% per lepton, respectively. The projections for 3000 fb^{-1} use different lepton efficiencies and misidentification rates to account for the higher pileup at the HL-LHC.

The uncertainties are in general dominated by statistical component, resulting in 3-4 times lower total uncertainties with respect to the ones from the Run I measurements for 300 fb^{-1}, and 5-10 times lower for the high luminosity measurements, with 3000 fb^{-1}. Further information on the projections made for top physics, heavy flavor physics, searches for dark matter and new heavy particles can be found in the CMS Detector Performance Summary [44].

5 Summary

As a conclusion from the presented results of the SM Higgs boson searches it follows that the precision of the $H \to \gamma\gamma$ and $H \to ZZ \to 4\ell$ analyses has been improved and the SM process $H \to \tau\tau$ has been observed. The sensitivity of the search for the production of the Higgs boson associated to top quarks has been increased and no significant deviations from the SM, neither extra Higgs bosons, nor a sign of the BSM physics have been yet found. As for the outlook, the updates of several important analyses with the full 2016 dataset are expected, looking forward for around 100 fb^{-1} of data to be collected by the end of the LHC Run II.

References

[1] CMS Collaboration, JINST 3 S08004 (2008)
[2] CMS Collaboration, Eur. Phys. J. **C75**, 5, 212 (2015)
[3] CMS and ATLAS Collaboration, JHEP, **08**, 045 (2016)
[4] CMS Collaboration, CMS-PAS-HIG-16-040 (2017), http://cds.cern.ch/record/2264515
[5] CMS Collaboration, CMS-PAS-HIG-17-015 (2017), http://cds.cern.ch/record/2257530
[6] LHC Higgs Cross Section Working Group, arXiv:1610.07922 (2016)

[7] CMS Collaboration, CMS-PAS-HIG-16-041 (2017), http://cds.cern.ch/record/2272260
[8] CMS Collaboration, JHEP 1604 (2016) 005
[9] Alioli, Simone and Nason, Paolo and Oleari, Carlo and Re, Emanuele, JHEP, **07**, 060 (2008)
[10] Nason, Paolo, JHEP, **11**, 040 (2004)
[11] Frixione, Stefano and Nason, Paolo and Oleari, Carlo, JHEP, **11**, 070 (2007)
[12] ATLAS and CMS Collaborations, Phys.Rev.Lett. 114 (2015) 191803
[13] CMS Collaboration, CMS-PAS-HIG-16-043 (2017), http://cds.cern.ch/record/2276465
[14] CMS Collaboration, CMS-PAS-HIG-16-038 (2016), http://cds.cern.ch/record/2231510
[15] CMS Collaboration, CMS-PAS-HIG-17-003 (2017), http://cds.cern.ch/record/2257067
[16] CMS Collaboration, CMS-PAS-HIG-17-004 (2017), http://cds.cern.ch/record/2256103
[17] CMS Collaboration, CMS-PAS-HIG-16-019 (2016), http://cds.cern.ch/record/2204925
[18] CMS Collaboration, CMS-PAS-HIG-17-005 (2017), http://cds.cern.ch/record/2264553
[19] CMS Collaboration, CMS-PAS-HIG-16-003 (2016), http://cds.cern.ch/record/2160154
[20] CMS Collaboration, CMS-PAS-HIG-16-037 (2016), http://cds.cern.ch/record/2231507
[21] Olive, K. A. and others, Chin. Phys., **C38**, 090001 (2014)
[22] Raidal, M. and others, Eur. Phys. J., **C57**, 13-182 (2008)
[23] Belanger, G. and Boudjema, F. and Cottrant, A. and Godbole, R. M. and Semenov, A., Phys. Lett., **B519**, 93-102 (2001)
[24] Giudice, Gian F. and Rattazzi, Riccardo and Wells, James D., Nucl. Phys., **B595**, 250-276 (2001)
[25] Shrock, Robert E. and Suzuki, Mahiko, Phys. Lett., **110B**, 250 (1982)
[26] Baek, Seungwon and Ko, P. and Park, Wan-Il and Senaha, Eibun, JHEP, **05**, 036 (2013)
[27] Servant, Geraldine and Tulin, Sean, Phys. Rev. Lett., **111**, 151601 (2013)
[28] Binoth, T. and van der Bij, J. J., Z. Phys., **C75**, 17-25 (1997)
[29] Branco, G. C. and Ferreira, P. M. and Lavoura, L. and Rebelo, M. N. and Sher, Marc and Silva, Joao P., Phys. Rept., **516**, 1-102 (2012)
[30] Fayet, Pierre, **B90**, 104-124 (1975)
[31] Agashe, Kaustubh and Davoudiasl, Hooman and Perez, Gilad and Soni, Amarjit, Phys. Rev., **D76**, 036006 (2007)
[32] Goertz, Florian and Papaefstathiou, Andreas and Yang, Li Lin and Zurita, Jose, JHEP, **04**, 167 (2015)
[33] CMS Collaboration, CMS-PAS-HIG-16-031 (2016), http://cds.cern.ch/record/2223865
[34] CMS Collaboration, CMS-PAS-HIG-16-036 (2017), http://cds.cern.ch/record/2242956
[35] CMS Collaboration, CMS-PAS-EXO-16-052 (2017), http://cds.cern.ch/record/2264690
[36] CMS Collaboration, JHEP, **02**, 135 (2017)
[37] CMS Collaboration, CMS-PAS-HIG-17-002 (2017), http://cds.cern.ch/record/2273798
[38] CMS Collaboration, CMS-PAS-HIG-17-006 (2017), http://cds.cern.ch/record/2278570
[39] CMS Collaboration, CMS-PAS-HIG-16-032 (2016), http://cds.cern.ch/record/2207960
[40] CMS Collaboration, CMS-PAS-HIG-16-026 (2017), http://cds.cern.ch/record/2209572
[41] CMS Collaboration, CMS-PAS-HIG-16-002 (2016), http://cds.cern.ch/record/2141024
[42] CMS Collaboration, CMS-PAS-B2G-16-026 (2017), http://cds.cern.ch/record/2264684
[43] Higgs PAG Summary Plots, http://twiki.cern.ch/twiki/bin/view/CMSPublic/SummaryResultsHIG
[44] CMS Collaboration, CMS-DP-2016-064 (2016), http://cds.cern.ch/record/2221747
[45] CMS Collaboration, CMS-PAS-HIG-16-020 (2016), http://cds.cern.ch/record/2205275
[46] CMS Collaboration, CMS-PAS-HIG-16-033 (2016), http://cds.cern.ch/record/2204926

EPJ Web of Conferences **158**, 01002 (2017) DOI: 10.1051/epjconf/201715801002
QFTHEP 2017

Top physics at the LHC *

Roberto Tenchini[1],**

[1] *INFN Sezione di Pisa, Largo Bruno Pontecorvo 3, 56127 Pisa, Italy*

Abstract. The status of top physics at the Large Hadron Collider is discussed, with particular attention on measurements of top production cross sections, including associate production of top quarks and electroweak bosons, and on measurements of the top mass.

1 Introduction

After seven years since the first beams, the Large Hadron Collider (LHC) can be called without any doubts a Top Factory. In year 2016, providing proton proton collisions at a centre-of-mass energy of 13 TeV, LHC has delivered about 30 millions top pairs to each general purpose experiment, ATLAS and CMS, corresponding to an integrated luminosity of about 36 fb^{-1}, one order of magnitude higher than the amount of data delivered in the previous year at the same energy. Data collected at lower centre-of-mass energies correspond to ≈ 5 fb^{-1} at 7 TeV (year 2010) and ≈ 20 fb^{-1} at 8 TeV (years 2011 and 2012). This impressive wealth of data has been used by the ATLAS and CMS experiments to study in detail top production and decay properties, as documented by about 150 journal publications [1, 2].

The most striking characteristics of top is its mass, 35 times larger than the mass of the other third generation quark. Thanks to the non-decoupling properties of the electroweak interactions, such a large mass makes the top quark the main actor in electroweak radiative corrections, which are very sensitive to the actual value of the top mass. In addition, because of its mass, top lifetime is extremely short ($\sim 10^{-25}$ s), actually shorter than the typical time involving strong interactions, allowing top properties to be studied as if it were a free quark.

The large mass also imply a large top-Higgs Yukawa coupling. Actually, measurements of top quark parameters and properties have an important impact to Higgs boson physics. The value of the top mass regulates the quartic Higgs couplings and, as a consequence, the stability of the electroweak vacuum: making a precise measurement of the top mass is of great importance for the Higgs sector. Top-pair production occurs at the LHC mainly through gluon-gluon fusion, the main production channel for the Higgs boson itself: measurements of inclusive and differential cross section allow improved determination of the gluon PDF to be performed and, as a consequence, improved understanding of Higgs production cross sections. Top is also playing an important role in searches for physics beyond the standard model; as an example searches for stop in supersymmetric scenarios or searches for vector-like quarks in composite Higgs models are often involving top-quarks in the final state.

**On behalf of the ATLAS and CMS Collaborations*

**e-mail: Roberto.Tenchini@cern.ch

In these short conference proceedings measurements for cross sections for various top-production mechanisms at the LHC are concisely reviewed. Associate production of top quarks and electroweak bosons is also mentioned. A discussion on the status of measurements of the top mass completes the proceedings, which are closed by a short outlook section. Top physics at LHC includes many additional subjects, which were not covered in the talk and corresponding proceedings. Among them, it is important to mention top decay measurements, such as measurement of the $t \to Wb$ branching fraction, searches for FCNC decays, studies of the structure of Wtb vertex including measurements of the W helicity and measurements of top polarisation and top-pair spin correlations. Other not covered subjects include the many searches for physics beyond the standard model with top quarks.

2 Top production cross sections

Top pair production at LHC occurs via gluon-gluon fusion and quark-antiquark annihilation. The first production mechanism is by far dominating at LHC centre-of-mass energies, contributing to about 90% of the inclusive cross section at 13 TeV. The cross section is computed at NNLO+NNLL [3] with a typical uncertainty of 3% due to scale variation and 2-3% related to PDF.

The top quark decays almost exclusively to Wb, other decays mode are suppressed by the values of the CKM elements. It is convenient to classify top decays with the related W decays, so top-pairs are classified in di-lepton, lepton+jet and all-hadron final states. Because of the different final state topology, decays to tau leptons are usually treated separately at hadron colliders. Di-lepton final states with electrons or muons (6% of the decays) are very clean, with main background originating from Drell-Yan production, usually suppressed by rejecting same-sign same-flavour leptons with invariant mass consistent with a Z boson decay. Muon+jet and electron+jet final states (34% of events in total) have also distinct signatures, but larger background mostly originating from W+jet events. The background can be suppressed by using b-tagging techniques and measuring the number of b-tagged or non-b-tagged jets. The all-hadron final state has the larger branching ratio (46%), but suffers from the large multi-jet generic QCD background. Finally, decays to tau leptons (14%) require tau reconstruction and are typically affected by W+jet and other top decay-mode backgrounds.

The most precise determinations of the top-pair production cross section are based on the analysis of di-lepton and lepton+jets top decays. The electron+muon di-lepton channel is particularly clean and a precise measurement of the cross section can be achieved without the use of b-tagging techniques, which are typically used to validate the methods employed. The lepton+jet channel requires the use of control regions where background are measured, typically in a common fit of backgrounds and signal, together with the main selection. Results of top-pairs cross section measurements at the centre-of-mass of 13 TeV are given in Fig. 1. The most precise measurements reach a precision comparable to the uncertainty on the theoretical cross section. All measurements are affected by knowledge of the absolute integrated luminosity, which can be known at LHC with a precision of 2-3% depending on the data-taking period. The cross section measured at various centre-of-mass energies is shown in Fig. 2: the measurements are in agreement with the expectations from NNLO+NNLL calculations.

Precise measurements of differential cross sections are also available, thanks to the high-statistics data provided by LHC. In order to facilitate comparisons with theory, and in particular to allow comparison with fixed-order calculations, yields are measured as a function of the observable of interest and then corrected by efficiency and bin-migration effects by means of unfolding procedures. As an example the differential production cross sections, as measured by the ATLAS and CMS experiments, are shown in Fig. 3 as a function of the transverse momentum of the top quark (left panel) and the invariant mass of the top-quark pair (right panel). The discrepancies observed in the past in

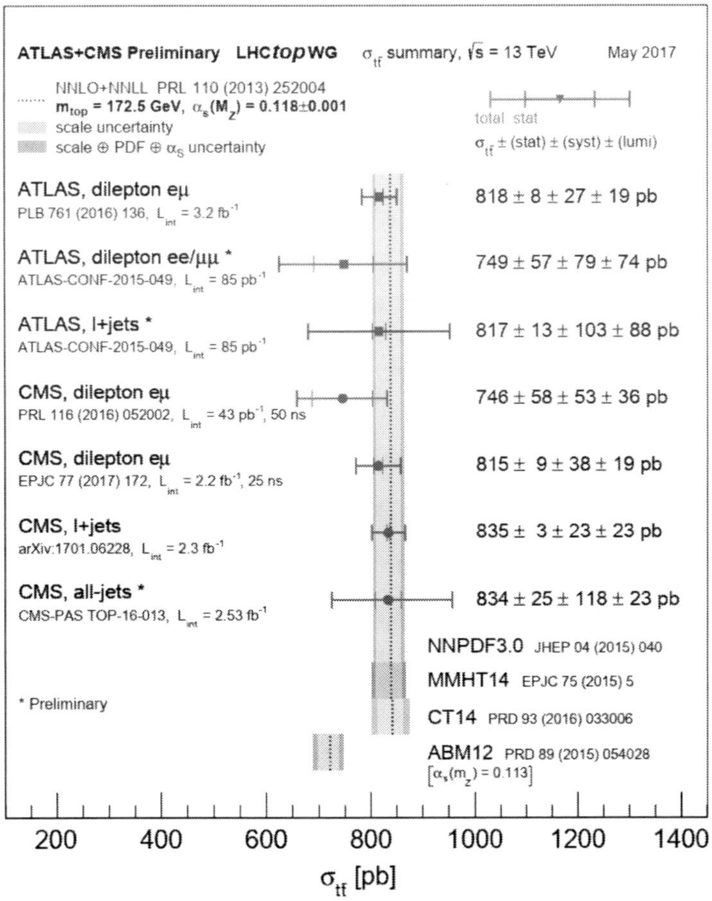

Figure 1. Summary of measurements of the top-pair production cross-section at 13 TeV compared to the exact NNLO QCD calculation complemented with NNLL resummation [3]. The theory band represents uncertainties due to renormalisation and factorisation scale, parton density functions and the strong coupling. The measurements and the theory calculation are quoted at m_{top}=172.5 GeV.

the first distribution, with data typically showing a softer spectrum with respect to NLO spectra, have essentially disappeared with the use recent NNLO+NNLL calculations.

Top quarks can be produced singly at hadron colliders via electroweak processes. The higher cross section is provided by t-channel production, where top is produced by a virtual W boson in presence of a spectator light quark. This channel has a specific signature related to the presence of a forward jet, originating from the spectator quark, which can be used in the selection of events. Associate top+W production has a lower cross section and a final state, which show similarities with top-pair production, from which is separated by counting the number of b-jets. The definition of this channel and the related signal requires also some care, since at NLO and higher orders it interferes with top-pair production. Top-quark+b-quark production in s-channel through a virtual W has an even lower cross section requiring a careful assessment of the top-pair-dominated background and the use

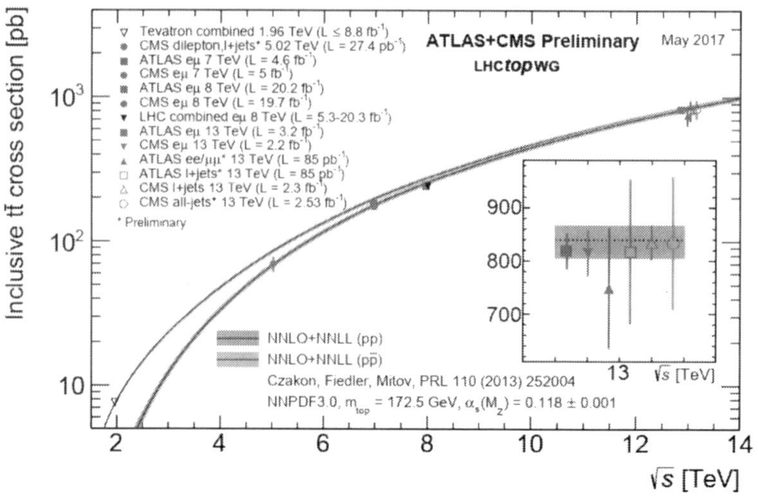

Figure 2. Summary of LHC measurements of the top-pair production cross-section as a function of the centre-of-mass energy compared to the NNLO QCD calculation complemented with NNLL resummation [3]. Results from $p\bar{p}$ collisions at Tevatron are also shown. The theory band represents uncertainties due to renormalisation and factorisation scale, parton density functions and the strong coupling. The measurements and the theory calculation are quoted at m_{top}=172.5 GeV. Measurements made at the same centre-of-mass energy are slightly offset for clarity.

of multivariate techniques. The measured inclusive cross sections are shown in Fig. 4 and compared with the expectations from recent calculations. Agreement between data and theory is observed also in the single top sector. Measurements of differential cross sections using single top data are also appearing [1, 2].

3 Top and electroweak bosons

Associate production of top quarks and electroweak bosons (vector or scalar) is of great interest, but production cross sections are typically lower than a picobarn. In order to separate these low signals from backgrounds, in particular from $t\bar{t}$ events, multilepton final states are often used; when this is done background originating from lepton misidentification needs to be treated carefully and measured with data using control samples.

Production of $t\bar{t}Z$ has been observed by both ATLAS and CMS experiments [4, 5]. Events with same-sign dileptons, three leptons and four leptons have been used in the measurements. In the analyses a common measurement of $t\bar{t}Z$ and $t\bar{t}W$ production is performed. Results based on data collected at the centre-of-mass energy of 8 TeV are shown in Fig. 5. The $t\bar{t}Z$ final state gives access to a direct measurement of the top-Z couplings, measurements have been used to constrain the Wilson coefficients of dim-6 operators. Associate production of single top and Z boson is also interesting, giving information complementary to $t\bar{t}Z$ and additional sensitivity to, e.g., FCNC top couplings. (Evidence for tqZ production has been shown by ATLAS and CMS after this conference.) Production of top-quark pairs and a direct photon ($t\bar{t}\gamma$ events) has also been measured at LHC [6, 7]; the differential cross section, as a function of the photon transverse momentum, is in agreement with standard model expec-

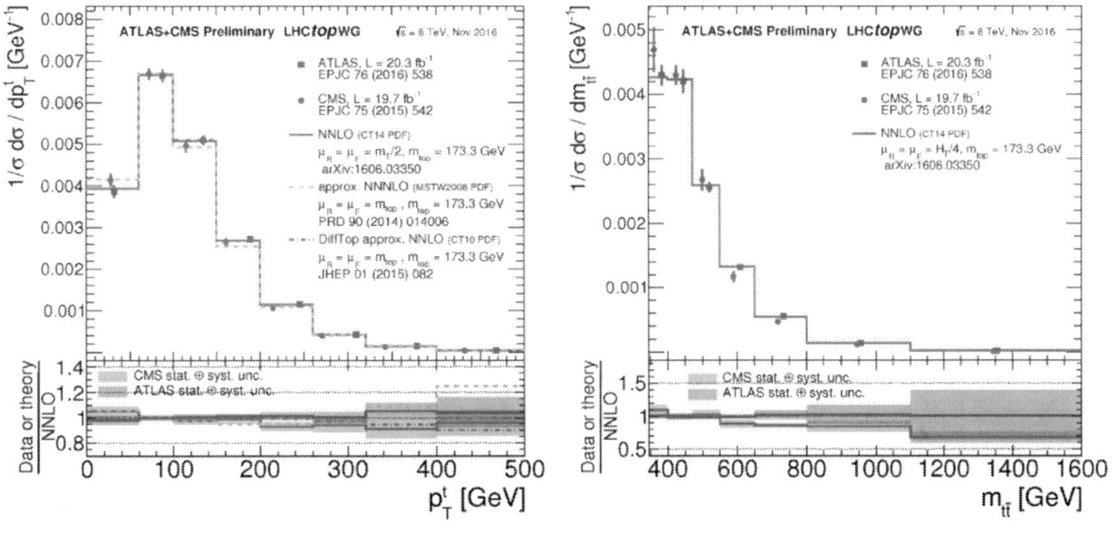

Figure 3. Full phase-space normalised differential ttbar cross-section as a function of the transverse momentum of the top quark (left panel) and the invariant mass of the top-quark pair (right panel). The CMS and ATLAS results are compared to NNLO and approximate NNLO calculations. The values for the top-quark mass (m_{top}), the renormalisation (μ_R) and factorisation (μ_F) scales, and the choice of the PDF set used in each calculation are provided. The variable m_T is defined as the square root of the sum of the squares of top-quark mass and the transverse momentum of the top quark. Both the CMS and ATLAS measurements are performed assuming a top-quark mass value of 172.5 GeV. The shaded bands show the total uncertainty on the data measurements in each bin. The lower panel shows the ratio of the data measurements and the approximate NNLO calculations to the full NNLO calculation.

tations. The $t\bar{t}\gamma$ process provides sensitivity to anomalous top electric dipole moments and deviations could indicate the presence of excited top states.

The $t\bar{t}H$ process allows a direct measurement of the top Higgs-Yukawa coupling to be performed. Once again multilepton final states, taking advantage of Higgs boson decays to Z or W pairs, give the higher sensitivity; other important channels are based on $H \to \gamma\gamma$ and $H \to b\bar{b}$ decays. The combination of ATLAS and CMS measurements in various channels, performed at the centre-of-mass energy of 8 TeV, gave evidence of this channel at 4.4σ level, with a moderate excess with respect to SM expectations [8]. First results at 13 TeV are available and a significant determination of the top Higgs-Yukawa coupling is expected soon. Associate production of single top and a Higgs boson is strongly sensitive to the sign of the Yukawa coupling (while $t\bar{t}H$ only depends on its square). Because of the interference between the two main tree-level diagrams, within the standard model the tH expected cross section is only 18 fb, however it can grow by more than one order of magnitude if the sign of the coupling is flipped: a detection of this process in the present data would be an evidence for physics beyond the standard model.

4 The top mass

Standard methods for top mass measurements are based on the identification of top decay products and determination of the corresponding invariant mass. The lepton+jet channel provides the best

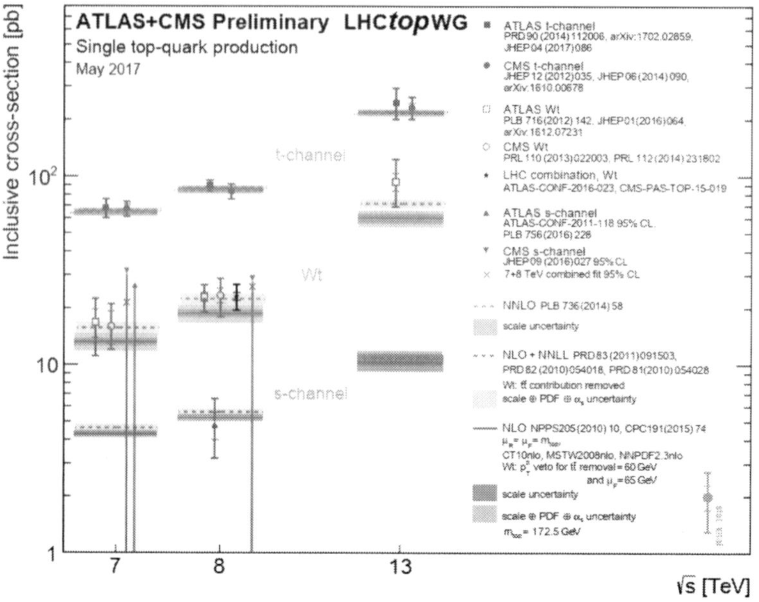

Figure 4. Summary of ATLAS and CMS measurements of the single top production cross-sections in various channels as a function of the centre-of-mass energy. The measurements are compared to theoretical calculations based on: NLO QCD, NLO QCD complemented with NNLL resummation and NNLO QCD (t-channel only).

precision because the only neutrino in the final state can be unambiguously associated to the missing transverse momentum, differently than the dilepton channel. In addition the distinct l+jet signature can be used, together with b-tagging, to keep backgrounds at low levels. B-tagging is also a powerful tool to reduce combinatorial in the association of jets to the top and antitop candidates.

Selection of lepton+jet events is typically based on requiring the presence of a single electron or single muon passing a transverse momentum threshold (a typical value for muons in 8 TeV analysis is 24 GeV), by requiring the presence of four reconstructed jets of which one or two are tagged as b-jets, and of significant missing transverse momentum. With such a simple selection the fraction of non-top events is reduced typically at the 5% level. Events are then subject to a kinematic fit procedure, where momentum conservation is imposed in the transverse plane. A major systematic uncertainty in the evaluation of the mass is related to the knowledge of jet energies; it is convenient to include jet energy scale (JES) correction factors in the kinematic fit and impose consistency of the invariant mass of the two light jets forming the hadronically decaying W boson candidate with the precisely known W mass. Dedicated b-jet JES correction factors can also be included by using variables sensitive to the b-jets/W-bosons global transverse momentum balance. Typically the mass of the top and antitop candidates are considered equal in the fit (this constraint is released for CPT invariance tests). The goodness of the kinematic fit is a powerful tool to reduce the number of combinations, since wrong jet-to-top-candidate associations are typically showing a poor χ^2.

Events reconstructed with these methods, or more precisely combinations reconstructed with these methods, can be used to build distributions providing sensitivity to the top mass (for example the distribution of the top mass parameter as measured by kinematic fitting, which is typically peaked around 170-175 GeV). The distributions however are distorted by many effects, as detector resolution, phase space constraints, background, pileup of other interactions in the same event, selection cuts,

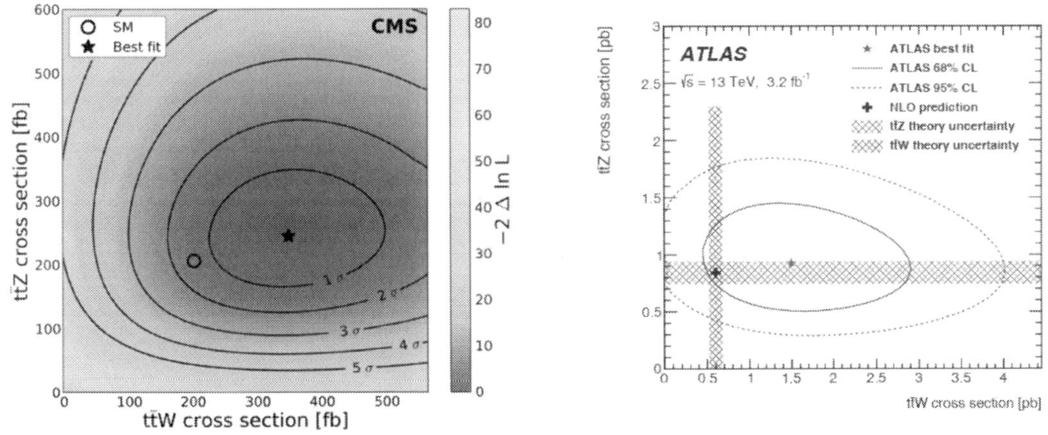

Figure 5. Measurements of the $t\bar{t}W$ and $t\bar{t}Z$ cross sections by the CMS and ATLAS experiments. Profile likelihoods as a function of $\sigma(t\bar{t}W)$ and $\sigma(t\bar{t}Z)$. Lines denote standard deviation (left) or CL contours (right).

etc., requiring a detailed Monte Carlo simulation to describe the distributions observed with data. The top mass and other parameters (such JES) can then be measured by generating simulations with different input parameters and use them to fit the data (for instance generating Monte Carlo *templates* for the distribution of interest). Results obtained with these methods at LHC and Tevatron, and their combinations, are displayed in Fig. 6. Measurements performed with di-lepton and all-hadron events, based on similar techinques, are also included.

Measurements obtained with standard methods, i.e. direct reconstruction of top decay products, provide high-precision: the combination of recent LHC measurements reaches a total uncertainty of only 0.3%. However, the interpretation of the measurements has caused an intense debate. As the standard methods are related to the invariant mass of top decay products, it is natural to associate the measured mass to the top pole mass. For a coloured particle as top, the pole mass has an intrinsic ambiguity related to renormalisation, the related uncertainty has been recently computed and it is in the 100-200 MeV range [9, 10]. More in general and beyond this purely technical issue top decays products cannot be unambiguously identified because top-quark colour charge is transferred to the b-quark in the $t \rightarrow Wb$ decay: in order to have colourless hadrons at the end of the decay chain colour needs to be transferred to the rest of the event (*colour reconnection effects*). Colour reconnection is a non-perturbative effect and cannot be calculated, its impact on the measurement is expected to be of the order of $\Lambda_{QCD} \approx 200$ MeV. In the measurements, uncertainties related to this effect are computed with heuristic Monte Carlo models, parameterised with multiplicity studies on generic proton-proton collisions.

Given the considerations of the previous paragraph, it is clear that a careful assessment of potential bias on present measurements, due to colour reconnection effects, is strongly desirable. Related studies are ongoing at LHC, as an example the top mass can be measured as a function of the top transverse momentum, of the minimal distance of two jets from the same or different top decay and of many other relevant observables. No bias has been observed with respect to the behaviour expected from simulation, with the current statistics.

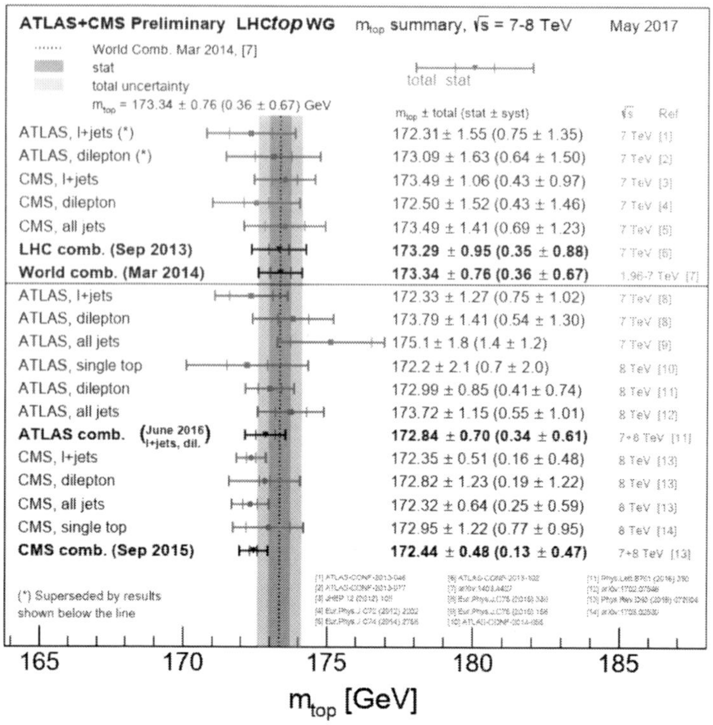

Figure 6. Summary of the ATLAS and CMS standard m_{top} measurements. The results are compared with the LHC and Tevatron+LHC m_{top} combinations. For each measurement, the statistical uncertainty includes the jet scale factor and b-jet scale factor contributions (when applicable), while the sum of the remaining systematic uncertainties is reported separately. The jet scale factors contributions are statistical in nature and apply to analyses performing in-situ (top quark pair based) jet energy calibration procedures. The results below the line are results produced after the LHC and Tevatron+LHC combinations were performed.

It also relevant to measure the top mass with alternative methods, based on very different techniques or based on events with a different colour structure, to check if results deviate with respect to standard methods. An important determination of the mass is based on the inclusive $t\bar{t}$ cross section, which is very sensitive to the top mass (and to the strong coupling constant, α_s). This interpretation of the cross section requires a careful assessment of the PDF uncertainties and it is limited by the knowledge of the accelerator luminosity. A measured based on the standard technique, but using events with a different colour structure is the top mass measurement with a sample enriched in single top events. An interesting measurement can be made by using dilepton events, using the dependence of kinematic endpoints on the top mass that can be analytically calculated. Other examples are measurements based on quasi-exclusive decays (as top decays to J/ψ) or on the b-jet or lepton energies measured in the reference frame where top is at rest (or in a reference frame close to it). Examples of these measurements are shown in Fig. 7: results are consistent with the standard measurements, albeit with larger uncertainties. These methods are very interesting in view of the large statistics expected from LHC in future years, since for many of these results the statistical errors (or systematic errors reducible with more statistics) are currently dominating.

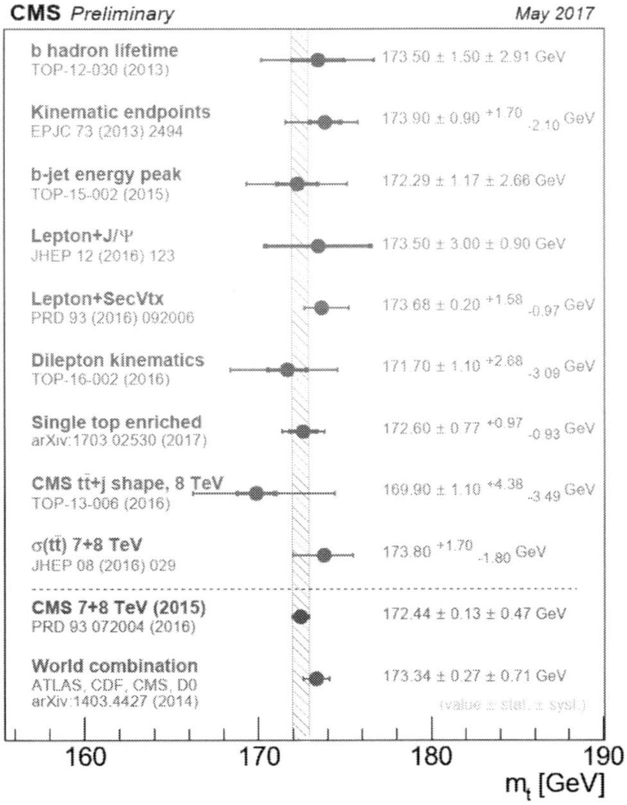

Figure 7. Summary of alternative measurements of the top mass, as made by the CMS experiment using data collected at 7 and 8 TeV centre-of-mass energies.

5 Outlook

In 2017 and 2018 additional 45 fb^{-1} per year at a centre-of-mass energy of 13 TeV are expected from the LHC. This additional statistics will complement the existing data and allow a detailed study of top production and decay to be performed. The first direct measurement of the top-Higgs Yukakwa coupling is expected, using these data. The accelerator will then undergo two years of shutdown (LS2) to allow upgrades of the experiments and of the collider to be installed. After LS2, 300 fb^{-1} at a centre-of-mass energy of 14 TeV are foreseen, followed by several ab^{-1} at the same centre-of-mass energy in the HL-LHC phase. The era of high-precision top physics has just started.

Acknowledgements

I would like to thank the conference organisers for the excellent organisation in a historical location. I am grateful to the LHC*top*WG working group for providing combinations of cross-section and mass results. Special thanks to Frederic Deliot, Rebeca Gonzalez-Suarez, Martijn Mulders, Paolo Nason, Reinhard Schwienhorst and Pedro Silva for providing material and for enlightening discussions.

References

[1] https://twiki.cern.ch/twiki/bin/view/AtlasPublic
[2] http://cms-results.web.cern.ch/cms-results/public-results/publications/
[3] M. Czakon, P. Fiedler, A. Mitov, Phys. Rev. Lett. **110**, 252004 (2013)
[4] ATLAS Collaboration, J. High Energy Phys. **11**, 172 (2015)
[5] CMS Collaboration, J. High Energy Phys. **01**, 096 (2016)
[6] ATLAS Collaboration, arXiv:1706.03046
[7] CMS Collaboration, arXiv:1706.08128
[8] ATLAS and CMS Collaborations, J. High Energy Phys. **08**, 045 (2016)
[9] M. Beneke, P. Marquard, P. Nason, M. Steinhauser, arXiv:1605.03609
[10] C. Lepenik, proceedings of this conference.

Nucleus-nucleus interactions in very-high-energy cosmic ray experiments

Anatoly Petrukhin* and *Aleksei* Bogdanov

National Research Nuclear University MEPhI (Moscow Engineering Physics Institute), 115409, Russian Federation, Moscow, Kashirskoe shosse, 31

Abstract. A review of unusual results of experiments in cosmic rays which cannot be explained in the frame of the existing models of hadron interactions is presented. Requirements to features of a new model which are necessary for explanation of all observed unusual events and phenomena are formulated. A model of hadron interactions with production of QGM blobs with large orbital momentum is considered. Its possibilities for explanation of various unusual events and phenomena are discussed.

1 Introduction

In collider physics, overwhelming majority of investigations (both experimental and theoretical) are connected with proton-proton interactions. It is clear that these interactions are more simple as for experimental analysis so for theoretical description then nucleus-nucleus interactions. In the last case various types of collective interactions of nucleons or quarks and gluons may appear.

In cosmic rays it is impossible to investigate proton-proton interactions since targets in the Earth's atmosphere are the nuclei of nitrogen and oxygen atoms. Additionally, most parts (about 60%) of cosmic rays are nuclei of various elements (up to iron) which come from cosmic space.

Above the energy of about 10 GeV primary cosmic ray (PCR) energy spectrum is described by a simple formula $dN/dE \sim E^{-\gamma}$, where $\gamma \approx$ const ≈ 2.7. In cosmic ray investigations, particles with energies up to about 10^{20} eV and some features in the behavior of the energy spectrum were observed. Possible classification of CR energy intervals is given in Table 1. Energy in the center-of-mass system (CMS) is usually calculated for nucleon-nucleon system $\sqrt{s_{NN}}$ irrespective of the type of interactions: nucleon-nucleon, nucleon-nucleus, nucleus-nucleus.

Composition of cosmic rays, or more exactly the relation between different components of cosmic rays, depends on the energy taken into account: energy per nucleon or energy per nucleus. Relation between various components combined in several groups is given in the Table 2.

* Corresponding author: AAPetrukhin@mephi.ru

Usually, the value <ln A> is used as a parameter of composition. For normal composition, this value is equal to ≈ 1.5. In simulations very often limiting assumptions on the composition are used: pure protons (ln A = 0) and pure iron (ln A = 4).

Table 1. Classification of CR energy intervals.

CR energy, eV	**Term**	**Energy in CMS, TeV**	**Main feature**
$< 10^{15}$	High	< 1.4	–
$10^{15} – 10^{17}$	Very High	1.4 – 14	the knee
$10^{17} – 10^{19}$	Ultra High	14 – 140	the ankle
$> 10^{19}$	Extremely High	> 140	GZK cutoff

Table 2. Relation between various nucleus.

Particles	**Z**	**<A>**	**Energy per nucleon**	**Energy per nucleus**
Protons	1	1	92%	42%
α-particles	2	4	7%	21%
Light nuclei	3-5	10	0.25%	1%
Medium nuclei	6-10	15	0.5%	18%
Heavy nuclei	≥ 11	32	0.25%	18%

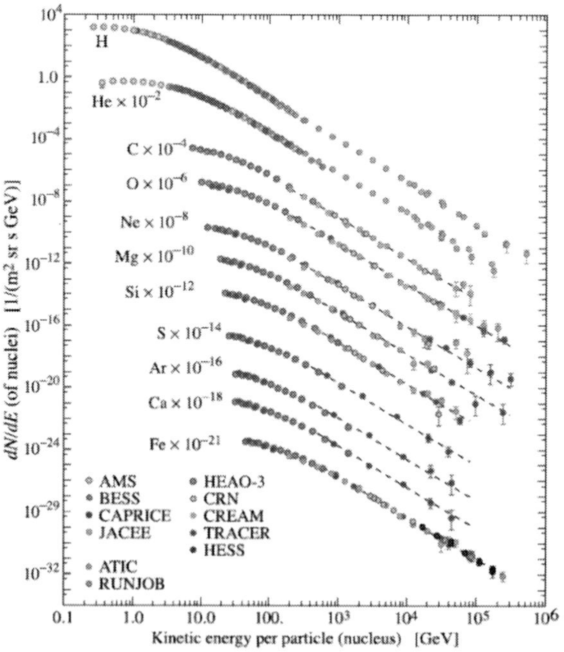

Fig. 1. Differential energy spectra of various cosmic ray nuclei obtained in direct measurements [1].

Investigations of cosmic rays are performed in space, at higher layers of the atmosphere, on the mountains, on the Earth's surface (near sea level), underground (under water and under ice). Outside of the Earth's atmosphere, direct measurements of primary particles (Z and E) by means of satellites are carried out. Taking into account a sharp decrease of CR flux with energy, such investigations can be carried out for energies less than 10^{15} eV only (Fig. 1). In the Earth's atmosphere, probability to detect primary particles is decreased very rapidly with increasing the depth.

Therefore on the ground level secondary particles can be investigated only. The most important phenomena generated by PCRs in the atmosphere are extensive air showers (EAS) which consist of various particles: hadrons, electrons, positrons, photons, muons and neutrinos, and they can reach the sea level starting from the energy near 10^{15} eV. EAS investigations are the only way to study CR energy region above 10^{15} eV.

For EAS investigations, systems of various detectors distributed over large area are applied. In addition to detectors of charged particles (mainly scintillation counters), detectors of different type radiations which are generated by EAS (Cherenkov light, fluorescence, radio, etc) are used.

The scheme of cosmic ray investigations by means of EAS observables is given in Fig. 2 [2].

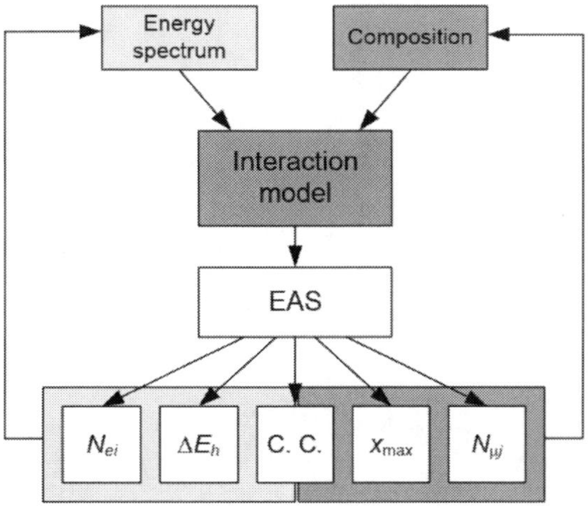

Fig. 2. Existing approach to EAS analysis.

Measured characteristics of EAS are: number of charged particles in each detector (N_{ei}), number of muons in each muon detector ($N_{\mu j}$), energy deposit of EAS core (ΔE_h), cascade curve (C.C., longitudinal profile of EAS development in the atmosphere) and maximum EAS development (X_{max}).

These experimental values are used for recalculation to primary energy spectrum and composition taking into account the interaction models which were checked at accelerator energies.

2 Results of cosmic ray experiments

Numerous experiments with various components of cosmic rays at primary particle energies below 10^{15} eV gave no deviations from a simple picture of CR energy spectrum (Fig. 1) and mass composition (Table 2). Though the results of several experiments showed unusual behavior of some components (especially muons), they gave no serious deviations from an assumed picture.

Above 10^{15} eV the picture is changed drastically. The first unusual phenomenon was observed in 1957 [3]. It was so-called "the break" in spectrum of charged particle number at energy about 3 PeV, later named "the knee". There can be two reasons of its origin: cosmophysical (changes in the energy spectrum and/or the composition of PCR, and nuclear physical (changes in the interaction model). After first discussions, most part of physicists agreed with the first possibility taking into account that it is in a good agreement with the Galactic model of cosmic ray origin.

At the further advancement to higher energies, other phenomena were observed: the second knee, the ankle, etc. All these changes in EAS characteristics were explained by changes in PCR energy spectrum and composition, results are shown in Fig. 3 and Fig. 4.

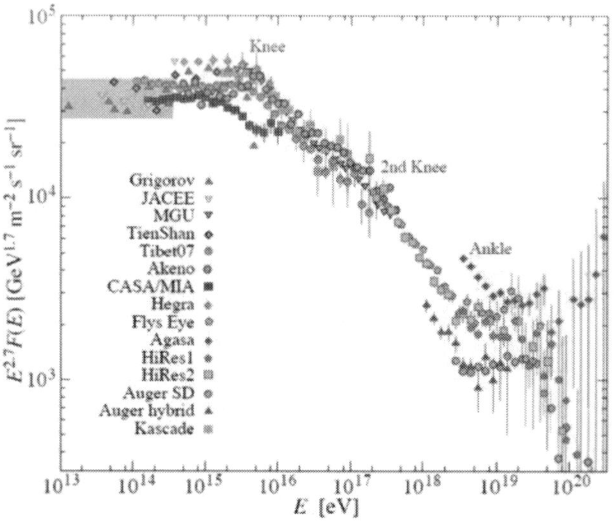

Fig. 3. All-particle energy spectrum of primary cosmic rays [1].

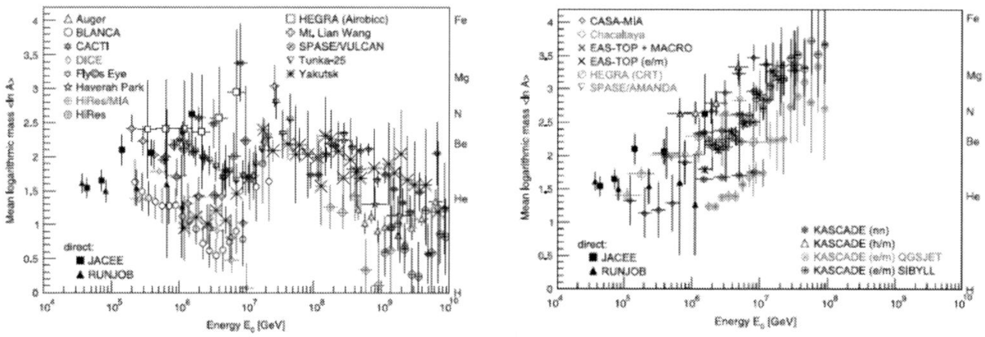

Fig. 4. Mean logarithmic mass of cosmic rays derived from the depth of the shower maximum (left), from the measurements of electrons, muons, and hadrons at ground level (right) [4].

Next serious deviations from the accepted picture of cosmic ray origin and existing model of their interactions were so-called unusual events which were detected at PeV energies and above in various experiments [5-7]. The list of such unusual events and phenomena includes:

- alignment (detection of three or more cascades which are located in a practically straight line; the probability of occasional generation of such events is negligibly small) – see Fig. 5;

- penetrating cascades (these cascades exhibit unusually slow absorption and correspondingly abnormally large attenuation length in comparison with the normal hadron cascade curve) – see Fig. 6;
- Centauros (observation of families with several tens of hadron cascades and practically without electromagnetic cascades) and Anti-Centauros (with opposite situation); these phenomena give evidence for the violation of isotopic invariance in secondary particle production;
- long-flying particles (these particles can penetrate a large thickness of matter and interact with the production of cascades);
- very young and very old EAS (with anomalous values of shower age), etc.

Possible reasons of their appearance were discussed in many papers, and various ideas of their explanation were proposed, see e.g. [8-12]. But no idea for their explanation from a single point of view was proposed.

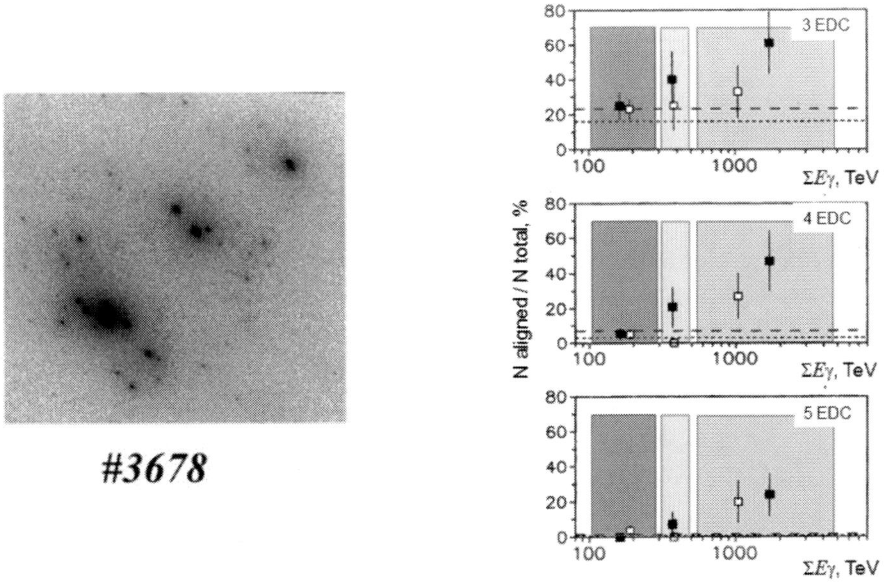

Fig. 5. Example of aligned event detected in Pamir experiment (left). Dependence of the fraction of families with alignment versus total γ-quanta energy of the event (right) [5].

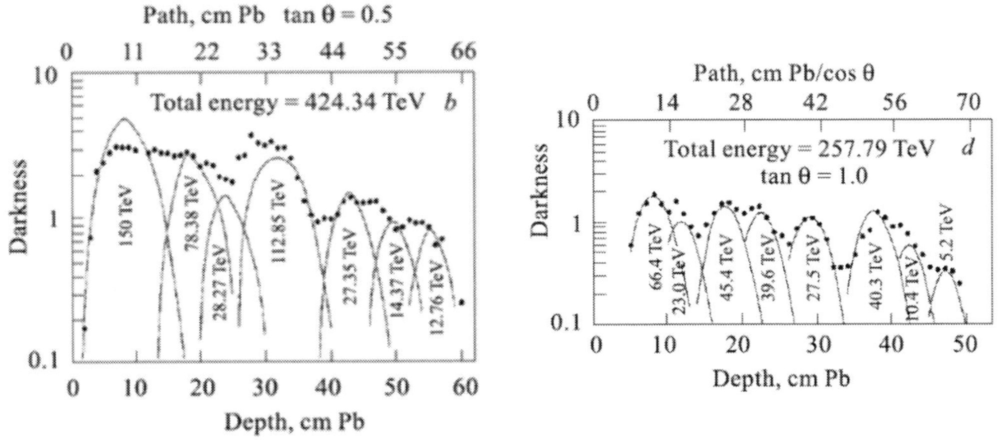

Fig. 6. Examples of events with penetrating cascades. Shower transition curve in spot darkness is shown [6].

3 Cosmic ray "muon puzzle"

During the last about 10 years, a new problem in cosmic ray muon investigations appeared. In several experiments, excess of muon bundles compared to simulations was observed. Firstly this excess was observed in LEP detectors ALEPH and DELPHI at muon multiplicity about 100 particles [13,14] (Fig. 7).

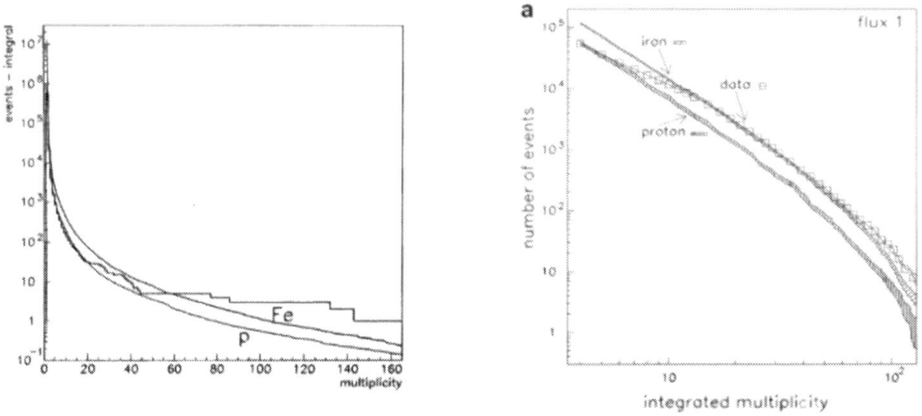

Fig. 7. Integral multiplicity distributions of muons in the TPC of ALEPH detector (left) [13] and in HAB of DELPHI detector (right) [14] compared to CORSIKA simulations for p and Fe.

However in these experiments there was no possibility to study the dependence of this excess on primary particle energies. This information was obtained in experiment NEVOD-DECOR in which muon bundles were measured at various zenith angles (Fig. 8 [15]).

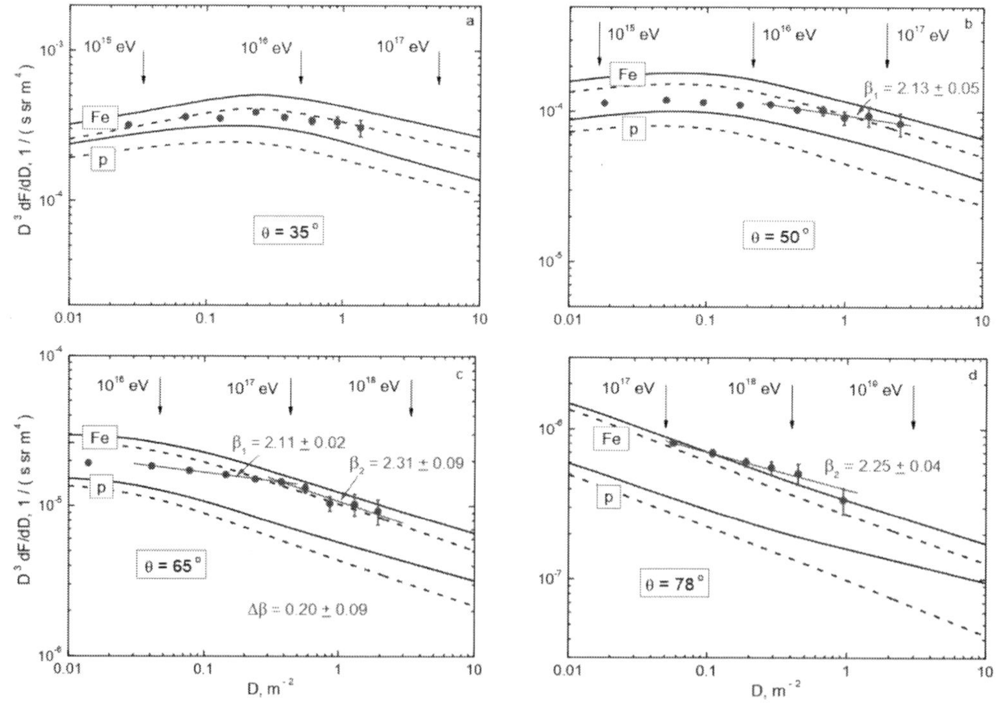

Fig. 8. Measured (points) and calculated (curves) differential local muon density spectra for 4 zenith angles [15].

For each zenith angle registered multiplicity m was re-calculated into local muon density (ratio of the number of muons that hit the setup to the detector area) and corresponding spectra were obtained, which are compared with results of simulations (Fig. 8). From Fig. 8a it is seen that at small zenith angles (35°) the experimental points are in a good agreement with calculations for normal energy spectrum and composition of cosmic rays, and even the knee in the interval 1-10 PeV is noticeable. Data for 50 and 65° (Fig. 8b,c) demonstrate a trend to a heavier composition and a hint for an increase of the slope near 10^{17} eV. However, at very high energies of primary particles around 10^{18} eV and above, which correspond to the last interval of zenith angles (78°, Fig. 8d), intensity of muon bundles is higher than predictions even for pure iron primary flux.

This result was confirmed in Auger experiment [16], where the number of muons exceeds the expected values for pure iron composition, too, and the excess increases with increasing energy of primary particles (Fig. 9). At the same time HiRes [17] and Auger [18] data on X_{max} distribution favor a light (predominantly proton) primary composition near 10^{18} eV.

The last measurements of muon energy spectrum above 100 TeV [19] also showed serious deviations from usual energy spectrum (Fig. 10).

The set of the listed results on the study of both the muon component of the EAS and single muons of cosmic rays is the so-called "muon puzzle" [20]. Obviously, in the frame of existing models of cosmic ray origin and their interaction it is impossible to find a solution of this muon puzzle, and introduction of a new model of interaction is required.

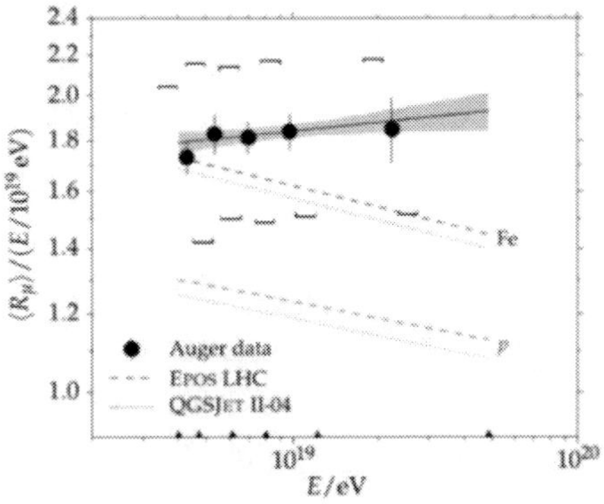

Fig. 9. Average muon content as a function of the shower energy [16]; shown for comparison are theoretical curves for proton and iron showers.

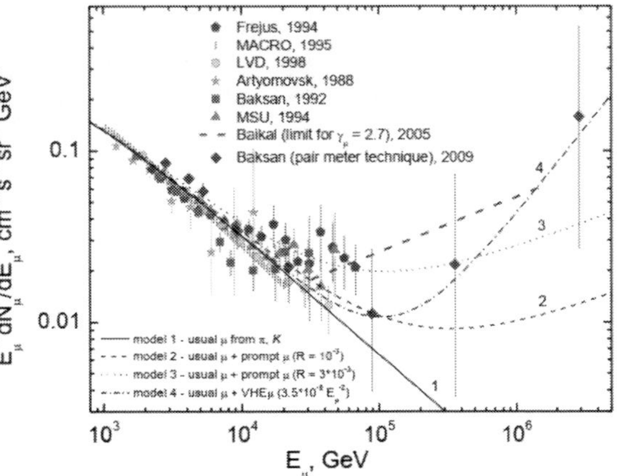

Fig. 10. Differential muon energy spectra for vertical direction measured in various experiments; the curves correspond to different spectrum models [19].

4 A new model of nucleus-nucleus interactions

What we need for explanation of all unusual data from a single point of view? A model of hadron interactions is required which gives:

1) threshold behavior (unusual events appear at several PeV only);
2) large cross section (to observe unusual events in CR);
3) large orbital momentum (to explain alignment);
4) large yield of HE leptons (excess of muon bundles, penetrating cascades);
5) the change of EAS development and, as a consequence, increasing N_μ/N_e ratio.

Which kind of model can be acceptable for this? The first, production of new heavy particles. But in this case the geometrical cross section will be very small

$$\sigma \sim \pi\lambda\!\!\!^{-2}, \quad \lambda\!\!\!^{-} \sim 1/m \qquad (1)$$

The second, production of blobs of quark-gluon plasma (QGP), though it is better to speak about quark-gluon matter (QGM), since usual plasma is a gas, while quark-gluon plasma behaves like a liquid. We will consider the last model, since it allows explain the inclusion of new interaction features observed in LHC and cosmic ray experiments.

Production of QGM blobs provides two main conditions:

- threshold behavior, since for that high temperature (energy) is required;
- large cross section, since the transition from quark-quark interaction to some collective interaction of many quarks and gluons occurs:

$$\sigma \sim \pi\lambda\!\!\!^{-2} \rightarrow \sigma \sim \pi R^2 \qquad (2)$$

where R is the size of quark-gluon blob.

But for explanation of other observed phenomena, a large value of orbital angular momentum is required. The question about orbital momentum which must appear in nucleus-nucleus interactions is usually not considered. For example in Fig. 11 the pictures from [21,22] are presented.

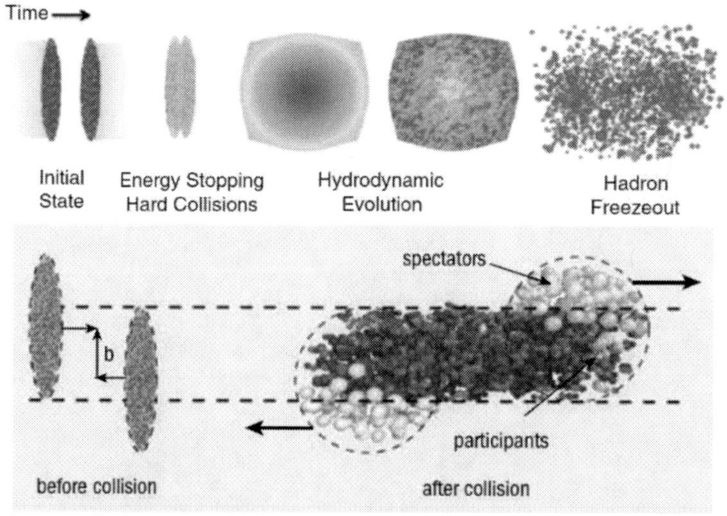

Fig. 11. The time evolution of a high-energy heavy-ion collision (top) [21]. Participants and spectators at the heavy-ion fireball (bottom) [22].

As one can see from this figure, QGM which is produced in peripheral collision, has no orbital momentum. Firstly, the importance of orbital momentum in heavy ion collisions was discussed in paper [23]. Corresponding scheme is presented in Fig. 12. Further investigations of the value of orbital momentum L were done in [24]. It was shown that the value of the orbital momentum for gold-gold collisions with energy 200A GeV (about 40

TeV) can reach $L \sim 10^4$. A blob of a globally polarized QGM with a large orbital angular momentum can be considered as a usual resonance with a large centrifugal barrier. The centrifugal barrier

$$V(L) = L^2/2mR^2 \tag{3}$$

will be large for light quarks but much less for top-quarks or other heavy particles. Though in interacting nuclei top-quarks are absent, the suppression of decays into light quarks gives time for the appearance of heavy quarks in boiling quark-gluon matter in the blob.

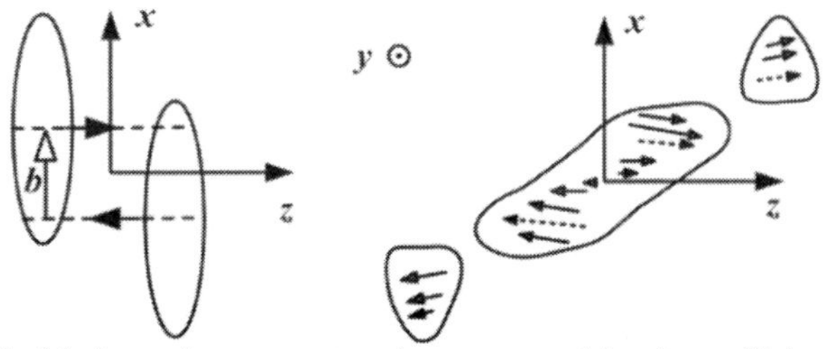

Fig. 12. Production of orbital angular momentum in non-central ion-ion collisions [22].

Simultaneous interactions of many quarks change the energy in the center of mass system drastically:

$$\sqrt{s} = \sqrt{2m_p E_1} \rightarrow \sqrt{2m_c E_1}\,, \tag{4}$$

where $m_c \approx n m_N$. At threshold energy, $n \approx 4$ (α-particle).

Produced $t\bar{t}$ -quarks take away from the QGM blob the energy $\varepsilon_t > 2m_t \approx 350$ GeV, and taking into account fly-out energy $\varepsilon_t > 4m_t \approx 700$ GeV in the center of mass system.

Top-quarks decay $t(\bar{t}) \rightarrow W^+(W^-) + b(\bar{b})$, W-bosons decay into leptons (~ 30%) and hadrons (~ 70%), b-quarks produce jets which generate multiple pions decaying into muons and neutrinos.

5 Explanation of CR and LHC results

In cosmic rays the new model allows explain all observed phenomena [25], in the fist turn "muon puzzle". Decays of W-bosons into muons and neutrinos explain excess of VHE muons with energy above 100 TeV (Fig. 13) and appearance of penetrating cascades [26]. Decays of W-bosons into hadrons (mainly pions, in average ~ 20) explain the increasing muon number (muon excess) with increasing energy.

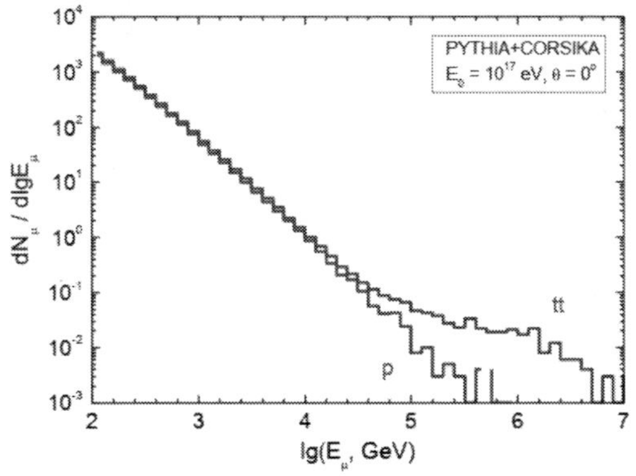

Fig. 13. The inclusive muon energy spectrum simulated by means of the CORSIKA code taking into account t-quarks pair production according to PYTHYA.

Let us consider the influence of the new model on interpretation of results of EAS investigations. Now the transition from measured data to the EAS energy does not take into account a missing energy which is carried away by VHE muons and three types neutrinos, and a change of EAS development due to a change of interaction model. A sharp increase of the missing energy leads to a change of EAS energy spectrum if this effect is not taken into account at reconstruction of EAS parameters from experimental data (Fig. 14).

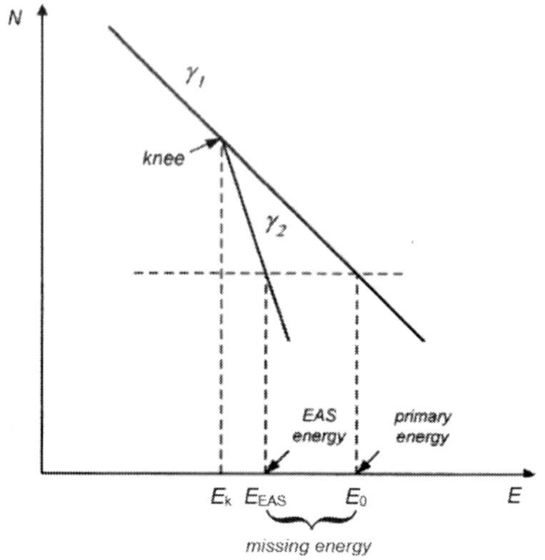

Fig. 14. Influence of missing energy on CR energy spectrum behavior.

A similar situation is connected with the change of the measured composition of CR. The threshold value of $\sqrt{s_{AA}}$ for iron nuclei will be attained at lower energy than for lighter nuclei. Therefore the number of registered EAS from iron and other heavy nuclei will be increased compared to EAS from more light nuclei (Fig. 15). If not to take into account the change of interaction model at transition to PCR energy spectrum and mass composition, this will lead to overestimation of heavy nuclei fraction in PCR.

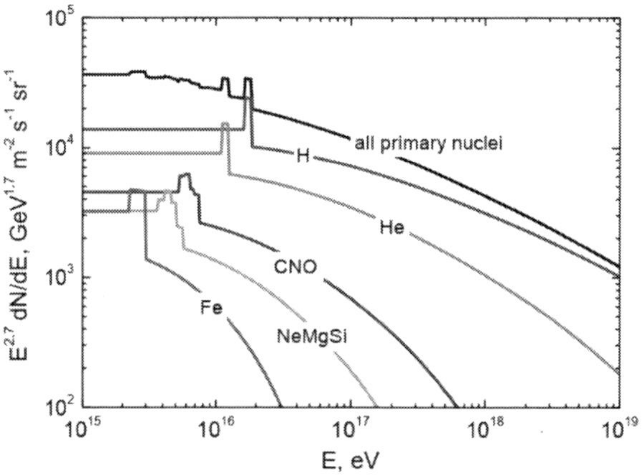

Fig. 15. Changes of various CR nuclei spectra in the frame of the considered interaction model.

The detailed discussion of explanation of other unusual events observed in CR experiments can be found in [27,28].

 DOI: 10.1051/epjconf/201715801003

To illustrate possibilities of the new model for explanation of unusual phenomena observed in nucleus-nucleus collisions at LHC, let's consider two examples.

The first example is observation of imbalance of jet energies at ATLAS in heavy ion collisions (Fig. 16) [29].

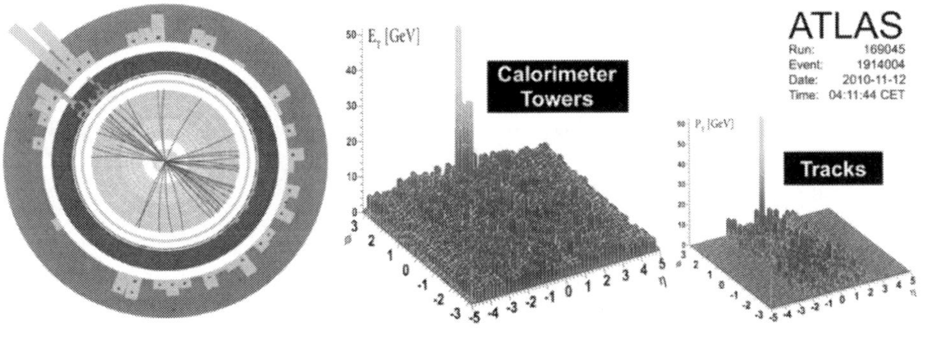

Fig. 16. Highly asymmetric dijet event in the ATLAS detector at Pb-Pb collision [29].

If to take into account a possibility of $t\bar{t}$ -quarks pair production, explanation will be the following: top-quark decays as $t \rightarrow W + b$, and kinetic energy of particles in the center-of-mass system will be equal to: $T_b \approx 65$ GeV, $T_W \approx 25$ GeV. If to take into account the fly-out energy, T_b may be more than 100 GeV which corresponds to jet energy in the observed event. In a case of decay W into pions (more than 10 particles), a picture observed by ATLAS will be obtained.

The second example is an excess of secondary particles observed in nucleus-nucleus (Pb-Pb) collisions compared to proton-proton collision at the same energy in the center-of-mass system $\sqrt{s_{NN}} \approx 2.76$ TeV [30]. If to take into account collective interactions, the average mass of intersecting parts of nuclei will be larger than nucleon mass, and corresponding value $\sqrt{s_{AA}}$ will be larger, too. Therefore, experimental point at the diagram must be moved to the right (Fig. 17).

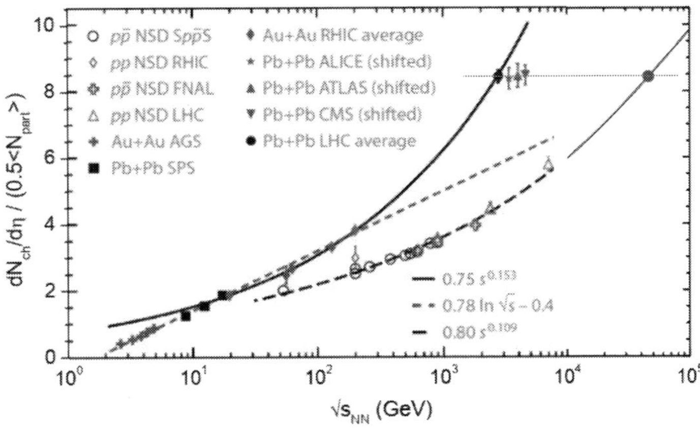

Fig. 17. Multiplicities of charged particles formed in nucleus–nucleus and proton–proton collisions as a function of collision energy [30].

If to take into account that $\sqrt{s_{AA}}$ cannot be less than $\sqrt{s_{NN}}$, it allows evaluate limiting average number of interacting nucleus (see Fig. 17): $\sqrt{n_N}$ < 50 TeV / 3.5 TeV ≈ 14. So in Pb-Pb interactions a blob of QGM can consist of up to 200 nucleons.

6 Conclusion

The results obtained in cosmic rays and in some experiments at the LHC evident in favor of a new state of matter existence. If this idea is correct, it can be investigated at LHC in detail. But it is necessary to search for the new state of matter at LHC not in proton-proton interactions but in nucleus-nucleus interactions. It is better to do this in interactions of light nuclei (nitrogen, oxygen), for which the multiplicity of secondary particles is not so large compared to Pb-Pb or Au-Au.

This work was performed with the financial support of the Ministry of Education and Science of the Russian Federation (government task and MEPhI Academic Excellence Project). Authors thank R.P. Kokoulin for fruitful discussions which appreciably improved this paper.

References

1. K. Nakamura et al. (Particle Data Group), J. Phys. G **37**, 075021 (2010).
2. A. Petrukhin, Measurements of EAS muon energy – the key to solution of primary cosmic ray energy spectrum problem, in Proc. 31st Int. Cosmic Ray Conf., 2009, Lodz, Poland (http://icrc2009.uni.lodz.pl/proc/pdf/icrc0884.pdf).
3. G.V. Kulikov, G.B. Khristiansen, Soviet Physics JETP **35**, 441 (1959).
4. J. Blumer, R. Engel, J.R. Horandel, Progress in Particle and Nuclear Physics **63**, 293 (2009).
5. S.A. Slavatinski, Nucl. Phys. B (Proc.Suppl.) **122**, 3 (2003).
6. M. Tamada, Nucl. Phys. B (Proc.Suppl.) **122**, 349 (2003).
7. V.I. Yakovlev, Nucl. Phys. B (Proc.Suppl.) **122**, 417 (2003).
8. S.I. Nikolsky, Nuclear Physics B (Proc. Suppl.) **60B**, 144 (1998).
9. Yu.V. Stenkin, Phys. of Atomic Nuclei **71**, 98 (2008).
10. I.M. Dremin, EPJ Web of Conferences **145**, 10003 (2017).
11. M. Tamada, N. Inoue, A. Misaki, A. Ohsawa, EPJ Web of Conferences **145**, 15001 (2017)
12. S.B. Shaulov, P.F. Beyl, R.U. Beysembaev et al., EPJ Web of Conferences **145**, 17001 (2017).
13. V. Avati, L. Dick, K. Eggert, J. Strom, H. Wachsmuth, S. Schmeling, T. Ziegler, A. Bruhl, C. Grupen, Astropart. Phys. **19**, 513 (2003).
14. J. Abdallah et al. (DELPHI Collab.), Astroparticle Physics **28**, 273 (2007).
15. A.G. Bogdanov, D.M. Gromushkin, R.P. Kokoulin, G. Mannocchi, A.A. Petrukhin, O. Saavedra, G. Trinchero, D.V. Chernov, V.V. Shutenko, I.I. Yashin, Phys. of Atomic Nuclei **73**, 1852 (2010).
16. A. Aab et al. (Pierre Auger Collaboration), Phys. Rev. D **91**, 032003 (2015).
17. P. Sokolsky, Nucl. Phys. B (Proc. Suppl.) **212-213**, 74 (2011).
18. A. Aab et al. (Pierre Auger Collaboration), Phys. Rev. D **90**, 122005 (2014).

19. A.G. Bogdanov, R.P. Kokoulin, Yu.F. Novoseltsev, R.V. Novoseltseva, V.B. Petkov, A.A. Petrukhin, Astropart. Phys. **36**, 224 (2012).
20. A.A. Petrukhin, Nucl. Instrum. Meth. A **742**, 228 (2014).
21. T.K. Nayak, Pramana **79**, 719 (2012).
22. A. Toia, CERN Courier **53** (N 4), 31 (2013).
23. Z.-T. Liang, X.-N. Wang, Phys. Rev. Lett. **94**, 102301 (2005).
24. J.-H. Gao, S.-W. Chen, W.-T. Deng, Z.-T. Liang, Q. Wang, X.-N. Wang, Phys. Rev. C **77**, 044902 (2008).
25. A. Petrukhin, Nucl. Phys. B (Proc. Suppl.) **212-213**, 235 (2011).
26. A.A. Petrukhin, V.I. Galkin, M.G. Kogan, R.P. Kokoulin, S.Yu. Matveev, V.S. Puchkov, Nucl. Phys. B (Proc. Suppl.) **196**, 165 (2009).
27. A. Petrukhin, Nucl. Phys. B (Proc. Suppl.) **175-176**, 125 (2008).
28. A. Petrukhin, Nuovo Cimento B **120**, 837 (2005).
29. G. Aad et al. (ATLAS Collaboration), Phys. Rev. Lett. **105**, 252303 (2010).
30. K. Aamodt et al. (ALICE Collaboration) Phys. Rev. Lett. **105**, 252301 (2010).

EPJ Web of Conferences **158**, 01004 (2017) DOI: 10.1051/epjconf/201715801004
QFTHEP 2017

Highlights from STAR heavy ion program

Vitalii Okorokov[1,*] (on behalf of the STAR Collaboration)

[1] *National Research Nuclear University MEPhI (Moscow Engineering Physics Institute), Kashirskoe avenue 31, 115409 Moscow, Russian Federation*

Abstract. Recent experimental results obtained in STAR experiment at the Relativistic heavy-ion collider (RHIC) with ion beams will be discussed. Investigations of different nuclear collisions in some recent years focus on two main tasks, namely, detail study of quark-gluon matter properties and exploration of the quantum chromodynamics (QCD) phase diagram. Results at top RHIC energy show clearly the collective behavior of heavy quarks in nucleus-nucleus interactions. Jet and heavy hadron measurements lead to new constraints for energy loss models for various flavors. Heavy-ion collisions are unique tool for the study of topological properties of theory as well as the magnetohydrodynamics of strongly interacting matter. Experimental results obtained for discrete QCD symmetries at finite temperatures confirm indirectly the topologically non-trivial structure of QCD vacuum. Finite global vorticity observed in non-central Au+Au collisions can be considered as important signature for presence of various chiral effects in sQGP. Most results obtained during stage I of the RHIC beam energy scan (BES) program show smooth behavior vs initial energy. However certain results suggest the transition in the domain of dominance of hadronic degrees of freedom at center-of-mass energies between 10–20 GeV. The stage II of the BES at RHIC will occur in 2019–2020 and will explore with precision measurements in the domain of the QCD phase diagram with high baryon densities. Future developments and more precise studies of features of QCD phase diagram in the framework of stage II of RHIC BES will be briefly discussed.

1 Introduction

Numerous experimental results have suggested that nucleus-nucleus collisions produce hot matter consisting of deconfined quark and gluons. This new state of matter behaves like (quasi)ideal quark-gluon liquid and the matter is called strongly-coupled quark-gluon plasma (sQGP). The sQGP is observed in wide energy domain, at least for $\sqrt{s_{NN}} \gtrsim 39$ GeV, where $\sqrt{s_{NN}}$ is the collision energy per nucleon-nucleon pair. The study of the transition from hadronic matter to the quark-gluon state as well as properties of sQGP is one of the most important and debated problems of modern physics of strong interactions. The final state created in nucleus-nucleus collision evolves through several stages [1]. The intensive investigations within relativistic nuclear physics during last decades provided the some general framework for space-time evolution of nuclear collisions called also "Standard Model" for (heavy) ion collisions (Fig. 1). According to this conception nucleus-nucleus collisions at relativistic energies can be considered as the collision of two sheets of strongly interacting matter appeared

*e-mail: VAOkorokov@mephi.ru;Okorokov@bnl.gov

DOI: 10.1051/epjconf/201715801004

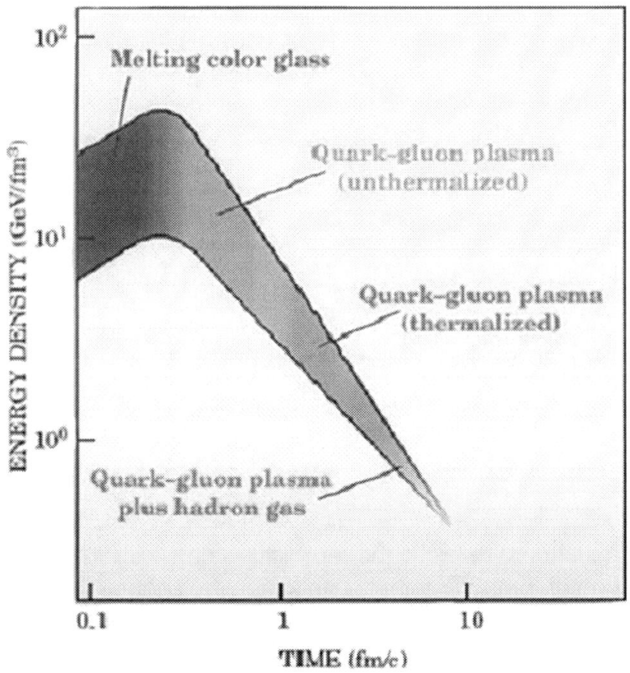

Figure 1. Cartoon for evolution of strongly interacting matter in nuclear collisions. Picture is taken from [1].

due to Lorentz contraction [2]. The initial state matter characterized by high energy density and high degree of coherence is the coherent limit of QCD at high energies. Thus this gluonic matter is called color glass condensate (CGC) [3]. At high energies, these sheets pass through one another, the fields become singular and CGC melts with production of the glasma [2, 4] which eventually materializes as quarks and gluons. It involves many strongly interacting and highly coherent color electric and color magnetic longitudinal flux lines at relatively low baryon densities. It is important to emphasize that because there are both longitudinal color electric and longitudinal color magnetic fields, there is a non-zero density of topological charge, $\vec{E}\cdot\vec{B}$ [5]. A non-zero average of such a quantity would signal the breaking of fundamental QCD symmetries ($\mathcal{P}$ and $\mathcal{CP}$). The one of the possible mechanisms for violation of these symmetries are the dynamical topological sphaleron transitions which may result in observable metastable $\mathcal{CP}$-violating domains. The glasma is a transient state of matter that connects the CGC to the QGP. As indicated above it has coherent fields, like the CGC, but it has properties similar to a plasma with strong background fields, and these fields evolve in time with the natural time scale $\tau_s \sim 1/Q_s$ which is also the proper formation time scale, τ^f_{qg}, for quarks and gluons. Here Q_s is the saturation scale and at RHIC $Q_s \sim 1$ GeV/c [6], $\tau_s \sim 0.2$ fm/c [7]. Then through some as yet not completely understood mechanism, the glasma eventually thermalizes into a sQGP [6] with a "thermalization" time, estimated between 0.15 fm/c and 1 fm/c after the nuclei cross. Thus the total duration of pre-equilibrium stage is estimated as $\Delta\tau_{p-e} \simeq 1$ fm/c for nucleus-nucleus collisions at RHIC energies, i.e. this is time for thermalized sQGP creation after the nuclei collision at $\tau_0 = 0$. Experimental results (see, for instance, [8]) confirm the short duration of the pre-equilibrium stage with respect to the later thermodynamically equilibrium stages which characterized by total duration $\Delta\tau_e \sim 20$ fm/c, where the high border of the last stage correspond to the kinetic freeze-out of the hadronic gas $\tau_{fr} \sim 20$ fm/c. Each of these principal periods, $\tau_0 < \tau \leq \tau_{p-e}$ and $\tau_{p-e} < \tau \leq \tau_{fr}$, contains several sub-stages and detailed description is model-dependent for evolution of strongly interacting matter during each of these two principal periods.

Table 1. Data samples obtained by STAR during RHIC runs.

Species	$\sqrt{s_{NN}}$, GeV
p+p [a]	22.0 [b], 62.4, 200, 410 [b], 500, 510
p+Al	200
p+Au	200
d+Au	19.6, 39.0, 62.4, 200
He^3+Au	200
Al+Au	4.9 [b,c],200
Cu+Cu	22.4 [b], 62.4, 200
Cu+Au	200
Au+Au	4.5 [b,c],7.7, 9.2 [b], 11.5, 14.5, 19.6, 27.0, 39.0, 54.4, 55.8 [b], 62.4, 130, 200
U+U	193

[a] with unpolarized ($\sqrt{s}$ = 62.4 GeV) and with longitudinal / transverse polarized beams;
[b] run with small integral luminosity
[c] run for STAR fixed target mode

Jet and leading hadron measurements are believed to probe the early stages of the dense medium while soft hadronic observables deliver information on the initial state and hydrodynamic evolution of the system. But the cleanest and most sensitive probes to the pre-equilibrium phase of the nucleus-nucleus collision are photons and leptons, since they do not interact with the concurrently forming QGP medium after their production [9]. The key questions discussed here are (*i*) properties of sQGP and (*ii*) features of QCD phase diagram.

The scientific complex with RHIC was designed and was built for investigations in QCD field specially. There were 17 successful physical runs since 2000 year. At present the only large experiment STAR continues to collect new data. The STAR setup is characterized by uniform and large acceptance, good particle identification and stable operation of the main subsystems, for instance, the main tracker – time projection chamber (TPC) – since first physics run in 2000 year [10]. Table 1 shows the data samples collected by STAR experiment during 17 runs at RHIC.

2 Hard physics

Jet quenching is considered as one of the most promising signatures of formation of sQGP and sensitive probe for transport properties of final- state matter. Study of jet quenching with various type triggers is considered as important tool for detailed investigation of transport properties of sQGP. There are two main experimental techniques for jet studies: with help of azimuthal correlations and via full jet reconstruction. The presence of a "trigger" particle, having p_T greater than some selected value, serves as part of the selection criteria to analyze the event for a hard scattering. Within the first approach the trigger particle with high p_T is selected in event, then pairs "trigger particle + associated particle" are formed and $\Delta x = x_t - x_a$ are calculated, where $x_{t(a)} \equiv \eta_{t(a)}, \phi_{t(a)}$ is the kinematic parameter (ϕ – azimuthal angle, η – pseudorapidity) of trigger (associated) particle. As expect, correlations appear at $\Delta\phi \approx 0$ the peak corresponds to jet in which the both trigger and associated particles are fragmentation products of one parton and at $\Delta\phi \approx \pi$ the peak corresponds to jet in which the trigger and associated particles are fragmentation products of different (back-to-back) partons. The following quantities are calculated for particle pairs:

$$C_2(\Delta\phi) = \frac{1}{N_{\rm trig}} \int d\Delta\eta N(\Delta\phi, \Delta\eta), \quad Y = \int d\Delta\phi C_2(\Delta\phi), \tag{1}$$

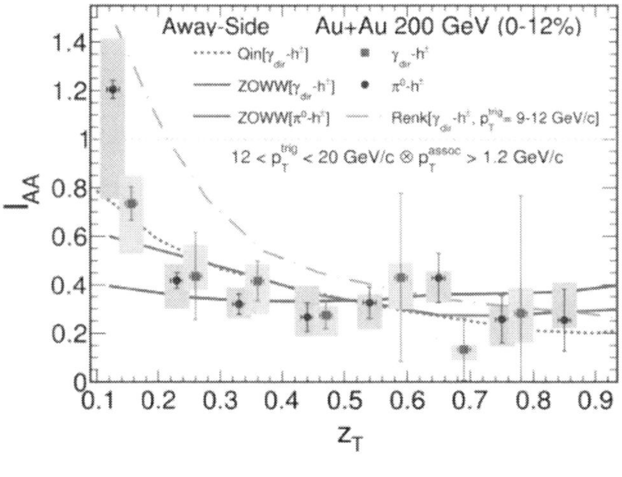

Figure 2. z_T–dependence of I_{AA} for π^0 and γ_{dir} triggers (the points are slightly shifted for visibility) in most central Au+Au collisions. The vertical lines are statistical errors, vertical extent of the boxes – systematic uncertainties. The curves represent the model predictions. Figure is taken from [13].

where $C_2(\Delta\phi)$ is the two-particle azimuthal correlation function, Y – the per-trigger yield of particle pairs. The development of experimental technique and data analysis results in the possibility for full reconstruction of hadron jets in nucleus-nucleus collisions. At present fully reconstructed jets are one of the main tools to constrain the medium transport parameters.

2.1 Neutral probes

It was pointed out that direct photons (γ_{dir}), real or virtual, are penetrating probes for the bulk matter produced in hadronic collisions, as they do not interact strongly and they have a large mean free path [11]. The recent STAR physics analyses are focused on azimuthal correlations with neutral triggers (γ_{dir} and π^0) produced in nucleus-nucleus collisions and on the γ_{dir} yield.

The azimuthal angular correlation of charged hadrons with respect to a direct-photon trigger was proposed as a promising probe to study the mechanisms of parton energy loss [12]. The ratio $I_{AA}(z_T) = Y^{\text{Au+Au}}(z_T)/Y^{p+p}(z_T)$ in Au+Au at $\sqrt{s_{NN}} = 200$ GeV has been studied with help of $\pi^0/\gamma_{dir} - h^{\pm}$ azimuthal correlations, where $z_T = p_T^a/p_T^t$ and it is expected $I_{AA} = 1$ in the absence of medium modification. Results are shown in Fig. 2. As seen $I_{AA}(z_T)$ for both π^0 and γ_{dir} triggers show indication on less suppression at low $z_T \in (0.1; 0.2)$ than at high z_T as well as similar levels of suppression for both trigger types [13]. There is no manifestation of the expected differences due to the color-factor effect and the difference in average path lengths between π^0 and γ_{dir} triggers (path-length dependence).

Also the γ_{dir} invariant yields are derived at midrapidity in the ranges $1 < p_T < 3$ GeV/c and $5 < p_T < 10$ GeV/c for Au+Au collisions at $\sqrt{s_{NN}} = 200$ GeV [14]. In Fig. 3 a clear excess in the γ_{dir} invariant yield is observed in comparison with the nuclear overlap function T_{AA} scaled $p + p$ reference in the first p_T. For $p_T > 6$ GeV/c the production follows T_{AA} scaling. Model calculations with contributions from thermal radiation and initial hard parton scattering are consistent within experimental uncertainties with the γ_{dir} invariant yield in 0–20%, 20–40% and 40–60% centrality bins. In 60–80% collisions there is systematic excess data over model curve at $2 < p_T < 3$ GeV/c.

2.2 Hadronic jets

In STAR the anti-k_T algorithm with $R = 0.2 - 0.5$ ($R^2 = (\Delta\phi)^2 + (\Delta\eta)^2$) and a low IR-cutoff ($p_T > 0.2$ GeV/c) is used for full jet reconstruction from charged tracks in Au+Au collisions. STAR measures

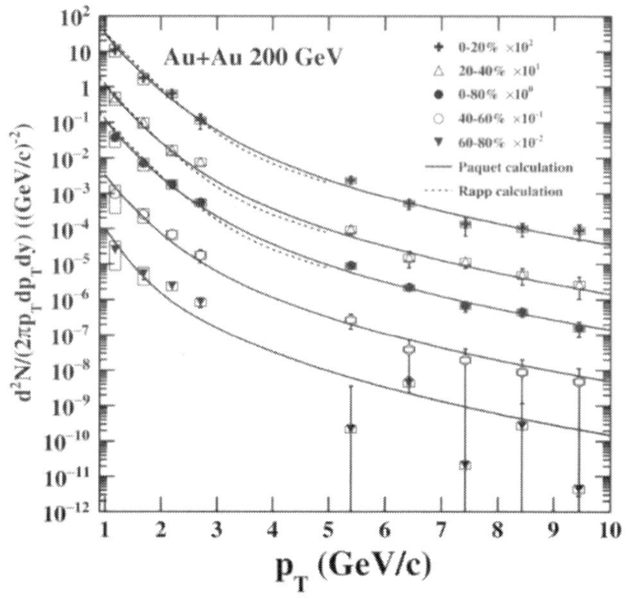

Figure 3. The direct γ invariant yield as a function of p_T in comparison with model predictions. The statistical and systematic errors are shown by bars and boxes, respectively. Figure is taken from [14].

the semi-inclusive distributions of reconstructed charged particle jets recoiling from a high p_T trigger hadron [15]

$$\frac{1}{N_{\text{trig}}} \frac{d^2 N_{\text{jet}}}{dp^{\text{ch}}_{T,\text{jet}}\, d\eta_{\text{jet}}} = \frac{1}{\sigma^{\text{A+A}\to h+X}} \frac{d^2\sigma^{\text{A+A}\to h+\text{jet}+X}}{dp^{\text{ch}}_{T,\text{jet}}\, d\eta_{\text{jet}}}, \tag{2}$$

where η_{jet} is the pseudorapidity of jet centroid, $p^{\text{ch}}_{T,\text{jet}}$ – transverse momentum of jet, $\sigma^{\text{A+A}\to h+X}$ is the cross-section to generate a trigger hadron, and $d^2\sigma^{\text{A+A}\to h+\text{jet}+X}/dp^{\text{ch}}_{T,\text{jet}}\, d\eta_J$ is the double differential cross-section for coincidence production of a trigger hadron and a recoil jet, $\text{A} + \text{A} \equiv p + p$, Au+Au. The combinatorial background is removed by mixed-event technique. This allows the access for low-p_T jets. The STAR has measured the I_{CP} ratio of jet yields in central to peripheral Au+Au events at $\sqrt{s_{NN}} = 200$ GeV [15]. Results for jet cone radius $R = 0.3$ is shown in Fig. 4 for invariant yield (upper panel) and for I_{CP} (lower panel) depends on $p^{\text{ch}}_{T,\text{jet}}$. The I_{CP} shows a clear suppression of yields in central events for $p^{\text{ch}}_{T,\text{jet}} > 10$ GeV/c for all cone radii $R = 0.2$, 0.3, 0.4 and 0.5 under study. The horizontal shifts in the range where the I_{CP} is flat ($10 < p^{\text{ch}}_{T,\text{jet}} < 20$ GeV/c) are $-6.3 \pm 0.6 \pm 0.8$ GeV/c and $-3.8 \pm 0.5 \pm 1.8$ GeV/c for $R = 0.3$ and $R = 0.5$, respectively. The suppression and shift, and their reduction for larger cones might indicate an out-of-cone energy transport by interaction of the jet with the medium, averaged over the recoil jet population.

The first dijet imbalance measurement at RHIC was made by STAR for central Au+Au collisions at $\sqrt{s_{NN}} = 200$ GeV [16], thus allowing a more direct comparison to jet quenching measurements at the Large Hadron Collider (LHC). The imbalance observable is defined as

$$A_J = \frac{p_{T,\text{lead}} - p_{T,\text{sublead}}}{p_{T,\text{lead}} + p_{T,\text{sublead}}} \tag{3}$$

where $p_{T,\text{lead}}$ and $p_{T,\text{sublead}}$ are the transverse momenta of the leading and subleading jet, respectively, in the dijets that are required to be approximately back to back [16]. In STAR analysis Au+Au events at $\sqrt{s_{NN}} = 200$ GeV were selected by an online high tower (HT) trigger, which required an uncorrected transverse energy of neutral hadrons $E_T > 5.4$ GeV in at least one barrel electromagnetic

DOI: 10.1051/epjconf/201715801004

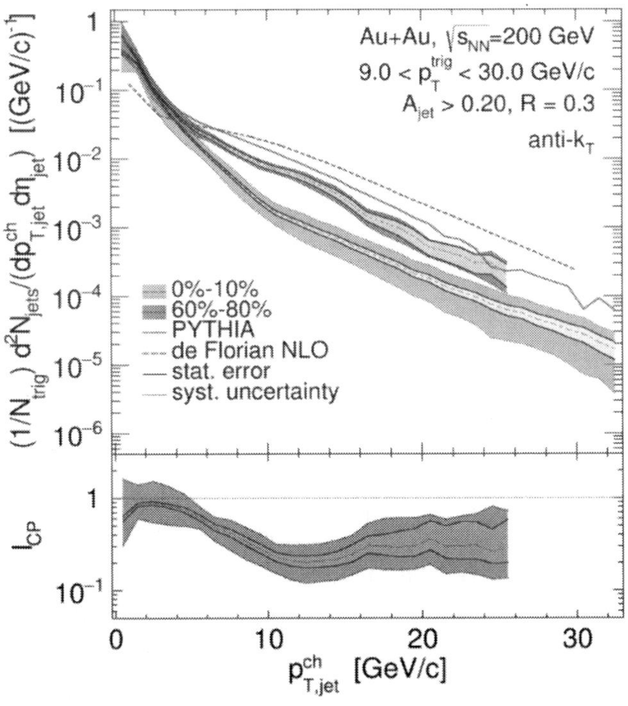

Figure 4. Fully corrected distributions of $Y(p_{T,jet}^{ch})$ (upper) and its ratio I_{CP} (lower) for central and peripheral Au+Au collisions, for anti-k_T jets with $R = 0.3$ and area $a_{jet} > 0.2$. The upper panels also show yield for $p + p$ collisions, calculated using PYTHIA at the charged-particle level and NLO pQCD transformed to the charged-particle level. The uncertainty of the NLO calculation is not shown. Figure is taken from [15].

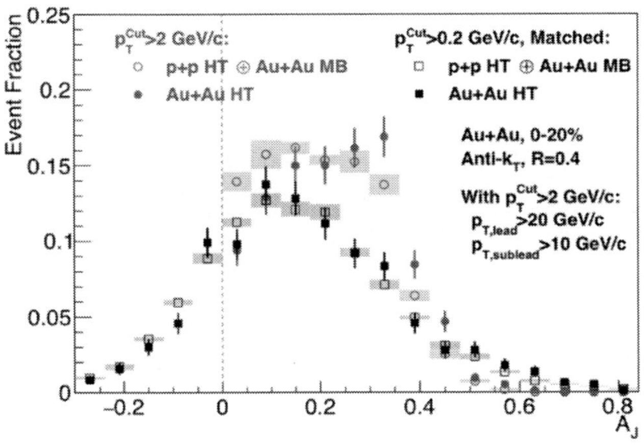

Figure 5. Normalized A_J distributions for Au+Au HT data (filled symbols) and $p + p$ HT ⊕ Au+Au MB (open symbols). The circles are for jets found using only constituents with $p_T^{Cut} > 2$ GeV/c and the squares for matched jets found using constituents with $p_T^{Cut} > 0.2$ GeV/c. In all cases $R = 0.4$. Statistical errors may be smaller than symbol size for $p + p$ HT ⊕ Au+Au MB. Figure is taken from [16].

calorimeter (BEMC) tower. It was shown that Au+Au HT leading jets are similar to $p + p$ HT leading jets embedded in an Au+Au minimum bias (MB) background [24]. In Fig. 5 the A_J distribution from central Au+Au collisions for anti-k_T jets with $R = 0.4$ (solid points) is compared to the $p + p$ HT embedding reference ($p + p$ HT ⊕ Au+Au MB, open symbols) for a jet constituents with $p_T^{Cut} > 2$ GeV/c. Dijets in central Au+Au collisions are significantly more imbalanced than the corresponding $p + p$ dijets. The calculation in accordance with a Kolmogorov–Smirnov test on the unbinned data [25] supports the hypothesis that the $Au + Au$ and $p + p$ HT ⊕ Au+Au MB data are not drawn from the same parent A_J distributions.

DOI: 10.1051/epjconf/201715801004

3 Soft physics

In relativistic nuclear physics the event shape is usually studied with help of the Fourier expansion of particle distribution [17]. Then with taking into account possible influence of QCD fundamental symmetries violation at finite temperature the invariant distribution of final state particles with certain type can be written as following

$$E\frac{d^3N}{d\vec{p}} = \frac{1}{2\pi}\frac{d^2N}{p_T dp_T dy}\left\{1 + 2\sum_{n=1}^{\infty}\sum_{j=1}^{2} k_n^j F_j\left[n\left(\phi - \Psi_{\rm RP}\right)\right]\right\}, \quad \{F\}_{j=1}^{2}(x) \equiv \begin{pmatrix}\sin\\ \cos\end{pmatrix}(x). \quad (4)$$

Here ϕ is an azimuthal angle of particle under study, $\Psi_{\rm RP}$ – azimuthal angle of reaction plane, $k_n^1 \equiv v_n = \langle\cos[n(\phi - \Psi_{\rm RP})]\rangle$ – collective flow of n-th order, the parameters $k_n^2 \equiv a_n = \langle\sin[n(\phi - \Psi_{\rm RP})]\rangle$ describe the possible effect of $\mathcal{P}/C\mathcal{P}$ violation in strong interactions.

3.1 Collective flows and QCD phase diagram

Up to the present time, the main efforts are focused on the study of the low-order v_n coefficients due to their sensitivity to the key features of the final-state matter.

The $v_2(\phi)/v_2(p)$ ratios are larger than unity at $p_T \sim 0.5$ GeV/c for most central 0–30% bin of the Au+Au collisions at $\sqrt{s_{NN}} = 200$ GeV showing an indication of breakdown of the expected mass ordering in that momentum range. This could be due to a large effect of hadronic rescattering on the proton v_2. The hypothesis has been verified by comparison with model results. The AMPT calculations show the increase of $v_2(\phi)/v_2(p)$ with growth hadronic cascade time. This is attributed to a decrease in the $v_2(p)$ due to an increase in hadronic rescattering, while the ϕ-meson v_2 remains unaffected. On the other hand the ratio under discussion from UrQMD is much smaller than unity because UrQMD model lacks partonic collectivity and therefore does not fully develop the ϕ-meson v_2 [18].

Similar to hadrons over the measured p_T range ($0.5 \lesssim p_T < 4$ GeV/c), light (anti)nuclei $v_2(p_T)$ show a monotonic rise with increasing p_T, mass ordering at low p_T, and a reduction for more central collisions. It is observed that v_2 of nuclei and antinuclei are of similar magnitude for Au+Au collisions in domain $\sqrt{s_{NN}} \geq 39$ GeV. In general the light-nuclei v_2 follow an atomic mass number scaling indicates that the coalescence of nucleons might be the underlying mechanism of light-nuclei formation in HIC. This fact is corroborated by model calculations [19].

For sufficiently central Au+Au collisions ($N_{\rm part} > 50$), the triangle flow v_3 persist down to the lowest energies studied. For more peripheral collisions, however, the correlation appears to be absent at low energies for $N_{\rm part} < 50$, in agreement with AMPT model. This non-QGP model predicts a smaller v_3 value than the data, suggesting that a QGP phase may exist in more central collisions at $\sqrt{s_{NN}} = 7.7$ GeV. When divided by multiplicity, v_3 shows a local minimum in the range $\sqrt{s_{NN}} = 15 - 20$ GeV in centrality range from 0% to 50% (Fig. 6). This feature has not been shown in any known models of HIC and could indicate an interesting trend in the pressure developed inside the system [20].

A finite difference in v_1 between positive and negative charged particles (Δv_1) was observed in the kinematic domain of $0.15 < p_T < 2$ GeV/c and $|\eta| < 1$. The Δv_1 seems to increase with p_T. The v_1 results from Au+Au collisions show much smaller values compared to those in Cu+Au. These results are consistent with the presumption of a strong, initial electric field in asymmetric collisions. The p_T dependence of Δv_1 is qualitatively described by the parton–hadron–string–dynamics (PHSD) model at $p_T < 2$ GeV/c. However, the magnitude of Δv_1 is smaller by a factor of 10 than the model predictions [21]. This may indicate that most of the quarks and antiquarks have not yet been created within the lifetime of the electric field ($t \lesssim 0.25$ fm/c).

DOI: 10.1051/epjconf/201715801004

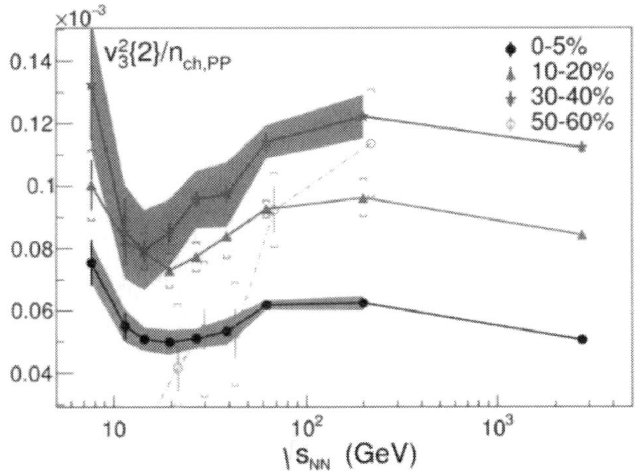

Figure 6. $v_3^2\{2\}$ divided by the midrapidity, charged-particle multiplicity density per participant pair in Au+Au and Pb+Pb ($\sqrt{s_{NN}}$ = 2.76 TeV) collisions. Systematic errors are shown either as a shaded band or as thin vertical error bars with caps. Figure is taken from [20].

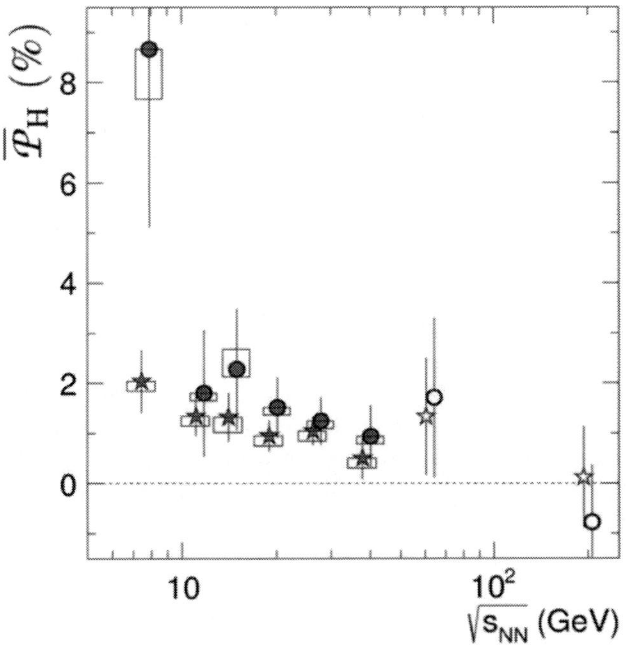

Figure 7. The average polarization for $H \equiv \Lambda(\bar{\Lambda})$ in Au+Au collisions. The results for Λ (stars) and $\bar{\Lambda}$ (circles) from 20—50% central collisions are plotted as a function of $\sqrt{s_{NN}}$. Error bars represent statistical uncertainties only, while boxes represent systematic uncertainties. The results of the present study ($\sqrt{s_{NN}} < 40$ GeV), indicated by filled symbols, are shown together with those reported earlier [23] for 62.4 GeV and 200 GeV collisions, indicated by open symbols and for which only statistical errors are plotted. Figure is taken from [22].

3.2 Hyperon polarization and rotating sQGP

Non-central nuclear collisions have angular momentum on the order of $10^3\hbar$, and shear forces generated by the interpenetrating nuclei may generate a clear vortical structure. The Λ and $\bar{\Lambda}$ are used to measure the global hyperon polarization in non-central Au+Au collisions because they are "self analyzing", i.e., for instance, in the decay $\Lambda \rightarrow p+\pi^-$ the p tends to be emitted along the spin direction of the parent Λ. At present due to the limited statistics collected by STAR, only the average projection of polarization on overall angular momentum was extracted. In Fig. 7 the average polarization for Λ ($\bar{\Lambda}$) hyperons is shown as a function of $\sqrt{s_{NN}}$ for semi-central Au+Au collisions [22]. At $\sqrt{s_{NN}}$ < 200 GeV, a positive polarization is observed for Λ ($\bar{\Lambda}$) hyperons. The data are statistically consistent with the hypothesis of energy-independent polarizations of 1.08 ± 0.15 (stat) ± 0.11 (sys) and 1.38 ± 0.30

(stat) ± 0.13 (sys) per cent for Λ and $\bar{\Lambda}$, respectively [22]. The $\sqrt{s_{NN}}$-averaged polarizations indicate a vorticity magnitude of $\omega = (9 \pm 1) \times 10^{21}$ s^{-1}, which is estimated using the hydrodynamic relations and takes into account the "feed-down" contributions. This extremely large estimation of ω evidences the sQGP is characterized by complex and developed vortical structure.

4 Heavy quarks in hot environment

Heavy quark masses are mostly uncharged by chiral symmetry restoration. Heavy flavor (HF) particles are produced early in collisions due to hard scattering processes. Therefore they allow the study of whole space-time evolution of sQGP. Study of heavy quark diffusion is essential for better understanding of transport properties of sQGP and underlying fluid of light partons. Precise measurements of c- and b-quarks energy loss are crucial for understanding of parton interactions with the hot environment.

4.1 Bottomonia in sQGP

The novel results have been obtained by STAR for production of Υ states in heavy ion collisions. The nuclear modification factor R_{AA} depends on number of participants N_{part} is shown in Fig. 8 for various Υ states, where

$$R_{AA} = \frac{1}{T_{AA}} \frac{d^2N^{A+A}/dp_T d\eta}{d^2\sigma^{p+p}/dp_T d\eta}. \tag{5}$$

The trend marked by the points $R_{AA}(N_{part})$ for Au+Au is confirmed by the U+U data. There is strong suppression of the Υ(2S+3S) states in U+U as well as in Au+Au collisions at highest available RHIC energies. There is neither a significant difference between the results in any of the centrality bins, nor any evidence of a sudden increase in suppression in central U+U compared to the central Au+Au data. But one can note the precision of the current measurement does not exclude a moderate drop in R_{AA}. These STAR data have been compared with several models for upsilon behavior in hot environment (Fig. 9). As seen the Rapp model included CNM effects, dissociation of bottomonia in the hot (T = 330 MeV) medium and regeneration for strongly bound scenario (SBS) for the internal-energy-based heavy quark potential describe the STAR data within uncertainties as well as other internal-energy-based models with an initial central temperature $428 < T < 442$ MeV (noted as "model B" in Fig. 3) and T = 340 MeV ("Liu-Chen model"). The "model A" with the free-energy-based potential corresponding to a more weakly bound scenario (WBS) tends to underpredict the R_{AA} especially for the Υ(1S).

4.2 Collectivity for heavy flavor hadrons

The heavy flavor tracker (HFT) is the newest member of the STAR detector subsystems. It is made using thinned monolithic active pixel sensor technology. STAR HFT includes pixel detector (PXL), intermediate silicon tracker (IST) and silicon strip detector (SSD). The HFT is characterized by low material budget and large acceptance coverage ($|\eta| < 1$, $0 < \phi < 2\pi$). The HFT makes it possible for the first time to directly track the decay products of hadrons comprised of charm and bottom quarks with high pointing resolution ($\sim$ 30 μm for particles with $p \geq 1.5$ GeV/c).

The clear mass ordering is obtained for elliptic flow in domain $p_T < 2$ GeV/c for various hadrons including D^0 mesons (Fig. 10). For larger $p_T > 2$ GeV/c, the D^0 meson v_2 follows that of other light mesons indicating a significant c-quark flow at highest RHIC energy. The elliptic flow for D^0 falls into the same universal trend as all other light hadrons, in particular, for the range $(m_T - m_0)/n_q < 1$

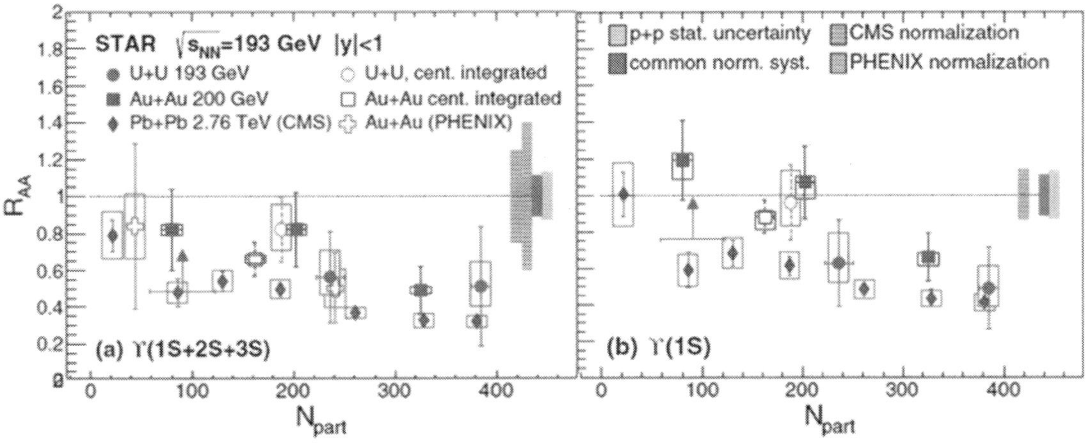

Figure 8. Υ(1S+2S+3S) (a) and Υ(1S) (b) nuclear modifi- cation factor R_{AA} as a function of N_{part} in relativistic heavy ion collisions. The data points in the 30—60% centrality bin have large statistical and systematical uncertainties, providing little constraint on R_{AA}. Therefore the 95% lower confidence bound is only indicated for the 30—60% centrality U+U data. Data are from [26].

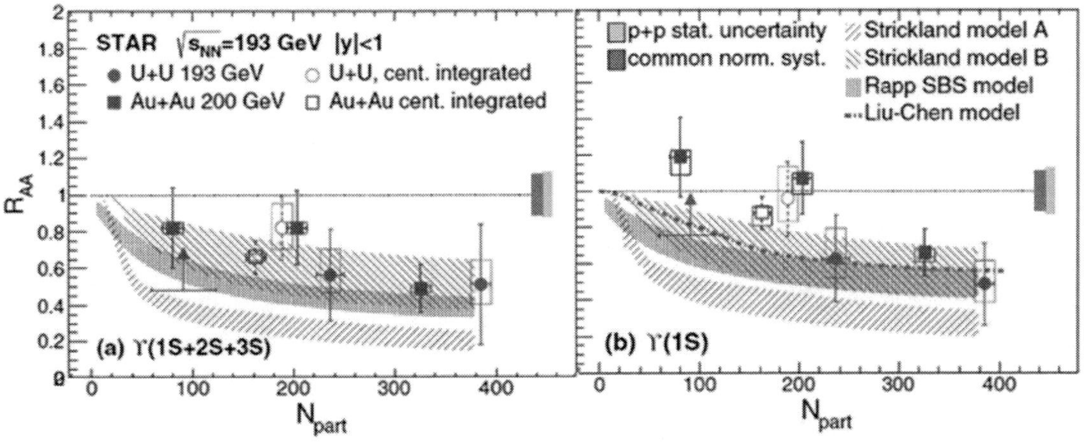

Figure 9. Υ(1S+2S+3S) (a) and Υ(1S) (b) R_{AA} depends on N_{part} in $\sqrt{s_{NN}}$ = 193 GeV U+U collisions, compared to different models. The 95% lower confidence bound is indicated for the 30–60% centrality U+U data. Data and model results are taken from [26].

GeV/c^2. This suggests that c-quarks have gained significant flow through interactions with the sQGP in the collisions under consideration. The comparison between STAR results and phenomenological calculations is shown in Fig. 11. The TAMU model describes the data only with no c-quark diffusion. A 3D viscous hydrodynamic simulation with $\eta/s = 0.12$ tuned to describe v_2 for light hadrons predicts elliptic flow values for D^0 that is consistent with data for $p_T < 4$ GeV/c. This suggests that c-quarks have achieved thermal equilibrium in these collisions. The statistical significance test was performed for the consistency between data and each model quantified by χ^2/ndf. The 3D viscous hydro models shows the best χ^2/ndf = 0.73 for the measured p_T region. The models that can describe both the R_{AA}

DOI: 10.1051/epjconf/201715801004

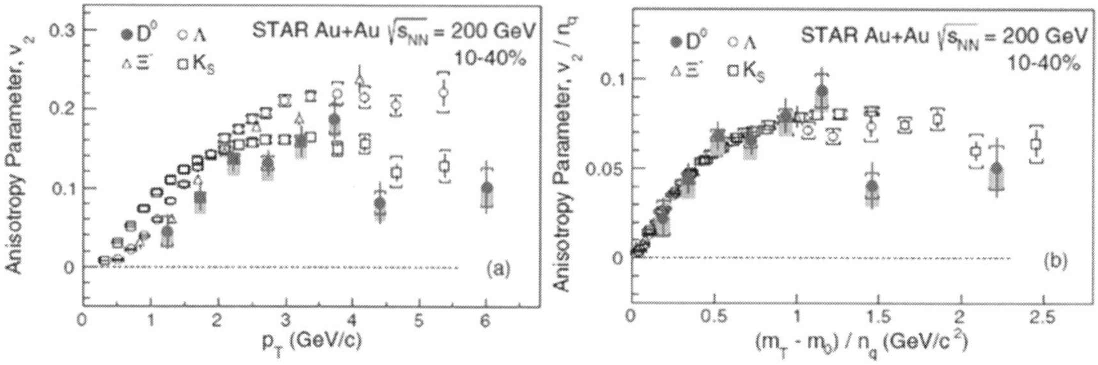

Figure 10. (a) v_2 as a function of p_T and (b) v_2/n_q dependence on $(m_T - m_0)/n_q$ for D^0 compared with earlier measurements for strange particles. The vertical bars and brackets represent statistical and systematic uncertainties, respectively, and the gray bands represent the estimated nonflow contribution. Data are from [27].

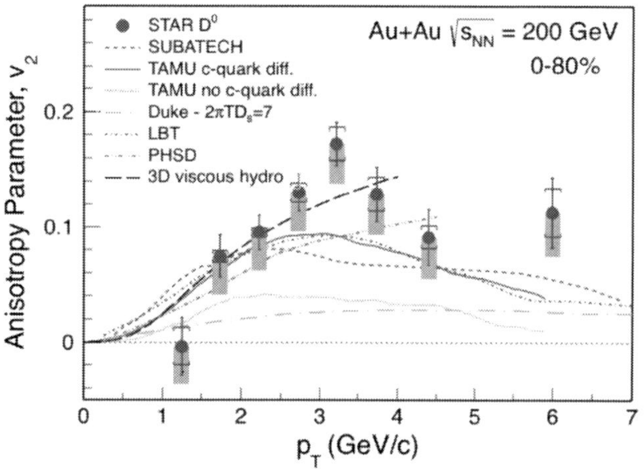

Figure 11. v_2 as a function of p_T for D^0 in minimum bias Au+Au collisions compared with model calculations. Figure is taken from [27].

and v_2 data include the temperature-dependent charm diffusion coefficient $\kappa = 2\pi T D_s$ in the range of $\sim 2-12$. The κ predicted by lattice QCD calculations fall in the same range [27].

5 Future plans

The STAR Collaboration works intensively under development the physics program and detector upgrades for nearest RHIC runs (2018–2020) as well as for future studies of strong interactions after 2020 year. Especially, STAR's highest scientific priority for run 18 is the successful realization of the isobaric collision program (Rb+Rb, Zr+Zr at $\sqrt{s_{NN}}$ = 200 GeV) for clarification the interpretation of measurements related to chiral effects, in particular chiral magnetic effect (CME).

5.1 Nearest upgrades for stage II of the BES program at RHIC

As expected there will important improvement of RHIC for stage II of the BES. Low energy electron cooling at RHIC allows the increase of luminosity by factors 3–10 for BES energy ranges in both the collider mode ($\sqrt{s_{NN}}$ = 7.7, 11.5, 14.5 and 19.6 GeV) and the fixed target mode ($\sqrt{s_{NN}}$ = 3.0, 3.5 and

 DOI: 10.1051/epjconf/201715801004

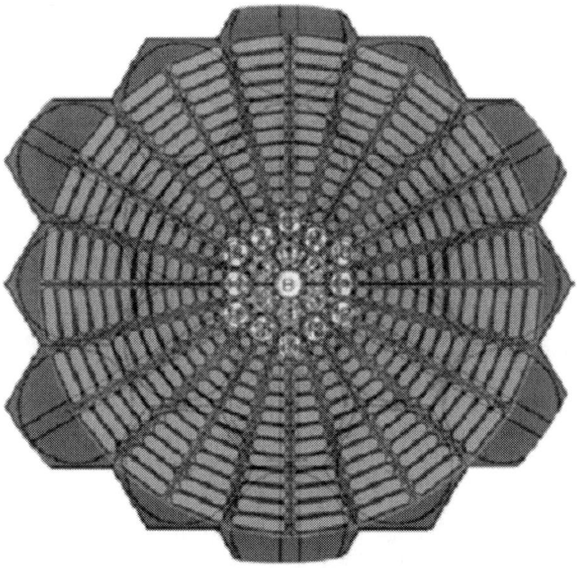

Figure 12. Schematic view of EPD with respect to the existing beam-beam counter (BBC) subsystem (central part). There will be 372 channels for EPD instead of 16 present channels for BBC.

4.5 GeV) for heavy (Au+Au) ion collisions. Upgrade for STAR promises several improvements for physics analyses like significant decrease of statistical and systematics uncertainties, advanced particle identification (PID) capability, broader kinematic coverage etc. The three detector subsystems will be made for stage II of BES at RHIC: upgrade of inner sectors of TPC (iTPC), event plane detector (EPD) and endcap for time-of-light (eTOF).

Replace all inner sector of TPC will improve the continues coverage (40 pad rows instead of 13 at present) and the dE/dx on 15–30%. Also coverage on $|\eta|$ will be extended from 1.0 to 1.5. As planed, iTPC will be characterized by better momentum resolution and lowers p_T cut from 125 MeV/c to 60 MeV/c. Beam-beam counter (BBC) will be replaced by new EPD with significantly higher azimuthal and radial segmentation (Fig. 12). The EPD is crucially important for study of Λ ($\bar{\Lambda}$) polarization in nuclear collisions as well as for BES physics due to independent event plane measurement. As expected, the EPD will provides better trigger and background reduction, extends $|\eta|$ coverage from $3.3 < |\eta| < 5.0$ for present BBC to $2.1 < |\eta| < 5.1$. The EPD will allows the measurement the centrality and event plane at forward rapidity and improvement for event plane resolution. During the Run 17 the first 1/8 part of EPD was installed and tested successfully. The eTOF is the joint project between STAR and CBM. This detector is important for fixed target program and complements the existing barrel TOF. The major improvements for eTOF installation are following: midrapidity particle identification in fixed target mode, extension of identification capabilities for π, p and K, additional for barrel TOF ($|\eta| < 0.9$) coverage: $-1.5 < \eta < -1.1$.

5.2 Towards cold QCD: STAR after 2020

The sQGP created in the final state of high energy nucleus-nucleus collision can be considered as the incoherent thermal limit of QCD matter at high temperatures while the CGC in the initial state is the coherent limit of QCD at high energies. Since the sQGP has to be created at modern colliders from the interaction of initial nuclear enhanced coherent chromo electric magnetic fields through glasma, both limiting forms of QCD matter as well as transient glasma state should be studied in detail. The forward heavy ion collisions provide the unique opportunity to study the various saturation regimes. The RHIC cold QCD plan requires an upgrade to the forward rapidity ($2.5 < \eta < 4.5$) detection

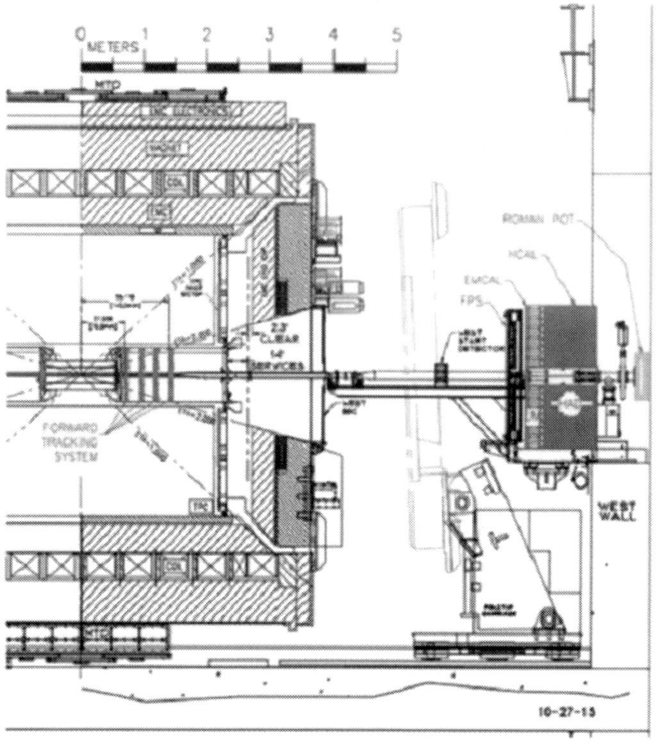

Figure 13. Scheme for location of the forward calorimeter system (FCS) at the west side of the STAR detector. Figure is taken from [28].

capabilities of STAR. According to the [28] it is suggested the development of the STAR calorimeter and tracking subsystems for forward region (Fig. 13). This upgrade allows the exploration of cold QCD physics in very high and low regions of Bjorken x_B as well as the longitudinal structure of the initial state and the temperature dependent transport properties of matter in nuclear collisions with help of measurements of azimuthal asymmetries, dihadron and h/γ-jet correlations, asymmetries in production of hadrons and jets, R_{pA} for γ_{dir} and Drell–Yan process.

6 Summary

During the last years many important results have been obtained within heavy ion program of the STAR experiment. Important progress was achieved for study of hard processes in nucleus-nucleus collisions, namely, azimuthal correlations with neutral triggers show the absence of path-length dependence for neutral pions and direct photons; semi-inclusive jet production demonstrate the possible reduction of medium-induced energy transport to large angles at RHIC with respect to the LHC. The STAR results for dijet imbalance in Au+Au collisions at $\sqrt{s_{NN}} = 200$ GeV are the first indication that at RHIC energies it is possible to select a sample of reconstructed dijets that clearly lost energy via interactions with the medium but whose lost energy reemerges as soft constituents accompanied with a small, but significant, broadening of the jet structure compared to $p + p$ fragmentation. The above observations are consistent with the qualitative expectations of perturbative QCD like radiative energy loss in the hot, dense medium created at RHIC. In soft physics field it was found $\Lambda(\bar{\Lambda})$ hyperons are polarized in nuclear collisions. The estimation of ω can be considered as evidence that quark-gluon matter produced at RHIC is most vortical fluid among known. Directed flow indicates on the presence of a strong, initial electric field in asymmetric nuclear collisions. Light (anti)nuclei and D^0 mesons show the collective behavior which agrees with general trends. Quarkonia measurements in heavy ion collisions consistent with the expectations from the sequential melting hypothesis.

Thus the present stage in studies of nuclear collisions in STAR experiment is the transition from the qualitative statements to the quantitative understanding in relativistic nuclear physics.

There are detailed plans for both the nearest runs (18 and 19) at RHIC and the future investigations after 2020 year. For the first case detector upgrades are in progress for precise study of sQGP within stage II of the BES. The future modification of the STAR setup after 2020 aims on the study of effects of QCD cold matter, longitudinal structure of initial conditions.

References

[1] T. Ludlam and L. McLerran, Phys. Today **56**, 48 (2003).

[2] A. Kovner, L.D. McLerran, H. Weigert, Phys. Rev. D**52**, 6231 (1995); 3809 (1995); A. Krasnitz, R. Venugopalan, Phys. Rev. Lett **84**, 4309 (2000); A. Krasnitz, R. Venugopalan, Nucl. Phys. B**557**, 237 (1999); A. Krasnitz, Y. Nara, R. Venugopalan, Phys. Rev. Lett. **87**, 192302 (2001).

[3] E. Iancu, R. Venugopalan, arXiv: hep-ph/0303204 (2003).

[4] A. Krasnitz, R. Venugopalan, Phys. Rev. Lett. **86**, 1717 (2001); T. Lappi, Phys. Rev. C**67**, 054903 (2003); T. Lappi, L. McLerran, Nucl. Phys. A**772**, 200 (2006).

[5] L. McLerran, arXiv: 0911.2987 [hep-ph] (2009).

[6] R. Baier et al., Phys. Lett. B**502**, 51 (2001); *ibid* **539**, 46 (2002).

[7] M. Gyulassy, L. McLerran, Nucl. Phys. A**750**, 30 (2005).

[8] I. Arsene et al. (BRAHMS Collaboration), Nucl. Phys. A**757**, 1 (2005); K. Adcox et al. (PHENIX Collaboration), 184; B.B. Back et al. (PHOBOS Collaboration), 28; J. Adams et al. (STAR Collaboration), 102.

[9] M. Martinez, M. Strickland, Phys. Rev. C**78**, 034917 (2008); Eur. Phys. J. C**61**, 905 (2009).

[10] K. H. Ackermann et al. (STAR Collaboration), Nucl. Inst. & Meth. A**499**, 624 (2003).

[11] E. L. Feinberg, Nuovo Cim. A**34**, 391 (1976).

[12] X.-N. Wang, Z. Huang, and I. Sarcevic, Phys. Rev. Lett. **77**, 231 (1996).

[13] L. Adamczyk et al. (STAR Collaboration), Phys. Lett. B**760**, 689 (2016).

[14] L. Adamczyk et al. (STAR Collaboration), Phys. Lett. B**770**, 451 (2017).

[15] L. Adamczyk et al. (STAR Collaboration), Phys. Rev. C**96**, 024905 (2017).

[16] L. Adamczyk et al. (STAR Collaboration), Phys. Rev. Lett. **119**, 062301 (2017).

[17] S. A. Voloshin and Y. Zhang, Z. Phys. C**70**, 665 (1996).

[18] L. Adamczyk et al. (STAR Collaboration), Phys. Rev. Lett. **116**, 062301 (2016).

[19] L. Adamczyk et al. (STAR Collaboration), Phys. Rev. C**94**, 034908 (2016).

[20] L. Adamczyk et al. (STAR Collaboration), Phys. Rev. Lett. **116**, 112302 (2016).

[21] L. Adamczyk et al. (STAR Collaboration), Phys. Rev. Lett. **118**, 012301 (2017).

[22] L. Adamczyk et al. (STAR Collaboration), Nature **548**, 62 (2017).

[23] B. I. Abelev et al. (STAR Collaboration), Phys. Rev. C**76**, 024915 (2007).

[24] V. Khachatryan et al. (CMS Collaboration), J. High Energy Phys. **02**, 156 (2016).

[25] I. M. Chakravarti, R. G. Laha, and J. Roy, *Handbook of Methods of Applied Statistics* (John Wiley & Sons, New York, 1967), V. **1**, pp. 392–394.

[26] L. Adamczyk et al. (STAR Collaboration), Phys. Rev. C**94**, 064904 (2016).

[27] L. Adamczyk et al. (STAR Collaboration), Phys. Rev. Lett. **118**, 212301 (2017).

[28] *The STAR forward calorimeter system and forward tracking system*. Proposal, May 2017.

EPJ Web of Conferences **158**, 01005 (2017) DOI: 10.1051/epjconf/201715801005
QFTHEP 2017

Charmless Hadronic Beauty Decays at LHCb

Timothy Williams,[1]* on behalf of the LHCb collaboration

[1] *School of Physics and Astronomy, University of Birmingham, Birmingham, B15 2TT, United Kingdom*

Abstract. A summary of six LHCb results on the topic of charmless hadronic b-hadron decays is presented. These are comprised of: a search for the decay $B_s^0 \to K_S^0 K^+ K^-$ and updated branching fraction measurements of $B_{(s)}^0 \to K_S^0 h^+ h'^-$ decays (h=K,π) [1]; the first observation of the decays $B^0 \to p\bar{p}\pi^+\pi^-$, $B_s^0 \to p\bar{p}K^+K^-$,$B_s^0 \to p\bar{p}K^+\pi^-$ and strong evidence for the decay $B^0 \to p\bar{p}K^+K^-$ [2]; the first observation of the decay $B_s^0 \to p\bar{\Lambda}K^-$ [3]; a search for the decay $B_s^0 \to \phi\eta'$ [4]; the first observation of the decay $\Xi_b^- \to pK^-K^-$ [5] and evidence for CP-violation in $\Lambda_b^0 \to p\pi^-\pi^+\pi^-$ decays [6].

1 Introduction

Charmless b-hadron decays provide fertile ground for studying CP violation and searching for the effects of new physics. These decays can typically proceed via either $b \to u$ tree level diagrams or $b \to s, d$ gluonic loop level diagrams. The former of these transisitons is heavily suppressed by the CKM matrix element V_{ub}, which means the tree and loop level diagrams often have similar amplitudes. The consequences of this are two fold; firstly the rates of these processes are sensitive to the presence of new physics entering into the loop diagrams and secondly CP-violation can arise from the interference of the tree and loop level diagrams. Therefore, by measuring branching fractions and *CP*-violation observables indirect searches for physics beyond the standard model can be performed. Furthermore, these measurements provide vital input for QCD calculations.

An overview of recent results from the LHCb experiment on the topic of charmless b- hadron decays is presented. These all make use of the Run I dataset collected by the LHCb detector during the years 2011 and 2012 which correspond to integrated luminosities of approximately 1 fb^{-1} and 2 fb^{-1} respectively. The LHCb detector is a single-arm forward spectrometer with a pseudorapidity coverage of $2 < \eta < 5$, which has an acceptance for $b\bar{b}$ pairs of $\sim 25\%$. The detector is specialised for studying particles containing beauty and charm quarks. The unique sub-detectors relevant for the studies presented here are: the silicon strip vertex locator (VELO) surrounding the beam pipe which gives an impact parameter resolution of $(15 \pm \frac{29}{p_T})$ μm [7] and the two ring imaging cherenkov sub-detectors (RICH) which provide excellent particle identification ability [8].

2 Updated Branching Fraction Measurements of $B_{(s)}^0 \to K_S^0 h^+ h'^-$ Decays

Several extensions to the standard model, which involve the presence of new particles, introduce additional weak phases to the amplitude of $b \to q\bar{q}s$ (q=s,d,u) decays [9]. As these new physics effects

*e-mail: timothy.williams@cern.ch

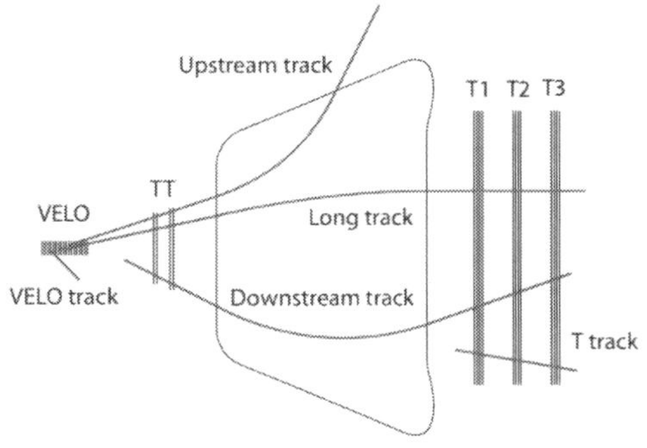

Figure 1. The different track types at LHCb. K_S^0 particles that decay inside the VELO produce Long tracks and K_S^0 particles that decay outside the VELO produce downstream tracks [12].

only enter via the loop diagrams which dominate the amplitudes of these charmless decays, the overall weak phase of $b \to c\bar{c}s$ decays would remain unchanged. Therefore, deviations in the values of weak phase measurements between those determined from $b \to q\bar{q}s$ (q=s,d,u) decays and from $b \to c\bar{c}s$ decays could be an indication of new physics. Mixing induced CP violation has been measured in charmed decays through channels such as $B^0 \to J/\psi K_S^0$ [10], and a time-dependent flavour tagged analysis of the three body Dalitz plot for $B^0 \to K_S^0\pi^+\pi^-$ and $B^0 \to K_S^0 K^+K^-$ decays would facilitate a comparison with charmless decays.

Unfortunately, the size of the data samples currently available at LHCb mean a flavour tagged time-dependent analysis of these decay modes is not feasible. In fact, the $B_s^0 \to K_S^0 K^+K^-$ is still unobserved. Therefore, the first aim is to observe all decay modes and develop selections that can later be used for Dalitz analyses. Previous LHCb results on these channels, making use of 1 fb^{-1} of 7 TeV data collected in 2011, made the first observation of the decay modes $B_s^0 \to K_S^0\pi^+\pi^-$, $B_s^0 \to K_S^0 K^\pm\pi^\mp$ and measured the branching fraction of all observed $B_{(s)}^0 \to K_S^0 h^+h'^-$ decays relative to the well measured channel $B^0 \to K_S^0\pi^+\pi^-$ [11]. Updated branching fraction results, which now include the 2 fb^{-1} of 8 TeV data collected in 2012, are presented below along with an updated search for the decay $B_s^0 \to K_S^0 K^+K^-$.

Due to the long lived nature of the neutral K_S^0 particle, two separate reconstruction categories are used in this measurement. If the K_S^0 decays inside the LHCb VELO the tracks from the daughter particles can be fully reconstructed. However, if the K_S^0 decays outside of the VELO there is less precise tracking information available and the mass resolution of the reconstructed K_S^0 is broader. Consequently, separate mass fits are required for each reconstruction category. As depicted in Figure 1, when the K_S^0 decays outside of the VELO it is labelled a downstream event and when the K_S^0 decays inside the VELO it is labelled as a long event.

A BDT classifier is used to separate signal and background events. The variables that enter into this classifier are deliberately chosen to avoid significant variations in efficiency over the phase space of the decay as such variations would be undesirable for future three body Dalitz plot analyses. There are two separate optimisations of the requirement on the BDT output for each final state; one for the favoured mode and one for the suppressed mode. It is possible for decays such as $\Lambda_b^0 \to K_S^0 p\pi^-$ to remain present as mis-ID backgrounds, therefore particle identification requirements are imposed to

Decay	Downstream Yield	Long Yield	Branching Fraction
$B^0 \to K^0_S\pi^+\pi^-$	2766 ± 66	1411 ± 45	
$B^0 \to K^0_S K^\pm\pi^\mp$	261 ± 24	160 ± 17	$(6.1 \pm 0.5 \pm 0.7 \pm 0.3) \times 10^{-6}$
$B^0 \to K^0_S K^+K^-$	1133 ± 39	685 ± 29	$(27.2 \pm 0.9 \pm 01.6 \pm 1.1) \times 10^{-6}$
$B^0_s \to K^0_S\pi^+\pi^-$	146 ± 19	74 ± 11	$(9.5 \pm 1.3 \pm 1.5 \pm 0.4) \times 10^{-6}$
$B^0_s \to K^0_S K^\pm\pi^\mp$	1100 ± 41	568 ± 28	$(84.3 \pm 3.5 \pm 7.4 \pm 3.4) \times 10^{-6}$
$B^0_s \to K^0_S K^+K^-$	12 ± 6	7 ± 4	$\in [0.4 - 2.5] \times 10^{-6}$ at 90% C.L.

Table 1. Signal yield and branching fraction results for each of the $B^0_{(s)} \to K^0_S h^+ h'^-$ decay modes

remove such backgrounds. There can be further backgrounds from B and Λ^0_b decays where the decay has proceeded through an intermediate charm (charmonium) resonance, for example $B^0 \to \bar{D}K^0_S$. These backgrounds are removed with explicit mass vetos of width 30(48) MeV around the world average mass of the charm (charmonium) resonance.

In order to extract the signal yields, the selected data samples are separated into 12 mutually exclusive categories; three separate final states, two different opitmisations and two different reconstruction categories. An unbinned simultaneous extended maximum likelihood fit to all categories is then performed. The signal decays are modelled with the sum of two Crystal Ball functions (CB) [13] and combinatorial background is modelled with a linear function. Cross-feed backgrounds, where events from one signal final state are mis-identified as another, are modelled with emperical shapes taken from simulated samples and Gaussian constraints are applied to the relative yield of these backgrounds. Partially reconstructed backgrounds from decays such as $B^0_s \to (K^{*0} \to K^0_S\pi^0)(\bar{K}^{*0} \to K^-\pi^+)$ are modelled with ARGUS functions [14]. The results of this fit for the favoured and suppressed optimisations are shown in Figure 2 and Figure 3 respectively. The resulting signal yields are shown in Table 1.

After correcting efficiencies for the variation over the phase space of the decay, the ratio of branching fractions (relative to the decay $B^0 \to K^0_S\pi^+\pi^-$) is determined separately for each reconstruction category. These are then combined for each decay mode by performing a minimum χ^2 fit where the convariance matrix includes the statistical and systematic uncertainties. The world average branching fraction measurement, with the previous LHCb result omitted, for the $B^0 \to K^0_S\pi^+\pi^-$ mode is then used to obtain the branching fraction results shown in Table 1.

The significance of the observed signal yield for the $B^0_s \to K^0_S K^+K^-$ mode is equivalent to 2.5 Gaussian standard deviations; this decay mode still remains unobserved. Therefore, a 90% confidence level interval is instead derived using the Feldman-Cousins method [15].

Despite not observing the $B^0_s \to K^0_S K^+K^-$ mode, these branching fraction measurements offer improved precision over, whilst still being consistent with, the previous LHCb results [11]. Work is now underway to perform Daltiz plot analyses of the dominant decay modes - $B^0 \to K^0_S\pi^+\pi^-$, $B^0_s \to K^0_S K^\pm\pi^\mp$ and $B^0 \to K^0_S K^+K^-$.

3 First Observation of the Decay $B^0_s \to p\bar{\Lambda}K^-$

The inclusive branching fraction of B decays to baryonic final states is $\sim 7\%$ of the total decay width; many experimental studies of baryonic B decays have been carried out. At previous e^+e^- collider experiments the observation of many baryonic decays of B^0 and B^+ mesons have been reported [16, 17]. More recently, the first observation of a baryonic B^+_c decay was reported by the LHCb collaboration [18]. However, the baryonic decay of a B^0_s meson has never previously been observed. The only

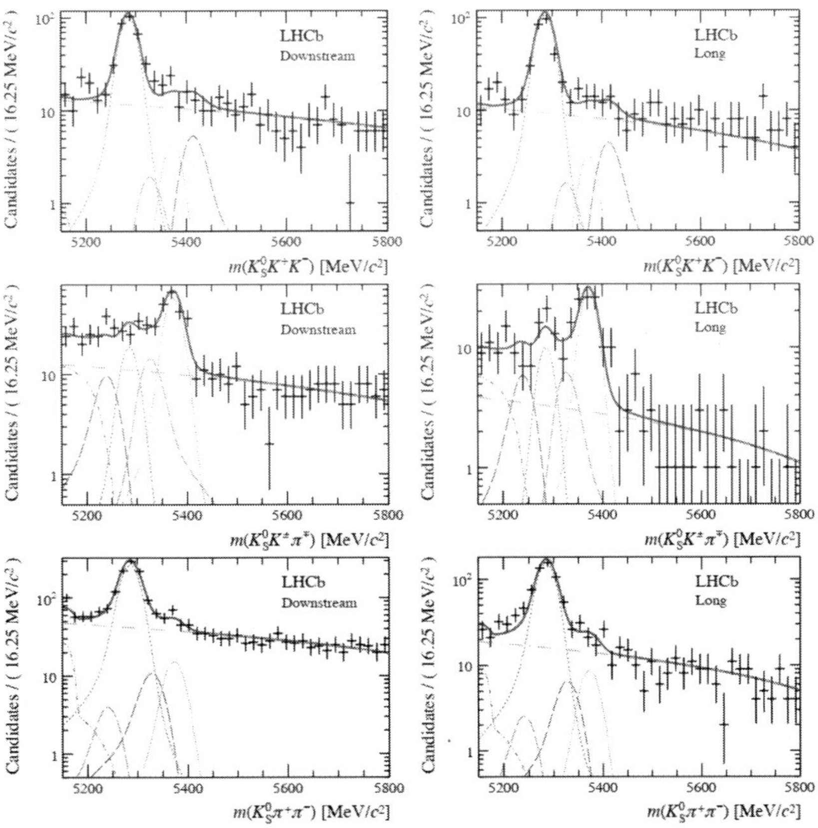

Figure 2. Invariant mass distributions for each possible final state with the favoured mode selection. The Downstream(Long) reconstruction category is shown on the right(left). The B^0 (B_s^0) signal candidates are shown by the magenta (orange) short dashed line, partially reconstructed backgrounds are shown by the red double dot dash line, the combinatorial background is shown by the grey dot dash line and the blue and green long dashed lines are mis-ID backgrounds. The full fit is shown by the solid blue line. [1]

evidence for such a decay is reported by the Belle collaboration, where 4.4σ evidence for the decay $B_s^0 \to \overline{\Lambda}_c^- \Lambda\pi^+$ was seen [19].

One interesting aspect of baryonic B decays is the hierachy of branching fractions; multi-body baryonic B decays generally have higher branching fractions than two body baryonic decays. It is predicted that the decay $B_s^0 \to p\overline{\Lambda}K^-$ has a branching fraction of the order of 10^{-6} which, based on similar charmless b decays, suggests it is likely to be accessible with the LHCb Run I dataset [20, 21]. It is also predicted that CP violating asymmetries in the decay $B^0 \to p\overline{\Lambda}\pi^-$ could be as high as $\sim 12\%$ [22], which provides extra motivation for studying the $B_{(s)}^0 \to p\Lambda h$ family of decays. For these reasons, a search for the decay $B_s^0 \to p\overline{\Lambda}K^-$ has been performed using the LHCb Run I dataset.

The branching fraction of the $B_s^0 \to p\overline{\Lambda}K^-$ is measured relative to the topologically very similar decay $B^0 \to p\overline{\Lambda}\pi^-$ in order to reduce systematic uncertainties. The branching fraction of this decay, $\mathcal{B}(B^0 \to p\overline{\Lambda}\pi^-) = (3.14 \pm 0.29) \times 10^{-6}$, has been previously measured by the BaBar and Belle collaborations [23]. For the same reasons as described in Section 2, the data is separated into two reconstruction categories, depending on whether the $\overline{\Lambda}$ is reoncstructed inside or outside of the LHCb

 DOI: 10.1051/epjconf/201715801005

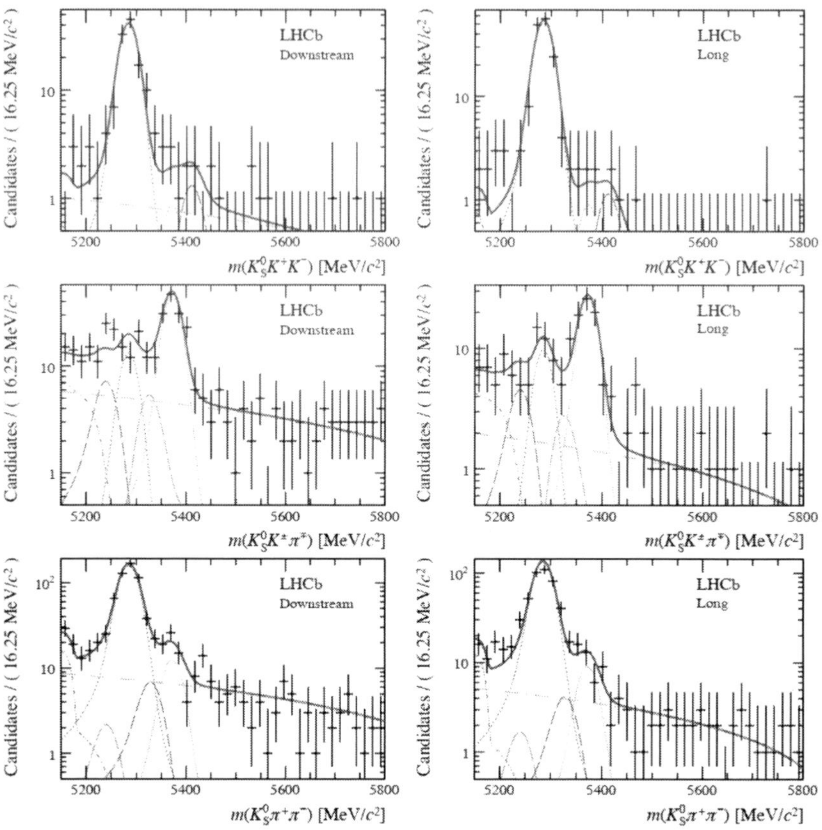

Figure 3. Invariant mass distributions for each possible final state with the suppressed mode selection. The Downstream(Long) reconstruction category is shown on the right(left). The individual fit components follow the same convention as Figure 2. [1]

VELO. After applying explicit mass vetoes to remove background events from the decay $B^0 \to \overline{\Lambda}_c p$, an MLP classifier is used to reduce combinatorial background. There are then further PID requirements applied to the p and $K^-/\,\pi^-$ from the B meson, but no PID requirements are placed on the p from the $\overline{\Lambda}$ because no evidence for contamination from $K^0_S \to \pi^+\pi^-$ decays has been found.

The signal and control channel yields are extracted by performing an extended maximum likelihood fit to the signal and control channel simultaneously. The results of these fits for both years of data taking and both reconstruction categories are shown in Figure 4. These mass fit models include components for: signal, combinatorial background and partially reconstructed background from the decays $B^0 \to p\overline{\Sigma}\pi^-$ and $B^0_s \to p\overline{\Sigma}K^-$.

It is clear from Figure 4 that signal events are unambiguously present in both channels; the resulting total yields are $N(B^0_s \to p\overline{\Lambda}K^-) = 234 \pm 29$ and $N(B^0 \to p\overline{\Lambda}\pi^-) = 519 \pm 28$. This is the first observation of a baryonic B^0_s decay, with a statistical significance of greater than 15 gaussian standard deviations. The branching fraction is subsequently measured to be,

$$\mathcal{B}(B^0_s \to p\overline{\Lambda}K^-) + \mathcal{B}(\bar{B}^0_s \to \bar{p}\Lambda K^+) = \left[5.46 \pm 0.61 \pm 0.57 \pm 0.50(\mathcal{B}) \pm 0.32(\frac{f_s}{f_d})\right] \times 10^{-6}$$

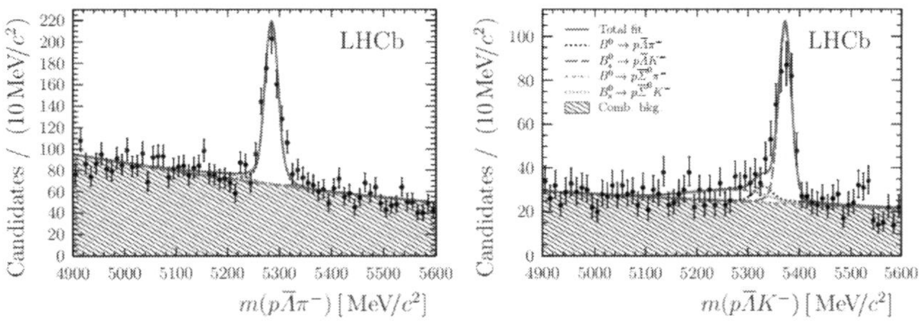

Figure 4. Left (Right): The invariant mass distribution of $B^0 \to p\bar{\Lambda}\pi^-$ ($B^0_s \to p\bar{\Lambda}K^-$) channel candidate events. The grey shaded area is combinatorial background, partially reconstructed backgrounds are represented by the dot and dot dash green and pink lines and signal is represented by black and red dashed lines. The full fit is shown by the solid blue line. [3]

where the first uncertainty is statistical, the second uncertainty is systematic, the third is due to the control channel branching fraction and the fourth is due to the knowledge of the ratio of B^0_s/B^0 b quark fragmentation fractions ($\frac{f_s}{f_d}$). The sum of the branching fractions from both B^0_s and $\bar{B}^0_s$ is quoted because the identical final states mean it is not possible to determine the flavour of the B^0_s meson without a flavour tagged analysis.

Another area of interest in the study of baryonic B decays is threshold enhancement in the baryon anti-baryon system. This effect has previously been observed by the BaBar and Belle collaborations in several baryonic B decays [17]. This effect is investigated in the $B^0 \to p\bar{\Lambda}\pi^-$ and $B^0_s \to p\bar{\Lambda}K^-$ decays by using the *sPlot* technique to perform a background subtraction and correct for efficiency variations across the phase space of the decay [24]. The efficiency corrected and background subtracted invariant mass distributions for the $p\,\bar{\Lambda}$ system are shown in Figure 5; clear threshold enhancement is observed.

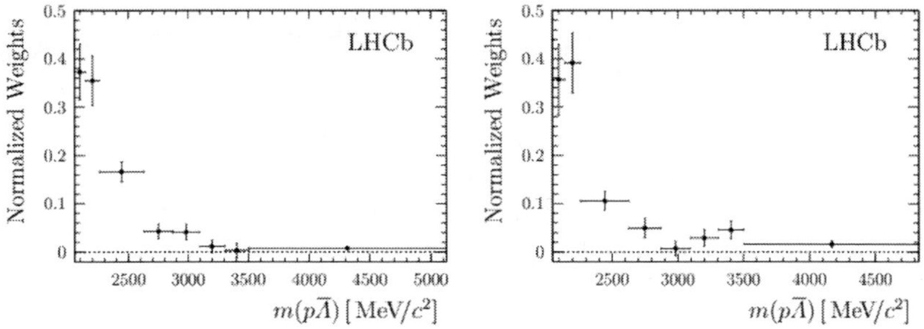

Figure 5. Left (Right): Normalised, background subtracted and efficiency corrected m($p\ \bar{\Lambda}$) invariant mass distributions for the $B^0 \to p\bar{\Lambda}\pi^-$ ($B^0_s \to p\bar{\Lambda}K^-$) candidates. [3]

4 Search for $B^0_{(s)} \to p\bar{p}h^+h^-$ decays

As described in Section 3 the baryonic decay of a B^0_s meson has now been observed. However, a four body baryonic decay of the B^0_s meson is still to be observed and previous results for the B^0 meson only set an upper limit $\mathcal{B}(B^0 \to p\overline{p}\pi^+\pi^-) < 2.5 \times 10^{-4}$ at 90% confidence level[25]. There is additional motivation in studying four body baryonic B meson decays because they are an ideal place to search for CP violation using triple-product correlations (TPCs); in contrast to the case of three body decays, they do not involve the spins of the final state particles. This motivation is strengthened by the first evidence of CP violation in baryonic B decays being seen in $B^+ \to p\overline{p}K^+$ decays [26] and predictions of CP violation as high as 20% in other baryonic B decays [27].

A search for the family of decays $B^0_{(s)} \to p\overline{p}h^+h^-$ is reported, where h is a kaon or pion. The branching fractions of these decays are measured relative to the decay $B^0 \to J/\psi K^{*0}$, where the resonances are reconstructed through the $J/\psi \to p\overline{p}$ and $K^{*0} \to K^+\pi^-$ decay channels. In order to remove charm and charmonium resonances in the signal channels the requirement $m(p\overline{p}) < 2.85$ GeV is imposed on the signal channels and specfic mass vetoes are applied to remove Λ_c^+ and D^0 decays. The selection further consists of a BDT, which makes use of 15 variables, to reduce combinatorial background and PID requirements remove mis-ID backgrounds.

The signal yields are extracted with an extended maximum likelihood fit which is simultaneous across the invariant masses of all 3 possible final states: $m(p\overline{p}K^+\pi^-)$, $m(p\overline{p}K^+K^-)$ and $m(p\overline{p}\pi^+\pi^-)$. A separate fit is performed for the $B^0 \to J/\psi K^{*0}$ control channel which is multi-dimensional for the $m(p\overline{p}K^+\pi^-)$, $m(p\overline{p})$ and $m(K^+\pi^-)$ invariant masses. These fits are shown in Figure 6 and the resulting yields and branching fractions are shown in Table 2. This is the first observation of the decays $B^0 \to p\overline{p}\pi^+\pi^-$ and $B^0_s \to p\overline{p}K^+K^-$ with significances of $> 25\sigma$. The decay $B^0_s \to p\overline{p}K^+\pi^-$ is also observed with a significance of 6.5σ, there is strong evidence for the decay $B^0 \to p\overline{p}K^+K^-$ at the level of 4.1σ and evidence at the level of 2.6σ is seen for the $B^0_s \to p\overline{p}\pi^+\pi^-$ decay. A limit is set on the $B^0_s \to p\overline{p}\pi^+\pi^-$ channel by integrating the likelihood with a uniform prior in the region of positive branching fraction.

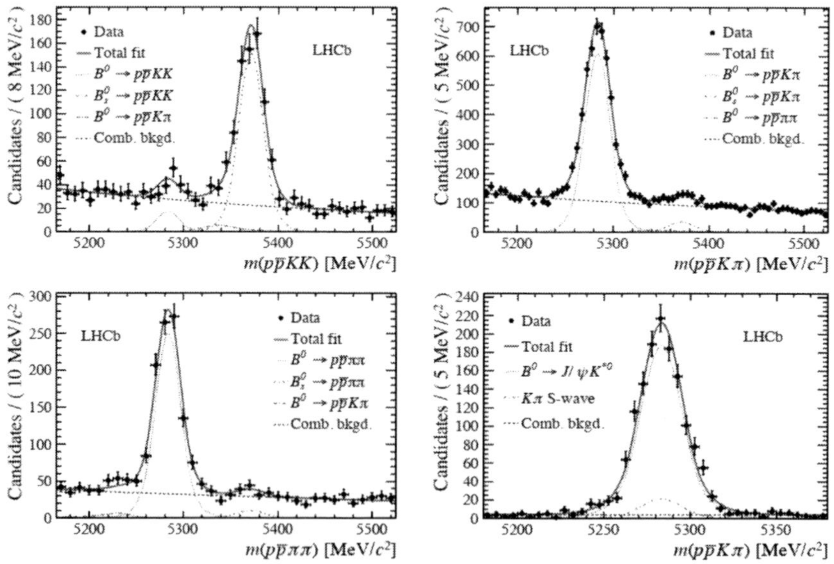

Figure 6. Top Left, Top Right and Bottom Left: Signal channel simultaneous mass fit to all three final states. B^0 signal is shown by the dotted green line, B^0_s signal by the dot dash red line and cross feed background is shown by the dot dash purple line. Bottom Right: Projection of the 3D control channel mass fit. [2]

DOI: 10.1051/epjconf/201715801005

Channel	Yield	Branching Fraction/10^{-6}
$B^0 \to p\bar{p}K^+K^-$	68 ± 17	$0.113 \pm 0.028 \pm 0.011 \pm 0.008$
$B^0 \to p\bar{p}K^+\pi^-$	4155 ± 83	$5.9 \pm 0.3 \pm 0.3 \pm 0.4$
$B^0 \to p\bar{p}\pi^+\pi^-$	902 ± 35	$2.7 \pm 0.1 \pm 0.1 \pm 0.2$
$B^0_s \to p\bar{p}K^+K^-$	635 ± 32	$4.2 \pm 0.3 \pm 0.2 \pm 0.3 \pm 0.3$
$B^0_s \to p\bar{p}K^+\pi^-$	246 ± 39	$1.3 \pm 0.21 \pm 0.11 \pm 0.09 \pm 0.08$
$B^0_s \to p\bar{p}\pi^+\pi^-$	39 ± 16	< 0.66 at 90% CL

Table 2. Signal yields and branching fraction results for the six $B^0_{(s)} \to p\bar{p}h^+h^-$ decays. The first uncertainty is statistical, the second systematic, the third is due to the uncertainty on the normalisation channel branching fraction and the fourth (in the case of B^0_s mesons) is due to the ratio of B^0_s/B^0 fragmentation fractions.

As described in Section 3, threshold enhancement has been observed in the baryon antibaryon system of several B meson decays. To investigate threshold enhancement in the $B^0 \to p\bar{p}K^+\pi^-$ decay, the *sPlot* technique is again used to perform a background subtraction and correct for the efficiency variation across the phase space of the decay. The normalised, background subtracted and efficiency corrected distribution of $m(p\bar{p})$ can be seen in Figure 7 [24]. As with the decay $B^0_s \to p\bar{\Lambda}K^-$ (see Figure 5), clear threshold enhancement is present.

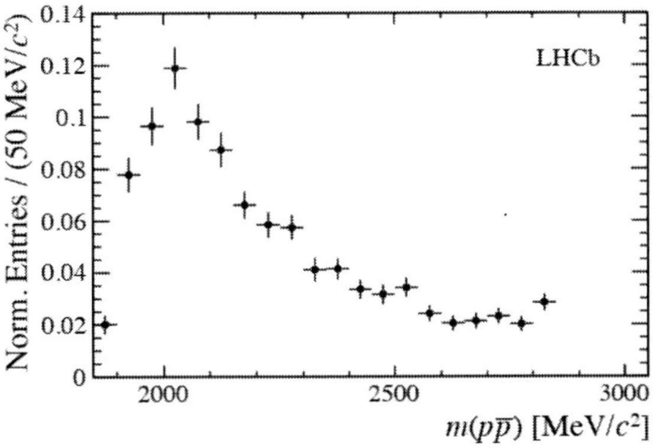

Figure 7. Normalised, background subtracted and efficiency corrected $m(p\bar{p})$ distribution for the $B^0 \to p\bar{p}K^+\pi^-$ decay channel. [2]

5 Search for the decay $B^0_s \to \eta'\phi$

Current experimental knowledge of B^0_s decays to a light pseudoscalar (P) meson and a vector meson (V) is very limited and theoretical predictions for the branching fractions of such decays have large uncertainties and cover a large range. For example, a QCD factorisation approach can yield the result $\mathcal{B}(B^0_s \to \eta'\phi) = (0.05^{1.18}_{-0.19}) \times 10^{-6}$ [28], whereas a perturbative QCD approach yields the prediction $\mathcal{B}(B^0_s \to \eta'\phi) = (20.0^{+16.3}_{-9.1}) \times 10^{-6}$ [29] (see Table 1 in Ref. [4] for a summary of theoretical predictions). These large uncertainties are a consequence of the currently limited knowledge of penguin contributions, the ω- ϕ mixing angle, the s-quark mass and form factors. Therefore, a measurement of $\mathcal{B}(B^0_s \to \eta'\phi)$ would be a useful input to the knowledge of B^0_s to ϕ form factors and would subsequently improve the accuracy of branching fraction predictions for $B^0_s \to P\phi$ decays.

A search for the decay $B_s^0 \to \eta'\phi$ is reported, where the η' is reconstructed through the channel $\eta' \to \pi^+\pi^-\gamma$ and the ϕ through the channel $\phi \to K^+K^-$. The well studied decay $B^+ \to K^+\eta'$ is used as a control channel due to its relatively large branching fraction $\mathcal{B}(B^+ \to K^+\eta') = (70.6 \pm 2.5) \times 10^{-6}$ and the presence of only combinatorial background [23, 30]. The selection makes use of a BDT with 9 variables to reject the majority of combinatorial background and particle identification requirements are imposed on the hadrons and the photon from the η'.

The signal yields are extracted with an extended maximum likelihood fit to the $B_s^0 \to \eta'\phi$ and $B^+ \to K^+\eta'$ channels simultaneously. The PDF used in the $B_s^0 \to \eta'\phi$ ($B^+ \to K^+\eta'$) channel is two dimensional for the $m(\eta'\phi)$ and $m(\pi^+\pi^-\gamma)$ ($m(\eta'K^+)$ and $m(\pi^+\pi^-\gamma)$) variables. In the case of the $B_s^0 \to \eta'\phi$ channel, there is an irreducible background from $B_s^0 \to \phi\phi$ decays where one of the ϕ mesons decays through $\phi \to K^+K^-$ and the other through $\phi \to \pi^+\pi^-\pi^0$ but one of the photons from the decay of the π^0 is not reconstructed. This background is therefore modelled in the fit with a two-dimensional Gaussian kernel function [31]. The fit results for the $B_s^0 \to \eta'\phi$ channel are shown in Figure 8.

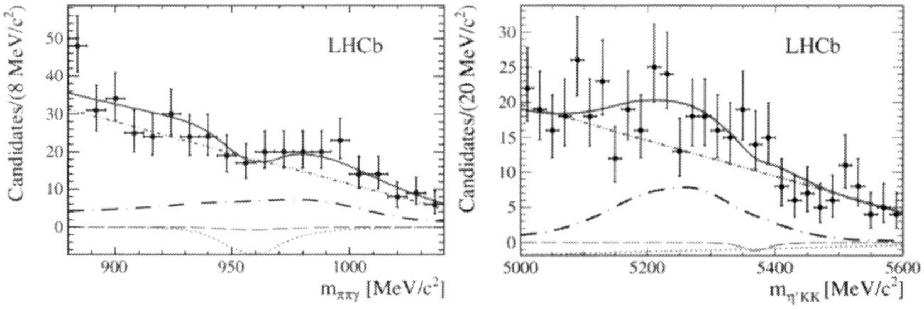

Figure 8. Left (Right): Distributions of m(π^+ π^- γ) (m(η' ϕ)) for the $B_s^0 \to \eta'\phi$ channel. The solid blue line represents the result of the two-dimensional simultaneous fit to both channels. The red dashed line shows the signal component, the green dashed line shows the combinatorial background with a real η', the blue dot-dash line shows combinatorial background without a real η' and the black dot-dash line is the irreducible $B_s^0 \to \phi\phi$ background described in the text. [4]

The yields from this fit are $N(B_s^0 \to \eta'\phi) = -3.2^{+5.0}_{-3.8}$ and $N(B^+ \to K^+\eta') = 11081 \pm 127$, therefore bayesian upper limits are set on $\mathcal{B}(B_s^0 \to \eta'\phi)$ using a uniform prior. The resulting limit is,

$$\mathcal{B}(B_s^0 \to \eta'\phi) < 0.82(1.01) \times 10^{-6} \text{ at } 90\% \ (95\%) \text{ CL},$$

which is the first upper limit to be set on $\mathcal{B}(B_s^0 \to \eta'\phi)$. This result is consistent with the lower end of the range of predictions available for this branching fraction; the central values of several predictions are inconsistent with this result, see for example Refs. [29, 32–34]. This result therefore provides very useful constraints on the theoretical models used to predict branching fractions and CP asymmetries of charmless B_s^0 decays to a pseudoscalar and vector meson final state.

6 Search for the decays $\Xi_b^-, \Omega_b^- \to ph^-h'^-$

Prior to the collection of the Run I dataset by LHCb, opportunities to study charmless decays of b-baryons were very limited. Consequently, the only observed decays were $\Lambda_b^0 \to p\pi^-$ and $\Lambda_b^0 \to pK^-$ by the CDF collaboration [35]. More recently, LHCb has observed the Λ_b^0 decays $\Lambda_b^0 \to pK_S^0\pi^-$ [36], $\Lambda_b^0 \to \Lambda\pi^+\pi^-$, $\Lambda_b^0 \to \Lambda K^+\pi^-$, $\Lambda_b^0 \to \Lambda K^+K^-$ [37] and $\Lambda_b^0 \to \Lambda\phi$ [38]. There has also been evidence at the level of 3σ seen for the decay $\Lambda_b^0 \to \Lambda\eta$ [21]. However a charmless decay of a Ξ_b^- or Ω_b^- baryon is yet to be observed, therefore a search for the decays $\Xi_b^-, \Omega_b^- \to ph^-h'^-$ where h=K^-, π^- is presented. It is not currently possible to measure the absolute branching fractions of Ξ_b^- and Ω_b^- decays because the fragmentation fractions, $f_{\Xi_b^-}$ and $f_{\Omega_b^-}$, are not known. This search uses the topologically similar and well measured decay $B^- \to K^+K^-K^-$ as a control channel, therefore the reported results are the product of branching fraction and ratio of fragmentation fractions, denoted $R_{ph^-h'^-}$.

The event selection firstly consists of a neural network, which makes use of 8 variables, to reduce combinatorial background. A tight particle identification requirement is then imposed on the proton to reject background from $B^- \to K^+h^-h'^-$ decays. Further particle identification requirements on the other hadrons are optimised in a manner that ensures mutually exclusive samples for each possible final state. There are then vetoes applied to remove charm backgrounds from decays such as $\Xi_b^- \to \Xi_c^0 h^- \to pK^-h^-$ and $B^- \to D^0K^- \to K^+K^-K^-$.

The signal yields are extracted using an extended unbinned maximum likelihood fit which is simultaneous across the invariant mass distributions of the three possible final states ($m(pK^-K^-)$, $m(pK^-\pi^-)$ and $m(p\pi^-\pi^-)$). The use of a simultaneous fit allows cross-feed background, where candidates from one final state are mis-identified as another, to be constrained to expected values. The control channel yield is extracted with a separate fit to the $m(K^+K^-K^-)$ distribution. There are also partially reconstructed backgrounds present that have to be modelled in the fit. In the $m(ph^-h'^-)$ distributions these arise from decays such as $\Xi_b^- \to N^+h^-h'^- \to p\pi^0h^-h'^-$ where the π^0 is not reconstructed. These backgrounds are modelled with an ARGUS function [14] convolved with a Gaussian function.

The projections of the simultaneous fit and the separate control channel fit are shown in Figure 9; the fitted yields and subsequent branching fraction times fragmentation fraction ratios are shown in Table 3. The significance of the $\Xi_b^- \to pK^-K^-$ signal yield is 8.7σ; this is the first observation of a charmless Ξ_b^- decay. There is also 3.4σ evidence for the decay $\Xi_b^- \to pK^-\pi^-$, but all other signal yields have a significance of $< 2.0\sigma$ meaning the charmless decay of an Ω_b^- baryon still remains unobserved. For these channels with signifiance $< 2.0\sigma$ an upper limit is set on $R_{ph^-h'^-}$ by integrating the likelihood after multiplying by a prior probability distribution which is uniform in the positive region.

Although it is not possible to measure absolute branching fractions, the observation of the $\Xi_b^- \to pK^-K^-$ decay means it is possible to measure the relative branching fractions of the Ξ_b^- decays. This cancels the dependance on fragmentation fractions, which is useful for comparisons with theoretical predictions. The measured relative branching fractions are,

$$\frac{\mathcal{B}(\Xi_b^- \to pK^-\pi^-)}{\mathcal{B}(\Xi_b^- \to pK^-K^-)} = 0.98 \pm 0.27(stat) \pm 0.09(syst),$$

$$\frac{\mathcal{B}(\Xi_b^- \to p\pi^-\pi^-)}{\mathcal{B}(\Xi_b^- \to pK^-K^-)} = 0.28 \pm 0.16(stat) \pm 0.13(syst) < 0.56(0.63),$$

where the upper limit quoted is at the 90%(95%) CL.

It is of interest to study the intermediate resonances that occur in the $\Xi_b^- \to pK^-K^-$ channel; it is hoped that when larger data samples are available an amplitude analysis to dis-entangle the intermediate resonances and to search for CP violation can be performed. Whilst this is not currently possible, it is still of interest to inspect the intermediate resonances. Figure 10 shows the background subtracted and efficiency corrected $m(pK^-)_{min}$ distribution for the $\Xi_b^- \to pK^-K^-$ channel; $m(pK^-)_{min}$ is the smaller of the two possible $m(pK^-)$ values for each candidate. This distribution shows clear peaks consistent with the $\Lambda(1520)$ and a combination of the $\Lambda(1670)$ and $\Lambda(1690)$ states. The less prominent peaks at higher mass suggest the possibility of further states being present.

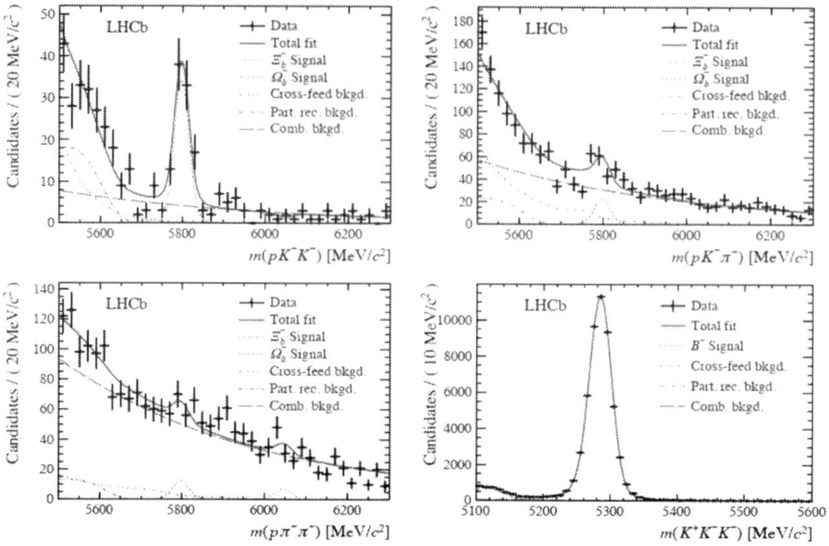

Figure 9. Top Left, Top Right and Bottom Left: Projections of the simultaneous fit to the $m(ph^-h'^-)$ final states. Bottom Right: The control channel fit to $m(K^+K^-K^-)$ mass distribution. The full fits are shown by the blue solid line, Ξ_b^- and B^- signal components are shown by the pink dashed line, Ω_b^- signal components are shown by the green dashed line, cross-feed backgrounds are shown by the red dot-dash line, partially reconstructed backgrounds are shown by the blue double dot-dash line and combinatorial background is shown by the grey long dashed line. [5]

Channel	Yield	$R_{ph^-h'^-}$ (10^{-5})
$\Xi_b^- \to pK^-K^-$	82.9 ± 10.4	$265 \pm 35 \pm 47$
$\Xi_b^- \to pK^-\pi^-$	59.6 ± 16.0	$259 \pm 64 \pm 49$
$\Xi_b^- \to p\pi^-\pi^-$	33.2 ± 17.9	$< 147(166)$
$\Omega_b^- \to pK^-K^-$	-2.8 ± 2.5	$< 18(22)$
$\Omega_b^- \to pK^-\pi^-$	-7.6 ± 9.2	$< 51(62)$
$\Omega_b^- \to p\pi^-\pi^-$	20.1 ± 13.8	$< 109(124)$
$B^- \to K^+K^-K^-$	50490 ± 250	

Table 3. Yield and $R_{ph^-h'^-}$ results for each of the decay channels studied, where $R_{ph^-h'^-}$ is the product of branching fraction and the ratio of Ξ_b^-/Ω_b^- and B^- fragmentation fractions $\frac{f_{\Xi_b^-/\Omega_b^-}}{f_{B^-}}$. The limits set are at the (90%)95% confidence level.

7 Search for CP violation in $\Lambda_b^0 \to p\pi^-\pi^+\pi^-$ decays

The levels of CP violation predicted by the standard model, through the CKM mechanism, are not enough to explain the matter-antimatter imbalance in the universe. It is therefore of great interest to search for additional sources of CP violation to explain the existence of a matter dominated universe. Thus far, CP violation has been observed in B_s^0 [39], B^+ [40] and B^0 [41] meson decays but has not been seen in the decay of a Λ_b^0 baryon. The LHC produces a copious amount of Λ_b^0 baryons; approximately 20% of all b hadrons prodcued are Λ_b^0 baryons. Therefore, LHCb exploits these unprecedented Λ_b^0 samples by searching for CP violation in $\Lambda_b^0 \to p\pi^-\pi^+\pi^-$ decays.

DOI: 10.1051/epjconf/201715801005

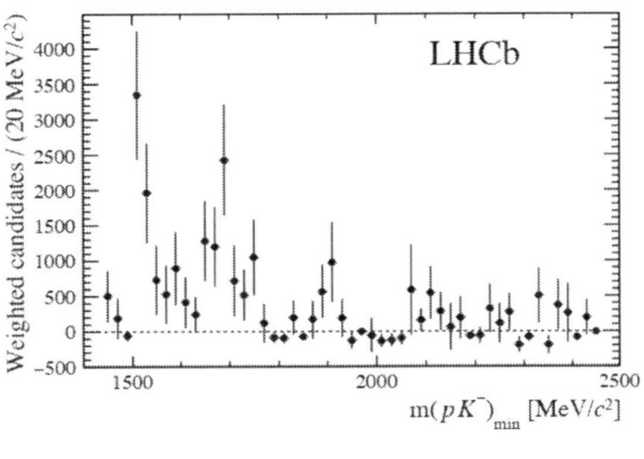

Figure 10. Background subtracted and efficiency corrected $m(pK^-)_{min}$ distribution for $\Xi_b^- \to pK^-K^-$ candidates. [5]

This search is performed by studying asymmetries in the $\hat{T}$ operator, which is the unitary operator that reverses both momentum and spin three-vectors [1]. Scalar triple products are defined,

$$C_{\hat{T}} = \vec{p}_p \cdot (\vec{p}_{\pi^-} \times \vec{p}_{\pi^+}), \qquad \bar{C}_{\hat{T}} = \vec{p}_{\bar{p}} \cdot (\vec{p}_{\pi^+} \times \vec{p}_{\pi^-}) \tag{1}$$

where $\vec{p}_h$ is the momentum of the final state hadron, h, in the Λ_b^0 centre-of-mass frame. The π^- with the largest momenta in the Λ_b^0 rest frame is used (π^-_{fast}). Asymmetries can then be defined as,

$$A_{\hat{T}}(C_{\hat{T}}) = \frac{N(C_{\hat{T}} > 0) - N(C_{\hat{T}} < 0)}{N(C_{\hat{T}} > 0) + N(C_{\hat{T}} < 0)}, \qquad \bar{A}_{\hat{T}}(\bar{C}_{\hat{T}}) = \frac{\bar{N}(-\bar{C}_{\hat{T}} > 0) - \bar{N}(-\bar{C}_{\hat{T}} < 0)}{\bar{N}(-\bar{C}_{\hat{T}} > 0) + \bar{N}(-\bar{C}_{\hat{T}} < 0)} \tag{2}$$

where $N(\bar{N})$ is the number of Λ_b^0 ($\overline{\Lambda}_b^0$) decays satisfying the given condition. Using these asymmetries, CP violation and P violation observables are defined:

$$a_{CP}^{\hat{T}-odd} = \frac{1}{2}(A_{\hat{T}} - \bar{A}_{\hat{T}}), \qquad a_{P}^{\hat{T}-odd} = \frac{1}{2}(A_{\hat{T}} + \bar{A}_{\hat{T}}). \tag{3}$$

Any deviation from zero in these operators would indicate the presence of CP/P violation. The advantage of using triple product asymmetries is that they are largely insensitive to prodcution asymmetries and charge asymmetries caused by the LHCb detector.

The event selection makes use of a BDT to reduce combinatorial background and optimised particle identification requirements. An extended unbinned maximum likelihood fit to the $m(p\pi^-\pi^+\pi^-)$ invariant mass distribution is used to extract the total signal yield of both Λ_b^0 and $\overline{\Lambda}_b^0$ decays. This fit is shown in Figure 11 and has several components: signal decays are modelled by a Gaussian core with power law tails [13]; combinatorial background is modelled with an exponential function; partially reconstructed backgrounds from five body Λ_b^0 decays where one particle is not reconstructed are modelled with an ARGUS [14] function convolved with a Gaussian and mis-ID backgrounds are modelled with shapes fixed from simulation. The signal yield from this fit is $N(\Lambda_b^0 \to p\pi^-\pi^+\pi^-) = 6646 \pm 105$ and is the first observation of the decay $\Lambda_b^0 \to p\pi^-\pi^+\pi^-$.

In order to extract the CP and P observables in Eqn. (3), the data sample is split into four categories by Λ_b^0 or $\overline{\Lambda}_b^0$ flavour and sign of $C_{\hat{T}}$ or $\bar{C}_{\hat{T}}$. An extended unbinned maximum likelihood fit is then

[1] The $\hat{T}$ operator is different to the time-reversal operator

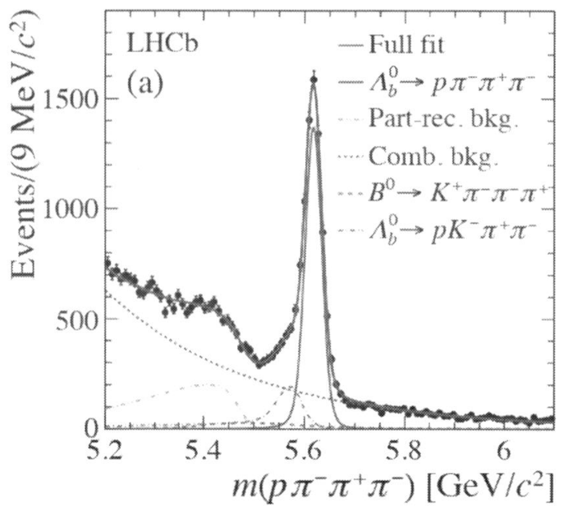

Figure 11. The invariant mass distribution of $\Lambda_b^0 \to p\pi^-\pi^+\pi^-$ candidates with the fit results overlayed. The fit has several components: the $\Lambda_b^0 \to p\pi^-\pi^+\pi^-$ signal component is shown by the solid red line, combinatorial background is shown by the blue dashed line, partially reconstructed backgrounds from 5 body decays are shown by the light blue double dot dashed line, mis-identified $\Lambda_b^0 \to pK^-\pi^+\pi^-$ events are shown by the purple dot dash line and mis-identified $B^0 \to K^+\pi^-\pi^-\pi^+$ candidates are shown by the brown dashed line. The full fit is shown by the solid blue line. [6]

performed simultaneously to all four categories to extract the asymmetries $A_{\hat{T}}$ and $\bar{A}_{\hat{T}}$ (defined in Eqn (2)), from which the CP and P observables $a_{CP}^{\hat{T}-odd}$ and $a_{p}^{\hat{T}-odd}$ are calculated (defined in Eqn (3)).

Many resonances are expected in the four body $\Lambda_b^0 \to p\pi^-\pi^+\pi^-$ decay and it is expected different resonances will exhibit different levels of CP violation. It is entirely possible that the CP violation from different resonances would cancel if a phase space integrated asymmetry was calculated, which would considerably reduce the sensitivity of this search. Therefore, different regions of the phase space are investigated by calculating the $a_{CP}^{\hat{T}-odd}$ and $a_{p}^{\hat{T}-odd}$ observables for two separate binning schemes of the phase space. The first binning scheme, denoted "Scheme A", makes use of two body invariant masses in order to exploit the strong resonant structure from decays such as $\Delta(1232)^{++} \to p\pi^+$. "Scheme B" consists of 10 bins in the angle Φ which is the angle between the p π^-_{fast} and π^-_{slow} π^+ decay planes, shown in Figure 12. The objective of this binning scheme is to probe the interference between different resonant contributions.

Figure 13 shows the results for $a_{CP}^{\hat{T}-odd}$ and $a_{p}^{\hat{T}-odd}$ in each binning scheme. A χ^2 test is used to assess the compatibility of each binning scheme with the null hypothesis of CP symmetry; the p-values are $4.9 \times 10^{-2}(2.0\sigma)$ and $7.1 \times 10^{-4}(3.4\sigma)$ for the binning schemes A and B respectively. The combined significance of the measured level of CP violation in $\Lambda_b^0 \to p\pi^-\pi^+\pi^-$ decays is determined using a permutation test where 40,000 pseudoexperiments are generated with randomly assigned $\Lambda_b^0/\overline{\Lambda}_b^0$ flavour. The product of p-values from both binning schemes is compared between these pseudoexperiments and the values obtained in data, this leads to a combined significance for CP violation in $\Lambda_b^0 \to p\pi^-\pi^+\pi^-$ decays of 3.3σ. This is the first evidence for CP violation in a baryonic decay and it will be of great interest to study this asymmetry further when larger data samples are available.

DOI: 10.1051/epjconf/201715801005

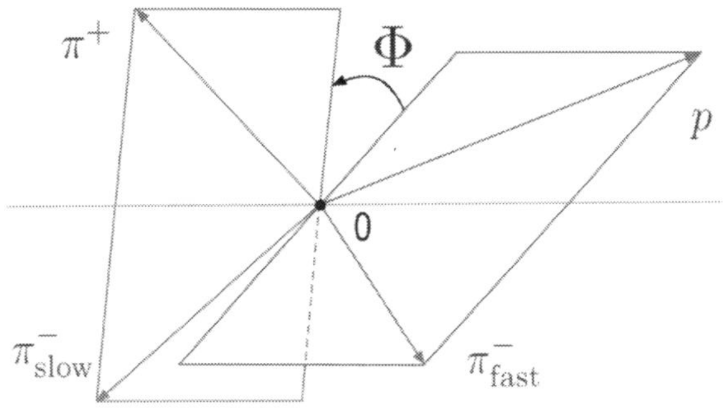

Figure 12. A diagram showing the defnition of the angle Φ which is used for binning scheme B. [6]

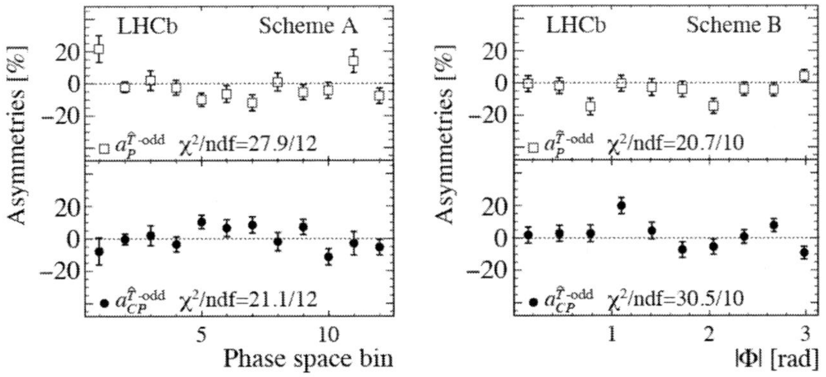

Figure 13. Left (Right): The CP(bottom) and P (top) violation asymmetries for binning scheme A(B). [6]

8 Summary

The study of charmless b hadron decays continues to yield many interesting results, and LHCb is leading the way in studying this sector. The branching fractions results for the $B^0_{(s)} \to K^0_S h^+ h'^-$ family of decays have been updated and the limits on the branching fraction of the $B^0_s \to K^0_S K^+ K^-$ have been narrowed. These results set the foundations for further Dalitz plot studies of these decay channels. Several baryonic decays of b mesons have been observed for the first time; the decay $B^0_s \to p\bar{\Lambda}K^-$ was the first baryonic decay of a B^0_s meson to be observed. The four body decays $B^0 \to p\bar{p}\pi^+\pi^-$, $B^0_s \to p\bar{p}K^+\pi^-$ and $B^0_s \to p\bar{p}K^+K^-$ have been observed for the first time, which open up opportunities for CP violation studies in four body baryonic b meson decays. A stringent upper limit has been set on the branching fraction of the decay $B^0_s \to \phi\eta'$, $\mathcal{B}(B^0_s \to \phi\eta') = 0.82(1.01) \times 10^{-6}$ at the 90%(95%) CL, which is a very useful input to the theoretical models used to predict the properties of $B^0_s \to PV$ decays. In the baryonic sector, the observation of the decay $\Xi^-_b \to pK^-K^-$ is the first observation of a charmless Ξ^-_b decay and opens up opportunities to search for CP violation in charmless Ξ^-_b decays. Lastly, 3.3σ evidence for CP violation in $\Lambda^0_b \to p\pi^-\pi^+\pi^-$ decays has been found; this is the first evidence for CP violation in a baryonic beauty decay. With further studies and larger data samples it is hoped these differences could contribute to understanding the absence of anti-matter in the universe.

References

[1] R. Aaij et al. (LHCb collaboration) (2017), submitted to JHEP, 1707.01665

[2] R. Aaij et al. (LHCb collaboration) (2017), submitted to Phys. Rev. D, 1704.08497

[3] R. Aaij et al. (LHCb collaboration), Phys. Rev. Lett. **119**, 041802 (2017), 1704.07908

[4] R. Aaij et al. (LHCb collaboration), JHEP **05**, 158 (2017), 1612.08110

[5] R. Aaij et al. (LHCb collaboration), Phys. Rev. Lett. **118**, 071801 (2017), 1612.02244

[6] R. Aaij et al. (LHCb collaboration), Nature Physics **13**, 391 (2017), 1609.05216

[7] R. Aaij et al., JINST **9**, P09007 (2014), 1405.7808

[8] M. Adinolfi et al., Eur. Phys. J. **C73**, 2431 (2013), 1211.6759

[9] M. Ciuchini, E. Franco, G. Martinelli, A. Masiero, L. Silvestrini, Phys. Rev. Lett. **79**, 978 (1997), hep-ph/9704274

[10] R. Aaij et al. (LHCb collaboration), Phys. Rev. Lett. **115**, 031601 (2015), 1503.07089

[11] R. Aaij et al. (LHCb collaboration), JHEP **10**, 143 (2013), 1307.7648

[12] R. Aaij et al. (LHCb collaboration), Int. J. Mod. Phys. **A30**, 1530022 (2015), 1412.6352

[13] T. Skwarnicki, Ph.D. thesis, Cracow, INP (1986), http://lss.fnal.gov/cgi-bin/find_paper.pl?other/thesis/skwarnicki.pdf

[14] H. Albrecht, H. Ehrlichmann, G. Harder, A. Krüger, A. Nau, A.W. Nilsson, A. Nippe, T. Oest, M. Reidenbach, M. Schäfer et al., Zeitschrift für Physik C Particles and Fields **48**, 543 (1990)

[15] G.J. Feldman, R.D. Cousins, Phys. Rev. **D57**, 3873 (1998), physics/9711021

[16] X. Fu, B. Nemati, S.J. Richichi, W.R. Ross, P. Skubic, M. Wood, M. Bishai, J. Fast, E. Gerndt, J.W. Hinson et al. (CLEO Collaboration), Phys. Rev. Lett. **79**, 3125 (1997)

[17] A.J. Bevan et al. (Belle, BaBar), Eur. Phys. J. **C74**, 3026 (2014), 1406.6311

[18] R. Aaij et al. (LHCb collaboration), Phys. Rev. Lett. **113**, 152003 (2014), 1408.0971

[19] E. Solovieva et al. (Belle), Phys. Lett. **B726**, 206 (2013), 1304.6931

[20] C.Q. Geng, Y.K. Hsiao, E. Rodrigues, Phys. Lett. **B767**, 205 (2017), 1612.08133

[21] R. Aaij et al. (LHCb collaboration), JHEP **09**, 006 (2015), 1505.03295

[22] C.Q. Geng, Y.K. Hsiao, Int. J. Mod. Phys. **A23**, 3290 (2008), 0801.0022

[23] C. Patrignani et al. (Particle Data Group), Chin. Phys. **C40**, 100001 (2016)

[24] M. Pivk, F.R. Le Diberder, Nucl.Instrum.Meth. **A555**, 356 (2005), physics/0402083

[25] C. Bebek, K. Berkelman, E. Blucher, J. Byrd, D.G. Cassel, E. Cheu, D.M. Coffman, T. Copie, G. Crawford, R. DeSalvo et al., Phys. Rev. Lett. **62**, 8 (1989)

[26] R. Aaij et al. (LHCb collaboration), Phys. Rev. Lett. **113**, 141801 (2014), 1407.5907

[27] C.Q. Geng, Y.K. Hsiao, J.N. Ng, Phys. Rev. Lett. **98**, 011801 (2007), hep-ph/0608328

[28] M. Beneke, M. Neubert, Nucl. Phys. **B675**, 333 (2003), hep-ph/0308039

[29] X.f. Chen, D.q. Guo, Z.j. Xiao (2007), hep-ph/0701146

[30] R. Aaij et al. (LHCb collaboration), Phys. Rev. Lett. **115**, 051801 (2015), 1503.07483

[31] K.S. Cranmer, Comput. Phys. Commun. **136**, 198 (2001), hep-ex/0011057

[32] W. Wang, Y.M. Wang, D.S. Yang, C.D. Lü, Phys. Rev. D **78**, 034011 (2008)

[33] H.Y. Cheng, C.W. Chiang, A.L. Kuo, Phys. Rev. D **91**, 014011 (2015)

[34] S.H. Zhou, Q.A. Zhang, W.R. Lyu, C.D. Lü, Eur. Phys. J. **C77**, 125 (2017), 1608.02819

[35] T. Aaltonen et al. (CDF), Phys. Rev. Lett. **103**, 031801 (2009), 0812.4271

[36] R. Aaij et al. (LHCb collaboration), JHEP **04**, 087 (2014), 1402.0770

[37] R. Aaij et al. (LHCb collaboration), JHEP **05**, 081 (2016), 1603.00413

[38] R. Aaij et al. (LHCb collaboration), Phys. Lett. **B759**, 282 (2016), 1603.02870

[39] R. Aaij et al. (LHCb collaboration), Phys. Rev. Lett. **110**, 221601 (2013), 1304.6173
[40] R. Aaij et al. (LHCb collaboration), Phys. Rev. **D90**, 112004 (2014), 1408.5373
[41] B. Aubert et al. (BaBar), Phys. Rev. Lett. **87**, 091801 (2001), hep-ex/0107013

 DOI: 10.1051/epjconf/201715801006

Exotic polyquark states and their properties in QCD

Dmitri Melikhov[1,2,3,★]

[1] *Institute for High Energy Physics, Austrian Academy of Sciences, Nikolsdorfergasse 18, A-1050 Vienna, Austria*
[2] *D. V. Skobeltsyn Institute of Nuclear Physics, M. V. Lomonosov Moscow State University, 119991, Moscow, Russia*
[3] *Faculty of Physics, University of Vienna, Boltzmanngasse 5, A-1090 Vienna, Austria*

Abstract. We formulate rigorous criteria for selecting diagrams to be taken into account in the analysis of potential polyquark (tetra or penta) poles in QCD. The central point of these criteria is the requirement that the Feynman diagrams for the relevant Green functions contain four- or five-quark intermediate states and the corresponding cuts. It is shown that some diagrams which "visually" seem to contain the four-quark cuts, turn out to be free of these singularities and therefore should not be taken into account when calculating the tetraquark properties. We then consider large-N_c QCD which in many cases provide qualitatively correct picture of hadron properties and discuss in detail the tetraquark states. For the "direct" and the "recombination" four-point Green functions, which may potentially contain the tetraquark poles, we formulate large-N_c consistency conditions which strongly restrict the behaviour of the tetraquark-to-ordinary meson transition amplitudes. In the end, these conditions allow us to obtain constraints on width of the potential tetraquark states at large N_c. We report that both flavor-exotic and cryptoexotic (i.e., flavor-nonexotic) tetraquarks, if the corresponding poles exist, have a width of order $O(1/N_c^2)$, i.e. they should be parametrically narrower compared to the ordinary $\bar{q}q$ mesons with a width of order $O(1/N_c)$. Moreover, for flavor-exotic states, the large-N_c consistency conditions require two narrow flavor-exotic states, each of these states coupling dominantly to one specific meson-meson channel.

1 Motivation

In the recent years, many candidates for exotic tetraquark mesons (most of the states with charmonium or bottomonium like structure, i.w. with the hidden charm or the hidden beauty) and a few candidates for pentaquark states have been reported (see [1, 2]). To be more precise, the experiments have observed the resonance-like structures in the region near the two-hadron thresholds tempted to be interpreted as the new hadron poles. The precise status of these structures is still not fully understood (whether they correspond to the poles or are just complicated kinematical near-threshold effects), and the interest in the theoretical understanding of these states is incredible. The two main questions concerning the exotic states to be answered are the following: (i) do the scattering amplitudes have new poles (in the complex plane), and (ii) if the new poles exist, what is the structure of the corresponding

★e-mail: dmitri_melikhov@gmx.de

hadrons, i.e. whether they are dominantly molecular-like states—bound states in the Van-der-Waals potential between colorless hadrons, or they are bound states of two or more colored objects in in a confined potential. Answering these questions starting from the QCD Lagrangian is an immensely difficult problem.

To provide the theoretical understanding of exotic states, one may consider QCD with a large number of colors N_c [i.e., SU(N_c) gauge theory for large N_c] with a simultaneously decreasing coupling $\alpha_s \sim 1/N_c$ [3, 4]: in the sector of the ordinary hadrons, large-N_c considerations are known to provide qualitative explanations of many phenomena of real strong interactions corresponding to $N_c = 3$. A specific feature of the large-N_c QCD, which simplifies the analysis of the QCD Green functions, is that at the leading order, large-N_c QCD Green functions receive contributions only from planar diagrams (with an infinite number of gluon exchanges) and as the result have only non-interacting mesons as intermediate states; if tetraquark bound states emerge at all, they may emerge only in subleading large-N_c diagrams [5]. For many years, this fact was believed to be an argument providing the theoretical explanation of the non-existence of exotic tetraquarks. However, it was noticed in [6], that even if the exotic tetraquark bound states appear in the subleading diagrams, the crucial question is the width of these objects: if narrow, they might still be well observed in nature. The conclusion in [6] obtained on the basis of the large-N_c counting, that if tetraquark states exist, they may be as narrow as the ordinary mesons; the latter are known to a width $\sim 1/N_c$. The question of the expected width of the exotic states (if such poles indeed exist) issue has been further addressed in [7], discussing the dependence of the width of tetraquark mesons on their flavor structure. Finally, [8] reported for the cryptoexotic tetraquarks an even smaller width of order N_c^{-3}.

A detailed analysis of the exotic states in large N_c was performed by our group in [9]. The main content of the present talk is based on these results.

2 Diagram selection criteria

In order to draw a conclusion about the width of a potential tetraquark pole, one has to understand which QCD diagrams (in the general case, not only at large-N_c) may lead to the appearance of this pole and thus should be included in the analysis.

The crucial property of such diagrams is in fact obvious: since we are going to study bound state containing four quarks, the appropriate QCD diagrams should contain four-quark intermediate states and the corresponding cuts in the variable s if we expect to observe a tetraquark pole in s [10]. An unambiguous way to establish the presence (or the absence) of such a four-particle s-cut is to apply the Landau equations [11] which allow one to find the location of all singularities of the Feynman diagrams. A detailed application of the Landau equations shows that some of the diagrams which visually seem to contain the four-quark intermediate state and which thus have been attributed to the tetraquark pole in previous analyses, in fact do not contain the necessary four-particle cut and should be therefore excluded from the analysis of the tetraquark properties. This observation has crucial consequences for the properties of the potential tetraquark poles at large-N_c: According to our findings [9], the tetraquark width at large N_c does not depend on its flavor structure: both flavor-exotic and flavor-nonexotic tetraquarks have the same width of order $1/N_c^2$.

Let us study the four-point Green functions of colorless bilinear quark currents (i.e. interpolating currents for ordinary mesons) of various flavor content.

The general four-point function depends on six kinematical variables: the four momenta squared of the external currents, p_1^2, p_2^2, $p_1'^2$, $p_2'^2$, $p = p_1 + p_2 = p_1' + p_2'$, and the two Mandelstam variables $s = p^2$ and $t = (p_1 - p_1')^2$. A bound state in the s-channel corresponds to a pole in the variable s in the four-point function.

Now, the diagrams which potentially may contribute to the tetraquark pole at $s = M_T^2$, should satisfy the following two criteria:

1. The diagram should have a nontrivial (i.e., non-polynomial) dependence on the variable s.

2. The diagram should have four-quark intermediate states in the s-channel and the corresponding s-cuts starting at $s = (m_1 + m_2 + m_3 + m_4)^2$, where m_i are the masses of the quarks which are suspected to form the tetraquark bound state. The presence or absence of this cut should be established by solving the Landau equations for the corresponding diagram.

Obviously, not all diagrams in the perturbative expansion of Green functions satisfy the above criteria. It is therefore convenient to decompose these diagrams into two sets: diagrams of the first set either do not depend on s or have no four-quark cut in the s-channel and thus cannot be related to the potential tetraquark poles; diagrams of the second set satisfy both above criteria and thus contribute to the potential tetraquark pole.

Having at hand the above criteria, we study separately two cases: tetraquarks of an exotic flavor content, i.e., built up of quarks of four different flavors, $\bar{q}_1 q_2 \bar{q}_3 q_4$, and cryptoexotic tetraquarks, i.e those of the flavor content $\bar{q}_1 q_2 \bar{q}_2 q_3$, carrying the same flavor as ordinary mesons. The necessity to treat flavor-exotic and cryptoexotic cases separately arises from the different topologies of the appropriate QCD diagrams in these two cases.

3 Flavour-exotic tetraquarks

Let us consider a bilinear quark current $j_{ij} = \bar{q}_i \hat{O} q_j$ producing a meson M_{ij} of flavor content $\bar{q}_i q_j$ from the vacuum, $\langle 0|j_{ij}|M_{ij}\rangle = f_{M_{ij}}$. Here, $\hat{O}$ is a combination of Dirac matrices corresponding to the meson spin and parity. We omit all Lorentz structures as they are not important for our analysis. Important for us is that at large N_c, the meson decay constants f_M scale as $f_M \sim \sqrt{N_c}$.

In the case of four-point functions of bilinear currents involving quarks of four different flavors, denoted by 1, 2, 3, 4, there are two types of Green functions: the "direct" functions $\Gamma_{\rm I}^{\rm (dir)} = \langle j_{12}^\dagger j_{34}^\dagger j_{12} j_{34}\rangle$ and $\Gamma_{\rm II}^{\rm (dir)} = \langle j_{14}^\dagger j_{32}^\dagger j_{14} j_{32}\rangle$, and the "recombination" functions $\Gamma^{\rm (rec)} = \langle j_{12}^\dagger j_{34}^\dagger j_{14} j_{32}\rangle$ and $\Gamma^{{\rm (rec)}\dagger}$. It turns out that the analysis of the four-quark singularities of the "direct" Green functions is rather straightforward, whereas the analysis of the "recombination" Green functions has subtleties.

3.1 Direct channel

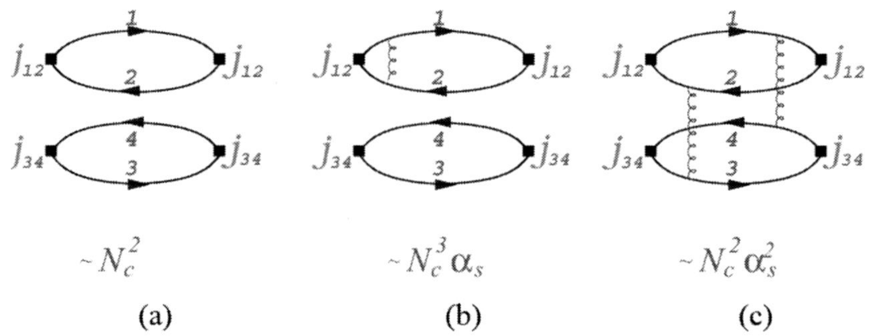

Figure 1. Perturbative expansion of the "direct" Green functions $\Gamma_I^{\rm (dir)}$.

DOI: 10.1051/epjconf/201715801006

Figure 1 shows the perturbative expansion of the direct correlator $\Gamma_{\rm I}^{(\rm dir)}$. Similar diagrams defined by evident flavor rearrangements describe the correlator $\Gamma_{\rm II}^{(\rm dir)}$. Obviously, not all these diagrams satisfy our above criteria for diagrams that potentially contain a tetraquark pole. For instance, the diagrams in Fig. 1(a,b) do not depend on s. The leading large-N_c diagram which depends on s and also has a four-quark s-cut is given by Fig. 1(c). The diagrams of this type are therefore the leading large-N_c diagrams of interest to us.

3.2 Recombination channel

The analysis of the recombination channel is more involved: among the diagrams in Fig. 2, the first two diagrams (a,b) do depend on s; however, in spite of their appearance, they have no four-quark s-cut (and in fact no four-quark cuts at all).

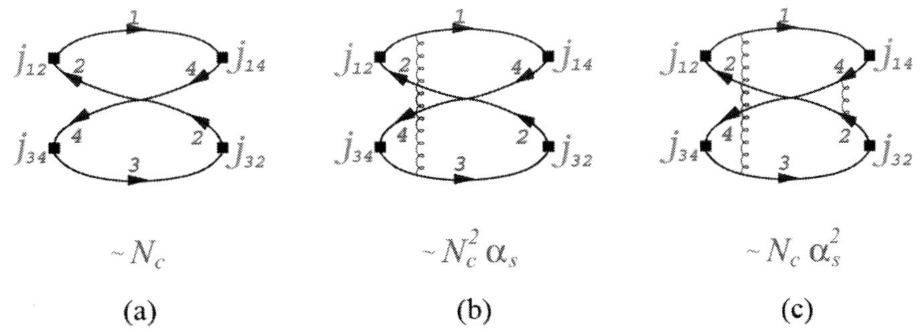

Figure 2. Diagrams for the recombination Green function $\Gamma^{(\rm rec)}$.

The easiest way to see this is to redraw these diagrams as the usual box diagram with gluon exchanges, see Fig. 3: The s-cut of the diagrams on the l.h.s. corresponds to the u-cut of the box diagrams on the r.h.s. For the r.h.s. diagrams (a) and (b), the Landau equations show the absence even of the two-particle u-cuts; the diagrams (a) and (b) have no four-particle singularities at all. Let us emphasize that planar gluon exchanges in the diagrams on the r.h.s. of (a) and (b) do not change this property. In order to develop a four-quark cut, one needs a non-planar gluon exchange: a non-planar diagram in the r.h.s. of 3 (c) develops a four-quark u-cut when the crossed propagators go on-shell. Respectively, this corresponds to the four-particle s-cut of the l.h.s. diagram in Fig. 3 (c). Finally, the N_c-leading diagram exhibiting the four-quark s-cut is the nonplanar diagram in Fig. 3(c).

Its four-quark cut with the threshold at $s = (m_1 + m_2 + m_3 + m_4)^2$ may be verified directly by solving the Landau equations.

Based on the established N_c-leading diagrams possessing four-particle intermediate states, we find the following N_c-leading behaviour of the Green functions Γ_T potentially involving a pole corresponding to some tetraquark T:

$$\begin{aligned}
\Gamma_{{\rm I},T}^{(\rm dir)} &= \langle j_{12}^\dagger j_{34}^\dagger j_{12} j_{34}\rangle = O(N_c^0),\\
\Gamma_{{\rm II},T}^{(\rm dir)} &= \langle j_{14}^\dagger j_{32}^\dagger j_{14} j_{32}\rangle = O(N_c^0),\\
\Gamma_{T}^{(\rm rec)} &= \langle j_{12}^\dagger j_{34}^\dagger j_{14} j_{32}\rangle = O(N_c^{-1}).
\end{aligned} \qquad (1)$$

These diagrams potentially contain the tetraquark pole, although the actual existence of this pole is still a conjecture. Now, let us assume that narrow resonances (i.e., resonances with widths vanishing

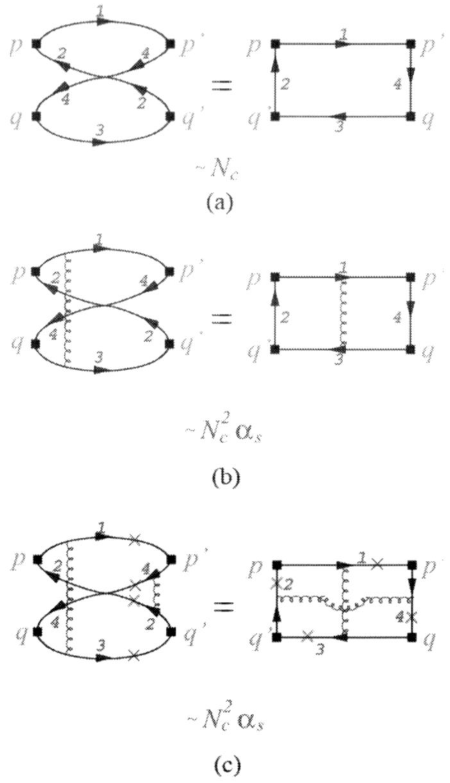

Figure 3. Diagrams for the recombination Green function $\Gamma^{(\mathrm{rec})}$ displayed as box diagrams.

for large N_c) show up at the lowest possible $1/N_c$ order and that the resonance mass M_T remains finite at large N_c. The fact that the "direct" and the "recombination" Green functions of flavour-exotic currents have different large-N_c behaviours, leads to the conclusion that a single pole cannot saturate the leading behaviour of both the direct and the recombination Green functions and we need at least two exotic poles, denoted by T_A and T_B: one state, T_A, couples stronger to the $M_{12}M_{34}$ channel, while another state, T_B couples stronger to the $M_{14}M_{32}$ channel.

Truncating the poles corresponding to the external mesons and retaining explicitly only the tetraquark poles, we get

$$\begin{aligned}
\Gamma^{(\mathrm{dir})}_{\mathrm{I},T} &= O(N_c^0) = f_M^4\left(\frac{|A(M_{12}M_{34}\to T_A)|^2}{p^2-M^2_{T_A}}+\frac{|A(M_{12}M_{34}\to T_B)|^2}{p^2-M^2_{T_B}}\right)+\cdots,\\
\Gamma^{(\mathrm{dir})}_{\mathrm{II},T} &= O(N_c^0) = f_M^4\left(\frac{|A(M_{14}M_{32}\to T_A)|^2}{p^2-M^2_{T_A}}+\frac{|A(M_{14}M_{32}\to T_B)|^2}{p^2-M^2_{T_B}}\right)+\cdots,\\
\Gamma^{(\mathrm{rec})}_{T} &= O(N_c^{-1}) = f_M^4\left(\frac{A(M_{12}M_{34}\to T_A)A(T_A\to M_{14}M_{32})}{p^2-M^2_{T_A}}\right.\\
&\qquad\left.+\frac{A(M_{12}M_{34}\to T_B)A(T_B\to M_{14}M_{32})}{p^2-M^2_{T_B}}\right)+\cdots.
\end{aligned}\tag{2}$$

At this point we have to take into account the scaling relation $f_M \sim \sqrt{N_c}$. Now, since we are interested in the properties of the potential tetraquark poles with **finite mass** at large N_c, the equations (2) have the following solution:

$$A(T_A \to M_{12}M_{34}) = O(N_c^{-1}), \qquad A(T_A \to M_{14}M_{32}) = O(N_c^{-2}),$$
$$A(T_B \to M_{12}M_{34}) = O(N_c^{-2}), \qquad A(T_B \to M_{14}M_{32}) = O(N_c^{-1}). \tag{3}$$

The widths $\Gamma(T_{A,B})$ of the states T_A and T_B are determined by the dominant channel. We therefore conclude that both states have parametrically the same widths, $\Gamma(T_{A,B}) = O(N_c^{-2})$.

So far we have ignored the mixing between T_A and T_B. Introducing their mixing parameter g_{AB}, we get additional contributions to the above Green functions. Most restrictive for g_{AB} is the recombination function, for which mixing provides the additional contribution

$$\Gamma_T^{(\text{rec})} = O(N_c^{-1}) = f_M^4 \left(\frac{A(M_{12}M_{34} \to T_A)}{p^2 - M_{T_A}^2} g_{AB} \frac{A(T_B \to M_{14}M_{32})}{p^2 - M_{T_B}^2} \right) + \cdots . \tag{4}$$

Equations (3) and (4) restrict the behaviour of the mixing parameter in the following way:

$$g_{AB} \le O(N_c^{-1}). \tag{5}$$

Thus, the two flavor-exotic tetraquarks of the same flavor content do not mix at large N_c.

4 Cryptoexotic tetraquarks

Let us now discuss tetraquarks with nonexotic flavor content, i.e., having the same flavor as the ordinary mesons. All confirmed tetraquark candidates observed experimentally belong to this class of states. The analysis proceeds along the same line as for the exotic states. There is however an important new ingredient compared to flavour-exotic states: diagrams of new topologies compared to the flavor-exotic case emerge.

4.1 Direct channel

For the direct Green functions $\Gamma^{(\text{dir})}_{(\text{I,II}),T}$ the new diagrams are shown in Fig. 4. Essentail for us is that the new diagrams do not change the leading large-N_c behaviour compared to the diagrams of the same topology as in the flavor-exotic case.

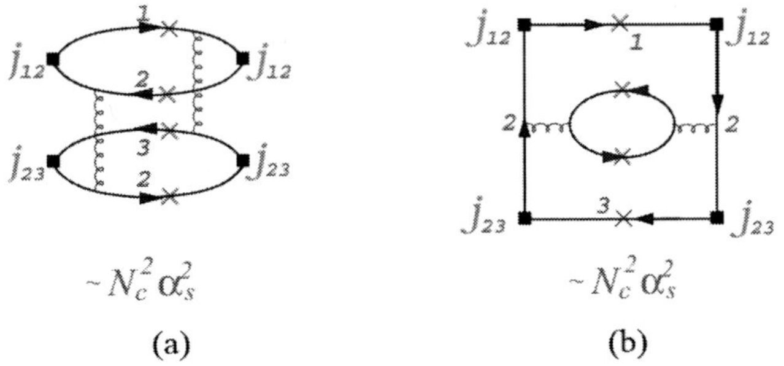

Figure 4. Diagrams contributing to $\Gamma^{(\text{dir})}_{\text{I},T}$. Crossed propagators denote on-shell particles contributing to the four-quark cut.

4.2 Recombination channel

For the recombination functions, however, the situation changes qualitatively: the new diagram, Fig. 5(b), dominates at large N_c and thus modifies the leading large-N_c behaviour of $\Gamma_T^{(\mathrm{rec})}$. We thus find

$$\begin{aligned}\Gamma_{\mathrm{I},T}^{(\mathrm{dir})} &= \langle j_{12}^\dagger j_{23}^\dagger j_{12} j_{23}\rangle = O(N_c^0),\\ \Gamma_{\mathrm{II},T}^{(\mathrm{dir})} &= \langle j_{13}^\dagger j_{22}^\dagger j_{13} j_{22}\rangle = O(N_c^0),\\ \Gamma_T^{(\mathrm{rec})} &= \langle j_{12}^\dagger j_{23}^\dagger j_{13} j_{22}\rangle = O(N_c^0).\end{aligned} \tag{6}$$

In contrast to the flavor-exotic case, now both the direct and the recombination Green functions have the same leading behaviour at large N_c. As a consequence, one exotic state T is already sufficient in order to satisfy the expected large-N_c behaviour of both Green functions. The couplings of this state to the meson-meson channels are

$$\begin{aligned}A(T \to M_{12}M_{23}) &= O(N_c^{-1}),\\ A(T \to M_{13}M_{22}) &= O(N_c^{-1}).\end{aligned} \tag{7}$$

The exotic state T is then expected to have coupling of the order to both possible decay channel, i.e. no preferrable decay channel is seen. The width of this single cryptoexotic state is of order $\Gamma(T) = O(N_c^{-2})$.

If all its quantum numbers allow, T can mix with the ordinary meson M_{13}. The restriction on the mixing parameter $g_{TM_{13}}$ may be obtained, e.g., from the direct amplitude

$$\Gamma_{\mathrm{I},T}^{(\mathrm{dir})} = O(N_c^0) = f_M^4\left(\frac{A(M_{12}M_{23} \to T)}{p^2 - M_T^2} g_{TM_{13}} \frac{A(M_{13} \to M_{12}M_{23})}{p^2 - M_{M_{13}}^2}\right) + \cdots . \tag{8}$$

Taking into account that $A(M_{13} \to M_{12}M_{23}) \sim 1/\sqrt{N_c}$ [3, 4], we obtain for the mixing parameter

$$g_{TM_{13}} \leq O(1/\sqrt{N_c}). \tag{9}$$

Thus, the cryptoexotic tetraquark and the ordinary meson with the same flavour do not mix at large N_c.

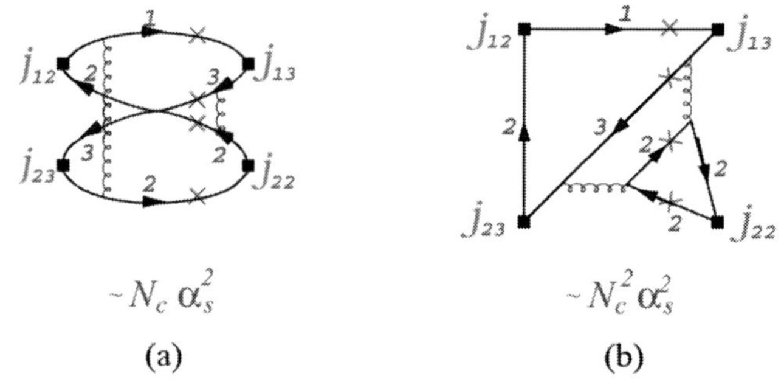

Figure 5. Diagrams contributing to the recombination Green function $\Gamma_T^{(\mathrm{rec})}$.

5 Conclusions and outlook

We have formulated a set of physically obvious and rigorous criteria for selecting Feynman diagram appropriate for the analysis of possible tetraquark (i.e. a bound state containing four quarks, independently of the structure of this bound state): namely, if one is going to study properties of the tetraquark, only those diagrams, which contain four-quark intermediate states and the corresponding four-quark cuts in the appropriate variable should be taken into account. Diagrams which do not contain the four-quark intermediate states cannot contribute to the appearance of the tetraquark pole and should therefore be neglected.

In particular, when one discusses four-point functions of the bilinear quark currents, describing the scattering of the ordinary $\bar{q}q$-mesons, one should consider only those contributions to the four-point Green functions which have a four-quark cut in the s-channel. The straightforward and rigorous way to establish these four-quark singularities is provided by the Landau equations. These simple criteria are quite general and are applicable in QCD for any N_c.

Now, these general criteria for the diagram selection, combined with the large-N_c limit, provide essential constraints on the properties of the possible exotic tetraquark mesons. Namely, using these criteria and requiring that the **narrow** poles contribute to the appropriate parts of the Green functions at the leading large-N_c order, we reported the following results:

1. One has important consistency conditions which arise from the large-N_c behaviour of the direct and the recombination four-point Green functions. These consistency conditions restrict the number of exotic states necessary to saturate the leading large-N_c behaviour of the four-point Green functions.

2. For the flavor-exotic four-point Green functions (all four quarks of different flavors $\bar{q}_1, \bar{q}_3, q_2, q_4$), their large-$N_c$ behaviour is not compatible with the existence of only one flavor-exotic tetraquark bound state; the consistency condition but requires two narrow states T_A and T_B with widths $\Gamma(T_A, T_B) = O(N_c^{-2})$. Each of these tetraquarks dominantly couples to one meson-meson channel; its coupling to the other meson-meson channel is suppressed: $A(T_A \to M_{12}M_{34}) = O(N_c^{-1})$, $A(T_A \to M_{14}M_{32}) = O(N_c^{-2})$, $A(T_B \to M_{14}M_{32}) = O(N_c^{-1})$, $A(T_B \to M_{12}M_{34}) = O(N_c^{-2})$. The parameter describing the mixing between the two tetraquarks vanishes for large N_c at least like $1/N_c$.

3. The large-N_c behaviour of four-point Green functions of nonexotic flavor content $(\bar{q}_1, \bar{q}_2, q_2, q_3)$ is compatible with the existence of a single narrow cryptoexotic tetraquark, T, with width $\Gamma(T) = O(N_c^{-2})$. This tetraquark couples parametrically equally to both two-meson channels: $A(T \to M_{12}M_{23}) = O(N_c^{-1})$, $A(T \to M_{13}M_{22}) = O(N_c^{-1})$. If quantum numbers allow, the cryptoexotic tetraquark $T = \bar{q}_1 q_3 \bar{q}_2 q_2$ mixes with the ordinary meson $M_{13} = \bar{q}_1 q_3$. The corresponding mixing parameter vanishes like $1/\sqrt{N_c}$, i.e., slower than that of the flavor-exotic tetraquarks.

4. The constraints on the width of the possible narrow exotic states at large N_c are general and universal, i.e. they do not depend on the structure of the tetraquark pole. In particular, the constraints should be satisifed by both molecular and confined tetraquarks.

5. We discussed the consequences of the large-N_c behaviour of the four-point Green functions for the narrow exotic states with the mass staying finite at large N_c. However, the constraints on the Green functions are quite general and cannot be violated by any exotic tetraquarks, including those with the masses which rise with N_c. In this case, however, one should take into account that the number of the open channels with many ordinary mesons in the final states also rises

with N_c. Therefore, obtaining constraints on the width of the exotic states with the mass rising with N_c needs a detailed treatment.

We would like to mention that, in principle, there is a possibility that narrow tetraquarks do exist but appear only in N_c-subleading diagrams with four-quark intermediate states, while they do not contribute to the N_c-leading topologies. This possibility seems rather unnatural to us: if such pole exists at all, there should be some special reason, not evident to us, why it does not appear in the set of the appropriate N_c-leading diagrams. Nevertheless, also in the N_c-subleading topologies one observes a difference in the large-N_c behaviour of the direct and the recombination diagrams of the flavor-exotic case: for any N_c-subleading topology, the direct diagrams are N_c-even, whereas the recombination diagrams are N_c-odd. Accordingly, if the latter scenario is realized in nature, one still encounters the necessity of two flavor-exotic poles, although with parametrically smaller widths at large N_c.

Acknowledgements. I have pleasure to thank W. Lucha and H. Sazdjian for a fruitful collaboration on the subject of this talk. Valuable discussions with V. Anisovich, T. Cohen, M. Knecht, B. Moussallam, O. Nachtmann, and B. Stech are gratefully acknowledged. The work was supported by the Austrian Science Fund (FWF) under project P29028.

References

[1] S. L. Olsen, Front. Phys. (Beijing) **10**, 101401 (2015).
[2] A. Ali, J. S. Lange, and S. Stone, arXiv:1706.00610 [hep-ph].
[3] G. 't Hooft, Nucl. Phys. **B72**, 461 (1974).
[4] E. Witten, Nucl. Phys. **B160**, 57 (1979).
[5] S. Coleman, Aspects of symmetry (Cambridge University Press, Cambridge, England, 1985), Chapter 8.
[6] S. Weinberg, Phys. Rev. Lett. **110**, 261601 (2013).
[7] M. Knecht and S. Peris, Phys. Rev. D **88**, 036016 (2013).
[8] L. Maiani, A. D. Polosa, and V. Riquer, JHEP **1606**, 160 (2016).
[9] W. Lucha, D. Melikhov, and H. Sazdjian, Phys. Rev. D **96**, 014022 (2017).
[10] T. D. Cohen and R. F. Lebed, Phys. Rev. D **90**, 016001 (2014).
[11] L. D. Landau, Nucl. Phys. **13**, 181 (1959).

DOI: 10.1051/epjconf/201715801007

Dark Matter Searches with the ATLAS detector

Alison Elliot[1,*] on behalf of the ATLAS Collaboration

[1] *University of Victoria*

Abstract. The presence of a non-baryonic dark matter component in the Universe is inferred from the observation of its gravitational interaction. If dark matter interacts weakly with the Standard Model it would be produced at the LHC, escaping the detector and leaving a large missing transverse momentum as its signature. The ATLAS detector has developed a broad and systematic search program for dark matter production in LHC collisions. The results of these searches on the first 13 TeV data, their interpretation, and the design and possible evolution of the search program will be presented.

1 Introduction

Dark Matter is an unsolved problem; there is nothing in the Standard Model that can explain it. However, something beyond this theory clearly exists. The astrophysical indicators of Dark Matter (DM) come primarily from the cosmic microwave background [1, 2], gravitational lensing [3], the motion of stars in galaxies [4], and the motion of galaxies in clusters [5].

In order to discover the nature of DM, some assumptions must be made about it. The most important assumption is that DM is a particle that interacts weakly with the Standard Model (SM) [6, 7]. Once this assumption is made, DM can be searched for in emission from galactic sources, from direct detection in nuclear recoils underground, or from particle production in colliders. The searches being performed using the 13 TeV dataset from the ATLAS detector [8] at the CERN Large Hadron Collider (LHC) [9] will be presented here.

2 The ATLAS Dataset

The data taken at the ATLAS detector at 13 TeV during 2015 and 2016 are used for the following searches. The recorded luminosity during these time periods can be seen in Figure 1. The searches presented use either the 2015 dataset with 3.2 fb^{-1}, the full 2015+2016 dataset with 36.1 fb^{-1}, or a 2015 plus a subset of 2016 data, amounting to 13.3 fb^{-1}.

3 Dark Matter Searches

The LHC DM Working Group is a collaboration between ATLAS, CMS, and the theory community in order to benchmark interesting models in the search for DM. The working group's recommendations

*e-mail: alison.elliot@cern.ch

 DOI: 10.1051/epjconf/201715801007

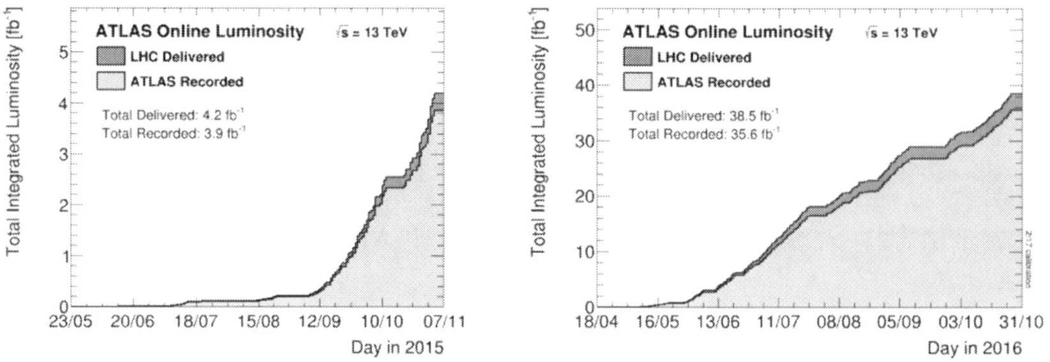

Figure 1. ATLAS recorded luminosity at 13 TeV by day in 2015 (left) and 2016 (right).

for the 13 TeV data collected at the LHC are found in Ref. [10]. These benchmark models include effective field theory (EFT) models which are defined by their interaction mass or energy scale (M^* or Λ). Other models included are the simple mediator models, often referred to as 'simplified models', whose parameters include the mass of the mediator (m_{med}) and the mass of the DM (m_χ). Both EFT and simplified models are measurable in the ATLAS detector when the DM particles are produced in conjunction with SM particles, which arise either as initial state radiation (ISR), or through direct contact at the interaction.

Further DM models being considered are 'two Higgs doublet models'. These models include either vector (Z') or scalar (h, A^0, S) particles that act as mediators to DM, as well as SM particles that interact in the detector. Recommendations and benchmarks for these models can also be found in Ref. [10].

3.1 Mono-X Searches

Searches for DM plus a detectable SM particle are referred to as 'Mono-X' searches, where 'X' is the SM particle being produced in conjunction with the DM particle(s). In Mono-X searches, the presence of DM is inferred from a momentum imbalance in the ATLAS detector. The key variable to these searches is the magnitude of the missing momentum, referred to as missing energy, which is in the plane transverse to the beam direction. The symbol for this variable is $E_{\text{T}}^{\text{miss}}$. The amount of $E_{\text{T}}^{\text{miss}}$ expected from the DM depends on the masses of both the DM and its mediator.

Other important variables and selections applied to the potential DM events in Mono-X include suppression of fake $E_{\text{T}}^{\text{miss}}$ through proxy on its uncertainty $\left(E_{\text{T}}^{\text{miss}} / \sqrt{\Sigma E_{\text{T}}}\right)$ and angular requirements, which could include a large separation $\Delta\phi$ between the missing momentum and the SM particle momentum or separation between the missing momentum and hadronic activity in the detector.

Mono-photon

One of the cleanest Mono-X signatures of new physics is the Mono-γ search. In this search, using 36.1 fb^{-1} of data collected at ATLAS, an ISR γ is produced along with at least 150 GeV of $E_{\text{T}}^{\text{miss}}$ [11], see Figure 2 (left). Events with leptons or more than one jet are vetoed. The background is predominantly Z + jets, where $Z \rightarrow \nu\bar{\nu}$. Normalization factors for the backgrounds are obtained from a simultaneous background-only fit.

DOI: 10.1051/epjconf/201715801007

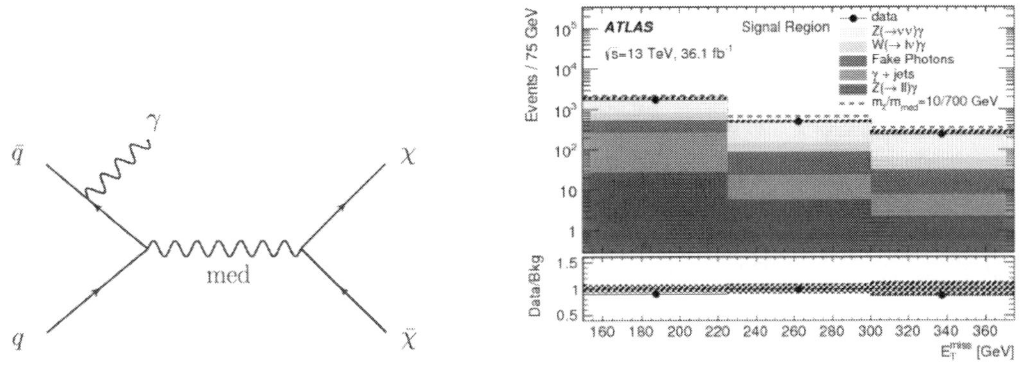

Figure 2. The Feynman diagram for a simplified model of dark matter with an ISR γ (left). The number of events in the backgrounds to the search as well as the data measurement are shown in bins of $E_{\mathrm{T}}^{\mathrm{miss}}$ up to 400 GeV (right) [11].

Using the results in Figure 2 (right), limits can be placed on the simplified models of DM that were analyzed. As seen in Figure 3, mediator masses up to 1.2 TeV can be excluded for the axial-vector mediator simplified model. As well, limits on the DM-nucleon cross-section for interaction are extremely competitive with direct detection searches at low and mid-range DM masses.

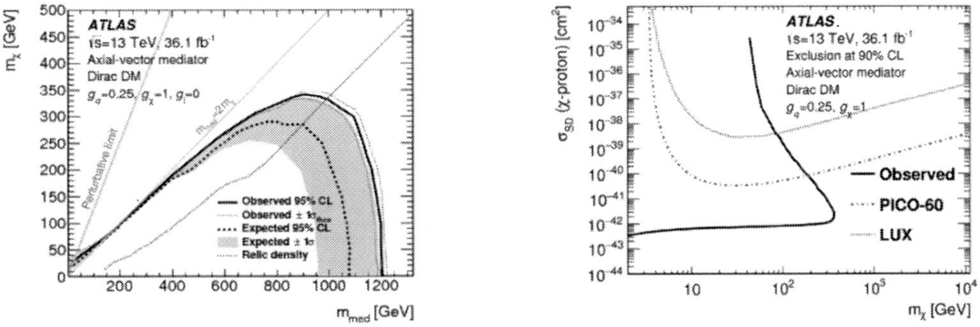

Figure 3. The 95% exclusion contours for axial-vector mediated simplified Mono-γ models of DM, in mass of the mediator, m_{med}, and mass of the DM, m_{χ} (left). Exclusion contours at 90% CL reinterpreted as spin-dependent DM-nucleon coupling strengths for various mediator masses, and compared to direct detection searches (right) [11].

Mono-jet

One of the first places Mono-X searches could find DM is the Mono-jet search [12], as ISR jets are a frequent occurrence and a distinct signature in the detector. In this search, using 3.2 fb^{-1} of data, an energetic ISR jet is produced along with at least 250 GeV of $E_{\mathrm{T}}^{\mathrm{miss}}$. An angular distance separation is applied to ensure the jets and $E_{\mathrm{T}}^{\mathrm{miss}}$ are well-separated, and any charged leptons are vetoed. The backgrounds, such as $Z \to \nu\bar{\nu}$ + jets, are normalized to data through Monte Carlo (MC) control regions. The results exclude the axial-vector mediator simplified model of DM out to 1 TeV, on masses of DM over 250 GeV. Cross section limits are competitive with the direct detection searches at low and mid-ranged DM masses, as seen in Figure 4.

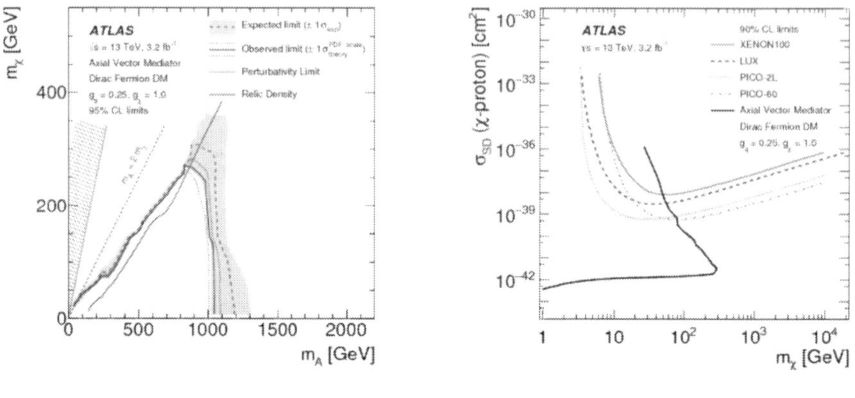

Figure 4. The 95% exclusion contours for axial-vector mediated simplified Mono-jet models of DM, by mass of the mediator, m_{med}, and mass of the DM, m_χ (left). Exclusion contours at 90% CL reinterpreted as spin-dependent DM-nucleon coupling strengths for various mediator masses and compared to direct detection searches (right) [12].

Mono-V

A search similar to the Mono-jet search is the Mono-V(hadronic) search [13]. This search for DM examines 3.2 fb^{-1} of data for an ISR W or Z boson in the form of an energetic jet containing substructure consistent with coming from a W or Z boson and at least 250 GeV of missing energy. Backgrounds, such as $Z \to \nu\bar{\nu}$ + jets, are estimated through MC and normalized to data through dedicated control regions. Both vector-mediated simplified models as well as EFT contact models are considered, and limits are set on the signal strength and couplings respectively. The EFT Feynman diagram and results can be seen in Figure 5.

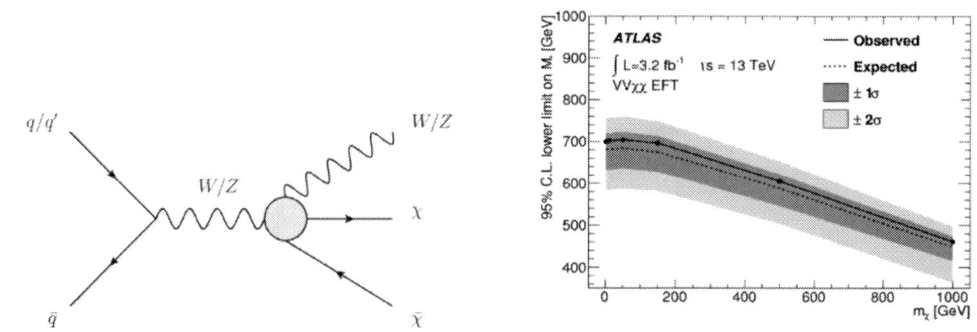

Figure 5. The Feynman diagram for an EFT model of dark matter in a contact interaction with a Z boson (left). 95% CL lower limits on EFT effective mass for various dark matter masses m_χ (right) [13].

DOI: 10.1051/epjconf/201715801007

An analysis with a similar structure is the Mono-$Z(\ell\ell)$ search that is performed with 13.3 fb^{-1} looking for the emission of an ISR Z boson where the Z boson decays into charged leptons [14]. This process has a smaller cross section (from the smaller branching fraction of the $Z \to \ell\ell$), but a cleaner signal than the Mono-V search. The background arises primarily from diboson processes, as well as $Z \to \ell\ell$ + jets, and is estimated through MC and normalized by dedicated control regions, see Figure 6.

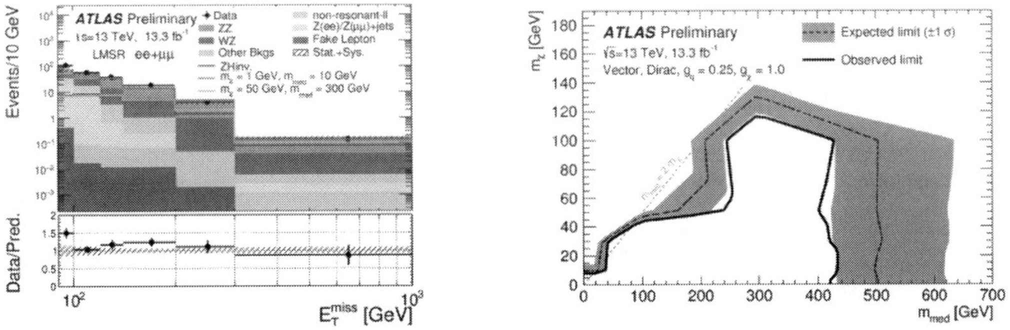

Figure 6. The number of events in the backgrounds to the Mono-$Z(\ell\ell)$ search as well as the data measurement are shown in bins of E_T^{miss} up to 1000 GeV (left). The 95% exclusion contours for axial-vector mediated simplified models of DM, by mass of the mediator, m_{med} and mass of the DM, m_χ (right) [14].

Mono-H

Another suite of analyses that have sensitivities to different models of DM are the Mono-H searches. They search for two Higgs doublet models, such as the one picture in Figure 7 (left). There are several searches for DM that are separated by the decay of the Higgs boson in the model. One such search is the Mono-H(bb) with 36.1 fb^{-1} of data [15], where the large branching fraction of the Higgs to b-quarks is exploited. The backgrounds, such as $Z \to \nu\bar{\nu}$ and t-quark backgrounds, are estimated through control regions in data. Limits are made on the mass of a Z' mediator up to 2.5 TeV, for pseudo scalar masses (m_A) up to 600 GeV, as seen in Figure 7 (right).

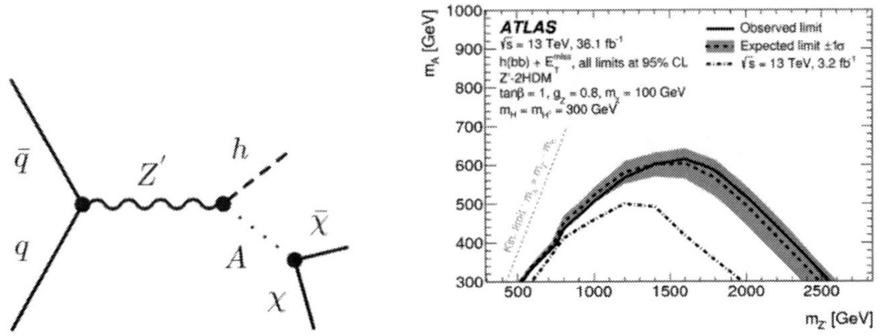

Figure 7. The Feynman diagram for a Mono-H model of dark matter (left). The 95% exclusion contours for by mass of the mediator, $m_{Z'}$, and mass of the pseudo scalar, m_A (right) [15].

Another Mono-H search for DM is the Mono-H($\gamma\gamma$) search with 36.1 fb^{-1} of data [16]. This search has a very clean signature, with the $\gamma\gamma$ non resonant spectrum estimated from sideband regions in data, and SM $H \to \gamma\gamma$ estimated from MC. Limits on a Z'_B mediator are placed up to 850 GeV for DM masses up to 200 GeV. These are reinterpreted as spin-independent cross section limits and compared to direct detection searches, see Figure 8. These limits are only competitive at the lowest masses of DM.

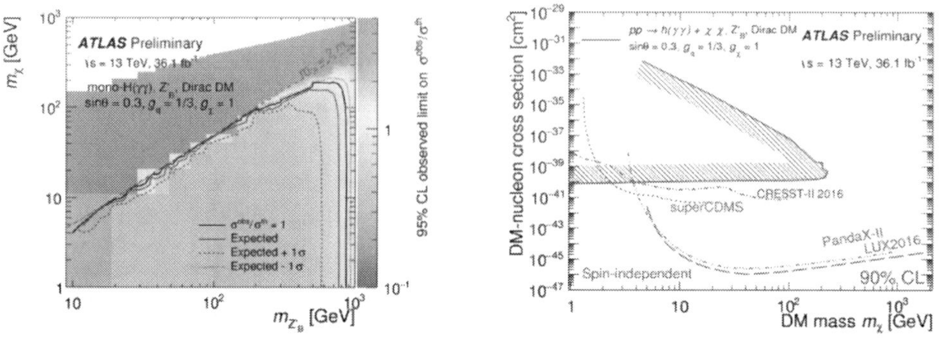

Figure 8. The 95% exclusion contours for Mono-H($\gamma\gamma$) models of DM, by mass of the mediator, $m_{Z'_B}$, and mass of the DM, m_χ (left). Exclusion contours at 90% CL reinterpreted as spin-independent DM-nucleon coupling strengths for various mediator masses (right) [16].

Finally, a Mono-H(4ℓ) search with 3.2 fb^{-1} of data was performed [17] . This has a clean signature with few background events, which are primarily from SM Higgs and diboson events. Upper limits were set on cross section times branching fraction for models with various masses of mediators and a DM mass of 1 GeV, see Figure 9.

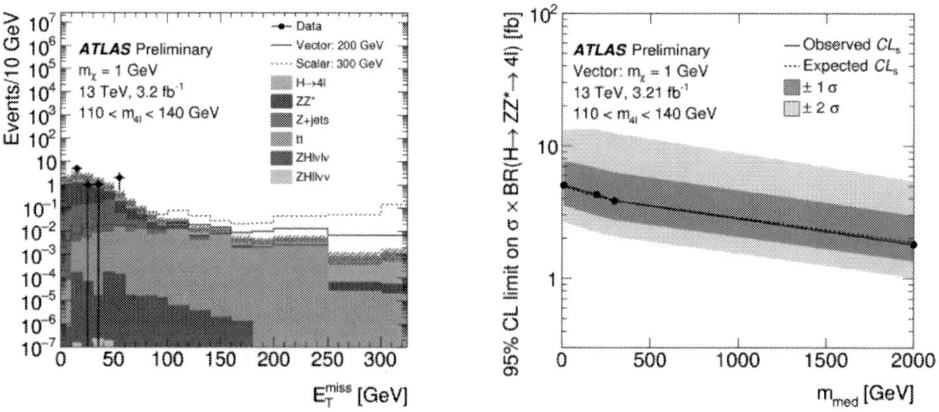

Figure 9. The number of events in the backgrounds to the search as well as the data measurement are shown in bins of E_T^{miss} up to 350 GeV (left). 95% CL limit on branching fraction of vector mediated DM, $m_\chi = 1$ GeV for various mediator masses (right) [17].

E_T^{miss} + quarks

The final category of Mono-X searches for DM is from $E_{\mathrm{T}}^{\mathrm{miss}}$ + quarks, in a simplified model such as in Figure 10 (left). This is where a $q\bar{q}$ pair is produced in conjunction with a scalar (ϕ) or pseudo scalar (a) mediator, which then decays into DM particles. The searches are being performed with 13.3 fb^{-1} where $q\bar{q}$ can be either $b\bar{b}$ [18] or $t\bar{t}$. In $t\bar{t}$ analyses, the decay products are split by number of leptons ℓ, and include 0ℓ [19], 1ℓ [20, 21], or 2ℓ [22].

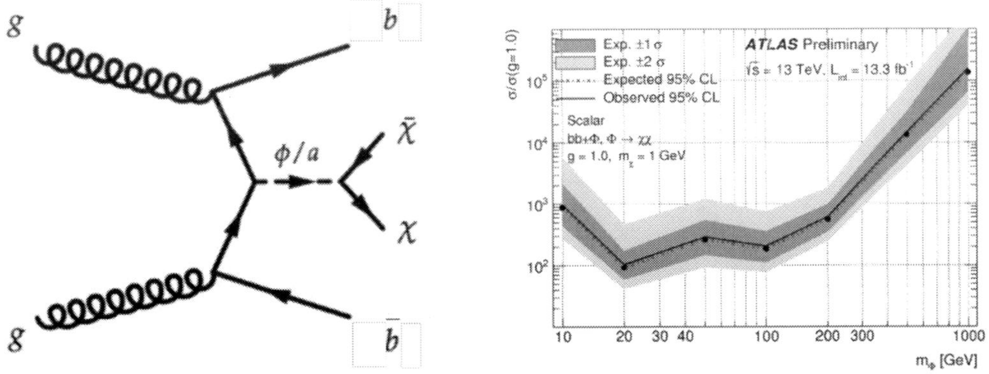

Figure 10. Model of DM produced in conjunction with a $b\bar{b}$ and a scalar or pseudo scalar mediator particle (left) and the limits set on the signal strength of the scalar mediator, for a range of masses (right) [18].

Both the $b\bar{b}$ and $t\bar{t}$ searches look for jets tagged as b-jets with between 150 GeV and 350 GeV of $E_{\mathrm{T}}^{\mathrm{miss}}$. Limits are set on the signals strength of the mediator particle (for $b\bar{b}$, see Figure 10 (right)), or in the m_χ–m_{med} plane up to ~ 350 GeV.

3.2 Mediator Searches

If the DM mediator can be produced at the LHC, then it must couple to quarks. Therefore, the mediator should also decay back to SM particles. This will show up as a resonance, such as a Z' resonance, in the invariant mass of the decay products. This invariant mass is the parameter of interest for these direct mediator searches, rather than the $E_{\mathrm{T}}^{\mathrm{miss}}$ for the Mono-X searches.

Dijet

If the mediator couples to quarks, then these would show up in the detector as two energetic jets, known as a dijet event. There are three separate searches for dijet events. The first is a search for at least two jets with a large transverse momentum [23]. This search used 37 fb^{-1} of data, and modelled the background with a fit to a smoothly falling QCD spectrum in m_{jj}. It is sensitive in the mass range 1.1–8.2 TeV. Another search with a very similar strategy is the 'Trigger Level Analysis', which probes a lower mass range (0.44–1.2 TeV), as it triggers to record events on a stream of partially rebuilt objects in the detector (tracking and muon information is thrown away), and searched 3.4 fb^{-1} of data [24]. Finally, the dijet + ISR analysis triggers on an event with an energetic photon or jet to look for lower mass dijets, but uses similar background estimations to the first two analyses, and used an intermediate 15.5 fb^{-1} of data to search for DM mediators [25]. This analysis is able to probe the region of m_{jj} in the range 0.17–1.5 GeV.

Each of these analyses uses y^*, which is half the difference in rapidity of the two jets, to reduce the QCD background. The analyses search ranges of the mass of the mediator ($m_{Z'}$) for various coupling strengths. The three searches described above are summarized in Figure 11. Finally, not pictured here, a dijet resonance search for Z' was done searching exclusively for t-quarks [26].

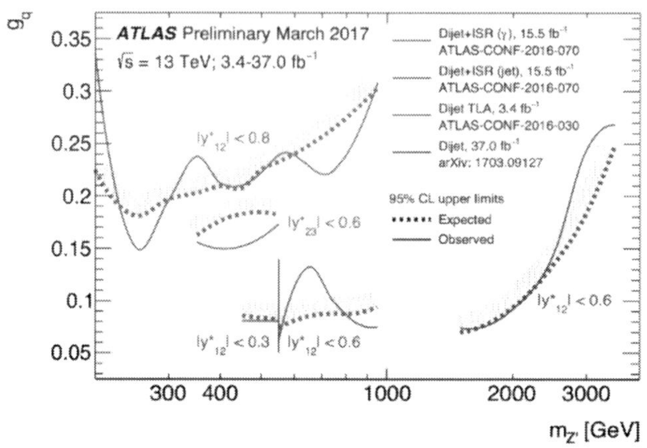

Figure 11. Dijet searches for DM mediator Z' for various couplings to quarks g_q and masses of Z' [23].

Dilepton

Under the assumption that the mediator also couples to charged leptons, the mediator could decay into pairs of electrons or muons. The search for a mediator decaying to charged leptons ($Z' \to \ell\ell$) was done with 36.1 fb^{-1} of data [27]. The main backgrounds from Drell-Yan, t-quarks, and dibosons are modelled with MC. Signal regions combined with ee and $\mu\mu$ exclude a mediator with a mass up to 4.1 TeV, see Figure 12.

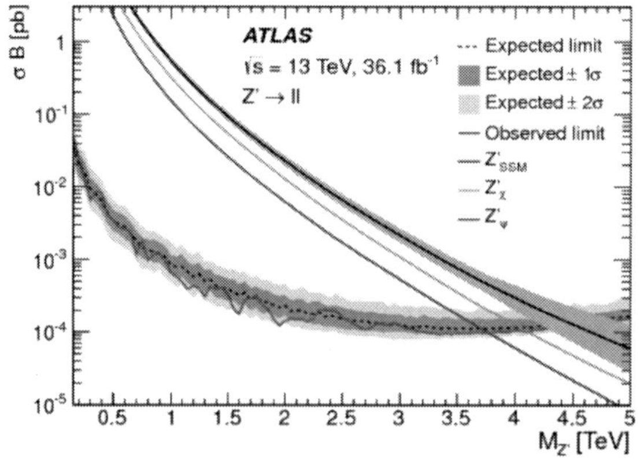

Figure 12. Dilepton exclusion limits on the DM mediator, Z' [27].

3.3 Combinations

A combination of the Mono-X and mediator searches can be seen in Figure 13, [28]. For the simplified models, a common set of parameters from the benchmark models were chosen in order to be able to consistently overlay the results from the different searches.

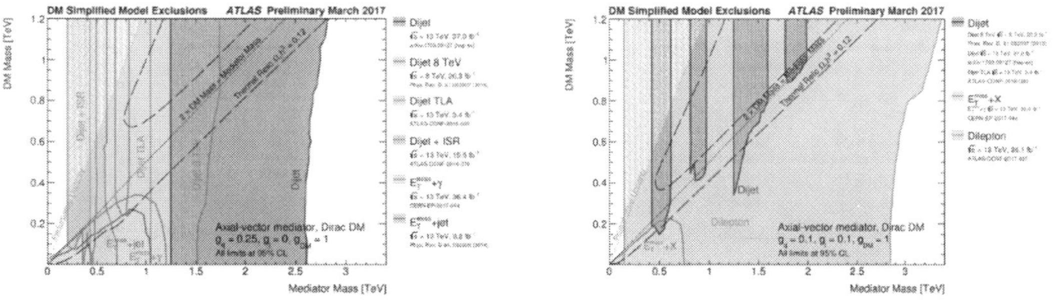

Figure 13. DM Combinations of Mono-X and mediator searches for the axial-vector mediator with no leptonic couplings (left) and small leptonic couplings (right) [28].

These searches can also be reinterpreted on DM–nucleon scattering cross section plots with direct detection searches, under the assumptions that all SM quarks couple equally to the mediator, and that there is only one variant of DM, which couples to this mediator, see Figure 14.

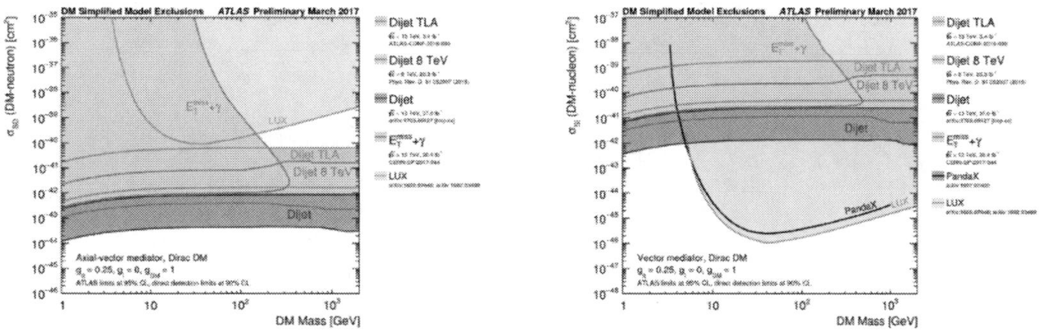

Figure 14. Comparison with direct detection searches with Mono-X and mediator searches for the axial-vector mediator with no leptonic couplings (left) and vector mediator with no leptonic couplings (right) [28].

3.4 Additional searches

In addition to the Mono-X and mediator searches presented, there are a large variety of searches at ATLAS to find particle DM. The ones presented previously are not the only ones that are competitive. Supersymmetry (SUSY) searches for the lightest supersymmetric partner, a DM candidate, and their mediators have excellent discrimination power. These searches follow methods similar to those in [29], and set competitive limits on the DM masses of charginos and neutrilinos, the SUSY candidates for DM.

4 Summary

The dark matter searches being performed in ATLAS are complimentary to each other and to direct detection searches. The recommendations have been to move from effective field theories to 'simplified' or mediator models in order to ensure the validity of the models at the large centre-of-mass energy of 13 TeV. This has also enabled searches for mediators themselves, providing additional constraints on the dark matter limits being published. There are many interesting 13 TeV results in both 2015 and 2016 data, and many more to come. The results presented here are not exhaustive, but show the cornerstone of the dark matter search program at ATLAS.

References

[1] C. Patrignani et al. (Particle Data Group), Chin. Phys. **C40**, 100001 (2016)
[2] R. Adam et al. (Planck), Astron. Astrophys. **594**, A1 (2016)
[3] J.A. Tyson, G.P. Kochanski, I.P. Dell'Antonio, Astrophys. J. **498**, L107 (1998)
[4] L. Bergström, Rept. Prog. Phys. **63**, 793 (2000)
[5] F. Zwicky, General Relativity and Gravitation **41**, 207 (2009)
[6] G. Bertone, D. Hooper, J. Silk, Phys. Rept. **405**, 279 (2005)
[7] J.L. Feng, Ann. Rev. Astron. Astrophys. **48**, 495 (2010)
[8] ATLAS Collaboration, JINST **3**, S08003 (2008)
[9] L. Evans, P. Bryant, Journal of Instrumentation **3**, S08001 (2008)
[10] D. Abercrombie et al. (2015), `arXiv:1507.00966[hep-ex]`
[11] ATLAS Collaboration, Eur. Phys. J. **C77**, 393 (2017)
[12] ATLAS Collaboration, Phys. Rev. **D94**, 032005 (2016)
[13] ATLAS Collaboration, Phys. Lett. **B763**, 251 (2016)
[14] ATLAS Collaboration, ATLAS-CONF-2016-056, `http://cds.cern.ch/record/2206138`
[15] ATLAS Collaboration (2017), `arXiv:1707.01302[hep-ex]`
[16] ATLAS Collaboration (2017), `arXiv:1706.03948[hep-ex]`
[17] ATLAS Collaboration, ATLAS-CONF-2015-059, `http://cds.cern.ch/record/2114825`
[18] ATLAS Collaboration, ATLAS-CONF-2016-086, `http://cds.cern.ch/record/2206279`
[19] ATLAS Collaboration, ATLAS-CONF-2016-077, `http://cds.cern.ch/record/2206250`
[20] ATLAS Collaboration, ATLAS-CONF-2016-050, `http://cds.cern.ch/record/2206132`
[21] ATLAS Collaboration, ATLAS-CONF-2017-037, `http://cds.cern.ch/record/2266170`
[22] ATLAS Collaboration, ATLAS-CONF-2016-076, `http://cds.cern.ch/record/2206249`
[23] ATLAS Collaboration (2017), `arXiv:1703.09127[hep-ex]`
[24] ATLAS Collaboration, ATLAS-CONF-2016-030, `http://cds.cern.ch/record/2161135`
[25] ATLAS Collaboration, ATLAS-CONF-2016-070, `http://cds.cern.ch/record/2206221`
[26] ATLAS Collaboration, ATLAS-CONF-2016-014, `http://cds.cern.ch/record/2141001`
[27] ATLAS Collaboration (2017), `arXiv:1707.02424[hep-ex]`
[28] ATLAS Collaboration, *Summary plots from the ATLAS Exotic physics group* (2017), `https://atlas.web.cern.ch/Atlas/GROUPS/PHYSICS/CombinedSummaryPlots/EXOTICS/index.html`
[29] ATLAS Collaboration, JHEP **09**, 175 (2016)

Non-perturbative effects of the electroweak interaction at the LHC

B.A. Arbuzov[1,⋆]

[1]*M.V. Lomonosov Moscow State University, 119991 Moscow, Russia*

Abstract. Assuming an existence of the anomalous triple electro-weak bosons interaction being defined by coupling constant λ we calculate its contribution to interactions of the Higgs with pairs of heavy particles. Bearing in mind experimental restrictions $-0.011 < \lambda < 0.011$ we present results for possible effects in processes $p\,p \to W^+W^-H$, $p\,p \to W^+ZH$, $p\,p \to W^-ZH$, $p\,p \to \bar{t}tH$, $pp \to \bar{b}bH$. Effects could be significant with negative sign of λ in associated heavy quarks t, b pairs production with the Higgs. In calculations we rely on results of the non-perturbative approach to a spontaneous generation of effective interactions, which defines the form-factor of the three-boson anomalous interaction.

1 Introduction

The totality of data nowadays confirms main features of the Standard Model, which consists of QCD, describing strong interactions, and the EW theory, describing electroweak interactions. This confirmation is essentially based on numerous perturbative calculations which describe corresponding data. However in QCD the inevitable introduction of non-perturbative effects is also evident. First of all the low momenta region of the strong interaction definitely can not be described in the framework of the perturbative calculations. Examples of non-perturbative quantities are well-known: vacuum averages the gluon condensate $< \frac{\alpha_s}{\pi} G^a_{\mu\nu} G^a_{\mu\nu} >$, the quark condensate $< \bar{q}\, q >$ *etc*. One of the most powerful methods of dealing with the non-perturbative effects is provided in the framework of approaches using the so-called effective interactions. The eldest and the most popular such effective interaction is the famous Nambu–Jona-Lasinio interaction [1, 2]. With application to quark structure of hadrons this approach adequately describes the low momenta region, see *e.g.* reviews [3–5].

The Nambu–Jona-Lasinio interaction deals with four-quark effective terms. However, the non-zero gluon condensate testifies for additional effective terms also in gluon interactions. There were also proposals for such terms. In particular, triple gluon interaction in the low momenta region of the following form

$$L_{eff} = -\frac{G}{3!} f_{abc} G^a_{\mu\nu} G^b_{\nu\rho} G^c_{\rho\mu}; \qquad (1)$$

where $G^a_{\mu\nu}$ is a gauge covariant gluon field and f_{abc} are structure constants of the color $SU(3)$, was proposed in work [6].

⋆e-mail: arbuzov@theory.sinp.msu.ru

In the electro-weak theory necessity of non-perturbative contribution is not nowadays so evident as in QCD. However the structure of gauge theories is the same for both cases. One might expect similar features in three-boson interactions. In particular, the following three weak boson interaction was introduced [7, 8]

$$L_{eff} = -\frac{G_W}{3!} F(p_i)\epsilon_{abc} W^a_{\mu\nu} W^b_{\nu\rho} W^c_{\rho\mu}; \quad G_W = \frac{g\,\lambda}{M_W^2}; \; W^3_{\mu\nu} = \cos\theta_W Z_{\mu\nu} + \sin\theta_W A_{\mu\nu}; \tag{2}$$

where g is the electro-weak gauge coupling and indices a, b, c take three values and the third boson $W^3_{\mu\nu}$ is a composition of neutral bosons Z and γ. Form-factor $F(p_i)$ in [7, 8] is postulated and it has to vanish for $|p_i^2| \gg \Lambda^2$, where Λ is a characteristic scale. Interaction (2) would lead to effects *e.g.* in electro-weak bosons pair production and was studied in experiments. The best limitations for parameter λ is provided by recent data of the CMS Collaboration [9]

$$-0.011 < \lambda < 0.011. \tag{3}$$

The ATLAS Collaboration presents similar results [10]

$$-0.013 < \lambda < 0.013. \tag{4}$$

Both the Nambu–Jona-Lasinio interaction and interaction (1) are supposed to act in a low momenta region. This means, that in both cases form-factors are present, which guarantee decreasing of intensity of the interactions for large momenta. In the original NJL [1, 2] interaction a cut-off was introduced for the purpose. Starting of fundamental gauge theories of interactions of the Standard Model we have to understand the origin of such effective cut-off. This can be done under assumption of these interactions being spontaneously generated. The notion of spontaneous generation is traced back to methods of the superconductivity theory. In application to superconductivity the conception of compensation principle was elaborated [11, 12] by N.N. Bogoliubov. This approach was applied to spontaneous generation of Nambu–Jona-Lasinio interaction in work [13] and of interaction (1) in work [14]. In the course of this application form-factors inherent to corresponding interactions are uniquely defined [1]. As an additional confirmation of applicability of the method to non-perturbative quantities, value V_2 of the gluon condensate was calculated [14] in agreement with its phenomenological value.

2 Additional interactions of the Higgs with electro-weak bosons

Electro-weak bosons $W^\pm$, Z due to their large masses interact with Higgs H significantly. Namely there are the following vertices for interaction of Higgs H with W^+W^- and ZZ respectively

$$\imath\, g M_W\, g_{\mu\nu}; \quad \imath \frac{g M_Z}{\cos\theta_W}\, g_{\mu\nu}. \tag{5}$$

Now let us assume, that in addition to usual three-boson gauge interaction, interaction (2) really exists. Then three-boson vertex takes the form

$$V(\mu, p; \nu, q; \rho, k) = -g\epsilon_{abc}(g_{\mu\nu}(q_\rho - p_\rho) + g_{\nu\rho}(k_\mu - q_\mu) + g_{\rho\mu}(p_\nu - k_\nu) + \frac{\lambda}{M_W^2} F(p, q, k) \times$$
$$(g_{\mu\nu}(q_\rho pk - p_\rho qk) + g_{\nu\rho}(k_\mu pq - q_\mu pk) + g_{\rho\mu}(p_\nu qk - k_\nu pq) + q_\mu k_\nu p_\rho - k_\mu p_\nu q_\rho)). \tag{6}$$

[1]Of course, in the framework of an approximation.

Here $F(p,q,k)$ is a form-factor, which is defined in the framework of the spontaneous generation of effective interaction (2) [15] in the compensation approach, which we have discussed in the Introduction. Then we have additional contribution to VVH vertex due to terms proportional to λ and to λ^2 according to diagrams presented in Fig. 1. As a consequence we have additional contribution to vertices $V_{VV'H}$, where V and V' correspond to electro-weak bosons W, Z, γ.

The combined account of both vertices (5, 6) leads to corrections for $VV'H$ vertex according to diagrams shown in Fig. 1.

We have for vertices of interactions of the Higgs with two electro-weak bosons $V(p,\mu)\,V'(q,\nu)$ instead of (5)

$$\begin{aligned}
&V_{W^+W^-H} = \imath g M_W (g_{\mu\nu} + G_{WW}(g_{\mu\nu}pq - p_\nu q_\mu));\\
&V_{ZZH} = \imath\frac{gM_Z}{\cos\theta_W}(g_{\mu\nu} + G_{ZZ}(g_{\mu\nu}pq - p_\nu q_\mu));\\
&V_{ZAH} = \imath\frac{gM_Z}{\cos\theta_W}G_{Z\gamma}\,(g_{\mu\nu}pq - p_\nu q_\mu);\quad V_{AAH} = \imath\frac{gM_Z}{\cos\theta_W}G_{\gamma\gamma}\,(g_{\mu\nu}pq - p_\nu q_\mu);\\
&G_{ZZ} = \cos^2\theta_W\, G_{W_0W_0};\quad G_{\gamma\gamma} = \sin^2\theta_W\, G_{W_0W_0};\quad G_{Z\gamma} = 2\cos\theta_W\sin\theta_W\, G_{W_0W_0};\\
&G_{W_0W_0} = \frac{2g}{M_W^2}\Big(\frac{3g\lambda}{8\pi^2}I_1 - \frac{\sqrt{2}\lambda^2}{\pi}I_2\Big);\\
&G_{WW} = \frac{g}{M_W^2}\Big(\frac{3g\lambda}{8\pi^2}I_{W1} - \frac{\sqrt{2}\lambda^2}{\pi}I_{W2} + \frac{3g\lambda}{8\pi^2}I_{Z1} - \frac{\sqrt{2}\lambda^2}{\pi}I_{Z2}\Big).
\end{aligned}\tag{7}$$

Here

$$\begin{aligned}
&I_1 = \int_0^{z_0}\frac{F(t)\,dt}{2(\sqrt{t}+\mu)^2};\ I_2 = \int_0^{z_0}\frac{2tF^2(t)\,dt}{(\sqrt{t}+\mu)^3};\quad I_{W1} = \int_0^{z_0}\frac{F(t)(\sqrt{t}+s\mu_Z)\,dt}{(\sqrt{t}+\mu_Z)^2(\sqrt{t}+\mu)};\\
&I_{W2} = \int_0^{z_0}\frac{F^2(t)\sqrt{t}(\sqrt{t}+s\mu_Z)\,dt}{(\sqrt{t}+\mu_Z)^2(\sqrt{t}+\mu)};\quad I_{Z1} = \int_0^{z_0}\frac{F(t)\sqrt{t}\,dt}{(\sqrt{t}+\mu_Z)^2(\sqrt{t}+\mu)(1-s)};\\
&I_{Z2} = \int_0^{z_0}\frac{F^2(t)\,t\,dt}{(\sqrt{t}+\mu_Z)^2(\sqrt{t}+\mu)(1-s)};\ \mu = \frac{g|\lambda|}{16\sqrt{2}\pi};\quad \mu_Z = \frac{g|\lambda|}{16\sqrt{2}\pi(1-s)};\\
&t = \frac{G_W^2(p^2)^2}{512\pi^2};\ s = \sin^2\theta_W.
\end{aligned}\tag{8}$$

We take form-factor $F(t) = F(p,-p,0)$ from results of work [15] in which the compensation approach was applied to the electro-weak interaction:

$$\begin{aligned}
&F(t) = \frac{1}{2}G_{15}^{31}\big(t|^{0}_{1,1/2,0,-1/2,-1}\big) - \frac{85\sqrt{2}g_0}{128\pi}G_{15}^{31}\big(t|^{1/2}_{1,1/2,1/2,-1/2,-1}\big) + C_1G_{04}^{10}\big(t|\tfrac{1}{2},1,-\tfrac{1}{2},-1\big) +\\
&C_2G_{04}^{10}\big(t|1,\tfrac{1}{2},-\tfrac{1}{2},-1\big);\ t = \frac{G_W^2\,(p^2)^2}{512\,\pi^2};\ F(t) = 0,\ t \geq t_0;\quad t_0 = 9.6175,\\
&g_0 = 0.6037,\ C_1 = -0.0351,\ C_2 = -0.0511.
\end{aligned}\tag{9}$$

Here g_0 is the value of gauge electro-weak coupling g at point $t = t_0$ and we use Meijer functions

$$G_{pq}^{mn}\big(t|^{a_1\ldots,a_p}_{b_1\ldots,b_q}\big),$$

for more details see, *e.g.* book [16].

With definitions (7,8,9) we calculate the couplings and show results in Table 1. Note, that additional interactions (7) were already considered *e.g.* in works [17, 18].

Table 1. Coupling constants $G_{VV'}$ GeV^{-2} of effective interactions HVV' in dependence on value of λ. All coupling values are multiplied by 10^7.

λ	G_{WW}	G_{ZZ}	$G_{Z\gamma}$	$G_{\gamma\gamma}$
0.01	3.10	4.40	4.71	1.26
0.006	2.20	3.10	3.32	0.89
0.003	1.33	1.86	2.00	0.54
0	0	0	0	0
−0.003	−3.51	−4.83	−5.17	−2.58
−0.006	−6.54	−9.01	−9.65	−2.58
−0.01	−10.3	−14.2	−15.2	−4.08

Then we calculate cross sections of pair weak boson production accompanied by the Higgs. In doing this we apply the CompHEP package [19]. In subsequent Tables we show results for values of couplings in (7) and cross sections of processes $p + p \to W^+W^-H + X$ ($\sigma(+-)$), $p + p \to W^+ZH + X$ ($\sigma(+0)$) and $p + p \to W^-ZH + X$ ($\sigma(-0)$). Results are shown in dependence on value of λ in admissible interval (3).

Table 2. Production LO cross sections of $VV'H$ for $\sqrt{s} = 8\,TeV$ at the LHC.

λ	$\sigma(+-)\,fb$	$\sigma(+0) fb$	$\sigma(-0) fb$
−0.01	5.18	1.21	0.499
−0.006	4.62	1.09	0.45
−0.003	4.22	1.02	0.43
0	3.86	0.98	0.42
0.003	3.75	0.98	0.42
0.006	3.69	0.98	0.42
0.01	3.62	0.98	0.42

Table 3. Production LO cross sections of $VV'H$ for $\sqrt{s} = 13\,TeV$ at the LHC.

λ	$\sigma(+-)\,fb$	$\sigma(+0) fb$	$\sigma(-0) fb$
−0.01	21.16	3.60	1.47
−0.006	17.06	2.71	1.20
−0.003	14.36	2.27	1.07
0	11.90	2.08	1.00
0.003	11.14	2.09	1.00
0.006	10.70	2.12	1.02
0.01	10.35	2.19	1.03

From Table 3 we see, that for $\sqrt{s} = 13\,TeV$ with negative λ the effect is noticable, especially for process $p + p \to W^+W^-H + X$, and *e.g* for $\lambda = -0.01$ the cross-section is almost two times more than the SM one. However the cross-section itself is presumably not sufficiently high for a productive study of the effect.

Let us note also, that $VV'H$ the additional interaction with $\lambda \neq 0$ might give effect for VBF Higgs production. However calculations show, that with couplings from Table 1 effects even for $\sqrt{s} = 13\,TeV$ are insignificant, as well as effects for branching ratios of the Higgs decays.

We have also calculated effects of the interaction for an associated single electro-weak boson production with the Higgs. While cross sections of processes are significant (few hundreds of *fb*), contributions of the additional interaction do not exceed few *per cent*. For example, for process $p\,p \to W^+\,H + X$ at $\sqrt{s} = 13\,TeV$ with $|\lambda| = 0.01$ ratio $\mu = 1.033$.

Possible manifestations of vertices (7) were studied in decays $H \to W^+W^-$, $H \to ZZ$ [18] . Results of this work give limitations, which definitely do not contradict values for couplings presented in Table 1.

In the next section we consider additional contributions of interactions (7) to interaction of the Higgs with quarks, especially with the heavy ones, which can lead to essentially more significant effects at the LHC.

3 Additional top and bottom quarks interaction with the Higgs

We use vertices (6) to define additional contribution for quarks interactions with Higgs. For the beginning we shall be interested in interactions of the most heavy top quarks.

Taking into account these vertices we calculate loop diagrams presented in Fig. 2 to obtain the following expression for $\bar{t}tH$ vertex, which corresponds to the first diagram in Fig.2

$$V_{\bar{t}tH} = -\frac{g}{2M_W}\,\bar{t}\,(M_t + 9\cos\theta_W M_Z M_W\; G_{WW} I_1\,(\hat{p}_1 - \hat{p}_2)(1+\gamma_5))\,tH;\;\; \hat{a} = a_\mu\gamma^\mu; \tag{10}$$

where G_{WW} is already defined in (6) and calculated in Table 1, p_1 and p_2 are respectively outcoming momenta of t and $\bar{t}$ quarks. Integral I_1 is defined in (8). For calculation of the integral we here use the same form-factor $F(t)$ (9).

Due to QCD gauge invariance we have to take into account also vertex for fourfold interaction involving also a gluon: $\bar{t}tHG_\mu$, which actually corresponds to the second diagram in Fig.2

$$V_{\bar{t}tHG} = 9 g g_s \cos\theta_W M_Z G_{WW} I_1\,\bar{t}\hat{G}(1+\gamma_5)tH; \tag{11}$$

where g_s is the QCD gauge coupling constant and of course usual structure of the QCD is used. Then we perform calculations for cross sections of process $p + p \to \bar{t}\,t\,H + X$ for two energies of the LHC: $\sqrt{s} = 8\,TeV$ and $\sqrt{s} = 13\,TeV$

Let us define for the same values of $\sqrt{s}$ ratios of cross-sections with nonzero λ in admissible interval (3) and its SM value for $\lambda = 0$

$$\mu_{\sqrt{s}} = \frac{\sigma_\lambda(pp \to \bar{t}tH)}{\sigma_0(pp \to \bar{t}tH)}; \tag{12}$$

where σ_0 is actually SM value for the cross section. Results of calculations with application of CompHEP package [19] are shown in Table 4.

Table 4. Production LO cross sections *fb* of $\bar{t}tH$ for $\sqrt{s} = 8\,TeV$, $13\,TeV$ and ratio $\mu_{\sqrt{s}}$ (12) at the LHC.

λ	$\sigma(13\,TeV)$	$\sigma(8\,TeV)$	μ_{13}	μ_8
−0.01	1628.2	460.6	3.14	3.15
−0.006	1212.9	342.6	2.34	2.34
−0.003	853.3	241.0	1.65	1.65
0	517.8	146.1	1.00	1.00
0.003	401.4	113.6	0.78	0.78
0.006	348.7	98.4	0.67	0.67
0.01	304.4	85.9	0.59	0.59

Values for cross-sections are calculated with the current value for the strong coupling [20]

$$\alpha_s(M_Z) = 0.1181 \pm 0.0011. \tag{13}$$

The uncertainty in (13) means 2% accuracy for calculated ross-sections. With taking into account of other sources of uncertainties we estimate overall accuracy to be around 10%.

The combination of the ATLAS and the CMS data, collected with $\sqrt{s} = 7\,TeV$ and $8\,TeV$, gives the following experimental result for ratio μ_8 [21]

$$\mu_8 = 2.3^{+0.7}_{-0.6}\,. \tag{14}$$

Due to significant uncertainties, result (14) does not mean convincing deviation from the SM value. With numbers from Table 4 we have from (14) the following prediction

$$\lambda = -0.0057^{+0.0028}_{-0.0039}\,. \tag{15}$$

The result is safely inside experimental limitations (3, 4). Let us note, that estimates for cut-off an energy scale also evidently do not contradict LHC results [9]. Of course we have in (15) again only two standard deviations effect, which undoubtedly needs further studies. We see that values of μ, practically, do not depend on $\sqrt{s}$ but with $\sqrt{s} = 13\,TeV$ cross sections are more than three times as much as those for conditions of result (14). One might hope to check the predictions in forthcoming experimental studies at the LHC with increased statistics. Emphasize, that in case of this study would give result $\lambda \neq 0$, we would come to the fundamental conclusion of non-perturbative effects in the electro-weak interaction to be necessarily present. Let us note, that recent NLO and NNLL SM calculations of $\bar{t}tH$ production cross section at 13 TeV are presented in works [22–24].

Let us consider also associative production of the Higgs with $\bar{b}\,b$ pairs. Unlike the t quark pairs case, for which the experimental studies were performed and have given results, *e. g.* (14), there were no dedicated studies. However, interactions (10,11) in our consideration also exist for other quarks. All the difference is connected only with value of the quark mass in (10). In particular, it is advisable to consider also process of b quark associative pair production with Higgs

$$p + p \rightarrow \bar{b}bH + X. \tag{16}$$

Results of calculations are shown in Table 5.

Table 5. Production LO cross sections fb of $\bar{b}bH$ for $\sqrt{s} = 8\,TeV$, $13\,TeV$ and ratio $\mu_{b\,\sqrt{s}}$ at the LHC.

λ	$\sigma(13\,TeV)$	$\sigma(8\,TeV)$	μ_{b13}	μ_{b8}
−0.01	1612.5	572.7	2.93	2.77
−0.006	1204.4	433.6	2.19	2.10
−0.003	903.0	329.1	1.64	1.59
0	550.8	206.7	1.00	1.00
0.003	446.0	151.8	0.81	0.82
0.006	396.4	151.8	0.72	0.73
0.01	354.6	137.1	0.64	0.66

Effects for the b-quarks are of the same order of magnitude as for the t-quark pairs. Unlike the t-quark case we have no datum for a comparison. As a matter of fact, analogous results are valid for light quarks u, d, c, s as well.

4 Conclusion

The problem of an existence of non-perturbative contributions in the electro-weak interaction is without doubt a fundamental one. Anomalous three-boson interaction (2) provides the crucial test for this problem. We have shown above, that there are promising processes

$$p + p \to \bar{t}tH + X \quad p + p \to \bar{b}bH + X; \tag{17}$$

for investigation of the problem at the LHC, and we can hope, that future results for these processes at $\sqrt{s} = 13\,TeV$ will confirm the existence of non-perturbative effects in the electro-weak interaction. Let us note, that we have studied how results for processes under the study depend on different cuts. It comes out, that for main process $p + p \to \bar{t}tH$ cuts $M(\bar{t}t) > M_0, M(tH) > M'_0, p_T(H) > p_{T0}$ *etc* lead, of course, to decreasing of cross sections, but, practically, do not change ratios μ for data. Thus introduction of cuts is to be defined by conditions of experiments, in particular by background considerations.

The important problem is, if predictions under the discussion could in any way contradict the present knowledge. We have already mentioned, that contributions of the additional interactions to branching ratios of the Higgs are negligible. The effects in process $p + p \to \bar{q}qH + X$, where we have to take into account all six flavors of quarks would lead to an additional contribution to the total Higgs production cross section. For example, for $\lambda = -0.006$, which actually is quite close to the central value in estimate (15) we have from Tables 4, 5 the following additional contributions $\Delta\sigma$ to the total cross section of the Higgs production

$$\begin{aligned} &\Delta\sigma(8TeV) = 1.36\,pb; \quad \sigma(8TeV) = 22.3\,pb; \\ &\Delta\sigma(13TeV) = 4.03\,pb; \; \sigma(13TeV) = 50.6\,pb; \end{aligned} \tag{18}$$

where we also show the SM values for the total cross-sections [20]. These additional contributions lead to a change in the global signal strength, which currently reads [20]

$$\mu = 1.09 \pm 0.07 \pm 0.04 \pm 0.03 \pm 0.07; \tag{19}$$

where two last errors are connected with uncertainties in the theoretical estimates. We easily see, that additional contributions (18) give the following changes for theoretical predictions for effective μ instead of unity

$$\mu(8TeV) = 1.061\,; \quad \mu(13TeV) = 1.080. \tag{20}$$

The results evidently do not contradict to value (19), which is based mostly on data collected with $\sqrt{s} = 7$ and $8\,TeV$.

In case of an existence of triple interaction (2), *e.g.* in processes (17), being confirmed, extensive studies of other possible non-perturbative effects will be desirable. Let us emphasize that simultaneous consideration of effects in reaction (17) and searches for anomalous interaction (2) are of the utmost importance. Provided these searches would give a positive and compatible results, we would obtain the strong evidence in favour of non-perturbative contributions of the electro-weak interaction.

Let us note also, that effects in top pair production in association with an electro-weak boson $W^{\pm}, Z$, which were discussed in work [25] under assumption of wouldbe existence of four-fold electro-weak bosons effective interaction [26, 27]. These effects might also indicate for the non-perturbative contributions.

In case the non-perturbative effects being discussed above would be discovered, there could appear new and really attractive possibilities. The approach [13–16, 25, 28] is connected with non-trivial solutions of compensation equations. Such solutions exist only for definite values of parameters inherent in a problem under a study. For example, the non-trivial solutions exist for values of coupling constants at the characteristic scale Λ for QCD $g_s(\Lambda_s) = 3.116$ [14] and for the EW theory $g_{EW}(\Lambda_{EW}) = 0.604$ [15] (see also (9)), that seems quite reasonable. Thus in case of the successful confirmation of the approach, we could hope to find a way for understanding of why physical values of couplings and maybe also of mass ratios are such as we do observe.

5 Acknowledgment

The author appreciates the collaboration with I.V. Zaitsev in obtaining results of the talk (see works [25, 28]). The studies are supported in part by grant NSh-7989.2016.2 of the President of Russian Federation.

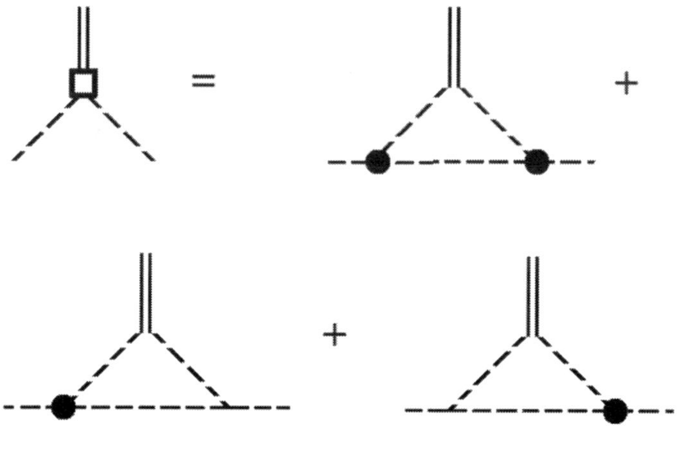

Figure 1. Diagram representation of non-perturbative contribution to VVH vertices depicted as square. Double lines represent Higgs boson, dotted lines represent electro-weak bosons. Black spots represent non-perturbative vertex (2, 6) and simple points corresponds to usual SM interaction.

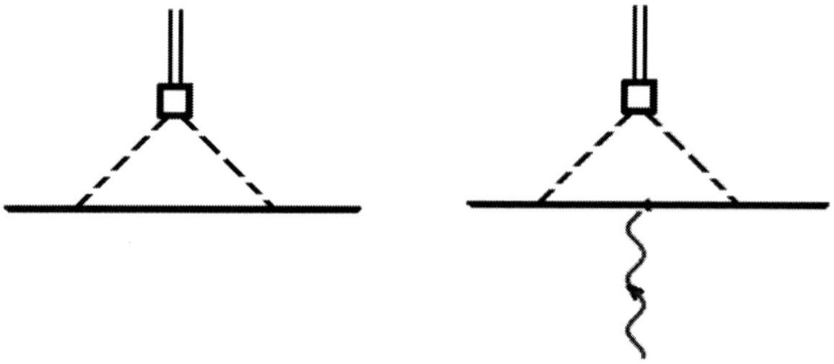

Figure 2. Diagram representation of non-perturbative contribution to $\bar{t}tH$ and $\bar{t}tHG$. Double lines represent Higgs boson, dotted lines represent electro-weak bosons. Thick lines represents the t quark, the wavy line corresponds to a gluon. Non-perturbative part of VVH vertex is defined in the previous section and simple points correspond to usual gauge interaction.

References

[1] Y. Nambu and G. Jona-Lasinio, Phys. Rev., 122 (1961), p. 345.
[2] Y. Nambu and G. Jona-Lasinio, Phys. Rev., 124 (1961), p. 246.
[3] M. K. Volkov, Annals Phys.,157 (1984), p. 282.
[4] D. Ebert, H. Reinhardt and M. K. Volkov, Prog. Part. Nucl. Phys., 33 (1994), p. 1.
[5] M. K. Volkov and A. E. Radzhabov , Phys. Usp., 49 (2006), p. 551.

[6] A. I. Alekseev, B. A. Arbuzov and V. A. Baikov, Theor. Math. Phys., 52 (1982), p. 739.
[7] K. Hagiwara, R. D. Peccei, D. Zeppenfeld and K. Hikasa, Nucl. Phys. B, 282 (1987), p. 253.
[8] K. Hagiwara, S. Ishihara, R. Szalapski and D. Zeppenfeld, Phys. Rev. D, 48 (1993), p. 2182.
[9] A.M. Sirunyan *et al.* (CMS Collaboration), Phys. Lett. B, 772 (2017) p. 21.
[10] M. Aabud *et al.* (ATLAS Collaboration), Eur. Phys. J. C, 77 (2017) p. 563.
[11] N. N. Bogoliubov, Soviet Phys.-Uspekhi, 67 (1959), p. 236.
[12] N. N. Bogoliubov, Physica Suppl. (Amsterdam), 26 (1960), p. 1.
[13] B. A. Arbuzov, M. K. Volkov and I. V. Zaitsev, Int. J. Mod. Phys. A, 21 (2006), p. 5721.
[14] B. A. Arbuzov and I. V. Zaitsev, Int. J. Mod. Phys. A, 28 (2013), Article 1350127.
[15] B. A. Arbuzov and I. V. Zaitsev, Phys. Rev. D, 85 (2012), Article 093001.
[16] B. A. Arbuzov, *Non-perturbative Effective Interactions in the Standard Model*, De Gruyter, Berlin, 2014.
[17] I. Anderson *et al.*, Phys. Rev. D, 89 (2014), Article 035007.
[18] V. Khachatryan *et al.* (CMS Collaboration) Phys.Rev. D, 92 (2015) Article 012004.
[19] E. E. Boos *et al.* (CompHEP Collaboration), Nucl. Instr. Meth. A, 534 (2004), p. 250.
[20] C. Patringnani *et al.* (Particle Data Group), Chin. Phys. C, 40 (2016), Article 100001.
[21] G. Aad *et al.* (ATLAS and CMS Collaborations), J. High Energy Phys., 1608 (2016), p. 045.
[22] A. Broggio *et al.*, JHEP, 1603 (2016) p. 124.
[23] A. Broggio *et al.*, JHEP, 1702 (2017) p. 126.
[24] A. Kulesza *et al.*, arXiv: 1704.03363 [hep-ph].
[25] B. A. Arbuzov and I. V. Zaitsev, Prog. Theor. Exp. Phys. 85 (2016), Article 093001.
[26] G. Belanger and F. Boudjema, Phys. Lett. B, 288 (1992), p. 201.
[27] G. Belanger *et al.*, Eur. Phys. J. C, 13 (2000), p. 283.
[28] B. A. Arbuzov and I. V. Zaitsev, Phys. Lett. B, 772 (2017), p. 184.

EPJ Web of Conferences **158**, 01009 (2017)
QFTHEP 2017
DOI: 10.1051/epjconf/201715801009

Extra dimensions hypothesis in high energy physics

Igor Volobuev[1,⋆], *Eduard* Boos[1,⋆⋆], *Viacheslav* Bunichev[1,⋆⋆⋆], *Maxim* Perfilov[1,⋆⋆⋆⋆], and *Mikhail* Smolyakov[1,†]

[1] *D.V. Skobeltsyn Institute of Nuclear Physics, Lomonosov Moscow State University, 119991 Moscow, Russia*

Abstract. We discuss the history of the extra dimensions hypothesis and the physics and phenomenology of models with large extra dimensions with an emphasis on the Randall-Sundrum (RS) model with two branes. We argue that the Standard Model extension based on the RS model with two branes is phenomenologically acceptable only if the inter-brane distance is stabilized. Within such an extension of the Standard Model, we study the influence of the infinite Kaluza-Klein (KK) towers of the bulk fields on collider processes. In particular, we discuss the modification of the scalar sector of the theory, the Higgs-radion mixing due to the coupling of the Higgs boson to the radion and its KK tower, and the experimental restrictions on the mass of the radion-dominated states.

1 Introduction

The hypothesis about the existence of extra space-time dimensions is discussed in theoretical physics for more than a century. There is a good reason to believe that this hypothesis grew from the ideas formulated by Bernhard Riemann in his famous lecture "On the Hypotheses which lie at the Bases of Geometry" [1]. In this lecture Riemann formulated the notion of n-dimensional space and suggested that the space-time geometry at microscopic distances may be very different from the macroscopic one.

The first attempt to construct a physical theory in a space-time with extra dimensions was made by Gunnar Nordström in 1914 in his work "On the possibility of unifying the electromagnetic and the gravitational fields" [2], where he tried to unify a relativistic scalar gravity theory, which he was developing at that time, with electrodynamics. In fact, the construction of Nordström was very similar to the present-day braneworld models: our four-dimensional world was embedded in a five-dimensional one as a submanifold. Though Nordström's scalar theory of gravity was a very important step in constructing Einstein's general theory of relativity, it was forgotten soon after the latter proved to be a success. Correspondingly, the idea of extra dimensions was abandoned for five years to be rediscovered by Theodor Kaluza and implemented in the framework of Einstein's general theory of relativity.

⋆e-mail: volobuev@theory.sinp.msu.ru
⋆⋆e-mail: boos@theory.sinp.msu.ru
⋆⋆⋆e-mail: bunichev@theory.sinp.msu.ru
⋆⋆⋆⋆e-mail: perfilov@theory.sinp.msu.ru
†e-mail: smolyakov@theory.sinp.msu.ru

2 Kaluza-Klein theory

The original theory of Kaluza [3] was a pure gravity in space-time $E = M^4 \times R^1$, where M^4 is four-dimensional space-time. If we denote the coordinates in E by $X^N = (x^\nu, y)$, $N = 0, 1, \cdots 4$, $\nu = 0, 1, \cdots 3$, the standard gravitational action in this space can be written as

$$S = \frac{1}{16\pi\hat{G}} \int_E \hat{R}\sqrt{-g}\, d^5X, \tag{1}$$

where $\hat{G}$ is the five-dimensional gravitational constant and $\hat{R}$ is the scalar curvature of the metric g_{MN}, $sign(g) = (-, + \cdots +)$. Kaluza suggested that this five-dimensional theory reproduced four-dimensional gravity theory coupled to electromagnetic field, if the fields of the original theory did not depend on the coordinate of the extra dimension.

The next step in theories with extra dimensions was made by Oskar Klein [4], who proposed to consider the vacuum configuration of the five-dimensional metric and to include scalar, spinor and vector fields in the multidimensional theory from the very beginning. He showed that, in this case, for every four-dimensional field there exists a tower of fields with the same quantum numbers and the masses of the order of the inverse size of the extra dimension, which are called nowadays the Kaluza-Klein modes.

Finally, in paper [5] by Einstein and Bergmann it was shown that the unobservability of the extra dimension could be explained by its compactness, i.e. that it is the circle S^1 instead of R^1, and extremely small size, - of the order of the Planck length $L_{Pl} = 1/M_{Pl}$.

Let us consider the Kaluza-Klein hypothesis in more detail. To isolate the physical degrees of freedom in this theory, it is useful to consider the linearized theory

$$g_{MN}(x, y) = \eta_{MN} + \kappa h_{MN}(x, y), \quad \kappa = \sqrt{16\pi\hat{G}}. \tag{2}$$

This theory is a gauge theory of tensor field with gauge transformations generalizing those of classical electrodynamics, which allows one to impose a convenient gauge condition and to obtain the linearized equations of motion. In the case, where the energy-momentum tensor of matter is equal to zero, these equations take the form

$$\partial^P \partial_P h_{MN}(x, y) = \left(\Box + \frac{\partial^2}{\partial y^2}\right) h_{MN}(x, y) = 0. \tag{3}$$

The corresponding eigenvalue problem for the mass spectrum of the gravitational field fluctuations looks as follows

$$\frac{\partial^2}{\partial y^2} h_{MN}(x, y) = -m^2 h_{MN}(x, y). \tag{4}$$

Thus, the mass spectrum can be found to be

$$m_n = \frac{2\pi n}{L}, \tag{5}$$

L denoting the size of the extra dimension, i.e. the circumference of S^1. It includes the massless zero modes, for which n = 0 and whose wave functions do not depend on the coordinate of the extra dimension, and the higher Kaluza-Klein modes that are very heavy, because the size of the extra dimension should be of the order of the Planck length to provide for the unobservability of the extra dimension.

The zero mode five-dimensional metric g_{MN} can be decomposed into the four-dimensional part and the extra components as

$$g_{MN} = \begin{pmatrix} \gamma_{\mu\nu} + \phi A_\mu A_\nu & \phi A_\mu \\ \phi A_\nu & \phi \end{pmatrix}, \qquad \phi = g_{44}, \phi A_\mu = g_{\mu 4}.$$

Since g_{MN} does not depend on y, which Kaluza called the "cylindricity condition", the scalar curvature $\hat{R}$ can be expressed as

$$\begin{aligned} \hat{R} &= R_{(4)} - \frac{1}{4} F_{\mu\nu} F^{\mu\nu} \phi^{-1} - \frac{1}{2} \phi^{-2} \partial_\mu \phi \partial^\mu \phi \\ F_{\mu\nu} &= \partial_\mu A_\nu - \partial_\nu A_\mu, \end{aligned}$$

$R_{(4)}$ being the scalar curvature of the four-dimensional space-time M^4 with metric $\gamma_{\mu\nu}$.

In the original paper by Kaluza the field A_μ was interpreted as electromagnetic field. But the subsequent studies showed that this interpretation failed.

The size L of the extra dimension being of the order of the Planck length, the Kaluza-Klein hypothesis leads to the conclusion that the observed fields (i.e. the fields of the Standard Model) should be the so called zero modes, i.e. should not depend on the coordinates of extra dimensions. Moreover, for each observed field there should exist a tower of fields with masses of the order of the Planck mass $M_{Pl} \sim 1/L_{Pl}$, which cannot be observed at the energies, which are available for experiments nowadays.

Nevertheless, even in this case there remain relations between the multidimensional and the four-dimensional theories, which can manifest themselves as certain symmetries of the reduced four-dimensional theories and the values of their coupling constants. In particular, an important point is the relation between the Planck mass in four dimensions M_{Pl} and the fundamental energy scale M of multidimensional space-time

$$M_{Pl}^2 = M^3 L, \tag{6}$$

which can be obtained for the zero modes of gravitational field from action (1) in space-time $E = M^4 \times S^1$ just by integrating it over the circle S^1.

3 Large extra dimensions

The interpretation of a theory in a space-time with extra dimensions in terms of a four-dimensional one was given the name of *dimensional reduction.* The presence of boson fields in the multidimensional theory led to the idea to explain the factorized structure of the multidimensional space-time dynamically as a result of interaction of gravity with gauge and scalar fields. This procedure was called *spontaneous compactification.* The attempts to construct the Standard Model along these lines showed that large extra dimensions were needed [6], which was obviously in contradiction with their unobservability.

A solution to this problem was found in 1983 by Rubakov and Shaposhnikov, who put forward a new scenario for Kaluza-Klein theories based on the idea of localization of states [7]. They considered a real Higgs field in the flat 5-dimensional space-time and found the exact solution connecting the two disjoint vacua. The energy density of the solution is localized in the vicinity of the 3-dimensional hypersurface $x^4 = 0$ and has a width of the order of the inverse mass of the Higgs field. Such an object is called a thick domain wall and serves as a potential well for the fluctuations of the Higgs field.

It has also been shown that there are only two modes of the fluctuations that are localized on the domain wall. Moreover, the interaction of these modes cannot produce unlocalized modes, if the center of mass energy is less than a certain energy scale proportional to the Higgs mass. Thus, for sufficiently small energies the theory describes a 4-dimensional world, although the whole world is, in fact, 5-dimensional.

This model suggests a possibility for extra dimensions to be unobservable, if the fields of the Standard Model are localized by a certain mechanism on a domain wall in multidimensional space-time. The corresponding localization mechanism turned out to be rather simple for spinors, but a plausible mechanism for gauge fields has not been found yet.

If the thickness of the domain wall goes to zero, it turns into a three-dimensional surface in the multidimensional space, on which the field of the Standard Model are presumably trapped; such an object is called a membrane, or just a "brane". There are indications that this scenario can be implemented in the theory of superstrings.

It is clear that within the framework of this scenario extra dimensions may be much larger, than the Planck length (or even infinite). Based on this assumption, in paper [8] a single brane without tension (i.e. energy density) in a space-time with an arbitrary number of compact extra dimensions was considered. In this case relation (6) between the fundamental multidimensional energy scale and the four-dimensional one, which for d extra dimensions looks like

$$M_{Pl}^2 = M^{d+2} L^d,$$

yields that the fundamental energy scale in multidimensional space (in the bulk) M may be much less than the Planck mass in four dimensions M_{Pl}, if the volume of the space of extra dimensions L^d is large. In other words, gravity in multidimensional space-time becomes "strong" not at the energies of the order of $10^{19}\,GeV$, but at much lower energies, maybe of the order of $1 - 10\,TeV$. Thus, the scenario provides a solution to the hierarchy problem: it gives a strong gravity in the multidimensional space-time and a weak gravity on the brane. The effects due to the interaction of this "strong" gravity with the fields of the Standard Model could be observable already at the energies, which are available at the existing colliders.

A flaw of this approach is the approximation of the zero brane tension, which is rather too rough: it turns out that the proper gravitational field of the brane cannot be taken into account perturbatively at all. In paper [9] an exact solution for two branes interacting with gravity in five-dimensional space-time was found, which allows one to estimate the influence of the proper gravitational field of the brane on these results. First of all, it turned out that, if extra dimensions are compact, there should exist at least two branes. A similar solution for two branes in space-time with two extra dimenions with a similar form of metric was later found in [10]. In what follows we will discuss in more detail the five-dimensional model, which turned out to be more interesting.

4 The Randall-Sundrum model

The model is called the Randall-Sundrum model and is similar to the original Kaluza-Klein theory in the sense that it is a gravity theory in 5-dimensional space-time $E = M^4 \times S^1$ with coordinates $X^M = (x^\mu, y)$, the circle S^1 now being of circumference $2L$. But unlike the original Kaluza-Klein theory the gravity interacts with two 3-branes. Thus, the action of the system is

$$S = \int d^4x \int_{-L}^{L} dy (2M^3 \hat{R} - \Lambda) \sqrt{-g} - \lambda_1 \int_{y=0} \sqrt{-\tilde{g}} d^4x - \lambda_2 \int_{y=L} \sqrt{-\tilde{g}} d^4x, \quad (7)$$

where M is the fundamental 5-dimensional energy scale, Λ is the cosmological constant in 5-dimensional space-time, $\tilde{g}_{\mu\nu}$ is the metric induced on the branes, $\tilde{g} = det\tilde{g}_{\mu\nu}$, and the subscripts 1,2 distinguish the branes. Moreover, the 5-metric g must be invariant under the reflection $(x^\mu, y) \to (x^\mu, -y)$, which is equivalent to identifying the points (x^μ, y) and $(x^\mu, -y)$ of S^1 and reducing it to the orbifold S^1/Z_2.

The ansatz for the metric, which preserves the Poincaré invariance on every hypersurface $y = const$, looks like

$$ds^2 = e^{-2\sigma(y)}\eta_{\mu\nu}dx^\mu dx^\nu + dy^2 \equiv \gamma_{MN}(y)dx^M dx^N, \tag{8}$$

When it is substituted into Einstein equations following from action (7), they reduce to

$$\begin{aligned}\left(\frac{d\sigma}{dy}\right)^2 &= -\frac{\Lambda}{24M^3}\\ 6\frac{d^2\sigma}{dy^2} &= \frac{1}{2M^3}\left(\lambda_1\delta(y) + \lambda_2\delta(y-L)\right).\end{aligned}$$

The solution to these equation with the orbifold symmetry is

$$\begin{aligned}\sigma(y) &= k|y| + c, \quad \Lambda = -24M^3k^2\\ \lambda_1 &= -\lambda_2 = 24M^3k,\end{aligned} \tag{9}$$

where k is a new parameter of the dimension of mass, which can be interpreted as the inverse effective thickness of the branes, and c is an integrations constant. The choice of this constant is equivalent to a choice of the coordinates $\{x^\mu\}$ on the branes, and we will keep it arbitrary for the moment.

It is worth noting that the space E with metric (8) and σ given by solution (9) is a piece of five-dimensional anti-de Sitter space in horospherical coordinates [11] and that brane 1 has a positive energy density, whereas brane 2 has a negative one.

Metric (8) with σ given by equation (9) is taken to be the vacuum of the theory, and the physical fields are fluctuations $h_{MN}(x, y)$ above this vacuum, which are treated perturbatively:

$$g_{MN}(x,y) = \gamma_{MN}(y) + \frac{1}{\sqrt{2M^3}}h_{MN}(x,y). \tag{10}$$

If we substitute this representation into action (7) and keep the terms of zero order in M, we get a free action for the fluctuations of the gravitational field. It is a common knowledge that linearized gravity is a gauge theory, and the corresponding gauge transformations can be used to isolate the physical degrees of freedom of the model. One can prove that there exist a gauge, where

$$h_{\mu 4} = 0, \; h_{44} = h_{44}(x) \equiv \phi(x). \tag{11}$$

The scalar field $\phi(x)$ is called the radion field. With the help of a substitution including this field one can extract a transverse-traceless tensor field $b_{\mu\nu}(x, y)$ from the tensor field $h_{\mu\nu}(x, y)$. The transverse-traceless tensor field $b_{\mu\nu}(x, y)$ and the scalar radion field $\phi(x)$ are the physical degrees of freedom of the Randall-Sundrum model and diagonalize its second variation Lagrangian. One can also prove that the radion field is massless.

To understand the physical meaning of the radion field let us calculate the distance between the branes along a geodesic $x = const$:

$$l = \int_0^L \sqrt{ds^2} \sim \int_0^L \left(1 + \frac{1}{2\sqrt{2M^3}}h_{44}\right)dy = L\left(1 + \frac{1}{2\sqrt{2M^3}}\phi(x)\right). \tag{12}$$

This formula means that the field $\phi(x)$ describes local variations of the inter-brane distance, i.e. oscillations of the branes with respect to each other.

The transverse-traceless tensor field $b_{\mu\nu}(x, y)$ can be expanded as

$$b_{\mu\nu}(x,y) = \sum_n b_{\mu\nu}^{(n)}(x)\psi_n(y), \tag{13}$$

where $b_{\mu\nu}^{(n)}(x)$ are four-dimensional fields with spin 2 and certain masses m_n. The latter and the wave functions of the modes $\psi_n(y)$ are defined by the equation

$$\left[\frac{1}{2}\frac{d^2}{dy^2} + 2k(\delta(y) - \delta(y-L)) - 2k^2\right]\psi_n(y) = \frac{m_n^2}{2}e^{2\sigma}\psi_n(y). \tag{14}$$

This equation can be solved exactly, the wave function of the zero mode being proportional to $\exp(-2\sigma)$ and those of massive modes being expressible in terms of Bessel and Neumann functions.

Now we are in a position to find the relation between the fundamental 5-dimensional energy scale M and the Planck mass. To this end we have to calculate the curvature of the metric

$$ds^2 = e^{-2\sigma(y)}\bar{g}_{\mu\nu}dx^\mu dx^\nu + dy^2 \tag{15}$$

with $\bar{g}_{\mu\nu} = \eta_{\mu\nu} + h_{\mu\nu}(x)$ and integrate it over the extra dimension. Since the wave function of the massless graviton is proportional to $\exp(-2\sigma)$, this metric corresponds to four-dimensional gravity only and integrating it over the extra dimension gives an effective action for 4-dimensional gravity

$$S_{eff} = 2M^3 \int_{-L}^{L} dy e^{-2(k|y|+c)} \int R(\bar{g})\sqrt{-\bar{g}}\,d^4x = 2M^3(1-e^{-2kL})\,e^{-2c}\int d^4x R(\bar{g})\sqrt{-\bar{g}}, \tag{16}$$

which has to coincide with the Hilbert-Einstein action for the metric $\bar{g}_{\mu\nu}$. But since we have two branes, we have to realize, to which brane this metric corresponds.

If we are on the brane at $y = 0$, the metric on the brane induced from metric (15) coincides with $\bar{g}_{\mu\nu}$ only if $c = 0$. This leads to the following relation between the energy scales

$$M_{Pl}^2 = M^3\,\frac{1-e^{-2kL}}{k}\,. \tag{17}$$

Thus, if we live on the brane at $y = 0$, relation (17) demands that the fundamental five-dimensional energy scale M and the parameter k should be of the order of M_{Pl}, $M \sim k \sim M_{Pl}$. This means that for the observer on brane 1 the five-dimensional gravity is as weak, as the four-dimensional one.

If we are on the brane at $y = L$, the metric on the brane induced from metric (15) coincides with $\bar{g}_{\mu\nu}$ only if $c = -kL$. In this case formula (16) gives us the following relation between the Planck mass M_{Pl} and the fundamental five-dimensional energy scale M:

$$M_{Pl}^2 = M^3\,\frac{e^{2kL}-1}{k}\,. \tag{18}$$

Obviously, both relations (17) - (18) reduce to (6) (with L replaced by $2L$) for $k \to 0$, i.e. when the effective thickness of the brane goes to infinity and the very notion of a brane is meaningless.

If our world is situated on the brane at $y = L$, we can put $kL \sim 35$ in formula (18), which gives that the funamental five-dimensional energy scale M and the parameter k may be of the order of $1TeV$. That is the hierarchy problem of the gravitational interaction is solved for the observer on brane 2: the energy scale of the five-dimensional gravity comes down to the electroweak energy scale, and the gravity on the brane becomes weak due to the exponential warp factor in relation (18). Thus, five-dimensional gravity in the Randall-Sundrum model looks different for observers on different branes.

In what follows we will assume that we live on the brane at $y = L$. If we consider the interaction of the physical fields with matter on this brane, we will find that the coupling of the massive tensor fields to the energy-momentum tensor of SM is of the order of inverse TeV. The coupling of the radion field $\phi(x)$ to the trace of the energy-momentum tensor is also of the same order, i.e. much stronger than that of the zero mode of the tensor field. Thus, a scalar gravity is realized on brane 2, which is phenomenologically unacceptable. A way to cure this problem without violating the solution of the hierarchy problem is to give a mass to the radion. It turns out that this is equivalent to stabilizing the distance between the branes.

5 Stabilized Randall-Sundrum model

Mechanisms for stabilizing the Randall-Sundrum model were put forward in papers [12, 13]. Both of them employ a five-dimensional scalar field to stabilize the inter-brane distance, but the mechanism of paper [13] seems to be preferable because it is based on an exact solution for gravity interacting with two branes and a scalar field in five-dimensional space-time. The physical degrees of freedom of models stabilized by this mechanism were isolated in paper [14]. They are the transverse-traceless tensor fields $b^n_\mu(x), n = 0, 1, \cdots$ with masses $m_n (m_0 = 0)$ and scalar fields $\varphi_n(x), n = 1, 2, \cdots$ with masses μ_n. The coupling constants to matter of the massive tensor fields remain essentially the same as in the unstabilized model, i.e. of the order of $1TeV^{-1}$. The couplings of the scalar fields are also defined by their wave functions and are essentially of the same order.

If we consider the extension of the Standard Model based on the stabilized Randall-Sundrum model, where the SM fields are localized on the brane at $y = L$, we see that the vacua of the stabilized Randall-Sundrum model and the SM are independent. If we assume that this brane world model is an indivisible theory, it is reasonable to believe that there should be a common interconnected vacuum solution for all these fields. To this end we modify action (7) so that it contains a quartic interaction of the stabilizing scalar field and of the Higgs field, which would connect the stabilization of the extra dimension size and the SM spontaneous symmetry breaking.

The modified action looks like

$$S = S_g + S_{\phi+SM}, \tag{19}$$

where S_g and $S_{\phi+SM}$ are given by

$$S_g = 2M^3 \int d^4x \int_{-L}^{L} dyR\sqrt{-g}, \tag{20}$$

$$\begin{aligned} S_{\phi+SM} &= \int d^4x \int_{-L}^{L} dy\left(-\frac{1}{2}g^{MN}\partial_M\phi\partial_N\phi - V(\phi)\right)\sqrt{-g} - \\ &- \int_{y=0} \sqrt{-\tilde{g}}V_1(\phi)d^4x + \int_{y=L} \sqrt{-\tilde{g}}(-V_2(\phi) + L_{SM-HP} + L_{int}(\phi, H))d^4x. \end{aligned} \tag{21}$$

Here M is the fundamental five-dimensional energy scale, $V(\phi)$ is a bulk stubilizing scalar field potential and $V_{1,2}(\phi)$ are brane scalar field potentials, $\tilde{g} = det\tilde{g}_{\mu\nu}$, and $\tilde{g}_{\mu\nu}$ denotes the metric induced on the branes. The SM fields are localized on the brane at $y = L$, and the Lagrangian L_{SM-HP} is the SM Lagrangian without the Higgs potential. The key point of the approach under consideration is the replacement of the Higgs potential by the interaction Lagrangian

$$L_{int}(\phi, H) = -\lambda(|H|^2 - \frac{\xi}{M}\phi^2)^2 \tag{22}$$

of the Higgs and Goldberger-Wise fields, ξ being a positive dimensionless parameter.

The background solutions for the metric and the scalar field, which preserve the Poincaré invariance in any four-dimensional subspace $y = const$, look like

$$ds^2 = e^{-2A(y)}\eta_{\mu\nu}dx^\mu dx^\nu + dy^2 \equiv \gamma_{MN}(y)dx^M dx^N, \tag{23}$$

$$\phi(x, y) = \phi(y), \tag{24}$$

$\eta_{\mu\nu}$ denoting the flat Minkowski metric, whereas the background (vacuum) solution for the Higgs field is standard

$$H_{vac} = \begin{pmatrix} 0 \\ \frac{v}{\sqrt{2}} \end{pmatrix}, \tag{25}$$

all the other SM fields being equal to zero.

If one substitutes this ansatz into the equations corresponding to action (19), one gets a system of differential equations for the functions $A(y)$ and $\phi(y)$, the brane scalar field potentials $V_{1,2}(\phi)$ and interaction Lagrangian (22) defining the boundary conditions for these equations on the branes. The potentials $V_{1,2}(\phi)$ are chosen so that they fix the values of the stabilizing field ϕ on the branes and stabilize the inter-brane distance [12, 13]. Interaction Lagrangian (22) does not affect this stabilization mechanism, if the relation

$$\phi(L)^2 = \frac{Mv^2}{2\xi}$$

between the values of the Higgs and stabilizing scalar field on the brane at $y = L$ is valid, which defines the vacuum value of the Higgs field. This means that in such a scenario the Higgs field vacuum expectation value, being proportional to the value of the stabilizing scalar field on the TeV brane, arises dynamically as a result of the gravitational bulk stabilization.

Now the linearized theory is obtained by expanding the metric, the scalar and the Higgs field in the unitary gauge around the background solution as

$$g_{MN}(x, y) = \gamma_{MN}(y) + \frac{1}{\sqrt{2M^3}}h_{MN}(x, y), \tag{26}$$

$$\phi(x, y) = \phi(y) + \frac{1}{\sqrt{2M^3}}f(x, y), \tag{27}$$

$$H(x) = \begin{pmatrix} 0 \\ \frac{v+\sigma(x)}{\sqrt{2}} \end{pmatrix}. \tag{28}$$

After substituting this representation into action (19) and keeping the terms of the second order in h_{MN}, f and σ one gets the Lagrangian of this action which is the standard free Lagrangian of the SM (i.e. the masses of all the SM fields are expressed in the same way as usually in terms of the vacuum value of the Higgs field and the coupling constants) together with the standard second variation Lagrangian of the stabilized RS model [14] supplemented by an interaction term of the scalar fields f and σ on the brane coming from interaction Lagrangian (22) [15].

Besides the fields f and σ, there are two more scalar fields in the linearized theory, – the fields $h_{44}(x, y)$ and $\gamma^{\mu\nu}h_{\mu\nu}(x, y)$. However, the fields $f(x, y)$, $h_{44}(x, y)$ and $\gamma^{\mu\nu}h_{\mu\nu}(x, y)$ are not independent: they are connected by the equations of motion of the linearized theory and a gauge condition [14, 16]. For this reason we can use any one of them to describe the scalar states.

It turns out to be convenient to use the field $h_{44}(x, y)$. This field can be expanded in KK modes, the lowest mode $\phi_1(x)$ is called the radion field and the modes $\phi_n(x)$, $n > 1$ belong to its KK tower. This expansion induces the corresponding expansion of the bulk scalar field $f(x, y)$. Substituting the latter expansion into the second variation Lagrangian and integrating over the extra dimension coordinate, one gets an effective four-dimensional Lagrangian. In case the Higgs and the radion masses are much lower than the masses of the radion excitations on can pass to a low energy approximation in the four-dimensional Lagrangian by integrating out the radion excitation fields. This gives an effective Lagrangian for the interactions of the Higgs and radion fields with the SM fields. However, due to the Higgs-radion mixing terms the fields $\sigma(x)$ and $\phi_1(x)$ are not mass eigenstates.

The physical mass eigenstate fields $h(x)$, $r(x)$ are, as usually, obtained by a rotation diagonalizing the mass matrix

$$\begin{aligned} h(x) &= \cos\theta\,\sigma(x) + \sin\theta\,\phi_1(x) \\ r(x) &= -\sin\theta\,\sigma(x) + \cos\theta\,\phi_1(x). \end{aligned} \tag{29}$$

The field $h(x)$ is the Higgs-dominated field and the field $r(x)$ is the radion-dominated field. Finally, one gets the effective interaction Lagrangian of the physical scalar fields $h(x)$ and $r(x)$ with the Standard Model fields in the following form [15]:

$$\begin{aligned} L_{h-r} &= -\frac{1}{2}\partial_\mu h(x)\partial^\mu h(x) - \frac{1}{2}m_h^2 h^2(x) - \frac{1}{2}\partial_\mu r(x)\partial^\mu r(x) - \frac{1}{2}\mu_r^2 r^2(x) \\ &- \frac{(c\cos\theta + \sin\theta)}{\Lambda_r} h(x)(T_\mu^\mu + \Delta T_\mu^\mu) + \frac{(c\sin\theta - \cos\theta)}{\Lambda_r} r(x)(T_\mu^\mu + \Delta T_\mu^\mu) - \\ &- \sum_f \frac{m_f}{v}\bar{\psi}_f\psi_f(\cos\theta\, h(x) - \sin\theta\, r(x)) - \frac{2M_W^2}{v}(W_\mu^- W^{\mu+})(\cos\theta\, h(x) - \sin\theta\, r(x)) + \\ &- \frac{M_Z^2}{v}(Z_\mu Z^\mu)(\cos\theta\, h(x) - \sin\theta\, r(x)) - \frac{M_W^2}{v^2}(W_\mu^- W^{\mu+})(\cos\theta\, h(x) - \sin\theta\, r(x))^2 + \\ &- \frac{M_Z^2}{2v^2}(Z_\mu Z^\mu)(\cos\theta\, h(x) - \sin\theta\, r(x))^2. \end{aligned} \tag{30}$$

Here m_h^2 and m_r^2 are the masses of the fields $h(x)$ and $r(x)$, Λ_r is the (inverse) coupling constant of the radion to the trace of the SM energy-momentum tensor, ΔT_μ^μ is the conformal anomaly of massless vector fields explicitly given by

$$\Delta T_\mu^\mu = \frac{\beta(g_s)}{2g_s} G_{\rho\sigma}^{ab} G_{ab}^{\rho\sigma} + \frac{\beta(e)}{2e} F_{\rho\sigma} F^{\rho\sigma} \tag{31}$$

with β being the well-known QCD and QED β-functions.

The parameter c accommodates the contributions of the integrated out heavy scalar modes and is expressed in terms of the physical masses and the mixing angle as follows:

$$c = \frac{(m_r^2 - m_h^2)\sin 2\theta}{m_r^2\cos^2\theta + m_h^2\sin^2\theta}\left(\sum_{n=2}^{\infty}\alpha_n^2\right). \tag{32}$$

It also depends on the sum of the coefficients α_n^2, where α_n is the ratio of the wave functions in the extra dimension of the modes ϕ_n and ϕ_1 taken at $y = L$. These ratios are, of course, model dependent and should fall off with n in order for the sum to be convergent. In certain models several first ratios may be of the order of unity [14]. Thus, one can conservatively estimate this sum to be also of the order of unity.

The effective four-dimensional interaction Lagrangian (30) expressed in terms of the physical Higgs-dominated $h(x)$ and radion-dominated $r(x)$ fields involves only five extra parameters in addition to those of the SM: the masses of the Higgs-dominated and radion-dominated fields m_h and m_r, the mixing angle θ, the (inverse) coupling constant of the radion to the trace of the energy-momentum tensor of the SM fields Λ_r and the parameter c.

Let us consider several interesting subspaces of the parameter space of the effective theory. If we put the parameters c and θ equal to zero, i.e. consider the case of the zero mixing, Lagrangian (30) becomes just the SM Higgs Lagrangian plus the usual Lagrangian of the radion interaction with the trace of the SM energy-momentum tensor. In paper [17] it was shown that this interaction Lagrangian leads to the same production and decay properties of a single radion as those of the Higgs boson. If we put $1/\Lambda_r = 0$, we get the effective Lagrangian of the real Higgs singlet extension of the Standard Model [18, 19]. If we formally put the parameter c equal to zero while keeping the mixing angle θ non-zero, we obtain an effective interaction Lagrangian which is very similar to the one of the unstabilized RS model with the Higgs-curvature term on the brane [20]. In particular, the couplings of the Higgs-dominated and the radion-dominated states to the conformal anomaly of massless vector fields, which turn to be very important for their production and decay to two photons, are the same. However, one should keep in mind that the observable parameters of the effective Lagrangian in different cases depend differently on the fundamental parameters of the models.

Thus, Lagrangian (30) is a very general effective Lagrangian of the interaction of two mixed scalar states with the fermion and vector fields of the Standard Model extended by a singlet scalar. The extra parameters of this Lagrangian are natural for considering the phenomenology of the mixed scalar states and allow one to compare easily the predictions of different models. In fact, it is unimportant, how they depend on the fundamental parameters of a particular model as long as the latter belong to the phenomenologically acceptable parameter subspace.

In the general case of a non-zero mixing, when all the parameters and θ, $1/\Lambda_r$, and c are not equal to zero, the additional terms in the Lagrangian, coming from the integrated out heavy modes and containing the parameter c that depends on the scalar state masses and the mixing angle, may lead to certain changes in the collider phenomenology of the Higgs-dominated and radion-dominated states as was demonstrated in [15], where also the Feynman rules for the model are explicitly given.

6 Restrictions on the radion mass

Since there are two light particles in the scalar sector, we have studied two a priori possible scenarios that the scalar state discovered at 125 GeV is either a Higgs-dominated state or a radion-dominated state [15]. In this analysis we used only two signal strengths corresponding to the channels $gg->\gamma\gamma$ and $gg->ZZ^*$. The analysis showed that the interpretation of the 125 GeV scalar state as a Higgs-dominated state is the preferred one. Then the restrictions on the mass of the radion-dominated state and the mixing angle can be demonstrated by Fig. 1

We see that the most probable values of the radion-dominated state masses are close to the mass of the Higgs-dominated state. Nevertheless, there is an allowed region of heavier masses of the radion-dominated state, where its production rate is decreased so that the corresponding state may be again allowed. Obviously, if one considers larger values of the parameter Λ_r, the cross section of the radion-dominated state gets smaller and the allowed region for such a state is increased. This is demonstrated in Fig. 2, where the parameter Λ_r is chosen to be 5 GeV with all the other parameters being the same as for the considered 3 TeV case.

Next we have studied the impact of the constraints from the measurements of the parameters of the observed 125 GeV Higgs boson and from the unconfirmed 750 GeV diphoton excess in the LHC experiments on the properties of a radion-dominated state [21]. It turned out that the allowed region of heavier masses of the radion-dominated state shrinks even more. In Fig. 3 it is shown that the allowed masses are now of the order of 1 TeV.

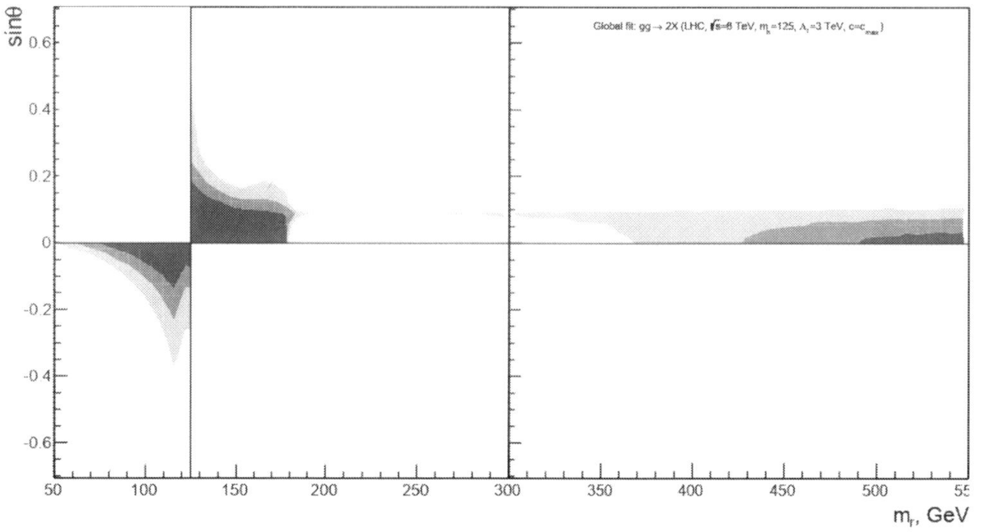

Figure 1. Exclusion contours for the combined χ^2 fit in the $(m_r, \sin\theta)$ plane that includes both channels $gg \to \gamma\gamma$ and $gg \to ZZ^*$ for the LHC at $\sqrt{s} = 8TeV$ and m_h=125, $\Lambda_r = 3TeV$, $c = c_{max}$. The dark, medium and light shaded areas correspond to CL of the fit 65%, 90% and 99% respectively.

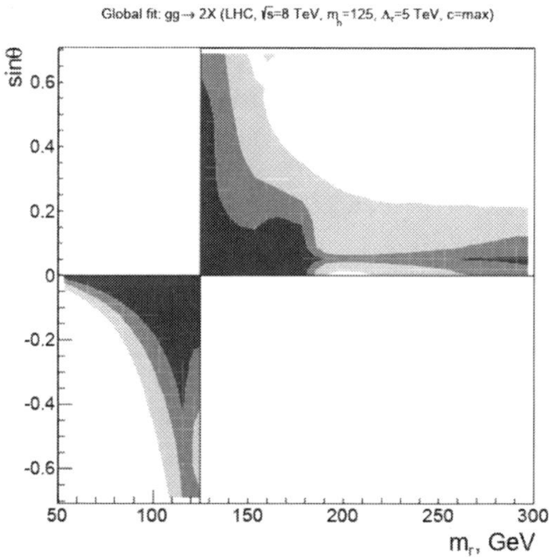

Figure 2. Exclusion contours for the combined χ^2 fit in the $(m_r, \sin\theta)$ plane that includes both channels $gg \to \gamma\gamma$ and $gg \to ZZ^*$ for the LHC at $\sqrt{s} = 8TeV$ and m_h=125, $\Lambda_r = 5TeV$, $c = c_{max}$. The dark, medium and light shaded areas correspond to CL of the fit 65%, 90% and 99% respectively.

However, there remain masses of the radion-dominated state below 125 GeV. The available data from the LHC also give restrictions on the model parameters in this region. We have studied in

 DOI: 10.1051/epjconf/201715801009

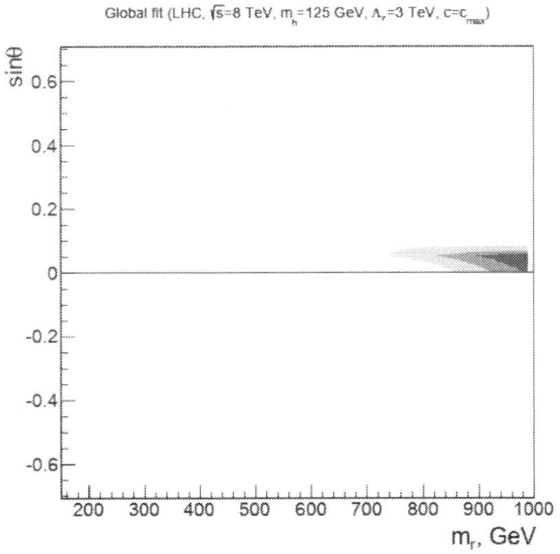

Figure 3. The region in the $(m_r, \sin\theta)$ still allowed by the common fit of the CMS and ATLAS exclusion limits for heavy Higgs searches and restictions from the fit of the influence of that parameters on the signal strengthes at the point 125 GeV. The dark, medium and light shaded areas correspond to CL of the fit 65%, 90% and 99% respectively. (The figure is taken from paper [15])

detail two possible values of masses below 125 GeV, - 30 GeV and 60 GeV that have already been discussed in the literature. The LHC data at 8 TeV give the restrictions on the model parameters for these radion-dominated state masses presented in Fig. 4

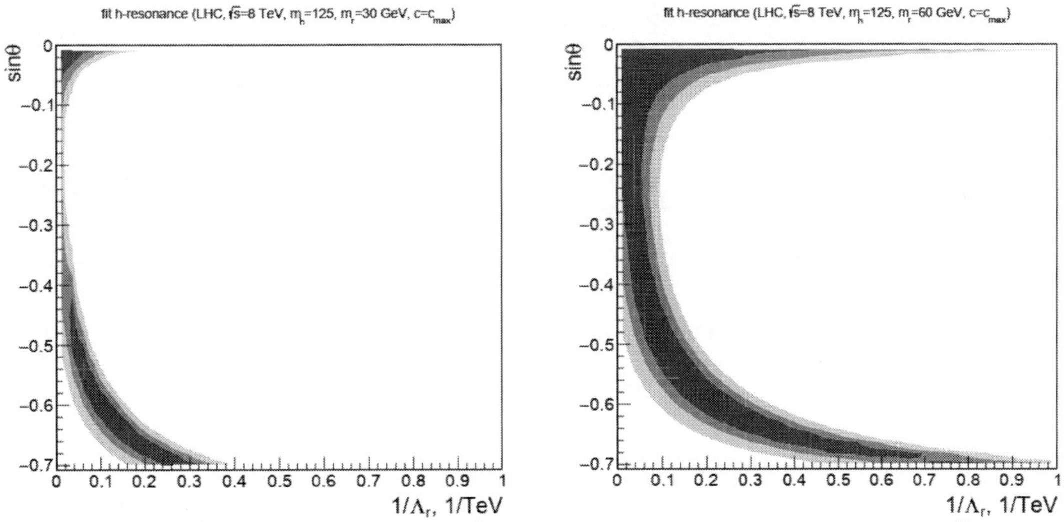

Figure 4. Exclusion contours for the partial χ^2 fit in the $(\sin\theta, 1/\Lambda_r)$ plane for the LHC at $\sqrt{s} = 8TeV$ and $m_r = 30GeV$ (left) and $m_r = 60GeV$ (right). The dark, medium and light shaded areas correspond to CL of the fit 65%, 90% and 99% respectively.

These restrictions become even stronger, if one takes into account the old LEP data. In fact, they almost exclude the existence of a light radion-dominated state for $\Lambda_r < 10TeV$.

7 Conclusion

Theories with extra dimensions have a long history characterized by alternating periods of oblivion and keen interest. The latest developments in this field have put an end to the old restrictions on the size of the extra dimensions, and now their predictions can be confronted with experiment.

We believe that nowadays the most consistent model with extra dimensions is the stabilized Randall-Sundrum model. If its fundamental parameters lie in the TeV energy range for the observer on the negative tension brane, the effects due to the massive modes can be observed in collider experiments. In particular, the higher excitations of the scalar radion field can influence the properties of the Higgs-dominated state at 125 GeV, whereas the restrictions on the model parameters imposed by the experimental study of the properties of 125 GeV scalar boson move the mass of the radion-dominated state to 1 TeV or higher.

References

[1] Bernhard Riemann, Nature **8** 14-17, 36, 37 (1873).

[2] G. Nordström, Physikalische Zeitschrift **15** 504-506 (1914) [physics/0702221].

[3] T. Kaluza, Sitzungsber. Preuss. Akad. Wiss, Berlin, Math. Phys. **H1**, 966-972 (1921).

[4] O. Klein, Z. Phys. **37**, 895-906 (1926).

[5] A. Einstein and P. Bergmann, Annals Math. **39** 683 (1938)

[6] Yu.A. Kubyshin, I.P. Volobuev, J. M. Mourao and G. Rudolph, Lecture Notes in Physics, **349** (1990).

[7] V.A. Rubakov, M.E. Shaposhnikov, Phys. Lett. **B125** 136 (1983).

[8] N. Arkani-Hamed, S. Dimopoulos, G. Dvali, **B429** 263 (1998) [hep-ph/9803315].

[9] L. Randall, R. Sundrum, Phys. Rev. Lett. **83** 3370 (1999) [hep-ph/9905221].

[10] Sean M. Carroll, Monica M. Guica, hep-th/0302067.

[11] M. N. Smolyakov, Class. Quantum Grav. **25** 238003 (2008) [gr-qc/0811.0299].

[12] W. D. Goldberger, M. B. Wise. Phys. Rev. Lett. **83** 4922 (1999) [hep-ph/9907447].

[13] O. DeWolfe, D. Z. Freedman, S. S. Gubser, A. Karch, Phys. Rev. D. V. **62** 046008 (2000) [hep-th/9909134].

[14] E. E. Boos, Yu. S. Mikhailov, M. N. Smolyakov, I. P. Volobuev, Mod. Phys. Lett. A **21**, 1431-1449 (2006) [hep-th/0511185].

[15] E. E. Boos, V. E. Bunichev, M. A. Perfilov, M. N. Smolyakov and I. P. Volobuev, Phys. Rev. D **92** (2015) no.9, 095010 [arXiv:1505.05892].

[16] C. Csaki, M. L. Graesser and G. D. Kribs, Phys. Rev. D **63**, 065002 (2001) [hep-th/0008151].

[17] E. Boos, S. Keizerov, E. Rahmetov and K. Svirina, Phys. Rev. D **90**, 095026 (2014) [arXiv:1409.2796].

[18] S. I. Godunov, A. N. Rozanov, M. I. Vysotsky and E. V. Zhemchugov, arXiv:1503.01618.

[19] T. Robens and T. Stefaniak, Eur. Phys. J. C **75** (2015) 104 [arXiv:1501.02234].

[20] G. F. Giudice, R. Rattazzi and J. D. Wells, Nucl. Phys. B **595**, 250 (2001) [hep-ph/0002178].

[21] E. E. Boos, V. E. Bunichev and I. P. Volobuev, J. Exp. Theor. Phys. **124** (2017) no.5, 722.

EPJ Web of Conferences **158**, 02001 (2017) DOI: 10.1051/epjconf/201715802001
QFTHEP 2017

Searches for new physics with heavy flavour at ATLAS

Semen Turchikhin*
On behalf of the ATLAS Collaboration
[1] *Joint Institute for Nuclear Research*

Abstract. The Flavour Changing Neutral Current processes are sensitive to New Physics contributions, in particular through additional electroweak loop amplitudes. The angular analysis of the decay of $B^0 \to \mu^+\mu^- K^{*0}$ is presented. A number of angular coefficients are measured as a function of the invariant mass squared of the dimuon system using data collected by the ATLAS experiment at the LHC. Comparison is made to theoretical predictions, including for the observable P'_5, for which there has been recent tension between theory and experiment. ATLAS result on the study of the $B^0_{(s)} \to \mu^+\mu^-$ rare decays is also presented.

1 Introduction

Studying heavy flavour decay processes at collider experiments provides an excellent opportunity for searches of indirect manifestations of new physics. Rare decays of b hadrons involving $b \to s\ell\ell$ transition belong to such kind of processes. They are mediated by flavour-changing neutral currents which are forbidden at tree level within the Standard Model (SM) and occur at higher orders via loop diagrams. Thus these processes are sensitive to contributions of heavy particles predicted by various models beyond the SM in the loops.

This proceeding presents two analyses of such decays recently performed in the ATLAS experiment [1] at Large Hadron Collider: measurement of angular distributions in the $B^0 \to \mu^+\mu^- K^{*0}$ semileptonic decay [2] and study of $B^0_{(s)} \to \mu^+\mu^-$ rare decay [3].

2 Angular analysis of $B^0 \to \mu^+\mu^- K^{*0}$ decay

Decay $B^0 \to \mu^+\mu^- K^{*0}$ occurs via loop diagrams and has a branching ratio of $(1.03 \pm 0.06) \times 10^{-6}$ [4]. Angular distributions of the decay products are observables of interest for such process as they are potentially sensitive to different types of new physics.

Analysis of the $B^0 \to \mu^+\mu^- K^{*0}$ decay is performed using the pp collisions data sample collected by the ATLAS detector at centre of mass energy $\sqrt{s} = 8$ TeV corresponding to an integrated luminosity of 20.3 fb^{-1} [2].

The kinematics of the $B^0 \to \mu^+\mu^- K^{*0}$ decay[1] (with subsequent $K^{*0} \to K^+\pi^-$) is described in terms of four variables: the dimuon invariant mass squared, q^2, and three angles describing the final state geometrical configuration, as illustrated in Fig. 1.

*e-mail: Semen.Turchikhin@cern.ch
[1]Charge conjugation is implied here and below unless stated otherwise.

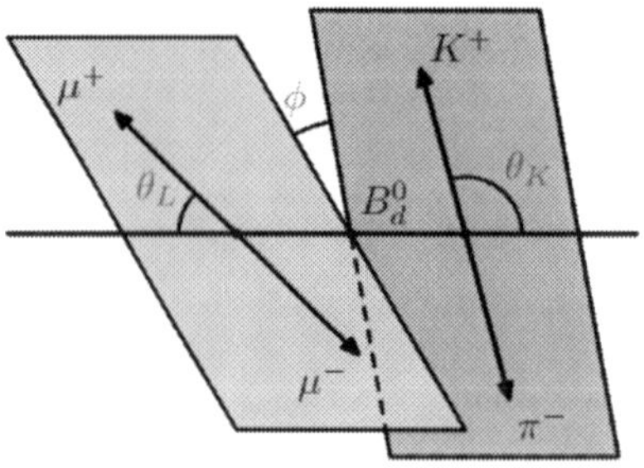

Figure 1. Illustration of the kinematic angles (ϕ, θ_L, θ_K) definition in $B^0 \to \mu^+\mu^- K^{*0}(K^+\pi^-)$ decay.

The differential decay rate as a function of the decay angles can be expressed as

$$\frac{1}{\mathrm{d}\Gamma/\mathrm{d}q^2}\frac{\mathrm{d}^4\Gamma}{\mathrm{d}\cos\theta_L\mathrm{d}\cos\theta_K\mathrm{d}\phi\mathrm{d}q^2} = \frac{9}{32\pi}\Bigg[\frac{3(1-F_L)}{4}\sin^2\theta_L + F_L\cos^2\theta_K + \frac{1-F_L}{4}\sin^2\theta_K\cos 2\theta_L$$
$$-F_L\cos^2\theta_K\cos 2\theta_L + S_3\sin^2\theta_K\sin^2\theta_L\cos 2\phi$$
$$+S_4\sin 2\theta_K\sin 2\theta_L\cos\phi + S_5\sin 2\theta_K\sin\theta_L\cos\phi$$
$$+S_6\sin^2\theta_K\cos\theta_L + S_7\sin 2\theta_K\sin\theta_L\sin\phi$$
$$+S_8\sin 2\theta_K\sin 2\theta_L\sin\phi + S_9\sin^2\theta_K\sin^2\theta_L\sin 2\phi\Bigg], \quad (1)$$

where F_L is the fraction of longitudinally polarised K^{*0}'s and S_i are the angular coefficients related to the decay transversity amplitudes. Extracting the full set of these parameters from a fit to data using the above equation requires sufficient statistics and purity of the data sample which are not available with the analysed dataset. To mitigate this, trigonometric transformations are used to simplify the fit so that terms in Eq. (1) drop out. The folding schema used in the analysis follows Ref. [5] and results in four sets of transformations, such that three parameters can be extracted from each of four fits: F_L, S_3 and one of the other S_i parameters. As a consequence, S_6 and S_9 cannot be extracted from the data.

The S_i parameters are sensitive to hadronic form factors, and to reduce the corresponding theoretical uncertainties, the following optimized set of parameters is used in the analysis:

$$P_1 = \frac{2S_3}{1-F_L}, \quad P_2 = \frac{1}{2}\frac{S_6}{1-F_L}, \quad P_3 = -\frac{S_9}{1-F_L}, \quad P'_{j=4,5,6,8} = \frac{S_{i=4,5,7,8}}{\sqrt{F_L(1-F_L)}}.$$

The B^0 decay candidates are reconstructed by fitting a common vertex of two oppositely charged muons and two other oppositely charged tracks forming a K^{*0} meson. Signal candidate selection mostly focuses on combinatorial background suppression. Selection criteria include a number of cuts on kinematic properties, quality of the vertex fit, displacement of the decay vertex and pointing of the candidate momentum towards the primary interaction point. Further signal fits are performed for the candidates with the q^2 value within the range [0.04, 6] GeV^2 excluding [0.98, 1.1] GeV^2 due to background from $\phi \to \mu^+\mu^-$ decays. In this q^2 range, 787 signal candidates are retained after the full selection.

Extended unbinned maximum likelihood fits of the distributions of B^0 candidate mass and angular variables $\cos\theta_K$, $\cos\theta_L$, and ϕ are performed on the data to extract the angular parameters. The fit model and procedure have been extensively validated using simulated pseudo-experiments. The fits

q^2 [GeV^2]	n_{signal}	$n_{\text{background}}$
[0.04, 2.0]	128 ± 22	122 ± 22
[2.0, 4.0]	106 ± 23	113 ± 23
[4.0, 6.0]	114 ± 24	204 ± 26
[0.04, 4.0]	236 ± 31	233 ± 32
[1.1, 6.0]	275 ± 35	363 ± 36
[0.04, 6.0]	342 ± 39	445 ± 40

Table 1. The values of fitted signal, n_{signal}, and background, $n_{\text{background}}$, yields obtained for different regions of q^2. The uncertainties indicated are statistical. Table taken from Ref. [2].

are done separately in three non-overlapping regions of q^2: [0.04, 2.0], [2.0, 4.0], and [4.0, 6.0] GeV^2, as well as in wider bins [0.04, 4.0], [1.1, 6.0], [0.04, 6.0] GeV^2, to facilitate comparison with the results of other experiments and theoretical predictions. The signal and background yields in these q^2 bins are shown in Table 1.

A number of sources of systematic uncertainty affecting the analysis are studied. The methods for determining these uncertainties are based on either a comparison of nominal and modified fit results, or on observed fit biases in modified pseudo-experiments. Examples of the most significant sources are: various background contributions producing peaks in the distributions of the angular variables; uncertainty of background angular shapes modelling; detector angular acceptance description; detector alignment and magnetic field calibration uncertainties; effect of intrinsic bias of the maximum likelihood estimator.

The distributions of F_L and $P_i^{(\prime)}$ parameters as a function of q^2 are shown in Fig. 2. The figure also shows results of theoretical predictions of SM by various groups: Ciuchini et al. (CFFMPSV) [6], Descotes-Genon et al. (DHMV) [7], and Jäger and Martin Camalich (JC) [8]. Results of measurements from LHCb [9], CMS [10], Belle [11, 12] and BaBar [13] experiments are also shown.

Theoretical predictions are in a good agreement with the ATLAS result, except for the P_4' and P_5' measurements in $q^2 \in [4.0, 6.0]$ GeV^2 and P_8' in $q^2 \in [2.0, 4.0]$ GeV^2. The observed deviations of P_4' and P_5' from the SM expectations are consistent with those reported by LHCb [9] and are 2.5 and 2.7 standard deviations from the DHMV predictions, respectively (with less significant difference for the other calculations). All measurements are found to be within three standard deviations of the range covered by the different predictions.

3 Study of $B^0_{(s)} \to \mu^+\mu^-$ rare decay

Decays of B^0_s or B^0 mesons into a muon pair represent another example of $b \to s\ell\ell$ transition. Measurement of their branching fractions is of particular interest because of their additional helicity suppression and since they can be predicted by the SM very precisely: $\mathcal{B}(B^0_s \to \mu^+\mu^-) = (3.65 \pm 0.23) \times 10^{-9}$ and $\mathcal{B}(B^0 \to \mu^+\mu^-) = (1.06 \pm 0.09) \times 10^{-10}$ [14]. Being a clear experimental signature, these decays are considered to be one of the most prominent probes for new physics in heavy flavour sector. The CMS and LHCb collaborations so far have reported the observation of $B^0_s \to \mu^+\mu^-$ [15, 16] and evidence of $B^0 \to \mu^+\mu^-$ decays, with combined values: $\mathcal{B}(B^0_s \to \mu^+\mu^-) = \left(2.8^{+0.7}_{-0.6}\right) \times 10^{-9}$ and $\mathcal{B}(B^0 \to \mu^+\mu^-) = \left(3.9^{+1.6}_{-1.4}\right) \times 10^{-10}$ [17].

ATLAS analysis of the $B^0_{(s)} \to \mu^+\mu^-$ decays presented below uses the full Run-1 pp collision dataset collected at centre of mass energies $\sqrt{s}$ = 7 and 8 TeV and corresponding to an integrated luminosity of 25 fb^{-1} [3].

 DOI: 10.1051/epjconf/201715802001

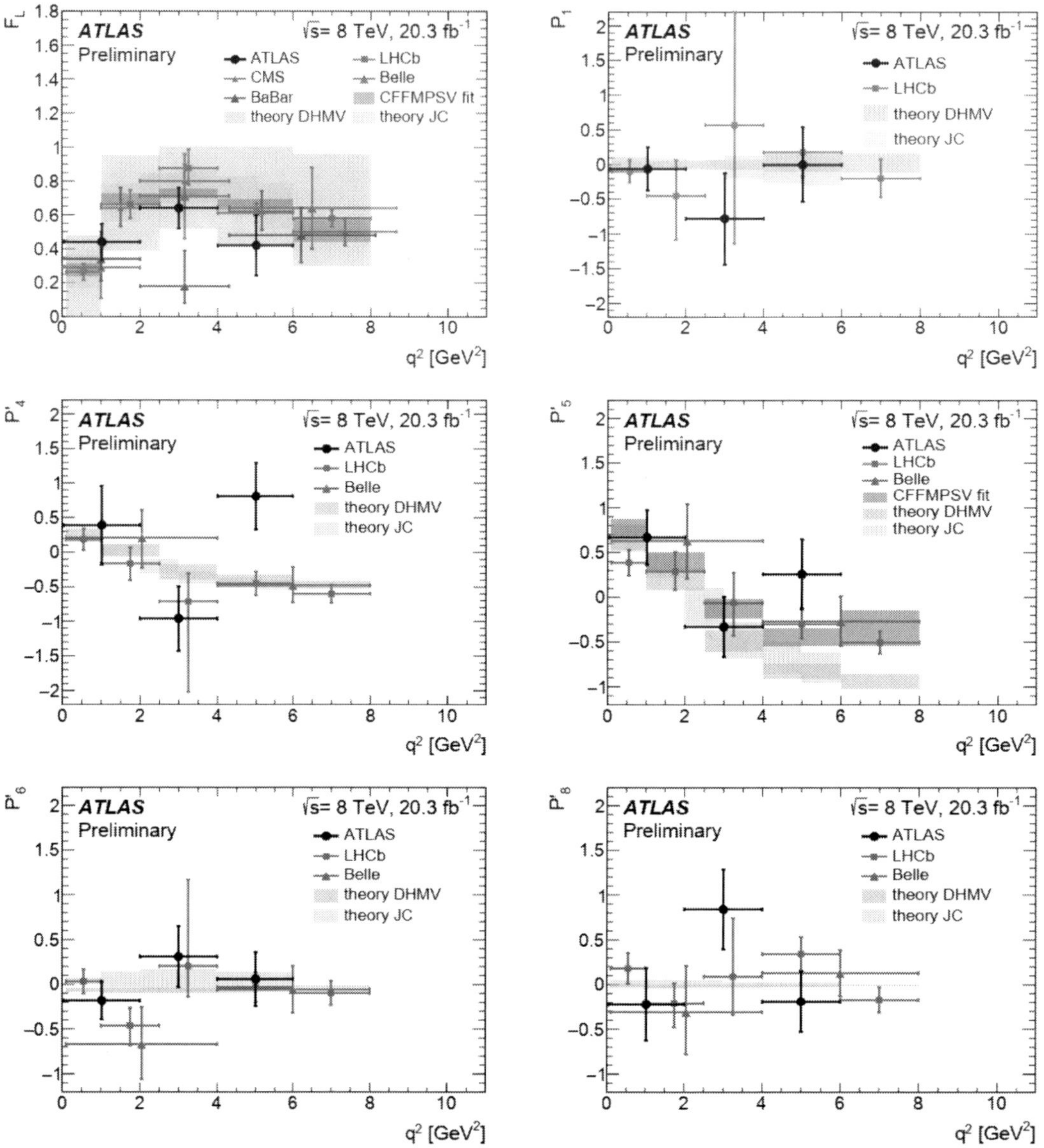

Figure 2. The measured values of F_L and $P_i^{(\prime)}$ compared with predictions from the theoretical groups and other experiment results. Figure taken from Ref. [2].

The branching fractions of the signal decay modes are measured relative to the normalisation decay $B^+ \to J/\psi(\mu^+\mu^-)K^+$ that has a known branching fraction [4]. This approach allows to substantially reduce systematic uncertainties related to the luminosity, detector acceptance, reconstruction and trigger efficiencies. The signal decay branching fraction is thus given by the following equation:

$$\mathcal{B}(B^0_{(s)} \to \mu^+\mu^-) = \frac{N_{d(s)}}{\varepsilon_{\mu^+\mu^-}} \times [\mathcal{B}(B^+ \to J/\psi K^+) \times \mathcal{B}(J/\psi \to \mu^+\mu^-)] \frac{\varepsilon_{J/\psi K^+}}{N_{J/\psi K^+}} \times \frac{f_u}{f_{d(s)}},$$

where N_d (N_s) is the $B^0 \to \mu^+\mu^-$ ($B_s^0 \to \mu^+\mu^-$) signal yield, $N_{J/\psi K^+}$ is the normalisation channel yield, $\varepsilon_{\mu^+\mu^-}$ and $\varepsilon_{J/\psi K^+}$ are the corresponding values of acceptance times efficiency evaluated from simulation, and f_u/f_d (f_u/f_s) is the known ratio of the hadronisation probabilities of a b-quark into B^+ and B^0 (B_s^0). The yield of the normalisation channel is obtained from a maximum likelihood fit of $J/\psi K^+$ invariant mass distribution in data.

A blind analysis is performed which means that data in the dimuon invariant mass region from 5166 to 5526 MeV are removed until the procedures for event selection and the signal yield extraction are completely defined.

Multivariate technique based on boosted decision trees (BDT) [18] is used for event selection of both signal and normalisation decay channels in the data. Two BDT classifiers are used. The first one referred to as "continuum BDT" is intended to separate the signal from a huge combinatorial background of pairs of muons coming from different sources (continuum background). It takes as input a number of variables characterising the decay candidate kinematics, properties of decay vertex, isolation of the decay candidate. The classifier is trained and tested using simulated samples of signal and background. The second classifier is used for suppression of mis-identified muons and exploits properties of their reconstructed tracks. This mis-identification can happen if a charged pion or kaon decays within the detector volume or punches through the calorimeters to the ATLAS muon spectrometer. If two hadrons from $B \to hh'$ (h, h' being charged pions or kaons) decays produce such fake muons, this signature mimics the signal decay thus being a source of irreducible background.

The yields of the decays $B_s^0 \to \mu^+\mu^-$ and $B^0 \to \mu^+\mu^-$ are extracted from data using a maximum likelihood fit of the dimuon invariant mass. To improve the sensitivity, the fit is performed simultaneously in three intervals of continuum BDT output designed to have equal signal efficiency of 18 %. These regions have different levels of background thus allowing to constrain its properties in the simultaneous fit. The likelihood includes various systematic uncertainties as nuisance parameters, but statistical uncertainties largely dominate. Total expected number of signal events according to the SM prediction is 41 and 5 for N_s and N_d, respectively, equally distributed among the three intervals.

The fit projections are shown in Fig. 3. Without applying any boundary on the values of the fitted yields, the values determined by the fit are $N_s = 16 \pm 12$, $N_d = -11 \pm 9$. The primary result of the analysis is obtained by applying the natural boundary of non-negative yields, for which the fit returns the values $N_s = 11$, $N_d = 0$.

The branching fraction $\mathcal{B}(B_s^0 \to \mu^+\mu^-)$ corresponding to the fitted yield with the constraint of non-negative values is $(0.9^{+1.1}_{-0.8}) \times 10^{-9}$ where the uncertainties are quoted for 68.3 % confidence interval. The observed significance of the $B_s^0 \to \mu^+\mu^-$ signal is determined using pseudo-experiments and found to be 1.4 standard deviations while the corresponding expected significance for the SM prediction is 3.1 standard deviations. Pseudo-experiments are also used to evaluate the compatibility of the simultaneous fit result with the SM expectations. This results in a p value of 0.048 ± 0.002, corresponding to 2.0 standard deviations.

Figure 4 shows the contours in the plane of $\mathcal{B}(B_s^0 \to \mu^+\mu^-)$ and $\mathcal{B}(B^0 \to \mu^+\mu^-)$ drawn for values of $-2\,\Delta\ln(L)$ equal to 2.3, 6.2 and 11.8, relative to the maximum of the likelihood L, allowing negative values of the branching fractions. The maximum within the physical boundary is shown with error bars indicating the 68.3% interval for the value of $\mathcal{B}(B_s^0 \to \mu^+\mu^-)$. Also shown are the corresponding contours obtained in the combination of the results of the CMS and LHCb experiments [17], and the prediction based on the SM.

An upper limit is placed on the $B_s^0 \to \mu^+\mu^-$ branching fraction using the $\mathrm{CL_s}$ method [19] at the 95% confidence level: $\mathcal{B}(B_s^0 \to \mu^+\mu^-) < 3.0 \times 10^{-9}$. The limit is obtained under the hypothesis of background only, with $\mathcal{B}(B^0 \to \mu^+\mu^-)$ left as a free parameter in the fit. The expected limit is $1.8^{+0.7}_{-0.4} \times 10^{-9}$.

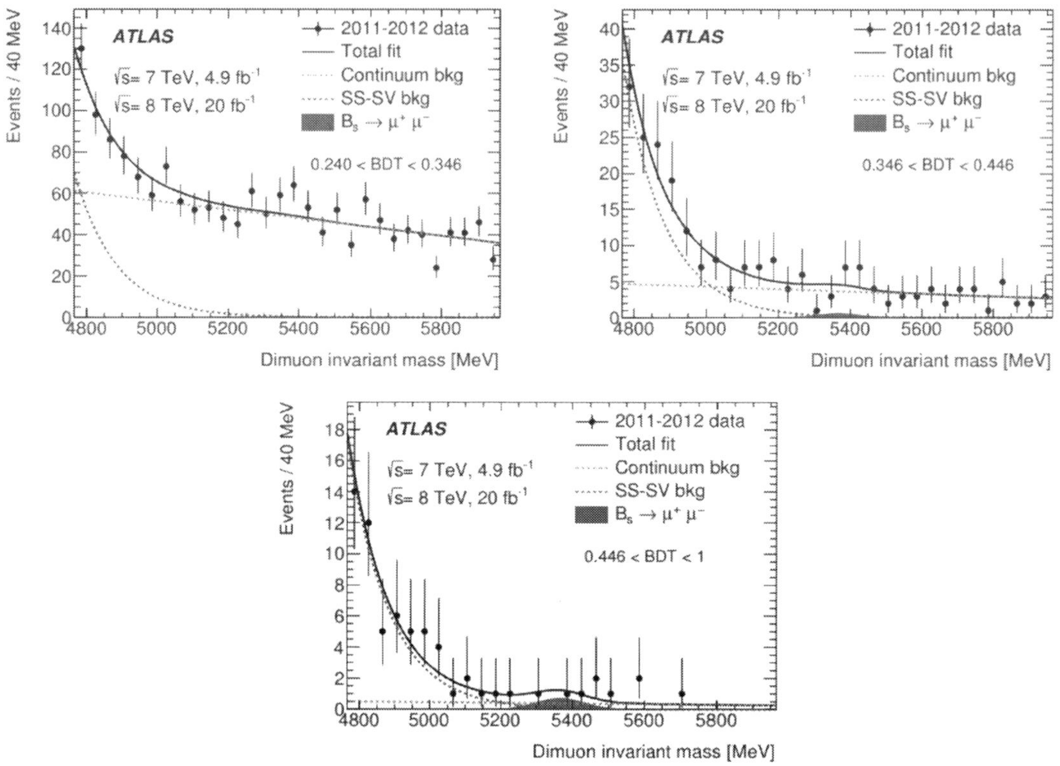

Figure 3. Dimuon invariant mass distributions in the unblinded data, in the three intervals of continuum-BDT output. Superimposed is the result of the maximum-likelihood fit, obtained imposing the boundary of non-negative signal contributions. The total fit is shown as a black continuous line, the filled area corresponds to the observed signal component, the blue and green dashed lines to background components. Figure taken from [3].

An upper limit is also set on the branching fraction of the B^0 decay: $\mathcal{B}(B^0 \to \mu^+\mu^-) < 4.2 \times 10^{-10}$ (95% CL). This upper limit is above the SM prediction and also covers the central value of the combination of the measurements by CMS and LHCb [17]. Tension in the value of this branching fraction is reduced with the most recent result of LHCb [20] not included in the combination, where the upper limit $\mathcal{B}(B^0 \to \mu^+\mu^-) < 3.4 \times 10^{-10}$ (95% CL) is set.

4 Conclusion

Two analyses of b hadron decays sensitive to new physics performed by ATLAS experiment are presented. The results of the angular analysis of $B^0 \to \mu^+\mu^- K^{*0}$ decay are found to agree with the SM predictions within three standard deviations across the whole studied range of q^2. In the rare $B^0_{(s)} \to \mu^+\mu^-$ decays study, although the expected sensitivity of ATLAS measurement is comparable to that of LHCb and CMS, the observed signals are lower than those in the other experiments. This result is compatible with the SM expectations at level of two standard deviations.

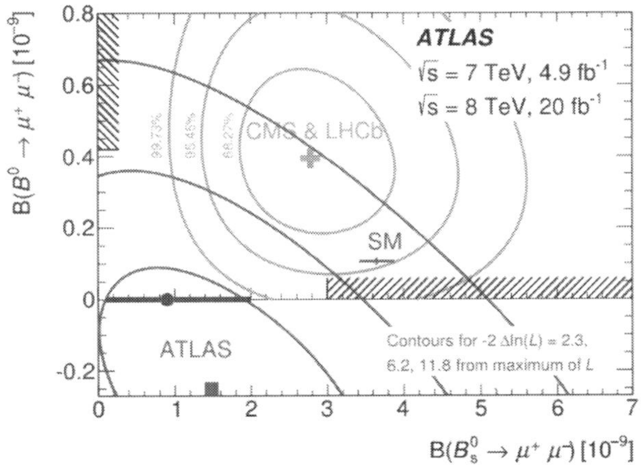

Figure 4. Contours in the plane $\mathcal{B}(B_s^0 \to \mu^+\mu^-)$, $\mathcal{B}(B^0 \to \mu^+\mu^-)$ for intervals of $-2\,\Delta\ln(L)$ equal to 2.3, 6.2 and 11.8 relative to the absolute maximum of the likelihood, without imposing the constraint of non-negative branching fractions. Also shown are the corresponding contours for the combined result of the CMS and LHCb experiments, the SM prediction, and the maximum of the likelihood within the boundary of non-negative branching fractions, with the error bars covering the 68.3 % confidence range for $\mathcal{B}(B_s^0 \to \mu^+\mu^-)$. Figure taken from [3].

References

[1] ATLAS Collaboration, JINST **3**, S08003 (2008)
[2] ATLAS Collaboration, ATLAS-CONF-2017-023, `http://cds.cern.ch/record/2258146`
[3] ATLAS Collaboration, Eur. Phys. J. C **76**, 513 (2016)
[4] C. Patrignani et al. (Particle Data Group), Chin. Phys. C **40**, 100001 (2016)
[5] LHCb Collaboration, Phys. Rev. Lett. **111**, 191801 (2013)
[6] M. Ciuchini, M. Fedele, E. Franco, S. Mishima, A. Paul, L. Silvestrini, M. Valli, JHEP **06**, 116 (2016)
[7] S. Descotes-Genon, L. Hofer, J. Matias, J. Virto, JHEP **12**, 125 (2014)
[8] S. Jäger, J. Martin Camalich, JHEP **05**, 043 (2013)
[9] LHCb Collaboration, JHEP **02**, 104 (2016)
[10] CMS Collaboration, Phys. Lett. B **753**, 424 (2016)
[11] J.T. Wei et al. (Belle Collaboration), Phys. Rev. Lett. **103**, 171801 (2009)
[12] S. Wehle et al. (Belle Collaboration), Phys. Rev. Lett. **118**, 111801 (2017)
[13] J.P. Lees et al. (BaBar Collaboration), Phys. Rev. D **93**, 052015 (2016)
[14] C. Bobeth, M. Gorbahn, T. Hermann, M. Misiak, E. Stamou, M. Steinhauser, Phys. Rev. Lett. **112**, 101801 (2014)
[15] CMS Collaboration, Phys. Rev. Lett. **111**, 101804 (2013)
[16] LHCb Collaboration, Phys. Rev. Lett. **111**, 101805 (2013)
[17] LHCb and CMS Collaborations, Nature **522**, 68 (2015)
[18] A. Hocker et al., PoS **ACAT**, 040 (2007)
[19] A.L. Read, J. Phys. G **28**, 2693 (2002)
[20] LHCb Collaboration, Phys. Rev. Lett. **118**, 191801 (2017)

EPJ Web of Conferences **158**, 02002 (2017)
QFTHEP 2017
DOI: 10.1051/epjconf/201715802002

Searches for Squarks and Gluinos with ATLAS

Oliver Ricken, on behalf of the ATLAS collaboration[1,*]

[1] *Physikaliches Institut, Universität Bonn, Nussallee 12, 53115 Bonn, Germany*

Abstract. One of the most versatile and attractive extensions to the successful yet incomplete Standard Model of particle physics is Supersymmetry – a theory the ATLAS experiment at the Large Hadron Collider is looking for in its recorded data. Due to the nature of proton-proton collisions, the recorded physics events are mainly produced via the strong force. This fact makes searches for the superpartners of the gluon and the quarks particularly promising.
This document provides an overview of searches for squarks and gluinos using the ATLAS experiment and describes two of the major analyses in detail. The analysis strategies are outlined, the results discussed and interpreted. Finally, an outlook onto other searches for strongly produced Supersymmetry with ATLAS is given.

1 Introduction

Being able to describe most phenomena observed in experimental particle physics, the Standard Model still leaves several open questions on both sides of particle physics, theoretical and experimental alike.
As an extended frame into which the Standard Model can be embedded, Supersymmetry (SUSY, for a pedagogical introduction see [1]) has been proposed as a theoretical model that is able to accommodate for many shortcomings of the Standard Model but for which experimental evidence is still to be found. This document reports on the latest results in the search for strong production of Supersymmetry using the ATLAS experiment [2] at the Large Hadron Collider LHC [3].
To illustrate the different approaches in the search for Supersymmetry with ATLAS, two different analyses following different strategies and searching for different models of Supersymmetry are introduced. After presenting the theoretical models being searched for, the analysis strategies and the obtained results are shown and discussed. In a brief outlook, further searches for strong production of SUSY are outlined.

2 Supersymmetry and the Standard Model

The most complete and successful theory in particle physics that is experimentally most thoroughly tested and confirmed is the Standard Model. It is able to describe interactions between elementary particles using three of the four fundamental forces of nature – gravity is not considered. This ability lets the Standard Model explain the formation of matter, the production and decay of heavier particles,

*e-mail: oliver.ricken@cern.ch

and the origin of the fundamental particles' properties. All in all, the Standard Model has been able to provide answers to most questions that arose from observed phenomena in the past and to predict observations before they were made, rendering itself a successful theory of particle physics. However, the successful model is yet neither complete nor able to explain all currently verified observations. Among other phenomena which the Standard Model is not able to describe is the existence of dark matter. Moreover, it is unable to predict the measured value of the Higgs boson mass in a way that does not exploit large, unnatural corrections – a problem referred to as the hierarchy problem. As a logical consequence of these flaws of the model, extensions to it and larger models to embed it into have been proposed, the most prominent being the theory of Supersymmetry (SUSY) [1].

As the name suggests, SUSY brings a new symmetry to the world of particle physics. In this larger model—of which the Standard Model would be a low-energy limit—a new fundamental symmetry between fermions and bosons is introduced. The Standard Model fermions receive bosonic superpartners while the bosons are accompanied by new fermions. These new particles allow, among many other effects, for cancellation of the huge corrections to the Higgs boson's mass introduced by couplings to Standard Model fermions, hence solving the aforementioned hierarchy problem.

If SUSY were to exist in the manner as described so far, the new particles were much likely already discovered—a bosonic electron-doppelgänger would have been visible in many experiments, for example—which is not the case until today. As a consequence, SUSY has got to be a broken symmetry. Since the minimal way of parametrising SUSY still needs more than 100 parameters[1], many approaches to restrict the available parameter space by making different assumptions and proposing different mechanisms of symmetry breaking have lead to a variety of different models of SUSY, none of them being found any evidene for as of now. As a consequence and due to the fact that nature is not expected to prefer one designated breaking mechanism of SUSY over others, the search strategies of particle physics experiments have changed.

In order to find a SUSY signature regardless of any particular model and to be able to cover as many different models when interpreting the results, Simplified Models are being looked for nowadays. These models usually contain one particular SUSY signature: only one process produces a fixed set of sparticles that, in turn, decay with fixed branching ratios to a set of stable and/or detectable particles. Simplified Models are not described by parameters that arise from the breaking mechanism of a designated theory but rather by pseudo-observables such as the masses of the sparticles involved. Depending on the outcome of an experimental analysis of such a model, either the discovery of a New Physics's signature can be claimed or limits on the describing parameters can be set – within the boundaries of the underlying assumptions and constraints of such a Simplified Model, of course. For model-builders on the theory side of particle physics, Simplified Models can help to narrow down the variety of prosperous new models to look for and are hence particularly worth searching for.

3 SUSY Searches at ATLAS – Strategies & Approaches

As outlined in section 2, the landscape of models of SUSY and Simplified Models is rich and not all of them are well testable at a collider such as the LHC. Due to the fact that the LHC collides protons, interactions driven by the strong force are more abundant than electro-weak processes. As a consequence, models producing gluinos and squarks via the strong interaction are expected to be well accessible and hence studied in various fashions. The ATLAS collaboration pursues two major approaches here:

- Analyses look for rather generic signatures of beyond-Standard-Model physics that are easily produced by the strong interaction to cover as much ground as possible in as many parts of the SUSY

[1]The Minimal Supersymmetric Standard Model comprises 104 parameters [1]

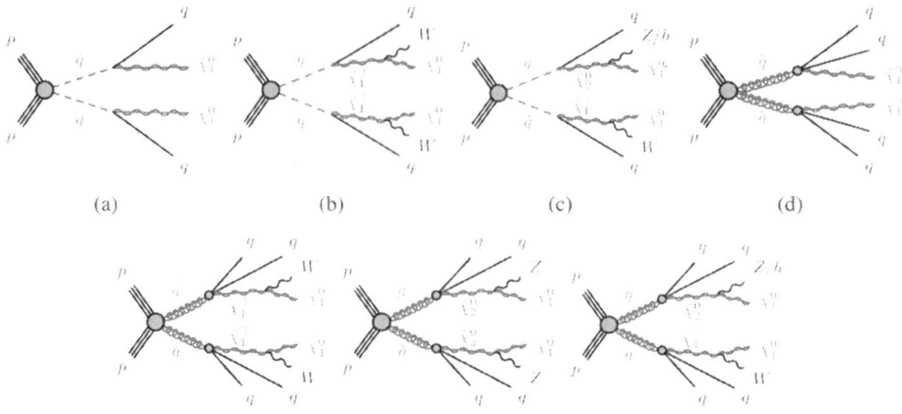

(a) (b) (c) (d)

(e) (f) (g)

Figure 1. Signal diagrams of the Simplified Models studied in [4], covering gluino and squark production.

phase space as possible. An example for such an analysis is the search for multiple jets and missing transverse energy along with the absence of any lepton[2] as it is presented in [4] and described in more detail in section 4.

- An alternative approach in the search for SUSY is to look for the production of the sparticles which are expected to be the easiest to find. Due to the nature of SUSY, the superpartners of the Standard Model's third generation fermions—the heaviest fermions—are expected to be the lightest sparticles and hence rather abundantly produced at lower energies. The expectation is thus to observe the presence of top and bottom squarks along with tau sleptons first. Searches tailored towards signatures of such third generation sparticles are another work horse of ATLAS. An example is the search for pair production of top squarks using signatures of two leptons as presented in [5] and described in section 5. The corresponding analyses and results by the CMS collaboration can be found in [6] and [7], and [8], respectively.

4 The Hunt for Gluinos and Squarks – Multiple Jets, Missing Energy and the Absence of Leptons as a Signature

The way an analysis in the search for physics beyond the Standard Model is designed strongly depends on the model(s) searched for. In the case of the search for signatures with multiple jets, missing transverse energy and the absence of leptons, the Simplified Models [4] searched for are models of direct squark[3] or gluino pair production via the strong force. After the production step, different decay chains are considered as can be inferred from Figure 1, all of them having in common the presence of multiple jets (here, originating from quarks), missing energy (introduced by the undetected Lightest Supersymmetric Particle (LSP), the neutralino) and the absence of any leptons. The last requirement limits the possible decays of W and Z bosons to pure hadronic ones.

[2]The term 'lepton' refers to electrons and muons exclusively. Since the ATLAS detector can only distinguish tau leptons from electrons and muons if they decay hadronically, taus are treated and referred to individually.

[3]In this context, the term 'squark' refers to the superpartners of first and second generation quarks. This terminology is chosen to separate these searches from endeavours to find third generation particles. Moreover, in the models investigated here all squarks regardless of their flavour and chirality are degenerate in mass.

4.1 Analysis Strategy

The approach chosen to access the rare and strongly suppressed SUSY signal in this analysis is to separate the available phase space into different parts to control and validate the predictions of Standard Model background and to extract the expected signal:

- **Signal Regions (SR):** As the name suggests, SRs are designed to ensure the presence of the signal searched for in an analysis. In order to be able to separate the desired signal from the potentially present Standard Model background, SRs are usually designed in such a way that they suppress any Standard Model process as strongly as possible. Design in this context refers to imposing selection criteria (cuts) that cut away Standard Model events and leave signal events mostly unaffected, leading to the desired separation effect. Having a signal-dominated region of phase space, extraction of results and their statistical interpretation can be addressed.
- **Control Regions (CR):** The aim of CRs is to isolate designated Standard Model backgrounds by imposing selection criteria that foster their presence and suppress other backgrounds and any signal of interest. Once a CR has reached the desired purity for one background, a fit of the prediction (usually Monte-Carlo simulations (MC), sometimes data-driven templates) to the actually observed data can be carried out. This way, an estimate of the fitted background is best-tailored to the observed data and expected to be reliable even in different parts of phase space (c. f. next point). It is important that CRs (for different backgrounds) do not overlap in phase space, neither among each other, nor with VRs (details see below) or SRs – a property referred to as orthogonality. Orthogonality is often achieved by inverting cuts between regions. Only then double-counting of events in the fits can be avoided to provide an unbiased result. A typical approach in an analysis like this one is to have CRs for all major backgrounds expected to contribute to the analysis.
- **Validation Regions (VR):** In contrast to CRs, VRs are not used for fitting background predictions to observed data but to study the quality of those very fits. To not overlap with the CRs and to probe the quality of extrapolation in variables defining SRs, VRs are usually located in the parts of phase space in between CRs and SRs. If the agreement between background and observed data in the SR-defining variables is good, the prediction is deemed trustworthy to be used in the SRs. In order to not be biased by signal contamination, VRs are sought to be signal-free – despite their vicinity to the SRs.

This analysis designs a total of 24 SRs probing different parts of model phase space, different models (c. f. Figure 1), and using different selection approaches – while here the m_{eff}-based approach is presented, the so-called 'Recursive Jigsaw Approach' is explained in detail in [4]. Depending on the SR their estimation is being used for, different CRs are used to obtain the correct normalisation of the contributing Standard Model backgrounds. Common to all sets of CRs are CRs for the processes W+Jets, Top, $Z \rightarrow \nu\nu$+Jets, and QCD Multijet events. While W+Jets and Top are estimated using the corresponding MC predictions directly, $Z \rightarrow \nu\nu$+Jets is estimated by fitting γ+Jets events and applying the obtained fit result to genuine $Z \rightarrow \nu\nu$+Jets in the SRs afterwards. This approach is justified by the difficulty to obtain a $Z \rightarrow \nu\nu$+Jets dominated CR and the similarities between $Z \rightarrow \nu\nu$ and γ-based signatures in the detector. The contribution of Multijet events arising from QCD is estimated in a semi-data-driven way by using the so-called 'Jet-Smearing'-method [4].
The results of an exemplary set of CRs—after application of the scaling factors obtained by fitting the normalisation of the Standard Model prediction in those very CRs to the observed data—are shown in Figure 2. The red arrows indicate the cut value which is chosen to define the actual CR where the fit is carried out. Beyond those arrows, the larger abundance of the background to which each CR is tailored with respect to the remaining other Standard Model contributions is visible. This way, confidence in the obtained background normalisations is ensured.

 DOI: 10.1051/epjconf/201715802002

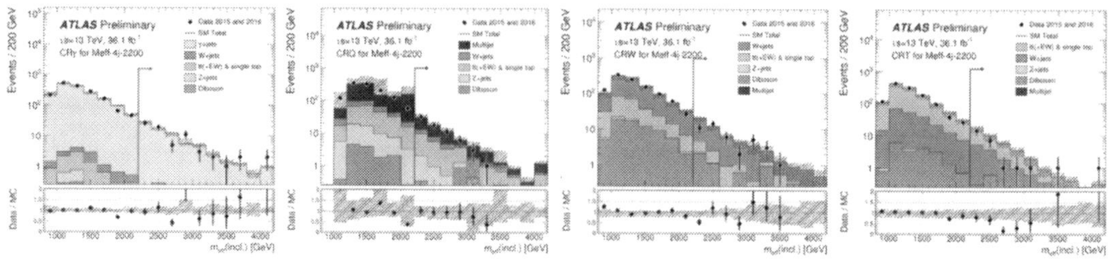

(a) γ+Jets CR (b) Multijet CR (c) W+Jets CR (d) Top CR

Figure 2. Distributions of the main discriminating variable m_{eff} before being cut on to define the set of control regions of one specific signal region. Such sets of CRs are designed for all SRs. The red arrow indicates the cut value that defines the final CR in which the fit of the background is performed. After application of that selection criterion, the CRs are rich in the background they are intended to probe. The distributions are shown after application of the normalisations obtained in the fit. The red lines indicate the total Standard Model prediction, while the red hatched band represents its total uncertainty (one standard deviation of combined statistical and systematic uncertainties). The last bin in each distribution contains the overflow bin of the histogram. The lower panels show the bin-wise ratios of data/MC with accordingly propagated uncertainties [4].

4.2 Analysis Results & Interpretations

The results of interest in such an analysis primarily lie in the signal regions. When applying the obtained fit results from the CRs to the considered backgrounds and overlaying the signal expectations, the SRs can be studied. An exemplary SR distribution is depicted in Figure 3(a); again, the red arrow indicates the cut value used to obtain the actual SR.

Summarising all 24 SRs of this analysis, Figure 3(b) shows the obtained yields for data and the predicted backgrounds integrated into one bin for each SR. The bottom plot shows the deviation between observed data and predicted background as a ratio with accordingly normalised uncertainty bands: looking over all SRs, no significant excess of data over background is visible, no discovery of a signal can be claimed.

Since no excess has been found, the degree to which the models of interest can be excluded is evaluated. This statistical interpretation is performed in two ways:

- **Model-Dependent Interpretation:** For every set of parameters—here, either $(m(\tilde{g}), m(\tilde{\chi}_1^0)$ or $(m(\tilde{q}), m(\tilde{\chi}_1^0)$—the corresponding CL of a hypothesis test of background-only versus signal+background hypothesis with respect to data is calculated. In a next step, a contour line is drawn along the values of CL =95% across the two-dimensional parameter plane. Since in every hypothesis test a certain signal model and its prediction in the studied SR is considered, such a set of results is model-dependent.
- **Model-Independent Interpretation:** In order to quantify how sensitive the performed analysis is towards finding any non-background signal, a model-independent interpretation is performed as well. In another hypothesis test the background-only hypothesis is tested against the observed data. The resulting level of (dis)agreement can be used to calculate upper limits on the production cross section of any model that is still compatible with the observed results.

Figures 4(a) and 4(b) show model-dependent results for two of the seven Simplified Models studied in this analysis, combining the most powerful results[4].

[4]For the best combination, the result from the signal region with the lowest value of CL_S is taken into account – this is equivalent to taking the SR with the strongest exclusion.

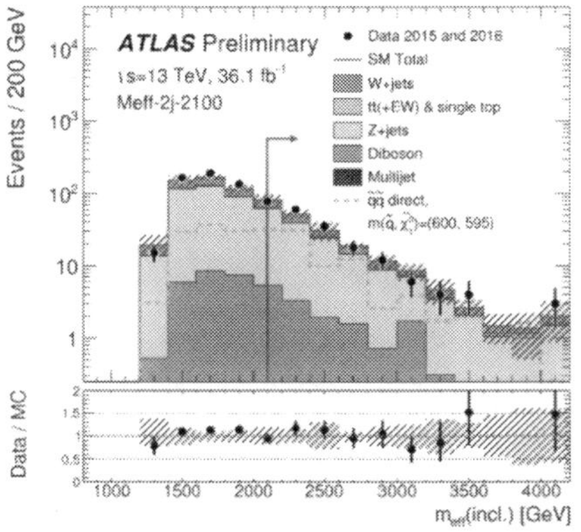

(a)

Distribution of the main discriminating variable m_{eff} before the application of the final selection criterion on it and before application of the fitted background normalisation. The cut value is again indicated by the red arrow. The growth of the signal contribution with respect to predicted background is visible. The red line and hatched band again represent the total Standard Model expectation and its combined uncertainty (one standard deviation of combined statistical and systematic uncertainty). The last bin contains the overflow bin of the histogram. The lower panel represents the ratio data/MC with accordingly propagated uncertainties [4].

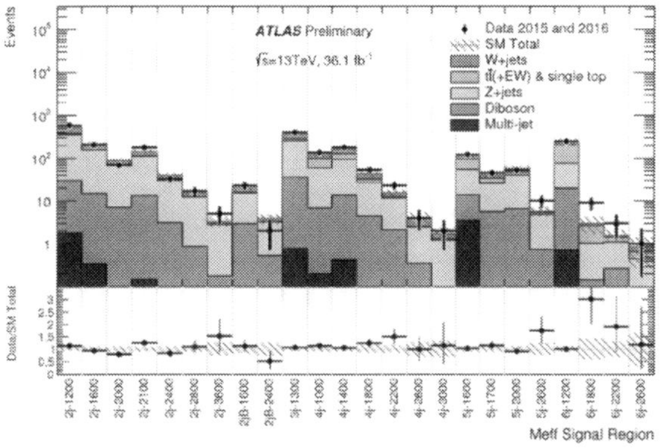

(b)

Summary of the agreement between observed data and the Standard Model expectation across all SRs after the fitted background normalisation is applied. While the upper panel shows the stacked Standard Model contributions and absolute yields, the lower panel shows the agreement as a ratio between data and simulation with accordingly propagated uncertainties [4].

Figure 3. Results in the signal regions of [4], (a) for an exemplary SR, (b) as an overview over all SRs.

In addition to the Simplified Models introduced in the beginning of section 4, a complete model of SUSY as introduced in section 2 is studied: the phenomenological MSSM (pMSSM) [9]. The 19 free parameters that describe the pMSSM are reduced to two—the mass of the gluino and the LSP—by imposing further assumptions. In a similar fashion as for the Simplified Model, the results are visualised in Figure 4(c) and 4(d): in this representation, the parameter plane is defined by the masses of the squark and the gluino while the mass of the LSP is fixed (massless case in Figure 4(c), massive case in Figure 4(d)). The case of the massless LSP reproduces the limit of the Simplified Model at low LSP masses and serves as a successful consistency check.

The model independent upper limits on production cross sections per signal region and breakdowns of the background contributions and systematic uncertainties can be found in [4].

5 The Search for a Top Superpartner – Two Leptons as an Indicator for the Presence of the Light Top Squark

As mentioned in section 3, the second major set of SUSY searches within the ATLAS collaboration is targeting third generation sfermions directly. One of the analyses driving the progress in the search for the light top squark is looking for its pair production and decays leading to the presence of Standard Model leptons [5]. Similar to the search presented in section 4, multiple Simplified Models are looked for. However, in this analysis, the motivation to study several models all at once is a different one: due to the fact that the light top squark is expected to be the Next-To-Lightest-Supersymmetric-Particle (NLSP) in many models [1], it can be similar in mass to its Standard Model partner, the top quark, or much heavier. In order to account for the wide range of potential top squark masses and

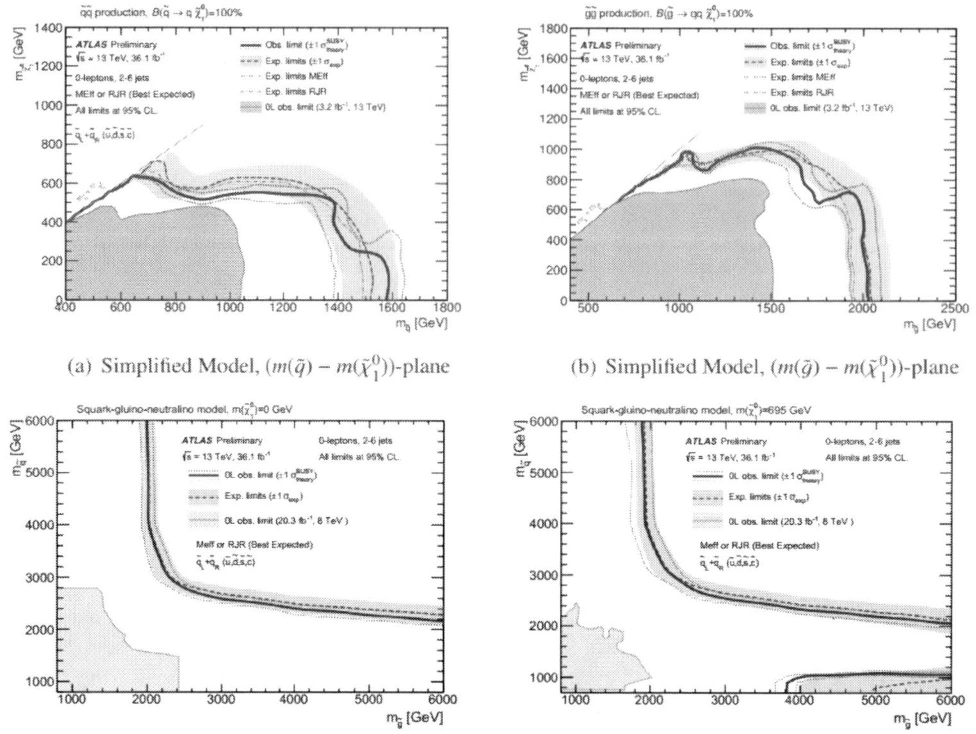

(a) Simplified Model, $(m(\tilde{q}) - m(\tilde{\chi}_1^0))$-plane

(b) Simplified Model, $(m(\tilde{g}) - m(\tilde{\chi}_1^0))$-plane

(c) pMSSM, massless LSP

(d) pMSSM, massive LSP

Figure 4. Exclusion contours at 95% CL computed for different signal models in their corresponding space of free parameters. The dashed blue line and the solid yellow band give the expected limit and its one-standard-deviation combined uncertainty. The corresponding observed results are visualised by the solid red line and the dotted contour around it. (a) and (b) show the results of the Simplified Models of squark production (a) and gluino production (b) plotted in the mass-plane of the corresponding particle produced in the collision and the LSP. Here, the blue solid area marks the exclusion obtained in an earlier iteration of this analysis. (c) and (d) show the findings interpreted in an constrained version of the pMSSM where all parameters but the squark, gluino and LSP masses are fixed. The results are given in the $(m(\tilde{g}) - m(\tilde{q}))$-plane. In (c) the LSP is set to be massless while (d) shows the limit with a massive LSP. The result of the massless LSP case in (c) reproduces the corresponding Simplified Model result shown in (b), serving as a sanity check for the consistency between the models [4].

their difference to Standard Model particles, different Simplified Models are proposed as depicted in Figure 5.

The Simplified Models cover the different mass hypotheses in the following way: the more particles participate in the decay of the top squark, the smaller the mass difference to its Standard Model partner. When the splitting between the two masses is large, enough energy is available to produce even another SUSY particle such as the lightest chargino (a) while a smaller difference could still allow for a decay into a top quark and a massive LSP (b). Compressing the mass spectrum further leads to the direct flavour changing decay into a b quark and a W boson at an effective vertex (c) while at lowest splittings the decay vertex is effective with the stop decaying into a b quark and a lepton-neutrino pair directly.

The strategy of this study is similar to what has been presented in section 4.1: for each dominant Standard Model background, a designated control region is designed. A simultaneous fit of the back-

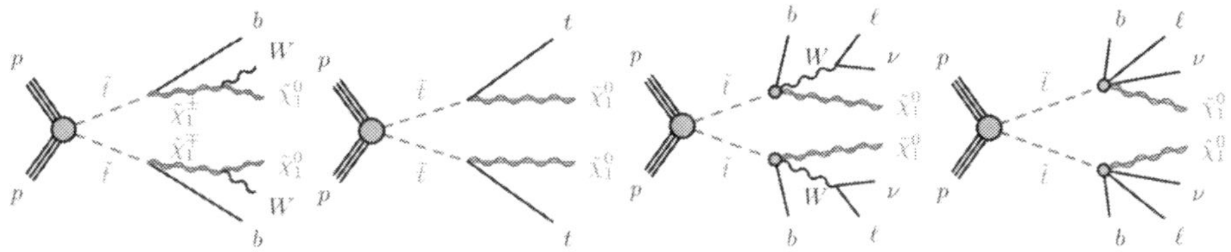

(a) 2-body decay, intermediate $\chi_1^\pm$ (b) 2-body decay to LSP and t (c) 3-body decay, effective vertex (d) 4-body decay, effective vertex

Figure 5. Signal diagrams of the Simplified Models as in [5]. Depending on the mass difference between the produced top squark and the Standard Model top quark, the decay process changes. The mass difference decreases from (a) to (d) prohibiting the decay to heavy particles such as a chargino (a), the top quark (b), or a real W-boson (c) until only light particles are produced (d) – the latter two cases make use of effective vertices.

grounds to the observed data in all CRs determines the normalisation of the Standard Model expectation to obtain a reliable prediction in the SRs. Exemplary control region distributions illustrating the agreement between the MC-based background predictions and data are shown in Figures 6(a) to 6(c). The observed agreement is good and can hence serve as a reliable prediction of the Standard Model backgrounds in the signal regions of interest.

Apart from the MC-based backgrounds, the most dominant background arises from fake and non-prompt leptons. This contribution is estimated in a data-driven way, described and validated in [5].

5.1 Analysis Results & Interpretations

The background fit results obtained in the CRs are applied in the SRs. Distributions of variables used to define an exemplary SR are shown in Figure 6(d) and 6(e). Again, the arrow indicates the final signal region that is obtained when cutting accordingly.

From the observed data and the predicted background and signals, again, model-dependent and model-independent limits can be computed in the same fashion as outlined in section 5.1. The model-dependent results for the different Simplified Models are visualised in the $(m(\tilde{t})_1, m(\tilde{\chi}_1^0))$-plane for all types of decays in Figure 7(a). The plot illustrates the power of this analysis: the gaps that were present after Run-I of the LHC (blue limit contours) are filled up and the overall picture is more completed – in addition to the general extension of the limit.

Similar to the analysis presented in section 4, this study, too, performs an interpretation of its results not only in the Simplified Models used for the search but also in the pMSSM. This analysis leaves one more degree of freedom and presents results for both possible signs of the Higgs mass parameter μ in the $(m(\tilde{t}_1, m(\tilde{\chi}_1^0))$-plane fixing the rest of the 16 model parameters as described in [5]. The obtained result is depicted in Figure 7(b).

DOI: 10.1051/epjconf/201715802002

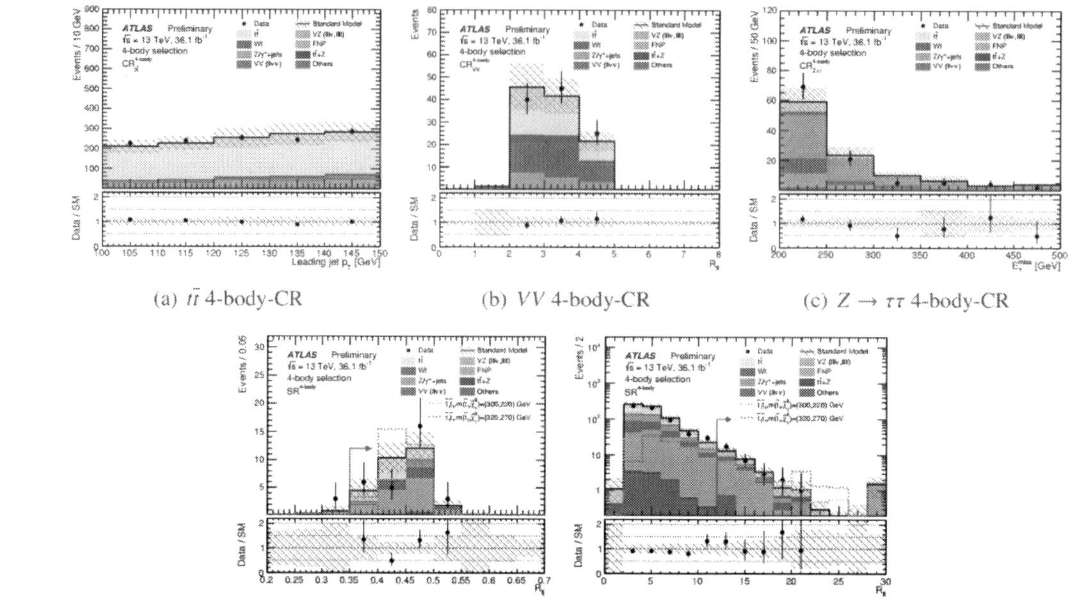

(a) $t\bar{t}$ 4-body-CR (b) VV 4-body-CR (c) $Z \to \tau\tau$ 4-body-CR

(d) R_{lj}, 4-body-SR (e) R_{ll}, 4-body-SR

Figure 6. Exemplary kinematic distributions after the fit of the background normalisations in the CRs. The black line and its blue hatched band represent the total Standard Model expectation and its combined uncertainty (one standard deviation combination of statistical and systematic uncertainties). (a) to (c) show the control regions associated to the 4-body Simplified Model and its SR. Each plot depicts the distribution of a variable characteristic for the CR. (d) and (e) show distributions of the variables R_{lj} and R_{ll} that characterise the 4-body SR before their application (definitions in [5]). Again, the red arrows illustrate the selection criterion which is imposed to obtain the final SR selection. Here, a gain in signal contributions with respect to background is visible. Again, the lower panels show the bin-wise ratio of data/MC with the corresponding uncertainties. The agreement between prediction and observation in the CRs is visible. The last bin in each distribution contains the overflow bin of the histogram [5].

6 Outlook & Conclusions

With two of the major analyses in the search program for strong production of SUSY within the ATLAS collaboration being described in detail, a brief look on even more specific analyses provides an idea of the full ATLAS scope on that topic.

While both analyses presented so far considered only R-parity conserving (RPC) models of SUSY, several studies searching for R-parity violating (RPV) SUSY are carried out, too. As an example the analysis targeting signatures of one lepton and multiple jets studying models of RPV SUSY can be considered [10]. Moreover, analyses using tau leptons such as [11] target the strong production of (RPC) SUSY. While all of the aforementioned studies consider only short-lived SUSY particles such as immediately decaying gluinos, few analyses try to look for longer-lived sparticles such as [12] that is able to probe areas of phase space which are not accessible using the more common signatures.

Concluding, a brief overview of the ATLAS collaboration's rich search program for strong production of Supersymmetry has been provided. The focus has been put on two of the main progress-driving analyses, their search strategies, the obtained results and their interpretation. Apart from the studies presented in detail, far more analyses address the topic of this work. Beyond the results compiled here, many analyses are still ongoing, relying on more recorded data due to rare signatures or better

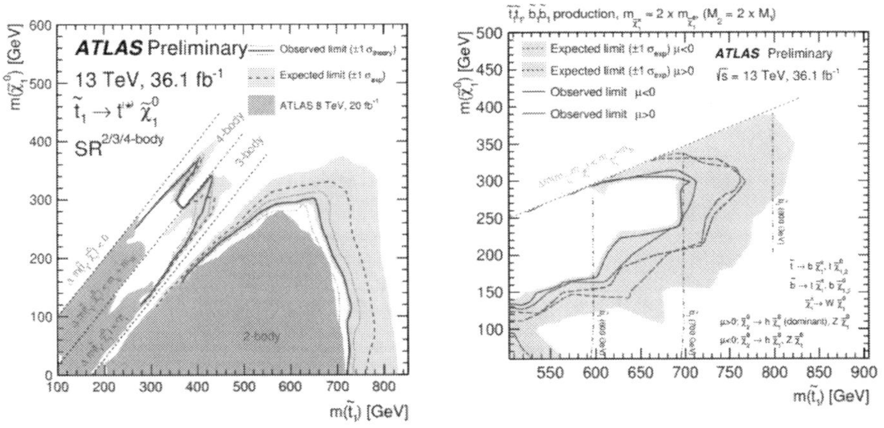

(a) Simplified Model (b) pMSSM

Figure 7. Exclusion contours at 95% CL computed for the combination of all Simplified Models (a) and an interpretation in the pMSSM phase space (b). The dashed line and the solid band give the expected limit and its one-standard-deviation combined uncertainty. The corresponding observed results are visualised by the solid line and the dotted contour around it (not computed in (b)). (a) displays the results of the Simplified Model findings in the $m(\tilde{t})_1) - m(\tilde{\chi}_1^0)$-plane. With the blue area depicting the Run-1 result of this search, both a filling of gaps in this plane and an extension of the exlcusion is visible. The still open gaps mark particularly difficult to access regions of phase space where the kinematics of the involved particles are hard to detect. (b) visualises the interpretation of the obtained results in the pMSSM model. In this interpretation the LSP mass and the top squark mass are left as floating parameters while all other parameters except the Higss mass parameter μ are fixed. Two interpretations for different values of μ are shown in red and blue, respectively. In addition, iso-lines for the light bottom squark are drawn to provide additional information for model builders and further interpretation [5].

calibrated data due to high sensitivity to the smallest of detector effects. With this in mind, ATLAS search for Supersymmetry continues and remains exciting.

References

[1] S.P. Martin (1997), adv. Ser. Direct. High Energy Phys.18,1(1998)
[2] ATLAS Collaboration, JINST **3**, S08003 (2008)
[3] L. Evans, P. Bryant, JINST **3**, S08001 (2008)
[4] ATLAS Collaboration, ATLAS-CONF-2017-022, `https://cds.cern.ch/record/2258145`
[5] ATLAS Collaboration, arXiv:1708.03247[hep-ex]
[6] CMS Collaboration, Phys. Rev. **D96**, 032003 (2017)
[7] CMS Collaboration, arXiv:1705.04650[hep-ex]
[8] CMS Collaboration, CMS-PAS-SUS-17-001, `http://cds.cern.ch/record/2256753`
[9] ATLAS Collaboration, JHEP **10**, 134 (2015)
[10] ATLAS Collaboration, arXiv:1704.08493[hep-ex]
[11] ATLAS Collaboration, Eur. Phys. J. C **76**, 683 (2016)
[12] ATLAS Collaboration, Phys. Rev. D **93**, 112015 (2016)

EPJ Web of Conferences **158**, 02003 (2017) DOI: 10.1051/epjconf/201715802003
QFTHEP 2017

ATLAS searches for resonances decaying to boson pairs

Andrey Ryzhov[1,⋆], on behalf of the ATLAS Collaboration

[1] *Institute for High Energy Physics of National Research Centre "Kurchatov Institute", Protvino, Russia*

Abstract. Many extensions to the Standard Model predict new particles decaying into two bosons (W, Z, photon, or Higgs bosons) making these important signatures in the search for new physics. Searches for such diboson resonances have been performed in final states with different numbers of leptons, photons, jets and b-jets where new jet substructure techniques to disentangle the hadronic decay products in highly boosted configuration are being used. This document summarizes recent ATLAS searches for resonances decaying to diboson final states, VV, VH and HH with LHC Run 2 data collected.

1 Introduction

The discovery of the Higgs boson [1] with a mass of approximately 125 GeV in 2012 finalizes the Standard Model (SM) in the description of particle interactions at energies up to a few hundred GeV. Despite the fact that a lot of precision measurements and searches with data from the Large Hadron Collider (LHC) show no significant deviations from the SM, some unanswered questions still exist. Therefore one of the main purposes of particle physics research at the LHC is to seek new intimations of physics Beyond the Standard Model (BSM). Many of BSM models, motivated by hierarchy and naturalness arguments [2], predict the existence of new heavy resonances decaying into diboson, such as composite Higgs models [3], warped extra dimensions [4], models with an extended Higgs sector [5] and Grand Unified Theories [6].

In order to evaluate the sensitivity, to optimise the event selections, and for comparison with the data, a number of models are used as benchmarks. Limits on cross-sections as a function of the particle mass are obtained in the framework of three representative benchmark models: an extended Higgs sector serves as benchmark model for spin-0 resonances, Heavy Vector Triplets (HVT) Model A (which has comparable branching ratios to fermions and gauge bosons) and Model B (which has enhanced couplings to gauge bosons and where the fermionic couplings are suppressed) for spin-1 in form of W' and Z' [7], and bulk Randall-Sundrum Gravitons (RSG) [8] for spin-2 resonances.

In this document, the results from recently performed searches for resonant production of VV, VH and HH are summarised. They are carried out using data taken with the ATLAS detector [9] during the 2015-2016 run of the LHC (Run 2).

The searches are performed by looking for localized excesses above the smoothly falling SM background. For this purpose, the mass of the diboson system is used as the final discriminant. This is the invariant mass unless Z boson decays to neutrinos (ν) are included, where only the transverse mass reconstruction is feasible.

⋆e-mail: Andrey.Ryzhov@ihep.ru

2 Experimental techniques

All models predict the existence of new particles with masses approximately few TeV, hence candidate decay products (especially W, Z or H bosons reconstructed through hadronic decays) must have high transverse momenta. In this case the distance in $\eta - \varphi$ space between direction of two quarks emerging from the boson decay can be approximated by $\Delta R = \sqrt{\Delta\eta^2 + \Delta\phi^2} \approx 2m/p_T$, where m and p_T are the mass and the transverse momenta of the parent boson (W, Z, H), respectively. When the resonance mass is significantly higher than the boson mass, the qq pair from the boson decay can be collimated. And hadrons from the two quarks overlap in the detector and are more efficiently reconstructed as a single large-radius jet (large-R jet, or J). Therefore two different reconstruction techniques for the hadronic decay are considered: resolved and merged. The resolved reconstruction attempts to identify boson decay products as two separate small-radius anti-k_t jets with a radius parameter of 0.4 (small-R jet, or j), while the merged reconstruction uses jet substructure techniques to identify the boson hadronic decay reconstructed as a large-R jet using the anti-k_t algorithm with $R = 1$. In order to remove a large contamination by soft radiation (pile-up, underlying event) a *trimming* algorithm [10] is applied to the large-R jets, retaining only components (anti-k_t jets with $R = 0.2$) carrying at least 5% of the original jet's energy. The large-R jet mass is calculated from combining calorimeter and tracking information, to improve jet mass resolution [11].

Vector boson tagging technique [12] using a variable named D_2 is applied to the large-R jets to select them from the $V \to qq$ decays. This D_2 variable is optimised to distinguish between jets originating from a single parton and those coming from the two-body decay of a heavy particle. The V boson jets are then selected by requiring that the large-R jet mass m_J should be in a p_T-dependent window centred around the expected value of the boson mass from simulations. Higgs bosons decaying to bottom quarks are identified by applying a dedicated Higgs-tagging technique [13] to the large-R jets. This algorithm proceeds by ghost-associating [14] anti-k_t jets ($R = 0.2$) reconstructed from tracks in the Inner Detector (ID) and applying a b-tagging algorithm [15]. The operating point of the b-tagging algorithm is chosen such that the b-tagging efficiency for jets containing b-hadrons is 77%. In addition, a mass window cut around the SM Higgs mass of 125 GeV is applied to the large-R jet.

3 Analyses overview and results

This section summarizes the individual searches presented in this article. Subsection 3.1 gives the results for the VV, VH, HH searches in fully hadronic decay modes, followed by the results for VV resonances in semileptonic decay modes in Subsect. 3.2. And finally a short summary is given in Table 1, which shows observed excluded resonance masses at the 95% confidence level (CL) for all analyses. The full description of each analysis can be found in the reference papers. These analyses are performed using data collected with the ATLAS detector at $\sqrt{s} = 13$ TeV with an integrated luminosity that differs for each particular analysis and that will be indicated in each subsection.

3.1 *VV*, *VH* and *HH* resonances in fully hadronic decay modes

The $\boldsymbol{VV \to qqqq}$ analysis [16] is performed only in the merged regime using 36.7 fb^{-1} of proton-proton collision data at a centre-of-mass energy of $\sqrt{s} = 13$ TeV recorded with the ATLAS detector at the Large Hadron Collider in 2015 and 2016. Events are required to contain at least two large-R jets with $|\eta| < 2.0$ (to guarantee a good overlap with the tracking acceptance) and mass $m_J > 50$ GeV. The leading (highest p_T) large-R jet must have $p_T > 450$ GeV and the subleading (second highest p_T) large-R jet must have $p_T > 200$ GeV. The invariant mass of the dijet system formed by these two jets must be $m_{JJ} > 1.1$ TeV to avoid inefficiencies due to the minimum jet-p_T requirements. Boson-jet candidates are identified by applying a boson-tagging procedure. Further requirements are applied on the number of tracks in the two jets, their rapidity difference and their p_T asymmetry. This selection strongly suppresses the large backgrounds due to SM multi-jet events. Other contributions to the background are small. The dijet invariant-mass spectrum of the background is estimated directly from data using a parametric form, and the search proceeds by seeking resonant structures in the falling spectrum of the background. The ability of the parametric shape to model distributions is tested with validation regions from collision data (see Figure 1(left)). No significant excess over the background is found in the signal regions, as shown for the WZ selection in Figure 1(right) and thus 95% CL exclusion limits are set on the production cross section times branching fraction in the HVT and bulk RS models as shown in Figure 2.

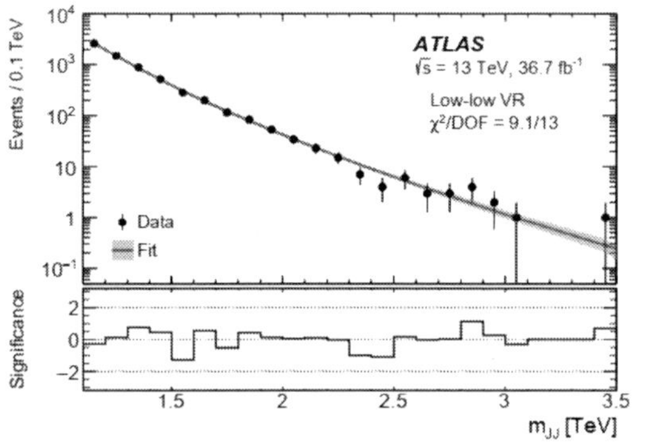

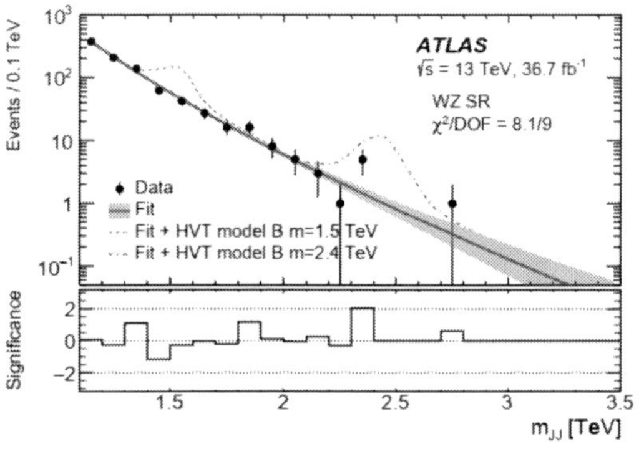

Figure 1. Left: Dijet mass distributions for data in the sideband validation regions. The shaded bands represent the uncertainty in the background expectation due to the maximum-likelihood fit's statistical uncertainty. The lower panels show the significance of the observed event yield relative to the background fits. **Right:** Dijet mass distributions for data in the WZ signal region. The red lines correspond to the result of the fit and the shaded bands represent the uncertainty in the background expectation. Expected signals are shown for the HVT model B and the bulk RS model. The predictions for G_{KK} production are multiplied by a factor of 10. Both figures are taken from ref. [16]

The $\boldsymbol{VH \to qqb\bar{b}}$ analysis [17] is carried out with 36.1 fb^{-1} of Run 2 data. There is only merged regime which uses two large-R jets selection with $p_T > 450$ (250) GeV for the leading (sub-leading) jet, $|\eta| < 2$ and an invariant mass larger than 50 GeV. The analysis is performed categorizing the signal in four regions, namely WH 1 b-tag, WH 2 b-tag, ZH 1 b-tag and ZH 2 b-tag. Signal and backgrounds from $t\bar{t}$ and W/Z+jets production are modelled with Monte Carlo (MC) simulation. While multijet MC events are used as a cross-check, the primary multijet background estimation is performed using data. This background shape is extracted from 0 b-tag "SR", which consists of $\approx 99\%$ multijet events. The normalization of each distribution is obtained using the sideband regions. The data are in agreement with the Standard Model expectations (see Figure 3(left)), with the largest excess observed at $m_{JJ} \approx$ 3.0 TeV in the ZH channel with a local significance of 3.3 σ. The global significance of this excess is 2.1 σ (see Figure 3(right)).

The search for $\boldsymbol{HH \to b\bar{b}b\bar{b}}$ final state [18] is done in both boosted and resolved regimes, covering high and low energy regimes, respectively. The results presented here benefit from an integrated luminosity of 13.3 fb^{-1}. The merged analysis selects events with at least two anti-k_t $R = 1.0$ jets

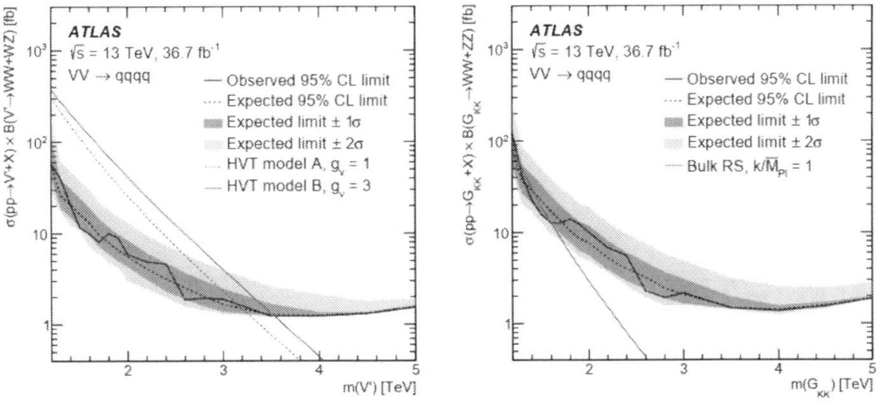

Figure 2. Upper limits at the 95% CL on the cross section times branching ratio for (**left**) $WW + WZ$ production as a function of V' mass, (**right**) $WW + ZZ$ production as a function of G_{KK} mass. The dotted and solid red lines show the predicted cross section times branching ratio as a function of resonance mass for the HVT Models A and B with $g_V = 1$ and $g_V = 3$, respectively, or the bulk RS model with $k/\overline{M}_{Pl} = 1$. Both figures are taken from ref. [16]

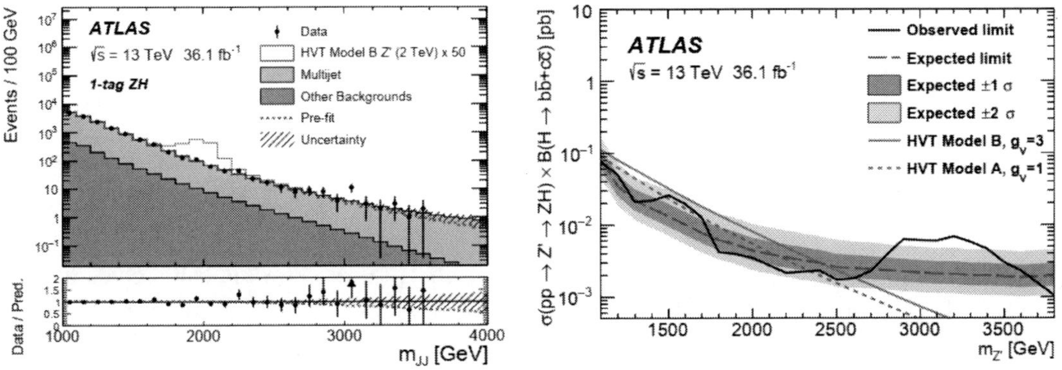

Figure 3. Left: The m_{JJ} distributions in the VH signal regions for data (points) and background estimate (histograms) after the likelihood fit for events in the 1-tag categories. The pre-fit background expectation is given by the blue dashed line. The expected signal distributions (multiplied by 50) for a HVT benchmark Model B V' boson with 2 TeV mass are also shown. **Right:** The observed and expected cross-section upper limits at the 95% confidence level on the cross section times branching ratio for HVT Model A and Model B in the ZH signal regions. The red and magenta curves show the predicted cross-sections as a function of resonance mass for the models considered. Both figures are taken from ref. [17]

with $p_T > 450$ (250) GeV for the leading (sub-leading) jet, $|\eta| < 2$ and an invariant mass larger than 50 GeV. Since high-mass resonances tend to produce jets that are more central than multijet background processes, the two large-R jets are required to have a separation $|\Delta\eta| < 1.7$. The resolved selection begins with the requirement that the event contains at least four b-tagged anti-k_t $R = 0.4$ jets with $p_T > 30$ GeV and $|\eta| < 2.5$. The leading jets are combined in pairs forming the Higgs candidates and requiring to pass ΔR and p_T cuts which depend on invariant mass of HH system. For both the boosted and the resolved analyses the dominant background is QCD multijet which is estimated in a side band region and validated in a control region in the m_{H_1}, m_{H_2} plane as shown in Figure 4(left), where m_H refers to the invariant mass of the Higgs candidate. Good agreement is observed between the data and the background prediction (see Figure 4(right)).

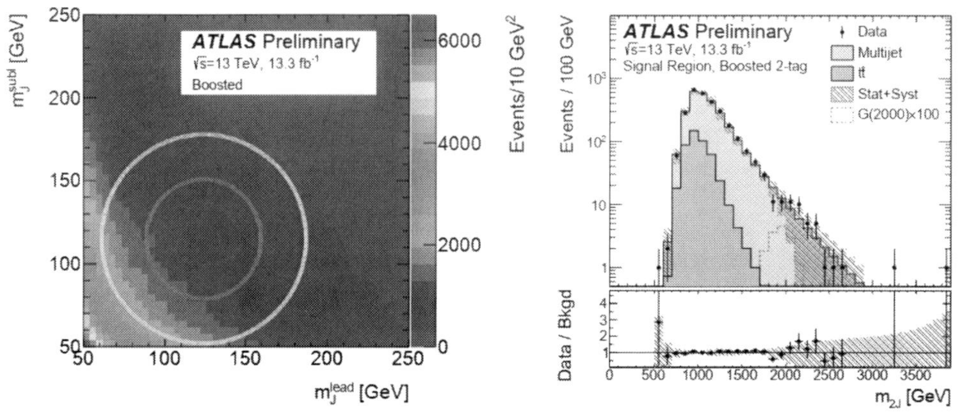

Figure 4. Left: The m_J^{subl} vs. m_J^{lead} distribution for the boosted analysis background model. The signal region is the area surrounded by the dashed red contour line, centred on (m_J^{lead} = 124 GeV, m_J^{subl} = 115 GeV). The control region is the area between the signal region and the orange contour line. The sideband region is the area between the control region and the yellow contour line. **Right:** Distributions of m_{2J} in the signal regions of the boosted analysis for the two-tag-split sample compared to the predicted backgrounds. The grey hatched bands represent the combined statistical and systematic uncertainties in the total background estimates. The expected signal shape for a G_{KK} resonance of mass 2.0 TeV is also shown. Both figures are taken from ref. [18]

3.2 *VV* resonances in semileptonic decay modes

All semileptonic analysis is carried out with 36.1 fb^{-1} of LHC proton-proton collision data collected at $\sqrt{s}$ = 13 TeV with the ATLAS detector in 2015 and 2016. One boson candidate is required to be formed by a $l\nu$, ll or $\nu\nu$ pair and the other boson by a qq pair or single jet, where both bosons should satisfy V mass window cuts. Events with two additional small-R jets (referred to as tag-jets) with large pseudorapidity separation and invariant mass are classified as vector-boson fusion (VBF) candidates, while absence of this topology is interpreted as gluon-gluon fusion/Drell-Yan (ggF/DY) production.

The $\boldsymbol{ZV \to llqq}$ search [19] explores the VBF and ggF production of a heavy Higgs boson H, the VBF and DY production of an HVT W' boson, and the ggF production of a bulk RS graviton G_{KK}. It also utilises both the merged and resolved reconstruction for the $V \to qq$ decay. The search begins with the identification of the $Z \to ll$ decay (the leading lepton must satisfy E_T (p_T) > 28 GeV and the dilepton invariant mass m_{ll} is required to be consistent with the Z boson mass), followed by classifying events into the VBF ($\eta_{tag_1} \cdot \eta_{tag_2} < 0$, $|\Delta\eta_{tag}| > 4.7$, $m_{tag} > 770$ GeV) or ggF/DY categories and finally the selection of either the $ZV \to llJ$ or $ZV \to lljj$ final states. Merged events are further required to have $p_T^J > 200$ GeV and $min(p_{ll}, p_J)/m_{llJ} > 0.3$ for the $H \to ZZ$ search and > 0.35 for the $W' \to ZW$ and $G_{KK} \to ZZ$ searches. Two signal regions are defined, high-purity (HP) and low-purity (LP) SR, that depend on boson tagging working point requirement. Events that have failed merged selection are required to have at least two small-R jets with $p_T > 60$ GeV. The kinematic quantity $\sqrt{(p_T^{ll})^2 + (p_T^{jj})^2}/m_{lljj}$ is required to be greater than 0.4 for $H \to ZZ$ and 0.5 for $W' \to ZW$

and $G_{KK} \to ZZ$. The dijet invariant mass must be in the window [70, 105] GeV for $Z \to qq$ and in the window [62, 97] GeV for $W \to qq$. Also the $ZV \to lljj$ candidates are divided into two signal regions: events with two b-tagged jets and events with fewer than two b-tagged jets. Multiple signal regions are defined to enhance the sensitivity of the search. The dominant backgrounds are the Z+jets, top quark and diboson processes. Their contributions are estimated from a combination of MC and data-driven techniques. The shapes of kinematic variables are taken from MC simulations. In order to extract the signal rate information, a simultaneous maximum-likelihood fit is performed to the observed distributions of the final discriminants in the signal regions, m_{llJ} or m_{lljj}. The data are found to be consistent with the SM background predictions and no evidence of heavy resonance production is observed (see Figure 5). Limits are presented in Figure 6.

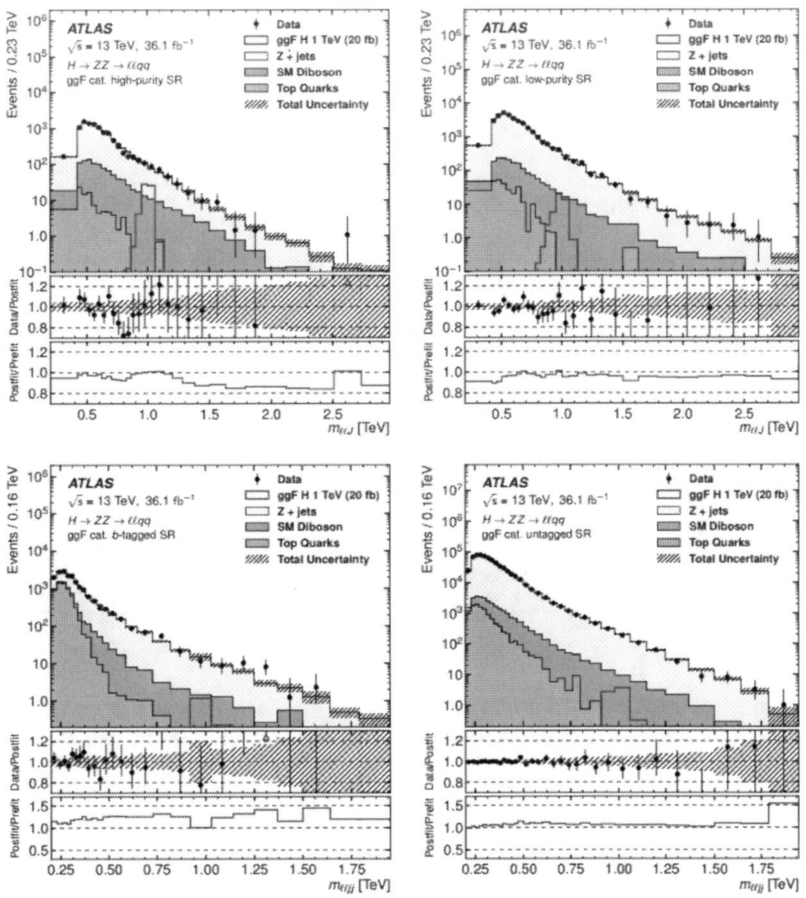

Figure 5. Comparisons of the observed data and expected background distributions of the final discriminants of the ggF category for the $H \to ZZ \to llqq$ search: m_{llJ} of (**top left**) high-purity and (**top right**) low-purity signal regions; m_{lljj} of (**bottom left**) b-tagged and (**bottom right**) untagged signal regions. For illustration, expected distributions from the ggF production of a 1 TeV Higgs boson with $\sigma \times B(H \to ZZ) = 20$ fb are also shown. The middle panes show the ratios of the observed data to the background predictions. The uncertainty in the total background prediction, shown as bands, combines statistical and systematic contributions. The bottom panes show the ratios of the post-fit and pre-fit background predictions. All figures are taken from ref. [19]

 DOI: 10.1051/epjconf/201715802003

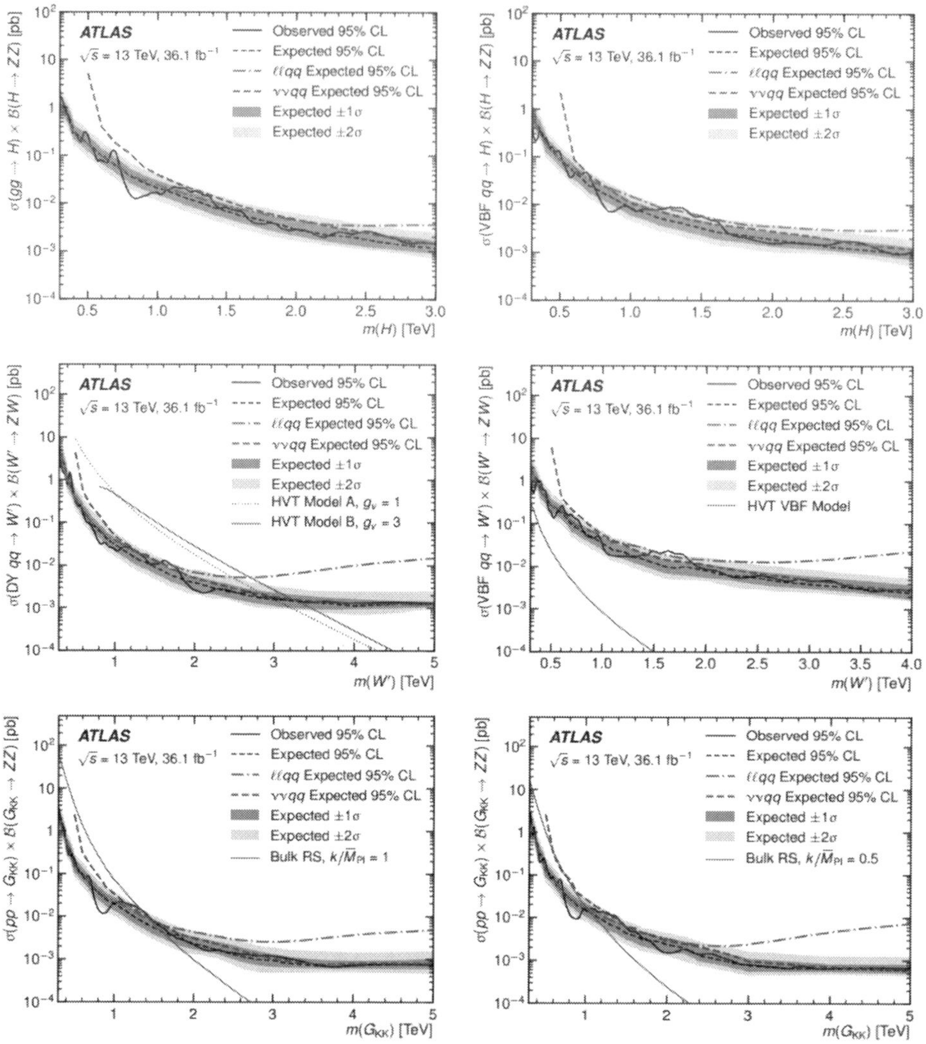

Figure 6. Observed (black solid curve) and expected (black dashed curve) 95% CL upper limits: on $\sigma \times B(H \to ZZ)$ for the (**top left**) ggF and (**top right**) VBF production of a heavy Higgs boson as a function of its mass; on $\sigma \times B(W' \to ZW)$ for the (**middle left**) DY and (**middle right**) VBF production of a W' boson in the HVT model as a function of its mass; on $\sigma \times B(G_{KK} \to ZZ)$ for the production of a G_{KK} in the bulk RS model with couplings of (**bottom left**) $k/\overline{M}_{Pl} = 1$ and (**bottom right**) $k/\overline{M}_{Pl} = 0.5$ as a function of the graviton mass, combining $llqq$ and $\nu\nu qq$ searches. Limits expected from individual searches (dashed curves in blue and magenta) are also shown for comparison. Limits are calculated in the asymptotic approximation below 2 TeV and are obtained from pseudo-experiments above that. All figures taken from ref. [19]

The $\boldsymbol{ZV \to \nu\nu qq}$ selection [19] closely follows the $VZ \to llqq$ search with additional requirements to remove multijet backgrounds but without the resolved selection. An initial selection is made by requiring $E_{miss} > 250$ GeV, and vetoing events with electrons or muons. The multijet background, originating primarily from the presence of mismeasured jets, and non-collision backgrounds are suppressed by using a track-based missing transverse momentum, $\overrightarrow{p}_T^{miss}$. Due to the presence of neutrinos in the final state, it is not possible to fully reconstruct the invariant mass of the $\nu\nu J$ system, so the transverse mass is used as the final discriminant: $m_T = \sqrt{(E_{T,J} + E_T^{miss})^2 + (\overrightarrow{p}_{T,J} + \overrightarrow{E_T}^{miss})^2}$, where $E_{T,J} = \sqrt{m_J^2 + p_{T,J}^2}$. The dominant backgrounds are Z+jets, W+jets and $t\bar{t}$ processes. Data control regions are defined to check the modelling of each contribution. As in the $llqq$ search, the shapes of kinematic variables, including the final discriminant m_T, are taken from MC simulations. Figure 7 shows that no significant excess over the background is found in the signal regions. Limits are presented in Figure 6.

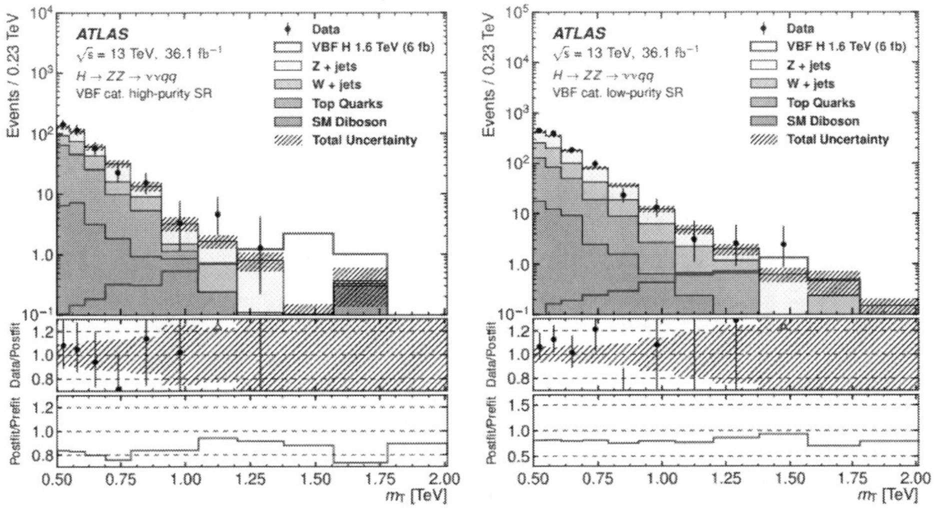

Figure 7. Comparisons of the observed data and expected background distributions of m_T in the VBF category of the $H \to ZZ \to \nu\nu qq$ search: **(left)** high-purity and **(right)** low-purity signal regions. For illustration, expected distributions from the VBF production of a 1.6 TeV Higgs boson with $\sigma \times B(H \to ZZ) = 6$ fb are also shown. The middle panes show the ratios of the observed data to the background predictions. The bottom panes show the ratios of the post-fit and pre-fit background predictions. Both figures taken from ref. [19]

The $\boldsymbol{WV \to l\nu qq}$ analysis [20] is very similar to the previous $ZV \to llqq$, also with resolved and merged regime, but some additional requirements are used to eliminate the multijet contamination ($E_T^{miss} > 100$ GeV, $p_T(l\nu) > 200$ GeV, $E_T^{miss}/p_T(l\nu) > 0.2$). The analysis selects events that contain exactly one charged signal lepton and no additional veto electrons or muons, then events are categorized to VBF or DY categories and are assigned to merged and resolved regions. Studies using simulated events show that W+ jets and $t\bar{t}$ production are the dominant background sources. The shapes of the mass distributions for events from SM production of W+jets and $t\bar{t}$ are modeled using simulated events. The results are extracted by performing a simultaneous binned maximum-likelihood fit to the $m(WV)$ distributions in the signal regions and the control regions for W+jets and $t\bar{t}$. The $m(WV)$ distributions are presented in Figure 8. The data are compatible with the Standard Model background hypothesis and no significant excess has been found.

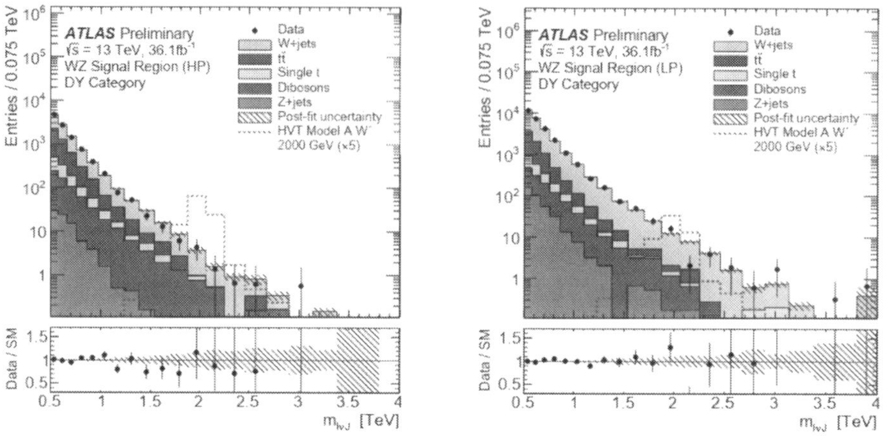

Figure 8. Post-fit signal region $m(WV)$ distributions in the DY category. Event failing the VBF selection are assigned to this category. The merged high-purity (HP) sample of WZ (**left**) events, the merged low-purity (LP) sample of WZ (**right**) events are presented. The background expectation is shown after the profile likelihood fit to the data and signal expectations are overlaid. The HVT Model A signal at 2000 GeV is presented. The band denotes the statistical and systematic uncertainty on the background after the fit to the data. The lower panels show the ratio of the observed data to the SM background estimation. In all regions, the overflow events are included in the last bin. Both figures are taken from ref. [20]

Table 1. Observed excluded resonance masses (at 95% CL) for the HVT and bulk RS models.

Analysis	Excluded mass range [TeV]		
	HVT W' Model A (Model B)	HVT Z' Model A (Model B)	RS G_{KK}, $k/\overline{M}_{Pl} = 1$ ($k/\overline{M}_{Pl} = 0.5$)
$VV \to qqqq$	1.1 – 3.0 (1.2 – 3.3)	1.2 – 2.2 (1.2 – 2.8)	1.3 – 1.45
$VV \to qqb\bar{b}$	1.1 – 2.4 (1.1 – 2.5)	1.1 – 1.48, 1.7 – 2.35 (1.1 – 2.6)	none
$HH \to b\bar{b}b\bar{b}$	none	none	0.36 – 0.86
$ZV \to llqq$, $\nu\nu qq$	< 2.9 (< 3.2)	none	< 1.3 (< 1.0)
$WV \to l\nu qq$	< 2.82 (< 2.98)	< 2.75 (< 3.09)	< 1.76

4 Conclusion

A brief summary of the most recent searches for resonances decaying to diboson final states, VV, VH and HH using ATLAS data from the 2015 and 2016 run at the LHC have been presented. No discrepancies with respect to the Standard Model expectations are observed and thus 95% CL exclusion limits are set on the production cross section times branching ratios in a number of benchmark scenarios. Observed excluded resonance masses at the 95% confidence level for the HVT and bulk RS models are shown in the table 1. The utilization of unprecedented luminosity of $\approx$ 85 fb^{-1} available at the end of the whole LHC Run 2 should give significant improvements on the exclusion limits.

References

[1] ATLAS Collaboration, *Observation of a new particle in the search for the Standard Model Higgs boson with the ATLAS detector at the LHC*, Phys. Lett. B 716 (2012) p. 1, arXiv: 1207.7214 [hep-ex]

[2] N. Arkani-Hamed, S. Dimopoulos and G. R. Dvali, *The Hierarchy problem and new dimensions at a millimeter*, Phys. Lett. B 429 (1998) 263, arXiv: hep-ph/9803315.

[3] M. J. Dugan, H. Georgi and D. B. Kaplan, *Anatomy of a Composite Higgs Model*, Nucl. Phys. B 254 (1985) 299.

[4] L. Randall and R. Sundrum, *A Large mass hierarchy from a small extra dimension*, Phys. Rev. Lett. 83 (1999) 3370, arXiv: 9905221.

[5] D. de Florian et al., *Handbook of LHC Higgs Cross Sections: 4. Deciphering the Nature of the Higgs Sector*, (2016), arXiv: 1610.07922.

[6] J. C. Pati and A. Salam, *Lepton Number as the Fourth Color*, Phys. Rev. D 10 (1974) 275.

[7] D. Pappadopulo, A. Thamm, R. Torre, A. Wulzer, *Heavy vector triplets: bridging theory and data*, JHEP 9 (2014) 60, arXiv:1402.4431 [hep-ph].

[8] K. Agashe, H. Davoudiasl, G. Perez, A. Soni, *Warped gravitons at the CERN LHC and beyond*, Phys. rev. D76 (2007) 036006, arXiv:hep-ph/0701186.

[9] ATLAS Collaboration, *The ATLAS Experiments at the CERN Large Hadron Collider*, JINST 3 (2008) S08003.

[10] D. Krohn, J. Thalr, L.-T. Wang, *Jet trimming*, JHEP 2 (2010) 84, arXiv:0912.1342 [hep-ph].

[11] ATLAS Collaboration, *ATLAS Combined Mass resolution and W mass reconstruction performance*, JETM-2017-002, https://atlas.web.cern.ch/Atlas/GROUPS/PHYSICS/PLOTS/JETM-2017-002.

[12] ATLAS Collaboration, *Identification of Boosted, Hadronically-Decaying W and Z Bosons in $\sqrt{s}$ = 13 TeV Monte Carlo Simulations for ATLAS*, ATL-PHYS-PUB-2015-033, https://cds.cern.ch/record/2041461.

[13] ATLAS Collaboration, *Expected Performance of Boosted Higgs ($\rightarrow b\bar{b}$) Boson Identification with the ATLAS Detector at $\sqrt{s}$ = 13 TeV*, ATL-PHYS-PUB-2015-035, https://cds.cern.ch/record/2042155.

[14] M. Cacciari, G. P. Salam, G. Soyez, *The Catchment Area of Jets*, JHEP 0804 (2008) 005, arXiv:0802.1188

[15] ATLAS Collaboration, *Expected performance of the ATLAS b-tagging algorithms in Run-2*, ATL-PHYS-PUB-2015-022, https://cds.cern.ch/record/2037697

[16] ATLAS Collaboration, *Search for diboson resonances with boson-tagged jets in pp collisions at $\sqrt{s}$ = 13 TeV with the ATLAS detector*, arXiv:1708.04445v1 [hep-ex].

[17] ATLAS Collaboration, *Search for heavy resonances decaying to a W or Z boson and a Higgs boson in the $q\bar{q}(')b\bar{b}$ final state in pp collisions at $\sqrt{s}$ = 13 TeV with the ATLAS detector*, arXiv:1707.06958v1 [hep-ex].

[18] ATLAS Collaboration, *Search for pair production of Higgs bosons in the $b\bar{b}b\bar{b}$ final state using proton-proton collisions at $\sqrt{s}$ = 13 TeV with the ATLAS detector*, ATLAS-CONF-2016-049, http://cds.cern.ch/record/2206131.

[19] ATLAS Collaboration, *Searches for heavy ZZ and ZW resonances in the llqq and vvqq final states in pp collisions at $\sqrt{s}$ = 13 TeV with the ATLAS detector*, arXiv:1708.09638v1 [hep-ex].

[20] ATLAS Collaboration, *Search for WW/WZ resonance production in lvqq final states in pp collisions at $\sqrt{s}$ = 13 TeV with the ATLAS detector*, ATLAS-CONF-2017-051, http://cds.cern.ch/record/2273867.

EPJ Web of Conferences **158**, 02004 (2017) DOI: 10.1051/epjconf/201715802004
QFTHEP 2017

On mass limits for vector leptoquarks from $K_L^0, B^0, B_s \to l_i^+ l_j^-$ decays with account of fermion mixing

Alexander D. Smirnov [1,*]

[1] *Yaroslavl State University, Russia*

Abstract. The contributions of the vector leptoquarks to branching ratios of $K_L^0, B^0, B_s \to l\,l'$ decays are calculated with account of fermion mixing in leptoguark currents of the general type and the corresponding mass limit for the vector leptoquarks resulting from the current data on these decays is obtained and analysed. It is found that the vector leptoquarks with masses $m_V > 78\ TeV$ are consistent with current data on these decays. The branching ratios of $K_L^0, B^0, B_s \to l\,l'$ decays at the possible lower mass $m_V = 78\ TeV$ are presented and the decays $B^0 \to \mu^+\mu^-$, $B_s \to e\mu$, $B_s \to \mu^+\mu^-$ are shown to be the most perspective ones for setting the new more stringent mass limits for the vector leptoquarks.

1 Introduction

The search for a new physics beyond the Standard Model (SM) is one of the aims of the current experiments at the LHC. There is a lot of models predicting new physics effects at LHC energies (such as models based on the supersymmetry, on the left-right symmetry, two Higgs models, extended dimension models, etc.).

One of such possible variants of new physics can be based on the well known idea of J.C. Pati and A. Salam on the possible four color symmetry between quarks and leptons regarding lepton number as the fourth color [1]. The four color symmetry can be unified with the SM in the minimal way by the gauge group [2, 3]

$$G_{MQLS} = SU_V(4) \times SU_L(2) \times U_R(1), \tag{1}$$

where $SU_V(4)$ is the vectorlike group of the four color symmetry [1–3], $SU_L(2)$ is the electroweak group for the left-handed fermions and $U_R(1)$ is the hypercharge factor for the right-handed ones (the minimal quark-lepton-symmetric model - $MQLS$-model [2, 3]). The four color symmetry of the vectorlike type predicts the new gauge particles - vector leptoquarks which form the color triplet $V_\alpha, \alpha = 1, 2, 3$ of the usual $SU_c(3)$ color group.

The lower mass limits for vector leptoquarks from their direct searches are of about 1 TeV. The essentially more stringent lower mass limits for vector leptoquarks are resulting from the rare decays of pseudoscalar mesons. The most stringent of them are given by data on the $K_L^0 \to e^\mp\mu^\pm$ decay and with neglect of fermion mixing in leptoquark currents achieve the order of 2 000 TeV [4–8].

* e-mail: asmirnov@uniyar.ac.ru

It should be noted however that the fermion mixing in leptoquark currents is natural. It is as natural as the fermion mixing in the week currents which are described by the well known matrices V_{CKM} and U_{PMNS} in the quark and lepton sectors respectively. The fermion mixing in the leptoquark currents and that in the week currents have a common origin and are resulting from the fact that the mass eigenstates of left- and right-handed quarks and leptons $Q^{L,R}_{pa\alpha}$, $l^{L,R}_{ia}$ can enter to interactions with gauge and scalar fields in general case through the superpositions

$$Q'^{L,R}_{pa\alpha} = \sum_q (A^{L,R}_{Q_a})_{pq}\, Q^{L,R}_{qa\alpha}, \qquad l'^{L,R}_{ia} = \sum_j (A^{L,R}_{l_a})_{ij}\, l^{L,R}_{ja},$$

where $A^{L,R}_{Q_a}$ and $A^{L,R}_{l_a}$ are unitary matrices describing the fermion mixing and diagonalizing the mass matrices of quarks and leptons, $p, q, i, j = 1, 2, 3, ...$ are the quark and lepton generation idexes, $a = 1, 2$ and $\alpha = 1, 2, 3$ are the $SU_L(2)$ and $SU_c(3)$ indexes, $Q_{q1} \equiv u_q = (u, c, t)$, $Q_{q2} \equiv d_q = (d, s, b)$ are the up and down quarks, $l_{j1} \equiv \nu_j$ are the mass eigenstates of neutrinos and $l_{j2} \equiv l_j = (e^-, \mu^-, \tau^-)$ are the charged leptons.

In the weak interaction the matrices $A^{L,R}_{Q_a}$ and $A^{L,R}_{l_a}$ form the CKM and PMNS matrices as

$$C_Q = (A^L_{Q_1})^+ A^L_{Q_2} = V_{CKM}, \qquad C_\ell = (A^L_{\ell_1})^+ A^L_{\ell_2} = U^+_{PMNS}.$$

In analogous way in the interaction of quarks and leptons with leptoquarks the matrices $A^{L,R}_{Q_a}$ and $A^{L,R}_{l_a}$ led to the four specific matrices

$$K^{L,R}_a = (A^{L,R}_{Q_a})^+ A^{L,R}_{\ell_a} \tag{2}$$

of the fermion mixing in leptoquark currents for up ($a = 1$), doun ($a = 2$) left (L)- and right (R)-handed fermions. The fermion mixing in leptoquark currents can lower the mass limits on leptoquark masses.

The responsible for the leptonic decays

$$P = K^0_L, B^0, B_s \to l^+_i l^-_j \tag{3}$$

of neutral pseudoscalar mesons interaction of the vector leptoquarks with down fermions in general case can be written as

$$L_{Vdl} = \frac{g_4}{\sqrt{2}}\{(\bar{d}_{p\alpha}[(K^L_2)_{pi}\gamma^\mu P_L + (K^R_2)_{pi}\gamma^\mu P_R]l_i)V_{\alpha\mu} + h.c.\}, \tag{4}$$

where $g_4 = g_{st}(M_c)$ is the $SU_V(4)$ gauge coupling constant related to the strong coupling constant at the mass scale M_c of the $SU_V(4)$ symmetry breaking, $P_{L,R} = (1 \pm \gamma_5)/2$ are the left and right operators of fermions and $K^{L,R}_2$ are the mixing matrices (2) for down fermions. It should be noted that in general case of $K^L_2 \neq K^R_2$ the interaction (4) of vector leptoquarks with quarks and leptons is not purely vectorlike. The mixing matrices $K^{L,R}_2$ in (4) can lower the mass limits on leptoquark masses resulting from the decays (3) and now the current experimental data on the decays (3) give the possibility to obtain new lower mass limits for the leptoquarks from these decays with account of the fermion mixing in leptoquark currents. The results of the first attempt to account the fermion mixing in leptoquark currents in the rare decays of type (3) was published in [9].

In the present paper the results of the calculations and analysis of the contributions of the vector leptoquarks into decays (3) with account of the fermion mixing in leptoquark currents of general form and the corresponding new lower mass limit for the vector leptoquarks are presented and discussed. With account of this lower mass limit the expected branching ratios of the decays (3) are also presented.

2 Branching ratios of $P \to l\,l'$ decays

Omitting the details of calculations (some details can be found in [7, 8]) the induced by the vector leptoquarks V branching ratios $Br_V(P \to l_i^+ l_j^-)$ of the decays of pseudoscalar meson P into lepton-antilepton pairs $l_i^+ l_j^-$ with $l_i^\pm = (e^\pm, \mu^\pm, \tau^\pm)$ can be written as

$$Br_V(P \to l_i^+ l_j^-) = B_P \, r_P(\mu_{l_i}, \mu_{l_j}) \, b_{P,ij}\,, \tag{5}$$

where

$$B_P = \frac{m_P \, \pi \, \alpha_{st}^2(M_c) \, f_P^2 \, \bar{m}_P^2 \, (R_P^V)^2}{2\, m_V^4 \, \Gamma_P^{tot}} \tag{6}$$

is the typical branching ratio of these decays, $r_P(\mu_{l_i}, \mu_{l_j})$ are the root factors

$$r_P(\mu_{l_i}, \mu_{l_j}) = \sqrt{[1 - (\mu_{l_i} + \mu_{l_j})^2][1 - (\mu_{l_i} - \mu_{l_j})^2]}$$

defined by the ratios $\mu_{l_i} = m_{l_i}/m_P$ of lepton m_{l_i} and meson m_P masses and $b_{P,ij}$ are the mixing factors depending on the matrices K_2^L, K_2^R. The enterring into (6) form factors f_P parametrize the matrix elements of the axial and pseudoscalar quark currents as

$$\langle 0|\bar{b}\gamma^\mu\gamma^5 d|P(p)\rangle = i f_P p^\mu, \quad \langle 0|\bar{b}\gamma^5 d|P(p)\rangle = -\, i\, \bar{m}_P f_P,$$

where p_μ is 4-momentum of the decaying meson, $\bar{m}_P = m_P^2/(m_{d_p} + m_{d_q})$, the factors $R_P^V = R_P(\mu_P, M_c)$ accounts the gluonic corrections to the pseudoscalar quark current and depend on mass scale M_c of the four color symmetry breaking and on mass scales μ_P at which the decays occur, Γ_P^{tot} is the total width of P meson and m_{d_p}, m_{d_q} are the masses of its valency quarks, m_V is the mass of the vector leptoquark.

The mixing factors $b_{P,ij}$ in (5) can be presented in the form [7, 8]

$$b_{P,ij} = \frac{1}{2}(|\beta_{P,ij}^L|^2 + |\beta_{P,ij}^R|^2)(1 - \mu_{l_i}^2 - \mu_{l_j}^2) + (\overset{*}{\beta}{}_{P,ij}^L \beta_{P,ij}^R + \beta_{P,ij}^L \overset{*}{\beta}{}_{P,ij}^R)\, \mu_{l_i} \mu_{l_j}, \tag{7}$$

where

$$\beta_{K_L^0,ij}^{L,R} = (\beta_{21,ij}^{L,R} + \beta_{12,ij}^{L,R})/\sqrt{2}, \qquad \beta_{B^0,ij}^{L,R} = \beta_{31,ij}^{L,R}, \qquad \beta_{B_s,ij}^{L,R} = \beta_{32,ij}^{L,R}$$

for $P = K_L^0 = ((\tilde{s}d) + (\tilde{d}s))/\sqrt{2}$, $P = B^0 = (\tilde{b}d)$, $P = B_s = (\tilde{b}s)$ respectively and

$$\beta_{pq,ij}^{L,R} = k_{pq,ij}^{L,R} - (\bar{\mu}_{l_j^-} k_{pq,ij}^{\prime L,R} + \bar{\mu}_{l_i^+} k_{pq,ij}^{\prime R,L})/2$$

with

$$k_{pq,ij}^{L,R} = (K_2^{L,R})_{pi} (\overset{*}{K}{}_2^{R,L})_{qj},$$
$$k_{pq,ij}^{\prime L,R} = (K_2^{L,R})_{pi} (\overset{*}{K}{}_2^{L,R})_{qj}$$

and $\bar{\mu}_{l_i^\pm} = m_{l_i^\pm}/(\bar{m}_P R_P^V)$.

Denoting the sums of the branching ratios of charge conjugated final states as

$$Br_V(P \to e\mu) = Br_V(P \to e^+\mu^-) + Br_V(P \to \mu^+ e^-), \tag{8}$$
$$Br_V(P \to e\tau) = Br_V(P \to e^+\tau^-) + Br_V(P \to \tau^+ e^-), \tag{9}$$
$$Br_V(P \to \mu\tau) = Br_V(P \to \mu^+\tau^-) + Br_V(P \to \tau^+\mu^-) \tag{10}$$

the branching ratios (5) can be rewritten as

$$Br_V(P \to l\,l') = B_P\; r_P(\mu_l, \mu_{l'})\; b_{P,ll'}\,, \tag{11}$$

with

$$b_{P,e^+e^-} = b_{P,11}, \qquad b_{P,\mu^+\mu^-} = b_{P,22}, \qquad b_{P,e\mu} = b_{P,12} + b_{P,21}, \tag{12}$$

$$b_{P,e\tau} = b_{P,13} + b_{P,31},\; b_{P,\mu\tau} = b_{P,23} + b_{P,32},\; b_{P,\tau^+\tau^-} = b_{P,33} \tag{13}$$

for $l\,l' = e^+e^-,\ \mu^+\mu^-,\ e\mu,\ e\tau,\ \mu\tau,\ \tau^+\tau^-$. With account of notations (8)–(10) the branching ratios (11)–(13) can be immediately compared with the experimental data on the decays (3).

3 Minimization of branching ratios of $P \to l\,l'$ decays

In the further analysis we use the general expressions

$$K_2^{L,R} = K_2^{L,R}(\theta_{12}^{L,R}, \theta_{13}^{L,R}, \theta_{23}^{L,R}, \delta^{L,R}, \varepsilon^{L,R}, \varphi_0^{L,R}, \varphi_1^{L,R}, \varphi_2^{L,R}, \varphi_3^{L,R}) \tag{14}$$

for the mixing matrices $K_2^{L,R}$. Each of the matrices (14) as the unitary 3×3 matrix depends on three angles $\theta_{12}^{L,R}$, $\theta_{23}^{L,R}$, $\theta_{13}^{L,R}$ and six phases $\delta^{L,R}$, $\varepsilon^{L,R}$, $\varphi_0^{L,R}$, $\varphi_1^{L,R}$, $\varphi_2^{L,R}$, $\varphi_3^{L,R}$.

In the case of neglecting the electron and muon masses

$$m_e, m_\mu \ll m_\tau, m_{K^0}, m_{B^0}, m_{B_s} \tag{15}$$

the mixing factors (12), (13) can be presented as the functions

$$b_{K_L^0,ll'} = b_{K_L^0,ll'}(\theta_{12}^{L,R}, \theta_{13}^{L,R}, \theta_{23}^{L,R}, \delta^{L,R}, \varepsilon^{L,R}), \tag{16}$$

$$b_{B^0,ll'} = b_{B^0,ll'}(\theta_{12}^{L,R}, \theta_{13}^{L,R}, \theta_{23}^{L,R}, \delta^{L,R}, \varepsilon^{L,R}, \chi_1^{L,R}, \chi_2^{L,R}), \tag{17}$$

$$b_{B_s,ll'} = b_{B_s,ll'}(\theta_{12}^{L,R}, \theta_{13}^{L,R}, \theta_{23}^{L,R}, \delta^{L,R}, \varepsilon^{L,R}, \chi_1^{L,R}, \chi_2^{L,R}) \tag{18}$$

with $b_{K_L^0,ll'}$ depending on the mixing angles $\theta_{12}^{L,R}$, $\theta_{23}^{L,R}$, $\theta_{13}^{L,R}$ and the phases $\delta^{L,R}$, $\varepsilon^{L,R}$ and with $b_{B^0,ll'}$, $b_{B_s,ll'}$ depending also on the phases $\varphi_0^{L,R}$, $\varphi_1^{L,R}$, $\varphi_2^{L,R}$, $\varphi_3^{L,R}$ through the phases

$$\chi_1^{L,R} = \varphi_0^{L,R} - \varphi_1^{L,R} - \varphi_2^{L,R}, \qquad \chi_2^{L,R} = \varphi_0^{L,R} + \varphi_3^{L,R}.$$

With fixed values of mixing angles $\theta_{12}^{L,R}$, $\theta_{23}^{L,R}$, $\theta_{13}^{L,R}$ and phases $\delta^{L,R}$ the mixing factors (17), (18) can be minimized over phases $\chi_1^{L,R}$, $\chi_2^{L,R}$, $\varepsilon^{L,R}$ by the conditions

$$\chi_1^L - \varepsilon^L - \chi_1^R + \varepsilon^R = 0, \qquad \chi_2^L - \chi_2^R = 0, \qquad \delta^L + \varepsilon^L - \delta^R - \varepsilon^R = 0. \tag{19}$$

Under conditions (19) the mixing factors (17), (18) depend only on the mixing angles $\theta_{12}^{L,R}$, $\theta_{23}^{L,R}$, $\theta_{13}^{L,R}$ and phases $\delta^{L,R}$ as

$$b_{B^0,ll'} = b_{B^0,ll'}(\theta_{12}^{L,R}, \theta_{13}^{L,R}, \theta_{23}^{L,R}, \delta^{L,R}), \tag{20}$$

$$b_{B_s,ll'} = b_{B_s,ll'}(\theta_{12}^{L,R}, \theta_{13}^{L,R}, \theta_{23}^{L,R}, \delta^{L,R}). \tag{21}$$

Keeping in mind that the most stringent lower mass limits for vector leptoquark are resulting from the experimental data $Br(K_L^0 \to l\,l')^{exp}$ on the branching ratios of the decays $K_L^0 \to l\,l'$ the mixing factors $b_{K_L^0,ll'}$ for the more small masses m_V must be very small (close to zero). In the further analysis the factors (16) for $l\,l' = e^+e^-$, $\mu^+\mu^-$, $e\mu$ can be assumed to be equal to zero

$$b_{K_L^0,ll'} = 0. \tag{22}$$

There are two solutions of the equations (22):

$$\text{solution A:} \qquad \theta^L_{23} = \theta^R_{23} = \pi/2, \qquad \theta^L_{13} = \theta^R_{13} \equiv \theta_{13}, \qquad \delta^L + \varepsilon^L + \delta^R + \varepsilon^R = \pi \tag{23}$$

and

$$\text{solution B:} \qquad \theta^L_{13} = \theta^R_{13} = \pi/2. \tag{24}$$

In the case (23) each of the matrices (14) K^L_2 (K^R_2) contains two angles θ_{13} and θ^L_{12} (θ^R_{12}) and the corresponding phases with $\delta^{L,R}$, $\varepsilon^{L,R}$ satisfying the last equation in (23). In the case (24) each of the matrices (14) K^L_2 (K^R_2) is defined by the unitary 2×2 matrix $U^L(2)$ ($U^R(2)$) and hence contains only one effective mixing angle. In the further analysis we use the matrices (14) of the case (23) as of the more general one because this case results in two mixing angles in each of the matrices K^L_2 (K^R_2).

In the case (23) the mixing factors (20), (21) take the form

$$b_{B^0,ll'} = b_{B^0,ll'}(\theta^L_{12}, \theta^R_{12}, \theta_{13}), \tag{25}$$

$$b_{B_s,ll'} = b_{B_s,ll'}(\theta^L_{12}, \theta^R_{12}, \theta_{13}) \tag{26}$$

of the functions on three angles θ^L_{12}, θ^R_{12}, θ_{13}.

The branching ratios (11) with mixing factors (22), (25), (26) have been numericaly analysed with account of experimental data on the decays (3) by varying the mixing angles θ^L_{12}, θ^R_{12}, θ_{13} to find the minimal vector leptoquark mass m_V satisfying these data.

4 Numerical results and discussion

In the numerical analysis we use from ref. [10] the data on the masses m_{l_i}, m_{d_i} of leptons and quarks, the data on the masses m_P, life times τ_P ($\tau_P \to \Gamma^{tot}_P$) and the form factors f_P

$$f_{K^0_L} = f_{K^-} = 155.72\, MeV, \; f_{B^0} = 190.9\, MeV, \; f_{B_s} = 227.2\, MeV$$

of mesons $P = (K^0_L,\, B^0,\, B_s)$ as well as the experimental data on the branching ratios $Br(P \to l\,l')^{exp}$ except the branching ratios

$$Br(B^0 \to \mu^+\mu^-)^{exp} < 3.4 \cdot 10^{-10} \quad [11],$$
$$Br(B^0 \to \tau^+\tau^-)^{exp} < 2.1 \cdot 10^{-3} \quad [12],$$
$$Br(B_s \to \mu^+\mu^-)^{exp} = (3.0 \pm 0.6^{+0.3}_{-0.2}) \cdot 10^{-9} \quad [11],$$
$$Br(B_s \to \tau^+\tau^-)^{exp} < 6.8 \cdot 10^{-3} \quad [12]$$

for which we use the current data of refs.[11, 12]. For the mass scale M_c of the $SU_V(4)$ symmetry we use the value $M_c = 100\, TeV$.

In the case of vectorlike mixing ($\theta^L_{12} = \theta^R_{12} \equiv \theta_{12}$) minimizing the vector leptoquark mass m_V over mixing angles θ_{12}, θ_{13} with satisfying the experimental values of $Br(P \to l\,l')^{exp}$ we find for vector leptoquark the mass limit

$$m_V > 78\; TeV. \tag{27}$$

The further analysis showed that the chiral mixing ($\theta^L_{12} \neq \theta^R_{12}$) does not change this mass limit for vector leptoquark. The additional analysis showed that in the case (24) the vector leptoquark mass limit exceeds the limit (27).

The mass limit (27) is essentially lower than that of order of 2 000 TeV obtained with neglect of the fermion mixing [4–8] at the same time it noticeably exceeds the mass limit of order of 1 TeV

resulting from the current direct searches for vector leptoquarks. It should be noted however that the mass limit (27) can be further lowered by the possible destructive interference of the vector leptoquark contributions discussed in this paper with those from the scalar leptoquarks which are also predicted (in addition to the vector leptoquarks) by the gauge group (1). The simultaneous account of the contributions of the vector and scalar leptoquarks to decays (3) needs the special consideration and we restrict ourselves here only by account of the vector leptoquark contributions. The mass limit (27) exceeds the mass limit $m_V > 38\ TeV$ declared in ref.[9] (mainly due to the difference of mixing factors (12), (13) from those of ref.[9]).

Keeping in mind the mass limit (27) we have calculated the contributions $Br_V(P \to l\,l')$ of vector leptoquarks to branching ratios of $P \to l\,l'$ decays with account of fermion mixing for $m_V = 78\ TeV$ with corresponding mixing angles. The results of calculations are presented in the second column of the table (1). In the third column of the table (1) we present for comparision the current experimental data on these decays.

Table 1. Contributions $Br_V(P \to l\,l')$ of vector leptoquarks to branching ratios of $P \to l\,l'$ decays with account of fermion mixing for $m_V = 78\ TeV$

$Br(P \to l\,l')$	$Br_V(P \to l\,l')$	$Br(P \to l\,l')^{exp}$
$Br(K_L^0 \to e^+e^-)$	0	$(9^{+6}_{-4}) \cdot 10^{-12}$
$Br(K_L^0 \to \mu^+\mu^-)$	0	$(6.84 \pm 0.11) \cdot 10^{-9}$
$Br(K_L^0 \to e\mu)$	0	$< 4.7 \cdot 10^{-12}$
$Br(B^0 \to e^+e^-)$	$3.3 \cdot 10^{-10}$	$< 8.3 \cdot 10^{-8}$
$Br(B^0 \to \mu^+\mu^-)$	$3.3 \cdot 10^{-10}$	$< 3.4 \cdot 10^{-10}$ *)
$Br(B^0 \to e\mu)$	$1.3 \cdot 10^{-9}$	$< 2.8 \cdot 10^{-9}$
$Br(B^0 \to e\tau)$	$2.0 \cdot 10^{-9}$	$< 2.8 \cdot 10^{-5}$
$Br(B^0 \to \mu\tau)$	$6.9 \cdot 10^{-9}$	$< 2.2 \cdot 10^{-5}$
$Br(B^0 \to \tau^+\tau^-)$	0	$< 2.1 \cdot 10^{-3}$
$Br(B_s \to e^+e^-)$	$2.9 \cdot 10^{-9}$	$< 2.8 \cdot 10^{-7}$
$Br(B_s \to \mu^+\mu^-)$	$2.9 \cdot 10^{-9}$	$(3.0 \pm 0.6^{+0.3}_{-0.2}) \cdot 10^{-9}$ *)
$Br(B_s \to e\mu)$	$1.09 \cdot 10^{-8}$	$< 1.1 \cdot 10^{-8}$ *)
$Br(B_s \to e\tau)$	$5.0 \cdot 10^{-10}$	-
$Br(B_s \to \mu\tau)$	$1.7 \cdot 10^{-9}$	-
$Br(B_s \to \tau^+\tau^-)$	0	$< 6.8 \cdot 10^{-3}$

As seen from the table (1) the contributions of vector leptoquarks with $m_V = 78\ TeV$ to branching ratios $Br(B^0 \to \mu^+\mu^-)$, $Br(B_s \to e\mu)$ are close their current experimental limits (marked by stars) and the corresponding contribution to $Br(B_s \to \mu^+\mu^-)$ is compatible with its current experimental value (also marked by the star). It means that the further search for these decays can immediately give new more stringent limits on the vector leptoquark mass m_V.

The contribution of the vector leptoquarks to branching ratio $Br(B^0 \to e\mu)$ is about twice lower than its current experimental limit and the search for this decay looks also as perspective for setting new mass limit for vector leptoquarks.

DOI: 10.1051/epjconf/201715802004

As concerns the branching ratios $Br(B^0 \to e^+e^-)$, $Br(B^0 \to e\tau)$, $Br(B^0 \to \mu\tau)$, $Br(B_s \to e^+e^-)$ the contributions of the vector leptoquarks to these decays are essentially (by order of $2 \div 4$) less than their current experimental limits, the corresponding contributions to branching ratios $Br(B^0 \to \tau^+\tau^-)$, $Br(B_s \to \tau^+\tau^-)$ are very small (under assumptions (22) they are equal to zero).

The contributions of the vector leptoquarks to branching ratios $Br(B_s \to e\tau)$ and $Br(B_s \to \mu\tau)$ are expected to be of order of 10^{-10} and 10^{-9} respectively, which can be kept in mind in the future experimental searches for these decays.

5 Conclusion

1. The contributions of the vector leptoquarks to branching ratios of $K_L^0, B^0, B_s \to l\,l'$ decays are calculated with account of fermion mixing in leptoguark currents of the general type and the corresponding mass limit for the vector leptoquarks resulting from the current experimental data on these decays are obtained and analysed.

2. It is found that the vector leptoquarks with masses

$$m_V > 78\ TeV$$

are consistent with current data on the branching ratios of $K_L^0, B^0, B_s \to l\,l'$ decays.

3. It is shown that the contributions of the vector leptoquarks with $m_V = 78\ TeV$ to branching ratios of the decays

$$B^0 \to \mu^+\mu^-, \quad B_s \to e\mu, \quad B_s \to \mu^+\mu^-$$

are close to the current experimental data on these decays and the further searches for these decays look as the most perspective ones for setting the new more stringent mass limits for the vector leptoquarks.

4. The contribution of the vector leptoquarks to branching ratio $Br(B^0 \to e\mu)$ is shown to be about twice lower than its current experimental limit and the search for this decay looks also as perspective for obtaining new mass limit for vector leptoquarks.

5. The contributions of the vector leptoquarks to branching ratios $Br(B_s \to e\tau)$ and $Br(B_s \to \mu\tau)$ are snown to be of order or less than 10^{-10} and 10^{-9} respectively.

6. The contributions of the vector leptoquarks to branching ratios of the others decays of $K_L^0, B^0, B_s \to l\,l'$ type are shown to be essentially lower (by order 2 or greater) than their current experimental limits.

References

[1] J. C. Pati, A. Salam, Phys. Rev. **D10**, 275 (1974)
[2] A. D. Smirnov, Phys. Lett. **B346**, 297 (1995)
[3] A. D. Smirnov, Phys. At. Nucl. **58**, 2137 (1995)
[4] G. Valencia, S. Willenbrock, Phys. Rev. **D50**, 6843 (1994)
[5] A. V. Kuznetsov, N. V. Mikheev, Phys. Lett. **B329**, 295 (1994)
[6] A. V. Kuznetsov, N. V. Mikheev, Phys. At. Nucl. **58**, 2228 (1995)
[7] A. D. Smirnov, Mod. Phys. Lett. **A 22**, 2353 (2007)

[8] A. D. Smirnov, Phys. At. Nucl. **71**, 1470 (2008)
[9] A. V. Kuznetsov, N. V. Mikheev, A. V. Serghienko, J. Mod. Phys. **A 27**, 1250062 (2012)
[10] C. Patrignani et al. (Particle Data Group), Chinese Physics **C 40**, 100001 (2016)
[11] R. Aaij et al. (LHCb Collaboration), Phys. Rev. Lett. **118**, 191801 (2017)
[12] R. Aaij et al. (LHCb Collaboration), Phys. Rev. Lett. **118**, 251802 (2017)

EPJ Web of Conferences **158**, 02005 (2017) DOI: 10.1051/epjconf/201715802005
QFTHEP 2017

Vacuum stability with the effective six-pointed couplings of the Higgs bosons in the heavy supersymmetry

Mikhail Dubinin[1,2,★] and *Elena* Petrova[1,2,★★]

[1]*Skobeltsyn Institute of Nuclear Physics, Lomonosov Moscow State University, 119991 Moscow, Russia*
[2]*Physics Department, Lomonosov Moscow State University, 119991 Moscow, Russia*

Abstract. Consistent and phenomenologically successful scenarios of the Minimal Supersymmetric Standard Model (MSSM) include the Higgs potential which is positively defined at large fields and has two minima, one of them is true and stable vacuum in which we live. However, the effective terms of the MSSM potential which appear at a loop level can spoil the vacuum stability leading to a metastable or unstable vacuum configurations. Such a situation is analysed and the general necessary conditions of vacuum stability in the MSSM parameter space are imposed. For a majority of the MSSM "benchmark scenarios" which are used for reconstruction of production channels at the LHC, the effective potential is found to be always stable, except *light stop* and τ-phobic scenarios.

1 Introduction

The discovery of the Higgs boson with mass 125.09±0.24 GeV at the LHC [1] confirmed the fundamental concept of symmetry breaking in the scalar sector. The properties of a new scalar, e.g. its spin, CP parity, and coupling strengths to the SM particles, are consistent within the statistical uncertainties with predictions of the Standard Model (SM). However, statistical uncertainties of the signal strength and significant signal strength errors in the production channels are not small and should be reduced by the future analyses of ATLAS and CMS Collaborations at the energy 13 TeV. Combined results [2] for the Higgs boson production and decay rates as well as constraints on its couplings to the SM particles leave a room for meaningful contributions of physics beyond the SM.

Precise measurements of the Higgs mass and the top quark mass are critical to determine the vacuum structure of the SM [3] which demonstrates a metastable vacuum. In the SM the scalar sector is single-field in the unitary gauge. In the SM extensions, the situation is more complicated because the Higgs sector includes several fields.

2 The MSSM effective Higgs potential at the one-loop

In the MSSM the Higgs sector consists of five Higgs bosons. In the CP-conserving limit, they are two CP-even neutral Higgs bosons h and H, one CP-odd neutral Higgs boson A and two charged scalars

★e-mail: dubinin@theory.sinp.msu.ru
★★e-mail: petrova@theory.sinp.msu.ru

$H^{\pm}$ [4]. As a rule, the lightest CP-even scalar h is identified with the observable Higgs boson. Large radiative corrections at loop level ensure the h-boson mass of approximately 125 GeV. The effective Higgs potential in the Coleman-Weinberg framework [5] expanded to all orders of perturbation theory is shown schematically in figure 1 and has a form

$$U(1-\text{loop}) = U^{(2)} + U^{(4)} + U^{(6)} + \dots \tag{1}$$

In the literature, it is usually supposed that all higher order potential terms of effective operators more

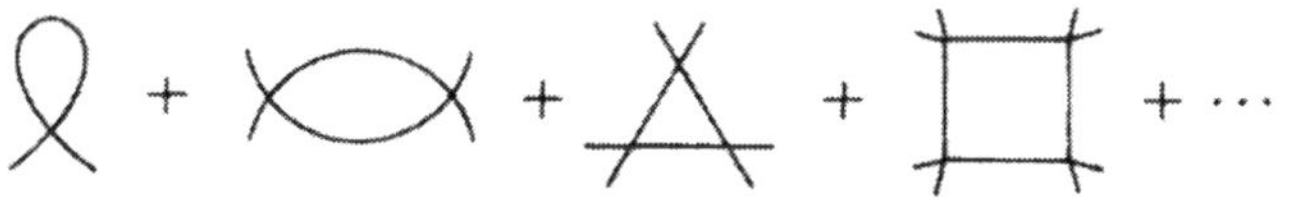

Figure 1. The one-loop approximation for the effective potential in the Coleman-Weinberg framework [5].

then dimension-four in the fields are negligibly small if the following conditions are respected [6]

$$2|m_t A_t| < M_S^2, \qquad 2|m_t \mu| < M_S^2, \tag{2}$$

where m_t =173.2 GeV is the top quark mass, A_t is the trilinear Higgs–t-squark interaction parameter, μ is the Higgs superfield mass parameter and M_S is the squark mass scale. This parametrization of the MSSM soft supersymmetry breaking sector [7] is the most common in the MSSM benchmark scenarios [8]. However, the dimension-six operators may play an important role in the A_t, μ parameter range about/of the order of 10^1 TeV and low and moderate M_S [9]. Such a situation is rather unusual in the most of MSSM scenarios but in general parameters A_t, μ may have so large values which respect the perturbative unitarity conditions (see e.g. [10]). Potential terms of the two-doublet scalar sector can be written as follows

$$\begin{aligned}
U^{(2)} &= -\mu_1^2(\Phi_1^\dagger\Phi_1) - \mu_2^2(\Phi_2^\dagger\Phi_2) - [\mu_{12}^2(\Phi_1^\dagger\Phi_2) + h.c.], & (3)\\
U^{(4)} &= \lambda_1(\Phi_1^\dagger\Phi_1)^2 + \lambda_2(\Phi_2^\dagger\Phi_2)^2 + \lambda_3(\Phi_1^\dagger\Phi_1)(\Phi_2^\dagger\Phi_2) + \lambda_4(\Phi_1^\dagger\Phi_2)(\Phi_2^\dagger\Phi_1) + & (4)\\
&+ [\lambda_5/2(\Phi_1^\dagger\Phi_2)(\Phi_1^\dagger\Phi_2) + \lambda_6(\Phi_1^\dagger\Phi_1)(\Phi_1^\dagger\Phi_2) + \lambda_7(\Phi_2^\dagger\Phi_2)(\Phi_1^\dagger\Phi_2) + h.c.], &\\
U^{(6)} &= \kappa_1(\Phi_1^\dagger\Phi_1)^3 + \kappa_2(\Phi_2^\dagger\Phi_2)^3 + \kappa_3(\Phi_1^\dagger\Phi_1)^2(\Phi_2^\dagger\Phi_2) + \kappa_4(\Phi_1^\dagger\Phi_1)(\Phi_2^\dagger\Phi_2)^2 + & (5)\\
&+ \kappa_5(\Phi_1^\dagger\Phi_1)(\Phi_1^\dagger\Phi_2)(\Phi_2^\dagger\Phi_1) + \kappa_6(\Phi_1^\dagger\Phi_2)(\Phi_2^\dagger\Phi_1)(\Phi_2^\dagger\Phi_2) + &\\
&+ [\kappa_7(\Phi_1^\dagger\Phi_2)^3 + \kappa_8(\Phi_1^\dagger\Phi_1)^2(\Phi_1^\dagger\Phi_2) + \kappa_9(\Phi_1^\dagger\Phi_1)(\Phi_1^\dagger\Phi_2)^2 + &\\
&+ \kappa_{10}(\Phi_1^\dagger\Phi_2)^2(\Phi_2^\dagger\Phi_2) + \kappa_{11}(\Phi_1^\dagger\Phi_2)^2(\Phi_2^\dagger\Phi_1) + \kappa_{12}(\Phi_1^\dagger\Phi_2)(\Phi_2^\dagger\Phi_2)^2 + &\\
&+ \kappa_{13}(\Phi_1^\dagger\Phi_1)(\Phi_1^\dagger\Phi_2)(\Phi_2^\dagger\Phi_2) + h.c.], &
\end{aligned}$$

where

$$\Phi_i = \begin{pmatrix} \phi_i^+(x) \\ \phi_i^0(x) \end{pmatrix} = \begin{pmatrix} -i\omega_i^+ \\ \frac{1}{\sqrt{2}}(v_i + \eta_i + i\chi_i) \end{pmatrix}, \qquad i = 1, 2 \tag{6}$$

– Higgs doublets with the $SU(2)$ field states and $v_1 = v\cos\beta$, $v_2 = v\sin\beta$ (v = 246 GeV) – vacuum expectation values of them. The real part of μ_{12}^2 is fixed by zero eigenvalue of the mass matrix (which ensures massless Goldstone boson state and defines the CP-odd scalar mass m_A^2, see details in ref. [9])

$$\begin{aligned}
\text{Re}\mu_{12}^2 &= s_\beta c_\beta[m_A^2 + \frac{v^2}{2}(2\text{Re}\lambda_5 + \text{Re}\lambda_6\cot\beta + \text{Re}\lambda_7\tan\beta)] + v^4\{\text{Re}\kappa_9 c_\beta^3 s_\beta & (7)\\
&+ \text{Re}\kappa_{10} c_\beta s_\beta^3 + \frac{1}{4}[\text{Re}\kappa_8 c_\beta^4 + \text{Re}\kappa_{12} s_\beta^4 + (9\text{Re}\kappa_7 + \text{Re}\kappa_{11} + \text{Re}\kappa_{13}) s_\beta^2 c_\beta^2]\}, &
\end{aligned}$$

where $s_\beta = \sin\beta$, $c_\beta = \cos\beta$ and so on. Tree level boundary conditions for Higgs self-couplings at the scale M_S are [11]

$$\lambda^{\text{tree}}_{1,2}(M_S) = \frac{g_1^2+g_2^2}{4}, \qquad \lambda^{\text{tree}}_{3}(M_S) = \frac{g_2^2-g_1^2}{4}, \qquad \lambda^{\text{tree}}_{4}(M_S) = \frac{g_2^2}{2}, \tag{8}$$
$$\lambda^{\text{tree}}_{5,6,7}(M_S) = 0, \qquad \kappa^{\text{tree}}_{1,\dots,13}(M_S) = 0,$$

where g_1 and g_2 are the standard gauge couplings, while at the loop-level and below M_S scale these couplings acquire radiative corrections

$$\lambda_{1,2}(M) = \lambda^{\text{tree}}_{1,2}(M_S) - \Delta\lambda_{1,2}(M)/2, \qquad \lambda_{3,\dots,7}(M) = \lambda^{\text{tree}}_{3,\dots,7}(M_S) - \Delta\lambda_{3,\dots,7}(M),$$
$$\kappa_{1,\dots 13}(M) = \Delta\kappa_{1,\dots 13}(M).$$

Radiative corrections to parameters λ_i (i=1,...7) in the effective field theory framework have been analysed in ref. [6, 7, 12, 13]. Radiative corrections to the parameters κ_i (i =1,...13) in the approximation of degenerate squark masses have been obtained in ref. [9]. An example of the one-loop renormalization group improved (RG-improved) threshold correction for λ_1 and the threshold correction for κ_1 can be written as

$$-\frac{\Delta\lambda_1^{\text{thr}}}{2} = \frac{3}{32\pi^2}\Big[h_b^4\frac{|A_b|^2}{M_S^2}\left(2-\frac{|A_b|^2}{6M_S^2}\right) - h_t^4\frac{|\mu|^4}{6M_S^4} + 2h_b^4 l + \frac{g_2^2+g_1^2}{4\,M_S^2}(h_t^2|\mu|^2 - h_b^2|A_b|^2)\Big] \tag{9}$$
$$+ \frac{1}{768\pi^2}\left(11g_1^4+9g_2^4-36\,(g_1^2+g_2^2)\,h_b^2\right)l,$$

$$\Delta\kappa_1^{\text{thr}} = \frac{h_b^6}{32M_S^2\pi^2}\left(2-\frac{3|A_b|^2}{M_S^2}+\frac{|A_b|^4}{M_S^4}-\frac{|A_b|^6}{10M_S^6}\right) - h_b^4\frac{g_1^2+g_2^2}{128M_S^2\pi^2}\left(3-3\frac{|A_b|^2}{M_S^2}+\frac{|A_b|^4}{2M_S^4}\right) \tag{10}$$
$$+ \frac{h_b^2}{512M_S^2\pi^2}\left(\frac{5}{3}g_1^4+2g_1^2g_2^2+3g_2^4\right)\left(1-\frac{|A_b|^2}{2M_S^2}\right) - h_t^6\frac{|\mu|^6}{320M_S^8\pi^2}$$
$$+ h_t^4\frac{(g_1^2+g_2^2)|\mu|^4}{256M_S^6\pi^2} - h_t^2\frac{(17g_1^4-6g_1^2g_2^2+9g_2^4)|\mu|^2}{3072M_S^4\pi^2} + \frac{g_1^2}{1024M_S^2\pi^2}(g_1^4-g_2^4),$$

where $l \equiv \ln\left(\frac{M_S^2}{\sigma^2}\right)$, σ is the renormalization scale, $h_t = \frac{g_2 m_t}{\sqrt{2}m_W\sin\beta}$ and $h_b = \frac{g_2 m_b}{\sqrt{2}m_W\cos\beta}$ are Yukawa couplings. One can notice inspecting these explicit forms that radiative corrections $\Delta\kappa^{\text{thr}}$ begin to play an important role if the following conditions are true

$$|\mu|m_t\cot\beta \approx M_S^2, \qquad |\mu|m_b\tan\beta \approx M_S^2, \qquad |A_t|m_t \approx M_S^2, \qquad |A_b|m_b \approx M_S^2, \tag{11}$$
$$|\mu A_t|m_t^2\cot\beta \approx M_S^4, \qquad |\mu A_b|m_b^2\tan\beta \approx M_S^4.$$

In this case, the corrections coming from the dimension-six operators of the one-loop potential can be appreciable even if the inequalities (2) are respected.

3 Vacuum stability with respect to $\Delta\kappa_{1,\dots 13}$ terms

The local minimum v =246 GeV of the Higgs potential exists if the following conditions are respected

$$\left.\frac{\partial U}{\partial\phi_1^0}\right|_{\phi_i^0=v_i} = 0, \qquad \left.\frac{\partial U}{\partial\phi_2^0}\right|_{\phi_i^0=v_i} = 0, \qquad \Delta = \left\|\begin{matrix} U''_{\phi_1^0\phi_1^0} & U''_{\phi_1^0\phi_2^0} \\ U''_{\phi_2^0\phi_1^0} & U''_{\phi_2^0\phi_2^0}\end{matrix}\right\|_{\phi_i^0=v_i} > 0, \qquad \left.U''_{\phi_1^0\phi_1^0}\right|_{\phi_i^0=v_i} > 0. \tag{12}$$

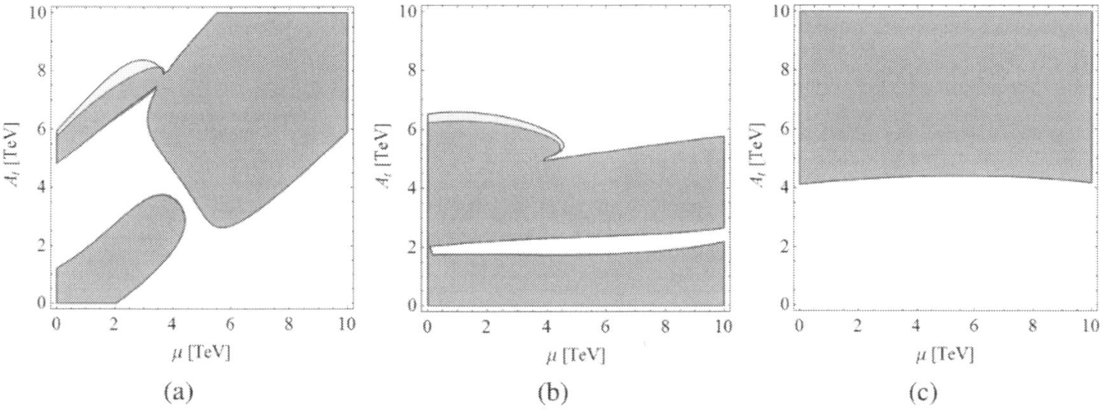

Figure 2. The allowed regions defined by Eq. (12) for m_h =125 GeV and (a) – $\tan\beta$=1, M_S=1.5 TeV, (b) – $\tan\beta$=10, M_S=1.5 TeV and (c) $\tan\beta$=10, M_S=5 TeV. Yellow and green areas correspond to an allowed parameter range for the Higgs potential $U = U^{(2)} + U^{(4)} + U^{(6)}$; yellow area corresponds to an allowed parameter range for the pottential $U^{(2)} + U^{(4)}$ in the one-loop expansion of the effective Higgs potential.

For example if κ_i=0, then

$$\begin{aligned} \Delta &= -[-2\text{Re}\mu_{12}^2 + v^2(3\lambda_6 c_\beta^2 + 3\lambda_7 c_\beta^2 + \lambda_{345} s_{2\beta})]^2 \\ &- [v^2(-\lambda_7 + 3\lambda_6 \cot^2\beta + 4\lambda_1 \cot^3\beta) + 2\text{Re}\mu_{12}^2 \csc^2\beta] \\ &\times [v^2(-4\lambda_2 - 3\lambda_7 \cot\beta + \lambda_6 \cot^3\beta) - 2\text{Re}\mu_{12}^2 \cot\beta \csc^2\beta] s_\beta^4 \tan\beta > 0, \end{aligned} \tag{13}$$

where $\text{Re}\mu_{12}^2$ is defined by eq. (7). Allowed domains on the plane (μ, A_t) for fixed values of M_S and $\tan\beta$ are shown in figure 2. They are reconstructed for the case when the lightest CP-even scalar mass m_h =125 GeV and masses of other Higgs bosons are positively defined. Several examples of the mass spectrum of scalars can be found in [9]. The combination of green and yellow regions includes an allowed A_t and μ range if the Higgs potential terms $U^{(2)} + U^{(4)}$ are extended by $U^{(6)}$ operators, eq. (5), and the yellow area represents the allowed region for $U^{(2)} + U^{(4)}$ terms only. Visible deviations of the contours for the dimension-four potential are observed if the values of A_t respect the conditions of eq. (11).

Configuration for $U^{(2)} + U^{(4)} + U^{(6)}$ in the case of "good" (A_t, μ) parameters for the effective potential decomposition is shown in figure 3 (a). The potential has two minima and is positively defined at a large ϕ_i^0. So, in this case, the vacuum in which we live is true and stable. But if A_t and μ are chosen from the yellow area of figure 2 (a), two minima degenerate to a gully which forms a saddle configuration in the origin, see figure 3 (b). The conditions defined by eq. (12) are not respected in this case and corrections coming from dimension-six operators spoil the vacuum stability. In the case $||U_{ij}||$ =0, the gully forms a flat direction when the stationary point at the origin is degenerate and nonisolated [14].

For the LHC analyses the MSSM multidimensional parameter space is reduced to specific benchmark parameter sets which demonstrate different collider phenomenology. Six benchmark sets which are specified in [8] are usually denominated as m_h^{max}, m_h^{mod+}, m_h^{mod-}, *light stop, light stau* and *τ-phobic* scenarios. In these benchmark scenarios M_S < 1.5 TeV, μ < M_S (except *τ-phobic* scenario) and $A_t = X_t + \mu/\tan\beta$, where $X_t \sim O(1)M_S$, so large values of A_t where the difference between various radiative corrections is appreciable can be achieved only for extremely small $\tan\beta$ region (see eq. 11), where one can expect an evidence of additional contributions. The minimum of the potential

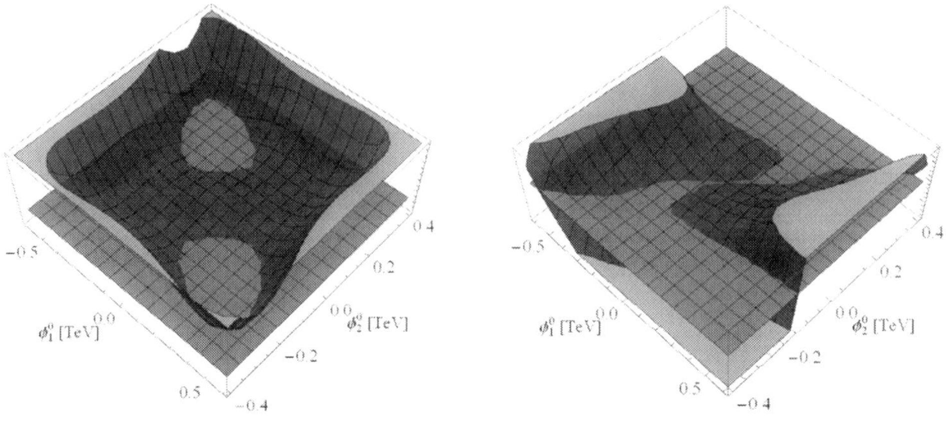

(a) (b)

Figure 3. Higgs potential forms, see (1), for M_S=1.5 TeV, tanβ=1 and (a) – A_t = μ=1.5 TeV, (b) – A_t=7.5 TeV, μ=2 TeV, see also figure 2 (a). The plane U =0 is shown in green colour.

disappears only at low tanβ for *light stop* and τ*-phobic* scenarios while other demonstrate remarkable stability with respect to radiative corrections.

4 Summary

The problem of the SM vacuum stability looks less alarming in the MSSM extension. The best enquiry would be if the beyond SM parameters can be chosen in such a way that the vacuum would be always stable. In the case of two-doublet Higgs sector, the potential has two identical minima, one of them is a vacuum in which we live. The conditions defined by eq. (12) guarantee the stable vacuum, but modification of Higgs potential demonstrates that some parameter sets respecting the minimization in the dimension-four version are not acceptable for the dimension-six extension. Instead of the stable minimum with vev's one can have a saddle configuration at the origin. New terms of dimension-six in the fields are significant if eq. (11) are respected. In the majority of the MSSM benchmark scenarios the local minimum is stable with respect to dimension-six terms, but for scenarios *light stop* and τ*-phobic* the low tanβ region is excluded.

The authors are grateful to I. Volobuev for useful remarks. This work was supported by Grant No. NSh-7989.2016.2.

References

[1] G. Aad et al. (ATLAS Collaboration), Phys. Lett. B **716**, 1(2012); S. Chatrchyan et al. (CMS Collaboration), Phys. Lett. B **716**, 30(2012)

[2] Aad G. et al. (ATLAS and CMS Collaborations), JHEP **1608**, 045 (2016); Aad G. et al. (ATLAS and CMS Collaborations), Phys. Rev. Lett. **114**, 191803 (2015)

[3] F. Bezrukov, M. Kalmykov, B. Kniehl, and M. Shaposhnikov, JHEP **1210**, 140 (2012); J. Elias-Miro et al, Phys. Lett. B **709**, 222 (2012); J. Casas, J. Espinosa, and M. Quiros, Phys. Lett. B **382**, 374 (1996); A. Linde, Phys. Lett. B **92**, 209 (1980); H. D. Politzer and S. Wolfram, Phys. Lett. B **82**, 242 (1979); *ibid,* B **83** 421 (1979); N. V. Krasnikov, Yad. Fiz., **28**, 549 (1978); L. Maiani, G. Parisi, and R. Petronzio, Nucl. Phys. B **136**, 115 (1978)

[4] A. Djouadi, Phys. Rep. **459**, 1 (2008); H. Haber and G. Kane, Phys. Rep. **117**, 75 (1985)
[5] S. Coleman and E. Weinberg, Phys. Rev. D **7**, 1888 (1973)
[6] M. Carena, J. R. Espinosa, M. Quiros, and C. E. M. Wagner, Phys. Lett. B **355**, 209 (1995)
[7] H. E. Haber and R. Hempfling, Phys. Rev. D **48**, 4280 (1993)
[8] M. Carena, S. Heinemeyer, O. Stal, C. E. M. Wagner, and G. Weiglein, Eur. Phys. J. C **73** 2552 (2013)
[9] M. N. Dubinin and E. Yu. Petrova, Phys. Rev. D **95**, 055021 (2017)
[10] S. Kanemura and K. Yagyu, Phys. Lett. B **751**, 289 (2015)
[11] K. Inoue, A. Kakuto, H. Komatsu, and S. Takeshita, Prog. Theor. Phys. **68**, 927 (1982); R. A. Flores and M. Sher, Ann. Phys. (N.Y.) **148**, 95 (1983)
[12] E. Akhmetzyanova, M. Dolgopolov, and M. Dubinin, Phys. Rev. D **71**, 075008 (2005); Phys. Part. Nucl. **37**, 677 (2006)
[13] S. Y. Choi, M. Drees, and J. S. Lee, Phys. Lett. B **481** 57 (2000)
M. Carena, J. Ellis, A. Pilaftsis, and C. E. M. Wagner, Nucl. Phys. B **586**, 92 (2000)
[14] M. N. Dubinin and E. Yu. Petrova, Theor. Math. Phys. **184**, 1170 (2015)

Resonant top pair production at NLO in QCD

Diogo Buarque Franzosi[1,⋆]

[1] *II. Physikalisches Institut, Universität Göttingen, Friedrich-Hund-Platz 1, 37077 Göttingen, Germany*

Abstract.
We compute the contribution of a scalar H or pseudo-scalar A resonance in top-quark pair production at NLO accuracy in QCD including the interference between pure QCD SM diagrams and the resonant $pp \to A(H) \to t\bar{t}$ process which drastically modifies the lineshape of the signal from a simple peak to peak-dip structure. We assume a point-like coupling between scalar and the gluons in a consistent effective field theory framework and improve the results with reweighting techniques in the case of resolved fermion loops in the production. The computation is set in an automatic framework and can be used to produce showered unweighted events. The detailed study is presented in Ref [1].

1 Introduction

Scalar resonances decaying into a $t\bar{t}$ system exist in many BSM scenarios. In order to correctly describe this process at the LHC, and to determine the constraints on the parameters of different models, it is important to get trustworthy and robust theoretical predictions.

It is now known that QCD corrections to this process have a large effect on the total cross section as well as the differential distributions, both for the SM QCD production and for (pseudo-)scalar production. It is also well known that the interference between resonant signal and the QCD background can drastically modify the lineshape of the signal, from a single peak to a dip-peak structure, and may be even larger than the pure signal. It appears to receive similar QCD corrections, as indicated by estimates in an effective field theory (EFT) approach in the soft gluon approximation [2], and in a simplified K-factor derived from the geometric mean of the background and signal K-factors [3].

In this talk, based on Ref [1], we provide the full perturbative calculation of the process $pp(\to A/H) \to t\bar{t}$, at NLO in QCD, including the interference, in the EFT approach, without any further approximation.

Our calculation is performed within the MadGraph5_aMC@NLO (MG5_aMC) [4] framework which allows us to perform automatic simulation of events with parton shower (PS) matching. An interesting feature of our calculation that has not been discussed before is that the gluon-scalar operator mixes into the chromo-magnetic dipole-moment operator, and therefore both operators must be taken into account in a consistent NLO calculation and put together in a coherent setup.

The gluon-gluon-scalar vertex can take two forms depending on the underlying model. If the vertex is dominated by heavy particle loops (e.g. vector-like fermions or scalar top partners), or induced

⋆e-mail: dbuarqu@gwdg.de

by strong dynamics at a high scale, this interaction will not be resolved at the scale of the scalar resonance. In this case the vertex can be simply represented by the dimension-five operators

$$O_{HG} = g_s^2 G^A_{\mu\nu} G^{A\mu\nu} H\,, \tag{1}$$

$$O_{A\tilde{G}} = g_s^2 G^A_{\mu\nu} \widetilde{G}^{A\mu\nu} A\,, \tag{2}$$

with $G^A_{\mu\nu}$ being the gluon field strength tensor and $\widetilde{G}^{A\mu\nu}$ its dual. On the other hand, if the vertex is loop-induced with lighter particles, such as top quarks running in the loop, it will be resolved at the resonance scale and the interaction will give rise to an absorptive phase. It is convenient to distinguish between these two cases, as they often lead to different lineshapes, and the resolved case is more difficult to compute at NLO accuracy. In practice, one may need to deal with a mixed scenario, if there are contributions from both light and heavy loop particles.

Our NLO computation of the interference is based on the EFT framework, which in the unresolved case provides accurate results (i.e. without using soft-gluon approximation), and in the resolved case, can be further improved by using reweighting techniques. These results will then be passed to PS simulation, to obtain more realistic predictions.

2 Theoretical setup

Including only the scalar-gluon operators (Eq. (1)-Eq. (2)) is not enough for a consistent calculation in the EFT. This is because operator mixing with the top-quark dipole moment becomes relevant in this process and including only these operators leads to uncancelled ultra-violet (UV) poles from non-factorisable contributions, whenever a top quark, a gluon and a scalar form a loop, as shown in Figure 1.

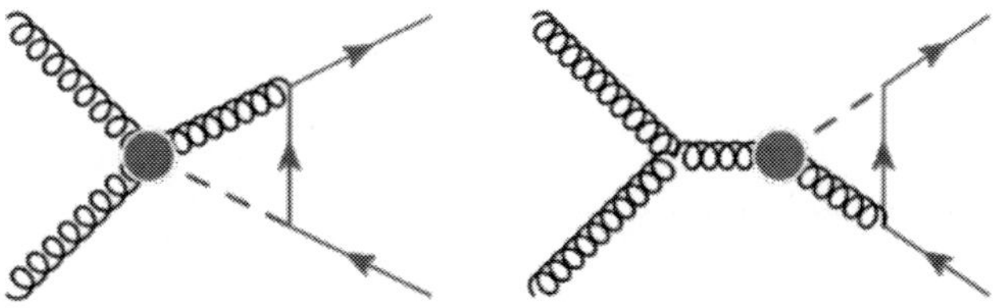

Figure 1. Selected UV divergent diagrams in non-factorisable contributions.

The reason for the UV divergence is not difficult to understand. It is well known that the EFT is renormalisable only if one considers the complete set of higher-dimensional operators up to a certain dimension. However, since so far only the operators O_{HG} and O_{AG} are added, it could mix into a different operator not included in the Lagrangian, leading to non-renormalisability of the theory. In fact, through the triangles shown in Figure 1, the O_{HG} (O_{AG}) induces a chromo-magnetic dipole operator of the top quark, leading to the following self-consistent Lagrangian,

$$L_{Eff} = L_{SM} + y_t \bar{t}tH + \tilde{y}_t \bar{t} i\gamma^5 tA + \frac{C_{HG}}{\Lambda} O_{HG} + \frac{C_{A\tilde{G}}}{\Lambda} O_{A\tilde{G}} + \frac{C_{tG}}{\Lambda} O_{tG}, \tag{3}$$

with the chromo-magnetic dipole operator given by

$$O_{tG} = g_s y_t \bar{t} \sigma^{\mu\nu} T^A t G^A_{\mu\nu}, \tag{4}$$

and O_{HG}, O_{AG} in Eq. (1)-Eq. (2). We include also Yukawa couplings between H, A and top quark.

If the full theory is known, one can determine the Wilson coefficient by explicitly computing the vertex in the full theory. For example, if this contribution is induced by a vector-like fermion, F, via the following Yukawa interactions:

$$L_{Yuk} = y_F \bar{F} F H + \tilde{y}_F F i\gamma^5 F A \,, \tag{5}$$

the matching of O_{HG} and O_{AG} coefficients can be performed in the usual way by identifying the triagle diagram with the vertex at LO and with an extra loop at NLO, and the O_{tG} can be matched with the two-loop Barr-Zee diagram [5], resulting in the following identification

$$\frac{C_{HG}(m_F)}{\Lambda} = -\frac{y_F}{48\pi^2 m_F} - \frac{11\alpha_s y_F}{192\pi^3 m_F} + O(\alpha_s^2) \,, \tag{6}$$

$$\frac{C_{A\tilde{G}}(m_F)}{\Lambda} = -\frac{\tilde{y}_F}{32\pi^2 m_F} + O(\alpha_s^2) \,, \tag{7}$$

$$\frac{C_{tG}(m_F)}{\Lambda} = -\frac{\alpha_s}{1152\pi^3 m_F}\left(8y_F - 9\tilde{y}_F \frac{\tilde{y}_t}{y_t}\right) + O(\alpha_s^2). \tag{8}$$

The matching procedure is obviously dependent on the specific underlying model.

On the other hand, the running and mixing of the effective coefficients are model-independent. We perform the calculation at the renormalisation scale μ, conventionally chosen at $\mu = m_H/2$ ($m_A/2$). In case the scale m_F and μ are well separated, one needs to run the EFT from m_F down to μ to resum the large logarithmic contributions. For $O(\alpha_s)$ mixing this is done by solving the RG equations

$$\frac{dC_i(\mu)}{d\log\mu} = \frac{\alpha_s(\mu)}{\pi}\gamma_{ij} C_j(\mu) \,, \tag{9}$$

where for the three operator coefficients $C_{HG}, C_{A\tilde{G}}, C_{tG}$, the matrix γ_{ij} is given by

$$\gamma = \begin{pmatrix} 0 & 0 & 0 \\ 0 & 0 & 0 \\ -1 & \tilde{y}_t/y_t & 1/3 \end{pmatrix}. \tag{10}$$

The coefficients at a given scale μ are thus given by

$$C_i(\mu) = \exp\left(\frac{-2}{\beta_0}\log\frac{\alpha_s(\mu)}{\alpha_s(m_F)}\gamma_{ij}\right) C_j(m_F) \,, \tag{11}$$

where $\beta_0 = 11 - 2/3 n_f$, and $n_f = 5$ is the number of running flavors.

The above procedures take into account the $O(\alpha_s)$ mixing from $O_{HG,A\tilde{G}}$ to O_{tG}. The mixing from O_{tG} to O_{HG} also exists, but is of order $O(y_t^2)$, and is negligible because $|C_{tG}| \ll |C_{HG}|$ (C_{tG} is two-loop induced).

Once $C_i(\mu)$ is known, NLO predictions based on the EFT can be computed. In this work we use the automated framework based on MG5_aMC. Calculations at NLO with higher-dimensional operators have been recently performed for the top-quark sector of the SMEFT [6–10].

The approach described above provides the exact NLO QCD prediction for the EFT, which is a good approximation if the ggH (ggA) vertex is induced by heavy particles with masses larger than $m_{H(A)}/2$. For lighter particles, in particular the top quark which always couples directly to the scalar, the EFT is not valid anymore, since it fails to capture the absorptive part of the loop, which is crucial for describing the lineshape. To partially improve this situation we perform an event-by-event reweighting [11] procedure using the ratio of Born amplitudes $|M_{\rm exact}|^2/|M_{\rm EFT}|^2$ as ratio, with complex effective coefficients to avoid divergent ratios for the interference.

3 Benchmarks and results

To study the lineshape, we will perform the calculation for several benchmark models, which cover both the resolved and unresolved cases. The impact of QCD corrections in our EFT approach will be illustrated by the first two benchmark models, where the $ggH(A)$ vertex is not resolved. In the first scenario, we consider a CP-even scalar that couples to a heavy vector-like fermion doublet, which induces the ggH vertex at one loop. The matching procedure discussed in the previous section can be explicitly carried out. In the second model, we consider a CP-odd scalar as a pseudo-Goldstone boson of new strong dynamics, whose decay into top-quarks may shed light on an eventual dynamical fermions mass generation [12, 13].

For our final benchmark scenarios we consider the CP-even and CP-odd Higgses of the 2HDM. In this case the top-loop induced $ggH(A)$ vertex is fully resolved as only top and bottom quarks run in the loops. For these benchmarks we construct the EFT and improve our predictions by Born-reweighting.

For the results shown here we fix our EFT scale at one half of the scalar resonance mass, while for renormalisation and factorisation scales we choose a dynamical scale equal to one half of the sum of the transverse masses in the final state. We vary the scales independently by a factor of two up and down to estimate the scale uncertainties. The Yukawa coupling between the H and the top quark is renormalised by $\overline{MS}$ scheme but fixed at the scale $m_H/2$. We use the LO and NLO NNPDF2.3 parton distribution functions [14] for the corresponding computation. We obtain results both at fixed-order and also matched to the parton shower with MC@NLO [15]. For parton shower we always use PYTHIA8 [16].

3.1 Benchmark A

Consider a vector-like quark doublet, F, with Yukawa coupling to a CP-even scalar H described by Eq. (5). We choose the following parameters:

$$m_H = 500 \text{ GeV}, \ \Gamma_H = 40 \text{ GeV}, \ m_F = 500 \text{ GeV}, \tag{12}$$

$$y_t = 0.4, \ y_F = 5. \tag{13}$$

The ggH vertex from the F fermion running in the loop is given by Eq. **??**, multiplied by a factor of two. For the top-quark loop, we simply replace m_F by m_t and y_F by $y_t/\sqrt{2}$. This does not fully capture the top-loop contribution, in particular the phase, however given that the dominant contribution is coming from the F fermion, the EFT is still a very good approximation.

Using Eqs. (6), (8), and (11), we find the following operator coefficients, at scale $\mu_{EFT} = m_H/2$:

$$\frac{C_{HG}}{\Lambda} = -5.11 \times 10^{-5} \text{ GeV}^{-1}, \tag{14}$$

$$\frac{C_{tG}}{\Lambda} = -1.40 \times 10^{-6} \text{ GeV}^{-1}, \tag{15}$$

where we have taken m_t to be 172.5 GeV. The above values define our EFT.

In the left panel of Figure 2 we show the fixed order $m_{t\bar{t}}$ distribution. Signal and interference lineshapes are displayed separately. The signal peaks at the resonant mass 500 GeV, and the radiative correction is quite large as expected. In particular, we observe an increasing K-factor as the $t\bar{t}$ invariant mass moves away from the resonance to lower values. On the right panel, the same results are shown but matched to the PS. Compared with the fixed order result, the size of the radiative corrections is now tamed below the resonance, mainly because the emission from decay products is partly taken into account by the PS. The K-factor for the signal is in general flat, with a small enhancement near

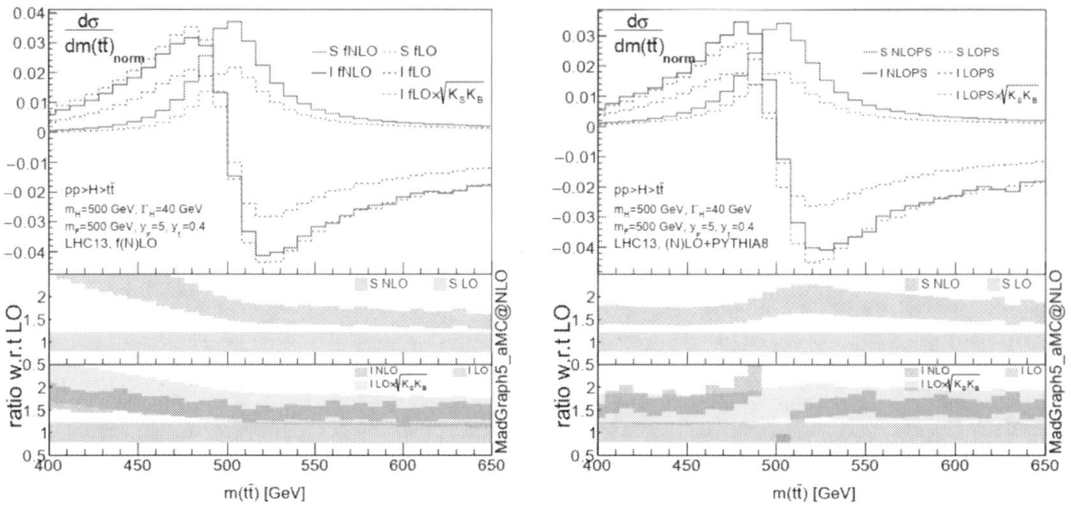

Figure 2. Signal and interference lineshapes at at (N)LO fixed order (*left*) and matched with PS (*right*) for LHC 13 TeV. Results are normalised to the SM fNLO lineshape. Lower panels show the K-factors and the scale uncertainties of signal and interference respectively.

the peak. In contrast, the K-factor for the interference shows a peak-dip structure, mainly because the zero point has been slightly shifted. This effect is not reproduced by a K-factor rescaling used in Ref. [3]. As one can see in the last panel, the inferred K-factor $\sqrt{K_S K_B}$ is flat and does not reproduce the peak-dip structure, and as a result this approximation will predict an interference lineshape that is shifted to the left compared with our full calculation. Finally, compared with the fixed-order results, the showered lineshape has a slightly lower peak due to smearing effects.

3.2 Benchmark B

As a benchmark of CP-odd state we consider the models of partial compositeness of Ref. [13], more specifically the a state of M3 model with $(n_\psi, n_\chi) = (-4, 2)$ in the η' decoupling limit, $\alpha = \zeta$. We choose the compositeness scale $f = 800$ GeV and the mass $m_a = 1$ TeV. The relevant couplings of $A \equiv a$ in Eq. (3) are [1]:

$$\frac{C_{A\tilde{G}}}{\Lambda} = -2.15308 \times 10^{-5}\ \mathrm{GeV}^{-1}\,, \tag{16}$$

$$\tilde{y}_t = -0.571406\,. \tag{17}$$

The $O_{A\tilde{G}}$ operator is generated through the axial-vector anomaly which couples the technipions to a pair of gauge bosons. We assume

$$C_{tG}(1\ \mathrm{TeV}) = 0\,. \tag{18}$$

The total width is dominated by the top decay and has some correction from di-gluon decay and is given by

$$\Gamma_A = 37.5\ \mathrm{GeV}\,. \tag{19}$$

[1]The conversion from parameters defined in Ref. [13] to our convention in Eq. (3) is given by: $\frac{C_{A\tilde{G}}}{\Lambda} = \frac{\kappa_g}{16\pi^2 f_\pi^2}$ and $\tilde{y}_t = C_t \frac{m_t}{f_\pi}$.

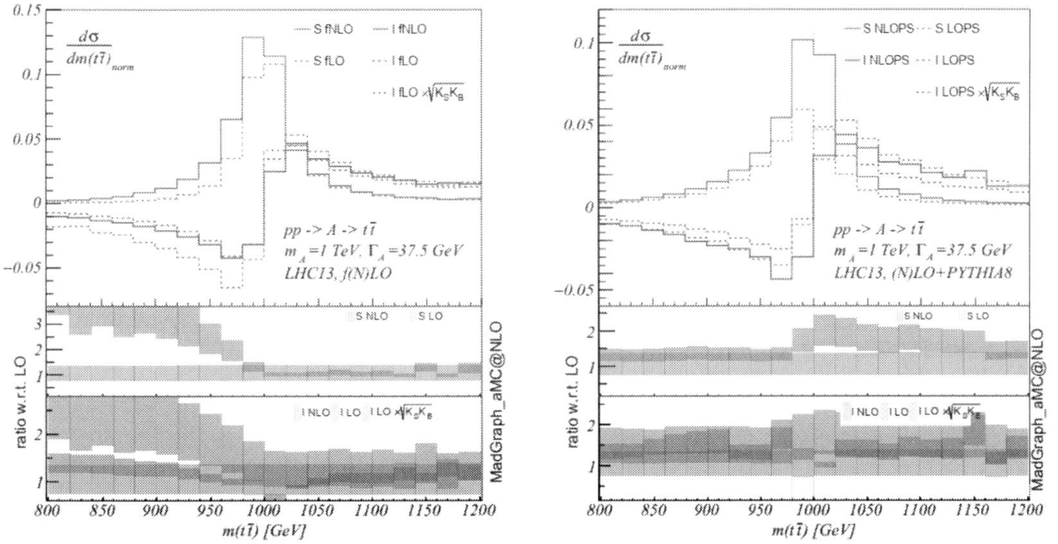

Figure 3. Signal and interference lineshapes at (N)LO fixed order (*left*) and matched with PS (*right*) for LHC 13 TeV. Results are normalised to the SM fNLO lineshape. Lower panels show the K-factors and the scale uncertainties of signal and interference respectively.

It is interesting to note that $C_{A\tilde{G}}$ and $\tilde{y}_t$ have the same sign and this implies that the top coupling has an opposite contribution to the gluon coupling, creating a dip-peak structure instead of the usual peak and dip. This opposite contribution also generates some strong cancellations if one approximates the full top loop with a heavy fermion, as we have done for benchmark A, thus for this benchmark we use only the effective coefficient generated by the high scale physics and neglect the top loop in the gluon coupling.

In Figure 3 we show the invariant mass distribution of the top pair system $m_{t\bar{t}}$ at fixed order in perturbation theory in the left and the result matched with PS in the right. In the upper panel the NLO and LO predictions are shown normalised bin-by-bin to the background QCD prediction.

3.3 Benchmarks C1 and C2

For our final BSM scenario we employ the 2HDM [17], which introduces a second $SU(2)_L$ doublet Φ_2 and gives rise to five physical Higgs bosons: one light (heavy) neutral, CP-even state h (H); one neutral, CP-odd state A; and two charged Higgs bosons $H^{\pm}$. In this work we take h to be the 125 GeV Higgs. The input parameters determining all properties of a 2HDM scenario are:$\tan\beta$, $\sin\alpha$, m_h , m_H , m_A , $m_{H^\pm}$, m_{12}^2 with the convention $0 \leq \beta - \alpha < \pi$ (with $0 < \beta < \pi/2$).

The ratio of the Yukawa couplings over the SM values are $\sin\alpha/\sin\beta$ and $\cot\beta$ for H and A respectively. We chose two benchmark scenarios whose parameters are listed in Table (1). Electroweak precision tests, the LHC Higgs results and searches for heavy scalar particles, along with unitarity, perturbativity and vacuum stability constrain the parameter space of the model. In the selection of 2HDM benchmarks, these constraints are taken into account.

	Type	$\tan\beta$	$\sin(\beta-\alpha)$	m_H	m_A	$m_{H^\pm}$	m_{12}^2
C1	I	2.0	1.0	300	450	450	20000
C2	II	0.9	1.0	450	600	620	10000

Table 1. Parameter choices for the 2HDM benchmarks used in our study. All masses are given in GeV. The light Higgs mass is fixed to $m_h = 125$ GeV.

In benchmark C1 only the pseudoscalar Higgs lies above the top–anti-top threshold. The corresponding Yukawa couplings (as rescalings of the SM Higgs top Yukawa) are: $g_t^H = -0.5$ and $g_t^A = 0.5$. The heavy scalar width is negligible while the pseudoscalar width is $\Gamma_A = 7.35$ GeV and the top quark decay branching fraction is approximately 65%.

Following Eqs. (6), (8), and (11), we find the following operator coefficients, at scale $\mu_{EFT} = m_A/2$:

$$\frac{C_{HG}}{\Lambda} = 4.73 \times 10^{-6}\ \text{GeV}^{-1}\,, \tag{20}$$

$$\frac{C_{AG}}{\Lambda} = -6.49 \times 10^{-6}\ \text{GeV}^{-1}\,, \tag{21}$$

$$\frac{C_{tG}}{\Lambda} = 9.56 \times 10^{-9}\ \text{GeV}^{-1}\,. \tag{22}$$

Where we have taken m_t to be 172.5 GeV.

The corresponding couplings and widths for benchmark C2 are: $g_t^H = -1.11$, $g_t^A = 1.11$, $\Gamma_H = 10.7$ GeV and $\Gamma_A = 38.7$ GeV. Both resonances decaying almost exclusively to top quarks with the relevant branching fractions being $Br(H \to t\bar{t}) = 0.995$ and $Br(A \to t\bar{t}) = 0.921$.

The operator coefficients at scale $\mu_{EFT} = m_A/2$ are:

$$\frac{C_{HG}}{\Lambda} = 1.05 \times 10^{-5}\ \text{GeV}^{-1}\,, \tag{23}$$

$$\frac{C_{AG}}{\Lambda} = -1.44 \times 10^{-5}\ \text{GeV}^{-1}\,, \tag{24}$$

$$\frac{C_{tG}}{\Lambda} = 4.11 \times 10^{-8}\ \text{GeV}^{-1}\,. \tag{25}$$

The above values define our EFT. For these scenarios, the EFT is not expected to accurately describe the lineshape as the top-loop is resolved. In order to improve the result we reweight event-by-event with the full one-loop LO amplitude $|M_{\text{exact}}|^2/|M_{\text{EFT}}|^2$. To carry out this procedure we must avoid division by zero and for that we incorporate a phase to the coefficients $\frac{C_{AG}}{\Lambda} \to (a+bi) \times \frac{C_{AG}}{\Lambda}$, where a and b are extracted by matching the exact and EFT amplitudes at leading order in particular at the point where they cross zero.

Differential results for the 2HDM benchmarks C1 and C2 are shown in Figure 4 left and right respectively. The invariant mass distribution of the top quark pair is computed at (N)LO+PS, with the signal and interference contributions and their corresponding K-factors displayed separately. The LO results are exact, i.e. they include the full top mass dependence, while for the NLO results we use the phase-improved EFT to compute the NLO corrections and then apply a Born-reweighting at an event-by-event level. As the reweighting is based on event generation we present only (N)LO+PS for these benchmarks. Results obtained using the approximate average K-factor, $\sqrt{K_S K_B}$, are also shown. For both benchmarks the signal K-factors peak around the resonant masses. The interference

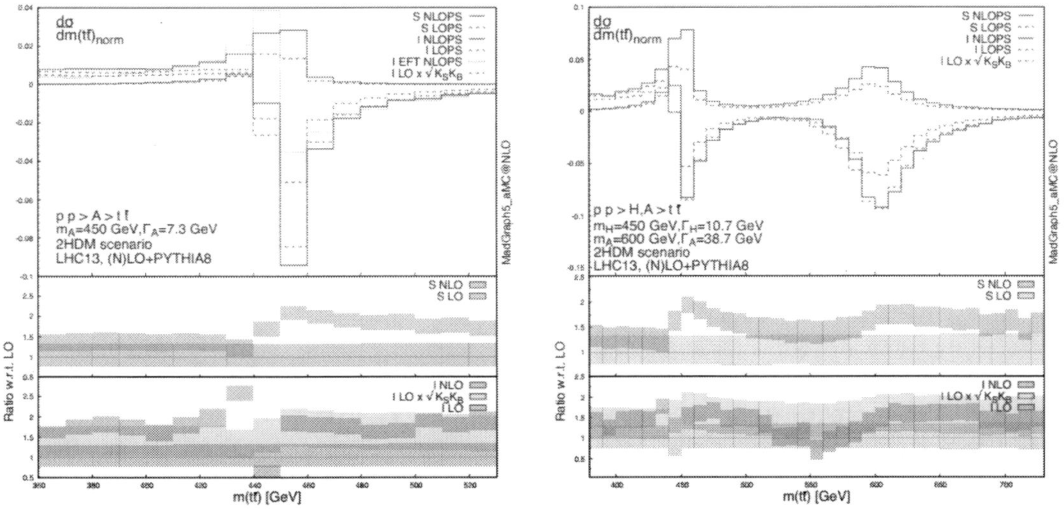

Figure 4. Signal and interference lineshapes at (N)LO matched with PS for LHC 13 TeV. Results are normalised to the SM fNLO lineshape. Lower panels show the K-factors and the scale uncertainties of signal and interference respectively.

K−factor shows a peak-dip structure as the NLO corrections shift the zero crossing point from its LO position. The approximate K-factor, $\sqrt{K_S K_B}$ is much flatter, with small peaks driven by the peaks in the signal K-factor at the resonance masses. Moreover, the uncertainties of that prediction are much larger as they are by construction LO.

For benchmark C1 we also show in yellow the prediction obtained for the interference in the EFT limit, i.e. without any reweighting. As expected the EFT prediction does not a provide a good description of the interference lineshape.

We stress that for scenarios where two resonances are present the experimental resolution will be crucial to establish whether one or two peaks are visible in the invariant mass distribution. For our benchmark the mass difference of 150 GeV should be sufficiently large. We note that here the interference distribution suffers from low statistics in the region between the two resonance masses, where the cross-section is very small.

4 Conclusions

The NLO QCD corrections to the interference between signal and background in resonant scalar or pseudoscalar top pair production have been computed using an EFT approach without any approximations. In this way we obtained accurate predictions for the $t\bar{t}$ lineshape for the case in which gluon fusion scalar production is dominated by heavy particles running in the loop or strong dynamics generating point-like interaction between the scalars and the gluon. The QCD corrections are important and cannot be correctly described by approximations.

A non-trivial feature of the calculation is the mixing between the gluon-scalar operators and the chromo-magnetic dipole operator, which implies that both operators and their mixings must be incorporated in the computation in order to extract meaningful physical results.

In addition, to treat the cases where loops of "light" fermions, like the top-quark itself, dominate the gluon fusion scalar production and thus the EFT approach is not immediately applicable, we have shown that our calculation can still be used. For that we introduced a phase to the Wilson coefficients to reproduce the absorptive part of the one-loop amplitude, and by reweighting the amplitudes using the exact LO amplitudes we obtained results beyond LO.

Our calculation is fully automated and can be matched to PS programs through the MG5_aMC program. We have presented results for a set of benchmark scenarios including pseudo-scalars and scalars in resolved and unresolved cases. We found that QCD corrections are important, increasing the yields and modifying the distribution with non-trivial K-factors, as well as reducing the scale uncertainties of the predictions.

We expect our predictions, combined with advances in experimental strategies, will improve the sensitivity in searches for BSM scalars in $t\bar{t}$, by reducing the systematical uncertainty from the theory side, and by optimising the experimental strategies according to an accurate lineshape description. Moreover, they will also allow the extraction of more reliable constraints on the parameters of various BSM models.

The reader is encouraged to read our full work in Ref. [1], where we give more details on the calculation and more extensive discussion, for instance about the impact of the uncertainty due to the treatment of the interference contribution in the matching to PS simulation and matching between underlying theory and EFT parameters.

References

[1] D. Buarque Franzosi, E. Vryonidou, C. Zhang (2017), 1707.06760

[2] W. Bernreuther, P. Galler, C. Mellein, Z.G. Si, P. Uwer, Phys. Rev. **D93**, 034032 (2016), 1511.05584

[3] B. Hespel, F. Maltoni, E. Vryonidou, JHEP **10**, 016 (2016), 1606.04149

[4] J. Alwall, R. Frederix, S. Frixione, V. Hirschi, F. Maltoni et al., JHEP **1407**, 079 (2014), 1405.0301

[5] S.M. Barr, A. Zee, Phys. Rev. Lett. **65**, 21 (1990), [Erratum: Phys. Rev. Lett.65,2920(1990)]

[6] C. Degrande, F. Maltoni, J. Wang, C. Zhang (2014), 1412.5594

[7] D. Buarque Franzosi, C. Zhang, Phys. Rev. **D91**, 114010 (2015), 1503.08841

[8] C. Zhang, Phys. Rev. Lett. **116**, 162002 (2016), 1601.06163

[9] O. Bessidskaia Bylund, F. Maltoni, I. Tsinikos, E. Vryonidou, C. Zhang, JHEP **05**, 052 (2016), 1601.08193

[10] F. Maltoni, E. Vryonidou, C. Zhang, JHEP **10**, 123 (2016), 1607.05330

[11] O. Mattelaer, Eur. Phys. J. **C76**, 674 (2016), 1607.00763

[12] T. Alanne, M.T. Frandsen, D. Buarque Franzosi, Phys. Rev. **D94**, 071703 (2016), 1607.01440

[13] A. Belyaev, G. Cacciapaglia, H. Cai, G. Ferretti, T. Flacke, A. Parolini, H. Serodio, JHEP **01**, 094 (2017), 1610.06591

[14] R.D. Ball, V. Bertone, S. Carrazza, L. Del Debbio, S. Forte, A. Guffanti, N.P. Hartland, J. Rojo (NNPDF), Nucl. Phys. **B877**, 290 (2013), 1308.0598

[15] S. Frixione, B.R. Webber, JHEP **06**, 029 (2002), hep-ph/0204244

[16] T. Sjostrand, S. Mrenna, P.Z. Skands, Comput. Phys. Commun. **178**, 852 (2008), 0710.3820

[17] G. Branco, P. Ferreira, L. Lavoura, M. Rebelo, M. Sher et al., Phys.Rept. **516**, 1 (2012), 1106.0034

The parameters of the fermion-mixing and search restriction on a mass leptoquark.

Aleksander Povarov[1,⋆]

[1] *Yaroslavl Higher Military Institute of the Air Defense*

Abstract. It is shown that the inclusion of fermion mixing in the framework of four-colour quark-lepton symmetric model can be greatly reduces the restrictions on the mass vector leptoquark. The lower limit for the mass of the vector leptoquark obtained from processes $K_L^0 \to l_i^+ l_j^-$, $B^0 \to l_i^+ l_j^-$ and $B_s \to l_i^+ l_j^-$ can be of the order of 85 TeV.

1 Introduction

The search for a new physics beyond the Standard Model (SM) is now one of the aims of the high energy physics. One of the possible variants of such new physics can be the variant induced by the possible four color symmetry [1] between quarks and leptons. This symmetry predicts the existence of leptoquarks. Leptoquarks are color triplet bosons of spin 0 or 1, carrying lepton (L) and barion (B) number and fractional electric charge. They couple to lepton-quark pairs and appear in many extensions of the Standart Model (SM) which unify lepton and quark in one multiplet in the framework Grand United Theory (GUT) on grounds that $SU(5)$ [2] and $SU(4)$ [1], superstring-inspired $E6$ model and other models. We study leptoquarks in the minimal extension SM with four color symmetry, that is Minimal Quark-Lepton Symmetry model (MQLS model)[3, 4]. The MQLS model predicted the existence of the vector leptoquarks with masses on the order of the mass scale, which $SU_c(4)$ symmetry is broken. The mass limits for the vector leptoquarks are well known and the most stringent of them are the indirect mass limits resulted from $K_L^0 \to e^\pm\mu^\mp$ decay and they are of order of 10^3 TeV [5–7]. However, these restrictions can be lowered if consider additional fermion mixing, in this case masses can be order 40 TeV [8, 9] This paper describes constraints on the masses vector leptoquarks from the current experimental data on the leptonic decays K_L^0-, B^0-, B_s- mesons.

2 Model

The MQLS model [3, 4] based on the group

$$G = SU_V(4) \times SU_L(2) \times U_R(1).$$

In the MQLS model the basic left- (L) and right- (R) handed quarks $Q'^{L,R}_{ia\alpha}$ and leptons $l'^{L,R}_{ia}$ form the fundamental quartets of $SU(4)$ color group, and can be written, in general, as superpositions of the

⋆e-mail: povarov@uniyar.ac.ru

quark and lepton mass eigenstates $Q^{L,R}_{ia\alpha}$ and $l^{L,R}_{ia}$

$$\psi^{L,R}_{ia\alpha} = Q'^{L,R}_{ia\alpha} = \sum_j \left(A^{L,R}_{Q_a}\right)_{ij} Q^{L,R}_{ja\alpha}, \qquad \psi^{L,R}_{ia4} = l'^{L,R}_{ia} = \sum_j \left(A^{L,R}_{l_a}\right)_{ij} l^{L,R}_{ja},$$

where i=1,2,3 are the generation indices, $a = 1, 2$ are the $SU_L(2)$ indices and $A = \alpha, 4 - SU_V(4)$ indices, $\alpha = 1, 2, 3$ are the $SU_c(3)$ color indices. The unitary matrices $A^{L,R}_{Q_a}$ and $A^{L,R}_{l_a}$ describe the fermion mixing and diagonalize the mass matrices of quarks and leptons. The combinations of these matrices give coresponding $C_Q = (A^L_{Q_1})^+ A^L_{Q_2}$ is the Cabibbo-Kobayashi-Maskawa matrix, $C_l = (A^L_{l_1})^+ A^L_{l_2}$ is the Pontecorvo-Maki-Nakagawa-Sakata matrix that is analogous to the preceding one in the lepton sector and is not diagonal, this evident from neutrino oscillation, and

$$K^{L,R}_a = (A^{L,R}_{Q_a})^+ A^{L,R}_{l_a}$$

unitary mixing matrices that are due to the possible distinctions between the quarks and leptons mixing matrices $A^{L,R}_{Q_a}$ and $A^{L,R}_{l_a}$.

The lagrangian interaction vector leptoquarks with down fermions can be written in the following form

$$L_{Vdl} = \frac{g_4}{\sqrt{2}} (\bar{d}_{p\alpha}[(K^L_2)_{pi}\gamma^\mu P_L + (K^R_2)_{pi}\gamma^\mu P_R] l_i) V_{\alpha\mu} + h.c.,$$

where $g_4 = g_{st}(M_c)$ is the $SU_V(4)$ gauge coupling constant, related to the strong coupling constant at the mass scale M_c of the $SU_V(4)$ symmetry breaking, $p, i = 1, 2, 3, ...$ are the quark and lepton generation indexes, $\alpha = 1, 2, 3 - SU(3)$ colour index, $P_{L,R} = (1 \pm \gamma_5)/2$.

3 The numerical results

3.1 Decays $K^0_L \to l^+_i l^-_j$

The the branching ratio of the decays $K^0_L \to l^+_i l^-_j$ with $i, j = 1, 2$, $l^\pm_i = e^\pm, \mu^\pm$ induced by the vector leptoquarks with neglect of electron and muon masses ($m_e, m_\mu << R^V_{K^0}\bar{m}_{K^0}$) in the case of the general fermion mixing can be written as [7]

$$BR(K^0_L \to l^+_i l^-_j) = \frac{m_{K^0}\pi\alpha^2_{st} f^2_{K^0} \bar{m}^2_{K^0} (R^V_{K^0})^2}{4 m^4_V \Gamma^{total}_K} |k^{(K^0)}_{ij}|^2,$$

where $\bar{m}_{K^0} = m^2_{K^0}/(m_s + m_d)$, m_s-, m_d- masses s and d quarks, $m_{K^0}-$ mass K^0_L-meson, m_V- mass vector leptoquark, Γ^{total}_K – total width decay K^0_L-meson, $f_{K^0} = 0.16$ GeV – form factor K^0_L-meson, factor $R^V_{K^0} = R_{K^0}(\mu_{K^0}, M_c)$ incorporate gluonic corrections, mixing factors incorporate fermionic mixing in general form

$$k^{(K^0)}_{ij} = \sqrt{(|\varkappa^L_{ij}|^2 + |\varkappa^R_{ij}|^2)/2},$$

$$\varkappa^{L,R}_{ij} = (K^{L,R}_2)_{2i} (\overset{*}{K}{}^{R,L}_2)_{1j} + (K^{L,R}_2)_{1i} (\overset{*}{K}{}^{R,L}_2)_{2j}.$$

Matrices additional fermion mixing $K^{L,R}_2$ can be parametrize as Cabibbo-Kobayashi-Maskawa matrix using three angles $\Theta_{12}, \Theta_{13}, \Theta_{23}$ and a phase δ for each matrix

$$K^{L,R}_2 = \begin{bmatrix} c^{L,R}_{12} c^{L,R}_{13} & s^{L,R}_{12} c^{L,R}_{13} & s^{L,R}_{13} e^{-i\delta_{L,R}} \\ s^{L,R}_{12} c^{L,R}_{23} - c^{L,R}_{12} s^{L,R}_{23} s^{L,R}_{13} e^{i\delta_{L,R}} & c^{L,R}_{12} c^{L,R}_{23} - s^{L,R}_{12} s^{L,R}_{23} s^{L,R}_{13} e^{i\delta_{L,R}} & s^{L,R}_{23} c^{L,R}_{13} \\ s^{L,R}_{12} s^{L,R}_{23} - c^{L,R}_{12} c^{L,R}_{23} s^{L,R}_{13} e^{i\delta_{L,R}} & c^{L,R}_{12} s^{L,R}_{23} - s^{L,R}_{12} c^{L,R}_{23} s^{L,R}_{13} e^{i\delta_{L,R}} & c^{L,R}_{23} c^{L,R}_{13} \end{bmatrix}.$$

where $s_{ij}^{L,R} = \sin\Theta_{ij}^{L,R}$, $c_{ij}^{L,R} = \cos\Theta_{ij}^{L,R}$.

For numerical calculations we used current experimental data [10]

$$\begin{aligned} BR(K_L^0 \to e^{\pm}\mu^{\mp}) &< 4.7\times 10^{-12}, \\ BR(K_L^0 \to \mu^{+}\mu^{-}) &= (6.8\pm 0.11)\times 10^{-9}, \\ BR(K_L^0 \to e^{+}e^{-}) &= (9^{+6}_{-4})\times 10^{-12}. \end{aligned}$$

Use this data can represent the lower limit mass of the vector leptoquark as function of the angles $\Theta_{ij}^{L,R}$ and phases $\delta^{L,R}$. Choosing these parameters we can obtain the minimal constraints on the masses of leptoquarks. So numerical calculations give the ranges of parameters where the contributions of the decays $K_L^0 \to l_i^+ l_j^-$ is zero. In particular, the choice of parameters in next values:

$$\Theta_{12}^{L} = \Theta_{13}^{L} = \Theta_{23}^{L} = \Theta_{12}^{R} = \Theta_{13}^{R} = \Theta_{23}^{R} = \pi/2, \quad \delta^{L} = \delta^{R} = 0$$

with correspondent matrices

$$K_2^L = K_2^R = \begin{bmatrix} 0 & 0 & 1 \\ 0 & -1 & 0 \\ 1 & 0 & 0 \end{bmatrix},$$

obtain $k_{21}^{(K^0)} = k_{11}^{(K^0)} = k_{22}^{(K^0)} = 0$, and experimental limits performed for all values m_V.

As shown in the case of fermion mixing into account only decays $K_L^0 \to l_i^+ l_j^-$ yields no constraints on the masses of vector leptoquarks.

3.2 Decays $B^0 \to l_i^+ l_j^-$ and $B_s \to l_i^+ l_j^-$

The branching ratio decays $B^0 \to l_i^+ l_j^-$ can be represented [7] in the form

$$BR(B^0 \to l_i^+ l_j^-) = \frac{m_{B^0}\pi\alpha_{st}^2 f_{B^0}^2 \bar{m}_{B^0}^2 (R_{B^0})^2}{4m_V^4 \Gamma_B^{total}} |k_{ij}^{(B^0)}|^2,$$

where $\bar{m}_{B^0} = m_{B^0}^2/(m_b + m_d)$, m_b-, m_d- masses b and d quarks, $f_{B^0} = 0.229$ GeV – form factor B^0-meson, factor $R_{B^0}^V = R_{B^0}(\mu_{B^0}, M_c)$ incorporate gluonic corrections, $\bar{\mu}_{l_i^\pm} = m_{l_i^\pm}/(\bar{m}_{B^0} R_{B^0}^V)$,

$$\begin{aligned} |k_{ij}^{(B^0)}|^2 &= \left[(|\beta_{ij}^L|^2 + |\beta_{ij}^R|^2)(1 - \mu_{l_i}^2 - \mu_{l_j}^2) + 2(\overset{*}{\beta_{ij}^L}\beta_{ij}^R + \beta_{ij}^L \overset{*}{\beta_{ij}^R})\mu_{l_i}\mu_{l_j}\right] \\ &\times \sqrt{[1 - (\mu_{l_i} + \mu_{l_j})^2][1 - (\mu_{l_i} - \mu_{l_j})^2]}, \end{aligned}$$

$$\beta_{ij}^{L,R} = k_{ij}^{L,R} - (\bar{\mu}_{l_j^-} k_{ij}'^{L,R} + \bar{\mu}_{l_i^+} k_{ij}'^{R,L})/2$$

$$k_{ij}^{L,R} = (K_2^{L,R})_{3i}(\overset{*}{K_2^{R,L}})_{1j}, \qquad k_{ij}'^{L,R} = (K_2^{L,R})_{3i}(\overset{*}{K_2^{L,R}})_{1j}$$

For numerical calculations, we used current experimental data [10–12]

$$\begin{aligned} BR(B^0 \to e^{\pm}\mu^{\mp}) &< 2.8\times 10^{-9}, \\ BR(B^0 \to e^{\pm}\tau^{\mp}) &< 2.8\times 10^{-5}, (*) \\ BR(B^0 \to \mu^{\pm}\tau^{\mp}) &< 2.2\times 10^{-5}, (*) \\ BR(B^0 \to \mu^{+}\mu^{-}) &< 3.4\times 10^{-10}, \\ BR(B^0 \to e^{+}e^{-}) &< 8.3\times 10^{-8}, \\ BR(B^0 \to \tau^{+}\tau^{-}) &< 2.1\times 10^{-3}. \end{aligned}$$

The joint fitting parameters of the decays of K_L^0 and B^0 mesons gives some ranges of parameters where the lower limit of the mass vector leptoquark is minimal.

For example if set parameters as

$$\Theta_{12}^L = \Theta_{12}^R = \pi/4, \quad \Theta_{13}^L = \Theta_{13}^L = \pi/2, \quad \Theta_{23}^L = \Theta_{23}^R = 0, \quad \delta^L = \delta^R = \pi/2,$$

then

$$K_2^L = K_2^R = \begin{pmatrix} 0 & 0 & -i \\ \frac{-i}{\sqrt{2}} & \frac{-i}{\sqrt{2}} & 0 \\ \frac{i}{\sqrt{2}} & \frac{-i}{\sqrt{2}} & 0 \end{pmatrix}$$

in this case $k_{ij}^{(K^0)} = 0$ and contribution K_L^0–meson to lower limit mass vector leptoquark negligible, but contribution B^0 meson are considerable (contribution give $B^0 \to e^\pm\tau^\mp$ and $B^0 \to \mu^\pm\tau^\mp$ decays)

$$k_{ij}^{(B^0)} = \begin{pmatrix} 0 & 0 & 0.61 \\ 0 & 0 & 0.62 \\ 0 & 0 & 0 \end{pmatrix}.$$

The restriction on the lower limit mass vector leptoquarks obtained from decays K_L^0 and B^0 mesons together are 10 TeV.

Include consideration of experimental data on the decays of B_s mesons [10–12]

$$\begin{aligned} BR(B_s \to e^\pm\mu^\mp) &< 1.1 \times 10^{-8}(*) \\ BR(B_s \to \mu^+\mu^-) &= (3.0^{+0.67}_{-0.63}) \times 10^{-10}(*) \\ BR(B_s \to e^+e^-) &< 8.3 \times 10^{-8} \\ BR(B_s \to \tau^+\tau^-) &< 6.8 \times 10^{-3} \end{aligned}$$

increases constraints on the masses vector leptoquarks.

For example if set parameters as

$$\Theta_{12}^L = \Theta_{12}^R = 9\pi/50, \quad \Theta_{13}^L = \Theta_{13}^R = \pi/2, \quad \Theta_{23}^L = \Theta_{23}^R = 0, \quad \delta^L = \delta^R = 0,$$

then

$$K_2^L = K_2^R = \begin{pmatrix} 0 & 0 & 1 \\ -0.54 & 0.84 & 0 \\ -0.84 & -0.54 & 0 \end{pmatrix}.$$

In this case

$$k_{ij}^{(K^0)} = \begin{pmatrix} 0 & 0 \\ 0 & 0 \end{pmatrix}$$

and contribution K_L^0 neglible,

$$k_{ij}^{(B^0)} = \begin{pmatrix} 0 & 0 & 0.87 \\ 0 & 0 & 0.35 \\ 0 & 0 & 0 \end{pmatrix}$$

and contribution B^0 small,

$$k_{ij}^{(B_s)} = \begin{pmatrix} 0.41 & 0.99 & 0 \\ 0.16 & 0.4 & 0 \\ 0 & 0 & 0 \end{pmatrix}$$

contribution B_s from $B_s \to e^{\pm}\mu^{\mp}$ and $B_s \to \mu^+\mu^-$ decays significant.

The restriction on the mass vector leptoquark obtained from decays K_L^0, B^0 and B_s mesons together are 85 TeV.

4 Conclusion

Accounting fermion mixing in leptonic decays of K_L^0, B^0 and B_s mesons leads to a significant weakening of restrictions on the lower limit mass vector leptoquark. The current limit is about 85 TeV, it is essentially weaker, than restriction without taking into account fermion mixing. This result is in good agreement with the result of prof. A.D. Smirnov presented in this seminar. The main result is to avoid strong constraints from decay $K_L^0 \to e^{\pm}\mu^{\mp}$ on the masses of vector leptoquarks. Also the result does not contradict the experimental constraints on the vector leptoquark type of Pati-Salam from $B_s \to e^{\pm}\mu^{\mp}$ and $B^0 \to e^{\pm}\mu^{\mp}$ decays [13].

Thank prof. A.D. Smirnov for useful discussions. Would like to express his deep gratitude to the organizers of the International Seminar for their warm hospitality.

References

[1] J. C. Pati, A. Salam, Phys. Rev. D **10**, 275 (1974).
[2] H. Georgi, S.L. Glashow Phys. Rev.Lett. 32, 438 - 441 (1974).
[3] A. D. Smirnov, Phys. Lett. B **346**, 297 (1995).
[4] A. D. Smirnov, YaF. **58**, 2252 (1995). (Physics of Atomic Nuclei **58**, (1995).)
[5] G. Valencia, S. Willenbrock, Phys. Rev. D **50** , 6843 (1994).
[6] A.V. Kuznetsov, N.V. Mikheev Phys.Lett.B **329**, 295 (1994).
[7] A. D. Smirnov, Mod.Phys. A **22**, 2353 (2007)
[8] A.V. Kuznetsov, N.V. Mikheev, A.V. Serghienko Int. J. Mod. Phys.Lett.A **27**, 1250062 (2012).
[9] A.V. Povarov and A.D. Smirnov, talk presented at the Conference "Physics of Fundamental Interactions", ITEP, Moscow, November 2011 (unpublished).
[10] C. Patrighani et al., Chin. Phys. C **40** 100001 (2016) and (2017)
[11] R. Aaij et al., Phys. Rev. Lett **118**, 191801 (2017).
[12] R. Aaij et al., Phys. Rev. Lett **118**, 251802 (2017).
[13] R. Aaij et al., Phys. Rev. Lett **111**, 141801 (2013) .

Missing energy signature for low scale supersymmetry breaking

S.V. Demidov[1,2,⋆] and I.V. Sobolev[1,3,4]

[1] *Institute for Nuclear Research of the Russian Academy of Sciences, 60th October Anniversary prospect 7a, Moscow 117312, Russia*
[2] *Moscow Institute of Physics and Technology, Institutsky per. 9, Dolgoprudny 141700, Russia*
[3] *M.V. Lomonosov Moscow State University, Vorobjevy Gory, 119991, Moscow, Russia*
[4] *DESY, Notkestraße 85, 22607 Hamburg, Germany*

Abstract. If the supersymmetry is broken at a scale about several TeVs particles from the sector responsible for the supersymmetry breaking - goldstino and sgoldstino - can reveal themselves already at the LHC experiments. We discuss bounds on supersymmetry breaking scale from LHC searches for jet and missing energy signature.

1 Introduction

Supersymmetry [1] still remains one of the attractive extensions of the SM and one of the main scenario of new physics which is extensively studied at the LHC experiments. If supersymmetry is indeed relevant to our Nature it should be spontaneously broken in some hidden sector with no direct interactions with the visible one. Transmission of this supersymmetry breaking can proceed in different ways. For example, it can be mediated by gravity or gauge interactions. This results in appearance of soft terms which break supersymmetry explicitly. According to supersymmetric analogue of Goldstone theorem there should exist massless fermionic degree of freedom which is called goldstino. In the simplest case goldstino is a part of a singlet chiral superfield $\left(\phi, \psi, F_\phi\right)$. Here ψ is a massless goldstino, ϕ is its scalar superpartner, sgoldstino, and F_ϕ is an auxiliary field. In the case of spontaneously broken supersymmetry the auxiliary field gets nonzero v.e.v., F and quantity $\sqrt{F}$ is called the scale of supersymmetry breaking. In supergravity theories goldstino becomes [4–6] a longitudinal part of spin-$\frac{3}{2}$ particle – gravitino, with the mass $m_{3/2} = \sqrt{\frac{8\pi}{3}} \frac{F}{M_{Pl}}$.

If gravitino is very heavy then the effective theory near the electroweak scale is just the MSSM endowed with the soft terms. However models with relatively light gravitino (i.e. with a mass of several keV) also exist and are phenomenologically viable (see e.g. models with gauge mediated SUSY breaking [7–12] or no-scale supergravity [13–15] where $\sqrt{F} \sim 10 - 100$ TeV). Here we consider the case when $\sqrt{F}$ is of order of several TeVs which is called low scale supersymmetry breaking. This can be realized in warped [16, 17] SUSY models and composite [3] models. According to the equivalence theorem [22–24] the effective lagrangian describing interactions of ultralight gravitino with the MSSM fields can be constructed using properties of goldstino.

Several aspects of collider phenomenology of low scale supersymmetry breaking have been studied for instance in [19, 25–30]. Gravitino production is one of the most distinct signatures of this

⋆e-mail: demidov@ms2.inr.ac.ru

setup [31–38]. Here we discuss missing energy signature of the process of gravitino pair production at the LHC. It has been recently studied in [39] where use ATLAS run-I data to obtain bounds on gravitino mass or equivalent supersymmetry breaking scale. However, in that work the limit of very heavy sgoldstino (with mass about 20 TeV) was considered. The goal of our work [40] is to include possibility light sgoldstinos.

2 Model

The model is described by the following lagrangian

$$\mathcal{L}_\Phi = \int d^2\theta\, d^2\bar{\theta} K(\Phi^\dagger, \Phi) + \left(\int d^2\theta F\Phi + \text{h.c.} \right), \tag{1}$$

where $\Phi = \phi + \sqrt{2}\theta\psi + \theta^2 F_\phi$ is goldstino chiral superfield and the Kahler potential $\tilde{K}$ is chosen as

$$\tilde{K}(\Phi^\dagger, \Phi) = \Phi^\dagger\Phi - \frac{m_s^2 + m_p^2}{8F^2}\left(\Phi^\dagger\Phi\right)^2 - \frac{m_s^2 - m_p^2}{12F^2}\Phi^\dagger\Phi\left(\Phi^2 + (\Phi^\dagger)^2\right). \tag{2}$$

Corresponding scalar potential has a local minimum where $\langle s \rangle = \langle p \rangle = 0$ and its expansion around the local minimum gives proper mass term for CP-even (s) and CP-odd (p) sgoldstinos

$$V(s,p) = F^2 + \frac{m_s^2 s^2}{2} + \frac{m_p^2 p^2}{2} + O\left(\frac{1}{F^2}\right) \tag{3}$$

Auxiliary fields F_ϕ and F_ϕ^* acquire non-zero vacuum expectation values F.

Here we are interested in interactions between invisible and QCD sectors of the MSSM. Spurion method [41] allows to describe this interaction by the following lagrangian

$$\begin{aligned}\mathcal{L}_{\Phi-\text{vis}} = &-\int d^2\theta d^2\theta^\dagger \frac{M^2_{\tilde{q}_L,ij}}{F^2}\Phi^\dagger\Phi Q^\dagger_{L,i} e^{2gV} Q_{L,j} - \int d^2\theta d^2\theta^\dagger \frac{M^2_{\tilde{u}_R,ij}}{F^2}\Phi^\dagger\Phi U^\dagger_{R,i} e^{2gV} U_{R,j} - \\ &- \int d^2\theta d^2\theta^\dagger \frac{M^2_{\tilde{d}_R,ij}}{F^2}\Phi^\dagger\Phi D^\dagger_{R,i} e^{2gV} D_{R,j} + \left(\frac{M_3}{2F}\int d^2\theta\, \Phi W^\alpha W_\alpha + \text{h.c.}\right).\end{aligned} \tag{4}$$

Apart from soft quark and gluino masses the above lagrangian describes interactions of goldstino with the MSSM fields. In particular it contains the following operators obtained by expansion of component field lagrangian in powers of $1/F$

$$\begin{aligned}\mathcal{L}_{\psi-\text{vis}} \supset &\frac{M_3}{4\sqrt{2}F}\bar{\psi}\left[\gamma^\mu,\gamma^\nu\right]\lambda^a F^a_{\mu\nu} - \frac{iM^2_{\tilde{d}_R,ij}}{F}\left(\tilde{d}^\dagger_{R,i}\bar{\psi}P_R d_j - \bar{d}_j P_L\psi\tilde{d}_{R,i}\right) - \\ &- \frac{iM^2_{\tilde{u}_R,ij}}{F}\left(\tilde{u}^\dagger_{R,i}\bar{\psi}P_R u_j - \bar{u}_j P_L\psi\tilde{u}_{R,i}\right) + \frac{iM^2_{\tilde{q}_L,ij}}{F}\left(\tilde{q}^\dagger_{L,i}\bar{\psi}P_L q_j - \bar{q}_j P_R\psi\tilde{q}_{L,i}\right) - \\ &- \frac{M^2_{\tilde{d}_R,ij}}{F^2}(\bar{\psi}P_R d_i)(\bar{d}_j P_L\psi) - \frac{M^2_{\tilde{u}_R,ij}}{F^2}(\bar{\psi}P_R u_i)(\bar{u}_j P_L\psi) - \frac{M^2_{\tilde{q}_L,ij}}{F^2}(\bar{\psi}P_L q_j)(\bar{q}_i P_R\psi),\end{aligned} \tag{5}$$

which we use in our study along with the following interactions of sgoldstinos

$$\mathcal{L}_{\phi-\psi} = \frac{m_s^2}{2\sqrt{2}F}s\bar{\psi}\psi + i\frac{m_p^2}{2\sqrt{2}F}p\bar{\psi}\gamma_5\psi - \frac{M_3}{2\sqrt{2}F}sF^{\mu\nu}_a F^a_{\mu\nu} + \frac{M_3}{2\sqrt{2}F}pF^{\mu\nu}_a \tilde{F}^a_{\mu\nu} \tag{6}$$

with gravitino and gluons.

3 Light gravitino production at the LHC: monojet signature

Here we describe main subprocesses which result in missing energy signature, discuss sgoldstino contribution and obtain bounds on parameter space of the model using the LHC run-I data. Our task is to extend the analysis of [39] and include possibility of light sgoldstinos. For the present study we use the CMS data at $\sqrt{s} = 8$ TeV and 19.6 fb^{-1} and perform leading order (LO) analysis at parton level.

Three main signal subprocesses contribute the process in question at the LO: 1) gravitino pair production in association with a parton (quark or gluon); 2) production of single gravitino with squark or gluino; 3) SUSY QCD pair production. For simplicity, we assume that squarks and gluino decay mainly as $\tilde{q}(\tilde{g}) \to q(g) + \psi$. Corresponding decay widths look as

$$\Gamma(\tilde{q}(\tilde{g}) \to q(g) + \psi) = \frac{M^5_{\tilde{q}(3)}}{16\pi F^2}. \quad (7)$$

In what follows we assume that $M_{\tilde{q}} = M_3$. In general, if sgoldstinos have masses around TeV scale they can decay into a pair of MSSM particles and corresponding decay widths are governed by relevant soft SUSY breaking parameters [41]. For the present analysis we consider only two dominant decay [1] channels: decay into a pair of gravitinos

$$\Gamma(s(p) \to \psi\psi) = \frac{m^5_{s(p)}}{32\pi F^2} \quad (8)$$

or a pair of gluons

$$\Gamma(s(p) \to gg) = \frac{M_3^2 m^3_{s(p)}}{4\pi F^2}. \quad (9)$$

Above expressions indicate that at fixed M_3 very heavy sgoldstino decays mostly into gravitino pair, while lighter sgoldstinos prefer to decay into gluons.

Let us start with direct gravitino pair production with a single jet

$$pp \to \psi\psi + \text{jet}\,.$$

Examples of relevant Feynman diagrams are presented on Fig. 1. In the heavy sgoldstino limit, the

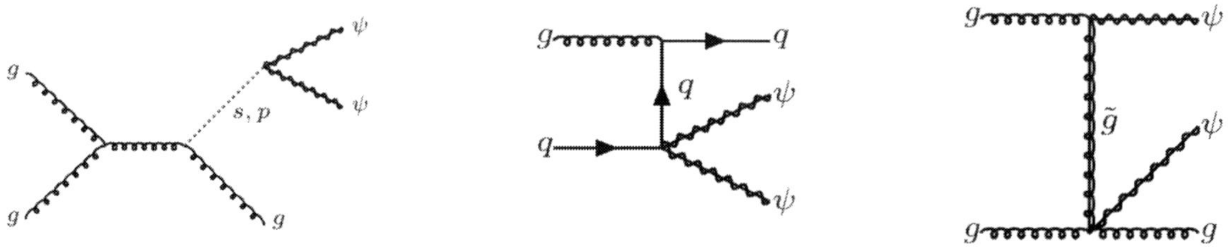

Figure 1. Examples of Feynman diagrams contributing to direct gravitino pair production with a jet.

cross section of this process scales as $1/F^4$. If sgoldstino contribution is dominant, which happens

[1] We disregard here possibility of sgoldstino mixing with the Higgs bosons which could result in a considerable changes of sgoldstino branching pattern [42, 43].

for very light sgoldstinos, corresponding cross section scales as $1/F^2$. Next subprocess is associated gravitino production with gluino or squark (see Fig. 2)

$$pp \to \tilde{q}\psi, \tilde{g}\psi \to \psi\psi + \text{jet}\,.$$

The cross section behaves as $1/F^2$ at large values of SUSY breaking scale. Finally, we have squark-

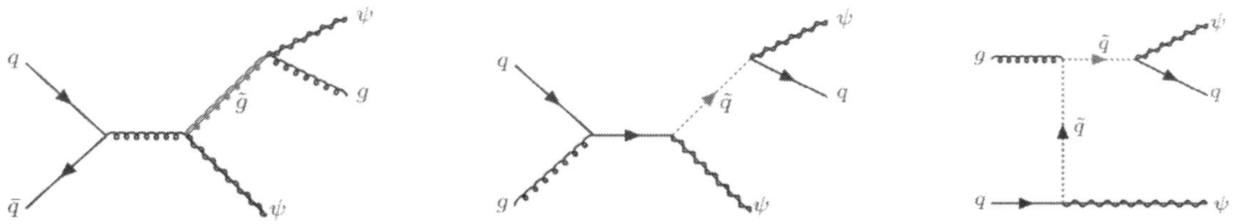

Figure 2. Examples of Feynman diagrams contributing to associated gravitino production with gluino or squark.

squark, gluino-gluino and squark-gluino production (or SUSY QCD pair production) with their subsequent decays into gravitino and gluons or quarks (see Fig. 3)

$$pp \to \tilde{q}\tilde{q}, \tilde{q}\tilde{g}, \tilde{g}\tilde{g} \to \psi\psi + 2\,\text{jets}\,.$$

The final state of these subprocesses contains in fact two jets but the event selection procedure used

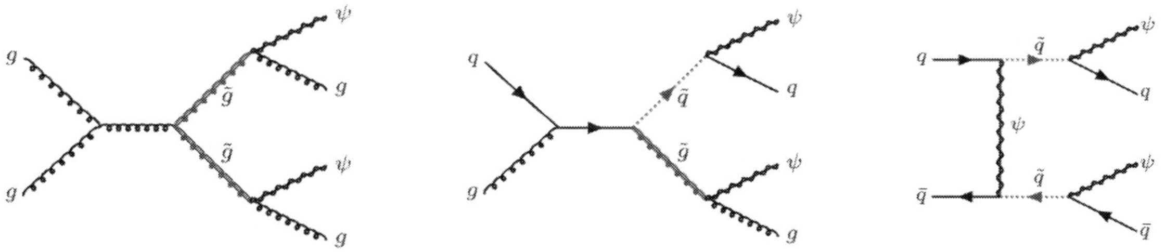

Figure 3. Examples of Feynman diagrams contributing to SUSY QCD pair (squark-squark, gluino-gluino and squark-gluino) production.

by ATLAS and CMS experiments for monojet searches actually admit more than one jet in a final state.

We implemented model (5)–(6) into `MadGraph` [44] and calculated cross section for the processes in question at the leading order applying all the relevant cuts from CMS monojet analysis with run-I data [45]. We decrease number of free parameters in the model by assuming that $M_{\tilde{q}_L,ii}$, $M_{\tilde{u}_R,ii}$, $M_{\tilde{d}_R,ii}$ are physical squark masses neglecting their mixing. Also we assume equal masses of squarks and gluinos and equal masses of sgoldstinos, i.e. $m_s = m_p$. For calculations of cross sections we use CTEQ6L1 PDFs [46] with relevant values of renormalization and factorization scales for different subprocesses contributing to the signal (see [40] for more details). We add cut on missing momentum, $\not{p}_T^{\text{min}} = 450$ GeV and in this case the upper limit on the visible cross section times acceptance times efficiency for non-SM production of events is about 7.8 fb, see [45].

It is known that for squark and gluino productions NLO corrections are quite large [47–49]. However, the whole theory which includes gravitinos and sgoldstinos is nonrenormalizable. Self-consistent NLO analysis would require knowledge of microscopic theory. This is beyond the scope of our work.

DOI: 10.1051/epjconf/201715802008

Given the fact the NLO corrections typically increase the production cross sections we expect that our bounds will be even stronger if we include them into our analysis.

On Fig. 4 we show cross sections of different subprocesses as functions of $\sqrt{F}$. One can see that

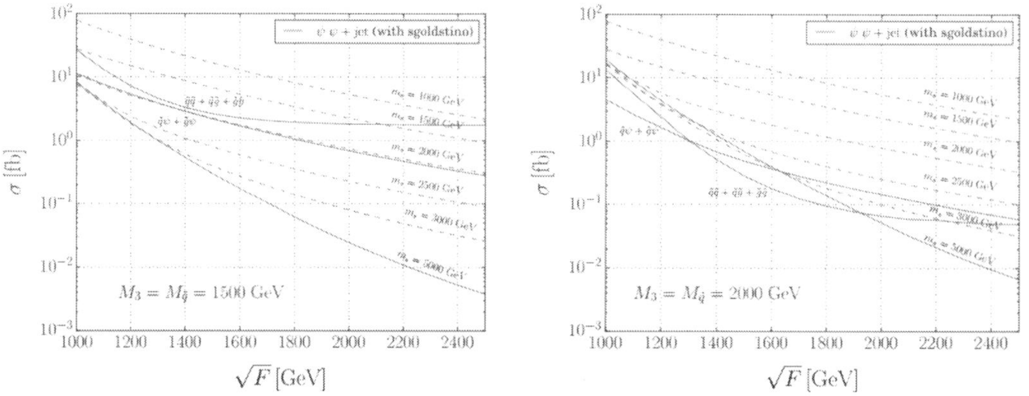

Figure 4. Cross sections of relevant subprocesses contributing to $pp \to \psi\psi + \text{jet} \to \not{p}_T + \text{jet}$ as functions of supersymmetry breaking scale for $M_3 = M_{\tilde{q}} = 1.5$ TeV (left panel) and $M_3 = M_{\tilde{q}} = 2$ TeV (right panel).

for sgoldstinos with masses of several TeVs corresponding cross sections can be comparable or even larger with production of superpartners. One can see that the slope of gravitino pair production in the heavy sgoldstino mass limit becomes more flat with light sgoldstino which respects changing of cross section scaling from $1/F^4$ to $1/F^2$ with the increase of light sgoldstino contribution.

On Fig. 5 we show the same cross sections but as functions of common mass of superpartners.

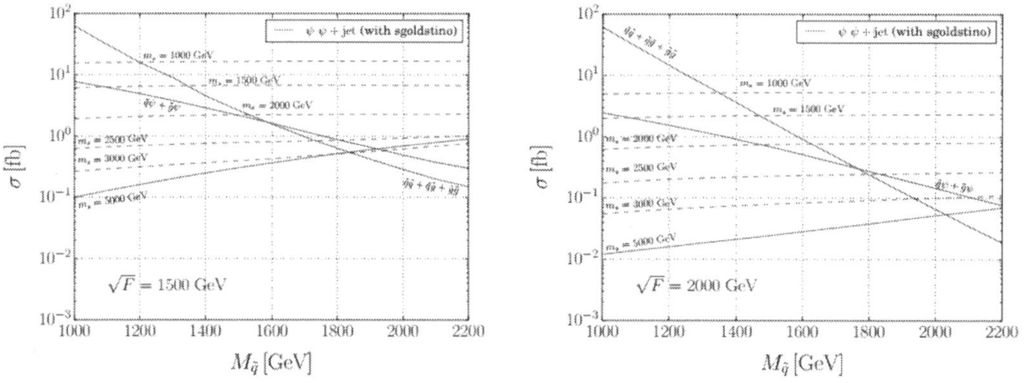

Figure 5. Cross sections of relevant subprocesses contributing to $pp \to \psi\psi + \text{jet} \to \not{p}_T + \text{jet}$ as functions of common mass of superpartners for $\sqrt{F} = 1.5$ TeV (left panel) and $\sqrt{F} = 2$ TeV (right panel).

Production cross section of superpartners decreases with increase of their masses. On the contrary, direct gravitino pair production increases and stays constant at large masses of squarks.

On Fig. 6 we show exclusion plots of $\sqrt{F}$ versus $M_3 = M_{\tilde{q}}$ at fixed values of sgoldstino masses. One can see that light sgoldstinos can change the bounds on $\sqrt{F}$ considerably. Flattening of the lines at large values of $\sqrt{F}$ is due to saturation of the total cross section by the pair production of

DOI: 10.1051/epjconf/201715802008

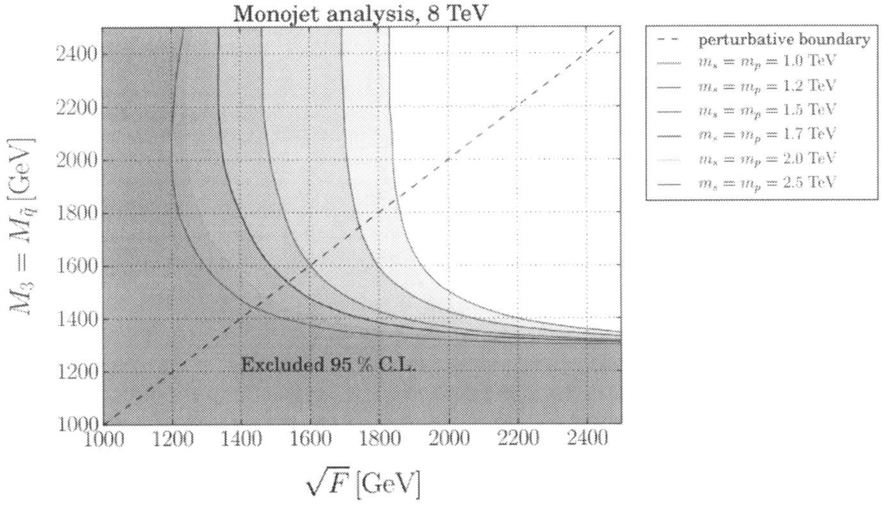

Figure 6. Exclusion plot of $\sqrt{F}$ vs $M_3 = M_{\tilde{q}}$ for different masses of sgoldstino.

superpartners, i.e. squarks and gluinos, in this limit. At large $M_3 = M_{\tilde{q}}$ the contour becomes vertical because in this limit the cross section is dominated by direct gravitino pair production in association with jet. The dashed line here is the boundary of applicability of perturbation theory for our model $m_{\text{soft}} < \sqrt{F}$. Above this line the theory is in the strong coupling regime.

On Fig. 7 we present another exclusion plot in different coordinates. Here we fix the mass scale of superpartners and vary common masses of sgoldstinos and SUSY breaking scale. One can see that sgoldstino contribution is the most important in the range of its mass from 1 TeV to 2.5 TeV where the bounds on SUSY breaking scale are strengthen by factor up to about 1.5 in comparison with those obtained in the heavy sgoldstino limit.

4 Direct sgoldstino production and dijet signature

TeV scale sgoldstinos can be produced directly in pp collisions mainly in gluon-gluon fusion [25] and after their decay into pair of gluons they can be observed as narrow dijet resonances. Here we compare the bounds from monojet analysis with the bounds which can be obtained from the the LHC searches of dijet resonance. For comparison we use both ATLAS [50] and CMS [51] dijet analyses with the data obtained at $\sqrt{s} = 8$ TeV.

To find excluded models we've compared calculated cross sections with the experimental upper 95 % C.L. limits on the dijet cross sections obtained by ATLAS [50] and CMS [51]. On Fig. 8 we present comparison of exclusion plots of $\sqrt{F}$ vs sgoldstino mass $m_s = m_p$ for different values of superpartners masses $M_3 = M_{\tilde{q}}$ obtained from dijet and monojet searches. We see that for relatively small masses of squarks and gluinos monojet analysis limits this models considerably stronger than dijets. In this case the monojet cross section is large due to contribution of these superpartners. With the increase of masses of superpartners direct production of sgoldstinos with their subsequent decays into pair of gluons becomes more constraining. But for heavy sgoldstinos it decreases because on the one hand direct production of sgoldstinos becomes suppressed by its mass and on the other in this

DOI: 10.1051/epjconf/201715802008

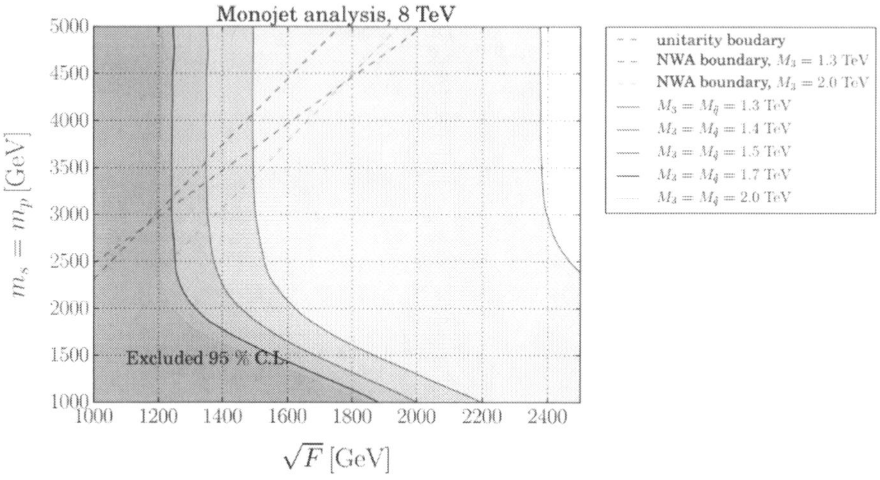

Figure 7. Exclusion plot of $\sqrt{F}$ vs $m_s = m_p$ for different scale of masses of superpartners. Above the dashed lines one of the corresponding conditions is violated: unitarity, condition for applicability of NWA. Here we depicted only lines which correspond to $M_3 = 1.3$ TeV (blue dashed) and $M_3 = 2.0$ TeV (orange dashed). All other lines which correspond to all intermediate scales lie between them.

case sgoldstinos decay dominantly a into pair of gravitinos. In this way searches for monojets and dijets are complimentary to each other.

On Fig. 9 we present similar exclusion plots but in different coordinates. In the upper left panel sgoldstino mass is equal to 1.2 TeV and here we see again that at small masses of superpartners the constraints from jet + missing transverse energy searches are stronger than from dijets, while for heavy superpartners it is vice versa. For sgoldstinos with mass more than ~ 1.7 TeV the bounds from dijets searches lie in a strong coupling regime of the theory $m_{\text{soft}} > \sqrt{F}$ and hence in this case constraints from monojet searches should be used.

5 Conclusions

To conclude we show that in models with low scale supersymmetry breaking contribution of light sgoldstino to gravitino pair production can be considerable. We calculate leading order cross sections of the processes contributing to jet and missing energy signal. We obtain bounds on the parameter space of the model within a simplified set of parameters using results of the CMS run-I searches for jet and missing transverse energy signature at $\sqrt{s} = 8$ TeV. We found that the bounds on supersymmetry breaking scale in the case of light sgoldstinos can be stronger by factor of 1.5 as compared to those in the heavy sgoldstino limit. We compare the bounds from monojet searches with those from dijets from ATLAS and CMS data of the same collision energy and statistics. We found them complimentary in different regions of parameter space.

Acknowledgments

The work was supported by the RSCF grant 14-22-00161. The numerical part of the work was performed on Calculational Cluster of the Theory Division of INR RAS.

DOI: 10.1051/epjconf/201715802008

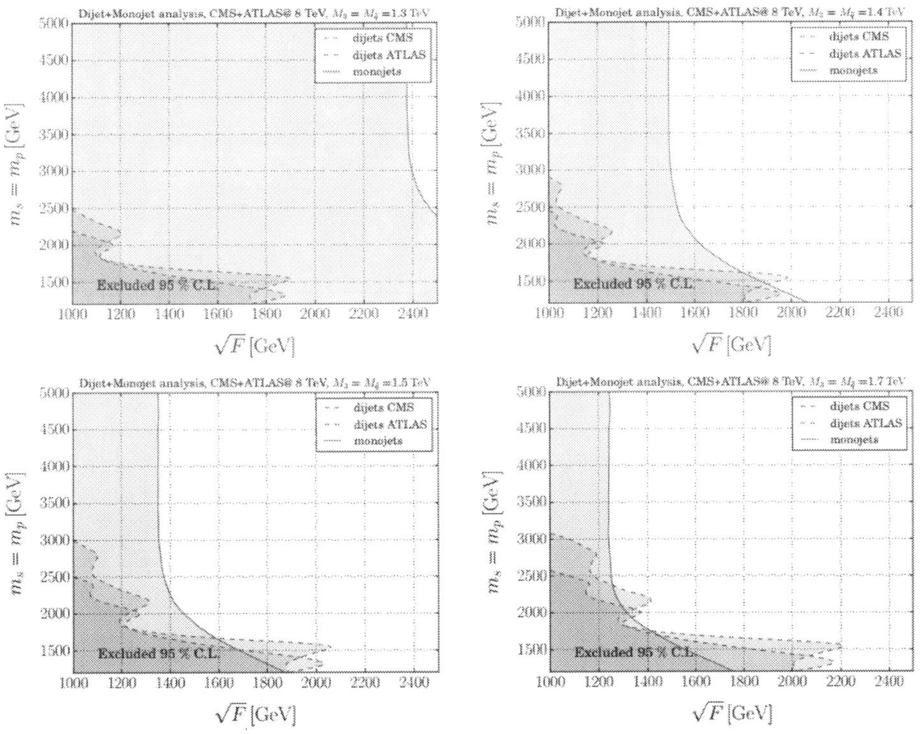

Figure 8. Exclusion plots of $\sqrt{F}$ vs $m_s = m_p$, comparison of bounds obtained from dijet resonances and missing energy searches for different values of $M_3 = M_{\tilde{q}}$: 1.3 TeV (upper left), 1.4 TeV (upper right), 1.5 TeV (lower left), 1.7 TeV (lower right).

References

[1] S. P. Martin, *Adv.Ser.Direct.High Energy Phys.* **21**, (2010) 1, [Adv. Ser. Direct. High Energy Phys. **18**, (1998) 1], [hep-ph/9709356].
[2] A. Brignole, F. Feruglio and F. Zwirner, *Nucl. Phys.* **B 501**, (1997) 332, [hep-ph/9703286].
[3] M. A. Luty, [hep-ph/9611387].
[4] D. V. Volkov and V. A. Soroka, JETP Lett. **18**, 312 (1973) [Pisma Zh. Eksp. Teor. Fiz. **18**, 529 (1973)].
[5] S. Deser and B. Zumino, *Phys. Rev. Lett.* **38**, 1433 (1977).
[6] E. Cremmer, B. Julia, J. Scherk, P. van Nieuwenhuizen, S. Ferrara and L. Girardello, *Phys. Lett.* **B 79**, 231 (1978).
[7] M. Dine, W. Fischler and M. Srednicki, *Nucl. Phys.* **B 189**, 575 (1981).
[8] S. Dimopoulos and S. Raby, *Nucl. Phys.* **B 192**, 353 (1981).
[9] M. Dine and W. Fischler, *Phys. Lett.* **B 110**, 227 (1982).
[10] M. Dine and W. Fischler, *Nucl. Phys.* **B 204**, 346 (1982).
[11] C. R. Nappi and B. A. Ovrut, *Phys. Lett.* **B 113**, 175 (1982).
[12] L. Alvarez-Gaume, M. Claudson and M. B. Wise, *Nucl. Phys.* **B 207**, 96 (1982).
[13] J. R. Ellis, K. Enqvist and D. V. Nanopoulos, *Phys. Lett.* **B 147**, 99 (1984).

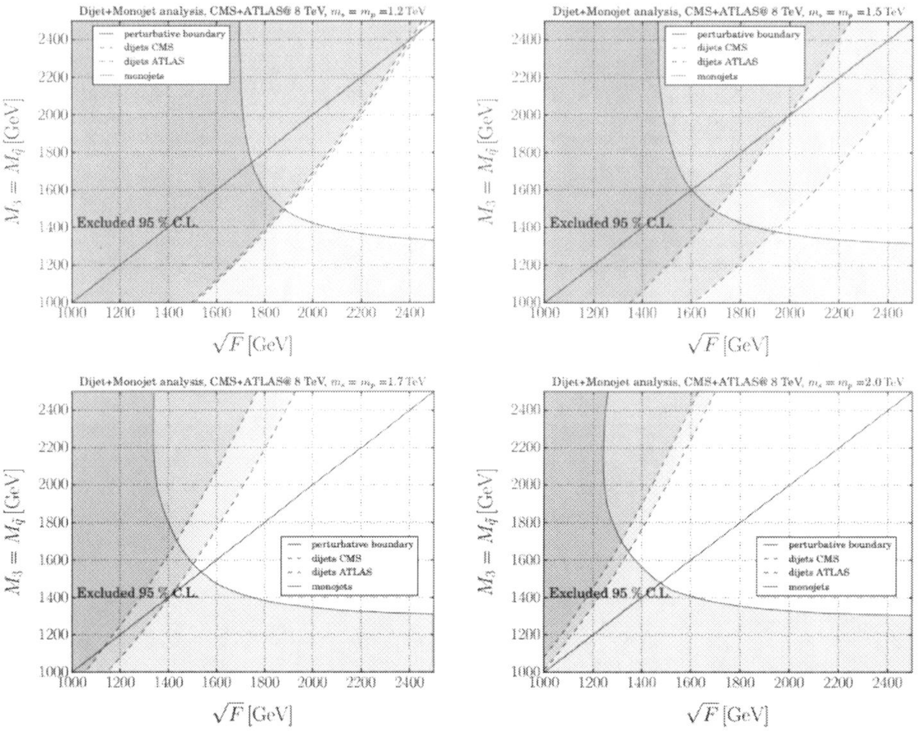

Figure 9. Exclusion plots of $\sqrt{F}$ vs $M_3 = M_{\tilde{q}}$, comparison of bounds obtained from dijet resonances and jet + missing energy searches for different values of $m_s = m_p$: 1.2 TeV (upper left), 1.5 TeV (upper right), 1.7 TeV (lower left), 2.0 TeV (lower right)

[14] J. R. Ellis, K. Enqvist and D. V. Nanopoulos, *Phys. Lett.* **B 151**, 357 (1985).
[15] A. B. Lahanas and D. V. Nanopoulos, *Phys. Rept.* **145**, 1 (1987).
[16] T. Gherghetta and A. Pomarol, *Nucl. Phys.* **B 586**, 141 (2000) [hep-ph/0003129].
[17] T. Gherghetta and A. Pomarol, *Nucl. Phys.* **B 602**, 3 (2001), [hep-ph/0012378].
[18] D. S. Gorbunov and N. V. Krasnikov, *JHEP* **07**, (2002) 043 [hep-ph/0203078].
[19] S. V. Demidov and D. S. Gorbunov, *Phys.Atom.Nucl.* **69**, (2006) 712, [hep-ph/0405213].
[20] K. O. Astapov and D. S. Gorbunov, *Phys. Rev.* **D 93**, (2016), no. 3, 035008, [arXiv:1511.05403 [hep-ph]].
[21] K. O. Astapov and D. V. Kirpichnikov, [arXiv:1612.02813 [hep-ph]].
[22] P. Fayet, *Phys. Lett.* **B 70**, 461 (1977).
[23] R. Casalbuoni, S. De Curtis, D. Dominici, F. Feruglio and R. Gatto, *Phys. Lett.* **B 215**, 313 (1988).
[24] R. Casalbuoni, S. De Curtis, D. Dominici, F. Feruglio and R. Gatto, *Phys. Rev. D* **D 39**, 2281 (1989).
[25] E. Perazzi, G. Ridolfi and F. Zwirner, *Nucl. Phys.* **B 590**, (2000) 287–305, [hep-ph/0005076].
[26] E. Perazzi, G. Ridolfi and F. Zwirner, *Nucl. Phys.* **B 574** (2000) 3 [hep-ph/0001025].
[27] E. Dudas, C. Petersson and P. Tziveloglou, *Nucl. Phys.* **B 870**, (2013) 353–383, [arXiv:1211.5609 [hep-ph]].

[28] D. Gorbunov, V. Ilyin and B. Mele, *Phys. Lett.* **B 502**, (2001) 181, [hep-ph/0012150].
[29] A. Brignole, J. A. Casas, J. R. Espinosa and I. Navarro, *Nucl. Phys.* **B 666**, (2003) 105–143, [hep-ph/0301121].
[30] C. Petersson, A. Romagnoni and R. Torre, *Phys. Rev.* **D 87**, (2013), no. 1, 013008, [arXiv:1211.2114 [hep-ph]].
[31] D. A. Dicus and P. Roy, *Phys. Rev.* **D 44**, 1593 (1991).
[32] S. Shirai and T. T. Yanagida, *Phys. Lett.* **B 680**, (2009) 351, [arXiv:0905.4034 [hep-ph]].
[33] K. Mawatari and B. Oexl, *Eur. Phys. J. C* **74**, no. 6, 2909 (2014), [arXiv:1402.3223 [hep-ph]].
[34] J. R. Ellis, J. L. Lopez and D. V. Nanopoulos, *Phys. Lett.* **B 394**, 354 (1997) [hep-ph/9610470].
[35] M. Klasen and G. Pignol, *Phys. Rev.* **D 75**, (2007) 115003, [hep-ph/0610160].
[36] A. Brignole, F. Feruglio, M. L. Mangano and F. Zwirner, *Nucl. Phys.* **B 526**, (1998) 136–152, Erratum: [*Nucl.Phys.* **B 582**, (2000) 759–761], [hep-ph/9801329].
[37] A. Brignole, F. Feruglio and F. Zwirner, *Nucl. Phys.* **B 516**, 13 (1998) *Erratum: [Nucl. Phys.* **B 555**, 653 (1999)] [hep-ph/9711516].
[38] C. Petersson, A. Romagnoni and R. Torre, *JHEP* **10**, (2012) 016 [arXiv:1203.4563 [hep-ph]].
[39] F. Maltoni, A. Martini, K. Mawatari and B. Oexl, *JHEP* **04**, (2015) 021 [arXiv:1502.01637 [hep-ph]].
[40] S. V. Demidov and I. V. Sobolev, [arXiv:1709.03830 [hep-ph]].
[41] D. S. Gorbunov and A. V. Semenov, *CompHEP package with light gravitino and sgoldstinos,* [hep-ph/0111291].
[42] K. O. Astapov and S. V. Demidov, *JHEP* **01**, (2015) 136, [arXiv:1411.6222 [hep-ph]].
[43] S. V. Demidov and I. V. Sobolev, *JHEP* **08**, (2016) 030, [arXiv:1605.08220 [hep-ph]].
[44] J. Alwall *et al.*, *JHEP* **07**, (2014) 079, [arXiv:1405.0301 [hep-ph]].
[45] **CMS Collaboration**, V. Khachatryan et al., *Eur. Phys. J. C* **75**, no. 5, 235 (2015), [arXiv:1408.3583 [hep-ex]].
[46] J. Pumplin, D. R. Stump, J. Huston, H. L. Lai, P. M. Nadolsky and W. K. Tung, *JHEP* **07**, (2002) 012, [hep-ph/0201195].
[47] W. Beenakker, R. Hopker, M. Spira and P. M. Zerwas, *Nucl. Phys.* **B 492**, 51 (1997), [hep-ph/9610490].
[48] W. Beenakker, S. Brensing, M. Kramer, A. Kulesza, E. Laenen and I. Niessen, *JHEP* **08**, (2010) 098, [arXiv:1006.4771 [hep-ph]].
[49] D. Gonçalves-Netto, D. López-Val, K. Mawatari, T. Plehn and I. Wigmore, *Phys. Rev.* **D 87**, no. 1, 014002 (2013), [arXiv:1211.0286 [hep-ph]].
[50] **ATLAS Collaboration**, G. Aad et al., *Phys. Rev.* **D 91**, no. 5, 052007 (2015), [arXiv:1407.1376 [hep-ex]].
[51] **CMS Collaboration**, V. Khachatryan et al., *Phys. Rev.* **D 91**, no. 5, 052009 (2015), [arXiv:1501.04198 [hep-ex]].

On mass limits for scalar color octet from the LHC data on $t\bar{t}$ invariant mass spectra

M. V. Martynov[1,*] and *A. D.* Smirnov[1,**]

[1] *Yaroslavl State University*

Abstract. The scalar color octet contribution to the resonance $t\bar{t}$-pair production at the LHC is calculated and analyzed with account of the one loop effective two gluon vertex. From current LHC data on total cross section and invariant mass spectrum of $t\bar{t}$ production we found the $(m_{F_2} - \sin\beta)$-plane exclusion regions.

1 Introduction

The search for new physics effects beyond the Standard Model (SM) is now one of the goals of the experiments at the LHC. There are many models predicting new physics effects. The most interesting of them look the models predicting new effects due to enlarging the symmetry of the SM because the search for such effects could help us to find the next symmetry in yet unknown hierarchy of the symmetries which possibly unify the known in the SM electroweak and strong interactions of quarks and leptons.

Extended color symmetries are attractive variants of the new physics. One of such variant of physics beyond the SM can be induced by the possible four color symmetry between quarks and leptons of Pati-Salam type [1]. The four color quark-lepton symmetry can be unified with the electroweak $SU_L(2)\times U(1)$ symmetry of the SM in the minimal way by the group

$$G_{MQLS} = SU_V(4) \times SU_L(2) \times U_R(1), \tag{1}$$

where the first factor is the vector-like group of the four color quark-lepton symmetry the second one is the usual SM electroweak symmetry group for the left-handed fermions and the third one is the corresponding hypercharge factor for the right-handed fermions (the minimal quark-lepton symmetry model – MQLS-model [2, 3]).

As a result of the Higgs mechanism of splitting the masses of quarks and leptons the four color symmetry in its minimal realization on the gauge group (1) predicts in addition to the SM Higgs doublet $\Phi^{(SM)}$ the existence of the new scalar $SU_L(2)$-doublets

$$\begin{pmatrix} \Phi'_1 \\ \Phi'_2 \end{pmatrix}; \begin{pmatrix} S^{(+)}_{1\alpha} \\ S^{(+)}_{2\alpha} \end{pmatrix}; \begin{pmatrix} S^{(-)}_{1\alpha} \\ S^{(-)}_{2\alpha} \end{pmatrix}; \begin{pmatrix} F_{1c} \\ F_{2c} \end{pmatrix} \tag{2}$$

*e-mail: martmix@mail.ru
**e-mail: asmirnov@uniyar.ac.ru

with electric charges

$$Q_{\Phi}^{em}: \quad \begin{pmatrix} 1 \\ 0 \end{pmatrix}; \begin{pmatrix} 5/3 \\ 2/3 \end{pmatrix}; \begin{pmatrix} 1/3 \\ -2/3 \end{pmatrix}; \begin{pmatrix} 1 \\ 0 \end{pmatrix}$$

respectively.

The fields (2) belong to the (1.2.1)+(15.2.1) representation of the group (1), here the fields Φ_1', Φ_2' form an additional colorless scalar doublet, $S_{1\alpha}^{(\pm)}$, $S_{2\alpha}^{(\pm)}$, $\alpha = 1,2,3$ are the color triplets forming two scalar leptoquark doublets and the fields F_{1c}, F_{2c}, $c = 1,2...8$ form the scalar doublet of the color octets (the scalar gluon doublet).

Because of their Higgs origin the coupling constants of the doublets (2) with the fermions occur to be proportional to the ratios m_f/η of the fermion masses m_f to the SM VEV η and are small for u-, d-, s-quarks are more significant for c-, b-quarks and are especially significant for t-quark ($m_t/\eta \sim 0.7$). As a result the scalar doublets (2) can manifest themselves more probably in the processes with t-quarks. In particular the scalar octet F_2 could manifest itself as a resonance in $t\bar{t}$-pair production at the LHC. It should be noted that the coupling constants of the doublets (2) with t-quark are known (up to the mixing parameters), which gives the possibility to estimate quantitatively the possible effects from these particles in dependence on their masses. The pair production of the scalar octets in pp-collisions at the LHC has been discussed in Refs [4–16].

In the present paper we calculate the contribution of the scalar octet to the cross section of the resonance $t\bar{t}$-pair production in pp-collisions and analyse the possibility of manifestation of the scalar gluon F_2 of the MQLS-model as the corresponding resonance peak in $t\bar{t}$-pair production at the LHC.

2 MQLS-model scalar color octets — scalar gluons

The details of interactions of the scalar doublets (2) with quarks and leptons can be found in [17, 18]. In particular the interaction of the scalar gluon F_2 with up- and down-quarks in the MQLS-model has the chiral form and can be written as

$$L_{F_2 u_i u_j} = \bar{u}_{i\alpha}\Big[(h_{1F_2}^L)_{ij} P_L\Big](t_c)_{\alpha\beta} u_{j\beta} F_{2c} + \text{h.c.}, \quad (3)$$

$$L_{F_2 d_i d_j} = \bar{d}_{i\alpha}\Big[(h_{2F_2}^R)_{ij} P_R\Big](t_c)_{\alpha\beta} d_{j\beta} F_{2c} + \text{h.c.}, \quad (4)$$

where $t_c, c = 1,2\ldots 8$ are the generators of the $SU_c(3)$ group, $P_{L,R} = (1 \pm \gamma_5)/2$ are the left and right projection operators and $(h_{1F_2}^L)_{ij}, (h_{2F_2}^R)_{ij}$ are the Yukawa coupling constants, $i,j = 1,2,3$ are the generation indices. As a result of the Higgs mechanism of generating the quark and lepton masses the Yukawa coupling constants $(h_{1F_2}^L)_{ij}, (h_{2F_2}^R)_{ij}$ are defined by the expressions

$$(h_{1F_2}^L)_{ij} = -\sqrt{3}\frac{1}{\eta \sin\beta}\Big[m_{u_i}(\delta)_{ij} - (K_1^R)_{ik} m_{\nu_k} (\overset{+}{K_1^L})_{kj}\Big], \quad (5)$$

$$(h_{2F_2}^R)_{ij} = -\sqrt{3}\frac{1}{\eta \sin\beta}\Big[m_{d_i}(\delta)_{ij} - (K_2^L)_{ik} m_{l_k} (\overset{+}{K_2^R})_{kj}\Big], \quad (6)$$

where m_{u_i}, m_{d_i} and m_{ν_k}, m_{l_k} are the masses of up- and down-quarks and of neutrinos and charged leptons, $K_1^{L,R}$, $K_2^{L,R}$ are the mixing matrices in leptoquark currents which are specific for the model with the four color quark-lepton symmetry and β is a mixing angle of two colorless scalar doublets of MQLS model. Among the coupling constants (5), (6) the largest is the constant $(h^L_{1F_2})_{33}$ which with neglect of the neutrinos masses takes the form

$$(h^L_{1F_2})_{33} = -\sqrt{3}\frac{m_t}{\eta \sin\beta}. \quad (7)$$

The interaction of the scalar gluon F_2 with t-quark can be written as

$$L_{F_2tt} = \bar{t}_\alpha(h^S_{F_2t\bar{t}} + h^P_{F_2t\bar{t}}\gamma_5)(t_c)_{\alpha\beta}t_\beta F_{2c} + \text{h.c.}, \quad (8)$$

where the correspondent scalar and pseudoscalar coupling constants with account of (7) take the form

$$h^S_{F_2t\bar{t}} = h^P_{F_2t\bar{t}} = -\frac{\sqrt{3}}{2}\frac{m_t}{\eta \sin\beta} \approx -0.61/\sin\beta. \quad (9)$$

The coupling constants (9) increase with decreasing $\sin\beta$ so that for $\sin\beta = 1, 0.7, 0.4$ the perturbation theory parameters take the values $(h^{S,P}_{F_2t\bar{t}})^2/4\pi \approx 0.03, 0.06, 0.18$ respectively. Below we restrict ourselves by the mixing angle region $0.4 \leq \sin\beta \leq 1$.

The interactions (3), (4) lead to the decays $F_2 \to u_i\bar{u}_i$, $F_2 \to d_i\bar{d}_i$ and in the case of $m_{F_2} > 2m_t$ the decay $F_2 \to t\bar{t}$ is dominant with the width [17, 18]

$$\Gamma(F_2 \to t\bar{t}) = m_{F_2}\frac{3}{32\pi}\left(\frac{m_t}{\eta}\right)^2\left(1 - 2\frac{m_t^2}{m_{F_2}^2}\right)\sqrt{1 - 4\frac{m_t^2}{m_{F_2}^2}\frac{1}{\sin^2\beta}}. \quad (10)$$

For the masses $m_{F_2} = 400 - 2000$ GeV the width (10) is of about $(2 - 30)/\sin^2\beta$ GeV and $\Gamma_{F_2}/m_{F_2} = (0.5 - 1.5)\%/\sin^2\beta$.

As seen from the expressions (5), (6) the coupling constants of the interaction of the scalar gluon F_2 with u- and d- quarks are of order of $m_u/\eta \sim m_d/\eta \sim 10^{-5}$ and the interactions of these quarks as the initial partons with the scalar gluon F_2 are negligibly small. On the other hand the Lagrangian (8), (9) can induce through the loop contribution of t-quark the more significant effective interaction of two initial gluons with the scalar gluon F_2, which should be taken into account. The analogous effective two gluon interaction is induced also with the colorless scalar Φ'_2.

3 Resonance contribution of scalar color octet to $t\bar{t}$ production

The calculation of the effective two gluon vertex of interaction with the scalar octet is like to that with the colorless scalar and we perform below these calculations simultaneously. For this purpose we write the flavour diagonal interactions of scalar octet and of the scalar color singlet with quarks in the model independent form as

$$L_{\Phi q\bar{q}} = \bar{q}_\alpha(h^S_{\Phi q\bar{q}} + h^P_{\Phi q\bar{q}}\gamma_5)\Phi_{\alpha\beta}q_\beta + \text{h.c.}, \quad (11)$$

where $\Phi_{\alpha\beta} = \Phi_0\delta_{\alpha\beta}$ for the colorless scalar particle Φ_0 and $\Phi_{\alpha\beta} = \Phi_{8c}(t_c)_{\alpha\beta}$ for the scalar octet Φ_8, t_c are the generators of the $SU_c(3)$ group ($c = 1, 2...8$), $h^S_{\Phi q\bar{q}}$ and $h^P_{\Phi q\bar{q}}$ are the corresponding scalar and pseudoscalar coupling constants. For the MQLS-model the scalars Φ_8 and Φ_0 correspond to F_2 and Φ'_2 respectively.

The effective vertex $\Gamma^{(q)\mu\nu}_{ab\Phi}(p, k_1, k_2)$ of interaction of two gluons with scalar field $\Phi = \Phi_0, \Phi_{8c}$ induced by the Lagrangian (11) with account of one loop contribution of quark q is described by the diagrams in the Fig. 1

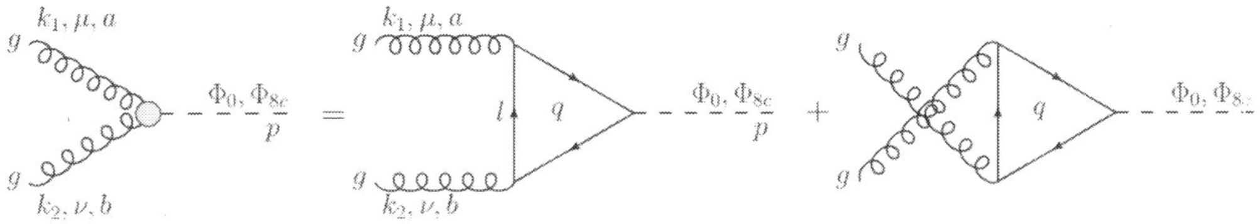

Figure 1. The effective vertex $\Gamma^{(q)\mu\nu}_{ab\Phi}(p, k_1, k_2)$ induced by one loop contribution of quark q for $\Phi = \Phi_0, \Phi_{8c}$.

With account of the contributions of all the quarks the resulted effective vertex $\Gamma^{\mu\nu}_{ab\Phi}(p, k_1, k_2)$ in the case of real gluons ($k_1^2 = 0,\ k_2^2 = 0,\ \hat{s} = p^2 = 2(k_1k_2)$) can be parametrized as [19]:

$$\Gamma^{\mu\nu}_{ab\Phi}(p, k_1, k_2) = \sum_q \Gamma^{(q)\mu\nu}_{ab\Phi}(p, k_1, k_2) =$$

$$= -C_{ab\Phi}\frac{\alpha_s\sqrt{\hat{s}}}{\pi}\left[\left(g^{\mu\nu} - \frac{2k_1^\nu k_2^\mu}{\hat{s}}\right)F^S_\Phi(\hat{s}) - 2i\varepsilon^{\mu\nu\rho\sigma}\frac{k_{1\rho}k_{2\sigma}}{\hat{s}}F^P_\Phi(\hat{s}) + \frac{2k_1^\mu k_2^\nu}{\hat{s}}G^S_\Phi(\hat{s})\right] \tag{12}$$

by the form factors

$$F^{S,P}_\Phi(\hat{s}) = \sum_q h^{S,P}_{\Phi q\bar{q}}\tilde{F}^{S,P}(\hat{s}, m_q^2), \qquad G^S_\Phi(\hat{s}) = \sum_q h^S_{\Phi q\bar{q}}\tilde{G}^S(\hat{s}, m_q^2), \tag{13}$$

where $C_{ab\Phi}$ is the color factor with

$$C_{ab\Phi_0} = \delta_{ab}/2 \equiv C_{ab}, \quad C_{ab\Phi_{8c}} = d_{abc}/4 \equiv C_{abc} \tag{14}$$

for the color singlet Φ_0 and for the color octet Φ_8.

For the form factors $\tilde{F}^{S,P}(\hat{s}, m_q^2)$, $\tilde{G}^S(\hat{s}, m_q^2)$ we have found the expressions

$$\tilde{F}^S(\hat{s}, m_q^2) = \frac{m_q}{\sqrt{\hat{s}}}\left[(\hat{s} - 4m_q^2)\,C_0(0,0,\hat{s},m_q^2,m_q^2,m_q^2) - 2\right] \equiv \tilde{F}^S(\rho_q), \tag{15}$$

$$\tilde{F}^P(\hat{s}, m_q^2) = m_q\sqrt{\hat{s}}\;C_0(0,0,\hat{s},m_q^2,m_q^2,m_q^2) \equiv \tilde{F}^P(\rho_q), \tag{16}$$

$$\tilde{G}^S(\hat{s}, m_q^2) =$$

$$= \frac{m_q}{\sqrt{\hat{s}}}\left[(\hat{s} + 4m_q^2)\,C_0(0,0,\hat{s},m_q^2,m_q^2,m_q^2) + 4B_0(\hat{s},m_q^2,m_q^2) - \frac{4A_0(m_q^2)}{m_q^2}\right] \equiv \tilde{G}^S(\rho_q), \tag{17}$$

where A_0, B_0, C_0 are the Passarino–Veltman (PV) scalar integrals [20, 21], with account of the explicit form of these integrals the form factors (15)–(17) depend only on the variable $\rho_q = \sqrt{\hat{s}}/m_q$. The form factors (15)–(17) account the one loop contribution of quark q in the model independent way, the specific features of the model are presented by the coupling constants $h^{S,P}_{\Phi q\bar{q}}$ in the form factors (13). Due to the gauge invariance the longitudinal component of the vertex (12) parametrized by the form factor $G^S_\Phi(\hat{s})$ do not enter to the observed variables.

We have calculated the cross section of the process $gg \to Q\bar{Q}$ of $Q\bar{Q}$ pair production in gluon fusion in QCD LO with account also of the effective vertex (12) in Ref. [19]. The diagrams of this process are shown in the Fig. 2.

 DOI: 10.1051/epjconf/201715802009

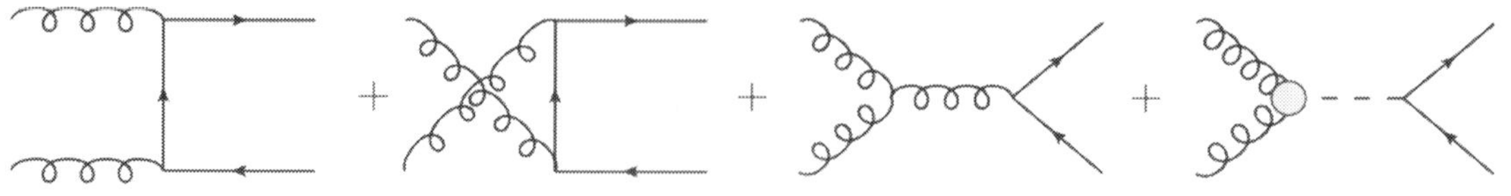

Figure 2. Diagrams of the process $gg \to Q\bar{Q}$ in the SM LO and with account of the effective vertex $\Gamma^{\mu\nu}_{ab\Phi}(p, k_1, k_2)$.

The total cross section of the process $gg \to Q\bar{Q}$ can be written as the sum

$$\sigma_0(gg{\to}Q\bar{Q}, \mu) = \sigma_0^{\rm SM}(gg{\to}Q\bar{Q}, \mu) + \Delta\sigma^{\Phi}(gg{\to}Q\bar{Q}, \mu) \tag{18}$$

of the well known QCD LO cross section

$$\sigma_0^{\rm SM}(gg{\to}Q\bar{Q}, \mu) = \frac{\alpha_s^2(\mu)\,\pi}{48\hat{s}}\left[(v^4 - 18v^2 + 33)\log\frac{1+v}{1-v} + v(31v^2 - 59)\right], \tag{19}$$

where $v = \sqrt{1 - 4m_Q^2/\hat{s}}$ is the velocity of quark Q in the center of mass frame, $\hat{s}$ is the squared energy in the center of momentum frame of the gluons, μ is a typical mass scale of the process, and the contribution $\Delta\sigma^{\Phi}(gg{\to}Q\bar{Q}, \mu)$ to this process from the scalar Φ.

For the contribution $\Delta\sigma^{\Phi}(gg{\to}Q\bar{Q}, \mu)$ we have found the expression

$$\Delta\sigma^{\Phi}(gg{\to}Q\bar{Q}, \mu) = \tag{20}$$

$$= \frac{\tilde{C}_{\Phi}^{(1)}}{64}\,\frac{\alpha_s^2(\mu) m_Q}{\pi\sqrt{\hat{s}}}\,\frac{{\rm Re}\left[(\hat{s} - m_{\Phi}^2 - i m_{\Phi}\Gamma_{\Phi})\left(-h^{S^*}_{\Phi Q\bar{Q}} v^2 F_{\Phi}^S(\hat{s}) - h^{P^*}_{\Phi Q\bar{Q}} F_{\Phi}^P(\hat{s})\right)\right]}{(\hat{s} - m_{\Phi}^2)^2 + m_{\Phi}^2\Gamma_{\Phi}^2}\log\frac{1{+}v}{1{-}v} +$$

$$+ \frac{\tilde{C}_{\Phi}^{(2)}}{1024}\,\frac{\alpha_s^2(\mu) v\hat{s}}{\pi^3}\,\frac{\left|h^S_{\Phi Q\bar{Q}}\right|^2 v^2 + \left|h^P_{\Phi Q\bar{Q}}\right|^2}{(\hat{s} - m_{\Phi}^2)^2 + m_{\Phi}^2\Gamma_{\Phi}^2}\left(|F_{\Phi}^S(\hat{s})|^2 + |F_{\Phi}^P(\hat{s})|^2\right) \tag{21}$$

where the form factors $F_{\Phi}^{S,P}(\hat{s})$ are given by the expressions (13),(15),(16) and $\tilde{C}_{\Phi}^{(1)}$, $\tilde{C}_{\Phi}^{(2)}$ are the color factors with

$$\tilde{C}_{\Phi_0}^{(1)} = C_{ab}C_{ab} = 2, \qquad \tilde{C}_{\Phi_0}^{(2)} = C_{ab}C_{ab}\, n_c = 6, \tag{22}$$

$$\tilde{C}_{\Phi_8}^{(1)} = C_{abc}C_{abc} = 5/6, \qquad \tilde{C}_{\Phi_8}^{(2)} = C_{abc}C_{abc}/2 = 5/12 \tag{23}$$

for the color singlet Φ_0 and for the color octet Φ_8 respectively, n_c is the number of colors of the $SU_c(n_c)$ group, the numerical values in (22), (23) correspond to the $SU_c(3)$ group.

The total cross section $\sigma_{tot}(pp \to t\bar{t})$ of the $t\bar{t}$ production in pp-collisions with account of the contribution of scalar octet F_2 can be written as the sum

$$\sigma_{tot}(pp \to t\bar{t}) = \sigma^{\rm SM}(pp{\to}t\bar{t}) + \Delta\sigma^{F_2}(pp{\to}t\bar{t}) \tag{24}$$

of the SM cross section $\sigma^{\rm SM}(pp{\to}t\bar{t})$ and the contribution $\Delta\sigma^{F_2}(pp{\to}t\bar{t})$ induced by scalar gluon F_2 via effective vertex (12)–(16).

Table 1. Summary of the most precise ATLAS and CMS measurements of the total $t\bar{t}$ cross sections at 7, 8 and 13 TeV.

Exp.	$\sqrt{s}$ [TeV]	$\mathcal{L}$ [fb^{-1}]	$\sigma^{\rm tot}(t\bar{t})$ [pb]	Ref.
ATLAS	7	4.6	182.9 ± 3.1 (stat) ± 4.2 (sys) ± 3.6 (lumi) ± 3.3 (bm)	[23]
	8	20.3	242.4 ± 1.7 (stat) ± 5.5 (sys) ± 7.5 (lumi) ± 4.2 (bm)	[23]
	13	3.2	818 ± 8 (stat) ± 27 (sys) ± 19 (lumi) ± 12 (bm)	[24]
CMS	7	5.0	173.6 ± 2.1 (stat)$^{+4.5}_{-4.0}$ (sys) ± 3.8 (lumi)	[25]
	8	19.7	244.9 ± 1.4 (stat)$^{+6.3}_{-5.5}$ (sys) ± 6.4 (lumi)	[25]
	13	2.3	835 ± 3 (stat) ± 23 (sys) ± 23 (lumi)	[26]

We obtain the total cross section from partonic cross sections (19), (21) by integrating the expression

$$\frac{d\sigma_{tot}(pp \to t\bar{t})}{dx_1 dx_2} = \sum_k F_k^{p\bar{p}}(x_1, x_2, \mu_f) K(s) \sigma_0^{SM}(\bar{q}_k q_k \to t\bar{t}, \mu) +$$
$$+ F_g^{pp}(x_1, x_2, \mu_f) K(s) \sigma_0^{SM}(gg \to t\bar{t}, \mu) + F_g^{pp}(x_1, x_2, \mu_f) \Delta\sigma^{F_2}(gg \to t\bar{t}, \mu) \quad (25)$$

over the variables $0 \leq x_1, x_2 \leq 1$, where x_1, x_2 are partonic parts of the momenta of protons, $\hat{s} = x_1 x_2 s$, $s = (P_1 + P_2)^2$, P_1, P_2 are the momenta of the colliding protons, $\sigma_0^{SM}(\bar{q}_k q_k \to t\bar{t}, \mu)$ is the well known SM LO cross section, $K(s)$ is the K-factor, which we use for the better agreement of the SM LO predictions of the cross section of $t\bar{t}$-pair production with the corresponding aNNNLO SM predictions [22]. The partonic functions in (25) are given as

$$F_k^{pp}(x_1, x_2, \mu_f) = f_{q_k}^p(x_1, \mu_f) f_{\bar{q}_k}^p(x_2, \mu_f) + f_{\bar{q}_k}^p(x_1, \mu_f) f_{q_k}^p(x_2, \mu_f), \quad (26)$$
$$F_g^{pp}(x_1, x_2, \mu_f) = f_g^p(x_1, \mu_f) f_g^p(x_2, \mu_f), \quad (27)$$

where $f_{q_k}^p(x, \mu_f)$, $f_{\bar{q}_k}^p(x, \mu_f)$, $f_g^p(x, \mu_f)$ are the parton distribution functions of quark q_k of flavor k, antiquark $\bar{q}_k$ and gluons in the proton, μ_f is the factorization scale.

For numerical calculations we use the analytical expressions for scalar Passarino-Veltman integrals A_0, B_0, C_0 from the Ref. [21] and we also perform the cross check with using numerical package LoopTools/FF [27, 28].

For calculations we use the parton distribution functions MMHT 2014 [29] (NNLO, $\mu = \mu_f = m_t$, $m_t = 173.21$ GeV), with the values of K-factors $K(s) = 1.6687, 1.6752, 1.6833$ for energies $\sqrt{s} = 7, 8, 13$ TeV respectively, in this case the cross section $\sigma^{\rm SM}(pp \to t\bar{t})$ reproduces well the aNNNLO SM predictions for the cross section of $t\bar{t}$ production [22] Also we perform cross check our partons integrations with use PDFs CT14 and METAv10 in the ManeParse (package for the Wolfram Mathematica for parsing various PDF functions) [30], the difference between these results was about 1%.

We have calculated the contributions $\Delta\sigma^{F_2}(pp \to t\bar{t})$ to the total cross section of the $t\bar{t}$ production from the scalar gluon F_2 defined by the last term in the expression (25) for $\sqrt{s} = 7, 8, 13$ TeV, $m_{F_2} = 400 \div 1000$ GeV and $\sin\beta = 0.4 \div 1$, $\Delta\sigma^{F_2}(pp \to t\bar{t})$ varies from 0.3 to 54.9 pb (depending on $\sqrt{s}, m_{F_2}, \sin\beta$) [19].

From current most precise ATLAS and CMS measurements of the total $t\bar{t}$ cross sections at 7, 8 and 13 TeV (Table 1) we found exclusion area at the $(m_{F_2} - \sin\beta)$-plane, the area is shown in the Fig. 3.

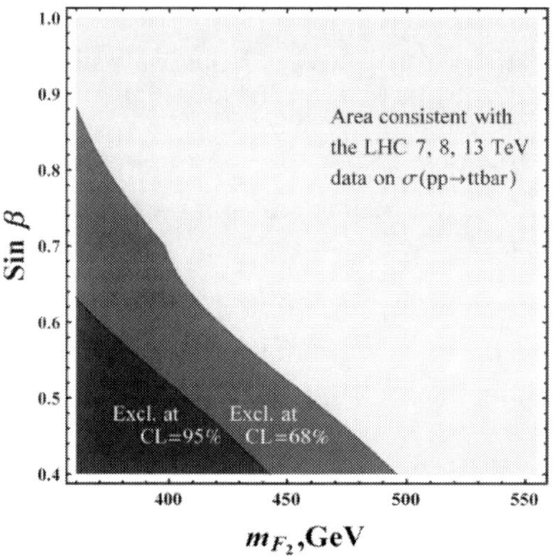

Figure 3. The $(m_{F_2} - \sin\beta)$-plane exclusion regions (at 95% and 68% probability level) resulting from the LHC data on total $t\bar{t}$ cross sections at 7, 8 and 13 TeV [23–26].

Using the known relations between the variables x_1, x_2 and the invariant mass $m_{t\bar{t}}$ of $t\bar{t}$-pair and the rapidity y of the final t-quark

$$m_{t\bar{t}}^2 = x_1 x_2 s, \quad y = \ln\frac{x_1}{x_2}, \qquad x_{1,2} = \frac{m_{t\bar{t}}}{\sqrt{s}}\, e^{\pm y/2} \tag{28}$$

and integrating the expression (25) over the rapidity y with account of the scalar gluon contribution $\Delta\sigma^{F_2}(gg{\to}t\bar{t}, \mu_f)$ we obtain the invariant mass spectrum $d\sigma(pp \to t\bar{t})/dm_{t\bar{t}}$ in the form

$$\frac{d\sigma_{tot}(pp \to t\bar{t})}{dm_{t\bar{t}}} = \frac{m_{t\bar{t}}}{s}\int_{-\ln(s/m_{t\bar{t}}^2)}^{+\ln(s/m_{t\bar{t}}^2)} \frac{d\sigma_{tot}(pp \to t\bar{t})}{dx_1 dx_2}\, dy. \tag{29}$$

In the same way but with neglect of the scalar gluon contribution $\Delta\sigma^{F_2}(gg{\to}t\bar{t}, \mu_f)$ we obtain the background invariant mass spectrum $d\sigma_b(pp \to t\bar{t})/dm_{t\bar{t}}$ which is in agreement with the theoretical predictions [31] and with the experimental results [32–34].

With using experimental data on the $t\bar{t}$ invariant mass spectrum (LHC ATLAS 8 TeV, L $= 20.3\,\text{fb}^{-1}$ [35]) we found exclusion area at the $(m_{F_2} - \sin\beta)$-plane, that is shown in the Fig. 4.

In conclusion, we summarize the results of this paper.

The effective vertex of interaction of the scalar color octet with two gluons is calculated with account of the one loop quark contribution. With account of this interaction the contribution of the scalar color octet to the partonic cross section of resonance $Q\bar{Q}$-pair production in the gluon fusion is calculated.

The total and differential cross sections of the $t\bar{t}$ production in pp-collisions at the LHC are calculated with account of the resonance contribution of scalar color octet F_2 predicted by the minimal model with the four color quark-lepton symmetry and analyzed in dependence on two parameters of the model, the F_2 mass m_{F_2} and mixing angle β.

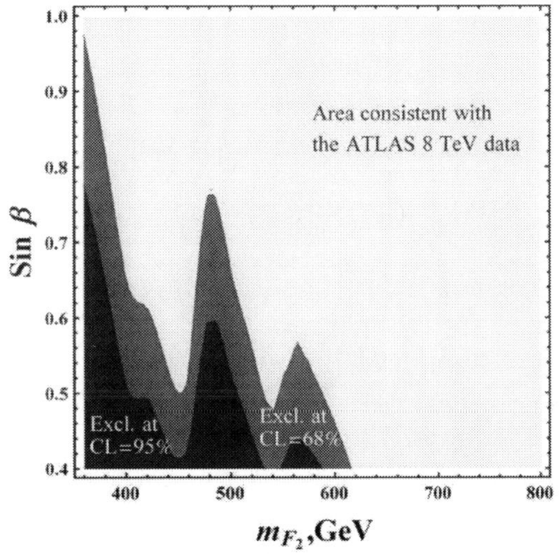

Figure 4. The $(m_{F_2} - \sin\beta)$-plane exclusion regions (at 95% and 68% probability level) resulting from the LHC ATLAS data on $t\bar{t}$ invariant mass spectrum at 8 TeV [35].

From the comparison with the LHC data on total (at $\sqrt{s} = 7, 8, 13$ TeV) and the differential cross sections (at $\sqrt{s} = 8$ TeV) of $t\bar{t}$ production it is shown that there is region of $(m_{F_2} - \sin\beta)$-plane of exclusion by these data. But for $\sin\beta = 1$ and for all the masses m_{F_2} the scalar color octet F_2 gives the contribution to this process of about a few percents and can not be visible in these data.

The work is done within "YSU Initiative Scientific Researches" (Project No. AAAA-A16-116070610023-3)

References

[1] J.C. Pati, A. Salam, Phys. Rev. D **10**, 275 (1974)
[2] A.D. Smirnov, Phys. Lett. B **346**, 297 (1995), hep-ph/9503239
[3] A.D. Smirnov, Phys. At. Nucl. **58**, 2137 (1995), [Yad. Fiz. 58, 2252 (1995)]
[4] A.V. Manohar, M.B. Wise, Phys. Rev. **D74,** 035009 (2006), hep-ph/0606172
[5] M.I. Gresham, M.B. Wise, Phys. Rev. **D76,** 075003 (2007), 0706.0909
[6] M. Gerbush, T.J. Khoo, D. Phalen, A. Pierce, D. Tucker-Smith, Phys. Rev. **D77**, 095003 (2008), 0710.3133
[7] A.R. Zerwekh, C.O. Dib, R. Rosenfeld, Phys. Rev. **D77**, 097703 (2008), 0802.4303
[8] P.F. Perez, H. Iminniyaz, G. Rodrigo, Phys. Rev. **D78,** 015013 (2008), 0803.4156
[9] S.Y. Choi, M. Drees, J. Kalinowski, J.M. Kim, E. Popenda, P.M. Zerwas, Phys. Lett. **B672**, 246 (2009), 0812.3586
[10] M.V. Martynov, A.D. Smirnov, Mod. Phys. Lett. **A23**, 2907 (2008), 0807.4486
[11] M.V. Martynov, A.D. Smirnov, Phys. At. Nucl. **73**, 1207 (2010)
[12] D. Goncalves-Netto, D. Lopez-Val, K. Mawatari, T. Plehn, I. Wigmore, Phys. Rev. **D85**, 114024 (2012), 1203.6358

[13] A. Idilbi, C. Kim, T. Mehen, Phys. Rev. **D82**, 075017 (2010), 1007.0865
[14] S. Calvet, B. Fuks, P. Gris, L. Valery, JHEP **04**, 043 (2013), 1212.3360
[15] J. Kubo, K.S. Lim, M. Lindner, Phys. Rev.Lett. **113**, 091604 (2014), 1403.4262
[16] G. Grilli di Cortona, E. Hardy, A.J. Powell, JHEP **08**, 014 (2016), 1606.07090
[17] P.Y. Popov, A.V. Povarov, A.D. Smirnov, Mod. Phys. Lett. **A20**, 3003 (2005), hep-ph/0511149
[18] A.V. Povarov, P.Yu. Popov, A.D. Smirnov, Phys. Atom. Nucl. **70**, 739 (2007), [Yad. Fiz.70,771(2007)]
[19] I.V. Frolov, M.V. Martynov, A.D. Smirnov, Mod. Phys. Lett. **A31**, 1650224 (2016), 1610.08409
[20] G. Passarino, M.J.G. Veltman, Nucl. Phys. **B160**, 151 (1979)
[21] A. Denner, Fortschr. Phys. **41**, 307 (1993), 0709.1075
[22] N. Kidonakis, Phys. Rev. **D90**, 014006 (2014), 1405.7046
[23] G. Aad et al. (ATLAS), Eur. Phys. J. **C74**, 3109 (2014), [Addendum: Eur. Phys. J.C76,no.11,642(2016)], 1406.5375
[24] M. Aaboud et al. (ATLAS), Phys. Lett. **B761**, 136 (2016), 1606.02699
[25] V. Khachatryan et al. (CMS), JHEP **08**, 029 (2016), 1603.02303
[26] A.M. Sirunyan et al. (CMS) (2017), 1701.06228
[27] T. Hahn, M. Perez-Victoria, Comput. Phys. Commun. **118**, 153 (1999), hep-ph/9807565
[28] G.J. van Oldenborgh, Comput. Phys. Commun. **66**, 1 (1991)
[29] L. Harland-Lang, A. Martin, P. Motylinski, R. Thorne, Eur.Phys.J. **C75**, 204 (2015), 1412.3989
[30] D.B. Clark, E. Godat, F.I. Olness, Comput. Phys. Commun. **216**, 126 (2017), 1605.08012
[31] V. Ahrens, A. Ferroglia, M. Neubert, B.D. Pecjak, L.L. Yang, JHEP **1009**, 097 (2010), 1003.5827
[32] G. Aad et al. (ATLAS), Phys. Rev. **D90**, 072004 (2014), 1407.0371
[33] S. Chatrchyan et al. (CMS), Eur.Phys.J. **C73**, 2339 (2013), 1211.2220
[34] V. Khachatryan et al. (CMS), Eur. Phys. J. **C75**, 542 (2015), 1505.04480
[35] G. Aad et al. (ATLAS), Eur. Phys. J. **C76**, 538 (2016), 1511.04716

The NMSSM dark sector constraints

Albina V. Gurskaya[1,⋆] and *Mikhail* V. Dolgopolov[1,⋆⋆]

[1] *Samara University, Academician Pavlov 1, 443011 Samara, Russia*

Abstract. In this paper authors present review of neutralino research results in LHC, cosmological observations and theoretical models. Nowdays there are constraints on the mass of dark matter particle from experiments. The neutralino as a candidate for the dark matter particle in the framework of supersymmetry is considered. There are some discuss advantages for it about the NMSSM model. The properties of neutralino are functions of free parameters of the model. In agreement with the foregoing, constraints on free parameters of the NMSSN are determined.

1 Introduction

Despite the fact that the problem of dark matter originates in the astronomical observation [1], this phenomenon can be explained in the framework of quantum field theory. There is the reason to believe that dark matter consists of new hypothetical particles. In particular, we can make this conclusion based on the paper [2] in which the cosmic microwave background (CMB) was investigated. At the time of formation of the CMB, the baryonic matter contribution to the full density of the Universe was one fifth. It means that 80% [3] was the dark matter at that time. It can be concluded that the phenomenon of dark matter is the cosmological only and a discussion on new particles is also relevant.

The most popular candidate is the supersymmetric neutralino. Supersymmetric models can also differ in composition of fields, and therefore, the neutralino properties may be different in models. The simplest extension of the standard model MSSM [4] has a very convincing basis. MSSM Superpotential contains the so-called μ-term ($\mu H_1 H_2$) including dimensional mass parameter of the Higgs field μ, the value of which cannot be accounted for and matched to masses scale of order TeV to ensure an acceptable range of Higgs bosons masses. Within NMSSM [5] μ-term is dynamically generated by spontaneous symmetry breaking ($\lambda S(H_1 H_2)$, where $< S >= \upsilon_3/\sqrt{2}$ and $\mu = \lambda\upsilon_3/\sqrt{2}$). This is the most reasonable, moreover, due to new Higgs field the model becomes more flexible to build new parametric scenarios.

Neutralino is not the only candidate for the role of dark matter particles. There is also a prediction of the axion in the NMSSM. It is necessary to note that an appearance of the axion can be realize in various symmetries (R-symmetry [6], Peccei-Quinn [7]). We can exclude this particle picking the appropriate components in superpotential Higgs. In particular, the component $\sim \kappa S^3$ [5] provides Peccei–Quinn violation.

⋆e-mail: a-gurska@yandex.ru
⋆⋆e-mail: mvdolg@yandex.ru

DOI: 10.1051/epjconf/201715802010

Previous papers were devoted to NMSSM with violation of CP invariance in the Higgs sector [8]. In the models describing elementary particles special place is prepared for the Higgs boson. It was shown that the Higgs field strongly associated with the problem of the baryon asymmetry[9].

Additional CP violating phases enrich scenarios for the physical parameters values of the model, however, it can be difficult to coexist in the case of CP-even and CP-odd state Higgs bosons mixing. The problem of the mass calculating in this case leads to the rigidly fine tuning parameters fixing.

2 Experimental data for dark matter particles

There is no conclusive evidence of supersymmetric particles up to now. The data of the Large Hadron Collider (LHC) give us constraints for supersymmetric particles masses. The main attention should be paid to quarks superpartners (squarks) and gluons (gluino). Such particles participate in the strong interaction and must be well created in protons collisions. Therefore gluino with the mass of 1-2 TeV can be discovered at the LHC. But here we have constraints above 1 TeV already. The figure 1 illustrate this situation.

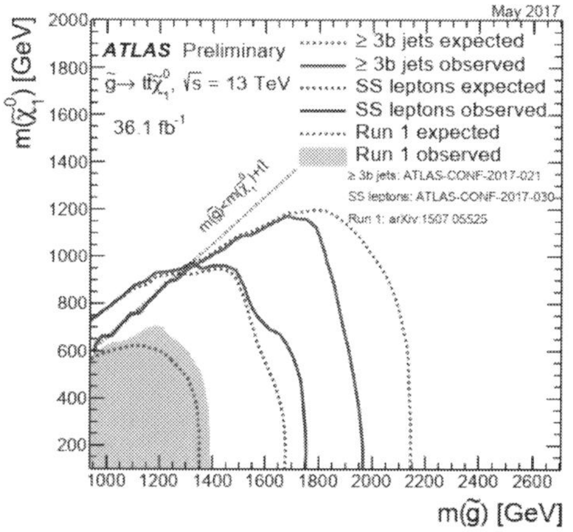

Figure 1. The neutralino mass dependence on the gluino mass in the decay to the top quark pair and an invisible neutralino [10].

The absence of a signal allows to exclude the areas bounded by solid curves. The gray area has been closed according to the results of Run 1. Along with the considered process also there are many other results.

Different supersymmetric models are checked on the LHC at this moment. For example, figure 2 shows fragment of data set of results of the ATLAS collaboration. There are no signals of neutralino and we have to conclude that constraints for this particle may be above 1-2 TeV and larger. Data which is presented in figures 1 & 2 is relevant for May 2017.

But some modified models give different predictions and can allow the light particle. We will discuss this below. In particular, the NMSSM has soft requirements of physical parameters values for neutralino. In addition, for example, the study of Higgs boson invisible decays to neutralino is one of the important and interesting problem in practice and in theory [11].

With regard to direct detection experiments they are designed to detect the nuclear recoil in the scattering of galactic WIMPs on nuclei target. The signal rate depends on the local density and

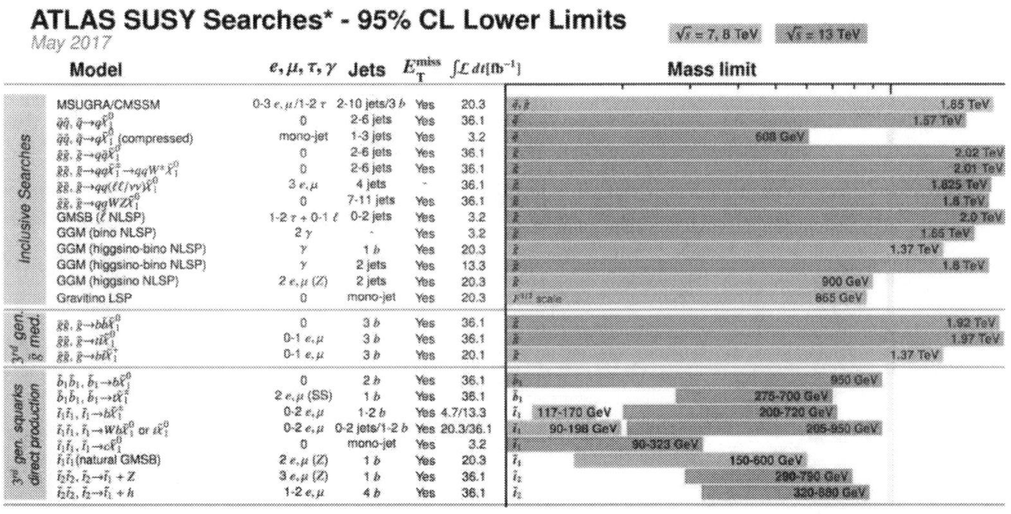

Figure 2. Mass reach of ATLAS searches for Supersymmetry. Fragment of the table [10].

velocity distribution of WIMPs in our Galaxy, the WIMP mass, and the interaction cross-section on the target nuclei [12]. We have no prediction on mass in this case yet, but the experiment gives us constraints on the WIMP–neutron scattering cross-section correlating dark matter particle mass. In the next figure 3 we can see data of 2017 year [13].

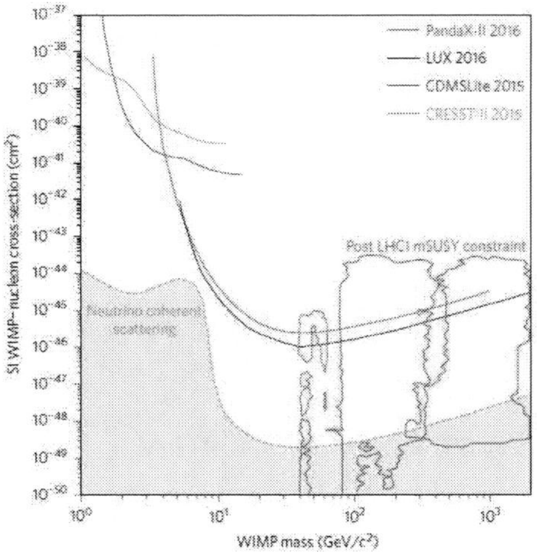

Figure 3. Upper limits on the spin-independent WIMP–nucleon scattering cross-section set by current leading experiments. The limit curves are from PandaX–II [14], LUX[15], SuperCDMS (CDMSLite)[16] and CRESST–II [17].

The sensitivity of detectors for the direct search of dark matter to the mass of the particle and to the cross section can achive the neutrino area. The scattering processes of neutrinos are background and represent the difficult in the registration of new particles. There is acceptable area for supersymmetric particles, but even here the neutrino background is the place to be.

There is amount of continuing attempts to observe the variety of products of annihilation or decay of dark matter: positrons, neutrinos, antiprotons, etc. However, there is indirect benefit to search them from photons. Photons are not deflected by magnetic fields and clearly indicate the source. In the paper [18] there is the description of methods for detecting of dark matter annihilations. It should be noted that the cross section of annihilation may not match the sensitivity of the telescopes. However in the past few years gamma-ray observations give some stringent constraints for annihilating dark matter. The lower limits of dark matter mass annihilating via e^+e^-, $\mu^+\mu^-$, $\tau^+\tau^-$, gg, $u\bar{u}$ and b^+b^- channels are 40 GeV, 28 GeV, 30 GeV, 57 GeV, 58 GeV and 66 GeV [19].

3 Theoretical aspects of constrains for physical properties of neutralino

From the theory point of view all physical properties of the models are functions of free parameters, in particular, the physical properties of the neutralino also. In this report we has interest in the NMSSN dark sector.

The definition of the basis in the neutralino sector is following:

$$\psi^0 = \left(-i\tilde{b}^0, -i\tilde{\omega}^3, \tilde{h}_1^0, \tilde{h}_2^0, \tilde{s}\right)^{\mathrm{T}}, \tag{1}$$

here $\tilde{b}^0$, $\tilde{\omega}^3$, $\tilde{h}_1^0$, $\tilde{h}_2^0$, $\tilde{s}$ – bino, wino, higgsino and singlino spinors. The mass part of the Lagrangian has the form:

$$\mathcal{L}_m = -\frac{1}{2}\,(\psi^0)^{\mathrm{T}}\, Y\, \psi^0 + \text{h.c.}, \tag{2}$$

where Y is:

$$Y = \begin{pmatrix} M_1 & 0 & -M_Z\, s_W\, c_\beta & M_z\, s_W\, s_\beta & 0 \\ 0 & M_2 & M_Z\, c_W\, c_\beta & -M_Z\, c_W\, s_\beta & 0 \\ -M_Z\, s_W\, c_\beta & M_Z\, c_W\, c_\beta & 0 & -\lambda v_3 & -\lambda v_2 \\ M_z\, s_W\, s_\beta & -M_Z\, c_W\, s_\beta & -\lambda v_3 & 0 & -\lambda v_1 \\ 0 & 0 & -\lambda v_2 & -\lambda v_1 & 2\kappa v_3 \end{pmatrix}. \tag{3}$$

Here M_1, M_2– U(1) and SU(2)– mass parameters of the soft violations supersymmetry fields $\tilde{b}^0$, $\tilde{\omega}^3$ accordingly; $s_W = \sin\theta_W$, $c_W = \cos\theta_W$; θ_W – the electroweak mixing angle (Weinberg angle); $s_\beta = \sin\beta$, $c_\beta = \cos\beta$, M_Z – mass of Z-boson.

The following parameters have no fixed values and also they have no predictions from the model and their value is unknown in the experiment [20, 21]:

$$1.0 < tg\beta \le 60, \quad M_1 = M_2, \quad 100\,\text{GeV} \le M_2 \le 2000\,\text{GeV}, \tag{4}$$
$$0.0001 \le \lambda \le 0.7, \quad 0 \le \kappa \le 0.65.$$
$$0\,\text{GeV} \le A_\lambda \le 1000\,\text{GeV}, \quad -100\,\text{GeV} \le A_\kappa \le -10\,\text{GeV}$$

These ranges of values were obtained by scanning of free parameters space on the mass superparticles scale. They were used by the authors of the given report to study the neutralino masses in the MSSM and NMSSM in the case of CP-invariance in the Higgs sector [22]. It is interesting to note that the values of the lightest neutralino NMSSM achives to 962.9 GeV, no larger, which is irrelevant to experimental limitations at the LHC, however, do not contradict the data of other methods for dark matter particles detection. But this value in the model obtained at fixed free parameters, therefore, the key theoretical restriction on the properties of dark matter are the allowing areas for free parameters of the model.

There is the scan of the free parameters in the model given the constraints on the neutralino mass used data of annihilation of dark matter particles (as the lower constrain we choose 28 GeV) on figure 4.

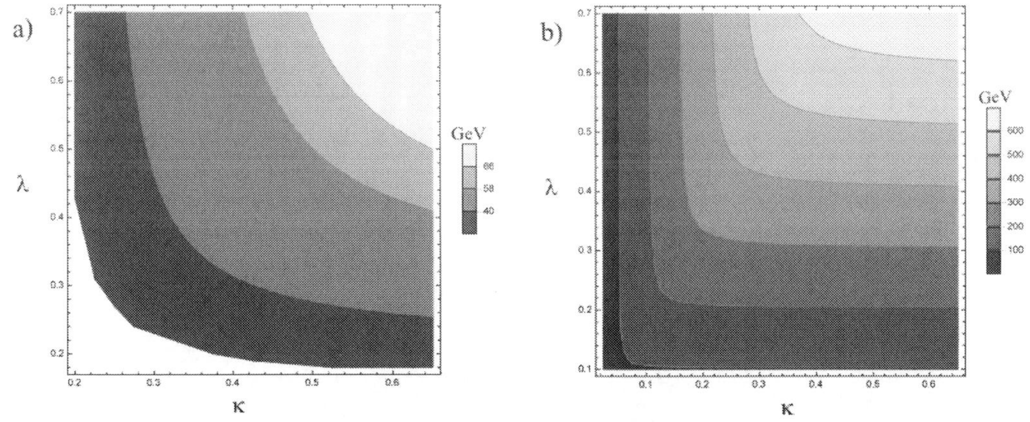

Figure 4. The space of parameters λ and κ correlating with low neutralino masses constrains. a) Two parameters are fixed: $M_2 = 2000$ GeV and $v_3 = 200$ GeV. The white area corresponds to forbidden values of λ and κ. b) It is the case for $M_2 = 2000$ GeV and $v_3 = 1000$ GeV.

The strong constraint on parameters may be the presence of CP-violating phases in the extended Higgs sector. We can define spontaneous violation of CP-invariance directly in the determination of doublet fields with use of phases θ and φ:

$$\Phi_1 = \begin{pmatrix} \phi_1^- \\ \frac{1}{\sqrt{2}}(v_1 + \phi_1^0 + i\chi_1) \end{pmatrix}, \quad \Phi_2 = e^{i\theta}\begin{pmatrix} \phi_2^+ \\ \frac{1}{\sqrt{2}}(v_2 + \phi_2^0 + i\chi_2) \end{pmatrix}, \quad S = \frac{1}{\sqrt{2}} e^{i\varphi}(v_3 + \phi_3^0 + i\chi_3) \tag{5}$$

We can see that vacuum expectation values (vevs) of Higgs fields are included into the neutralino mass matrix. In this case there are more possibilities for construct different scenarios of investigation. But if we will add mixing of CP-odd and CP-even Higgs bosons states the number of possibilities will decrease radically. This way CP violation was considered in the work [8]. In this case there is the problem to calculate the mass of the Higgs boson, therefore, it remains limited number of scenarios studies [23] in accordance with local minimum physical condition. Consequently, there are only two values for the particle candidate of dark matter. Also there are constrains on parameters from Electric Dipole Moments [24].

4 Conclusion

We have reviewed the existing limits on masses of dark matter particles in the various experimental data. It should be noted that supersymmetric particle neutralino is still relevant for the study. Moreover, non-minimal supersymmetric model can quite well describe the available results and predict new scenarios of research. However, it is necessary to consider changing the ranges of possible values of free parameters. So, we can conclude that the mass of the lightest neutralino does not exceed 1 TeV. In this case, the value υ_3 has a strong influence on the mass range. As can be seen from the results of the scan region of NMSSM free parameters the mass of the neutralino reaches close to 1 TeV at values of $\upsilon_3 = 1000$ GeV. We also indicated that when considering CP-violation free parameters become larger, but this does not always lead to the valuable possibilities for theoretical predictions. In addition, there are some other ways to restrict values of the parameters, for example, the Electric Dipole Moments calculation.

References

[1] F. Zwicky, ApJ **V.86**, 217 (1937).
[2] D. N. Spergel, L. Verde, H. V. Peiris, E. Komatsu, M. R. Nolta, C. L. Bennett, M. Halpern, G. Hinshaw, N. Jarosik, A. Kogut, et al., ApJS **148**, 175 (2003), astro-ph/0302209.
[3] D. S. Gorbunov, V. A. Rubakov, Introduction to the theory of the early Universe (Lenand, 2016).
[4] J.Rosiek, Physical Review D, **V.41**, 3464–3501 (1990).
[5] M.Maniatis, IJMP A., **V.25**, 3505–2602 (2010).
[6] B.A. Dobrescu, K.T. Matchev, JHEP, **V.2000**, (2000).
[7] D.J. Miller, R. Nevzorov, arXiv:hep-ph/0309143v1
[8] Gurskaya A.V., Dolgopolov M.V., Vestn. Samar. Gos. Tekhn. Univ. Ser. Fiz.-Mat. Nauki **20:4**, 581–588 (2016).
[9] A. D. Sakharov, JETP Lett. **5**, 24 (1967).
[10] https://twiki.cern.ch/twiki/bin/view/AtlasPublic/SupersymmetryPublicResults
[11] Mendez M.del R.A. et al., Journal of Physics:Conference Series, **V.468**, 012006 (2013).
[12] Lewin, J. D., Smith, P. F., Astropart. Phys. **6**, 87–112 (1996).
[13] J.Liu, X.Chen, X.Ji, Nature Physics, **13**, 212–216 (2017).
[14] Tan, A. et al. (PandaX-II Collaboration)Phys. Rev. Lett. **117**, 121303 (2016).
[15] Akerib, D. S. et al. (LUX Collaboration) Phys. Rev. Lett. **118**, 021303 (2017).
[16] Agnese, R. et al. (SuperCDMS Collaboration) Phys. Rev. Lett. **116**, 071301 (2016).
[17] Angloher, G. et al. (CRESST Collaboration) Eur. Phys. J. C **76**, 25 (2016).
[18] G.Bertone, D.Hooper, J.Silk, Phys. Rep. **405**, 279 (2005).
[19] Chan, M.H. Astrophys Space Sci (2017) 362: 147.
[20] D.A.Vasquez, G.Belanger, C.Boehm, A.Pukhov, J.Silk Phys. Rev. D **82**, 115027 (2010).
[21] M. M. Almarashi, S. Moretti Phys. Rev. D **83**, 035023 (2011).
[22] Gurskaya A.V., Filippov Yu.P. Teoreticheskaya fizika [Theoretical Physics], **12**, 69–81 (2013).
[23] Gurskaya A.V., Dolgopolov M.V., EPJ Web of Conferences, **V.125**, 02011 (2016)
[24] S.F.King, M.Muhlleitner, R.Nevzorov, K.Walz, Nuclear Physics B **901**, 526-555 (2015).

Supersymmetric models with broken Lorentz invariance.

Arthur Marakulin[1,⋆] and *Sergey* Sibiryakov[1,2,3,⋆⋆]

[1] *Institute for Nuclear Research of the Russian Academy of Sciences, Moscow, Russia*
[2] *CERN Theory division, Geneva, Switzerland;*
[3] *Institut de Physique, EPFL, Lausanne, Switzerland*

Abstract. Several supersymmetric theories with broken Lorentz invariance are considered. We study at the component level Lorentz violating representations of the supersymmetry algebra and construct Lagrangians for the scalar and vector supermultiplets with broken Lorentz invariance. Lorentz violating model for the gravitational supermultiplet is constructed using the superfield formalism as supersymmetric extension of the linearized Einstein-aether theory. The most general Lagrangian of the linearized Einstein-aether supergravity is constructed. We show that the Lagrangian for this model is unique and obtain its bosonic part in components. The constraints imposed by supersymmetry on the parameters of the theory are obtained. The phenomenological consequences of the model are discussed.

1 Introduction

Incompatibility of the principles of quantum field theory with general relativity is one of the most important problems in modern theoretical physics. A potential way out of this situation consists in restricting the domain of validity of some basic postulates of general relativity and abandoning them at high energy scales.

There is a number of motivations for abandoning Lorentz invariance at small scales. The first motivation to study models with Lorentz breaking is their rich phenomenology[1]. Another motivation comes from the theory. A number of Lorentz violating models of gravity demonstrate better quantum behavior than general relativity. A promising approach to quantization of gravity based on abandoning the Lorentz invariance has been proposed by Hořava [2] and extended by several authors [3, 4]. Hořava gravity is closely related[5, 6] to the so-called Einstein-aether model [1]. In the latter model violation of the Lorentz invariance is described by the time-like vector field u^m with unit norm ($u^m u_m = -1$) called aether that minimally couples to the Einstein-Hilbert action for gravity,

$$S = S_{GR} - \frac{1}{2}\int d^4x\,\sqrt{-g}\left(c_1\left(\nabla_n u_m\right)^2 + c_2\left(\nabla_m u_m\right)^2 + c_3\nabla_n u_m \nabla^m u^n - c_4 u^r u^s \nabla_r u_m \nabla_s u^m\right). \qquad (1)$$

The theory contains four dimensionless parameters $c_i, i = 1, 2, 3, 4$. In general case of the theory there are strong constraints on these parameters, because the PPN-parameters depends on the constants in the Einsten-aether model [1]

⋆e-mail: marakulin@physics.msu.ru
⋆⋆e-mail: sergey.sibiryakov@cern.ch

$$\alpha_1 = -8\frac{c_3^2 + c_1c_4}{2c_1 - c_1^2 + c_3^2}, \tag{2}$$

$$\alpha_2 = \frac{\alpha_1}{2} - \frac{(c_1 + 2c_3 - c_4)(2c_1 + 3c_2 + c_3 + c_4)}{(c_1 + c_2 + c_3)(2 - c_1 - c_4)}, \tag{3}$$

and these are constrained from observations in the Solar System [7]:

$$|\alpha_1| \lesssim 10^{-4}; \ |\alpha_2| \lesssim 4 \times 10^{-7} \tag{4}$$

A phenomenologically acceptable theory with Lorentz violation in the ultraviolet must incorporate a mechanism that ensures emergence of Lorentz invariance at low energies. One of the possible mechanisms is based on non-relativistic supersymmetry [8, 9]. As remarked, supersymmetry does not necessarily require Lorentz invariance. SUSY enforces emergence of Lorentz symmetry at low energies in the Standard Model, even if the high-energy theory is not Lorentz invariant. It is the natural motivation to consider supersymmetric models with Lorentz breaking for better theoretical understanding of their structure.

The main purpose of the present work is to consider the existence and uniqueness of the supersymmetric Lorentz violating field theories. We study at the component level Lorentz violating representations of the supersymmetry algebra and construct Lagrangians for the scalar and vector supermultiplets with broken Lorentz invariance. Lorentz violating model for the gravitational supermultiplet is constructed using the superfield formalism as supersymmetric extension of the linearized Einstein-aether theory. We consider the coupling of supersymmetric aether to supergravity. We use the superfield formulation of the linearized non-minimal supergravity to construct a Lagrangian for the aether field coupled to linearized gravity and prove that it is unique, up to the choice of a single parameter.

2 Supersymmetry and Lorentz breaking

Apart from the paper[9], we start from the on-shell field theories in components. Let's consider massless supermultiplet (A, Ψ^α) consisting of the scalar and spinor fields. To analyze the existence and unity of the scalar supermultiplet theory with broken Lorentz invariance we consider the relevant representation for the supersymmetry algebra with Lorentz invariance broken by the timelike vector u_m with unit norm. It is possible to obtain after straightforward calculations, that supersymmetry algebra closen on-shell as follows: $[\delta_\xi\delta_\eta]X = 2i\left(\eta\sigma_m\bar{\xi} - \xi\sigma_m\bar{\eta}\right)\partial_m X$, takes the form

$$\begin{cases} \delta_\xi A = \xi\Psi \\ \delta_\xi\Psi_\alpha = 2i\left(\sigma_m\bar{\xi}\right)_\alpha \partial_m A + C_1 i u_m \xi_\alpha \partial_m \bar{A} + C_2 i u_m \left(\sigma_{mn}\xi\right)_\alpha \partial_n \bar{A} \end{cases} \tag{5}$$

It is the most general form of the supersymmetric transformations in the relevant case. The requirement to the theory to be set by Lagrangian, consistent with the equations of motion and invariant under the supersymmetric transformations, leads us to the constraints

$$C_1 = C_2 = 0, \tag{6}$$

and to the consequence that there is no non-trivial Lorentz violating supersymmetric model containing the most simple set of fields.

Thus, there is no any new possibilities for the constructing the supersymmetic model with scalar and spinor without Lorentz invariance for the scalar supermultiplet. Another analysis is required if we consider several scalar and several spinor fields $\left(A^i, \Psi^i_\alpha\right)$. In this case all of the coefficients will have two indices (C^{ij}), and the transformations from the supersymmetry algebra closed on-shell will be matrices.

$$\begin{cases} \delta_\xi A^i = \xi\Psi^i \\ \delta_\xi \Psi^i_\alpha = 2i\left(\sigma_m\bar{\xi}\right)_\alpha \partial_m A^i + C_1^{ij} i u_m \xi_\alpha \partial_m \bar{A}^j + C_2^{ij} i u_m \left(\sigma_{mn}\xi\right)_\alpha \partial_n \bar{A}^j \end{cases} \tag{7}$$

The requirement of the Lagrangian existence leads us to the constraints

$$C_1^{ij} = -C_1^{ji}, \tag{8}$$

$$C_2^{ij} = 0, \tag{9}$$

and to the non-trivial supersymmetric model with Lorentz breaking, containing a set of spinor and scalar fields

$$L = i\Psi^i\sigma_m\partial_m\bar{\Psi}^i + \bar{A}^i\Box A^i - \frac{i}{4}C_1^{ij}u_m\left(\Psi^i\partial_m\Psi^j + \text{h. c.}\right) - \frac{1}{2}\left(C_1^2\right)^{ij} u_m u_n \partial_m A^i \partial_n \bar{A}^j \tag{10}$$

Thus, existence of Lorentz violating models for the scalar supermultiplets depends on the number of scalars and spinors. This fact is important for the case of one scalar and one spinor, because the model in this case can be only trivial.

By the analogical calculations one can obtain there is no non-trivial Lorentz-violating supersymmetric models for the vector supermultiplet (v_m, λ_α). The existence of the models for the scalar supermultiplet depends on the number of scalars and spinors, and also on the fact if we require the existence of the Lagrangian of the theory. These results agree with the superfield formalism consequences[9], and may indirectly confirm the correctness of the superfield formalism in the Lorentz violating case.

The next natural question is if non-trivial theories for the gravitational multiplet exist. The first step in answering this question was made in [8]. The authors of the paper cosidered the supersymmetric extension of the aether model in terms of chiral vector superfiels:

$$U^m = u^m\left(x_L\right) + \sqrt{2}\theta\eta^m\left(x_L\right) + \theta^2 G^m\left(x_L\right) \tag{11}$$

where $\left(x_L^m = x^m + i\theta\sigma^m\bar{\theta}, \theta^\alpha\right)$. The unit norm constraint on the super-aether is similar to the ordinary case:

$$U_m U^m = -1 \tag{12}$$

The authors showed that the most general super-aether action

$$S = \int d^8 z f\left(U_m\bar{U}^m\right) + \int d^6 z \Lambda\left(U_m U^m + 1\right) \tag{13}$$

leads to the constraints:

$$c_2 + c_3 = 0, \tag{14}$$

$$c_4 = 0. \tag{15}$$

It has been also shown that in the supersymmetric extension of the Einstein-aether theory based on chiral vector super-aether field interactions between the aether and the fields of the Standard Model are suppressed.

This cannot be realized within the minimal $N = 1$ supergravity which does not allow to define a chiral aether vector superfield ($\bar{\nabla}_{\dot{\alpha}} U^a = 0$): anti-commutator of two supercovariant derivatives with spinor indices acting on a vector superfield does not vanish prohibiting to impose the chirality condition. This problem is resolved in the non-minimal formulation of the $N = 1$ supergravity.

3 Supergravity with broken Lorentz invariance

Non-minimal linearized supergravity is formulated [10] in terms of real superfield H^m and so-called linear compensator Γ ($\bar{D}^2\Gamma = 0$) as a gauge theory with the action

$$S_{SG} = \frac{1}{\varkappa^2} \int d^8 z \Big[\frac{1}{4}\Big((\partial_k H_m)^2 - (\Delta_k H_m)^2\Big) + \frac{n+1}{2n}(\partial_m H^m)^2 + \frac{n+1}{2}(\Delta_m H^m)^2 - \\ - i\frac{3n+1}{2n}\partial_m H^m \left(\Gamma - \bar{\Gamma}\right) + \frac{3n+1}{2}\Delta_m H^m \left(\Gamma + \bar{\Gamma}\right) + \frac{9n^2-1}{8n}(\Gamma^2 + \bar{\Gamma}^2) + \frac{(3n+1)^2}{4n}\Gamma\bar{\Gamma}\Big]. \quad (16)$$

where $\Delta_k H_m = \frac{1}{4}\bar{\sigma}_k^{\dot{\alpha}\alpha}[\bar{D}_{\dot{\alpha}}, D_\alpha]H_m$. The action is invariant under super-gauge transformations

$$\delta H_{\alpha\dot{\alpha}} = \bar{D}_{\dot{\alpha}} L_\alpha - D_\alpha \bar{L}_{\dot{\alpha}} \quad (17)$$

$$\delta\Gamma = -\frac{n+1}{4(3n+1)}\bar{D}^2 D^\alpha L_\alpha + \frac{1}{4}\bar{D}^{\dot{\alpha}} D^2 \bar{L}_{\dot{\alpha}} \quad (18)$$

We construct linearized Einstein-aether supergravity theory in terms the latter superfields and the chiral vector super-aether field which is considered as consisting of the real vacuum and small perturbation:

$$U^a = w^a + V^a \quad (19)$$

The super-aether field gauge transformation

$$\delta V^a = w^b M_b{}^a \quad (20)$$

where

$$M_{ab} = \frac{1}{4}(\sigma_{ab})_\beta{}^\alpha D_\alpha \bar{D}^2 L^\beta + \frac{1}{4}(\bar{\sigma}_{ab})^{\dot{\alpha}}{}_{\dot{\beta}} \bar{D}^{\dot{\beta}} D^2 \bar{L}_{\dot{\alpha}} \quad (21)$$

is consistent with the chirality condition,

$$\bar{D}_{\dot{\alpha}} V^c = -w^b \Phi_{\dot{\alpha}b}{}^c\ . \quad (22)$$

where

$$\Phi_{\dot{\alpha}bc} = -\frac{1}{4}(\sigma_{bc})_\alpha{}^\beta \bar{D}^2 D^\alpha H_{\beta\dot{\alpha}} - (\bar{\sigma}_{bc})^{\dot{\beta}}{}_{\dot{\alpha}} \bar{D}_{\dot{\beta}}\Gamma \quad (23)$$

is superconnection.

To carry out the Lagrangian of the theory one have to write down supersymmetric transformations of all possible independent terms quadratically dependent on the aether or gravitational superfields.

Coefficients in these transformations define the system of linear equation to be solved. A straightforward but tedious calculation yields [11] the most general and unique Lagrangian invariant under gauge transformations with one free parameter C:

$$S_{\ae} = \frac{C}{2\kappa^2}\int d^8z\Big[V_a\bar{V}^a + iw^a w^b\partial_a H_b(\Gamma - \bar{\Gamma}) + w^a w^b\Delta_a H_b(\Gamma + \bar{\Gamma}) +$$
$$+\frac{1}{4}(\Delta_k H_m\Delta^k H^m - \partial_k H_m\partial^k H^m - (\Delta_m H^m)^2 + (\partial_m H^m)^2) +$$
$$+\frac{i}{4}\partial_m H^m(\Gamma - \bar{\Gamma}) + \frac{1}{4}\Delta_m H^m(\Gamma + \bar{\Gamma}) + \frac{3}{8}(\Gamma^2 + \bar{\Gamma}^2)\Big], \quad (24)$$

4 From superfields to component fields

In the real world supersymmetry must be broken. As shown in [8], a soft breaking of supersymmetry gives mass to the imaginary component of the super-aether, whereas its real part remains massless. At low energies, the theory is described by the action (1), where the parameters c_i, $i = 1,2,3,4$, however, are constrained by supersymmetry. To find these constraints, we write the bosonic part of the Lagrangian (24) in terms of component fields. We use the bosonic parts of the gravitational superfield in the Wess-Zumino gauge [10] and use the local Lorentz transformations to make the tetrad symmetric $\left(e_{mn} = e_{nm} = \frac{1}{2}h_{mn}\right)$,

$$H_m\left(x,\theta,\bar{\theta}\right)|_{\text{bos}} = \frac{1}{2}\theta\sigma^n\bar{\theta}e_{mn}(x) + \theta^2\bar{\theta}^2 d_m(x); \quad (25)$$

We use the linear compensator in terms of the "left" coordinates $\left(x_L^m = x^m + i\theta\sigma^m\bar{\theta}, \theta^\alpha\right)$

$$\Gamma\left(x_L,\theta,\bar{\theta}\right)|_{\text{bos}} = \frac{n+1}{2(3n+1)}h(x_L) + \theta^2 B(x_L) + \theta\sigma^a\bar{\theta}(v_a(x_L) + i\mu_a(x_L)); \quad (26)$$

and the aether superfield

$$V_b\left(x_L,\theta,\bar{\theta}\right)|_{\text{bos}} = \upsilon_b(x_L) + \theta^2 G_b(x_L) + \theta\sigma^k\bar{\theta}f_{bk}(x_L), \quad (27)$$

where $\upsilon_a = \upsilon_a^R + i\upsilon_a^I$ is the first order aether perturbation, and

$$f_{bk} = w_c d_p \text{tr}\left(\sigma_{bc}\sigma_{pk}\right) + \frac{i}{2}w_c\partial_m h_{np}\text{tr}\left(\sigma_k\bar{\sigma}_p\sigma_{bc}\sigma_{nm}\right) + w_c\left(v_p + i\mu_p\right)\text{tr}\left(\bar{\sigma}_{bc}\bar{\sigma}_{kp}\right). \quad (28)$$

Expanding in the Grassmann coordinates we obtain the bosonic part of the Lagrangian in components. Integrating out the auxiliary fields under the assumption $C \ll 1$ through order $O\left(\sqrt{C}\right)$ we reduce to the linearized Lagrangian of the Einstein-aether supergravity through order $O(C)$

$$L = \frac{1}{2\varkappa^2}\Big\{\frac{1}{4}h_{km}\Box h^{km} + \frac{1}{2}\partial^k h_{km}\partial_l h^{lm} - \frac{1}{2}\partial_k h^{km}\partial_m h +$$
$$+\frac{1}{4}\partial_m h\partial^m h - \partial_m\hat{\upsilon}_a^R\partial^m\hat{\upsilon}^{R,a} - \partial_m\hat{\upsilon}_a^I\partial^m\hat{\upsilon}^{I,a} + \sqrt{C}\,\hat{\upsilon}^{R,a}w^b(\partial_b\partial^k h_{ka} - \partial_a\partial^k h_{kb}) -$$
$$-\frac{C}{4}w^a w^b(\partial_a h_{mn} - \partial_m h_{na})(\partial_b h^{mn} - \partial^m h^n{}_b) - \frac{C}{2}w^a w^b\partial_a\hat{\upsilon}^{I,m}\partial_b\hat{\upsilon}_m^I +$$
$$+\frac{C}{2}(\partial_a\hat{\upsilon}^{I,a})^2 - Cw^b w^c\epsilon_{bkam}\partial^k\hat{\upsilon}^{R,a}\partial_c\hat{\upsilon}^{I,m} + O(C^{3/2})\Big\}. \quad (29)$$

In this expression we have canonically normalized the aether perturbations, so their leading kinetic term became of order 1: $\upsilon_a^{R,I} \mapsto \hat{\upsilon}_a^{R,I} = \sqrt{C}\,\upsilon_a^{R,I}$ [11].

5 Discussion

Apart from the usual gravitational coupling, the theory is described by a single dimensionless constant C. When restricted to the case of real aether, $\hat{v}_a^I = 0$, the Lagrangian (29) coincides with the quadratic part of the Einstein-aether Lagrangian 1 for the choice of couplings: $c_1 = C$, $c_2 = c_3 = c_4 = 0$. Thus we conclude that SUSY reduces the number of free parameters in the Einstein-aether model from four down to one.

Modern experimental tests provide tight bounds on the parameters of Lorentz violation. In general, the Einstein-aether model gives rise to non-zero values of the post-Newtonian parameters $\alpha_{1,2}$ that can be expressed in terms of c_i (2, 3). In the supersymmetric model $c_{2,3,4} = 0$ and

$$\alpha_1 = 0\,, \quad \alpha_2 = -\frac{2C}{2-C}\,, \tag{30}$$

so the constraints translate into a bound on C:

$$C \lesssim 10^{-7}. \tag{31}$$

More strong constraints to the parameter can be found from the pulsar astronomy[12]. This gives the bound,

$$C < 1.6 \times 10^{-9}, \tag{32}$$

but note that these observations are strictly relevant to the strong field case and should be used for the non-linear model. We'll briefly outline it at the end of the paper.

Another phenomenological consequence is the connection between structure of the supersymmetric algebra and the small perturbations velocity in the theory. It is curious that in supersymmetric theory of gravity the velocities of small perturbations is only luminal or superluminal[11]. In the Einstein-aether supergravity several modes with different helicities have the unit velocity, and for the others, including graviton,

$$s^2 = 1 + C, \tag{33}$$

where s describes the linear dispersion relation in the theory:

$$E = s \cdot p\,. \tag{34}$$

Due to the model properties [1] superluminal velocities don't lead to any disturbances in the theory. The fact that graviton velocity in the theory is over the speed of light can be used in the possible experimental tests of the model from the gravitational waves observations [13].

Recent extension of the model will be its generalization to the full non-linear supergravity case. It can be simply done in the superfield formalism, using notification of[14, 15], but the full analysis of the model and the reduction from theory in superspace to component fields are technically difficult and tedious. Nevertheless, this work will open the way to study possible manifestations of the model in cosmology, in particular, the effects of the additional fermionic and bosonic fields present in the model on the dynamics of the early universe. Other possible extensions of the model are the study of the different helicity mode velocities in different supersymmetric models with Lorentz breaking and its connection to the superalgebra structure, the effects of spontaneous supersymmetry breaking in the theory, etc. We leave it to the future works.

The work has been supported by the Russian Science Foundation grant 14-22-00161.

References

[1] T. Jacobson, PoS **QG-PH**, 020 (2007).
[2] P. Horava, Physical Review D, **79**, 8, 084008 (2009).
[3] D. Blas, O. Pujolas, O., and S. Sibiryakov, (2009). JHEP, **0910**, 029 (2009)
[4] D. Blas, O. Pujolas, O., and S. Sibiryakov, Phys. Rev. Lett. **104**, 181302, (2010)
[5] T. Jacobson, Phys. Rev. D **81**, 101502 (2010) Erratum: [Phys. Rev. D **82**, 129901 (2010)], Phys. Rev. D **89**, 081501 (2014)
[6] D. Blas, O. Pujolas and S. Sibiryakov, JHEP **1104**, 018 (2011)
[7] B. Z. Foster and T. Jacobson. Physical Review D, **73**, 6, 064015, 1-9 (2006).
[8] O. Pujolas and S. Sibiryakov, JHEP, **1201**, 062 (2012).
[9] S. Groot Nibbelink and M. Pospelov, Phys. Rev. Lett. **94**, 081601 (2005).
[10] I. L. Buchbinder and S. M. Kuzenko, *Ideas and methods of supersymmetry and supergravity, or A walk through superspace*. (Institute of Physics Pub., 1998).
[11] A. Marakulin and S. Sibiryakov, arXiv:1610.07805.
[12] K. Yagi, D. Blas, N. Yunes and E. Barausse, Phys. Rev. Lett. **112**, no. 16, 161101 (2014) Phys. Rev. D **89**, no. 8, 084067 (2014) Erratum: [Phys. Rev. D **90**, no. 6, 069902 (2014)] Erratum: [Phys. Rev. D **90**, no. 6, 069901 (2014)]
[13] J. Ellis, N. E. Mavromatos and D. V. Nanopoulos, Mod. Phys. Lett. A **31**, 1650155 (2016)
[14] G. Girardi, R. Grimm, M. Muller and J. Wess, Z. Phys. C **26**, 123 (1984).
[15] G. Girardi, M. Muller and J. Wess, Z. Phys. C **26**, 427 (1984).

Rare radiative leptonic B-decays as a tool to study New Physics

Anastasiia Kozachuk[1,2,⋆]

[1] *Faculty of Physics, M.V. Lomonosov Moscow State University, Leninskie gory 1(2), 119991 Moscow, Russia*
[2] *D. V. Skobeltsyn Institute of Nuclear Physics, M. V. Lomonosov Moscow State University, Leninskie gory 1(2), 119991 Moscow, Russia*

Abstract. We present results for rare radiative leptonic B-decays $\bar{B}_{d,s} \to \gamma\ell^+\ell^-$. We calculate non-perturbative QCD part via relativistic quark model, which is a dispersion approach based on the constituent quark picture. We include the long-distance contribution, represented by light and $c\bar{c}$ vector resonances, as an additional effective term to the Wilson coefficient C_9. We give predictions for the branching ratios, as well as distributions of the differential rates and the forward-backward asymmetry. Finally, we consider experimental limits on the Wilson coefficient $C_{7\gamma}$ and the corresponding sensitivity of the zero-point of the forward-backward asymmetry.

Introduction

Rare radiative leptonic B-decays proceed through flavor-changing neutral currents (FCNCs), which are forbidden at tree level in the Standard Model (SM). These currents are represented by penguin and box diagrams, which are loop diagrams, and thus lead to small probabilities of the considered decays. Because of possible contribution of new particles to the loops these decays are believed to be sensitive to physics beyond the Standard Model. There are several discrepancies between the predictions of the Standard Model and the experimental results in this sector of FCNC $b \to \{d, s\}$ transitions, which are widely discussed [1]. A well-known one is the quantity

$$\mathcal{R}_K = \frac{\mathcal{B}(B^+ \to K^+\mu^+\mu^-)}{\mathcal{B}(B^+ \to K^+e^+e^-)} = 0.745^{+0.090}_{-0.074}\,(\text{stat}) \pm 0.036\,(\text{syst})$$

in the region $q^2 \in [1;\, 6]\,\text{GeV}^2$, q^2 being the squared invariant mass of the lepton pair [2]. This differs from unity (the SM prediction) at $2.6\,\sigma$. A separate analysis of both of the modes showed that the deviation is rather in muons than in electrons. Another deviation comes from similar decays into K^{*0} mesons:

$$\mathcal{R}_{K^{*0}} = \frac{\mathcal{B}(B^0 \to K^{*0}\mu^+\mu^-)}{\mathcal{B}(B^0 \to K^{*0}e^+e^-)} = 0.69^{+0.11}_{-0.07}\,(\text{stat}) \pm 0.05\,(\text{syst}),$$

for the region $q^2 \in [1.1;\, 6]\,\text{GeV}^2$ [3]. This gives a deviation of $2.5\,\sigma$.

Similar situation is observed for $B \to \phi\ell^+\ell^-$ decays [4]. On the other hand, the famous result for $\mathcal{B}(B_s^0 \to \mu^+\mu^-)$ obtained by LHCb and CMS is in agreement with the SM at $1\,\sigma$ level [5].

⋆e-mail: anastasiia.kozachuk@cern.ch

In this work we give theoretical predictions for rare decays $\bar{B}_{d,s} \to \gamma\ell^+\ell^-$ in the framework of the Standard Model. In Sections 1 and 2 we consider the effective Hamiltonian and the diagrams contributing to the decays. In Section 3 we discuss our calculations of the form factors within the framework of relativistic quark model. Section 4 represents our numerical results. In Section 5 we consider experimental limits of the Wilson coefficient C_7 and how they effect the zero-point of the forward-backward asymmetry. Section 6 is our conclusions.

1 Effective Hamiltonian

On the quark level rare decays $\bar{B}_{d,s} \to \gamma\ell^+\ell^-$ correspond to FCNC $b \to q$, $q = \{d, s\}$ transitions. The latter are described by an effective Hamiltonian - the Wilson operator expansion at the scale $\mu \approx 5\,\mathrm{GeV}$, which is relevant for B-mesons[1] [6]

$$H_{\rm eff}^{b\to q} = \frac{G_F}{\sqrt{2}} V_{tq}^* V_{tb} \sum_i C_i(\mu) O_i^{b\to q}(\mu), \tag{1}$$

here G_F being the Fermi constant, C_i being the Wilson coefficients and O_i being the operators of the expansion. Wilson coefficients contain the contributions of massive virtual particles, such as W and Z bosons, and t-quarks. These heavy degrees of freedom, which contribute to the loop FCNC diagrams (e.g. to mentioned earlier penguin and box diagrams) are integrated out and become encoded in the Wilson coefficients. Lighter virtual particles, such as u and c quarks also contribute to the loop diagrams and should be taken into account separately as an addition to the effective Hamiltonian. O_i are four-fermion operators which only include light quarks and leptons.

2 Contributions to the amplitude of $\bar{B}_{d,s} \to \gamma\ell^+\ell^-$ decays

In this section, we consider the contributions to the $\bar{B}_{d,s} \to \gamma\ell^+\ell^-$ amplitude induced by the effective Hamiltonian (1). The transition form factors are defined as follows [8, 9]:

$$\begin{aligned}
\langle\gamma^*(k,\,\epsilon)|\bar{s}\gamma_\mu\gamma_5 b|B_s(p)\rangle &= i\,e\,\epsilon_\alpha^*\left(g_{\mu\alpha}\,k'k - k'_\alpha k_\mu\right)\frac{F_A(k'^2,k^2)}{M_{B_s}},\\
\langle\gamma^*(k,\,\epsilon)|\bar{s}\gamma_\mu b|B_s(p)\rangle &= e\,\epsilon_\alpha^*\,\epsilon_{\mu\alpha k'k}\frac{F_V(k'^2,k^2)}{M_{B_s}},\\
\langle\gamma^*(k,\,\epsilon)|\bar{s}\sigma_{\mu\nu}\gamma_5 b|B_s(p)\rangle\, k'^\nu &= e\,\epsilon_\alpha^*\left(g_{\mu\alpha}\,k'k - k'_\alpha k_\mu\right)F_{TA}(k'^2,k^2),\\
\langle\gamma^*(k,\,\epsilon)|\bar{s}\sigma_{\mu\nu} b|B_s(p)\rangle\, k'^\nu &= i\,e\,\epsilon_\alpha^*\epsilon_{\mu\alpha k'k}F_{TV}(k'^2,k^2).
\end{aligned}\tag{2}$$

We define the form factors as functions of two variables, $F_i(k'^2, k^2)$, where k' is the momentum emitted from the FCNC $b \to q$ vertex, and k is the momentum of the photon emitted from the valence quark of the B-meson.

2.1 Lepton pair couples to the effective $b \to q$ vertex

We define $A_{(1)}^{B\to\gamma\ell\ell}$ as the contribution to the $\bar{B}_{d,s} \to \gamma\ell^+\ell^-$ amplitude, when the lepton pair is produced from the FCNC $b \to q$ vertex, while the photon is emitted from the valence quark of the B-meson.

[1] Our notations and conventions are: $\gamma^5 = i\gamma^0\gamma^1\gamma^2\gamma^3$, $\sigma_{\mu\nu} = \frac{i}{2}[\gamma_\mu, \gamma_\nu]$, $\varepsilon^{0123} = -1$, $\epsilon_{abcd} \equiv \epsilon_{\alpha\beta\mu\nu}a^\alpha b^\beta c^\mu d^\nu$, $e = \sqrt{4\pi\alpha_{\rm em}}$.

Diagrams, contributing to $A_{(1)}^{B\to\gamma\ell\ell}$ are shown in Figure 1. In this case we denote the momenta as $k' = q$, $k = p - q$, $k'^2 = q^2$ and $k^2 = 0$, thus obtaining $F_i(q^2, 0)$:

$$A_{(1)}^{B\to\gamma\ell\ell} = \langle\gamma(k,\epsilon), \ell^+(p_1), \ell^-(p_2)|H_{\rm eff}^{b\to d\ell^+\ell^-}|B(p)\rangle = \frac{G_F}{\sqrt{2}} V_{tb}V_{tq}^* \frac{\alpha_{\rm em}}{2\pi} e\,\epsilon_\alpha^* \tag{3}$$

$$\times\Big[\epsilon_{\mu\alpha k'k}A_V^{(1)}(q^2)\bar\ell(p_2)\gamma_\mu\ell(-p_1) - i\left(g_{\mu\alpha}\,k'k - k'_\alpha k_\mu\right)A_A^{(1)}(q^2)\bar\ell(p_2)\gamma_\mu\ell(-p_1)$$

$$+\epsilon_{\mu\alpha k'k}A_{5V}^{(1)}(q^2)\bar\ell(p_2)\gamma_\mu\gamma_5\ell(-p_1) - i\left(g_{\mu\alpha}\,k'k - k'_\alpha k_\mu\right)A_{5A}^{(1)}(q^2)\bar\ell(p_2)\gamma_\mu\gamma_5\ell(-p_1)\Big], \tag{4}$$

$k' = q$, $k = p - q$ and

$$A_{V(A)}^{(1)}(q^2) = \frac{2m_b\,C_{7\gamma}(\mu)}{q^2}F_{TV(TA)}(q^2,0) + C_{9V}(\mu)\frac{F_{V(A)}(q^2,0)}{M_B}, \tag{5}$$

$$A_{5V(5A)}^{(1)}(q^2) = C_{10AV}(\mu)\frac{F_{V(A)}(q^2,0)}{M_B}. \tag{6}$$

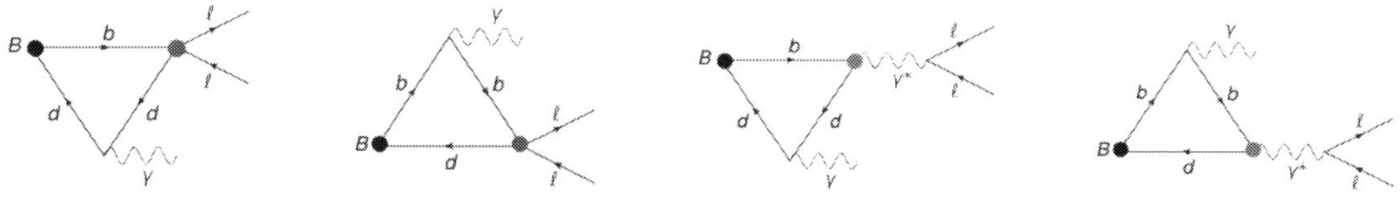

Figure 1. Diagrams, contributing to $A_{(1)}^{B\to\gamma\ell\ell}$ amplitude

2.2 Effective $b \to q$ vertex emits a real photon

Another contribution to the amplitude, which we denote as $A_{(2)}^{B\to\gamma\ell\ell}$, is relevant for the process when the real photon is emitted from the penguin FCNC vertex, while the valence quark of the B-meson emits a virtual photon.The diagrams, contributing to $A_{(2)}^{B\to\gamma\ell\ell}$ are shown on Figure 2.2.

The amplitude $A_{(2)}^{B\to\gamma\ell\ell}$ has the same Lorentz structure as the $C_{7\gamma}$ part of $A^{(1)}$ but now the momenta are defined as $k = q$, $k' = p - q$, $k'^2 = 0$ and $k^2 = q^2$. The amplitude thus includes the form factors $F_{TA,TV}(0,q^2)$, with $F_{TA}(0,q^2) = F_{TV}(0,q^2)$:

$$A_{(2)}^{B\to\gamma\ell\ell} = \frac{G_F}{\sqrt{2}} V_{tb}V_{tq}^* \frac{\alpha_{\rm em}}{2\pi} e\,\epsilon_\mu^*\bar\ell(p_2)\gamma_\alpha\ell(-p_1) \tag{7}$$

$$\times\ \Big[\varepsilon_{\mu\alpha k'k}A_V^{(2)}(q^2) - i\left(g_{\mu\alpha}k'k - k'_\alpha k_\mu\right)A_A^{(2)}(q^2)\Big],\quad k = q,\quad k' = p - q,$$

where

$$A_{V(A)}^{(2)}(q^2) = \frac{2m_bC_{7\gamma}(\mu)}{q^2}F_{TV(TA)}(0,q^2). \tag{8}$$

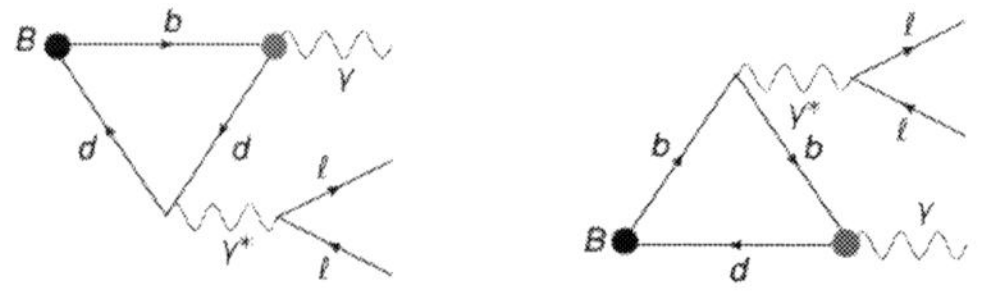

Figure 2. Diagrams, contributing to $A_{(2)}^{B\to\gamma\ell\ell}$ amplitude

In addition to the amplitudes described in Sections 2.1 and 2.2 we also include the contribution of bremsstrahlung and weak annihilation. We do not discuss the details here in order not to overload this paper.

3 Parametrization for the F_i form factors

In this section we discuss the parametrization for the form factors defined in (2).

3.1 Form factors $F_i(q^2, 0)$

We calculate the form factors $F_i(q^2, 0)$ in the framework of relativistic quark model which is a dispersion approach based on the constituent quark picture [7]. Each form factor is represented as a dispersion integral over the mass variable. The integrand in each case is a complicated function, for some details of the calculation see, for e.g. [10]. In case of radiative leptonic decays $\bar{B}_{d,s} \to \gamma\ell^+\ell^-$ we obtain form factors as functions of q^2, where q^2 is the squared invariant mass of the lepton pair. It is convenient to introduce every form factor as a sum of two parts

$$F_i(q^2) = Q_b F^{(b)}(q^2)_i + Q_{d,s} F^{(d,s)}(q^2). \tag{9}$$

Finally, we use the following parametrization:

$$F_i^{(\alpha)}(q^2) = \frac{F_i^{(\alpha)}(0)}{1 - q^2/M_{R_i}^2}, \tag{10}$$

where the parameters $F_i^{(\alpha)}(0)$ and M_{R_i} are presented in Tables 1 and 2.

Table 1. Parameters $F_i^{(\alpha)}(0)$ and M_{R_i} for the $\bar{B} \to \gamma\ell\ell$ form factors

	$F_V^{(d)}$	$F_V^{(b)}$	$F_A^{(d)}$	$F_A^{(b)}$	$F_{TV}^{(d)}$	$F_{TV}^{(b)}$	$F_{TA}^{(d)}$	$F_{TA}^{(b)}$
$F(0)$	-0.295	-0.037	0.259	0.042	-0.313	-0.04	-0.313	-0.04
M_R	5.6	5.55	5.9	5.45	5.6	5.6	6.0	6.2

Table 2. Parameters $F_i^{\alpha}(0)$ and M_{R_i} for the $\bar{B}_s \to \gamma\ell\ell$ form factors

	$F_V^{(s)}$	$F_V^{(b)}$	$F_A^{(s)}$	$F_A^{(b)}$	$F_{TV}^{(s)}$	$F_{TV}^{(b)}$	$F_{TA}^{(s)}$	$F_{TA}^{(b)}$
$F(0)$	-0.289	-0.043	0.257	0.049	-0.305	-0.046	-0.305	-0.046
M_R	5.6	5.6	5.9	5.5	5.6	5.6	6.0	6.2

3.2 Form factors $F_i(0, q^2)$

For the form factors $F_{TV,TA}(0, q^2)$ we use the VMD formula, which includes decay constants f_V of light vector resonances and $B \to V$ transition form factors:

$$F_{TV,TA}(0, q^2) = F_{TV,TA}(0, 0) - \sum_V 2 f_V^{\text{e.m.}} g_+^{B\to V}(0) \frac{q^2/M_V}{q^2 - M_V^2 + iM_V\Gamma_V}, \tag{11}$$

where M_V and Γ_V are the mass and the width of the vector meson resonance, $g_+^{B\to V}(0)$ are the $B \to V$ transition form factors, defined according to the relations

$$\langle V(q,\varepsilon)|\bar{d}\sigma_{\mu\nu}b|B(p)\rangle = i\varepsilon^{*\alpha}\,\epsilon_{\mu\nu\beta\gamma}\left[g_+^{B\to V}(k^2)g_{\alpha\beta}(p+q)^\gamma + g_-^{B\to V}(k^2)g_{\alpha\beta}k^\gamma + g_0^{B\to V}(k^2)p_\alpha p^\beta q^\gamma\right].$$

The form factors $g_+^{B\to V}(0)$ were calculated via relativistic quark model in [7]. The e.m. leptonic decay constant of a vector meson is given by

$$\langle 0|j_\mu^{\text{e.m}}|V(\varepsilon, p)\rangle = \varepsilon_\mu M_V f_V^{\text{e.m.}}. \tag{12}$$

Table 3. Values of the mass and the width of neutral vector mesons [12], decay constant f_V (with isotopic factors omitted) extracted from the leptonic width neglecting the meson-mixing effects, and the $g_+^{B\to V}(0)$ calculated via dispersion quark model.

V	M_V, MeV	Γ_V, MeV	f_V, MeV	$g_+^{B\to V}(0)$, GeV
ρ^0	775	149	220	0.29
ω	783	8.49	195	0.24
ϕ	1019	4.27	226	0.27

The electromagnetic decay constants which enter the VMD formula (11) are related to f_V as follows: $f_{\rho^0}^{\text{e.m.}} = \frac{1}{\sqrt{2}} f_{\rho^0}$, $f_\omega^{\text{e.m.}} = \frac{1}{3\sqrt{2}} f_\omega$, $f_\phi^{\text{e.m.}} = -\frac{1}{3} f_\phi$.

4 Numerical Results

In this section we present our numerical results for the decays $\bar{B}_{d,s} \to \gamma l^+ l^-$. We give two estimates for the branching ratios: the first one for the range $q^2 \in [1; 6]\,\text{GeV}^2$, which does not include any resonances, and the second one for the whole range of the invariant mass of the lepton pair. We include the long-distance contribution, represented by light and $c\bar{c}$ vector resonances, using a simple Breit-Wigner formula [11]. The numerical predictions are given in Table 4. In Figure 3 we present out results for the distributions of the differential branching ratios and the forward-backward asymmetry for $\bar{B}_{d,s} \to \mu^+\mu^-\gamma$ decays.

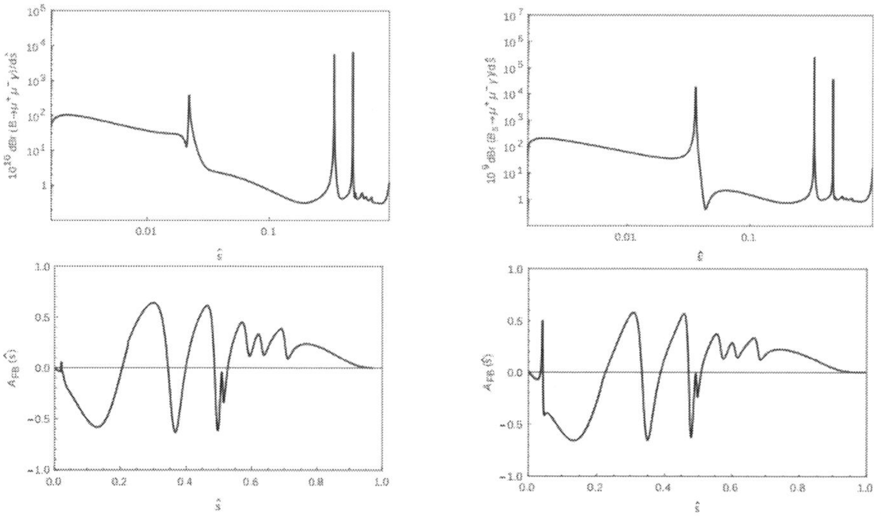

Figure 3. Differential branching ratios and forward-backward asymmetries for $\bar{B}_d \to \gamma\mu^+\mu^-$ (left) and $\bar{B}_s \to \gamma\mu^+\mu^-$ (right) decays, $\hat{s} = q^2/M_B^2$

DOI: 10.1051/epjconf/201715803001

Table 4. Our results for the branching ratios of $\bar{B}_{d,s} \to l^+l^-\gamma$ decays for different ranges of q^2: for $q^2 \in [1;6]\,\text{GeV}^2$ and for the full range. The value of the photon energy cut is 500 MeV

q^2-range	$[1;6]\,\text{GeV}^2$	Full
$Br(B \to e^+e^-\gamma) \times 10^{10}$	0.12	4.6
$Br(B \to \mu^+\mu^-\gamma) \times 10^{10}$	0.12	1.5
$Br(B_s \to e^+e^-\gamma) \times 10^{9}$	0.76	18.4
$Br(B_s \to \mu^+\mu^-\gamma) \times 10^{9}$	0.73	11.6

5 Experimental limits on C_7 and the forward-backward asymmetry

After obtaining results within the Standard Model, we started to look at different possibilities of including the effects of New Physics into our calculations. One of the simplest opportunities was to try to change the values of the Wilson coefficients within available experimental limits and to look at the corresponding sensitivity of the observables. So far we implemented this idea in comparing the distributions of the forward-backward asymmetry for different values of the Wilson coefficient $C_{7\gamma}$. We expect the forward-backward asymmetry to be almost independent from the model of the form factors description and thus sensitive to possible NP effects. In Figure 4 [13] one can see all the range of possible values for NP contributions to the Wilson coefficient $C_{7\gamma}$ (we assume that $C_{7\gamma} = C_{7\gamma}^{SM} + C_{7\gamma}^{NP}$).

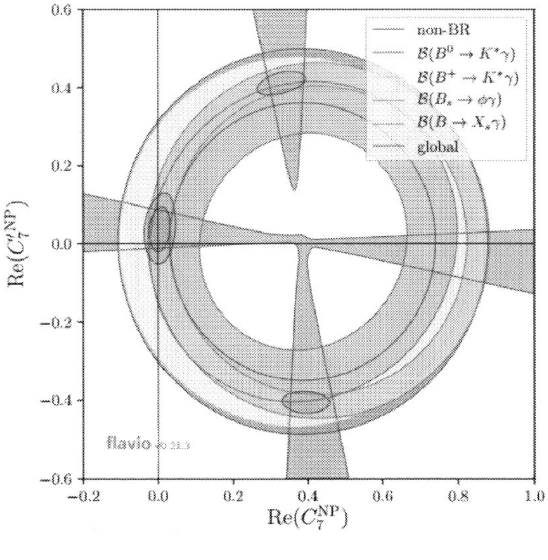

Figure 4. $C_7 - C_7'$ plot: experimental limits [13]

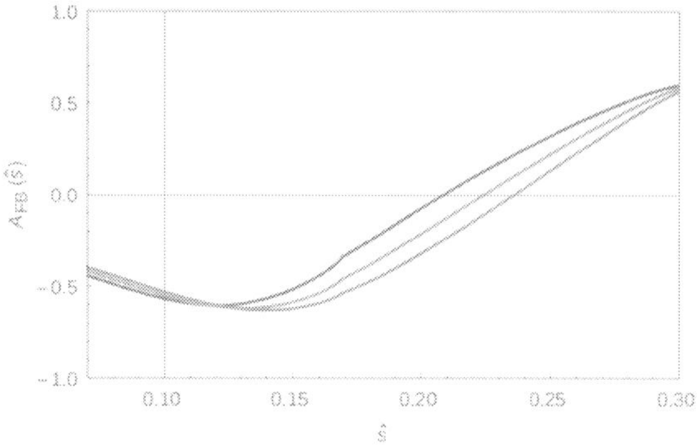

Figure 5. Sensitivity of the zero-point of the forward-backward asymmetry to the change of $C_{7\gamma}$ Wilson coefficient within its experimental limits, $\hat{s} = q^2/M_B^2$

We have plotted the forward-backward asymmetry for three of possible values of $C_{7\gamma}$: 0.276, 0.308 and 0.335 (see Figure 5). This range of values correspond to the case $C_7' = 0$, which is in agreement with our calculations. One can see that such a small change of $C_{7\gamma}$ has almost no impact on the position of the zero-point of the forward-backward asymmetry.

6 Conclusions

We gave predictions for rare radiative leptonic decays $\bar{B}_{d,s} \to \gamma \ell^+ \ell^-$. We calculated the form factors $F_i(q^2, 0)$ via relativistic quark model. Form factors $F_i(0, q^2)$ were taken into account via vector meson dominance. We included the long-distance contribution: light and $c\bar{c}$ vector resonances via simple Breit-Wigner formula. We obtained numerical predictions for the branching ratios and distributions for the branching fractions and the forward backward asymmetry. We found very small sensitivity of the zero-point of the forward-backward asymmetry to the change of the Wilson coefficient $C_{7\gamma}$ within its experimental limits. We are now working on a better way of including the $c\bar{c}$-resonances and the continuum above the $D\bar{D}$-threshold.

Acknowledgments

The author would like to thank Diego Guadagnoli, Dmitri Melikhov, Nikolai Nikitin and Peter Strangl for useful discussions. The work was supported by grant RNF-16-12-10280 of the Russian Science Foundation.

References

[1] D. Guadagnoli, D. Melikhov and M. Reboud, Phys. Lett. B **760**, 442 (2016)
[2] LHCb Collaboration, Phys. Rev. Lett. **113**, 151601 (2014)
[3] LHCb Collaboration, JHEP **08**, 055 (2017)
[4] LHCb Collaboration, JHEP **09**, 179 (2015)
[5] CMS and LHCb Collaborations, Nature **522**, 68 (2015)
[6] B. Grinstein, M. J. Savage and M. B. Wise, Nucl. Phys. B **319**, 271 (1989)
[7] D. Melikhov, Phys. Rev. D **53**, 2460 (1996); D. Melikhov, Phys. Rev. D **56**, 7089 (1997)
[8] F. Kruger and D. Melikhov, Phys. Rev. D **67**, 034002 (2003)
[9] D. Melikhov and N. Nikitin, Phys. Rev. D **70**, 114028 (2004)
[10] A. Kozachuk, D. Melikhov and N. Nikitin, Phys. Rev. D **93**, 014015 (2016)
[11] A. Ali, T. Mannel and T. Morozumi, Phys. Lett. B **273**, 505 (1991)

[12] Particle Data Group, Chin. Phys. C **40**, 100001 (2016)
[13] A. Paul and D. M. Straub, JHEP **1704**, 027 (2017); P. Strangl, "Constraints on the Wilson coefficients C7 and C7" ' June, 2017, https://doi.org/10.5281/zenodo.804453

EPJ Web of Conferences **158**, 03002 (2017) DOI: 10.1051/epjconf/201715803002
QFTHEP 2017

Semileptonic decays of B_c mesons into charmonium states

Aidos Issadykov[1,2,⋆], *Mikhail A.* Ivanov[1], and *Guliya* Nurbakova[3]

[1] *Joint Institute for Nuclear Research, Dubna, Russia*
[2] *The Institute of Nuclear Physics,Ministry of Energy of the Republic of Kazakhstan, Almaty,Kazakhstan*
[3] *Al-Farabi Kazakh National University, Almaty, Kazakhstan*

Abstract. In this work we study the semileptonic decays of B_c meson. We evaluated $B_c \to D(D^*)$, $B_c \to D_s(D_s^*)$ and $B_c \to \eta_c(J/\psi)$ transitions form factors in the full kinematical region within the covariant quark model. The calculated form factors are used to evaluate the semileptonic decays of B_c meson and it was defined ratios (R_{η_c}, $R_{J/\psi}$, R_D ,R_{D^*}) of the branching ratios, which will be hopefully tested on LHC experiments.We compare the obtained results with the results from other theoretical approaches.

Model

The covariant quark model was developed by G.V.Efimov and M.A.Ivanov [1–3].

The effective Lagrangian describing the transition of a meson $M(q_1\bar{q}_2)$ to its constituent quarks q_1 and $\bar{q}_2$ in model looks like

$$\begin{aligned} \mathcal{L}_{\rm int}(x) &= g_M M(x)\cdot J_M(x) + \text{h.c.}, \\ J_M(x) &= \int dx_1 \int dx_2 F_M(x,x_1,x_2)\bar{q}_2(x_2)\Gamma_M q_1(x_1), \end{aligned} \tag{1}$$

with Γ_M a Dirac matrix which projects onto the spin quantum number of the meson field $M(x)$. The vertex function F_M characterizes the finite size of the meson. Translational invariance requires the function F_M to fulfill the identity $F_M(x+a, x_1+a, x_2+a) = F_M(x, x_1, x_2)$ for any four-vector a. A specific form for the vertex function is adopted

$$F_M(x,x_1,x_2) = \delta(x - w_1x_1 - w_2x_2)\Phi_M((x_1-x_2)^2), \tag{2}$$

where Φ_M is the correlation function of the two constituent quarks with masses m_{q_1}, m_{q_2} and the mass ratios $w_i = m_{q_i}/(m_{q_1}+m_{q_2})$.

A simple Gaussian form of the vertex function $\bar{\Phi}_M(-k^2)$ is selected

$$\bar{\Phi}_M(-k^2) = \exp\left(k^2/\Lambda_M^2\right) \tag{3}$$

with the parameter Λ_M linked to the size of the meson. The minus sign in the argument is chosen to indicate that we are working in the Minkowski space. Since k^2 turns into $-k_E^2$ in the Euclidean space, the form (3) has the appropriate fall-off behavior in the Euclidean region. Any choice for Φ_M

⋆e-mail: issadykov@jinr.ru

is appropriate as long as it falls off sufficiently fast in the ultraviolet region of the Euclidean space to render the corresponding Feynman diagrams ultraviolet finite. We choose a Gaussian form for calculational convenience.

The fermion propagators for the quarks are given by

$$S_i(k) = \frac{1}{m_{q_i} - \not k} \tag{4}$$

with an effective constituent quark mass m_{q_i}.

The so-called *compositeness condition* [4, 5] is used to determine the value of the coupling constants g_M. It means that the renormalization constant Z_M of the elementary meson field $M(x)$ is to be set to zero, i.e.,

$$Z_M = 1 - \frac{3g_M^2}{4\pi^2}\tilde{\Pi}'_M(m_M^2) = 0, \tag{5}$$

where $\tilde{\Pi}'_M$ is the derivative of the meson mass operator. Its physical meaning in Eq. (5) becomes clear when interpreted as the matrix element between the physical and the corresponding bare state: $Z_M = 0$ implies that the physical state does not contain the bare state and is appropriately described as a bound state. The interaction makes the physical particle dressed, i.e. its mass and wave function have to be renormalized. The condition $Z_M = 0$ also effectively excludes the constituent degrees of freedom from the space of physical states. It thereby guarantees the absence of double counting for the physical observable under consideration, the constituents exist only in virtual states. The tree-level diagram together with the diagrams containing self-energy insertions into the external legs (i.e. the tree-level diagram times $Z_M - 1$) give a common factor Z_M which is equal to zero.

The mass functions for the pseudoscalar meson (spin $S = 0$) and vector meson (spin $S = 1$) are defined as

$$\Pi_P(x-y) = +i\langle T\{J_P(x)J_P(y)\}\rangle_0, \tag{6}$$

$$\Pi_V^{\mu\nu}(x-y) = -i\langle T\{J_V^\mu(x)J_V^\nu(y)\}\rangle_0. \tag{7}$$

Herein we use the updated values of the model parameters from [6] which are shown in Eq. (8,9).

$m_{u/d}$	m_s	m_c	m_b	λ	
0.241	0.428	1.67	5.05	0.181	GeV

(8)

Λ_{B_c}	Λ_{η_c}	$\Lambda_{J/\psi}$	Λ_D	Λ_{D^*}	Λ_{D_s}	$\Lambda_{D_s^*}$	Λ_B	Λ_{B^*}	Λ_{B_s}	
2.73	3.97	1.74	1.6	1.53	1.75	1.56	1.96	1.8	2.05	GeV

(9)

Semileptonic decays

We give the necessary definitions of the leptonic decay constants, invariant form factors and helicity amplitudes. The leptonic decay constants are defined by

$$M(H_{12} \to \bar{l}\nu) = \frac{G_F}{\sqrt{2}} V_{q_1q_2} \mathcal{M}^\mu_H(p)\, \bar{u}_l(k_l)\, O^\mu\, u_\nu(k_\nu),$$

$$\mathcal{M}^\mu_H(p) = -3\, g_{12} \int \frac{d^4k}{(2\pi)^4 i} \widetilde{\Phi}_{12}\left(-k^2\right) \mathrm{tr}\left[\Gamma_H\, \widetilde{S}_2(k - c^2_{12}\, p)\, O^\mu\, \widetilde{S}_1(k + c^1_{12}\, p)\right],$$

$$\Gamma_P = i\gamma^5, \quad \Gamma_V = \varepsilon_V \cdot \gamma,$$

$$\mathcal{M}^\mu_P(p) = -if_P p^\mu, \qquad \mathcal{M}^\mu_V(p) = f_V m_V \varepsilon^\mu_V. \tag{10}$$

The semileptonic decays of the B_c-meson may be induced by a b-quark transition.

$$M(H_{13} \to H_{23} + \bar{l}\nu) = \frac{G_F}{\sqrt{2}} V_{q_1q_2} \mathcal{M}^\mu_{12}(p_1, p_2)\, \bar{u}_l(k_l)\, O^\mu\, u_\nu(k_\nu),$$

$$\mathcal{M}^\mu_{12} = -3\, g_{13}\, g_{23} \int \frac{d^4k}{(2\pi)^4 i} \widetilde{\Phi}_{13}\left(-(k + c^3_{13}\, p_1)^2\right) \widetilde{\Phi}_{23}\left(-(k + c^3_{23}\, p_2)^2\right)$$
$$\times \mathrm{tr}\left[i\gamma^5\, \widetilde{S}_3(k)\, \Gamma_{32}\, \widetilde{S}_2(k + p_2)\, O^\mu\, \widetilde{S}_1(k + p_1)\right],$$

$$\times \mathrm{tr}\left[i\gamma^5\, \widetilde{S}_3(k - p_1)\, O^\mu\, \widetilde{S}_2(k - p_2)\Gamma_{21}\, \widetilde{S}_1(k)\right], \tag{11}$$

where $q_1 \equiv b$ and $q_3 \equiv c$ whereas q_2 denotes either of c, u, d, s.

The invariant form factors for the semileptonic B_c-decay into the hadron with spin $S = 0, 1$ are defined by

$$\mathcal{M}^\mu_{S=0} = P^\mu F_+(q^2) + q^\mu F_-(q^2), \tag{12}$$

$$\mathcal{M}^\mu_{S=1} = \frac{1}{m_1 + m_2} \epsilon^\dagger_\nu \left\{ -g^{\mu\nu} Pq\, A_0(q^2) + P^\mu P^\nu A_+(q^2) + q^\mu P^\nu A_-(q^2) + i\,\varepsilon^{\mu\nu\alpha\beta} P_\alpha\, q_\beta\, V(q^2) \right\}, \tag{13}$$

$$P = p_1 + p_2, \qquad q = p_1 - p_2.$$

It is convenient to express all physical observables through the helicity form factors H_m. The helicity form factors H_m can be expressed in terms of the invariant form factors in the following way [7]:

(a) Spin $S = 0$:

$$H_t = \frac{1}{\sqrt{q^2}} \left\{ (m_1^2 - m_2^2)\, F_+ + q^2\, F_- \right\},$$
$$H_\pm = 0, \tag{14}$$
$$H_0 = \frac{2\, m_1\, |\mathbf{p_2}|}{\sqrt{q^2}}\, F_+ .$$

(b) Spin $S = 1$:

$$\begin{aligned}
H_t &= \frac{1}{m_1+m_2}\frac{m_1\,|\mathbf{p_2}|}{m_2\,\sqrt{q^2}}\left\{(m_1^2-m_2^2)\,(A_+-A_0)+q^2A_-\right\},\\
H_\pm &= \frac{1}{m_1+m_2}\left\{-(m_1^2-m_2^2)\,A_0\pm 2\,m_1\,|\mathbf{p_2}|\,V\right\},\\
H_0 &= \frac{1}{m_1+m_2}\frac{1}{2\,m_2\,\sqrt{q^2}}\left\{-(m_1^2-m_2^2)\,(m_1^2-m_2^2-q^2)\,A_0+4\,m_1^2\,|\mathbf{p_2}|^2\,A_+\right\}.
\end{aligned} \tag{15}$$

where $|\mathbf{p_2}| = \lambda^{1/2}(m_1^2, m_2^2, q^2)/(2\,m_1)$ is the momentum of the outgoing particles in the B_c rest frame. The semileptonic B_c-decay widths are given by

$$\begin{aligned}
\Gamma(B_c^-\to M_{\bar{c}c}\,l\bar{\nu}) &= \frac{G_F^2}{(2\,\pi)^3}|V_{cb}|^2\int\limits_{m_l^2}^{q_-^2} dq^2\frac{(q^2-m_l^2)^2\,|\mathbf{p_2}|}{12\,m_1^2\,q^2}\\
&\times\left\{\left(1+\frac{m_l^2}{2\,q^2}\right)\sum_{i=\pm,0}\left(H_i^{B_c\to M_{\bar{c}c}}(q^2)\right)^2+\frac{3\,m_l^2}{2\,q^2}\left(H_t^{B_c\to M_{\bar{c}c}}(q^2)\right)^2\right\},\\
\Gamma(B_c^-\to \overline{D}^0\,l\bar{\nu}) &= \frac{G_F^2}{(2\,\pi)^3}|V_{ub}|^2\int\limits_{m_l^2}^{q_-^2} dq^2\frac{(q^2-m_l^2)^2\,|\mathbf{p_2}|}{12\,m_1^2\,q^2}\\
&\times\left\{\left(1+\frac{m_l^2}{2\,q^2}\right)\sum_{i=\pm,0}\left(H_i^{B_c\to\overline{D}^0}(q^2)\right)^2+\frac{3\,m_l^2}{2\,q^2}\left(H_t^{B_c\to\overline{D}^0}(q^2)\right)^2\right\},
\end{aligned}$$

where $q_-^2 = (m_1 - m_2)^2$, $m_1 \equiv m_{B_c}$, and $m_2 \equiv m_f$. Note that $M_{\bar{c}c}$ and $\overline{D}^0$ denote both the pseudoscalar and vector cases.

Numerical results

We take the following values (16) of the meson masses and the B_c-meson's lifetime from the PDG [8].

m_{B_c}	m_{η_c}	$m_{J/\psi}$	m_D	m_{D^*}	m_{D_s}	$m_{D_s^*}$		τ_{B_c}
6.275	2.984	3.097	1.869	2.010	1.968	2.112	GeV	$0.507ps$

(16)

The calculation of the semileptonic decay widths is straightforward. For the CKM-matrix elements we use

$\lvert V_{ud}\rvert$	$\lvert V_{us}\rvert$	$\lvert V_{cd}\rvert$	$\lvert V_{cs}\rvert$	$\lvert V_{cb}\rvert$	$\lvert V_{ub}\rvert$
0.974	0.225	0.220	0.995	0.0405	0.00409

(17)

DOI: 10.1051/epjconf/201715803002

The value of the decay constant f_{η_c} was calculated from the branching ratio for the η_c meson decay into two photons using the last data [8]. The quality of the fit may be assessed from the entries in Table 1.

The form factors are calculated in the full kinematical region of momentum transfer squared and are shown in Table 2. The curves are depicted in Fig. 1 and Fig. 2.

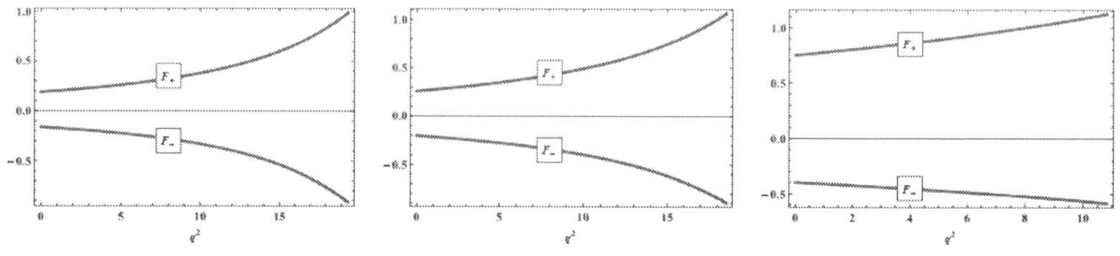

Figure 1. The $F_+(q^2)$ and $F_-(q^2)$ form factors for $B_c \to D, B_c \to D_s$ and $B_c \to \eta_c$ transitions, respectively.

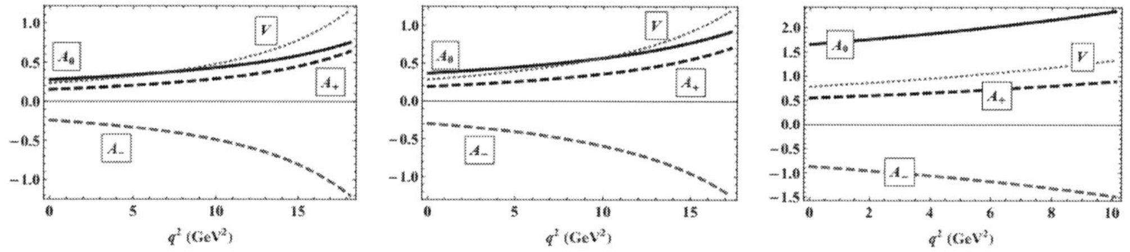

Figure 2. The A_0, A_-, A_+ and V form factors for $B_c \to D^*$, $B_c \to D_s^*$ and $B_c \to J/\psi$ transitions, respectively.

Table 1. Leptonic decay constants f_H (MeV).

	This work	Other	Ref.
f_{B_c}	489	$489 \pm 4 \pm 3$	LAT [9]
		395 ± 15	[10]
f_D	206	$222.6 \pm 16.7^{+2.8}_{-3.4}$	CLEO [11]
		$201 \pm 3 \pm 17$	MILC LAT [12]
		$235 \pm 8 \pm 14$	LAT [13]
		$210 \pm 10^{+17}_{-16}$	UKQCD LAT [14]
		$211 \pm 14^{+2}_{-12}$	LAT [15]
		204.6 ± 5.0	PDG [16]
f_{D^*}	244	$245 \pm 20^{+3}_{-2}$	LAT [15]
		$278 \pm 13 \pm 10$	LAT [17]
		$252.2 \pm 22.3 \pm 4$	QCD SR [18]
f_{D_s}	257	257.5 ± 4.6	PDG [16]
		$249 \pm 3 \pm 16$	MILC LAT [12]
		$266 \pm 10 \pm 18$	LAT [13]
		$290 \pm 20 \pm 29 \pm 29 \pm 6$	LAT [19]
		$236 \pm 8^{+17}_{-14}$	UKQCD LAT [14]
		$231 \pm 12^{+8}_{-1}$	LAT [15]
$f_{D_s^*}$	272	311 ± 9	LAT [17]
		$272(16)^{+3}_{-20}$	LAT [15]
		$305.5 \pm 26.8 \pm 5$	QCD SR [18]
$\frac{f_{D_s}}{f_D}$	1.25	1.258 ± 0.038	PDG [16]
		$1.24 \pm 0.01 \pm 0.07$	MILC LAT [12]
		$1.13 \pm 0.03 \pm 0.05$	LAT [13]
		$1.13 \pm 0.02^{+0.04}_{-0.02}$	UKQCD LAT [14]
		1.10 ± 0.02	LAT [15]
f_{η_c}	628	420 ± 52	[20]
		337.7 ± 18.2	pQCD[21]
$f_{J/\psi}$	415	405 ± 14	pQCD[22]
		416.2 ± 7.4	pQCD[21]

DOI: 10.1051/epjconf/201715803002

Table 2. Form factors for $B_c \to D(D^*)$, $B_c \to D_s(D_s^*)$ and $B_c \to \eta_c(J/\psi)$ transitions. Form factors are *approximated* by the form $F(q^2) = F(0)/(1 - a\,\hat{s} + b\,\hat{s}^2)$ with $\hat{s} = q^2/m_{B_c}^2$.

	$B_c \to D(D^*)$	$B_c \to D_s(D_s^*)$	$B_c \to \eta_c(J/\psi)$
$F_+(0)$	0.186	0.254	0.74
$F_-(0)$	-0.160	-0.202	-0.39
$A_0(0)$	0.276	0.365	1.65
$A_+(0)$	0.151	0.190	0.55
$A_-(0)$	-0.236	-0.293	-0.87
$V(0)$	0.230	0.282	0.78

Table 3. Branching ratios (in %) of semileptonic B_c decays into ground state charmonium states.

Mode	This work	[23]	[7]	[24, 25]	[26]	[27]	[28]
$B_c^- \to \eta_c \ell \nu$	0.95	0.81	0.98	0.75	0.97	0.59	0.44
$B_c^- \to \eta_c \tau \nu$	0.24	0.22	0.27	0.23		0.20	0.14
$B_c^- \to J/\psi \ell \nu$	1.67	2.07	2.30	1.9	2.35	1.20	1.01
$B_c^- \to J/\psi \tau \nu$	0.40	0.49	0.59	0.48		0.34	0.29
$B_c^- \to \overline{D}^- \ell \nu$	0.0033	0.0035	0.018		0.004	0.006	0.0032
$B_c^- \to \overline{D}^- \tau \nu$	0.0021	0.0021	0.0094	0.002			0.0022
$B_c^- \to \overline{D}^{*-} \ell \nu$	0.006	0.0038	0.034		0.018	0.018	0.011
$B_c^- \to \overline{D}^{*-} \tau \nu$	0.0034	0.0022	0.019	0.008			0.006

Table 4. Ratios of semileptonic decays of the B_c meson

Decay rate	This work	[23]	[28]
$R_{\eta_c} = \frac{B_c^- \to \eta_c \ell \nu}{B_c^- \to \eta_c \tau \nu}$	3.96	3.68	3.2
$R_{J/\psi} = \frac{B_c^- \to J/\psi \ell \nu}{B_c^- \to J/\psi \tau \nu}$	4.18	4.22	3.4
$R_D = \frac{B_c^- \to D^- \ell \nu}{B_c^- \to D^- \tau \nu}$	1.57	1.67	1.42
$R_{D^*} = \frac{B_c^- \to D^{*-} \ell \nu}{B_c^- \to D^{*-} \tau \nu}$	1.76	1.72	1.66

The results of our evaluation of the branching ratios of the semileptonic B_c decays appear in Table 3, which contains our predictions for the semileptonic B_c decays into ground state charmonium states and charm meson states. We compare our ratios of semileptonic decays of the B_c meson with those of other models in Table 4.

Acknowledgment

Authors A. Issadykov, M.A. Ivanov and G.S. Nurbakova acknowledge the partial support by the Ministry of Education and Science of the Republic of Kazakhstan, grant 3092/GF4, state registration No. 0115RK01040.

Author A. Issadykov is grateful for the support by the JINR, grant number 17-302-03.

References

[1] G. V. Efimov and M. A. Ivanov, Int. J. Mod. Phys. A **4** (1989) 2031. doi:10.1142/S0217751X89000832

[2] A. Faessler, T. Gutsche, M. A. Ivanov, J. G. Korner and V. E. Lyubovitskij, Eur. Phys. J. direct **4** (2002) no.1, 18 doi:10.1007/s1010502c0018 [hep-ph/0205287].

[3] T. Branz, A. Faessler, T. Gutsche, M. A. Ivanov, J. G. Korner and V. E. Lyubovitskij, Phys. Rev. D **81** (2010) 034010 doi:10.1103/PhysRevD.81.034010 [arXiv:0912.3710 [hep-ph]].

[4] S. Weinberg, Phys. Rev. **130** (1963) 776. doi:10.1103/PhysRev.130.776

[5] A. Salam, Nuovo Cim. **25** (1962) 224. doi:10.1007/BF02733330

[6] S. Dubnička,A.Z. Dubničková, A. Issadykov, M. A. Ivanov, A. Liptaj and S. K. Sakhiyev, Phys. Rev. D **93** (2016) no.9, 094022 doi:10.1103/PhysRevD.93.094022 [arXiv:1602.07864 [hep-ph]].

[7] M. A. Ivanov, J. G. Korner and P. Santorelli, Phys. Rev. D **63** (2001) 074010 doi:10.1103/PhysRevD.63.074010 [hep-ph/0007169].

[8] C. Patrignani *et al.* [Particle Data Group], Chin. Phys. C **40** (2016) no.10, 100001. doi:10.1088/1674-1137/40/10/100001

[9] T. W. Chiu *et al.* [TWQCD Collaboration], Phys. Lett. B **651** (2007) 171 doi:10.1016/j.physletb.2007.06.017 [arXiv:0705.2797 [hep-lat]].

[10] V. V. Kiselev, Central Eur. J. Phys. **2**, 523 (2004) [arXiv:hep-ph/0304017].

[11] M. Artuso *et al.* [CLEO Collaboration], Phys. Rev. Lett. **95** (2005) 251801 [arXiv:hep-ex/0508057].

[12] C. Aubin *et al.*, Phys. Rev. Lett. **95** (2005) 122002 [arXiv:hep-lat/0506030].

[13] T. W. Chiu, T. H. Hsieh, J. Y. Lee, P. H. Liu and H. J. Chang, Phys. Lett. B **624**, 31 (2005) [arXiv:hep-ph/0506266].

[14] L. Lellouch and C. J. D. Lin [UKQCD Collaboration], Phys. Rev. D **64**, 094501 (2001) [arXiv:hep-ph/0011086].

[15] D. Becirevic, P. Boucaud, J. P. Leroy, V. Lubicz, G. Martinelli, F. Mescia and F. Rapuano, Phys. Rev. D **60**, 074501 (1999) [arXiv:hep-lat/9811003].

[16] K. A. Olive *et al.* [Particle Data Group], Chin. Phys. C **38** (2014) 090001. doi:10.1088/1674-1137/38/9/090001

[17] D. Becirevic, V. Lubicz, F. Sanfilippo, S. Simula and C. Tarantino, JHEP **1202** (2012) 042 doi:10.1007/JHEP02(2012)042 [arXiv:1201.4039 [hep-lat]].

[18] W. Lucha, D. Melikhov and S. Simula, Phys. Lett. B **735** (2014) 12 doi:10.1016/j.physletb.2014.06.007 [arXiv:1404.0293 [hep-ph]].

[19] M. Wingate, C. T. H. Davies, A. Gray, G. P. Lepage and J. Shigemitsu, Phys. Rev. Lett. **92**, 162001 (2004) [arXiv:hep-ph/0311130].

[20] D. S. Hwang and G. H. Kim, Z. Phys. C **76**, 107 (1997) [arXiv:hep-ph/9703364].

[21] J. Sun, Y. Yang, N. Wang, J. Huang and Q. Chang, Phys. Rev. D **95** (2017) no.3, 036024 doi:10.1103/PhysRevD.95.036024 [arXiv:1703.00155 [hep-ph]].

[22] Z. Rui and Z. T. Zou, Phys. Rev. D **90** (2014) no.11, 114030 doi:10.1103/PhysRevD.90.114030 [arXiv:1407.5550 [hep-ph]].

[23] M. A. Ivanov, J. G. Korner and P. Santorelli, Phys. Rev. D **73** (2006) 054024 doi:10.1103/PhysRevD.73.054024 [hep-ph/0602050].

[24] V. V. Kiselev, A. E. Kovalsky and A. K. Likhoded, Nucl. Phys. B **585**, 353 (2000) [arXiv:hep-ph/0002127]; arXiv:hep-ph/0006104.

[25] V. V. Kiselev, arXiv:hep-ph/0211021.

[26] C. H. Chang and Y. Q. Chen, Phys. Rev. D **49**, 3399 (1994).

[27] A. Y. Anisimov, I. M. Narodetsky, C. Semay and B. Silvestre-Brac, Phys. Lett. B **452**, 129 (1999) [arXiv:hep-ph/9812514]; A. Y. Anisimov, P. Y. Kulikov, I. M. Narodetsky and K. A. Ter-Martirosian, Phys. Atom. Nucl. **62**, 1739 (1999) [Yad. Fiz. **62**, 1868 (1999)] [arXiv:hep-ph/9809249].

[28] W. F. Wang, Y. Y. Fan and Z. J. Xiao, Chin. Phys. C **37** (2013) 093102 doi:10.1088/1674-1137/37/9/093102 [arXiv:1212.5903 [hep-ph]].

Annihilation-Type Semileptonic B-Meson Decays

Anna Kuznetsova[1] and *Alexander* Parkhomenko[1,⋆]

[1] *P. G. Demidov Yaroslavl State University, Yaroslavl, Russia*

Abstract. The rare semileptonic decay $B^0 \to \phi\ell^+\ell^-$, where ℓ is an electron or muon, is considered in the QCD Factorization approach. The differential branching fraction, partially integrated over the interval $q^2 = \left[1~\text{GeV}^2, 8~\text{GeV}^2\right]$ of dilepton invariant mass squared, is estimated. The systematic uncertainty up to 10% resulting from the choice of the B-meson distribution amplitude is determined. The purely perturbative contribution to the $B^0 \to \phi\ell^+\ell^-$ total branching fraction is found to be $\mathcal{B} \sim 10^{-12}$ which suggests searches of the $B^0 \to \phi\mu^+\mu^-$ decay in the four-muon events at the LHC collider.

1 Introduction

Heavy quark physics, in particular related to the bottom (or b) quark, has been an important aspect of the high energy physics research carried out in experiments at hadron colliders at CERN and Fermilab and at the so-called B-meson factories operated at SLAC and KEK. Currently, the LHCb Collaboration working at the CERN proton-proton collider LHC is at the experimental forefront in flavor physics. These researches will be set forth at the Super-KEKB factory at KEK which is close to start its operation now. From a theoretical point of view, one of basic research goals in B-meson physics is a search for physics beyond the Standard Model (BSM) in rare B-meson decays. Here, the main question is whether all experimental observations can be consistently described in the context of the Standard Model. This requires a controlled theoretical framework to understand the strong interaction aspects of flavor physics.

In the heavy quark limit, $m_b \to \infty$, matrix elements of the four-quark operators entering the Effective Weak Hamiltonians [1–4] can be factorized in terms of theoretically more tractable hadronic quantities, then the problem can, in principle, be solved. Of course, one has still to calculate the factorized parts of the decay amplitudes using the QCD perturbation theory (see, for example, [5, 6]) and non-perturbative techniques, such as the QCD Sum Rules and Lattice QCD. In some cases, factorization theorems have been rigorously proved, and in these cases one has, in principle, a well-defined and controlled theory for heavy-hadron decays [1–4].

Phenomenologically, the factorization framework works remarkably well in accounting for the observed non-leptonic decays involving the current-current operators, inducing the $b \to c$ transition [7]. The $B \to h_1 h_2$ decays induced by the charged $b \to u$ and neutral $b \to s$ and $b \to d$ transitions [1–4] are more complex as they involve, in general, in addition to the current-current contributions, also the loop-induced QCD- and electroweak-penguin amplitudes. Here, h_1 and h_2 are light hadrons, such as π-, K-, η-, η'-, ρ-, ω-, ϕ-, and K^*-mesons. During two last decades, these decays have been studied

⋆e-mail: parkh@uniyar.ac.ru

in two specific factorization methods, the so-called perturbative QCD (pQCD) approach by Keum, Li and Sanda [8, 9] and the generalized QCD factorization approach by Beneke, Buchalla, Neubert and Sachrajda (BBNS) [10, 11]. A difference between the pQCD and the BBNS approaches is now understood in terms of the so-called Sudakov behavior of the non-leptonic decay amplitudes, required to alleviate some of the technical problems from the end-point regions of momentum distributions in the perturbative QCD kernels. An outgrowth of this work is the Soft-Collinear Effective Theory (SCET) [12–16], which is developed to deal with problems having different scales, such as obtained by separating the soft, semi-hard and hard regions in decay processes.

Precise data from the B-factory experiments BELLE and BaBar in the decays $B \to \pi\pi$, πK, KK, ηK, $\eta' K$, and related decays in which one of the pseudoscalar particles is replaced by a vector particle, show several mismatches between the predictions of the theories based on the QCD factorization and measurements (see, for example, [17]). This has been the outcome of several theoretical investigations using data from the B-factories.

Within the QCD Factorization approach, the phenomenological analysis of the $B^\pm \to \pi^\pm \ell^+ \ell^-$ decays, where ℓ is an electron or muon, has been completed [18, 19]. These decays are very similar to the exclusive $B \to (K, K^*)\,\ell^+\ell^-$ decays which have received a lot of theoretical attention in the past and are under intensive experimental and phenomenological analysis at present (see [20, 21]). The theory of the $B \to (K, K^*)\,\ell^+\ell^-$ processes — decay rates and a number of distributions — has been developed to the next-to-next-to-leading logarithmic (NNLL) accuracy in the SM [20, 21]. The decay rates have recently been measured by the experiments at the B-factories and the LHC, and are found to be basically in agreement with the corresponding SM-based theoretical estimates. However, the CKM-suppressed semileptonic decays involving the $b \to d\ell^+\ell^-$ transition at the quark level, which can be measured in the exclusive processes such as $B \to (\pi, \eta, \rho, \omega)\,\ell^+\ell^-$ and their $SU(3)_F$-symmetric analogues $B_s \to (K, K^*)\,\ell^+\ell^-$, have not yet been studied to the same theoretical accuracy. On the experimental front, there exist a measurement of the $b \to d\ell^+\ell^-$ transition in the $B^\pm \to \pi^\pm\mu^+\mu^-$ decay presented by the LHCb collaboration first as the total branching ratio $\mathcal{B}(B^\pm \to \pi^\pm\mu^+\mu^-) = (2.3 \pm 0.6 \pm 0.1) \times 10^{-8}$ [22] which was updated on the larger statistics to the smaller but consistent with the previous one value $\mathcal{B}(B^\pm \to \pi^\pm\mu^+\mu^-) = (1.83 \pm 0.24 \pm 0.05) \times 10^{-8}$ [23] and the first measurement of the direct CP-asymmetry $\mathcal{A}_{\rm CP}(B^\pm \to \pi^\pm\mu^+\mu^-) = -0.11 \pm 0.12 \pm 0.01$ [23] was also presented. Our prediction for the branching fraction of the $B^\pm \to \pi^\pm\mu^+\mu^-$ decay coincides with the LHCb measurement [22] within large experimental errors. The $B^+ \to \pi^+\mu^+\mu^-$ dilepton invariant-mass spectrum was measured by the LHCb collaboration recently [23] and is also in a good agreement with theoretical predictions [18, 19] except the first bin $q^2 < 1$ GeV^2. This part of the spectrum is sizably enhanced by the long-distance contributions from the two-body B-meson decays like $B \to \pi(\rho, \omega, \phi)$ and was worked out in [24] which correct our predictions. The predictions for the $B^0 \to \pi^0\ell^+\ell^-$ decay also presented by us [18, 19] can be precisely tested at the forthcoming Super-KEKB factory at KEK.

There exist more rare Semileptonic and radiative decays of B-mesons which are of the annihilation type and consequently they are not observed at present. Examples of pure annihilation-type decays are as follows: $B^0 \to \phi\ell^-\ell^+$, $B_s^0 \to \rho^0\ell^-\ell^+$, $B_s^0 \to \omega\ell^-\ell^+$, $B^0 \to \phi\gamma$, $B_s^0 \to \rho^0\gamma$, and $B_s^0 \to \omega\gamma$. On the best of our knowledges, there are experimental restrictions on one decay mode only: $\mathcal{B}(B^0 \to \phi\gamma) < 8.5 \times 10^{-7}$ at 90% CL, by the BaBar Collaboration [25] which has been superseded by the more stringent upper limit: $\mathcal{B}(B^0 \to \phi\gamma) < 1.0 \times 10^{-7}$, obtained by the Belle Collaboration [26] recently.

In this paper, we present some results concerning the analysis of the rare semileptonic $B^0 \to \phi\ell^-\ell^+$ decays obtained within the QCD Factorization approach. This is the extended version of [27] and the details will be published elsewhere [28]. This decay is can be experimentally searched by the LHCb Collaboration at the LHC collider and by the Belle-II Collaboration at the forthcoming Super-KEKB factory at KEK.

2 Theory of Rare Semileptonic Annihilation-Type B-Meson Decays

From theoretical side, the description of such decays is convenient to do within the method of Effective Weak Hamiltonians [1] when the heavy degrees of freedom — top-quark, W- and Z-boson and Higgs boson are integrated out as well as photons and gluons of energies larger than the W-boson mass. The corresponding effective Lagrangian consists of the usual QED and QCD Lagrangians responsible for the electromagnetic and strong interactions in the effective theory and the effective Lagrangian of the neutral $b \to s$ or $b \to d$ transitions [1]:

$$\mathcal{L}_{\rm eff} = \mathcal{L}_{\rm QCD\times QED} + \mathcal{L}_{\rm weak}^{b\to s} + \mathcal{L}_{\rm weak}^{b\to d}. \quad (1)$$

For us, the later one is of interest for its application to the $B \to \phi\ell^+\ell^-$ decay [1]:

$$\mathcal{L}_{\rm weak}^{b\to d} = -\frac{4G_F}{\sqrt{2}} \sum_{p=u,c} \lambda_p^{(d)} \sum_j C_j(\mu)\mathcal{P}_j(\mu) + \text{h. c.}, \quad (2)$$

where G_F is the Fermi constant, $\lambda_p^{(d)} = V_{p,d}^* V_{p,b}$ is the product of the elements of the Cabibbo-Kobayashi-Maskawa matrix, and $O_j(\mu)$ is a set of local operators accompanied by the corresponding Wilson coefficients $C_j(\mu)$. The local operators can be divided into several groups [1]: tree, strong- and electroweak-penguins, electromagnetic and chromomagnetic dipole and semileptonic operators. For the decay $B \to \phi\ell^+\ell^-$, we are interested in the strong-penguin, electromagnetic dipole and semileptonic operators which give non-trivial contributions to the decay amplitude

In calculations we adopt the following set of operators [29]:

Tree operators ($p = u, c$):

$$\mathcal{P}_1^{(p)} = (\bar{d}\gamma_\mu L T^a p)(\bar{p}\gamma^\mu L T^a b), \quad \mathcal{P}_2^{(p)} = (\bar{d}\gamma_\mu L p)(\bar{p}\gamma^\mu L b), \quad (3)$$

where $L = (1 - \gamma_5)/2$ is the left-handed projection of the fermionic fields and T^a ($a = 1, \dots, 8$) are the generators of the color $SU(3)_C$-group in the fundamental representation [6].

QCD-penguin operators:

$$\mathcal{P}_3 = (\bar{d}\gamma_\mu L b) \sum_q (\bar{q}\gamma^\mu q), \quad \mathcal{P}_4 = (\bar{d}\gamma_\mu L T^a b) \sum_q (\bar{q}\gamma^\mu T^a q),$$
$$\mathcal{P}_5 = (\bar{d}\gamma_\mu\gamma_\nu\gamma_\rho L b) \sum_q (\bar{q}\gamma^\mu\gamma^\nu\gamma^\rho q), \quad \mathcal{P}_6 = (\bar{d}\gamma_\mu\gamma_\nu\gamma_\rho L T^a q) \sum_q (\bar{q}\gamma^\mu\gamma^\nu\gamma^\rho T^a q). \quad (4)$$

Electromagnetic and chromomagnetic dipole operators:

$$\mathcal{P}_{7\gamma} = \frac{e}{16\pi^2}\left[\bar{d}\sigma^{\mu\nu}(m_b R + m_d L)\, b\right] F_{\mu\nu}, \quad \mathcal{P}_{8g} = \frac{g_{\rm st}}{16\pi^2}\left[\bar{d}\sigma^{\mu\nu}(m_b R + m_d L)\, T^a b\right] G_{\mu\nu}^a, \quad (5)$$

where $F_{\mu\nu}$ and $G_{\mu\nu}^a$ are the electromagnetic and gluonic field strength tensors, respectively.

Semileptonic operators:

$$\mathcal{P}_{9\ell(10\ell)} = \frac{\alpha}{2\pi}(\bar{d}\gamma_\mu L b) \sum_\ell (\bar{\ell}\gamma^\mu(\gamma^5)\ell), \quad (6)$$

where $\alpha = e^2/(4\pi)$ is the fine-structure constant.

As for the Wilson coefficients, they are scale-dependent quantities and their evolution is determined by the renormalization group [29]:

$$C_j(\mu_W) = \sum_{k=0}^{\infty} \left[\frac{\alpha_s(\mu_W)}{4\pi}\right]^k C_j^{(k)}(\mu_W), \quad (7)$$

Table 1. The values of the Wilson coefficients accompanied the strong- and electroweak-penguin operators as well as the electromagnetic and chromomagnetic dipole operators at the scale of the b-quark pole mass $\mu_b = m_b = 4.8$ GeV.

$C_1(m_b)$	1.080	$C_3(m_b)$	0.011	$C_7(m_b)$	4.9×10^{-4}
$C_2(m_b)$	−0.177	$C_4(m_b)$	−0.033	$C_8(m_b)$	4.6×10^{-4}
$C_{7\gamma}(m_b)$	−0.317	$C_5(m_b)$	0.010	$C_9(m_b)$	-9.8×10^{-3}
$C_{8g}(m_b)$	0.149	$C_6(m_b)$	−0.040	$C_{10}(m_b)$	1.9×10^{-3}

where $\mu_W \sim m_W$ is a typical scale of order of the W-boson mass where the effective theory is matched into the exact theory — the Standard Model. Their values at the scale of the b-quark mass are presented in Table 1.

3 $B^0 \to \phi \ell^+ \ell^-$ Decay: Differential Branching Fraction

We are discussing the B-meson decay into the ϕ-meson and charged lepton pair, where the lepton ℓ is either an electron or muon. The calculations of the branching fraction were done in the QCD Factorization approach and the ϕ-meson is assumed to be an energetic particle. Kinematics of the process is restricted by the condition $m_\phi^2 \lesssim q^2 \lesssim m_{J/\psi}^2$, where q^2 is the invariant mass squared of the lepton pair and $m_\phi = 1019.46$ MeV and $m_{J/\psi} = 3096.9$ MeV [7] are the ϕ- and $J/\psi(1S)$-meson masses, respectively. This q^2-region is free from the long-distance contributions from the two-body B-meson decays and dominates by the perturbative contribution. To the leading order in the b-quark mass, the differential branching fraction can be presented in the form [27]:

$$\begin{aligned} \frac{d\mathcal{B}}{dq^2} &= \tau_B \frac{G_F^2 |V_{td}^* V_{tb}|^2 \alpha^2}{216\pi} M_B f_B^2 f_\phi^2 Q_d^2 \lambda^3(1, m_\phi/M_B, \sqrt{q^2}/M_B) |C_3 + 4C_4|^2 \\ &\times \left[\left| \frac{1}{\lambda_-^B(q^2)} \right|^2 + \frac{m_\phi^2}{q^2 \left(1 - q^2/M_B^2\right)^2} \left| \frac{1}{\lambda_+^B(q^2)} \right|^2 \right], \end{aligned} \tag{8}$$

where M_B and τ_B are the B-meson mass and life-time, respectively, m_ϕ is the ϕ-meson mass, $\alpha \simeq 1/137$ is fine-structure constant, $Q_d = -1/3$ is the relative charge of the d-quark, and $\lambda^2(a, b, c) = (a + b + c)(a - b + c)(a + b - c)(a - b - c)$ is the kinematical function. The differential branching fraction also depends on several non-perturbative quantities of which f_B and f_ϕ are the leptonic decay constants of the B- and ϕ-meson, respectively, and $\lambda_\pm^B(q^2)$ are the first inverse moments of the B-meson distribution amplitudes. These inverse moments should be known as functions of the momentum squared, q^2, and are dependent on the choice of the B-meson distribution amplitudes $\tilde{\varphi}_\pm^B(t)$ [30]. In the position space, they are defined as scalar functions entering the transition matrix element from the B-meson state to the vacuum one [30]. In the limit of the static heavy quark and massless light quark, the later quark is situated on the light cone (the separation z^μ between the quarks is assumed to be light-like, $z^2 = 0$) and the matrix element can be parameterized as follows [30, 31]:

$$\langle 0 | q_\alpha(z)\, E(0, z)\, h_{v\beta}(0) | \bar{B}(v) \rangle = -\frac{i f_B M_B}{4} \left[(1 + \hat{v}) \left\{ \tilde{\varphi}_+^B(t) - \left[\tilde{\varphi}_+^B(t) - \tilde{\varphi}_-^B(t) \right] \frac{\hat{z}}{2t} \right\} \gamma_5 \right]_{\beta\alpha}. \tag{9}$$

Here, v^μ is the four-velocity of the heavy quark and $t = (zv)$ is the time in the rest frame of the B-meson. To calculate the first inverse moments, one needs the Fourier transforms of the distribution

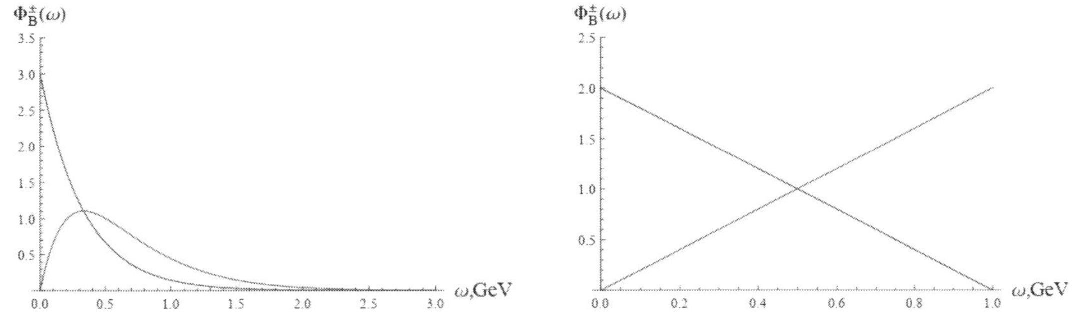

Figure 1. The exponential (the left panel) and linear (the right panel) model of the B-meson distribution amplitudes.

amplitudes defined above in the momentum space, $\varphi_\pm^B(\omega)$. Note that these two distribution amplitudes are not independent but can be related to each other by the Wandzura-Wilczek-type equation [30, 31]:

$$\phi_-^B(\omega) = \int_\omega^\infty \frac{\phi_+^B(\omega')}{\omega'}\, d\omega', \tag{10}$$

which is applicable when the three-particle contribution in the B-meson wave-function is neglected. This approximation is accepted in calculations. From the existing models suggested for the B-meson distribution amplitudes, $\varphi_\pm^B(\omega)$, we consider only two: the exponential one by Grozin and Neubert [30] and the linear model suggested by Kawamura et al. [32]. The explicit shapes of the distribution amplitudes are presented in Fig. 1.

4 First Inverse Moments

First inverse moments (ordinary and logarithmic) are determined as follows [30, 31, 33]:

$$\lambda_B^{-1}(\mu) = \int_0^\infty \frac{d\omega}{\omega}\, \phi_+^B(\omega;\mu), \qquad \frac{\sigma_{B,n}(\mu)}{\lambda_B(\mu)} = \int_0^\infty \frac{d\omega}{\omega}\, \phi_+^B(\omega;\mu) \ln^n \frac{\mu}{\omega}. \tag{11}$$

The former one can be generalized to include the q^2-dependence [30, 31, 34]:

$$\lambda_{B,\pm}^{-1}(q^2) = \int_0^\infty \frac{\phi_\pm^B(\omega)\, d\omega}{\omega - q^2/M_B - i\epsilon}. \tag{12}$$

Note that $\lambda_{B,-}^{-1}(q^2)$ has the logarithmic divergence at $q^2 \to 0$.

Within the representative models selected by us, the moments (12) can be calculated analytically. For the exponential model, the first inverse moments (ordinary and logarithmic with $n = 1$) have the forms [30, 34]:

$$\lambda_B(\mu) = \omega_0(\mu) = \frac{2}{3}\,\bar{\Lambda}; \quad \sigma_B(\mu) = \ln \frac{\mu}{\lambda_B(\mu)} + \gamma_E, \tag{13}$$

where $\bar{\Lambda} = M_B - m_b$ is the effective B-meson mass with m_b being the b-quark mass, $\gamma_E \simeq 0.577$ is the Euler constant. The q^2-dependent first inverse moments (12) can be also calculated rather easily [27, 34]:

$$\lambda_{B,+}^{-1}(q^2) = \lambda_B^{-1} + \zeta\, \lambda_{B,-}(q^2), \quad \lambda_{B,-}^{-1}(q^2) = \lambda_B^{-1}\, e^{-\zeta} \left[-\mathrm{Ei}(\zeta) + i\pi\right], \tag{14}$$

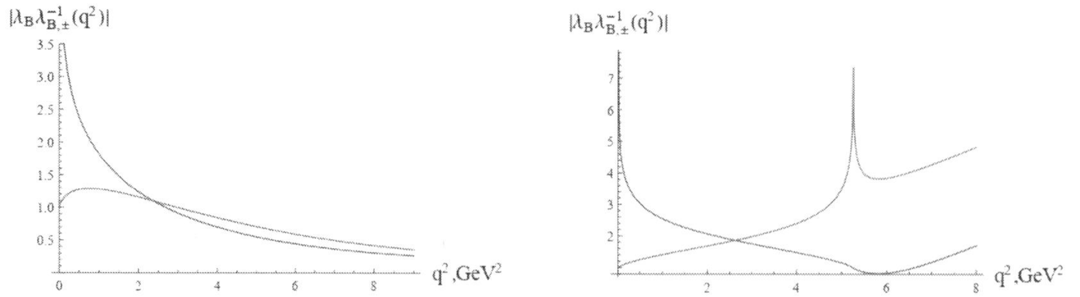

Figure 2. The q^2-dependent first inverse moments in the interval $q^2 \in [1\ \text{GeV}^2, 8\ \text{GeV}^2]$. The moments in the exponential model are shown on the left panel and corresponding curves in the linear model are presented on the right panel.

where $\zeta = q^2/(M_B \lambda_B)$ is the reduced momentum squared and $\text{Ei}(x)$ is the exponential integral function.

For the linear model [32], one can make similar calculations with the following results for first inverse (ordinary and $n = 1$ logarithmic) moments [27, 32]:

$$\lambda_B(\mu) = \bar{\Lambda} = m_B - m_b(\mu), \qquad \sigma_B(\mu) = \ln \frac{\mu}{2\lambda_B(\mu)} + 1, \tag{15}$$

and for the q^2-dependent first inverse moments [27]:

$$\begin{aligned} \lambda^{-1}_{B,+}(q^2) &= \lambda_B^{-1} \left[\xi \ln |1/\xi - 1| + 1 + i\pi\xi\, \Theta\,(1-\xi)\right], \\ \lambda^{-1}_{B,-}(q^2) &= \lambda_B^{-1} \left[(1-\xi) \ln |1/\xi - 1| - 1 + i\pi\,(1-\xi)\,\Theta\,(1-\xi)\right], \end{aligned} \tag{16}$$

where $\xi = q^2/(2M_B\lambda_B)$ is the reduced momentum squared and $\Theta(x)$ is the unit-step function. The dependence of $|\lambda_B \lambda^{-1}_{B,\pm}(q^2)|$ in the interval $q^2 \in [1\ \text{GeV}^2, 8\ \text{GeV}^2]$ is shown in Fig. 2: the exponential- and linear-model curves are presented in the left and right panels, respectively.

5 Numerical results

For experimentalists, the partially-integrated branching fractions over some q^2-interval are of interest:

$$\Delta\mathcal{B}(q^2_{\text{min}} < q^2 < q^2_{\text{max}}) = \int_{q^2_{\text{min}}}^{q^2_{\text{max}}} \frac{d\mathcal{B}}{dq^2}\, dq^2 \tag{17}$$

The differential branching fractions (8) involves the q^2-dependent first inverse moments. To work out the dependence on the choice of the B-meson distribution amplitudes, the branching fraction is evaluated in the interval $q^2 \in [1\ \text{GeV}^2, 8\ \text{GeV}^2]$, where the perturbative calculations are applicable. For the exponential (GN) and linear (KKQT) models, the branching fractions are [27]:

$$\begin{aligned} &\Delta\mathcal{B}^{\text{GN}}(1\ \text{GeV}^2 < q^2 < 8\ \text{GeV}^2) = 5.70 \times 10^{-13}, \\ &\Delta\mathcal{B}^{\text{KKQT}}(1\ \text{GeV}^2 < q^2 < 8\ \text{GeV}^2) = 5.25 \times 10^{-13}. \end{aligned} \tag{18}$$

One can see, that the branching fraction obtained is approximately 5×10^{-13} and the choice of the B-meson model results into the 10% error. This uncertainty can be accumulated into the total theoretical error with a sizable impact. Note that a dominant contribution into the total error is from the uncertainty in the energy scale $\mu \sim m_b$ which is the typical scale of the process considered. It is difficult to quantify all uncertainties numerically at present but we plan to perform such an analysis soon [28]. The last comment is about a possibility to observe such a decay. Assuming that this decay is dominated by the perturbative contribution, the total branching fraction can be estimated as [27]:

$$\mathcal{B}\left(B^0 \to \phi\ell^+\ell^-\right) \sim 10^{-12}. \tag{19}$$

With such a probability, the $B^0 \to \phi\ell^+\ell^-$ decay can not be observed experimentally at the LHC at present but after several years of data taking, it could be possible to get an evidence or strong limitation on this process.

6 Conclusions

The annihilation-type semileptonic decay $B \to \phi\ell^-\ell^+$ is considered in the QCD Factorization approach. Its branching fraction is sensitive to the choice of the B-meson distribution amplitude in the form of the momentum-dependent first inverse moments. Two models for the B-meson distribution amplitudes are considered and their q^2-dependent first inverse moments are calculated. The partially integrated branching fraction is calculated in interval $q^2 \in [1\ \mathrm{GeV}^2, 8\ \mathrm{GeV}^2]$, where the perturbative contribution dominates, and the dependence on the choice of the distribution amplitude model results in the uncertainty of order 10%. The total branching fractions of the $B^0 \to \phi\ell^-\ell^+$ decays, where $\ell = e,\ ,\mu$, are too small to be observed experimentally at the LHC at present but the LHCb Collaboration can, at least, get a strong limit on this decay probability after several years of the LHC run.

The paper is partially supported by the *Russian Foundation for Basic Research* (Project No. 15-02-06033-a) and by the *YSU Initiative Scientific Research Activity* (Project No. AAAA-A16-116070610023-3).

References

[1] G. Buchalla, A.J. Buras, M.E. Lautenbacher, Rev. Mod. Phys. **68**, 1125 (1996), arXiv:hep-ph/9512380
[2] A.V. Manohar, M.B. Wise, *Heavy quark physics*, Vol. 10 of *Cambridge Monographs on Particle Physics, Nuclear Physics and Cosmology* (Cambridge University Press, 2000)
[3] A.G. Grozin, *Heavy quark effective theory*, Vol. 201 of *Springer Tracts in Modern Physics* (New York: Springer-Verlag, 2004)
[4] T. Mannel, *Effective Field Theories in Flavor Physics*, Vol. 203 of *Springer Tracts in Modern Physics* (New York: Springer-Verlag, 2004)
[5] V.B. Berestetsky, E.M. Lifshitz, L.P. Pitaevsky, *Course of Theoretical Physics. Vol. 4. Quantum Electrodynamics* (Pergamon, Oxford, 1982)
[6] M.E. Peskin, D.V. Schroeder, *An Introduction to Quantum Field Theory* (Addison-Wesley, Reading, Massachusetts, 1995)
[7] C. Patrignani et al. (Particle Data Group), Chin. Phys. **C40**, 100001 (2016)
[8] Y.Y. Keum, H.n. Li, A.I. Sanda, Phys. Lett. **B504**, 6 (2001), hep-ph/0004004
[9] Y.Y. Keum, A.I. Sanda, Phys. Rev. **D67**, 054009 (2003), hep-ph/0209014
[10] M. Beneke, G. Buchalla, M. Neubert, C.T. Sachrajda, Phys. Rev. Lett. **83**, 1914 (1999), hep-ph/9905312
[11] M. Beneke, M. Neubert, Nucl. Phys. **B675**, 333 (2003), hep-ph/0308039
[12] C.W. Bauer, S. Fleming, D. Pirjol, I.W. Stewart, Phys. Rev. **D63**, 114020 (2001), hep-ph/0011336

[13] C.W. Bauer, I.W. Stewart, Phys. Lett. **B516**, 134 (2001), hep-ph/0107001
[14] C.W. Bauer, D. Pirjol, I.W. Stewart, Phys. Rev. **D65**, 054022 (2002), hep-ph/0109045
[15] M. Beneke, M. Neubert, Nucl. Phys. **B651**, 225 (2003), hep-ph/0210085
[16] M. Beneke, T. Feldmann, Phys. Lett. **B553**, 267 (2003), hep-ph/0211358
[17] M. Antonelli et al., Phys. Rept. **494**, 197 (2010), 0907.5386
[18] A. Ali, A.Ya. Parkhomenko, A.V. Rusov, Phys. Rev. **D89**, 094021 (2014), 1312.2523
[19] A. Ali, A.Ya. Parkhomenko, A.V. Rusov, Phys. Atom. Nucl. **78**, 436 (2015), [Yad. Fiz. 78, 466 (2015)]
[20] A. Ali, Int. J. Mod. Phys. **A31**, 1630036 (2016), 1607.04918
[21] A. Ali, Int. J. Mod. Phys. **A32**, 1741015 (2017)
[22] R. Aaij et al. (LHCb), JHEP **12**, 125 (2012), 1210.2645
[23] R. Aaij et al. (LHCb), JHEP **10**, 034 (2015), 1509.00414
[24] C. Hambrock, A. Khodjamirian, A. Rusov, Phys. Rev. **D92**, 074020 (2015), 1506.07760
[25] B. Aubert et al. (BaBar), Phys. Rev. **D72**, 091103 (2005), hep-ex/0501038
[26] Z. King et al. (Belle), Phys. Rev. **D93**, 111101 (2016), 1603.06546
[27] A.L. Kuznetsova, A.Ya. Parkhomenko, Phys. Part. Nucl. **48**, 851 (2017)
[28] A.L. Kuznetsova, A.Ya. Parkhomenko, in preparation (2018)
[29] K.G. Chetyrkin, M. Misiak, M. Munz, Phys. Lett. **B400**, 206 (1997), [Erratum: Phys. Lett.B425,414(1998)], hep-ph/9612313
[30] A.G. Grozin, M. Neubert, Phys. Rev. **D55**, 272 (1997), arXiv:hep-ph/9607366
[31] M. Beneke, T. Feldmann, Nucl. Phys. **B592**, 3 (2001), hep-ph/0008255
[32] H. Kawamura, J. Kodaira, C.F. Qiao, K. Tanaka, Phys. Lett. **B523**, 111 (2001), hep-ph/0109181
[33] V.M. Braun, D.Y. Ivanov, G.P. Korchemsky, Phys. Rev. **D69**, 034014 (2004), hep-ph/0309330
[34] M. Beneke, T. Feldmann, D. Seidel, Eur. Phys. J. **C41**, 173 (2005), hep-ph/0412400

EPJ Web of Conferences **158**, 03004 (2017) DOI: 10.1051/epjconf/201715803004
QFTHEP 2017

Associated production of J/ψ and Υ mesons and the prospects to observe a new hypothetical tetraquark state

Sergey Baranov[1,*]

[1] *P.N. Lebedev Institute of Physics, 53 Lenin Avenue, Moscow 119991, Russia*

Abstract. We propose a new mechanism for prompt simultaneous production of J/ψ and Υ mesons in high energy hadronic collisions. The process is considered as a perturbative production of $B_c^{(*)}$ mesons $g+g \to B_c^{(*)}+\bar{B}_c^{(*)}$ followed by a long-distance final state interaction that rearranges the quarks to form J/ψ and Υ mesons. Passing from $B_c^{(*)}+\bar{B}_c^{(*)}$ configuration to $J/\psi+\Upsilon$ configuration may proceed via a hypothetical resonance state, the tetraquark. The goal of this work is to examine whether the respective cross section is large enough to encourage a direct search for the tetraquark at the LHC conditions (yes), and whether this hypothesis can help to explain recent D0 data without assigning an unusually low value to $\sigma_{\rm eff}$ in the double parton scattering mechanism (no).

1 Introduction

After a number of discoveries revealing the existence of tetra-quark mesonic states and penta-quark baryonic states, one should not be surprised with finding yet more particles in these families. For example, it would not be unreasonable to suggest the existence of a $b\bar{b}c\bar{c}$ state composed of four heavy quarks. Its inner structure can be described as a hadron molecule $B_c+\bar{B}_c$, or a hadro-quarkonium $\Upsilon+J/\psi$, or, most probably, the quantum superposition of the two. Such a state must then be naturally connected to the $B_c+\bar{B}_c$ and $\Upsilon+J/\psi$ states, so that these quark configurations could easily transform into one another if the invariant energy of the system is close to the $b\bar{b}c\bar{c}$ resonant mass. The hypothetical Tetraquark can thus be observed experimentally via its four-lepton decay in the $\Upsilon + J/\psi$ channel. The goal of this note is to examine whether the respective cross section is large enough to encourage a direct detection of the tetraquark at the LHC conditions, and whether this hypothesis can help to explain the D0 data without assigning an unusually low value to $\sigma_{\rm eff}$ in the double parton scattering (DPS) mechanism.

2 The model

We understand the production mechanism as a two-step process. The first step is a purely perturbative one and refers to the production of a $B_c\bar{B}_c$ or $B_c^*\bar{B}_c^*$ pair in a short-distance interaction. At the second step, the four heavy quarks get reshuffled in a long-distance final state interaction thus forming the

*e-mail: baranov@sci.lebedev.ru

J/ψ and Υ mesons. We start with calculating the production of B_c pairs in the gluon-gluon fusion subprocess at $O(\alpha_s^4)$:

$$g + g \to B_c^{(*)} + \bar{B}_c^{(*)}. \quad (1)$$

Here we follow the standard QCD Feynman rules and basically repeat our earlier work [1]. The only innovation is in using the k_t-factorization approach [2]. We see the advantages of the latter in the ease of including higher-order corrections, which can be taken into account in the form of k_T-dependent parton densities. These corrections significantly modify the event kinematics; in particular, they give non-zero transverse momentum to the B_c pair and destroy the back-to-back azimuthal correlations. Technically, the computation of Feynman diagrams implies a modified gluon spin density matrix, which is now taken in the form [2–4] $\overline{\epsilon_g^\mu \epsilon_g^{*\nu}} = k_T^\mu k_T^\nu/|k_T|^2$, where k_T is the component of the gluon momentum perpendicular to the beam axis. In the collinear limit, when $k_T \to 0$, this expression converges to the ordinary $\overline{\epsilon_g^\mu \epsilon_g^{*\nu}} = -\frac{1}{2}\, g^{\mu\nu}$. In all other respects the calculations are standard.

Throughout this note, the calculations are based on the following parameter setting. The charmed and bottom quark masses are set to one half of the J/ψ and Υ masses, respectively: $m_c = m_\psi/2 =$ 1.55 GeV and $m_b = m_\Upsilon/2 = 4.8$ GeV (we also have to obey $m_{B_c} = m_b + m_c$); the factorization and renormalization scales are set equal to each other and $\mu_F^2 = \mu_R^2 = \hat{s}/4$ where $\hat{s}$ is the invariant energy of the partonic subprocess (this choice is Lorentz-invariant and is symmetric with respect to both final state particles); we use $A0$ parametrization for the unintegrated (k_T-dependent) gluon density taken from Ref. [5]; and we use radial wave functions of $B_c^{(*)}$ mesons $|\mathcal{R}_{B_c}(0)|^2 = 1.2$ GeV^3 taken from the potential model of Ref. [6]. Variations in the scales μ_F^2 and μ_R^2 or using the $A+$ or $A-$ gluon parametrizations instead of $A0$ may increase or decrease the estimated production rate by a factor of 2 or 3. Note however, that the aim of this note is not in giving exact predictions for an unknown particle, but rather in giving some reasonable hints that would help to make a decision on whether a dedicated search for a new particle is worth starting.

For an illustrative example we consider the kinematics of LHCb experiment, as it has no complications and no reduction of the cross section coming from the cuts on the muon transverse momentum. Note also that the k_t-factorization approach was able to perfectly describe the recent LHCb data on double J/ψ production [7]; the matrix element of the latter process can be obtained from that of $B_c\bar{B}_c$ case by just replacing m_b with m_c.

The $B_c\bar{B}_c$ invariant mass spectrum calculated for $\sqrt{s} = 13$ TeV is shown in Fig. 1. We see that the cross section mostly concentrates at nearby threshold, where the relative quark velocities are not large, thus giving a good chance for final state interactions. As a consequence, the quarks can reshuffle and form J/ψ and Υ mesons:

$$B_c + \bar{B}_c \to b\bar{b}c\bar{c} \to J/\psi + \Upsilon. \quad (2)$$

To calculate the transition amplitude of (2) we, strictly speaking, need to know the overlap integrals for the initial and intermediate, and for the intermediate and final state wave functions. For the lack of the information on the tetraquark wave function, we use a semi-qualitative approximation based on the $1/N_c$ topological expansion and quark-hadron duality.

The diagram of Fig. 1 with crossing quark lines is a typical nonplanar diagram in the $1/N_c$ topological expansion [8–10]. Diagrams of this kind violate Okubo-Zweig-Iizuka (OZI) rule and are suppressed by an approximate factor of $1/N_c^2$ in comparison with OZI non-violating planar diagrams of the same quark content (the latter refer to the B_c+$\bar{B}_c$ final states). The suppression factor is easily calculable in our case by taking the projection of the B_c+$\bar{B}_c$ color state $\delta^{b\bar{c}}\delta^{c\bar{b}}$ onto the J/ψ+Υ color state $\delta^{b\bar{b}}\delta^{c\bar{c}}$; taking the trace of color indices yields a factor of 1/8.

This factor gives the ratio of the production cross sections for two different color configurations, but it tells nothing on whether the two color singlets are bound or unbound. We can appeal here to

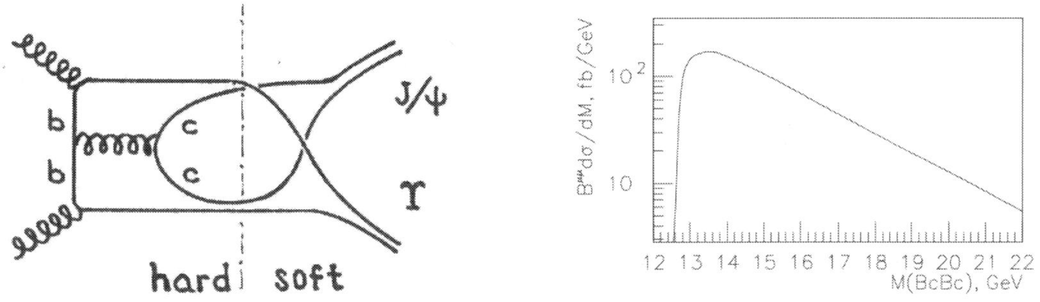

Figure 1. A diagram representing the J/ψ+Υ production mechanism discussed in the text (left panel) and our predictions on the $B_c\bar{B}_c$ invariant mass ditribution at LHCb at $\sqrt{s}$ = 13 TeV (right panel).

the ideas of quark-hadron duality [11] that the cross section integrated over a continuous spectum must be the same as the cross section summed over a discrete number of resonances. That is, the estimate $\sigma(J/\psi$+$\Upsilon) \simeq \sigma(B_c$+$\bar{B}_c)/N_c^2$ shows the correct number of J/ψ+Υ states, and if the hypothetical tetraquark does exisit, this estimate corresponds to the tetraquark production cross section; otherwise, it corresponds to the unbound J/ψ 's and Υ 's.

Now we turn to numerical results. The integral production cross section times muon branching fraction amounts to $\sigma(pp \to B_c\bar{B}_c)Br^{\Upsilon\to\mu\mu}Br^{\psi\to\mu\mu} \simeq 550$ fb. By including the muon branchings we also mean that the muons fit the LHCb kinematics $2 < \eta(\mu) < 4.5$. Then, applying the topological suppression factor we find for the J/ψ+Υ production rate

$$\sigma(\Upsilon+J/\psi)\, Br^{\Upsilon\to\mu\mu}Br^{\psi\to\mu\mu} \simeq 60 \text{ fb}, \tag{3}$$

where the estimated [12–14] four-lepton decay branching is not in contradiction with the naive guess $Br^{b\bar{b}c\bar{c}\to\mu\mu\mu\mu} \simeq Br^{\Upsilon\to\mu\mu}Br^{\psi\to\mu\mu} \simeq 1.5\ 10^{-3}$. So, the expected production rate is small, though not discouragingly small, and is at least not smaller than the integral Υ + J/ψ cross section measured by the collaboration D0.

The tetraquark signal should be seen over a background formed by both single (SPS) and double (DPS) parton scattering. First, we briefly discuss the SPS contributions. Contrary to the production of same-flavor meson pairs, the associated production of J/ψ and Υ mesons in the color-singlet mode is not possible at $O(\alpha_s^4)$. The leading SPS mechanism refers to a mixed singlet-octet scheme, with Υ formed in the color-singlet mode $g + g \to \Upsilon + g^*$ and J/ψ coming from the virtual (color-octet) gluon fragmentation $g^* \to c\bar{c}[^3S_1^8] \to J/\psi$. The fully octet modes are suppressed because of typically much smaller values of the color-octet matrix elements (as compared to the color-singlet wave functions). These contributions can only become important at high p_T, that is not the main region of the LHCb phase space. Simulataneous color-singlet production of P-wave states $g + g \to \chi_b + \chi_c$ is possible at $O(\alpha_s^4)$, but is suppressed by lower values of the P-state wave functions and, further, by the subsequent radiative transition branching fractions $\chi_c \to J/\psi$+γ, $\chi_b \to \Upsilon$+γ. Simultaneous production of S-wave color-singlets is possible at $O(\alpha_s^6)$; it is suppressed by two extra powers of α_s and, especially, by the color algebra, as is explained in detail in Ref. [15]. Besides, one can consider a non-prompt mechanism, i.e., the production of $\Upsilon + b + \bar{b}$ states at $O(\alpha_s^4)$ followed by the decays $b, \bar{b} \to J/\psi + X$. This contribution is suppressed because of small decay branching fraction $Br(B \to J/\psi + X) \simeq 3\ 10^{-3}$ [16]. All these contributions taken together can hardly bring more than $\sim$ 2 fb, even with the most optimistic choice for the color-octet matrix elements. The dominant background is certainly brought by the double parton scattering. Using the simple factorization formula [17]

$$\sigma_{\rm DPS}(\Upsilon{+}J/\psi) = \sigma_{\rm incl}(\Upsilon)\,\sigma_{\rm incl}(J/\psi)\;/\;\sigma_{\rm eff} \tag{4}$$

along with the previously measured single inclusive J/ψ [18] and Υ [19] cross sections at $\sqrt{s}$= 7, 8, and 13 TeV

$$\sigma_{\rm incl,\,13\,TeV}(J/\psi) Br^{\psi\to\mu\mu} \simeq 380 \text{ nb}, \tag{5}$$

$$\sigma_{\rm incl,\,7\,TeV}(\Upsilon) Br^{\Upsilon\to\mu\mu} \simeq 2.5 \text{ nb}, \quad \sigma_{\rm incl,\,8\,TeV}(\Upsilon) Br^{\Upsilon\to\mu\mu} \simeq 3.3 \text{ nb}, \tag{6}$$

extrapolating the Υ results to $\sqrt{s}$=13 TeV as

$$\sigma_{\rm incl,\,13\,TeV}(\Upsilon) Br^{\Upsilon\to\mu\mu} \simeq 6 \text{ nb}, \tag{7}$$

and taking $\sigma_{\rm eff}$ =15 mb we arrive at

$$\sigma(\Upsilon{+}J/\psi)\, Br^{\Upsilon\to\mu\mu} Br^{\psi\to\mu\mu} \simeq 150 \text{ fb}. \tag{8}$$

With smaller values of $\sigma_{\rm eff}$ suggested by some other measurements ($\sigma_{\rm eff}$ = 5 mb [20] or even $\sigma_{\rm eff}$ = 2.2 mb [21]) we arrive at accordingly larger estimations of the background. The signal is smaller than the integral background, but can be distinguished from the latter due to different kinematic properties. The resonant peak in the four-lepton invariant mass distribution contrasts with the broad and flat mass spectrum generated by double parton scattering.

Finally, we turn to the data [21] provided by the Collaboration D0 at the Tevatron. The Collaboration reported on the simultaneous production of J/ψ and Υ mesons in pp collisions at $\sqrt{s}$ = 1.96 TeV. Both mesons were seen in the muon decay channel, where all muons were required to have transverse momentum $p_T^\mu > 2$ GeV and pseudorapidity $|\eta^\mu| < 2.0$. The production cross section measured with these requirements was found to be

$$\sigma(\Upsilon{+}J/\psi)\, Br^{\Upsilon\to\mu\mu} Br^{\psi\to\mu\mu} = 27 \pm 9(\text{stat}) \pm 7(\text{syst}) \text{ fb}. \tag{9}$$

This number has to be compared with the hypothetical effect of a $c\bar{c}b\bar{b}$ tetraquark. Following the method described above we find for the integral production rate at the D0 conditions $\sigma(pp \to B_c\bar{B}_c)\, Br^{\Upsilon\to\mu\mu} Br^{\psi\to\mu\mu} \simeq 22$ fb. This is comparable with the experimental result Eq.(9), but one has to keep in mind that only a fraction of all $B_c^{(*)}$ pairs can convert into $\Upsilon{+}J/\psi$ states. After applying the topological suppression factor we are left with only 1 or 2 fb. The flatness of the $\Delta\phi$ distribution seen in the data is yet not incompatible with the tetraquark hypothesis (if the tetraquark mass is around 14 GeV; a result of the calculation not shown in the present note), but the absolute production rate probably is. So we conclude that the considered mechanism cannot take credit for the $\Upsilon{+}J/\psi$ events observed by D0.

3 Conclusions

We have considered a new mechanism for prompt simultaneous production of J/ψ and Υ mesons. This mechanism consists in a perturbative production of $B_c^{(*)}$ mesons $g{+}g \to B_c^{(*)}{+}\bar{B}_c^{(*)}$ followed by a long-distance interaction that rearranges the quarks to form J/ψ and Υ mesons. Passing from $B_c^{(*)}{+}\bar{B}_c^{(*)}$ configuration to $J/\psi{+}\Upsilon$ configuration may proceed via a hypothetical resonance state with $b\bar{b}c\bar{c}$ quark content.

We find that the considered mechanism may be strong enough to encourage a direct search for the $b\bar{b}c\bar{c}$ tetraquark at the LHC conditions. At the same time, the estimated production rate at the D0 conditions is insufficient to provide a reasonable explanation for the observed $\Upsilon{+}J/\psi$ events. Of course, the conclusive evidence could only be provided by an explicit peak in the four-muon mass spectrum.

Acknowledgments

This work was supported by the DESY directorate in the framework of Moscow-DESY project on Monte-Carlo implementations in Physics.

References

[1] S.P. Baranov, Phys. Rev. D **55**, 2756 (1997).

[2] L.V. Gribov, E.M. Levin, M.G. Ryskin, Phys. Rep. **100**, 1 (1983);
E.M. Levin, M.G. Ryskin, Yu.M. Shabelsky, A.G. Shuvaev, Sov. J. Nucl. Phys. **53**, 657 (1991).

[3] S. Catani, M. Ciafaloni, F. Hautmann, Phys. Lett. B **242**, 97 (1990); Nucl. Phys. **B366**, 135 (1991).

[4] J.C. Collins, R.K. Ellis, Nucl. Phys. **B360**, 3 (1991).

[5] H. Jung, arXiv:hep-ph/0411287;
H. Jung *et al.*, Eur. Phys. J. C **70**, 1237 (2010);
H. Jung, http://www.desy.de/~jung/cascade/updf.html

[6] E. Bagan, H.G. Dosch, P. Gosdzinsky, S. Narison, J.-M. Richard, Z. Phys. C **64**, 57 (1994).

[7] R. Aaij *et al.* (LHCb Collab.), arXiv:1612.07451.

[8] Chan Hong-Mo *et al.*, Nucl. Phys. B **86**, 470 (1975); Nucl. Phys. B **92**, 13 (1975).

[9] C.F. Chew, C. Rosenzweig, Nucl. Phys. B **104**, 290 (1976); Phys. Rev. C **41**, 263 (1978).

[10] G. Veneziano, Phys. Lett. B **52**, 220 (1974); Nucl. Phys. B **117**, 519 (1976).

[11] E.C. Poggio, H.R. Quinn, S. Weinberg, Phys. Rev. D **13**, 1958 (1976).

[12] M. Karliner, S. Nussinov, J.L. Rosner, arXiv:1611.00348.

[13] Y. Bai, S. Liu, J. Osborne, arXiv:1612.00012.

[14] Z.-G. Wang, arXiv:1701.04285.

[15] S.P. Baranov, W. Schäfer, A.M. Snigirev, A. Szczurek, N.P. Zotov, Phys. Rev. D **87**, 034035 (2013).

[16] C. Patrignani *et al.* (Particle Data Group), Chin. Phys. **C40**, 100001 (2016).

[17] P. Bartalini *et al.*, arXiv:1111.0469,
H. Abramowicz *et al.*, arXiv:1306.5413,
S. Bansal *et al.*, arXiv:1410.6664.

[18] R. Aaij *et al.* (LHCb Collab.), JHEP **1510**, 172 (2015).

[19] R. Aaij *et al.* (LHCb Collab.), JHEP **1511**, 103 (2015).

[20] V.M. Abazov *et al.* (D0 Collab.), Phys. Rev. D **90**, 111101(R) (2014).

[21] V.M. Abazov *et al.* (D0 Collab.), Phys. Rev. Lett. **116**, 082002 (2016).

EPJ Web of Conferences **158**, 03005 (2017) DOI: 10.1051/epjconf/201715803005
QFTHEP 2017

Rare four-leptonic B-decays with light leptons in the final state in the Standard Model

Anna Danilina[1,2,3,⋆] and *Nikolai* Nikitin[1,2,3,⋆⋆]

[1] *M.V.Lomonosov Moscow State University, Physics Faculty*
[2] *D.V.Skobeltsyn Institute of Nuclear Physics*
[3] *Institute for Theoretical and Experimental Physics*

Abstract. We present predictions for the four-leptonic decays of B^--mesons with three light charged leptons and one neutrino in the final states. We obtain the branching ratio for $B^- \to \mu^+\mu^-e^-\bar{\nu}_e$ decay at the level 1.3×10^{-7} taking into account the photon emission from the u- and b- valence quarks of B^-–meson and the Bremsstrahlung. The contribution of the virtual photon was calculated using the Vector Meson Dominance model (VMD). As a next step we present prediction for branching ratio of the decay $B^- \to \mu^-\mu^+\mu^-\bar{\nu}_\mu$ and the estimation for $Br(B_s \to \mu^+\mu^-e^+e^-)$.

1 Introduction

The investigation of rare four leptonic decays of the B-mesons provides an important test of the Standard Model (SM) and its possible extensions in the high order of the perturbation theory. The branching ratios of these decays are sufficiently sensitive to the contribution of the "New Physics". Therefore it is very important to precisely predict different characteristics of the B-mesons decays. Four-lepton B-decays are investigated at the LHCb, CMS and ATLAS experiments at LHC. By now, only upper limits on the branching ratios for the decays $B_{d,s} \to \mu^+\mu^-\mu^+\mu^-$ were obtained at the LHCb [1, 2].

2 Total Hamiltonian for the decays $B^- \to \ell^+\ell^-\ell'^-\bar{\nu}_{\ell'}$

The total Hamiltonian for four-leptonic decays $B^- \to \ell^+\,\ell^-\,\ell'^-\,\bar{\nu}_{\ell'}$ in the Standard Model is given by following expression:

$$\mathcal{H}_{tot}(x) = \mathcal{H}_{ub}(x) + \mathcal{H}_{em}(x) + \mathcal{H}_{VMD}(x). \tag{1}$$

The Hamiltonian relevant for the transition of interest $b \to uW^- \to u\ell^-\bar{\nu}_\ell$ has the form

$$\mathcal{H}_{ub}(x) = \frac{G_F}{\sqrt{2}}\,V_{ub}\,\left(\bar{u}(x)\,\gamma^\mu(1-\gamma^5)\,b(x)\right)\left(\bar{\ell}(x)\,\gamma_\mu(1-\gamma^5)\nu_\ell(x)\right) + h.c.,$$

with G_F the Fermi constant, V_{ub} the CKM–matrix element, and γ^5 defined as $\gamma^5 = i\gamma^0\gamma^1\gamma^2\gamma^3$.

⋆e-mail: anna.danilina@cern.ch
⋆⋆e-mail: nnikit@mail.cern.ch

The electromagnetic Hamiltonian is given by

$$\mathcal{H}_{em}(x) = e \sum_f Q_f \left(\bar{f}(x)\gamma^\mu f(x)\right) A_\mu(x),$$

where $e = |e| > 0$ is the elementary charge, $e^2 = 4\pi\alpha_{em}$, $\alpha_{em} \approx 1/137$ the fine structure constant, Q_f the charge of a fermion with flavour f in units of elementary charge.

The Hamiltonian describing the Vector Meson Dominance model has the form

$$\mathcal{H}_{VMD}(x) = \sum_V \frac{e\,M_V^2}{f_V} V^\mu(x) A_\mu(x),$$

M_V – is the mass of corresponding vector meson, $V^\mu(x)$ – is the vector meson field. The absolute value of the VMD-constant f_V is obtained from

$$\Gamma(V \to \ell^+\ell^-) = \frac{4\pi}{3}\left(\frac{\alpha_{em}}{|f_V|}\right)^2 M_V. \quad (2)$$

It is assumed that VMD–constants f_V for ρ and ω mesons are real and positive.

3 The $B^- \to \mu^+\mu^- e^-\bar{\nu}_e$ amplitude of the decay

The amplitude of the decays $B^-(p) \to \ell^+(k_1)\,\ell^-(k_2)\,\bar{\nu}_{\ell'}(k_3)\,\ell'^-(k_4)$ involves several different contributions. Firstly, it is the photon emission from the valence u-quark of the B^--meson; the corresponding diagram is given in Fig. 1.

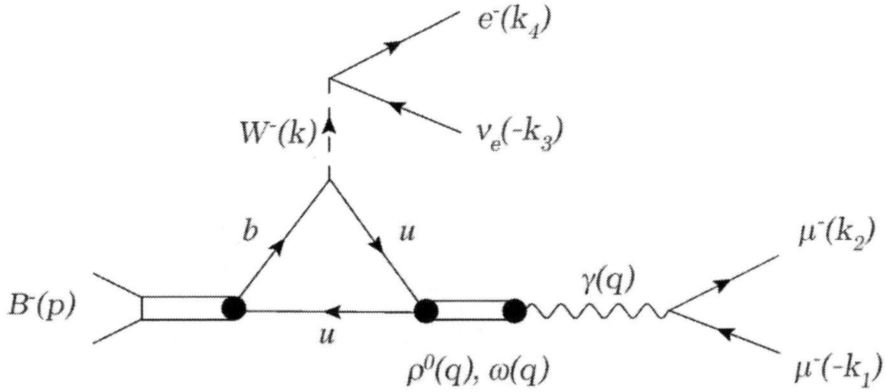

Figure 1. Emission of the virtual photon from u-quark in the VMD approach.

In this case we described the photon emissions via VMD includes the intermediate $\rho^0(770)$ and $\omega(782)$ mesons. The amplitude corresponding to the diagram of Fig. 1 can be parametrized as follows

$$F^{(u)}\left(q^2, k^2\right) = \sum_i \frac{M_{2i}^2}{q^2} \frac{M_1^2}{q^2 - M_{2i}^2 + i\Gamma_{2i} M_{2i}} G^{(u)}\left(q^2, k^2, M_{2i}^2\right), \quad (3)$$

where M_{2i} and Γ_{2i} – are the mass and the full width of the intermediated vector meson that contains $u\bar{u}$-pair, $G^{(u)}\left(q^2, k^2, M_{2i}^2\right)$ – the smooth function without poles, $q = k_1 + k_2$ and $k = k_3 + k_4$.

The function $F^{(u)}\left(q^2, k^2\right)$ has the poles at the $q^2 = M^2_{2i}$.

The leftmost singularity in the k^2–variable lies at $k^2 = M^2_{B^{*-}}$. This singularity was excluded from the kinematic region which was taken into account in the transition form factors $B \to \rho$ and $B \to \omega$ parameterization.

Before calculating the decay rate, it is useful to estimate the contribution of all diagrams. The contribution of the diagram Fig. 1 can be estimated roughly considering the conrtibution of $\rho^0(770)$ and $\omega(782)$ intermediate mesons:

$$\begin{aligned} Br^{(VMD)}(B^- &\to \mu^+\mu^- e^-\bar{\nu}_e) \approx \\ &\approx \left|\sqrt{Br(B^- \to \rho^0 e^-\bar{\nu}_e) * Br(\rho^0 \to \mu^+\mu^-)} + \right. \\ &+ \left. \sqrt{Br(B^- \to \omega e^-\bar{\nu}_e) * Br(\omega \to \mu^+\mu^-)}\right|^2 \approx 0,4 \times 10^{-7}. \end{aligned}$$

An alternative way to estimate of the contribution from the diagram Fig. 1 is to represent $B^- \to \mu^+\mu^- e^-\bar{\nu}_e$ decay as a chain of processes $B^- \to \gamma^* e^-\bar{\nu}_e \to \mu^+\mu^- e^-\bar{\nu}_e$.

The branching ratio of $B^- \to e^-\bar{\nu}_e$ can be written as

$$Br(B^- \to e^-\bar{\nu}_e) = \tau_{B^-} \frac{{G_F}^2|V_{ub}|^2 {f_B}^2 {M_1}^3}{8\pi} \left(\frac{m_e}{M_1}\right)^2 \approx 10^{-11},$$

Using the previous formula, the branching ratio for decay $B^- \to \mu^+\mu^- e^-\bar{\nu}_e$ is given by

$$Br^{(B^*)}(B^- \to \mu^+\mu^- e^-\bar{\nu}_e) \approx Br(B^- \to e^-\bar{\nu}_e) \times {\alpha_{em}}^2 * \left(\frac{M_1}{m_e}\right)^2 \approx 0,5 \times 10^{-7},$$

where the helicity suppression is reduced by emission of photon from light quark of B^-–meson.

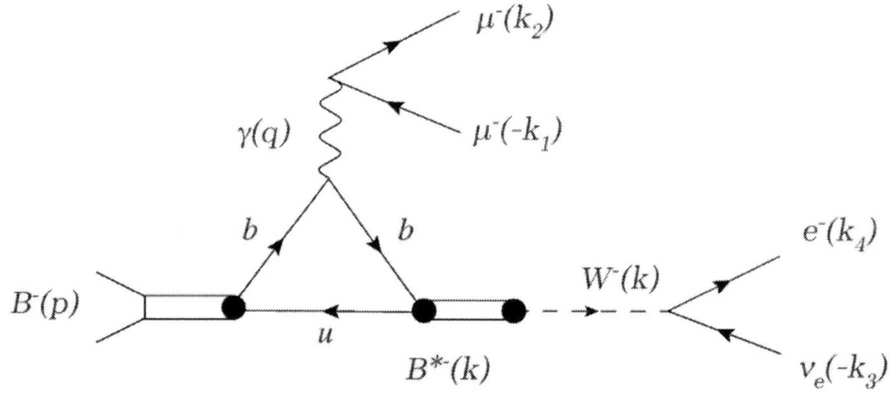

Figure 2. Emission of the virtual photon from b-quark in the VMD approach.

The second contribution to the amplitude describes the photon emission from the valence b–quark of the B^--meson (see Fig. 2). In this case, VMD is used to calulate only B^{*-} intermediate resonance contribution. The amplitude corresponding to the diagram of Fig. 2 has no singularity in the physical region of k^2, because the nearest singularity lies at $k^2 = M^2_{B^{*-}}$ and $M_{B^{*-}} > M_1$.

In addition, the Feynman graph in Fig. 2 contains the pole in q^2-variable coming from the photon propagator. The next singularity in q^2 is at the point $q^2 = M^2_{\Upsilon(1S)}$, where $M_{\Upsilon(1S)}$ is the mass of $\Upsilon(1S)$ –

meson. This pole is far beyond the kinematical decay region. Contribution of the $\Upsilon(1S)$ – meson was included in the parametrization of transition form factor [3–5]. Therefore the part of the amplitude corresponding to the diagram of Fig. 2 can be presented as

$$F^{(b)}\left(q^2, k^2\right) = \frac{M_{B^{*-}} f_{B^{*-}}}{q^2} G^{(b)}\left(q^2, k^2, M^2_{B^{*-}}\right). \qquad (4)$$

Let us note that the diagram of Fig. 2 is suppressed compared to the diagram of Fig. 1 by the suppression factor λ/m_b, where $\lambda \approx 0.5$ GeV and m_b is the b-quark mass.

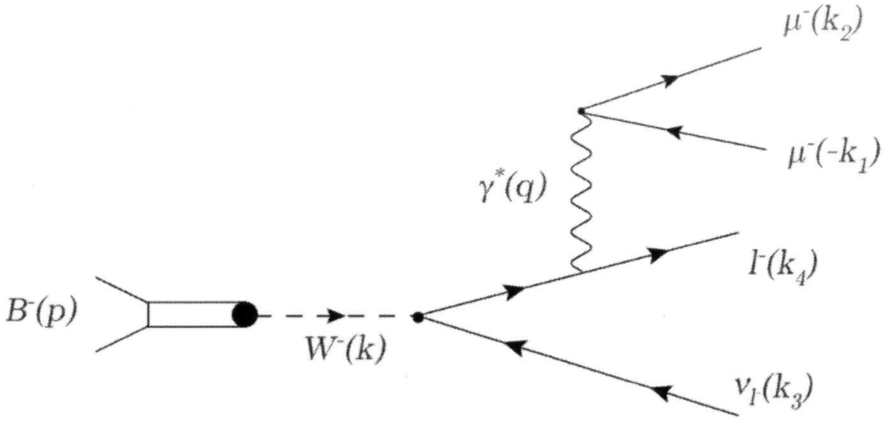

Figure 3. Bremstraulung contribution to the decay $B^- \to \mu^+\mu^- e^- \bar{\nu}_e$.

The diagram Fig. 3 corresponds to Bremsstrahlung. This diagram contains the pole from photon propagator at q^2–variable and it is significant only at small q^2–region.

Therefore, contribution of the Fig. 1 is most significant, but the other diagrams are important and have to be taken into account.

4 Calculation of the matrix elements for the decays $B^- \to \ell^+\ell^-\ell'^-\bar{\nu}_{\ell'}$

Contribution to the amplitude of the decay $B^-(p) \to \ell^+(k_1)\,\ell^-(k_2)\,\bar{\nu}_{\ell'}(k_3)\,\ell'^-(k_4)$ from diagram in the Fig. 1 one can calculate using the VMD model. We set $m_\ell = m_{\ell'} = 0$ and use the expression for the total Hamiltonian (1). For the matrix element we find

$$\begin{aligned} i\,M^{(VMD)}_{fi} &= -i\,\frac{G_F}{\sqrt{2}}\,\frac{4\pi\,\alpha_{em}}{q^2}\,V_{ub}\left[\sum_{i=\rho^0,\,\omega}\frac{M^2_{2i}}{f^i_V}\,\frac{1}{q^2 - M^2_{2i} + i\Gamma_{2i}M_{2i}}\mathcal{F}^{(i)}_{\mu\nu}(q,k)\right] \\ & \left(\bar{\ell}(k_2)\,\gamma^\nu\,\ell(-k_1)\right)\left(\bar{\ell'}(k_4)\,\gamma^\mu(1-\gamma^5)\,\nu_{\ell'}(-k_3)\right). \end{aligned} \qquad (5)$$

Taking into account the equation of motion of the lepton fields for $\mathcal{F}^{(i)}_{\mu\nu}(q,k)$ we find

$$\mathcal{F}^{(i)}_{\mu\nu}(q,k) = \frac{2\,V^{(i)}(k^2)}{M_1 + M_{2i}}\,\epsilon_{\mu\nu\alpha\beta}\,k^\alpha q^\beta - i\,(M_1 + M_{2i})A^{(i)}_1(k^2)g_{\mu\nu} + 2i\,\frac{A^{(i)}_2(k^2)}{M_1 + M_{2i}}\,q_\mu k_\nu.$$

For the diagram of Fig. 2 we have the folloving amplitude

$$i\,M_{fi}^{(b)} = -i\,\frac{G_F}{\sqrt{2}}\,\frac{4\pi\,\alpha_{em}}{q^2}\,V_{ub}\,\frac{M_{B^*}f_{B^*}}{k^2-M_{B^*}^2+i\Gamma_{B^*}M_{B^*}}\,\frac{2}{3}\,\frac{V_b(q^2)}{M_1+M_{B^*}}\,\epsilon_{\mu\nu pq}$$
$$\left(\bar{\ell}(k_2)\,\gamma^{\nu}\,\ell(-k_1)\right)\left(\bar{\ell'}(k_4)\,\gamma^{\mu}(1-\gamma^5)\,\nu_{\ell'}(-k_3)\right), \qquad (6)$$

and for Fig. 3 we obtained the amplitude

$$i\,M_{fi}^{(brem)} = -i\,\frac{G_F}{\sqrt{2}}\,\frac{4\pi\,\alpha_{em}}{q^2}\,V_{ub}\,i\,f_{B_u}\,g_{\mu\nu}\left(\bar{\ell}(k_2)\gamma^{\nu}\ell(-k_1)\right)\left(\bar{\ell'}(k_4)\gamma^{\mu}(1-\gamma^5)\nu_{\ell'}(-k_3)\right). \qquad (7)$$

5 Numerical results for the branching ratio and differential distributions of the decay $B^- \to \ell^+\ell^-\ell'^-\bar{\nu}_{\ell'}$

After the calculation of the square of the matrix elemant and integration over the variables $x_{12} = q^2/M_1^2$ and $x_{34} = k^2/M_1^2$, we got the numerical prediction for the $B^- \to \mu^+\mu^-e^-\bar{\nu}_e$ branching ratio:

$$Br(B^- \to \mu^+\mu^-e^-\bar{\nu}_e) \approx 1,3 \times 10^{-7}.$$

Let us mention that all leptons are taken massless in the calculations of the matrix element. However, the integration over $\sqrt{q^2}$–variable startes from double muon mass in order to avoid the singularity related to the photon propagator. Thus the integration over x_{12} started from the point $x_{12}^{min} = (2m_\mu/M_1)^2 \approx 0.0016$.

The double differencial distribution for the decay rates are shown in Fig. 4 and Fig. 5, where we can see peaks corresponding to the virtual photon emission and the contributions of $\rho^0(770)$ and $\omega(782)$ resonances.

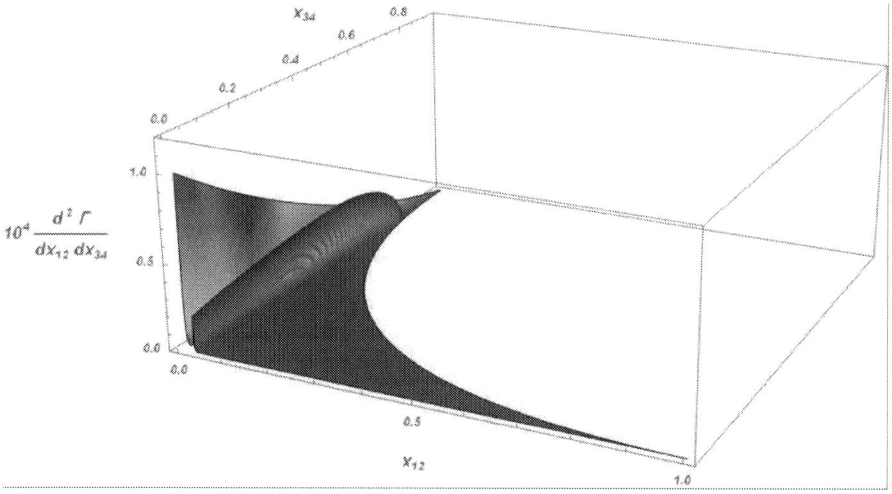

Figure 4. Double differential distributions for $\dfrac{d^2\Gamma(B^- \to \mu^+\mu^-e^-\bar{\nu}_e)}{dx_{12}\,dx_{34}} \times 10^4$.

Finally, we estimate the branching ratio of the decay of $B^- \to \mu^+\mu^-\mu^-\bar{\nu}_\mu$. Similarly to the decay $B^- \to \mu^+\mu^-e^-\bar{\nu}_e$ we take into account the contributions of the photon emission from u- and b- quarks

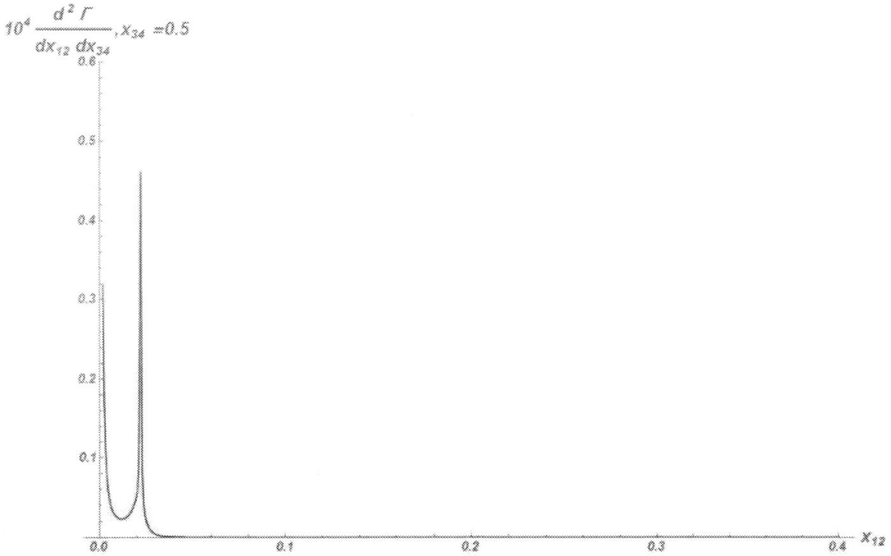

Figure 5. One-dimensional face of the double differencial distribution $\frac{d^2\Gamma(B^- \to \mu^+\mu^-e^-\bar{\nu}_e)}{dx_{12}\,dx_{34}}$ at $x_{34} = 0.5$.

and Bremsstrahlung contribution. But in this case the number of diagramms doubles because of the identical fermions in the final state. So we obtain addition interferecne terms in the expression of the square matrix element.

The basic problem in estimation of branching ratio of $B^- \to \mu^+\mu^-\mu^-\bar{\nu}_\mu$ was calculation of interference part in the square matrix element. The similar processes contributed to the decay rations of $B^- \to \mu^+\mu^-e^-\bar{\nu}_e$ and $B^- \to \mu^+\mu^-\mu^-\bar{\nu}_\mu$. Therefore branching ratio for the $B^- \to \mu^+\mu^-\mu^-\bar{\nu}_\mu$ decay we can represent as

$$Br(B^- \to \mu^+\mu^-\mu^-\bar{\nu}_\mu) = Br(B^- \to \mu^+\mu^-e^-\bar{\nu}_e) - Br^{(interf)}(B^- \to \mu^+\mu^-\mu^-\bar{\nu}_\mu).$$

Numerically $Br^{(interf)}(B^- \to \mu^+\mu^-\mu^-\bar{\nu}_\mu) \approx 2.4 \times 10^{-8}$. Therefore

$$Br(B^- \to \mu^+\mu^-\mu^-\bar{\nu}_\mu) \approx 10^{-7}.$$

It means that the measurements of branching ratios for the decays $B^- \to \mu^+\mu^-e^-\bar{\nu}_e$ and $B^- \to \mu^+\mu^-\mu^-\bar{\nu}_\mu$ are possible at LHCb and at Belle II.

6 Estimations of the branching ratio for the decay $B_s \to \mu^+\mu^-e^+e^-$

Using the previous result for $Br(B^- \to \mu^+\mu^-e^-\bar{\nu}_e)$, we can evaluate the branching ratio of the decay $B_s \to \mu^+\mu^-e^+e^-$. We suppose, that nonperturbative contribution of strong interactions have the same magnitude for both decays. But the difference between these decays cosists in the diffrent contributions of weak and electromagnetic processes. This difference contanning by coefficients multiplied by the hadronic parts of the matrix elements. After calculation of this coefficient we obtain the following result:

$$\frac{Br(B_s \to \mu^+\mu^-e^-e^+)}{Br(B^- \to \mu^+\mu^-e^-\bar{\nu}_e)} \approx \frac{\tau_{B_s}}{\tau_{B^-}}\frac{1}{8\pi^2}{\alpha_{em}}^2|C_{10A}|^2\frac{|V_{tb}V_{ts}^*|^2}{|V_{ub}|^2},$$

 DOI: 10.1051/epjconf/201715803005

where $|C_{10A}| = 4.64$ – the one of the Wilson coefficients for rare semileptonic B–decays [6]. Using the $Br(B^- \to \mu^+\mu^-e^-\bar{\nu}_e)$ and numerical data from [7] gives the following estimation

$$Br(B_s \to \mu^+\mu^-e^-e^+) \approx 2 \times 10^{-10}.$$

Such estimation is in good agreement with the same result given by [8], where the branching ratio of $B_s \to \mu^+\mu^-e^+e^-$ decay was determined by taking into account only electromagnetic effects

$$Br(B_s \to \mu^+\mu^-e^+e^-) \approx 1.1 \times 10^{-10}.$$

Also our result is in good agreement with theoretical estimates from $Br(B_s \to \ell^+\ell^-\gamma)$, using the result from [9–11], we have

$$Br(B_s \to \mu^+\mu^-e^+e^-) \sim \alpha_{em}\, Br(B_s \to e^+e^-\gamma) \approx 1.8 \times 10^{-10}.$$

7 Conclusion

We predicted the branching ratios of the decays $B^- \to \mu^+\mu^-e^-\bar{\nu}_e$ and $B^- \to \mu^+\mu^-\mu^-\bar{\nu}_\mu$. Taking into account the processes of the photon emission from u– and b–quarks of the B-meson and the Bremsstrahlung. The following predictions are obtained:

$$Br(B^- \to \mu^+\mu^-e^-\bar{\nu}_e) \approx 1,3 \times 10^{-7};$$

$$Br(B^- \to \mu^+\mu^-\mu^-\bar{\nu}_\mu) \approx 10^{-7};$$

Based on the results for the $Br(B^- \to \mu^+\mu^-e^-\bar{\nu}_e)$ we get the estimation for the branching ratio of the decay $B_s \to \mu^+\mu^-e^+e^-$

$$Br(B_s \to \mu^+\mu^-e^+e^-) \sim 10^{-10}.$$

We would like to express our deep gratitude to D. Melikhov and A. Kozachuk for multiple helpful discussions which improved the paper significantly. The work was supported by grant 16-12-10280 of the Russian Science Foundation.

References

[1] R. Aaij *et al.* [LHCb Collaboration], Phys. Rev. Lett. **110**, 211801 (2013).
[2] R. Aaij *et al.* [LHCb Collaboration], JHEP **1703**, 001 (2017).
[3] D. Melikhov and B. Stech, Phys. Rev. D **62**, 014006 (2000).
[4] D. Melikhov, Phys. Rev. D **53**, 2460 (1996).
[5] D. Melikhov, Phys. Rev. D **56**, 7089 (1997).
[6] G. Buchalla, A. J. Buras and M. E. Lautenbacher, Rev. Mod. Phys. **68**, 1125 (1996).
[7] C. Patrignani *et al.* (Particle Data Group), Chin. Phys. C **40**, 100001 (2016).
[8] Y. Dincer and L. M. Sehgal, Phys. Lett. B **556**, 169 (2003).
[9] F. Kruger and D. Melikhov, Phys. Rev. D **67**, 034002 (2003).
[10] D. Melikhov and N. Nikitin, Phys. Rev. D **70**, 114028 (2004).
[11] I. Balakireva, D. Melikhov, N. Nikitin and D. Tlisov, Phys. Rev. D **81**, 054024 (2010).

EPJ Web of Conferences **158**, 03006 (2017) DOI: 10.1051/epjconf/201715803006
QFTHEP 2017

Production and spectroscopy in heavy flavour

Artem Maevskiy[1,⋆]
On behalf of the ATLAS Collaboration

[1] *Skobeltsyn Institute of Nuclear Physics (SINP MSU), 1(2), Leninskie gory, GSP-1, Moscow 119991, Russian Federation*

Abstract. A wide programme of studies of heavy flavour production at the LHC is performed with the ATLAS detector. This paper covers recent results in measurements of pair b-quark production, di-J/ψ production cross-section and effective cross-section from double parton scattering. Studies on the production of $\psi(2S)$ and $X(3872)$ are presented as well. Paper also covers the measurement of J/ψ-meson non-prompt production fraction and $B^{\pm}$-meson mass reconstruction.

1 Introduction

Studies of heavy flavour production in pp-collisions at the LHC provide the means for both verifying predictions of perturbative quantum chromodynamics (pQCD) and Monte-Carlo (MC) models tuning. Understanding the mechanisms of heavy flavour production is also crucial for accurate description of background channels in Standard Model (SM) measurements and new physics searches. This paper briefly covers several results in this area obtained by the ATLAS experiment [1] in pp collisions at $\sqrt{s} = 8$ TeV (sections 2 to 4) and $\sqrt{s} = 13$ TeV (sections 5 and 6).

2 Measurement of b-hadron pair production

The mass of the b-quark introduces an additional scale for QCD calculations, which can be introduced at different stages to the factorisation into parton distribution functions, hard matrix elements and softer parton shower components. Therefore, the optimal settings must be determined by comparisons to data. Previous measurements of $b\bar{b}$ pair production have highlighted differences between theoretical predictions, especially in the small-angle $b\bar{b}$ production region, motivating further study.

Pair production of b-hadrons was measured in the ATLAS experiment using a 11.4 fb^{-1} pp data sample collected at $\sqrt{s} = 8$ TeV [2]. Events were selected where one b-hadron decayed to $J/\psi(\rightarrow \mu\mu) + X$ and the other to $\mu + Y$, resulting in three muons in the final state. To probe b-hadron production, several differential cross sections were measured, based on the kinematics of the J/ψ and the third muon. These are:

- the azimuthal separation between the J/ψ and third muon, $\Delta\phi(J/\psi, \mu)$;
- the transverse momentum of the three-muon system, $p_{\mathrm{T}}(J/\psi, \mu)$;

⋆e-mail: artem.maevskiy@cern.ch

- the separation between the J/ψ and the third muon in the azimuth-rapidity plane, $\Delta R(J/\psi, \mu)$;
- the separation in rapidity between the J/ψ and the third muon, $\Delta y(J/\psi, \mu)$;
- the magnitude of the average rapidity of the J/ψ and the third muon, y_{boost};
- the mass of the three-muon system, $m(J/\psi, \mu)$;
- the ratio of the transverse momentum of the three-muon system to the invariant mass of the three-muon system, p_{T}/m, and its inverse.

The measured cross sections as a function of ΔR and $\Delta\phi$ are shown in Fig. 1 (for other variables see [2]).

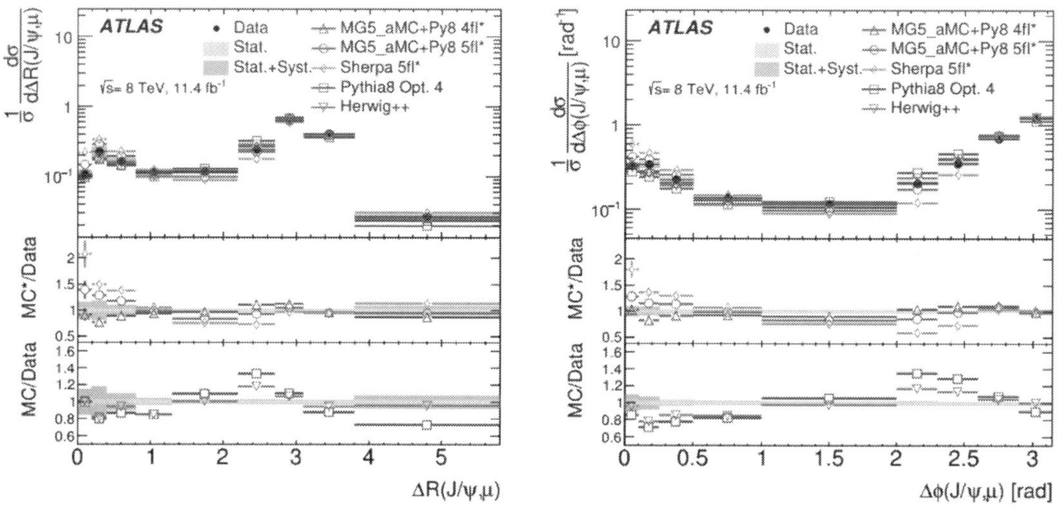

(a) (b)

Figure 1: Measured normalised differential cross sections as a function of $\Delta R(J/\psi, \mu)$ (a) and $\Delta\phi(J/\psi, \mu)$ (b) [2].

ATLAS results were compared with PYTHIA8 predictions, exploring several different options for the $g \to b\bar{b}$ splitting kernel, as this process dominates the region of particular interest: small-angle b-hadron production. The best agreement was obtained with a p_{T}-based scale splitting kernel [2], which is shown in Fig. 1. Figure 1 also shows predictions of other generators: HERWIG++, MADGRAPH5_aMC@NLOv2.2.2 at leading order in QCD interfaced to the PYTHIA8.186 parton shower model and SHERPA2.1.1. The best overall agreement is obtained with the 4-flavour MADGRAPH5_aMC@NLO+PYTHIA8, although none of the models can well-describe all the kinematic properties.

3 Measurement of the prompt J/ψ pair production

The study of simultaneous production of two prompt J/ψ mesons offers an opportunity to test our understanding of non-perturbative quantum chromodynamics. Di-J/ψ events can be produced from a single gluon-gluon collision via a single parton scattering (SPS) or from two independent parton-parton scatters in a single proton-proton collision, double parton scattering (DPS). Differential production cross section for pairs of prompt-prompt J/ψ-mesons as a function of various kinematic variables was measured by the ATLAS experiment using 11.4 fb^{-1} of $\sqrt{s}$ = 8 TeV data collected in 2012 [3]. A data-driven method was used to extract SPS-weighted and DPS-weighted distributions of the kinematic variables, allowing the extraction of SPS and DPS components from the measured cross sections. Measured total differential cross sections along with the DPS components are shown in Fig. 2.

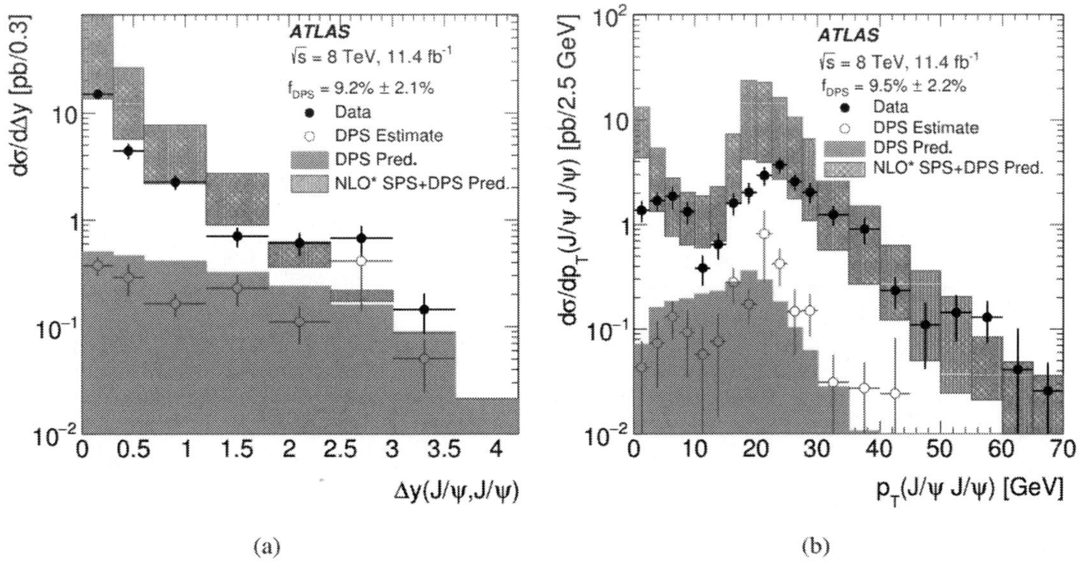

Figure 2: The DPS and total differential cross sections as a function of rapidity difference between the two J/ψ-mesons (a) and the transverse momentum of the di-J/ψ (b) [3].

In Fig. 2, the DPS components and the total distributions are compared to LO DPS and sum of the LO DPS and NLO* SPS predictions [3]. One can see that the shape of the data-driven DPS distribution approximately agrees with the shape of DPS predictions, while there is disagreement between the total data distribution and the total theory predictions at large Δy and low-p_T region that corresponds to di-J/ψ production in an away topology.

With the measured inclusive di-J/ψ cross section $\sigma_{J/\psi J/\psi}$ and the fraction of DPS events, f_{DPS} = (9.2 ± 2.1 (stat) ± 0.5 (syst))%, as well as the prompt J/ψ cross-section, $\sigma_{J/\psi}$, the effective cross-section can be derived:

$$\sigma_{\text{eff}} = \frac{1}{2}\frac{\sigma^2_{J/\psi}}{\sigma^{J/\psi,J/\psi}_{\text{DPS}}} = \frac{1}{2}\frac{\sigma^2_{J/\psi}}{f_{\text{DPS}} \times \sigma_{J/\psi J/\psi}}. \quad (1)$$

This value was measured to be:

$$\sigma_{\text{eff}} = 6.3 \pm 1.6\,(\text{stat}) \pm 1.0\,(\text{syst}) \pm 0.1\,(\text{BF}) \pm 0.1\,(\text{lumi})\ \text{mb}.$$

This value is compared with such values obtained in various processes by various experiments in Fig. 3, as a function of centre-of-mass energy, $\sqrt{s}$. One can see that σ_{eff} obtained in di-J/ψ and

$J/\psi\Upsilon$ measurements are lower than that obtained in other processes, which gives us a hint of possible process-dependence of DPS rates.

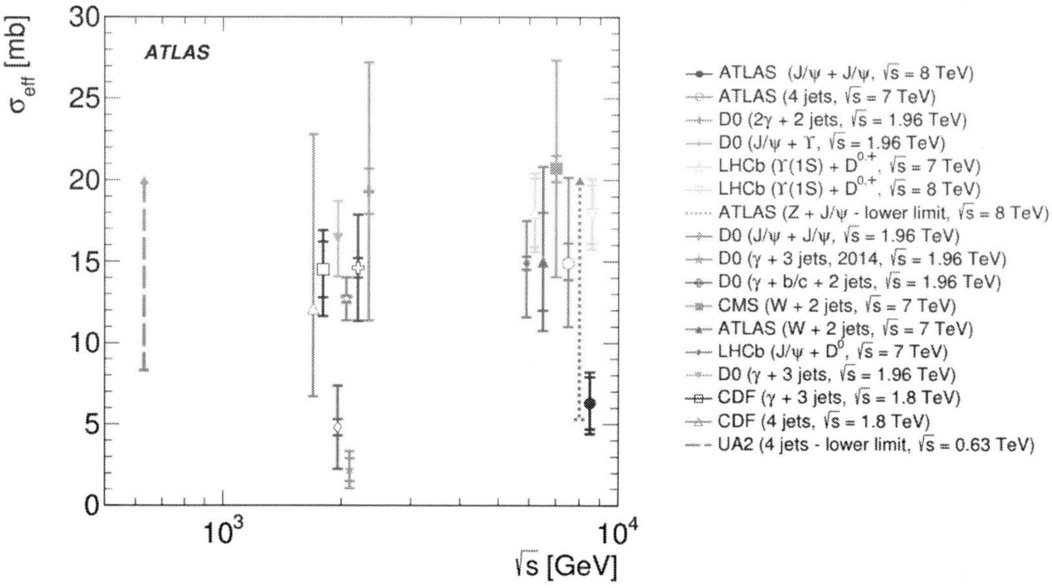

Figure 3: The effective cross section of DPS as a function of $\sqrt{s}$ [3].

4 Production measurements of $\psi(2S)$ and $X(3872)$

The hidden-charm state X(3872) was discovered by the Belle Collaboration in 2003 [4]. Due to its mass being close to the $D^0\bar{D}^{*0}$ threshold, it was hypothesised that this state is a $D^0\bar{D}^{*0}$-molecule with a very small binding energy. However, the non-relativistic QCD (NRQCD) prediction for its production cross section, assuming the $D^0\bar{D}^{*0}$-molecule hypothesis held, was too high, although the shape of the p_T dependence was described fairly well. Later, a NLO NRQCD model fitted to CMS data interpreting $X(3872)$ as a mixed $\chi_{c1}(2P) - D^0\bar{D}^{*0}$ state, where production is dominated through the $\chi_{c1}(2P)$ component, showed good agreement with data.

Measuring prompt and non-prompt components of $X(3872)$ production cross section, as well as their ratios to those of $\psi(2S)$, which is fairly close in mass, may give further insight into the nature of this state and mechanisms of its production.

Differential cross sections were measured in the ATLAS experiment for the production of $\psi(2S)$ and $X(3872)$ states in the decay channel $J/\psi\pi^+\pi^-$, using 11.4 fb^{-1} of pp collisions at $\sqrt{s}$ = 8 TeV [5]. The prompt and non-prompt contributions for both states are separated by analysing the displacement of the production vertex, and non-prompt production fractions are extracted, as well as $X(3872)/\psi(2S)$ production ratios separately for prompt and non-prompt components.

The measured $\psi(2S)$ cross section times branching fraction as a function of p_T of the candidate is shown in Fig. 4, where the prompt component is compared to NLO NRQCD, k_T factorisation model and the NNLO* Colour-Singlet Model (CSM) [5], while non-prompt one is compared with the predictions of FONLL calculation. One can see that NLO NRQCD and k_T factorisation models describe the prompt data fairly well, although there are some discrepancies at the high-p_T region. The non-prompt component is well-described by the FONLL model.

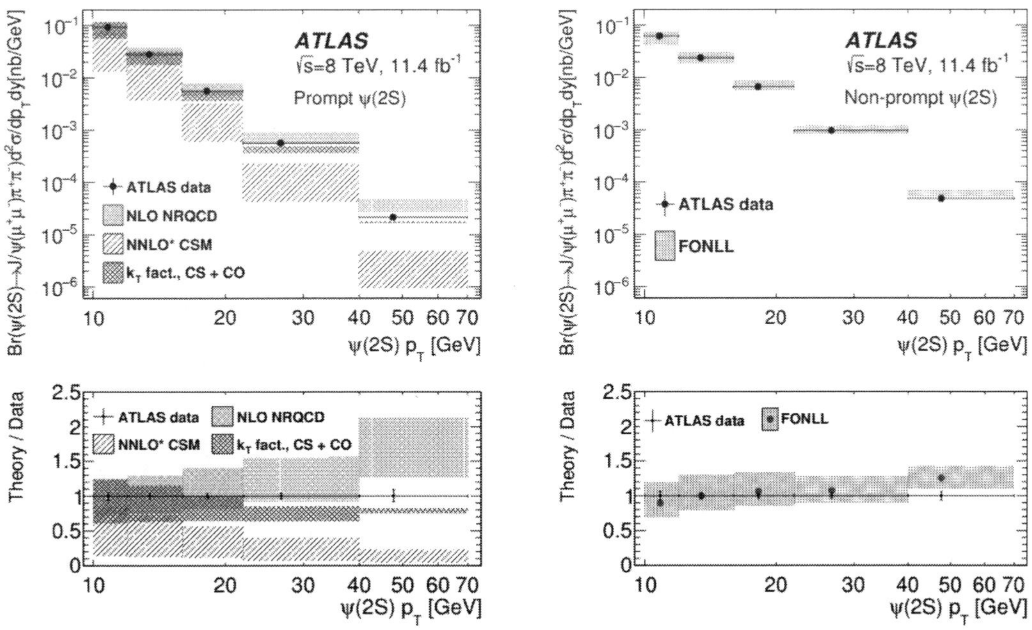

(a) (b)

Figure 4: Measured cross section times branching fractions as a function of p_T for (a) prompt and (b) non-prompt $\psi(2S)$ production [5].

Figure 5 shows the measured cross section times branching fractions as a function of p_T for prompt and non-prompt $X(3872)$ components. The prompt component, shown in Fig. 5a, is described within the theoretical uncertainty by the prediction of the NRQCD model which, in this case, considers $X(3872)$ to be a mixture of $\chi_{c1}(2P)$ and $D^0\bar{D}^{*0}$ molecular state, with the production being dominated by the $\chi_{c1}(2P)$ component and the normalisation fixed through the fit to CMS data. For the non-prompt component the FONLL model prediction for $\psi(2S)$ was used, rescaled using simulated kinematic template for the non-prompt $X(3872)/\psi(2S)$ ratio and the effective value of the product of branching fractions $\mathscr{B}(B \to X(3872))\mathscr{B}(X(3872) \to J/\psi\pi^+\pi^-)$ estimated based on the Tevatron data (for references see [5]).

The measured non-prompt production fractions for $\psi(2S)$ and $X(3872)$ are shown in Fig. 6, compared with CMS result. Both results agree between the two experiments within errors.

5 Non-prompt J/ψ production fraction

The non-prompt J/ψ production fraction $\sigma(pp \to b + X \to J/\psi + X')/\sigma(pp \to J/\psi + X')$ was measured in pp collisions at $\sqrt{s} = 13$ TeV with a data sample corresponding to an integrated luminosity of 6.4 pb^{-1} [6]. The J/ψ candidates were reconstructed in $\mu^+\mu^-$ channel and their invariant mass and

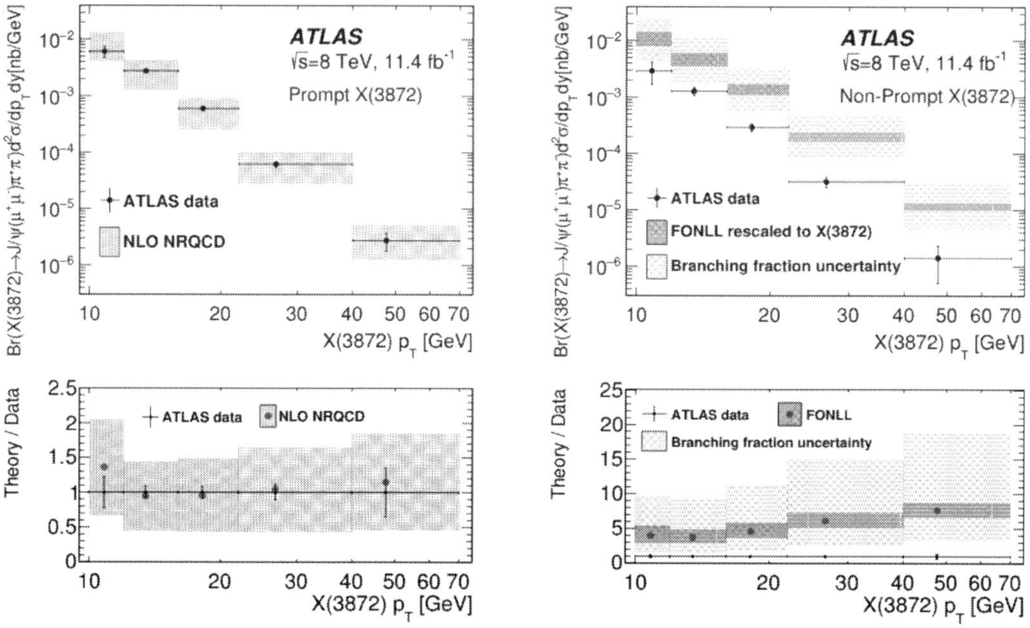

(a) (b)

Figure 5: Measured cross section times branching fractions as a function of p_T for (a) prompt and (b) non-prompt $X(3872)$ production [5].

pseudo-proper decay time[1] distributions were simultaneously fitted using an unbinned maximum likelihood fit. The extracted non-prompt fraction values are shown in Fig. 7. One can see that the fraction grows with p_T of the J/ψ candidates, while no rapidity dependence can be observed at current level of precision.

6 $B^\pm$ mass reconstruction

Performance of the ATLAS detector in reconstructing $B^\pm \to J/\psi(\mu\mu)K^\pm$ events was tested using 3.2 fb^{-1} of pp collision data at $\sqrt{s}$ = 13 TeV [7]. Events with two oppositely charged muons and one charged track were selected to form $B^\pm$ candidates. Their invariant mass distribution is shown in Fig. 8a. This distribution was fitted using an unbinned maximum likelihood fit with one signal and three background components. The three background components are: combinatorial, from partially reconstructed decays, and the resonant component from $B^\pm \to J/\psi\pi^\pm$ decays. Resonant background was described by a shape derived from MC, while all other components were represented by analytical functions with free parameters. It was decided to perform fits independently in the intervals of rapidity of the $B^\pm$ candidates to account for varying mass resolution and to test calibration of the tracker system. The mass values extracted from these fits are shown in Fig. 8b as a function of $B^\pm$ rapidity. One can see that deviation of each measurement from the weighted mean value of the mass is less that 0.1%, which indicates good momentum calibration.

[1]Pseudo-proper decay time is defined as $\tau = L_{xy}(\mu\mu) \cdot \frac{m_{J/\psi}}{p_T(\mu\mu)}$, where L_{xy} is the transverse decay length.

DOI: 10.1051/epjconf/201715803006

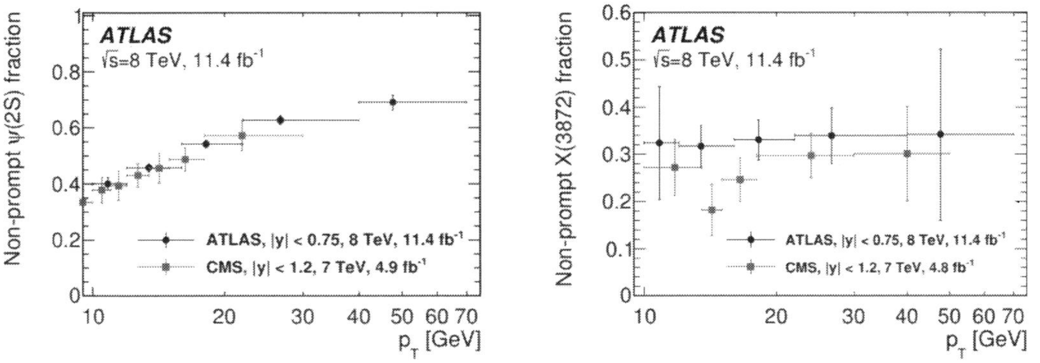

(a) (b)

Figure 6: Measured non-prompt fractions for (a) $\psi(2S)$ and (b) $X(3872)$ production [5], compared to CMS results at 7 TeV.

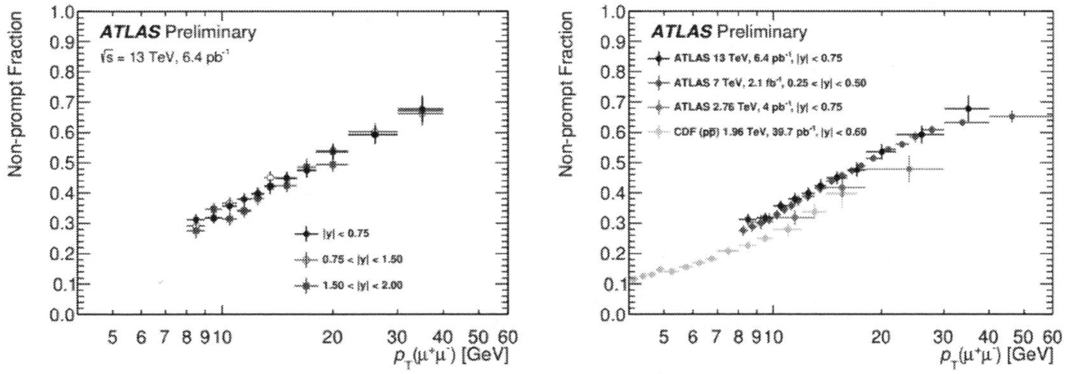

(a) (b)

Figure 7: Measured non-prompt J/ψ production fraction as a function of $p_T(J/\psi)$ for three rapidity regions (a) and for the central rapidity region (b) compared with earlier measurements by ATLAS and CDF (for complete list of references see [6]).

An additional test of the fit was performed by introducing a cut on transverse decay length (L_{xy}) of the candidates in order to modify the background composition. The $B^{\pm}$ mass values measured by the two versions of the fit are shown in table 1 compared with a measurement by LHCb [8] and with the world average [9]. One can see that all the four numbers are in good agreement. The systematic uncertainty of the ATLAS measurement is partially evaluated to be 0.25 MeV, while that value does not yet include the uncertainty related to the momentum scale of the ATLAS Inner Detector.

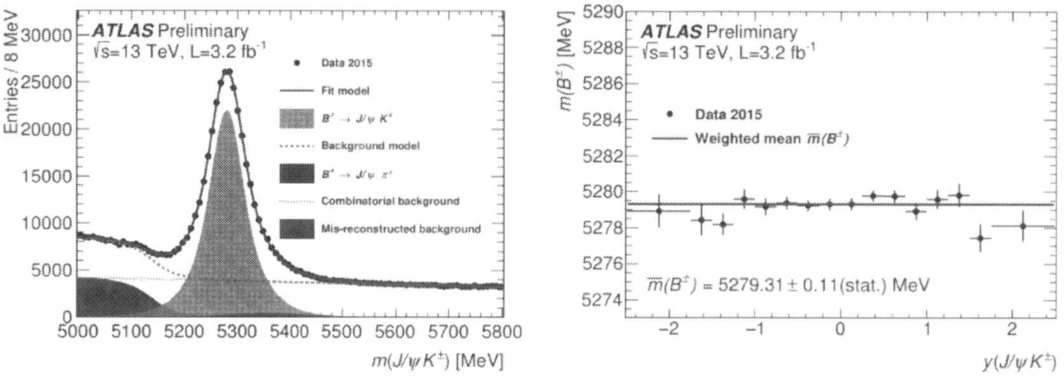

(a) (b)

Figure 8: Invariant mass distribution of $B^{\pm}$ candidates with the maximum likelihood fit components (a) and $B^{\pm}$ mass values extracted from the fits as a function of rapidity (b) [7].

Table 1: $B^{\pm}$ mass value measured in two ways [7] compared with the world average [9] and LHCb [8].

Fit	$B^{\pm}$ mass [MeV]	Fit error [MeV]
Default Fit	5279.31	0.11 (stat.)
$L_{xy} > 0.2$ mm	5279.34	0.09 (stat.)
World Average fit	5279.29	0.15
LHCb	5279.38	0.11 (stat.) $\pm$ 0.33 (syst.)

7 Summary

The paper covers a number of recent results in measurements of heavy flavour production. The measured pair b-quark differential production cross section provides new constraints to MC models parameter tuning, especially in the nearby b-hadron pair production region, which is of big importance for many LHC measurements. Di-J/ψ production cross-section was measured at $\sqrt{s} = 8$ TeV and fraction of DPS events was extracted, as well as effective cross-section from double parton scattering, which shows hints of possible process dependence of DPS. Studies on the production of $\psi(2S)$ and $X(3872)$ were presented as well, showing good agreement between ATLAS and CMS. The non-prompt production fraction of J/ψ-meson and $B^{\pm}$-meson mass were measured with $\sqrt{s} = 13$ TeV data collected in 2015, showing good detector performance in new Run-2 operation conditions.

Acknowledgements

Participation in the conference was supported by the Russian Foundation for Basic Research, grants 15-02-08133 and 17-02-20235.

DOI: 10.1051/epjconf/201715803006

References

[1] ATLAS Collaboration, JINST **3**, S08003 (2008).
[2] ATLAS Collaboration, arXiv:1705.03374 [hep-ex].
[3] ATLAS Collaboration, Eur. Phys. J. **C77**, 76 (2017).
[4] S.K. Choi et al. (Belle), Phys. Rev. Lett. **91**, 262001 (2003).
[5] ATLAS Collaboration, JHEP **01**, 117 (2017).
[6] ATLAS Collaboration (2015), ATLAS-CONF-2015-030, http://cds.cern.ch/record/2037967.
[7] ATLAS Collaboration (2015), ATLAS-CONF-2015-064, http://cds.cern.ch/record/2114830.
[8] R. Aaij et al. (LHCb), Phys. Lett. **B708**, 241 (2012).
[9] K. Olive et al. (Particle Data Group), Chinese Physics C **38**, 090001 (2014).

Recent measurements of $K^{\pm}_{l3}$ form factors at NA48

Sergey Shkarovskiy[1,★] on behalf of the NA48/2 Collaboration

[1] *Joint Institute for Nuclear Research, Joliot-Curie, 6 Dubna, Moscow region, Russia, 141980*

Abstract. The NA48/2 experiment presents a final result of the charged kaon semileptonic decays form factors measurement based on 4.28 million $K^{\pm}_{e3}$ and 2.91 million $K^{\pm}_{\mu 3}$ selected decays collected in 2004. The result is competetive with other measurements in $K^{\pm}_{\mu 3}$ mode and has a smallest uncertainty for $K^{\pm}_{e3}$, that leads to the most precise combined $K^{\pm}_{l3}$ result and allows to reduce the form factor uncertainty of $|V_{US}|$.

1 Introduction

The main purpose of the NA48/2 experiment at the CERN SPS was a search for the direct CP violation in $K^{\pm}$ decay to three pions [1]. The experiment used simultaneous K^{+} and K^{-} beams with momenta of 60 GeV/c propagating through the detector along the same beam line. Data were collected in 2003-2004, providing the large samples of reconstructed $K^{\pm} \rightarrow 3\pi$ decays and a high precision data for many rare kaon decay studies. The layout of beams and detectors is shown in Fig. 1.

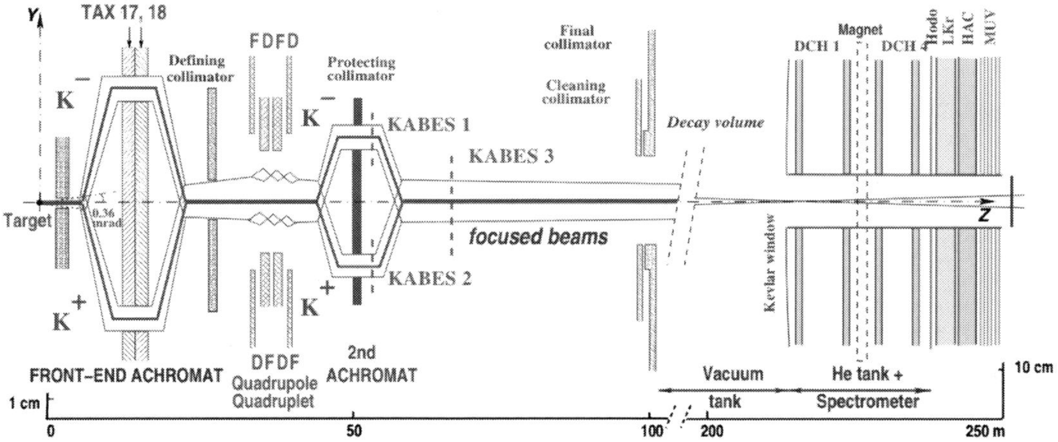

Figure 1. Schematic side view of the NA48/2 beamline and detectors.

Apart from that, a large statistics of $K^{\pm} \rightarrow \pi^{0} l^{\pm} \nu$ (K_{l3}) events has been collected during a special data taking period of 2004 (l means e or μ lepton). Semileptonic kaon decays (K_{l3}) offer the most

★e-mail: Sergey.Shkarovskiy@cern.ch

precise determination of the CKM matrix element $|V_{US}|$ [2], that require both a branching ratio and a formfactors experimental measurement. The K_{l3} precision form factors measurement results based on the NA48/2 data analysis are presented in this paper.

The K_{l3} decay width in the absence of electromagnetic effects can be represented by the Dalitz plot density depending on the lepton and pion energies in kaon rest frame E_l and E_π respectively [3]:

$$\frac{d^2\Gamma_0(K_{l3})}{dE_l dE_\pi} = N(Af_+^2(t) + Bf_+(t)f_-(t) + Cf_-^2(t)), \tag{1}$$

where $t = (P_K - P_\pi)^2 = m_K^2 + m_\pi^2 - 2m_K E_\pi$, N is a normalization constant and $f_-(t) = (f_+(t) - f_0(t))(m_K^2 - m_{\pi+}^2)/t$. Here $f_+(t)$ and $f_0(t)$ are the so called vector and scalar K_{l3} form factors, respectively. The m_K is a mass of charged kaon m_π is a mass of neutral pion and $m_{\pi+}$ is a mass of charged pion.

Kinematical factors in the (1) expression are:

$$\begin{aligned} A &= m_K(2E_l E_\nu - m_K(E_\pi^{max} - E_\pi)) + m_l^2((E_\pi^{max} - E_\pi)/4 - E_\nu) \\ B &= m_l^2(E_\nu - (E_\pi^{max} - E_\pi)/2) \\ C &= m_l^2(E_\pi^{max} - E_\pi)/4, \end{aligned}$$

where $E_\pi^{max} = (m_K^2 + m_\pi^2 - m_l^2)/(2m_K)$ (m_l is the lepton mass), and $E_\nu = m_K - E_l - E_\pi$ is a neutrino energy in the kaon rest frame. For K_{e3} the terms B and C are negligible due to the small electron mass, so in this case in fact only the vector form factor participates in the Dalitz plot description.

	$f_+(t)$	$f_0(t)$
Quadratic	$1 + \lambda'_+ t/m_\pi^2 + \frac{1}{2}\lambda''_+(t/m_\pi^2)^2$	$1 + \lambda'_0 t/m_\pi^2$
Pole	$\frac{M_V^2}{(M_V^2 - t)}$	$\frac{M_S^2}{(M_S^2 - t)}$
Dispersive	$\exp\left(\frac{(\Lambda_+ + H(t))t}{m_\pi^2}\right)$	$\exp\left(\frac{(ln[C] - G(t))t}{(m_K^2 - m_\pi^2)}\right)$

Table 1. Definitions of the form factor parameterizations used in the analysis

Definitions of the implemented parameterizatios are shown in the Table 1: the Quadratic [4] parameterization (fit parameters $\lambda'_+, \lambda''_+, \lambda'_0$), the Pole [5] (fit parameters M_V, M_S) and the Dispersive [6] one (fit parameters $\Lambda_+, ln[C]$).

2 Beams and detectors

A detailed descriptions of the detector elements and beam line are available in [1, 7]. The NA48/2 experiment used two simultaneous beams produced by 400 GeV/c protons impinging on a berillium target. Particles of opposite charge with a central momentum of 60 GeV/c and a momentum band of ±3.8% (RMS) were selected by a system of dipole magnets, focusing quadrupoles, muon sweepers and collimators. Two beams with an opposite charges were split in the vertical plane and than recombined on a common axis. The decay volume was a 114 m long vacuum tank.

Charged particles from $K^\pm$ decays were measured by a magnetic spectrometer (DCH) that included four drift chambers (DCH1–DCH4) and a dipole magnet between DCH2 and DCH3. The spectrometer momentum resolution was $\sigma(P)/P = 1.02\% \oplus 0.044\%P$, where P is a charged particle momentum in GeV/c. The magnetic spectrometer was followed by a scintillator hodoscope (HOD) consisting of two planes segmented into horizontal and vertical strips and arranged in four quadrants.

 DOI: 10.1051/epjconf/201715803007

A liquid Krypton calorimeter (LKr) [8] was used to reconstruct $\pi^0 \to \gamma\gamma$ decays. It was an almost homogeneous ionization chamber with an active volume of $\sim$ 10 m^3 of liquid krypton, segmented transversally into 2 cm $\times$ 2 cm projective cells. The calorimeter was 27 X_0 thick and has an energy resolution $\sigma(E)/E = 0.032/\sqrt{E} \oplus 0.09/E \oplus 0.0042$ (E in GeV). The space resolution for a single electromagnetic shower can be parameterized as $\sigma_x = \sigma_y = 0.42/\sqrt{E} \oplus 0.06$ cm both for X and Y coordinates.

The LKr was followed by a hadronic calorimeter with a total iron thickness of 1.2 m. A muon detector (MUV) was located further downstream. It consisted of three planes of scintillator strips, each preceded by a 80 cm thick iron wall. The strips were aligned horizontally in the first and last planes, and vertically – in the second plane. They were 2.7 m long and 2 cm thick, and read out by photomultipliers at both ends.

3 Events reconstruction and selection

The data used for the form factor (FF) analysis were collected in 2004 during a dedicated run with a special trigger setup which required at least one charged track crossing the Hodoscope HOD and an energy deposit of at least 10 GeV in the electromagnetic calorimeter. Nearly 480 $\times$ 10^6 triggered events have been recorded during the special short period.

3.1 π^0 selection

The event is considered as a preliminary candidate to K_{l3} decays, if it contains at least 2 LKr clusters consistent with a photons of reconstructed energy above 3 GeV (good photons). Fiducial cuts on the minimum distance between the good photon and LKr edges or centre are applied in order to avoid elecromagnetic showers energy loss. In addition, a minimum distance between the good photon and the nearest LKr cell with a known readout problems (dead cell) is required to be at least 2 cm. The minimum distance from the selected photon to any in-time (within 10 ns) charged track impact point at LKr front face is 15 cm, and the minimum distance to any other in-time (within 5 ns) cluster is 10 cm.

Each pair of the close in time (within 5 ns) good photons forms a π^0 candidate, if there is no extra good photons in $\pm$5 ns vicinity of the two π^0 photons average time. This extra-photons cut suppresses $\pi^\pm\pi^0\pi^0$ background. A distance between π^0 photons on LKr is required to be more than 20 cm, and the minimum sum of two photon energies is 15 GeV. So high energy threshold ensures a high efficiency of $E_{LKr} > 10$ GeV trigger requirement.

Longitudinal K_{l3} decay position Z_n (neutral vertex Z coordinate) is defined as a longitudinal position of π^0 decay, reconstructed from LKr data assuming PDG [9] value for π^0 mass.

3.2 Charged leptons selection

A candidate to K_{l3} decay is required to contain, apart from the reconstructed π^0, also at least one reconstructed DCH track of charged particle with a minimum momentum of 5 GeV/c. A harder muon momentum cut $P\mu >= 10$ GeV/c is applied in the case of muon positive identification in order to ensure high MUV efficiency. A distance from the track impact point on the LKr front face to the closest dead cell is required to be above 2 cm, and a minimum distance from the nominal beam axis to the reconstructed track at each DCH plane is 15 cm. A track is required to be in-time with the reconstructed π^0 within 10 ns, and no extra good tracks are allowed to be close in time to the considered track (within 8 ns).

Track with $2.0 > E/P > 0.9$ is identified as an electron from K_{e3} decay. For experimental data, if $E/P < 0.9$ and there is a MUV muon candidate associated to the track, it is identified as a muon from $K_{\mu 3}$ decay.

3.3 Kaon momentum

Kaon momentum is measured ignoring the possible real radiative photons, so this radiative effect is treated just as an extra source of measurement error.

For the kaon momentum measurement we direct Z axis along the beam average position in space, measured from $3\pi^{\pm}$ data. In the assumptions of zero neutrino mass and kaon flight along the beam axis (that means availability of the measured neutrino transverse momentum $P_T(\nu) = -P_t$), two solutions of quadratic equation for kaon momentum P_K exist:

$$P_K = P_{1,2} = (\phi P_Z \pm \sqrt{d})/(E^2 - P_Z^2), \tag{2}$$

where

$$\phi = 0.5(M_K^2 + E^2 - P_t^2 - P_Z^2);$$
$$d = \phi^2 P_Z^2 - (E^2 - P_Z^2)(M_K^2 E^2 - \phi^2)),$$

and E, P_t, P_Z are the total energy and total momentum of all the registered particles π^0, l.

An average beam momentum P_B is known from $3\pi^{\pm}$ decays data. For each event, both for K_{e3} and $K_{\mu 3}$ selections a combination with the minimum $\Delta P = |P_K - P_B|$ is chosen as the best candidate. Finally, a cut on the reconstructed kaon momentum is applied: $-7.5\ GeV/c < (P_K - P_B) < 7.5\ GeV/c$.

3.4 Background suppression cuts

For the clean K_{e3} selection one need to reject the $K^{\pm} \to \pi^{\pm}\pi^0$ decays with a $\pi^{\pm}$ misidentified as e due to the rare occasional high enegry deposit in LKr, resulting in $E/P > 0.9$. Reconstructed transversal momentum for these background events doesn't depend on the charged particle mass and should be close to zero. So a requirement on the reconstructed neutrino transverse momentum is applied: $P_t(\nu) >= 0.03\ GeV/c$, that takes into account both the effect of resolution and the beam angular spread influence.

For $K_{\mu 3}$ selection, an essential background may come from $K^{\pm} \to \pi^{\pm}\pi^0$ decays with a subsequent $\pi^{\pm} \to \mu^{\pm}\bar{\nu}$ process. In order to suppress this background, we remove the cases, when a kaon mass can be well reconstructed in the assumption of 2π kaon decay – by means of the requirement of $m(\pi^{\pm}\pi^0) < 0.47\ GeV/c^2$. Additionally, an empirically found cut $m(\pi^{\pm}\pi^0) < (0.6 - P_t(\pi^0))\ GeV/c^2$ (P_t in GeV/c) improves the suppression of 2π background further.

Also for $K_{\mu 3}$, we suppress the possibility of successfull $\pi^{\pm} \to \mu^{\pm}\bar{\nu}$ reconstruction by the requirement on the minimum dilepton invariant mass $m(\mu\bar{\nu}) > 0.18\ GeV/c^2$, that is well above $m(\pi^+)$. The majority of background events rejected by this cut are also rejected by the above empirical $P_t(\pi^0)$)-dependent cut, that makes its nature more clear – in both cases we reject the $K^{\pm} \to \pi^0(\pi^{\pm} \to \mu^{\pm}\bar{\nu})$ background events with a muon that conserves approximately the momentum of its parent pion. Nevertheless, these two cuts suppress the tails of background distribution in a somewhat different ways, so we apply both.

For both $K_{\mu 3}$ and K_{e3} samples the $K^{\pm} \to \pi^{\pm}\pi^0\pi^0$ decays can contribute to the background, if two γ clusters from π^0 are not detected and if the charged pion is misidentified as electron (for K_{e3}), or if the pion decays into muon and neutrino (for $K_{\mu 3}$). It has been found, that for this background source a difference between two possible solutions (2) of quadratic equation for P_K is relatively large. So for the $K_{\mu 3}$ events selection we apply a requirement of $|P_2 - P_1| < 60GeV$. For K_{e3} events such a cut also can suppress the 3π source of background, but even without this suppression our K_{e3} selection is rather clean. The loss of the signal statistics caused by this cut is not justified in K_{e3} case, so we don't apply this cut for K_{e3} selection.

3.5 $P_L(\nu)^2$ cut

One can evaluate from kinematics of K_{l3} decay reconstruction, that the physical reason of two-fold P_K uncertainty is in fact the unknown sign of true longitudinal neutrino momentum $p_L(\nu)$ in the kaon center of mass system.

When the measured value of $P_L(\nu)^2 = E(\nu)^2 - P_T(\nu)^2$ is negative (and when it is smaller than its resolution) a sign of $P_L(\nu)$ is uncertain. In this case the choice of the best P_K value from two close solutions of quadratic equation becomes arbitrary. As a result, the region of small and negative measured values of $P_L(\nu)^2$ is dominated by the events with an essentially mismeasured $P_L(\nu)^2$ values.

The measurement of $P_L(\nu)^2$ depends on reconstructed $P_T(\nu)^2$, that essentially depends on the assumed direction of kaon filght as well as on the transverse coordinates of the reconstructed decay vertex. As a consequence, the accurate modelling of the kaon momentum choice at small and negative $P_L(\nu)^2$ is sensitive to the fine details of beams geometry, that is problematic for exact simulation.

A comparison between the reconstructed experimental and Monte Carlo $P_L(\nu)^2$ distributions is shown in the Fig. 2. One can see, that there is some discrepance between the data and simulation at small and negative $P_L(\nu)^2$ values. It is not so well visible on the normalized overlapped plots, but the ratios of these distributions show the clear difference of the negative tails, and this difference is larger for K_{e3} decays.

Due to the inevitable residual discrepance between the simulated and real beam geometry, the result of kaon momentum solution choise for the events from the negative $P_L(\nu)^2$ tail essentially depends on the vertex transverse coordinates measurement procedure (CDA or neutral vertex). And even for the neutral vertex we have for K_{e3} a considerable dependence of our results on the minimum $P_L(\nu)^2$ as long as we don't cut at $P_L(\nu)^2 > 0.0014\ (GeV/c)^2$. For a higher cut values all the results remain stable.

For the $K_{\mu 3}$ case no essential variations of our results versus $P_L(\nu)^2$ cut have been observed, and no considerable systematic uncertainty may be linked to the sources of MC/Dtata discrepance. Together with the fact of smaller discrepance in comparison with K_{e3} case, it leads us to the decision not to apply $P_L(\nu)^2$ cut for $K_{\mu 3}$ in order to save signal statistics.

So a special cut has been applied to exclude the K_{e3} events subsample, that introduces a large sensitivity of the results to the precision of beam simulation: $P_L(\nu)^2 > 0.0014\ (GeV/c)^2$. This cut leads to the considerable loss of statistics: K_{e3} one decreases from 6.05 millions to 4.28 millions of events (about 30% are lost). It makes the statistical error larger, but a large systematic uncertainty of complex nature is diminished.

The selected event Dalitz plots with a binning of $5\times 5\ MeV$ are shown in the Fig. 3. They are used for all the further fits after correction for residual background. The total statistics of selected data is 4.278×10^6 events for K_{e3}, and 2.907×10^6 events for $K_{\mu 3}$ selection.

4 Monte Carlo simulation and form factor fits

Semileptonic radiative Monte Carlo samples have been simulated with the KLOE generator [10]. It is based on the Dalitz plot density (1) with a linear approximation for the vector form factor $f_+(t) =$

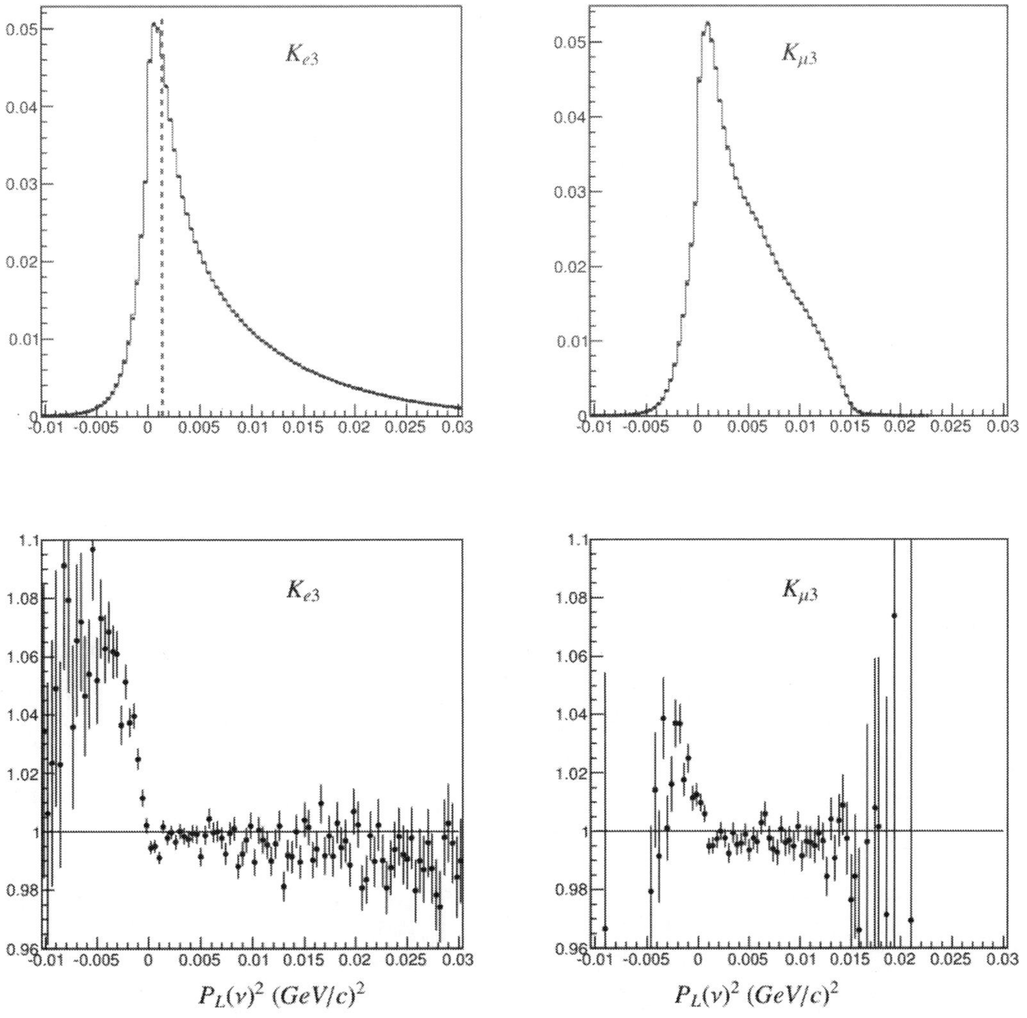

Figure 2. Normalized $P_L(\nu)^2$ distributions for the data (dots) and background-corrected Monte Carlo (histograms). Upper left plot – K_{e3}, upper right – $K_{\mu 3}$. Lower plots – the corresponding MC/Data ratios. Vertical dashed line shows the cut applied for the K_{e3} events selection.

$1 + 0.0296 \cdot t/m_{\pi^+}^2$ and without negative form factor ($f_- = 0$) for the both cases of semileptonic decay. The simulation codes for two semileptonic modes differ only in the lepton mass value. It corresponds to $f_+(t) = f_0(t)$ assumption for $K_{\mu 3}$ simulation. Apart from the Dalitz plot radiative correction, one real radiative photon is simulated for each K_{e3} event as well as for some fraction of $K_{\mu 3}$ decays.

We will denote here the Dalitz plot density expression (1) with $N = 1$ and with a form factors written in terms of Quadratic parameterization as $Q_l(E_\pi, E_l, \lambda'_+, \lambda''_+, \lambda'_0)$ for the corresponding K_{l3} case ($l = \mu, e$).

For each fit iteration with the current changed form factor parameters the fitting code reads a special file of selected events data. For each event, a probability to simulate this event with the

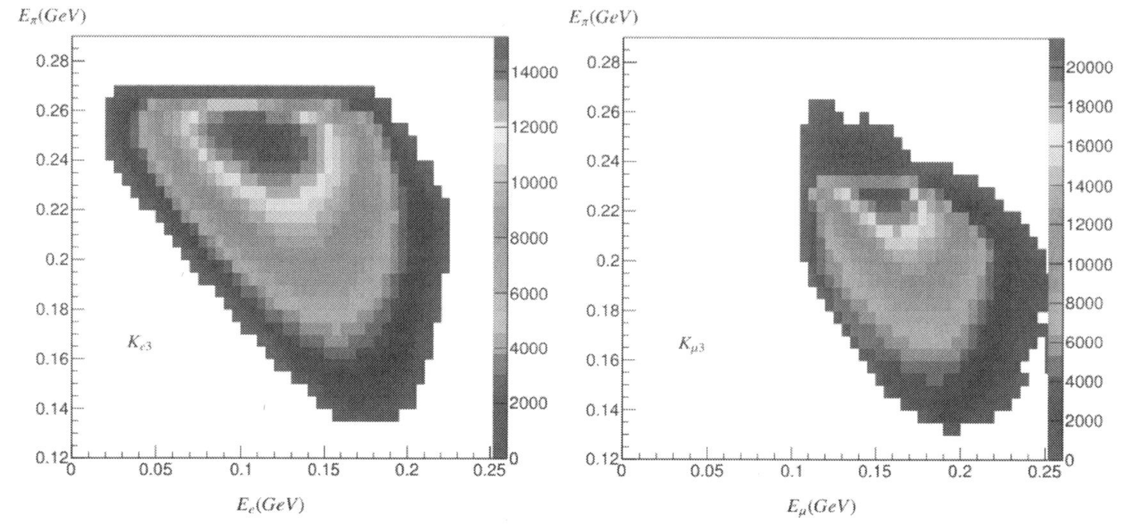

Figure 3. Reconstructed Dalitz plots for K_{e3} (left) and $K_{\mu3}$ (right) selections of experimental data

Monte Carlo generator [10] $\rho_0^{gen} = Q_l(E_\pi, E_l, 0.0296, 0.0, 0.0296)$is calculated. Than a probability to simulate the same event with the current fit parameters **ff** is also obtained from one of the considered form factor parameterisation $\rho_0^{ff}(E_\pi, E_l, \mathbf{ff})$. The individual event weight $W_{evt} = \frac{\rho_0^{ff}}{\rho_0^{gen}}$ is used in order to fill the simulated Dalitz plot according to the current form factor parameters.

Than, for each fit iteration, χ^2 is calculated from the experimental background-corrected Dalitz plot $D_{i,j}$ and the current Monte-Carlo simulated one $MC_{i,j}$:

$$\chi^2 = \Sigma_{i,j} \frac{(D_{i,j} - MC_{i,j})^2}{(\delta D_{i,j})^2 + (\delta MC_{i,j})^2}, \tag{3}$$

where i, j indices correspond to the cell of Dalitz plot with a center, lying inside the kinematically allowed region for K_{l3} decays without radiative γ [11] and containing at least 20 reconstructed data events. MINUIT [12] package called from the ROOT [13] interface minimizes χ^2 by means of parameters variation, and in such a way the resulting fit parameter values, their errors and correlation coefficients are found.

5 Background

Four kaon decay modes have been considered as a possible sources of residual background for the both semileptonic decays and have been simulated for the present analysis (see Table 2). Additionally, the effect of $K_{\mu3}$ misidentification as K_{e3} (due to the $\mu \to e\nu$ decay) has been taken into account.

Inner bremsstrahlung part of $\pi^\pm\pi^0\gamma$ decay was simulated separately for the kaon rest frame kinetic energy of the charged pion $T^*_{\pi^\pm} < 90MeV$, its probability estimation is taken from [14].

The experimental two-dimensional Dalitz plot is corrected for background by subtracting of the estimated background contributions.

Process	Notation	Br	N_g	F_e	F_μ
$K^\pm \to \pi^\pm(\pi^0 \to 2\gamma)$	2π	20.66	393.2	0.270	0.264
$K^\pm \to \pi^\pm 2(\pi^0 \to 2\gamma)$	3π	1.761	62.5	0.286	1.833
$K^\pm \to \pi^\pm(\pi^0 \to e^+e^-\gamma)$	$2\pi D$	1.174	1.5	0.049	0.000
$K^\pm \to \pi^\pm\gamma(\pi^0 \to 2\gamma)$	$2\pi\gamma$	0.0275	35.3	0.004	0.044
$K^\pm \to \pi^0\mu^\pm\nu(\mu \to e\nu)$	$K^e_{\mu3}$	0.03353	174.3	0.004	0.000

Table 2. Simulated background processes, their probabilities Br (in %), generated MC statistics N_g (in 10^6 events) and the estimated fractions F_e and F_μ (both in units of per mill) in K_{e3} and $K_{\mu3}$ samples for the present selection.

6 Form factor results

Fit results and contributions to systematic uncertainty for Quadratic, Pole and Dispersive Parameterisation are shown in the Table 3.

Quadratic	λ'_+	λ''_+	λ_0
$K_{\mu3}$	$23.32 \pm 3.08_{stat} \pm 3.50_{syst}$	$2.14 \pm 1.06_{stat} \pm 0.96_{syst}$	$14.33 \pm 1.11_{stat} \pm 1.25_{syst}$
K_{e3}	$23.52 \pm 0.78_{stat} \pm 1.29_{syst}$	$1.60 \pm 0.30_{stat} \pm 0.39_{syst}$	
K_{l3}	$23.35 \pm 0.75_{stat} \pm 1.23_{syst}$	$1.73 \pm 0.29_{stat} \pm 0.41_{syst}$	$14.90 \pm 0.55_{stat} \pm 0.80_{syst}$
Pole	m_V	m_S	
$K_{\mu3}$	$879.1 \pm 8.1_{stat} \pm 13.5_{syst}$	$1196.4 \pm 18.1_{stat} \pm 28.8_{syst}$	
K_{e3}	$896.8 \pm 3.4_{stat} \pm 7.6_{syst}$		
K_{l3}	$894.3 \pm 3.2_{stat} \pm 5.4_{syst}$	$1185.5 \pm 16.6_{stat} \pm 35.5_{syst}$	
Dispersive	Λ_+	$ln[C]$	
$K_{\mu3}$	$23.55 \pm 0.50_{stat} \pm 0.97_{syst}$	$186.68 \pm 5.12_{stat} \pm 9.23_{syst}$	
K_{e3}	$22.54 \pm 0.20_{stat} \pm 0.62_{syst}$		
K_{l3}	$22.67 \pm 0.18_{stat} \pm 0.55_{syst}$	$186.12 \pm 4.91_{stat} \pm 11.09_{syst}$	

Table 3. Fit results for the Quadratic ($\times 10^3$), Pole (MeV/c^2) and Dispersive ($\times 10^3$) Parameterisation

The NA48/2 is the first experiment measuring the FF using both K^+ and K^-. In $K_{\mu3}$ the result is dominated by the statistical error, for K_{e3} by the systematic. The NA48/2 K_{e3} and $K_{\mu3}$ in agreement within each other and our combined results are competitive with the current world average.

In order to avoid the problem of partially correlated systematic uncertainties in the K_{e3} and $K_{\mu3}$ results averaging, we just repeated the complete analysis considering the two decay modes information as the joint data set, containing two Dalitz plots that should be simultaneousely fitted with a common form factor parameters.

The final results of the fit for quadratic, pole and dispersive parametrizations are listed in Table 3. The comparison between K_{l3} quadratic fit results by recent experiments is shown in Fig. 6. The 68% confidence level contours are displayed for both K^0_{l3} (KLOE, KTeV and NA48) and charged kaon decays (ISTRA+ studied K^-_{l3} only). The final NA48/2 results presented here are the first high precision measurements done with both K^+ and K^- decays. All the measured parameters are in good agreement with the measurements done by the other experiments.

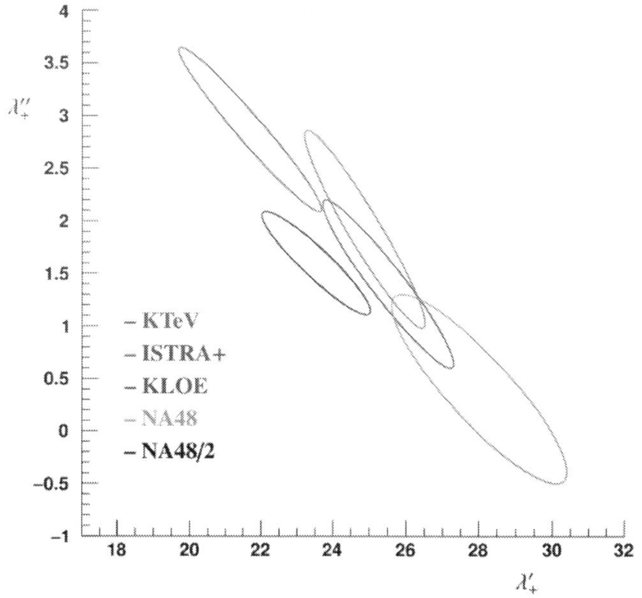

Figure 4. (Color online) 1σ confidence contours for measurements of K_{e3} vector form factor parameters. NA48/2 : result of the present work.

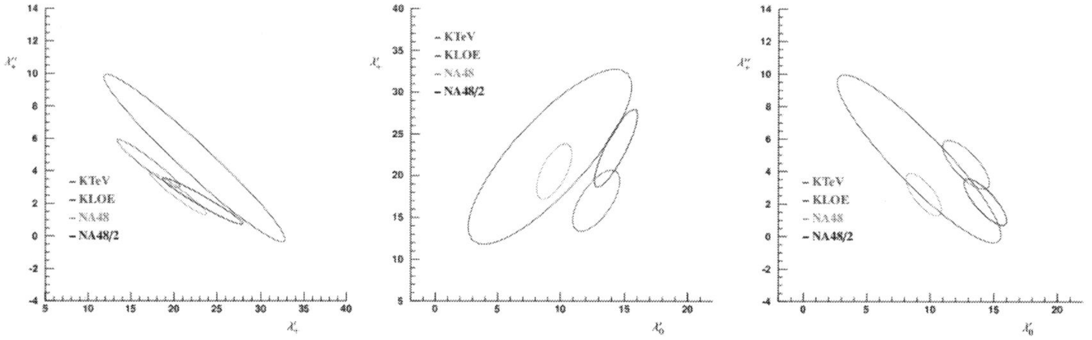

Figure 5. (Color online) 1σ confidence contours for measurements of $K_{\mu3}$ λ'_+, λ''_+ and λ'_0 form factor parameters

Conclusion

K_{l3} form factors measurement is performed by NA48/2 experiment on the basis of 2004 run data. Result is competetive with the other ones in $K^{\pm}_{\mu3}$, and a smallest error in $K^{\pm}_{e3}$ has been reached, that gives us also the combined result with the smallest error.

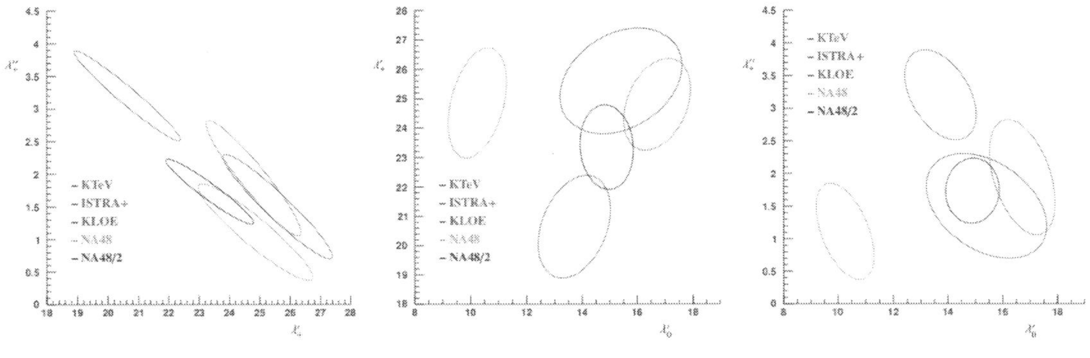

Figure 6. Joint K_{l3} results for the λ'_+, λ''_+ and λ'_0 form factor in comparison with the K_{l3} from another experiments [2].

References

[1] J.R. Batley et al. (NA48/2), Eur. Phys. J. **C52**, 875 (2007), 0707.0697
[2] M. Antonelli et al. (FlaviaNet Working Group on Kaon Decays), Eur.Phys.J. **C69**, 399 (2010), 1005.2323
[3] L.M. Chounet, J.M. Gaillard, M. Gaillard, Phys.Rept. **4**, 199 (1972)
[4] K. Olive et al. (Particle Data Group), Chin.Phys. **C38**, 090001 (2014)
[5] P. Lichard, Phys.Rev. **D55**, 5385 (1997), hep-ph/9702345
[6] V. Bernard, M. Oertel, E. Passemar, J. Stern, Phys.Rev. **D80**, 034034 (2009), 0903.1654
[7] V. Fanti et al. (NA48), Nucl. Instrum. Meth. **A574**, 433 (2007)
[8] G.D. Barr et al. (NA48), Nucl. Instrum. Meth. **A370**, 413 (1996)
[9] C. Patrignani et al. (Particle Data Group), Chin. Phys. **C40**, 100001 (2016)
[10] C. Gatti, Eur.Phys.J. **C45**, 417 (2006), hep-ph/0507280
[11] E.S. Ginsberg, Phys.Rev. **162**, 1570 (1967)
[12] F. James, M. Roos, Comput. Phys. Commun. **10**, 343 (1975)
[13] R. Brun, F. Rademakers, Nucl. Instrum. Meth. **A389**, 81 (1997)
[14] C. Amsler et al. (Particle Data Group), Phys.Lett. **B667**, 1 (2008)

EPJ Web of Conferences **158**, 03008 (2017) DOI: 10.1051/epjconf/201715803008
QFTHEP 2017

Strange Charmed Baryons Spectroscopy

Elena Solovieva[1 2,⋆]

[1]*Moscow Institute of Physics and Technology, 9 Institutskiy pereulok, Dolgoprudny, Moscow Region, 141700, Russia*
[2]*P.N. Lebedev Physical Institute of the Russian Academy of Sciences, 53 Leninskiy Prospekt, Moscow, 119991, Russia*

Abstract. A present status of experimentally observed states of strange charmed baryons is given. The classification of charmed baryons is presented, a quark model for ground states is brie fly described, and the energy levels of excited states are analyzed.

1 Introduction

The spectroscopy of charmed baryons is beautiful and intricate. With three quarks there are numerous degrees of freedom, giving rise to many more states than in the charmed meson sector. At the same time, the large difference in mass between the charm quark and the light quarks provides a natural way to classify and understand these states: Heavy Quark Effective Theory (HQET). The spectrum of known singly-charmed states can be thought of in three broad regimes: the ground states, which are a vindication of the constituent quark model; the low-lying excited states, which are described well by HQET; and higher excited states, where the situation is murkier.

2 Quark Model for Ground States

In the constituent quark model [1], baryons composed of u, d, s, c quarks can be classified into $SU(4)$ multiplets according to the symmetry of their flavor, spin, and spatial wavefunctions. All states in a given $SU(4)$ multiplet have the same angular momentum J, and parity P, but can have different quark flavors. For excited states with multiple units of orbital angular momentum the number of possible multiplets becomes large, but for the ground states the picture is much simpler. This framework is not exact — different states with the same conserved quantum numbers will mix, and baryons are not pure three-quark objects — but it works remarkably well for the ground states.

Quarks are fermions, so the baryon wavefunction must be overall antisymmetric under quark interchange[1]. Baryons are color singlets, and so have an antisymmetric color wavefunction. In the ground state, the orbital angular momentum L is zero (S-wave) and the spatial wavefunction is symmetric. Therefore, the product of the spin and flavor wavefunctions must also be symmetric for ground-state

⋆e-mail: lena@lebedev.ru

[1]Strictly, it only needs to be antisymmetric under interchange of equal-mass quarks, but in order to build the model we assume $SU(4)$ is a good symmetry.

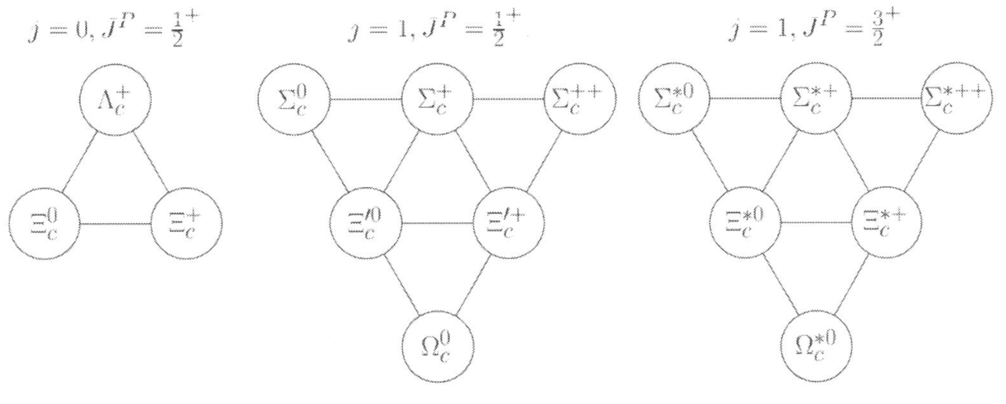

Figure 1. The $SU(3)$ multiplets containing the ground state baryons, grouped according to the spin j of the light diquark and the spin-parity J^P of the baryon.

baryons. There are two ways this can be accomplished: both wavefunctions can be fully symmetric, or both can have mixed symmetry with the product being symmetric.

In concrete terms, we can consider a singly-charmed baryon to consist of a heavy c quark and a light diquark with spin-parity j^P. Assuming isospin symmetry and letting q denote a u or d quark, there are four possibilities for the flavor content of the diquark:

- qq with isospin 0 (flavor antisymmetric);
- qq with isospin 1 (flavor symmetric);
- sq with isospin $1/2$ (either);
- ss with isospin 0 (flavor symmetric).

These correspond to the Λ_c, Σ_c, Ξ_c, and Ω_c states, respectively. The diquark wavefunction must be antisymmetric under quark interchange. Its color wavefunction is antisymmetric and in the ground state its spatial wavefunction is symmetric, so it may be either flavor-symmetric and spin-symmetric ($j^P = 1^+$) or flavor-antisymmetric and spin-antisymmetric ($j^P = 0^+$). Combining the diquark with the charm quark gives rise to the possible states illustrated in figure 1, where the multiplets of the full SU(3) symmetry (formed by the u-, d-, and s-quarks) are shown. Those with $J^P = 1/2^+$ are all members of the same multiplet as the proton, and those with $J^P = 3/2^+$ are all members of the same multiplet as the Δ and Ω. Note that there is a second isospin doublet of Ξ_c states with $J^P = 1/2^+$, denoted Ξ_c'.

The constituent quark model predicts relations between the masses of these states as well as their existence and quantum numbers.

3 Excited States

Baryons can be given orbital (l) or radial (k) excitations. Since they are three-body systems there are two degrees of freedom in each case (denoted ρ, λ). For baryons with one heavy quark (mass M) and two light quarks (mass m), a natural way to specify these is to divide the system into a light diquark and the heavy quark. Taking a simple potential model based on the harmonic oscillator, the energy

 DOI: 10.1051/epjconf/201715803008

levels are given by ([2]):

$$E = \sqrt{\frac{K}{m}}\left(3 + 2l_\rho + 4k_\rho\right) + \sqrt{\frac{K}{\mu}}\left(3 + 2l_\lambda + 4k_\lambda\right), \tag{1}$$

where K is a constant describing the potential and $\mu = (2/3M + 1/3m)^{-1} \approx 3m$ in the heavy quark limit. Thus, the ρ excitations (within the diquark) require roughly three times as much energy as the corresponding λ excitations (between quark and diquark). Therefore the lowest-lying excitations are those with $l_\lambda = 1$ and the other quantum numbers zero, i.e. $L = 1$. (Within this band there will be further splitting, e.g. due to spin-spin and spin-orbit couplings.) The second band will consist of two groups of states that have comparable energy: those with $l_\lambda = 2$ ($L = 2$) and those with $k_\lambda = 1$ ($L = 0$), with the other quantum numbers being zero. Beyond the second band the degeneracy grows further, but we lack useful experimental data in this region in any case.

All this said, it is important to bear in mind that states which share all conserved, external quantum numbers (J, P, I, C, S) can mix. Therefore we should be careful when interpreting observed resonances as specific expected states, particularly for higher excitations.

4 Ξ_c Family

Since all three quark flavors are different for the Ξ_c, there are many allowed configurations. These may be divided into states for which the light diquark wavefunction is flavor-antisymmetric (analogous to Λ_c) or flavor-symmetric (analogous to Σ_c).

The two ground states are the only members of the group that decay weakly, and their masses, lifetimes, and many of their decay modes have been measured [3]. The Ξ_c states also exist in many angular momentum configurations of the constituent quarks, each as an isospin pair (Table 1).

The Ξ_c' and $\Xi_c(2645)$ states form a doublet analogous to the $\Sigma_c(2455)$ and $\Sigma_c(2520)$ with expected (j^P, J^P) of $(1^+, 1/2^+)$ and $(1^+, 3/2^+)$, respectively. The former is too light to decay strongly, but the electromagnetic transition $\Xi_c' \to \Xi_c\gamma$ is allowed. BaBar performed an angular analysis of $\Xi_c'^0 \to \Xi_c^0(\Xi^-\pi^+)\gamma$ in the helicity formalism and found the data to be consistent with $J = 1/2$ [4]. However, due to the inclusive production environment higher spins could not be ruled out.

The low-lying excited states $\Xi_c(2790)$ and $\Xi_c(2815)$ are analogous to the $\Lambda_c(2595)$ and $\Lambda_c(2625)$, and their decays follow a corresponding pattern: $\Xi_c(2790) \to \Xi_c'\pi$, $\Xi_c(2815) \to \Xi_c(2645)\pi$. They were therefore identified as the $1/2^-, 3/2^-$ doublet with $j^P = 1^-$ and the diquark in a flavor-antisymmetric configuration [5]. Recently Belle collaboration significantly improved measured mass differences (between excited and ground states) and widths (where applicable) for Ξ_c', $\Xi_c(2645)$, $\Xi_c(2790)$, $\Xi_c(2815)$, and $\Xi_c(2970)$ states [6]. The $\Xi_c(2930)$ was seen in $B^- \to \Lambda_c^+\bar{\Lambda}_c^- K^-$ decays [7]. The Dalitz plot was clearly not flat and the $\Lambda_c^+K^-$ projection was consistent with a single resonance with the parameters given in table 1. However, given the small sample size and the inability to rule out other explanations (such as two overlapping Ξ_c resonances or a complicated interference pattern between Ξ_c and charmonium resonances) this is considered unconfirmed.

The remaining resonances were all seen in the $\Lambda_c^+\overline{K}\pi^+$ isodoublet of final states (and, in the case of the $\Xi_c(2970)$, in $\Xi_c(2645)\pi$). The $\Xi_c(2970)$ and $\Xi_c(3077)$ were discovered by Belle in $\Lambda_c^+\overline{K}\pi$ and confirmed by BaBar. Since this is a three-body decay it could proceed via an intermediate Σ_c. BaBar tested this by fitting a two-dimensional PDF in $M(\Lambda_c^+\pi)$, $M(\Lambda_c^+\overline{K}\pi)$. It was found that approximately half of the $\Xi_c(2970)$ decays to this final state proceed through an intermediate $\Sigma_c(2455)$ with the rest non-resonant. By contrast, most if not all of the $\Xi_c(3077)$ decays to this final state proceed via $\Sigma_c(2455)$ or $\Sigma_c(2520)$ with approximately equal branching fractions to each. Because the $\Xi_c(2970)$ is

DOI: 10.1051/epjconf/201715803008

Table 1. Summary of excited Ξ_c baryons.

State	Decay mode	Mass, MeV$/c^2$	Width, MeV	J^P
$\Xi_c'^+$	$\Xi_c^+\gamma$	2577.4 ± 1.2		$\frac{1}{2}^+$
$\Xi_c'^0$	$\Xi_c^0\gamma$	2578.8 ± 0.5		$\frac{1}{2}^+$
$\Xi_c(2645)^+$	$\Xi_c^0\pi^+$	2645.53 ± 0.31	2.14 ± 0.19	$\frac{3}{2}^+$
$\Xi_c(2645)^0$	$\Xi_c^+\pi^-$	2646.32 ± 0.31	2.35 ± 0.22	$\frac{3}{2}^+$
$\Xi_c(2790)^+$	$\Xi_c'^0\pi^+$	2792.0 ± 0.5	8.9 ± 1.0	$\frac{1}{2}^-$
$\Xi_c(2790)^0$	$\Xi_c'^+\pi^-$	2792.8 ± 1.2	10.0 ± 1.1	$\frac{1}{2}^-$
$\Xi_c(2815)^+$	$\Xi_c^+\pi^+\pi^-$, $\Xi_c(2645)^0\pi^+$, $\Xi_c^{0\prime}\pi^+$	2816.67 ± 0.31	2.43 ± 0.26	$\frac{3}{2}^-$
$\Xi_c(2815)^0$	$\Xi_c^0\pi^+\pi^-$, $\Xi_c(2645)^+\pi^-$, $\Xi_c^{+\prime}\pi^-$	2820.22 ± 0.32	2.54 ± 0.25	$\frac{3}{2}^-$
$\Xi_c(2930)^0$	$\Lambda_c^+K^-$	2931 ± 6	36 ± 13	
$\Xi_c(2970)^+$	$\Lambda_c^+K^-\pi^+$, $\Sigma_c^{++}K^-$, $\Xi_c(2645)^0\pi^+$, $\Xi_c^{0\prime}\pi^+$	2969.4 ± 0.8	$20.9^{+2.4}_{-3.5}$	
$\Xi_c(2970)^0$	$\Xi_c(2645)^+\pi^-$, $\Xi_c^{+\prime}\pi^-$	2967.8 ± 0.8	$28.1^{+3.4}_{-4.0}$	
$\Xi_c(3055)^+$	$\Sigma_c^{++}K^-$, ΛD^+	3055.9 ± 0.4	7.8 ± 1.9	
$\Xi_c(3055)^0$	ΛD^0	3059.0 ± 0.8	6.4 ± 2.4	
$\Xi_c(3080)^+$	$\Lambda_c^+K^-\pi^+$, $\Sigma_c^{++}K^-$, $\Sigma_c(2520)^{++}K^-$, ΛD^+	3077.2 ± 0.4	3.6 ± 1.1	
$\Xi_c(3080)^0$	$\Lambda_c^+K_S^0\pi^-$, $\Sigma_c^0K_S^0$, $\Sigma_c(2520)^0K_S^0$	3079.9 ± 1.4	5.6 ± 2.2	

close to threshold on the scale of its natural width, especially with an intermediate Σ_c, the available phase space changes significantly across the resonance. Different handling of this threshold behavior is the reason for the mild tension in the fitted $\Xi_c(2970)$ masses between [8] and [9]. The masses measured in the $\Xi_c(2645)\pi^+$ final state [10], which is far from threshold, are consistent with the BaBar treatment, although the widths are smaller than either experiment saw in $\Lambda_c^+\overline{K}\pi$. Requiring an intermediate Σ_c reduces the background levels, and by doing this BaBar was able to identify two further candidate states, the $\Xi_c(3055)$ and $\Xi_c(3123)$. The latter had a limited statistical significance (3σ), and later was not confirmed by Belle with twice as much statistics [11].

Till recently all measurements of excited Ξ_c baryons were performed using decay modes in which the charm quark is contained in the final state baryon. Belle reported studies of Ξ_c baryons decaying to the ΛD^+ and ΛD^0 final states [12]. Significant signals were found for $\Xi_c(3055)^+$ and $\Xi_c(3080)^+$ decays into ΛD^+. In the ΛD^0 final state, observation of the $\Xi_c(3055)^0$ was reported.

As a final point it should be noted that no direct measurements of the J^P of any of the excited Ξ_c states are available. Mild constraints on the quantum numbers can be inferred from the decay pattern. For example, the observation of the $\Xi_c(3077)$ in $\Sigma_c(2455)\overline{K}$ and $\Sigma_c(2455)\overline{K}$ excludes states with diquark $j^p = 0^-$. However, many quantum numbers are still allowed for these states and there is a range of opinions on the best match to the data.

5 Ω_c Family

In contrast to the Λ_c, Σ_c, and Ξ_c families available experimental data on the Ω_c ground states were limited before the B Factories. The weakly-decaying Ω_c^0 ($J = 1/2^+$) had been seen in a number of different decay modes and production environments, but with only limited statistics (typically samples of $O(10)$ events in a given decay mode, and never more than 100); and the $J = 3/2^+$ Ω_c^{*0} had not been observed. Belle carried out a precise measurement of the Ω_c^0 mass using a sample of 725 decays to the $\Omega^-\pi^+$ final state [13], obtaining $(2693.6 \pm 0.3^{+1.8}_{-1.5})$ MeV$/c^2$.

DOI: 10.1051/epjconf/201715803008

Table 2. Summary table of excited Ω_c baryons. For each parameter, the first uncertainty is statistical and the second systematic. The asymmetric uncertainty arising from the Ξ_c^+ mass is given separately.

State	Decay mode	Mass, MeV/c^2	Width, MeV	J^P
Ω_c^{*0}	$\Omega_c^0\gamma$	2765.9 ± 2.0		$\frac{3}{2}^+$
$\Omega_c(3000)^0$		$3000.4 \pm 0.2 \pm 0.1^{+0.3}_{-0.5}$	$4.5 \pm 0.6 \pm 0.3$	
$\Omega_c(3050)^0$		$3050.2 \pm 0.1 \pm 0.1^{+0.3}_{-0.5}$	$0.8 \pm 0.2 \pm 0.1$	
$\Omega_c(3066)^0$	$\Xi_c^+ K^-$	$3065.6 \pm 0.1 \pm 0.3^{+0.3}_{-0.5}$	$3.5 \pm 0.4 \pm 0.2$	
$\Omega_c(3090)^0$		$3090.2 \pm 0.3 \pm 0.5^{+0.3}_{-0.5}$	$8.7 \pm 1.0 \pm 0.8$	
$\Omega_c(3119)^0$		$3119.1 \pm 0.3 \pm 0.9^{+0.3}_{-0.5}$	$1.1 \pm 0.8 \pm 0.4$	

The short list of known doubly strange charmed baryons is summarized in Table 2. The Ω_c^{*0} is too light to undergo strong decay and so decays purely to $\Omega_c^0\gamma$. It was discovered by BaBar [14] and confirmed by Belle [13]. Both measured the mass difference $m(\Omega_c^{*0}) - m(\Omega_c^0)$, and the two results are in excellent agreement with each other as well as with most theoretical predictions [15].

Recently LHCb Collaboration performed a study of the $\Xi_c^+ K^-$ spectrum, observing five new narrow excited Ω_c^0 states [16] and thus presenting first excited doubly strange charmed baryons, that decay strongly. The observation of these states in an inclusive process through a two-body decay does not allow the determination of their quantum numbers. More information can be obtained from the study of possible three-body decays or when reconstructing these states in decays of heavy baryons.

6 Summary

Recently observed excited Ω_c states present a unique opportunity to test and further improve theoretical models, that predict properties of heavy hadrons. More accurate Ξ_c mass values is of both practical and theoretical interest, and knowing their widths (where applicable) can then lead to measurements of the matrix elements of their decays. These matrix elements are also applicable to other excited charmed and bottom baryons.

Interesting feature of the charmed baryons spectrum is that highly excited charmed baryons can decay to a charmed meson and a non-charmed baryon.

It should be noted that no direct measurements of the J^P of any of the excited strange charmed baryons are available. Constraints on the quantum numbers can be inferred only from the decay pattern.

Acknowledgements

This work was supported by the Russian Foundation for Basic Research grant No. 16-32-60134.

References

[1] M. Gell-Mann, Phys. Lett. **8**, 214 (1964);
G. Zweig, CERN-TH-401 (1964);
G. Zweig, Developments in the Quark Theory of Hadrons **1**, 22 (1980).
[2] E. Klempt and J.-M. Richard, Rev. Mod. Phys. **82**, 1095 (2010).
[3] C. Patrignani *et al.* (Particle Data Group), Chin. Phys. C **40**, 100001 (2016), and 2017 update.
[4] B. Aubert *et al.* (BaBar Collaboration), hep-ex/0607086 (2006).

[5] C. E. Csorna *et al.* (CLEO Collaboration), Phys. Rev. Lett. **86**, 4243 (2001);
J. P. Alexander *et al.* (CLEO Collaboration), Phys. Rev. Lett. **83**, 3390 (1999).
[6] J. Yelton *et al.* (Belle Collaboration), Phys. Rev. D **94**, 052011 (2016).
[7] B. Aubert *et al.* (BaBar Collaboration), Phys. Rev. D **77**, 031101 (2008).
[8] R. Chistov *et al.* (Belle Collaboration), Phys. Rev. Lett. **97**, 162001 (2006).
[9] B. Aubert *et al.* (BaBar Collaboration), Phys. Rev. D **77**, 012002 (2008).
[10] T. Lesiak *et al.* (Belle Collaboration), Phys. Lett. B **665**, 9 (2008).
[11] Y. Kato, T. Iijima *et al.* (Belle Collaboration), Phys. Rev. D **89**, 052003 (2014).
[12] Y. Kato, T. Iijima *et al.* (Belle Collaboration), Phys. Rev. D **94**, 032002 (2016).
[13] E. Solovieva, R. Chistov *et al.* (Belle Collaboration), Phys. Lett. B **672**, 1 (2009).
[14] B. Aubert *et al.* (BaBar Collaboration), Phys. Rev. Lett. **97**, 232001 (2006).
[15] R. Roncaglia, B. Lichtenberg, and E. Predazzi, Phys. Rev. D **52**, 1722 (1995);
L. Rosner, Phys. Rev. D **52**, 6461 (1995);
J. Savage, Phys. Lett. B **359**, 189 (1995);
B. Lichtenberg, R. Roncaglia, and E. Predazzi, Phys. Rev. D **53**, 6678 (1996);
E. Jenkins, Phys. Rev. D **54**, 4515 (1996);
Y. Glozman and D. Riska, Nucl. Phys. A **603**, 326 (1996);
A. Zalewska and K. Zalewski, hep-ph/9608240 (1996);
L. Burakovsky, T. Goldman, and L. Horwitz, Phys. Rev. D **56**, 7124 (1997);
N. Mathur, R. Lewis, and M. Woloshyn, Phys. Rev. D **66**, 014502 (2002).
[16] R. Aaij *et al.* (LHCb Collaboration), Phys. Rev. Lett. **118**, 182001 (2017).

 DOI: 10.1051/epjconf/201715803009

Z-boson production in association with heavy flavor in the k_T-factorization

Maxim Malyshev[1,⋆], *Sergei* Baranov[2,⋆⋆], and *Artem* Lipatov[1,3,⋆⋆⋆]

[1] *Skobeltsyn Institute of Nuclear Physics, Moscow State University, 119991 Moscow, Russia*
[2] *P.N. Lebedev Physics Institute, 119991 Moscow, Russia*
[3] *Joint Institute for Nuclear Research, Dubna 141980, Moscow region, Russia*

Abstract. We present the calculations of associated production of Z bosons and heavy (charm or beauty) quarks at the LHC energies in the framework of k_T-factorization QCD approach. Our consideration is mainly based on the $O(\alpha\alpha_s^2)$ off-shell gluon-gluon fusion subprocess $g^*g^* \to ZQ\bar{Q}$, where produced Z boson subsequently decays into a lepton pair. Several subleading contributions from the $O(\alpha\alpha_s^2)$ and $O(\alpha\alpha_s^3)$ subprocesses are taken into account. Contributions from the double parton scattering mechanism are estimated. The transverse momentum dependent (or unintegrated) gluon densities in a proton are determined using the Catani-Ciafaloni-Fiorani-Marchesini evolution equation. We achieve reasonably good agreement of our predictions and latest experimental data taken by the CMS and ATLAS Collaborations, discuss the theoretical uncertainties of our calculations and demonstrate the importance of subleading quark contributions in description of the LHC data in the whole kinematical region.

Very recently, the CMS Collaboration has presented the measurements of Z boson production in association with charm [1] quarks at the LHC energy $\sqrt{s} = 8$ TeV. Also measurements of angular correlations of B-hadrons at $\sqrt{s} = 7$ TeV were presented [2] and the ATLAS Collaboration has reported the experimental data on the total and differential cross sections of Z boson production in association with beauty quark jets at $\sqrt{s} = 7$ TeV [3]. Such processes involve both strong and weak interactions, so are important as global test of the Standard Model (SM). The $Z + b$-jets production is an important background for studies of the associated production of Higgs and Z bosons, where the Higgs boson decays into $b\bar{b}$ pairs [4–6]. Many physics scenarios beyond the Standard Model (SM), for example, new generations of heavy quarks (b', t') decaying into Z bosons and b-quarks [7], supersymmetric Higgs bosons produced in association with beauty quarks [8] and some SM extentions with additional $SU(2)$ doublets with enhanced $Zb\bar{b}$ coupling [9], predict final states with b-quarks and Z bosons. In addition, $Z + b/Z + c$ cross sections ratio is highly sensitive to the charm content of the proton [10]. Finally, such processes may serve as potential indicators of the Double Parton Scattering (DPS) mechanism [11–13].

In the present work we analyse recent CMS [1, 2] and ATLAS [3] data using so called k_T-factorization QCD approach [14]. This approach is based on the famous Balitsky-Fadin-Kuraev-Lipatov (BFKL) or Ciafaloni-Catani-Fiorani-Marchesini (CCFM) gluon evolution equations and pro-

⋆e-mail: malyshev@theory.sinp.msu.ru
⋆⋆e-mail: baranov@sci.lebedev.ru
⋆⋆⋆e-mail: lipatov@theory.sinp.msu.ru

vides solid theoretical grounds for the effects of gluon radiation in initial state and intrinsic gluon transverse momentum. The certain advantages are connected with the fact that, even with the LO partonic amplitudes, one can include a large piece of higher-order corrections (namely, part of NLO + NNLO + ... terms containing leading $\log 1/x$ enhancement of cross sections due to real initial state gluon emissions) taking them into account in the form of CCFM-evolved transverse momentum dependent (TMD) gluon densities. In particular, it gives better agreement with Tevatron data on the associated production of prompt photons and charm or beauty quarks compared to the NLO pQCD predictions [15], that is an additional motivation of our present study.

Let us start from a short review of calculation steps. Our consideration is mainly based on the off-shell gluon-gluon fusion subprocess:

$$g^*(k_1) + g^*(k_2) \to Z(p) + Q(p_1) + Q(p_2), \tag{1}$$

where Q denotes the produced charm or beauty quark, and four-momenta of the particles are given in parentheses. The corresponding gauge-invariant off-shell amplitude was calculated earlier [16, 17]. To fully reproduce the experimental setup [1–3] we simulate the subsequent decay of the produced Z boson into lepton pair according to the electroweak theory, that has not been made in the previous calculations [16, 17]. So, the off-shell gluon-gluon fusion gives the $O(\alpha\alpha_s^2)$ contribution to the production cross-section. Additionally, we take into account several subprocesses involving quarks in the initial state, namely

$$q(k_1) + Q(k_2) \to Z(p) + q(p_1) + Q(p_2), \tag{2}$$

$$q(k_1) + \bar{q}(k_2) \to Z(p) + Q(p_1) + \bar{Q}(p_2), \tag{3}$$

$$q(k_1) + g(k_2) \to Z(p) + q(p_1) + Q(p_2) + \bar{Q}(p_3), \tag{4}$$

where the produced Z bosons also decay to lepton pairs. These subprocesses give $O(\alpha\alpha_s^2)$ (subprocesses (2) and (3)) and $O(\alpha\alpha_s^3)$ (subprocess (4)) contributions. Subprocess (4) is suppressed by the additional degree of strong coupling QCD constant, however, we include it into the consideration because it can give sizeble contribution to the production cross section due to large gluon flux at the LHC energies, especially at low $\eta - \phi$ distances between the produced heavy quarks. Moreover, while the gluon-gluon fusion (1) contributes mainly at low and moderate transverse momenta, the subprocesses (2) — (4) become important at high transverse momenta, which correspond to the region of relatively large x, where the standard Dokshitzer-Gribov-Lipatov-Altarelli-Parisi (DGLAP) parton evolution works perfectly and contribution of the $\log 1/x$-enhanced terms is negligible. Therefore, the subprocesses (2) — (4) can be safely taken into account in the usual collinear QCD factorization. Thus, we rely on a combination of two techniques with each of them being used where it is most suitable. For the subprocesses (2) and (3) we use here the on-shell limit of formulas obtained earlier [15] supplementing them with the Z boson decays.

According to the collinear QCD factorization theorem, to calculate the contributions of subleading subprocesses (2) — (4) one has to convolute the corresponding partonic cross sections $d\hat{\sigma}_{ab}$ and conventional parton distribution functions $f_a(x_1, \mu^2)$ in a proton:

$$\sigma = \int dx_1 dx_2 d\hat{\sigma}_{ab}(x_1, x_2, \mu^2) f_a(x_1, \mu^2) f_b(x_2, \mu^2), \tag{5}$$

where indices a and b denote quark and/or gluon, x_1 and x_2 are the fractions of longitudinal momenta of colliding protons and μ^2 is the hard scale. In the case of the leading off-shell gluon-gluon fusion subprocess (1) we employ the k_T-factorization approach:

$$\sigma = \int dx_1 dx_2 d\mathbf{k}_{1T}^2 d\mathbf{k}_{2T}^2 d\hat{\sigma}_{gg}^*(x_1, x_2, \mathbf{k}_{1T}^2, \mathbf{k}_{2T}^2, \mu^2) f_g(x_1, \mathbf{k}_{1T}^2, \mu^2) f_g(x_2, \mathbf{k}_{2T}^2, \mu^2), \tag{6}$$

where $\mathbf{k}_{1T}^2$ and $\mathbf{k}_{2T}^2$ are the transverse momenta of the initial off-shell gluons, and $f_g(x_1, \mathbf{k}_T^2, \mu^2)$ is the TMD distribution in a proton.The essential point of our consideration is that we use the numerical solution of the CCFM evolution equation to determine the distribution. Numerically, we adopt the latest JH'2013 set, where two density functions, namely JH'2013 set 1 and JH'2013 set 2, were released [18]. The input parameters of corresponding initial gluon distributions were fitted from the best description of the precision DIS data on the inclusive F_2 data (set 1) or both F_2 and F_2^c data (set 2). Both these fits are based on the TMD matrix elements (which are directly related with the resummation of DIS coefficient functions) and include two-loop strong coupling constant, kinematic consistency constraint [19, 20] and non-singular terms in the CCFM gluon splitting function [21]. However, the inclusive structure function F_2 receives significant contributions from quark channels, whereas charm production is dominated by the gluon distribution. Therefore, below we will use JH'2013 set 2 gluon density as a default choice. For the conventional quark and gluon distributions in a proton we apply the LO MSTW'2008 set [22].

To calculate the DPS contributions one commonly makes use of a simple factorization formula (for details see the reviews [11–13] and references therein):

$$\sigma_{\rm DPS}(Z + Q + \bar{Q}) = \frac{\sigma_{\rm SPS}(Z)\sigma_{\rm SPS}(Q + \bar{Q})}{\sigma_{\rm eff}}, \quad (7)$$

where $\sigma_{\rm eff} \simeq 15$ mb is a normalization constant which incorporates all "DPS unknowns" into a single phenomenological parameter. The numerical value of $\sigma_{\rm eff}$ was earlier obtained from fits to pp and $p\bar{p}$ data. This will be taken as default value throughout the paper. Deriving the formula (7) relies on two simplifying approximations: the double parton distribution functions can be decomposed into longitudinal and transverse components and the longitudinal component reduces to the diagonal product of independent single parton densisites. The inclusive SPS cross sections for the individual partonic subrocesses can be derived in a usual way (in the collinear QCD approximation or k_T-factorization). To calculate them below we strictly follow the approaches described earlier [23–25].

We now are in a position to present our numerical results. First we describe the input parameters and kinematic conditions. The predicted cross sections depend on the renormalization (μ_R) and factorization (μ_F) scales. As it is often done in the collinear QCD factorization, which we apply for the subprocesses (2) — (4), we set both these scales to be equal to transverse mass of produced Z bosons. In the k_T-factorization approach, employed for the gluon-gluon fusion subprocess (1), we set $\mu_F^2 = \hat{s} + \mathbf{Q}_T^2$ with $\hat{s}$ and $\mathbf{Q}_T^2$ being the energy of scattering subprocess and transverse momentum of the incoming off-shell gluon pair, respectively. The special choice of μ_F is connected with the CCFM evolution [18]. We set charm and beauty masses $m_c = 1.4$ GeV and $m_b = 4.75$ GeV, Z boson mass $m_Z = 91.1876$ GeV, its total decay width $\Gamma_Z = 2.4952$ GeV and $\sin^2\theta_W = 0.23122$. For the subprocesses (2) — (4) we used LO formula for the strong coupling constant with $n_f = 4$ massless quark flavours and $\Lambda_{QCD} = 200$ MeV, so that $\alpha_s(m_Z^2) = 0.1232$. According to the fit [18], we apply two-loop strong coupling constant for the off-shell gluon-gluon fusion subprocess. The multidimensional integration everywhere was performed by means of a Monte Carlo technique, using the VEGAS routine [26].

Let us consider first the associated production of Z bosons and b-jets [27]. The ATLAS Collaboration collected the data [3] at $\sqrt{s} = 7$ TeV. Both leptons originating from the Z boson decay were required to have $p_T^l > 20$ GeV and $|\eta^l| < 2.4$, the lepton pair invariant mass lied in the interval $76 < M_{ll} < 106$ GeV, the beauty jets were required to have $p_T^b > 20$ GeV and $|\eta^b| < 2.4$. We confront our predictions with the available data in Figs. 1 and 2. To estimate the theoretical uncertainties in the quark-involving subprocesses (2) — (4), calculated using the collinear QCD factorization, we have varied the scales μ_R and μ_F by a factor of 2 around their default values. In the k_T-factorization approach, employed for off-shell gluon-gluon fusion subprocess (1), the scale uncertainties have been

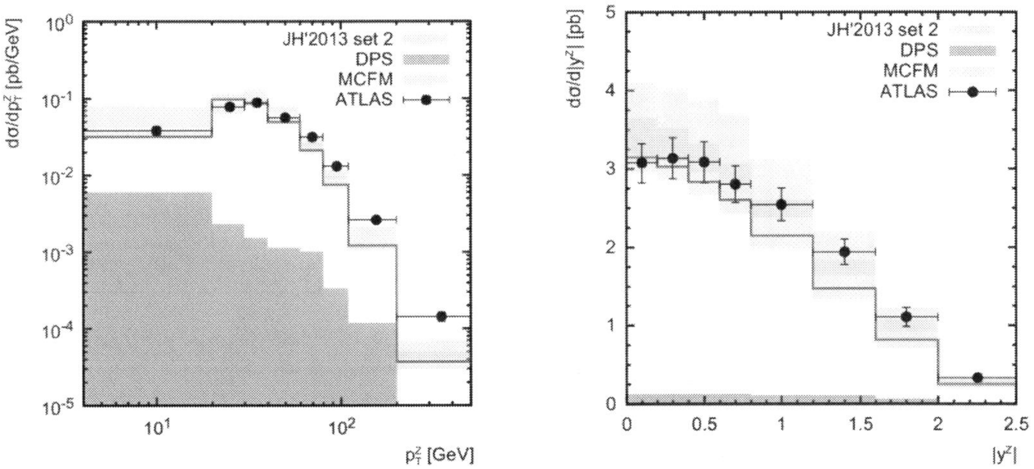

Figure 1. Associated $Z + b$ production cross section at $\sqrt{s} = 7$ TeV presented as a function of the Z boson transverse momentum (left panel) or rapidity (right panel). Solid histograms show our predictions at the default scale while shaded bands correspond to scale variations described in the text. The estimated DPS contributions and MCFM [28] predictions (taken from [3]) are shown additionally. The data are from ATLAS [3]

estimated by using the gluon densities JH'2013 set 2+ and JH'2013 set 2- instead of default density JH'2013 set 2. These two sets refer to the varied hard scales in the strong coupling constant α_s in the off-shell amplitude: JH'2013 set 2+ stands for $2\mu_R$, while JH'2013 set 2- refers to $\mu_R/2$ (see [18] for more information). The estimated scale uncertainties are shown as shaded bands. As one can see, we achieve reasonably good agreement with the ATLAS data [3] within the experimental and theoretical uncertainties, although we observe some underestimation of these data at high p_T^Z and slight overestimation at small transverse momenta. The slight overestimation of the data at low p_T^Z can probably be attributed to the TMD gluon density used, since the region $p_T^Z < 100$ GeV is fully dominated by off-shell gluon-gluon fusion, as it is demonstrated in Fig. 2. The rapidity distribution is well described practically everywhere. The NLO pQCD calculations, performed using MCFM routine [28], tend to slightly overestimate our predictions and better decribe the data at large transverse momenta.

We find that the quark-initiated subprocesses (2) — (4) become important only at high transverse momenta, where the typical x values are large, and that supports using of the DGLAP quark and gluon dynamics for these subprocesses (see Fig. 2). The subprocesses (2) — (4) are important to achieve an adequate description of the data in the whole p_T^Z region. The estimated DPS contributions are found to be small in the considered kinematic region. Some reasonable variations in $\sigma_{\rm eff} \simeq 15 \pm 5$ mb would affect DPS predictions, though without changing our basic conclusion. We note also that scale uncertainties of the CCFM-based predictions are comparable with the ones of NLO pQCD calculations.

Now we turn to the associated production of Z bosons and two beauty jets [27]. The data provided by the ATLAS Collaboration [3] refer to the same energies and kinematic restrictions as in the previous subsection. The observables shown by the ATLAS Collaboration are the Z boson transverse momentum p_T^Z and rapidity y^Z, invariant mass of the b-jet pair M^{bb} and angular separation in $\eta - \phi$ plane between the jets ΔR^{bb}. The latter is useful to identify the contributions where scaterring amplitudes are dominated by terms involving gluon splitting $g \to Q + \bar{Q}$.

 DOI: 10.1051/epjconf/201715803009

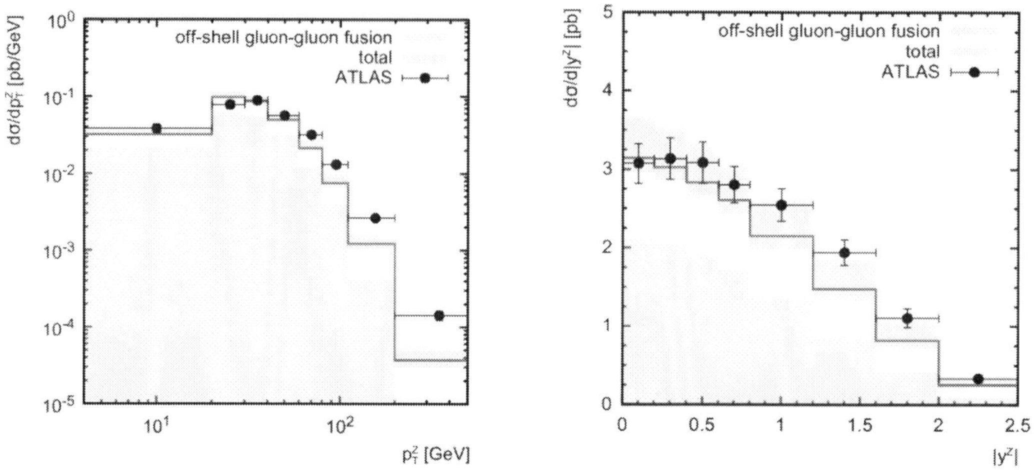

Figure 2. The off-shell gluon-gluon fusion contribution to the associated $Z + b$ production at $\sqrt{s} = 7$ TeV. The data are from ATLAS [3]

The results of our calculations are shown in Fig. 3 in comparison with the ATLAS data [3]. As one can see, our results describe the data reasonably well within the experimental and theoretical uncertainties, although some tendency to slightly underestimate the data at high transverse momentum p_T^Z and large M^{bb} can be seen. The role of off-shell gluon-gluon fusion subprocess is a bit enhanced here compared to the case of $Z + b$ production because the quark-antiquark annihilation subprocess (3) gives a negligible contribution and gluon splitting subprocess (4) populates mainly at low $\eta - \phi$ distances ΔR^{bb}. The estimated DPS contribution is small and can play a role at low p_T^Z only. The NLO pQCD calculations, performed using MCFM program, tend to slightly underestimate the ATLAS data at low ΔR^{bb} and M^{bb}, although provide better description of the data at large transverse momentum p_T^Z and invariant mass M^{bb}.

In the measurements reported by CMS Collaboration [2], both b-hadrons were identified explicitly by their full decay reconstruction. This data sample allows to study the production properties of a $Zb\bar{b}$ system even in the region of small angular seperation between the b quarks (where the usual jet analysis is not possible as the jets would overlap). In a specific subsample, an additional cut on the Z boson transverse momentum is applied, $p_T^Z > 50$ GeV. The CMS Collaboration described the angular configuration of the $Zb\bar{b}$ system in terms of spatial (in $\eta - \phi$ plane) and azimuthal separation between the b-hadrons ΔR^{bb} and $\Delta\phi^{bb}$, spatial separation $\min \Delta R^{Zb}$ between the Z boson and closest b-hadron and the asymmetry in the $Zb\bar{b}$ system defined as

$$A^{Zbb} = \frac{\max \Delta R^{Zb} - \min \Delta R^{Zb}}{\max \Delta R^{Zb} + \min \Delta R^{Zb}}, \tag{8}$$

where $\max \Delta R^{Zb}$ is the distance between the Z boson and remote b-hadron. The correlation observables are useful to identify the different production mechanisms (or specific higher-order corrections). For example, low $\min \Delta R^{Zb}$ identifies Z bosons in the vicinity of one of the b-hadrons (Z bosons promptly radiated from b-quarks), small $\Delta\phi^{bb}$ indicates gluon to quark splitting $g \to Q + \bar{Q}$. Moreover, while the configurations where the two b-hadrons are emitted symmetrically with respect to the Z directions leads to a zero value of A^{Zbb} assymetry, the additional final-state gluon radiation results in a non-zero one, that provides us with the possibility to test the high-order pQCD corrections.

 DOI: 10.1051/epjconf/201715803009

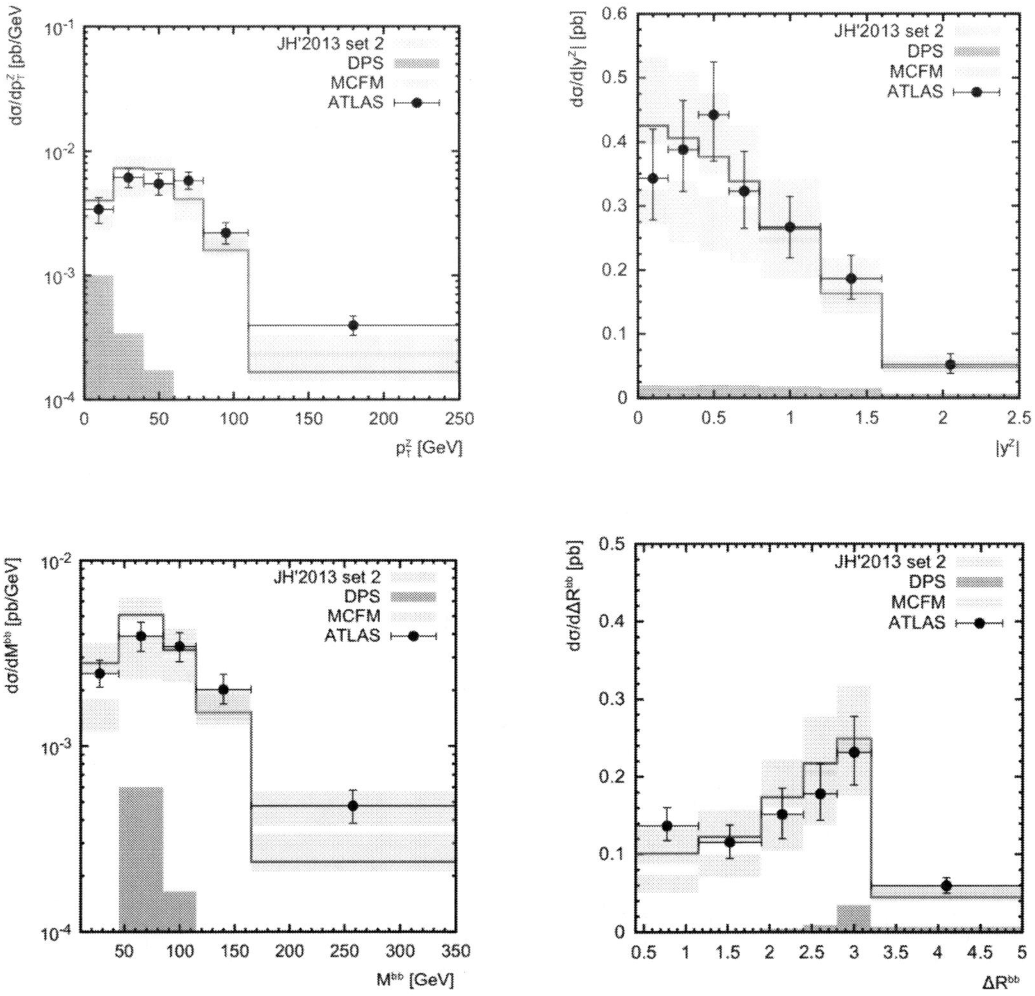

Figure 3. Associated production of a Z boson with two beauty jets at $\sqrt{s} = 7$ TeV calculated as a function of the Z boson transverse momentum, rapidity, invariant mass of the b-jet pair and angular separation between the jets. Notation of the histograms is the same as in Fig. 1. The data are from ATLAS [3]. The MCFM [28] predictions are taken from [3].

Our predictions [27] are shown in Figs. 4 and 5 in comparison with the CMS data [2]. As one can see, our results with default b-quark fragmentation parameters reasonably well describe the data within the theoretical and experimental uncertainties. To estimate an additional uncertainty coming from the b-quark fragmentation, we repeated our calculations with varied shape parameter $\epsilon_b = 0.003$ (not shown), which is often used in NLO pQCD calculations. We find that the predicted cross sections (in the considered p_T region) are larger for smaller ϵ_b values. However, the typical dependence of numerical predictions on the fragmentation scheme is much smaller than the scale uncertainties of our calculations. The NLO pQCD predictions, obtained using the aMC@NLO [29] event generator, are rather close to our results.

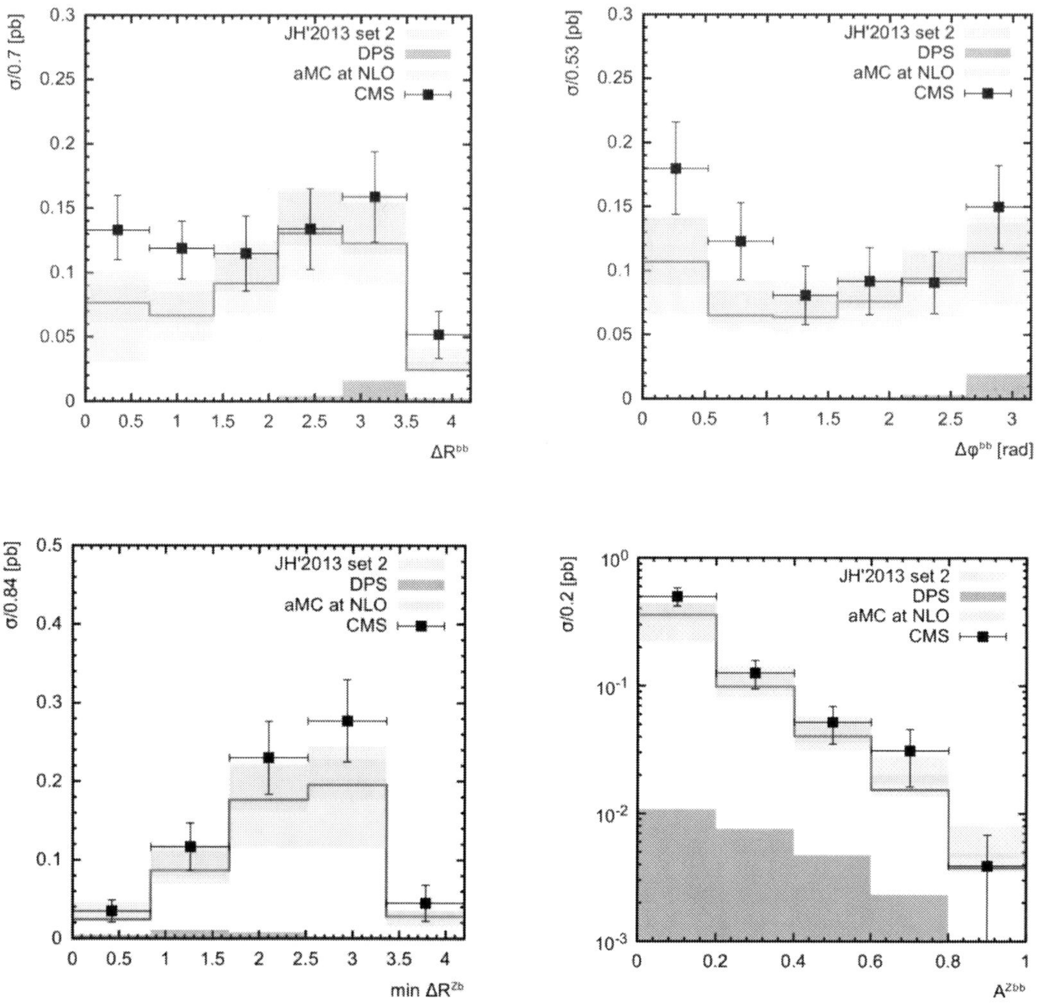

Figure 4. Associated production of a Z boson and two b-hadrons at $\sqrt{s} = 7$ TeV. Notation of the histograms is the same as in Fig. 1. The data are from CMS [2]. The aMC@NLO [29] predictions are taken from [2].

Finally, let us consider measurements of the associated $Z+c$-jet production performed by the CMS Collaboration at $\sqrt{s} = 8$ TeV [1]. Similar to $Z + b$ production, leptons originating from the Z boson decay have transverse momenta $p_T > 20$ GeV and pseudo-rapidities $|\eta| < 2.1$. The invariant mass of the lepton pair must lie within the 71 — 111 GeV interval and c-jets are required to have $p_T > 25$ GeV and $|\eta| < 2.5$. The cross sections ratios $\sigma(Z+c)/\sigma(Z+b)$ was also measured as a functions of Z boson and jet transverse momenta. Our predictions are shown in Figs. 6 and 7 in comparison with the CMS data. We have achieved good description of the Z boson transverse momentum distribution, whereas the measured cross section as a function of leading c-jet is above the predicted one at $p_T^{c-\text{jet}} < 40$ GeV. Taking into account the DPS contributions only slightly increases the calculated cross sections. The NLO pQCD predictions, obtained using the MCFM program [28], are lower than the data in the first and second Z boson p_T bins and in the first $p_T^{c-\text{jet}}$ bin. The LO pQCD predictions matched with parton

DOI: 10.1051/epjconf/201715803009

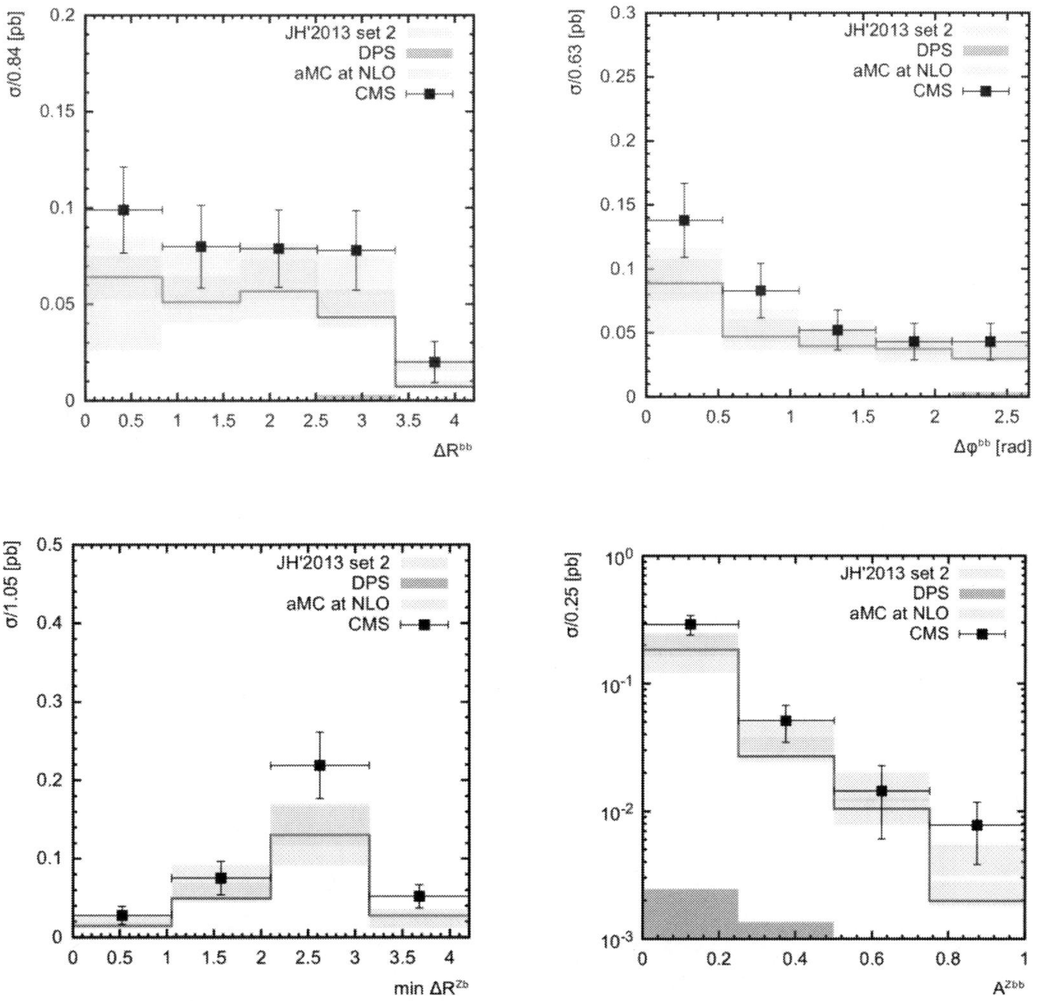

Figure 5. Associated production of a Z boson and two b-hadrons at $\sqrt{s} = 7$ TeV under additional kinematical cut on the Z boson transverse momentum $p_T^Z > 50$ GeV. Notation of the histograms is the same as in Fig. 1. The data are from CMS [2]. The aMC@NLO [29] predictions are taken from [2].

showering, obtained with MadGraph event generator, are in better agreement with the data than the NLO ones, although they still slightly underestimate the CMS data. The differences between these predictions and the data are reduced in the $\sigma(Z + c)/\sigma(Z + b)$ ratios. The results of our calculations underestimate these ratios at high Z-boson and jet transverse momenta. However, they are rather close to the CMS data within the large experimental uncertainties.

To conclude, we have considered the associated production of Z bosons and charmed or beauty quarks at the LHC energies in the framework of the k_T-factorization approach. Our consideration has been based on the $O(\alpha\alpha_s^2)$ off-shell gluon-gluon fusion subprocess $g^*g^* \to ZQ\bar{Q}$, where the produced Z boson subsequently decays into the lepton pair. Several subleading $O(\alpha\alpha_s^2)$ and $O(\alpha\alpha_s^3)$ contributions from the quark-involved subprocesses, which come into play at high transverse momenta, were

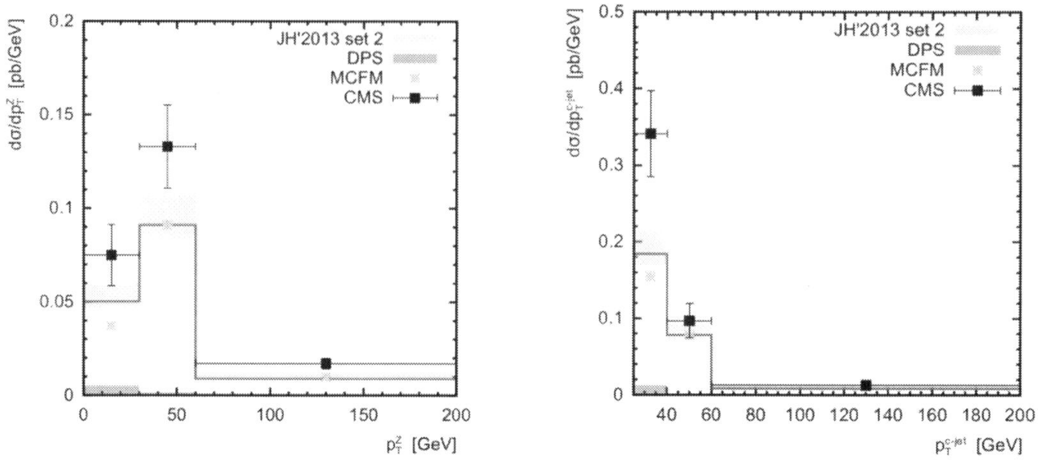

Figure 6. The associated $Z + c$ production calculated as a function of Z boson and c-jet transverse momenta at $\sqrt{s} = 8$ TeV. Notation of histograms is the same as in Fig. 1. The experimental data are from CMS [1].

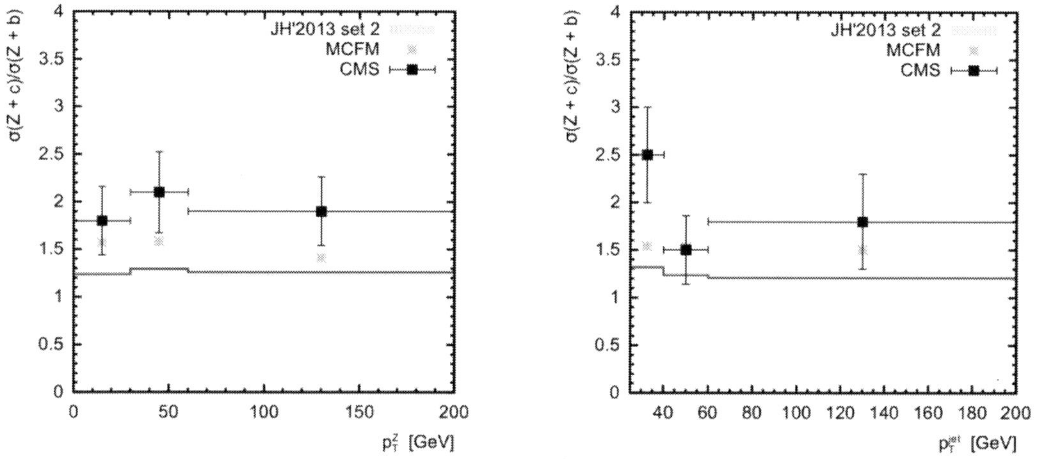

Figure 7. The ratio $\sigma(Z+c)/\sigma(Z+b)$ calculated as a function of Z boson and c-jet transverse momenta at $\sqrt{s} = 8$ TeV. Notation of histograms is the same as in Fig. 1. The experimental data are from CMS [1].

taken into account in the collinear approximation of QCD. Additionally, we have estimated the contributions from the double parton scattering mechanism. Using the TMD gluon densities in a proton derived from the CCFM evolution equation, we have analyzed the latest experimental data taken by the ATLAS and CMS Collaborations at different energies $\sqrt{s} = 7$ and 8 TeV. We have achieved reasonably good agreement between our predictions and the LHC data in the whole kinematical region and demonstrated that the subleading quark contributions are important to describe the LHC data, especially at high transverse momenta. The double parton scattering contributions are estimated to be small.

M.A.M. would like to thank the QFTHEP team for the excellent organization. The investigation of the $Z + b$ production was made in collaboration with H. Jung. We thank F. Hautmann, G.I. Lykasov and S. Turchikhin for very useful discussions and remarks. This research was supported in part by RFBR grant 16-32-00176-mol-a and grant of the President of Russian Federation NS-7989.2016.2. We are grateful to DESY Directorate for the support in the framework of Moscow — DESY project on Monte-Carlo implementation for HERA — LHC. M.A.M. was also supported by a grant of the foundation for the advancement of theoretical physics "Basis" 17-14-455-1.

References

[1] CMS Collaboration, CMS PAS SMP-15-009.
[2] CMS Collaboration, JHEP **1312**, 039 (2013).
[3] ATLAS Collaboration, JHEP **1410**, 141 (2014).
[4] J.-M. Gerard, M. Herquet, Phys. Rev. Lett. **98**, 251802 (2007).
[5] S. de Visscher et al., JHEP **0908**, 042 (2009).
[6] R. Dermisek, J.F. Gunion, Phys. Rev. D **79**, 055014 (2009).
[7] B. Holdom et al., PMC Phys. A **3**, 4 (2009).
[8] L.J. Hall, D. Pinner, J.T. Ruderman, JHEP **1204**, 131 (2012).
[9] D. Choudhury, T.M.P. Tait, C.E.M. Wagner, Phys. Rev. D **65**, 053002 (2002).
[10] A.V. Lipatov et al., Phys. Rev. D **94**, 053011 (2016).
[11] P. Bartalini et al., arXiv:1111.0469 [hep-ph].
[12] H. Abramowicz et al., arXiv:1306.5413 [hep-ph].
[13] S. Bansal et al., arXiv:1410.6664 [hep-ph].
[14] Small-x Collaboration, Eur. Phys. J. C **25**, 77 (2002); **35**, 67 (2004); **48**, 53 (2006).
[15] A.V. Lipatov, M.A. Malyshev, N.P. Zotov, JHEP **1205**, 104 (2012).
[16] S.P. Baranov, A.V. Lipatov, N.P. Zotov, Phys. Rev. D **78**, 014025 (2008).
[17] M. Deak, F. Schwennsen, JHEP **0809**, 035 (2008).
[18] F. Hautmann, H. Jung, Nucl, Phys. B **883**, 1 (2014).
[19] J. Kwiecinski, A.D. Martin, P. Sutton, Z. Phys. C **71**, 585 (1996).
[20] B. Andersson, G. Gustafson, J. Samuelsson, Nucl. Phys. B **467**, 443 (1996).
[21] M. Hansson, H. Jung, proceedings of DIS 03, arXiv:hep-ph/0309009.
[22] A.D. Martin et al., Eur. Phys. J. C **63**, 189 (2009).
[23] S.P. Baranov, A.V. Lipatov, N.P. Zotov, Phys. Rev. D **89**, 094025 (2014).
[24] H. Jung et al., JHEP **1101**, 085 (2011).
[25] H. Jung et al., Phys. Rev. D **85**, 034035 (2012).
[26] G.P. Lepage, J. Comput. Phys. **27**, 192 (1978).
[27] S.P. Baranov et al., arXiv:1708.07079 [hep-ph].
[28] J.M. Campbell, R.K. Ellis, Phys. Rev. D **60**, 113006 (1999); J.M. Campbell, R.K. Ellis, C. Williams, JHEP **1107**, 018 (2011); J.M. Campbell, R.K. Ellis, W. Giele, Eur. Phys. J. C **75**, 246 (2015).
[29] J. Alwal et al., JHEP **1407**, 079 (2014).

Angular correlations in $B\bar{B}$ pair production at the LHC in the parton Reggeization approach

Anton Karpishkov[1,⋆], *Maxim* Nefedov[1,⋆⋆], and *Vladimir* Saleev[1,⋆⋆⋆]

[1] *Samara National Research University - 34, Moskovskoe Shosse, Samara, 443086, Russia*

Abstract. We calculate angular correlation spectra between beauty (B) and anti-beauty ($\bar{B}$) mesons in proton-proton collisions in the leading order approximation of the parton Reggeization approach consistently merged with the next-to-leading order corrections from the emission of additional hard gluon (NLO⋆ approximation). To describe b-quark hadronization we use the universal scale-depended parton-to-meson fragmentation functions extracted from the combined e^+e^- annihilation data. The Kimber-Martin-Ryskin model for the unintegrated parton distribution functions in a proton is implied. We have obtained good agreement between our predictions and data from the CMS Collaboration at the energy $\sqrt{S} = 7$ TeV for $B\bar{B}$ angular correlations within uncertainties and without free parameters.

1 Introduction

Production of bottom-flavored mesons (B) and $b-$quarks in the high energy $pp-$collisions is the object of an intensive experimental study at the CERN LHC. Physics of beauty quarks is very interesting for test of higher-order radiative corrections in Quantum Chromodynamics (QCD), for search of new exotic bound states of heavy quarks and for precise study of parameters of the Standard Model in the flavor sector. Measurements of $b\bar{b}$ angular and momentum correlations provide a test of dynamics of hard interactions, which is sensitive to the next-to-leading order (NLO) corrections of QCD. There are two ways to study this $b\bar{b}$ correlations. The first one is based on a reconstruction of pairs of $b-$jets [1, 2], in the second case we get an information on dynamics of hard production of $b\bar{b}$-pair using data on $B\bar{B}-$pair production. In turn, long-lived $B-$mesons are reconstructed via their semileptonic decays. One advantage of this method is the unique capability to detect $B\bar{B}-$pairs even at small opening angles, in which case the decay products of the B hadrons tend to be merged into a single jet and the standard $b-$jet tagging techniques are not applicable [3]. The conventional formalism of the Collinear Parton Model (CPM) is most suitable for calculation of single-scale variables, such as inclusive production cross section, transverse momentum (p_T) or rapidity spectra of $B-$mesons. Correlation spectra, such as the spectrum of azimuthal angle difference between B and $\bar{B}$ transverse momenta and their rapidity differences spectrum are multi-scale observables. To describe these multi-scale quantities in perturbative QCD one should take into account significant radiative corrections

⋆e-mail: karpishkov@ssau.ru
⋆⋆e-mail: nefedovma@gmail.com
⋆⋆⋆e-mail: saleev@samsu.ru

from the emission of hard partons with large transverse momentum. The aim of our paper is to describe $B\bar{B}$ angular correlation spectra in the Parton Reggeization Approach (PRA) of QCD, as it has been done early for dijets [4], $b\bar{b}-$jets [5, 6], prompt photon pairs [7], $D\bar{D}$ and DD pairs [8, 9]. The novelty of our present results is the merging of Leading-Order (LO) PRA contribution and real NLO* correction from the emission of additional hard gluon.

2 Basic formalism

The PRA is gauge-invariant scheme of high-energy QCD calculations which extends the CPM for description of hard multi-scale processes at the high-energy regime, when we deal with the processes in the multi-Regge kinematics (MRK). In the MRK, energy of collision ($\sqrt{S}$) is sufficiently larger than transverse momenta of final particles ($\sqrt{S} \gg |\mathbf{k}_{Ti}|$) and invariant masses of their pairs ($M_{ij} = \sqrt{(k_i + k_j)^2}$). The main ingredient of the PRA are the high-energy or k_T-factorization of unintegrated parton distribution functions (unPDFs) and gauge-invariant partonic cross sections with *off-shell* partons in the initial state, where these off-shell partons are considered as Reggeized gluons (R) and Reggeized quarks (Q).

The following LO parton subprocesses are relevant for our study of $B\bar{B}-$pair production in the PRA [10]:

$$R + R \to b + \bar{b}, \tag{1}$$

$$Q + \bar{Q} \to b + \bar{b}. \tag{2}$$

Here, we explore the main mechanisms of $B-$meson production via $b-$quark fragmentation into $B-$meson. At high energy and moderate transverse momenta of $B-$mesons, the contribution of the subprocess of quark-antiquark annihilation (2) is negligible in comparison to the gluon-gluon fusion contribution [5] and we will consider contribution of subprocess (1) only. Taking into account the experimental conditions for $B\bar{B}-$pair production data measured by the CMS Collaboration [3], which we study, we should add contribution of NLO subprocess

$$R + R \to b + \bar{b} + g, \tag{3}$$

which generates events with leading jet originating from gluon but not from final b or $\bar{b}$ quarks, see the discussion in the Sec. 5 below.

The general factorization formula for the cross-section in PRA reads:

$$\begin{aligned} d\sigma^{PRA}(pp \to b\bar{b} + X) &= \int \frac{dx_1}{x_1} \int \frac{d^2\mathbf{q}_{1T}}{\pi} \int \frac{dx_2}{x_2} \int \frac{d^2\mathbf{q}_{2T}}{\pi} \Phi_g(x_1, t_1, \mu^2) \times \\ &\times \Phi_g(x_2, t_2, \mu^2) \frac{\overline{|\mathcal{A}_{PRA}|^2}}{2S x_1 x_2} d\Phi(k_3, k_4), \end{aligned} \tag{4}$$

where $q^{\mu}_{1,2} = x_{1,2}P^{\mu}_{1,2} + q^{\mu}_{1,2T}$, $t_{1,2} = -q^2_{1,2T} = \mathbf{q}^2_{1,2T}$, $P^{\mu}_{1,2}$ are the four-momenta of initial-state protons, $(P_1 + P_2)^2 = S$ and $\overline{|\mathcal{A}_{PRA}|^2}$ is the modulus squared of the matrix element of the corresponding PRA subprocess, which is discussed in more detail in the Sec. 3. The unintegrated gluon PDF, first introduced by Kimber, Martin and Ryskin (KMR) [11] is defined as

$$\Phi_g(x, t, \mu^2) = T_g(t, \mu^2) \frac{\alpha_s(t)}{2\pi \cdot t} \int\limits_x^{1-\Delta} dz\, P_{gg}(z) \frac{x}{z} f_g\left(\frac{x}{z}, t\right), \tag{5}$$

where $\Delta(t,\mu^2) = \frac{\sqrt{t}}{\mu+\sqrt{t}}$ and the Sudakov form factor is equal to

$$T_g(t,\mu^2) = \exp\left[-\int_t^{\mu^2} \frac{\alpha_s(t')}{2\pi}\frac{dt'}{t'} \int_0^{1-\Delta(t',\mu^2)} dz' P_{gg}(z')\right].$$

The unintegrated PDF, defined by Eq. (5) approximately satisfies the normalization condition,

$$\int_0^{\mu^2} dt\, \Phi_g(x,t,\mu^2) = x f_g(x,\mu^2). \tag{6}$$

The relation between MRK limit of QCD scattering amplitudes and KMR unPDF is explained, and the derivation of factorization formula (4) is presented in the Ref. [12]. In a stage of numerical calculation we use the LO PDFs from the MSTW-2008 set [13] proton PDFs as collinear input.

3 Reggeized amplitudes

To obtain Reggeized amplitudes of the subprocesses (1) and (3) we exploit our in-house implementation of the Feynman rules of the Lipatov's Effective Theory of Reggeized gluons and Reggeized quarks [14, 15] as a model file for the popular automatic Feynman amplitude generator `FeynArts` [16]. This model file can be used to generate $2 \to 2$ and $2 \to 3$ tree-level scattering amplitudes with Reggeized quarks and Reggeized gluons in the initial state and with Yang-Mills gluons, quarks and photons in the final state. This instrument has been developed as a continuation of the work done in the Ref. [17] and will be publicly released soon, under the acronym `ReggeQCD`. The generated amplitudes can be further processed using `FeynCalc` [18] or `FormCalc` [19] programs.

The generated sets of Feynman diagrams for subprocesses (1) and (3) are shown in Fig. 1 and Fig. 2, correspondingly. The exact analytical formula for Reggeized amplitude $\mathcal{A}_{PRA}(RR \to b\bar{b})$ as well as its squared modulus $\overline{|\mathcal{A}_{PRA}(RR \to b\bar{b})|^2}$ has been obtained earlier, see the Appendix of Ref. [4]. The squared amplitude $\overline{|\mathcal{A}_{PRA}(RR \to b\bar{b}g)|^2}$ is too large for the analytical presentation in the paper. The `FORTRAN` code for this squared amplitude can be obtained by the request from authors.

We have checked the gauge invariance of the obtained results w. r. t. the choice of the gauge for external gluon legs and internal gluon propagators. Also it has been verified, that in the collinear limit $\mathbf{q}_{(1,2)T} \to 0$, squared amplitudes (1) and (3) transform to the squared amplitudes of the corresponding parton subprocesses in the CPM after averaging over the azimuthal angles. Additionally, we have compared our results for the both squared amplitudes numerically with the results obtained from the Monte-Carlo generator `Katie` [20], which is based on the recently suggested scheme to obtain gauge-invariant matrix elements in the k_T−factorization by exploiting the spinor-helicity representation and recursion relations for the tree-level amplitudes [21, 22].

4 Cross sections

The hadronic cross section of subprocess (1) in the PRA can be written as follows:

$$\begin{aligned}\frac{d\sigma(pp \to b(k_1)\bar{b}(k_2) + X)}{dy_1 dy_2 dk_{1T} dk_{2T}} &= \frac{k_{1T}k_{2T}}{16\pi^3}\int d\phi_1 \int d\Delta\phi \int dt_1 \times \\ &\times \Phi_g(x_1,t_1,\mu^2)\Phi_g(x_2,t_2,\mu^2)\frac{\overline{|\mathcal{A}_{PRA}(RR \to b\bar{b})|^2}}{(x_1x_2S)^2},\end{aligned} \tag{7}$$

where $\Delta\phi$ is the azimuthal angle between $\mathbf{k}_{1T}$ and $\mathbf{k}_{2T}$, and $k_{Ti} = |\mathbf{k}_{Ti}|$ in the Eqns. 7 and 8.

The $2 \to 3$ differential cross section for subprocess (3) includes additional integration on phase space of final gluon $g(k_3)$ relatively $2 \to 2$ cross section (7) and reads

$$\frac{d\sigma(pp \to b(k_1)\bar{b}(k_2)g(k_3) + X)}{dy_1 dy_2 dk_{1T} dk_{2T}} = \frac{k_{1T}k_{2T}}{(2\pi)^6} \int dy_3 \int d\phi_1 \int d\phi_2 \int dk_{3T} \times$$
$$\times \int d\Delta\phi \int dt_1 \int dt_2 \, \Phi_g(x_1, t_1, \mu^2)\Phi_g(x_2, t_2, \mu^2) \frac{\overline{|\mathcal{A}_{PRA}(RR \to b\bar{b}g)|^2}}{8(x_1 x_2 S)^2}. \quad (8)$$

To describe the hadronization stage we use the fragmentation approach. The transition from the produced $b(\bar{b})$-quark to the $B(\bar{B})$ meson is described by the corresponding fragmentation function (FF) $D_{b,g\to B}(z, \mu^2)$. According the factorization theorem of the fragmentation model, the master formula for the $B\bar{B}$–pair production via subprocess (1) reads:

$$\frac{d\sigma(pp \to B\bar{B} + X)}{dp_{BT}dy_B dp_{\bar{B}T}dy_{\bar{B}}} = \int_0^1 \frac{dz_1}{z_1} \int_0^1 \frac{dz_2}{z_2} D_{b\to B}(z_1, \mu^2) D_{\bar{b}\to\bar{B}}(z_2, \mu^2) \times$$
$$\times \frac{d\sigma(pp \to b(k_1) + \bar{b}(k_2) + X)}{dk_{1T}dy_1 dk_{2T}dy_2}, \quad (9)$$

where $D_{b\to B}(z_1, \mu^2) = D_{\bar{b}\to\bar{B}}(z_2, \mu^2)$ are the FFs for the transitions of $b(\bar{b})$–quark into $B(\bar{B})$ meson with momentum fraction $z_1(z_2)$ at the scale μ^2, $p_B^\mu = z_1 k_1^\mu$, and $p_{\bar{B}}^\mu = z_2 k_2^\mu$.

In our calculations we use the LO FFs from Ref. [23], where the fit of non-perturbative $b(\bar{b})$ and gluon FF's to data for the $B(\bar{B})$ production from the ALEPH [24] and OPAL [25] Collaborations at the CERN LEP1 collider has been performed. These data were taken at the energy of Z-boson peak, that strongly suppresses the finite-m_b effects which are of relative order m_b^2/m_Z^2, giving the internal consistency of resulting FFs with the zero-mass scheme [26] which we keep throughout our analysis. These FFs satisfy two desirable properties: at first, their scaling violations are ruled by DGLAP evolution equations [27]; at second, they are universal. In our analysis of CMS data [3] we should take the sum of all B–mesons as it is was performed in the experiment.

5 Results

We apply PRA to describe $B\bar{B}$–pair production data from CMS Collaboration at the CERN LHC [3]. The data were collected at the $\sqrt{S} = 7$ TeV. There are three samples of data which correspond to different values of leading jet transverse momentum in the event: $p_{TL} > 56$ GeV (A), $p_{TL} > 84$ GeV (B) and $p_{TL} > 120$ GeV (C). The leading jet is used to define the energy scale of the event and it is required to be within $|\eta(jet)| < 3.0$. This leading jet can be initiated either by b–quark or $\bar{b}$–quark production, which we describe by the LO hard subprocess (1), or by the hard gluon jet. We take into account the latter possibility by the NLO subprocess (3). For the jet reconstruction, the anti-k_T jet algorithm [28] with a distance parameter $R_{k_T} = 0.5$ was used. The visible kinematic range for the measurements was defined at the B–meson level by the requirements $|\eta(B, \bar{B})| < 2.0$ and $p_T(B) > 15$ GeV for both of the B–mesons. Originally, data consist sum of events with B–mesons and B–baryons, but we can ignore the small contribution from bottom baryons, which is about several percents of summed contribution.

The variables used for the characterization of the angular correlations between the two hadrons are the difference in azimuthal angles ($\Delta\phi$) and the difference in polar angles, usually expressed in terms of pseudorapidity ($\Delta\eta$), or the combined separation variable $\Delta R = \sqrt{\Delta\eta^2 + \Delta\phi^2}$.

In the PRA, the parton emission in hard subprocess and during evolution are not ordered in transverse momentum and we should use cuts on the transverse momenta of initial Reggeized gluons $\max(\sqrt{t_1}, \sqrt{t_2}) < p_{TL}$ to ensure, that the leading jet can be produced only in hard subprocess. After this, we consider LO contribution of subprocesses (1) requiring $\max(p_{Tb}, p_{T\bar{b}}) = p_{TL}$, and NLO subprocess (3) taking into account that $p_{Tg} = p_{TL}$ and with additional cut $\max(p_{Tb}, p_{T\bar{b}}) < p_{TL}$. If the ΔR distance between b and $\bar{b}$ is smaller than R_{k_T} they are considered as one jet and we compare the sum of their transverse momenta with the threshold value of leading jet transverse momentum p_{TL}. It is important for calculation in region of first two bins, where $0 \leq \Delta\phi \leq 1$, $\Delta R \leq 0.5$.

Experimental uncertainties, related to the shape of $\Delta\phi$ and ΔR distributions are relatively small ($\sim 20-30\%$). They are indicated by the error-bars in the Fig. 3. However, an additional uncertainty in the absolute normalization of the cross-sections $\simeq \pm 47\%$ is reported in the Ref. [3]. It is not included into the error bars of the experimental points in the Fig. 3, as well as in the plots presented in the experimental paper. Taking this large uncertainty into account, it is reasonable to consider the overall normalization of the cross-section to be a free parameter, which is also the case in MC simulations presented in the Ref. [3]. Following this route we find, that to obtain a very good agreement of the central curve of our predictions both with the shape and normalization of all experimental spectra we have to multiply all our predictions by the universal factor $\simeq 0.4$. Since the major part of the reported normalization uncertainty is due to the uncertainty in the efficiency of identification of B-mesons, our finding seems to support the assumption, that the B-meson reconstruction efficiency is largely independent from the kinematics of the leading jet, and in particular, from the value of p_{TL}. In the plots below, we show theoretical predictions multiplied by the above-mentioned factor, however our default result is also compatible with experiment, if one takes into account full experimental uncertainties and the scale-uncertainty of our predictions.

In the Fig. 3 we present the comparison of the predictions of our model with $\Delta\phi$ and ΔR spectra from the Ref. [3]. Apart from the above-mentioned overall normalization uncertainty, our model does not contain any free parameters. In both LO (1) and NLO (3) contributions we set the renormalization and factorization scales to be equal to the p_T of the leading jet: $\mu_R = \mu_F = \xi p_{TL}$, where $\xi = 1$ for the central lines of our predictions, and we vary $1/2 < \xi < 2$ to estimate the scale-uncertainty of our prediction, which is shown in the following figures by the gray band. All numerical calculations has been performed using the adaptive MC integration routines from the `CUBA` library [29], mostly using the `SUAVE` algorithm, but with the cross-checks against the results obtained by `VEGAS` and `DIVONNE` routines.

We have obtained very good agreement between our predictions and experimental data both for $\Delta\phi$ and ΔR spectra at all values of p_{TL}. We have found that LO and NLO contributions are dominating in the different ranges of $\Delta\phi$ and ΔR. The LO contribution dominates at the large $\Delta\phi, \Delta R \geq 2$, but NLO contribution is extremely important for regions of small and moderate values of $\Delta\phi$ and ΔR.

6 Conclusions

We have calculated the angular spectra between beauty and anti-beauty mesons in proton-proton collisions in the LO approximation of the parton Reggeization approach consistently merged with the NLO corrections from the emission of additional hard gluon. We have obtained good agreement between our predictions and data from the CMS Collaboration at the energy $\sqrt{S} = 7$ TeV for $B\bar{B}$ angular correlations within uncertainties and without free parameters. The NLO contribution from the emission of additional real gluon for $B\bar{B}$-pair production cross section in the k_T–factorization scheme has been calculated in the first time.

7 Acknowledgements

We are grateful to A. van Hameren for help in comparison between amplitudes obtained in the PRA and ones obtained using Monte-Carlo generator `Katie`.

This work was supported by the Ministry of Education and Science of Russia under Competitiveness Enhancement Program of Samara University for 2013-2020, project 3.5093.2017/8.9.

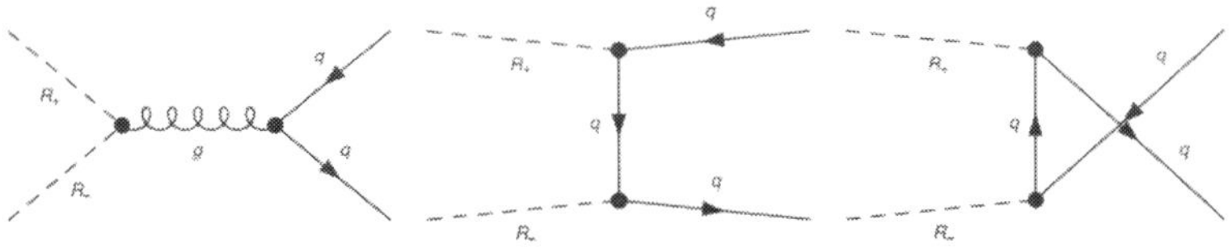

Figure 1. The Feynman diagrams of Lipatov's Effective Theory for the Reggeized amplitude of subprocess $R + R \to b + \bar{b}$.

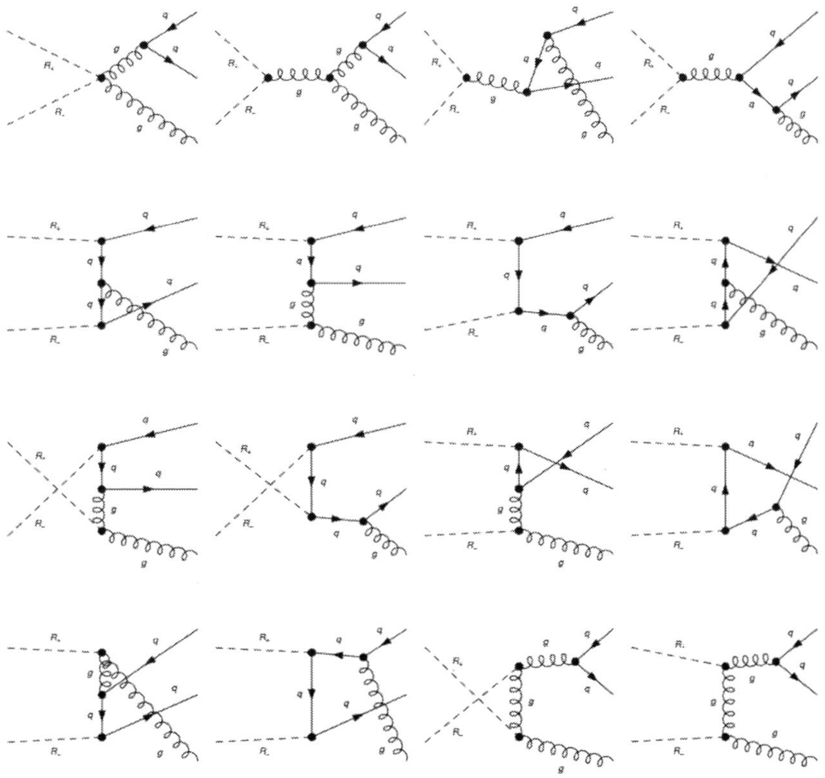

Figure 2. The Feynman diagrams of Lipatov's Effective Theory for the Reggeized amplitude of subprocess $R + R \to b + \bar{b} + g$.

DOI: 10.1051/epjconf/201715803010

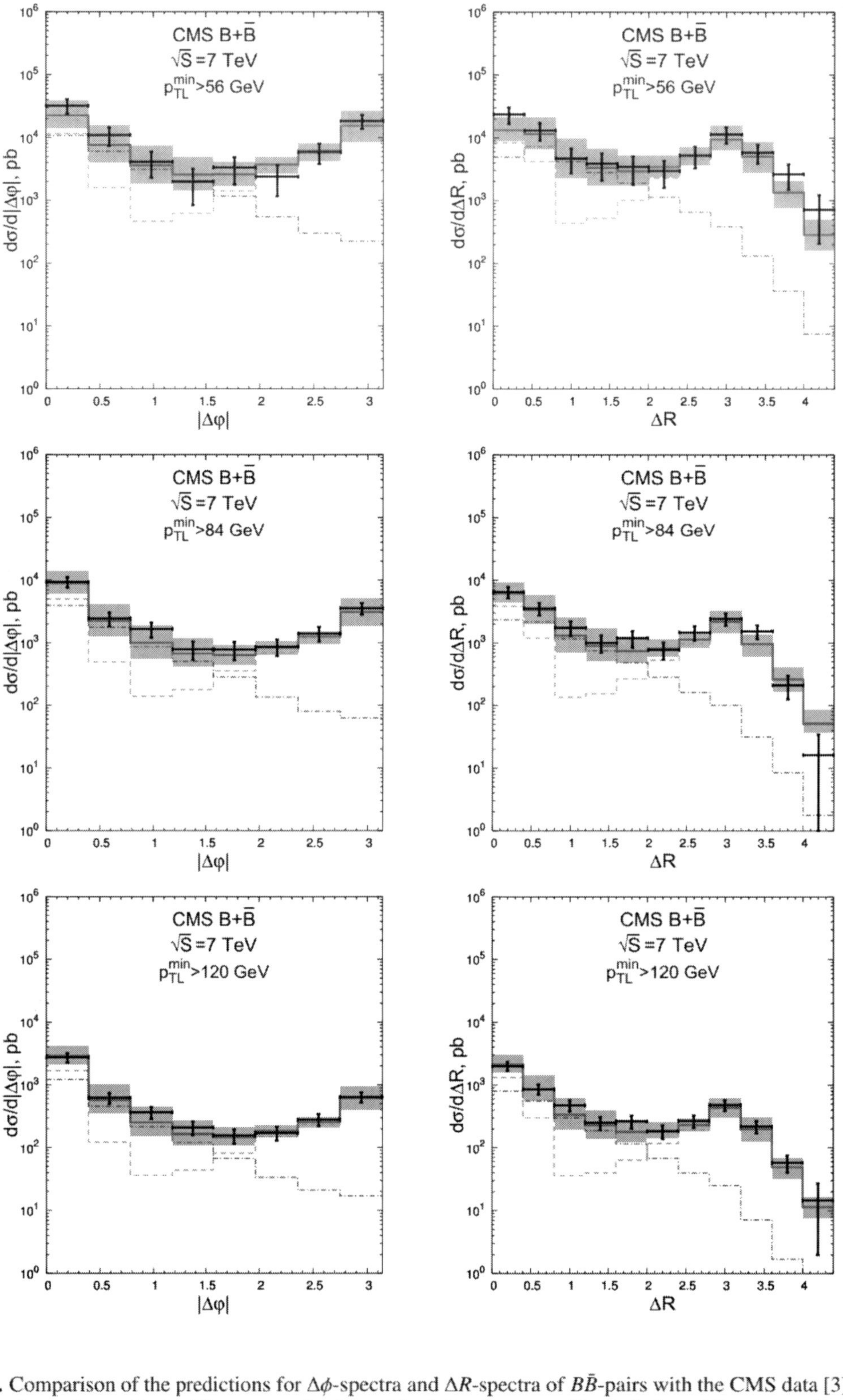

Figure 3. Comparison of the predictions for $\Delta\phi$-spectra and ΔR-spectra of $B\bar{B}$-pairs with the CMS data [3] at the different values of p_{TL}. Dashed line – contribution of the LO subprocess (1), dash-dotted line – contribution of the NLO subprocess (3), solid line – sum of LO and NLO contributions.

References

[1] CDF Collaboration, T. Aaltonen et al., CDF note 8939, (2007).
[2] ATLAS Collaboration, G. Aad et al., Eur. Phys. J. C **71**, 1846 (2011).
[3] CMS Collaboration, JHEP **03**, 136 (2011).
[4] M. A. Nefedov, V. A. Saleev, A. V. Shipilova, Phys. Rev. D **87**, 094030 (2013).
[5] B. A. Kniehl, A. V. Shipilova, and V. A. Saleev, Phys. Rev. D **81**, 094010 (2010).
[6] V. A. Saleev, A. V. Shipilova Phys. Rev. D **86**, 034032 (2012).
[7] M. Nefedov and V. Saleev, Phys. Rev. D **92**, 094033 (2015).
[8] R. Maciula, V. A. Saleev, A. V. Shipilova and A. Szczurek, Phys. Lett. B **758**, 458 (2016).
[9] A. Karpishkov, V. Saleev and A. Shipilova, Phys. Rev. D **94**, 114012 (2016).
[10] A. V. Karpishkov, M. A. Nefedov, V. A. Saleev and A. V. Shipilova, Int. J. Mod. Phys. A **30**, 1550023 (2015).
[11] M. A. Kimber, A. D. Martin, and M. G. Ryskin, Phys. Rev. D **63**, 114027 (2001).
[12] A. Karpishkov, M. Nefedov and V. Saleev, arXiv:1707.04068 [hep-ph], (2017).
[13] A.D. Martin, et. al., Eur. Phys. J. C **63**, 189 (2009).
[14] L. N. Lipatov, Nucl. Phys. B **452**, 369 (1995).
[15] L. N. Lipatov and M. I. Vyazovsky, Nucl. Phys. B **597**, 399 (2001).
[16] T. Hahn, Comput. Phys. Commun. **140**, 418 (2001).
[17] M. Nefedov and V. Saleev, Phys. Rev. D **92**, 094033 (2015).
[18] R. Mertig, M. Bohm, and A. Denner, Comput. Phys. Commun. **64**, 345 (1991).
[19] T. Hahn and M. Perez-Victoria, Comput. Phys. Commun. 118, 153 (1999).
[20] A. van Hameren, arXiv:1611.00680 [hep-ph], (2016).
[21] A. van Hameren and M. Serino, JHEP **07**, 010 (2015); K. Kutak, A. Hameren and M. Serino, JHEP **02**, 009 (2017).
[22] A. van Hameren, J. High Energy Phys. **07**, 138 (2014).
[23] B. A. Kniehl, G. Kramer, I. Schienbein, and H. Spiesberger, Phys. Rev. D **77**, 014011 (2008).
[24] ALEPH Collaboration, A. Heister et al., Phys. Lett. B **512**, 30 (2001).
[25] OPAL Collaboration, G. Abbiendi et al., Eur. Phys. J. C **29**, 463 (2003).
[26] J. Binnewies, B. A. Kniehl, and G. Kramer, Phys. Rev. D **58**, 034016 (1998).
[27] V. N. Gribov and L. N. Lipatov, Sov. J. Nucl. Phys. **15**, 438 (1972) [*Yad. Fiz.* **15**, 781 (1972)]; Yu. L. Dokshitzer, Sov. Phys. JETP **46**, 641 (1977) [Zh. Eksp. Teor. Fiz. **73**, 1216 (1977)]; G. Altarelli and G. Parisi, Nucl. Phys. **B126**, 298 (1977).
[28] M. Cacciari, G.P. Salam and G. Soyez, JHEP **04**, 063 (2008).
[29] T. Hahn, Comput. Phys. Commun. **168**, 78 (2005); J. Phys. Conf. Ser. **608**, 012066 (2015).

DOI: 10.1051/epjconf/201715803011

DIS structure functions in the NLO approximation of the Parton Reggeization Approach

Maxim Nefedov[1,⋆] and *Vladimir* Saleev[1,⋆⋆]

[1] *Samara National Research University, Moskovskoe Shosse, 34, 443086, Samara, Russia*

Abstract. The main ideas of the NLO calculations in Parton Reggeization Approach are illustrated on the example of the simplest NLO subprocess, which contributes to DIS: $\gamma^\star + R \to q + \bar{q}$. The double counting with the LO contribution $\gamma^\star + Q \to q$ is resolved. The problem of matching of the NLO results for single-scale observables in PRA on the corresponding NLO results in Collinear Parton Model is considered. In the developed framework, the usual NLO PDFs in the $\overline{MS}$-scheme can be consistently used as the collinear input for the NLO calculation in PRA.

1 Introduction

Multiscale and correlational observables, related with the hard processes in the hadronic or lepton-hadronic collisions, still present a challenge for the state-of-the-art calculational methods in perturbative QCD. By such observables we mean, for example, angular correlations of pairs of vector bosons, jets or reconstructed hadrons, or observables related with the polarization of virtual photon/Z-boson in the Drell-Yan process. All these observables are highly sencitive to QCD radiation over wide range of scales – from soft and collinear to the hard emissions with transverse momenta of the order of the hard scale. Parton showers (PS) are most accurate in modelling of the soft and collinear emissions, while fixed-order calculations in QCD demonstrate good convergence only for single-scale quantities. The aim of the Parton Reggeization Approach (PRA) is to improve our understanding of the transition between soft/collinear and hard regimes, using the information from the Multi-Regge limit of scattering amplitudes in QCD, and thus to reduce the gap between fixed-order and PS Monte-Carlo (MC) techniques.

In the present contribution we continue the development of formalism of the Next-to-Leading Order (NLO) calculations in PRA, concentrating on the classic example of *single-scale* observable: $F_2(x_B, Q^2)$-structure function of the electron-proton Deep Inelastic Scattering (DIS). For the single-scale observables, the NLO results of PRA should be consistent with the corresponding NLO results in the Collinear Parton Model (CPM). We exploit this relation to fix the formalism of NLO calculations in PRA in such a way, that the Parton Distribution Functions (PDFs) from NLO global fits, which are conventionally defined in the $\overline{MS}$-scheme, can be used as an input to the NLO calculations in PRA.

The discussion of loop corrections in PRA has been started by us in the Ref. [1]. In the present contribution we concentrate exclusively on tree-level corrections, and issues related with the double-counting and scheme-dependence of PDFs. The present paper has the following structure. In the

⋆e-mail: nefedovma@gmail.com
⋆⋆e-mail: saleev@samsu.ru

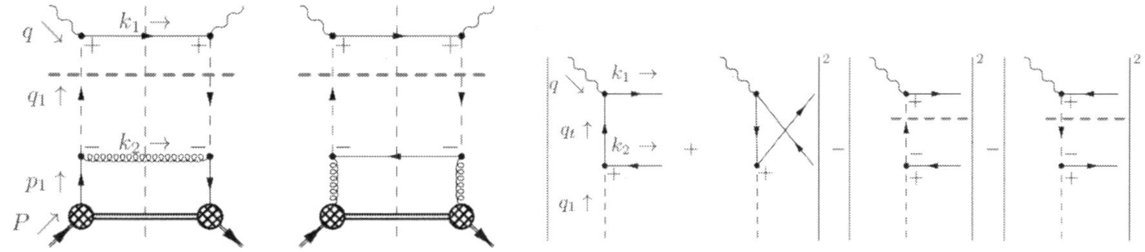

(a) (b)

Figure 1. Panel (a): contributions to the DIS structure functions in the LO of PRA. Panel (b): Contribution $\gamma^\star + R \to q + \bar{q}$ and double-counting subtraction terms. The horizontal thick dashed line denotes the mMRK-approximation, i.e. the "small" q_1^- and q_t^- light-cone components of t-channel momenta do not propagate to the upper part of the graph.

Sec. 2 the LO-formalism of PRA is reviewed and the main ideas beyond the NLO calcualtions in PRA are formulated. In the Sec. 3 the first $O(\alpha_s)$ contribution, which comes from the dependence of the LO hard-scattering coefficient (HSC) in PRA on the transverse momentum of initial-state parton is isolated. In the Sec. 4 the contribution of the NLO subprocess $\gamma^\star + R \to q + \bar{q}$ and the corresponding double-counting subtraction is introduced. Finally, in the Sec. 5 the formulated scheme of NLO calculations is tested numerically.

2 DIS at LO and NLO framework in PRA

We will consider the process of lepton-proton DIS, which is traditionally described as a process of desintegration of the proton with the four-momentum P_μ (we put $P^2 = 0$, since $E_p \gg M_p$) by the virtual photon with the four-momentum q_μ ($q^2 = -Q^2 < 0$). We will work in the center-of-mass frame of the proton and virtual photon, where the Sudakov(light-cone) components[1] of momentum of the photon are:

$$q^+ = -x_B P^+, \quad q^- = \frac{Q^2}{x_B P^+}, \quad \mathbf{q}_T = 0,$$

where $x_B = Q^2/(2qP)$ is the usual Bjorken variable. Inclusive DIS is fully described by the hadronic tensor, with the following standard parametrization in terms of structure functions [2, 3]:

$$W^{\mu\nu} = \left(-g_{\mu\nu} + \frac{q_\mu q_\nu}{q^2}\right) F_1(x_B, Q^2) + \left(P_\mu - \frac{q_\mu (Pq)}{q^2}\right)\left(P_\nu - \frac{q_\nu (Pq)}{q^2}\right) \frac{F_2(x_B, Q^2)}{(Pq)}, \tag{1}$$

where the function F_1 is related with the commonly used structure function $F_L(x_B, Q^2)$ as follows: $F_1 = (F_2 - F_L)/(2x_B)$.

In the CPM, the hadronic tensor is represented as a convolution of the partonic tensor of hard interaction of the virtual photon with the partonic target and the PDF $f_i(x, \mu^2)$. To derive the factorization formula of PRA for DIS, we will start with the specific NLO CPM hard subprocesses:

$$\gamma^\star(q) + q(p_1) \to q(k_1) + g(k_2), \tag{2}$$
$$\gamma^\star(q) + g(p_1) \to q(k_1) + \bar{q}(k_2). \tag{3}$$

[1]We use the following Sudakov decomposition for the four-momentum k: $k^\mu = (n_+^\mu k^- + n_-^\mu k^+)/2 + k_T^\mu$, $n_\pm^\mu = (n^\pm)^\mu = (1, 0, 0, \mp 1)^\mu$, $n_\pm k_T = 0$, $k^\pm = k_\pm = (n^\pm k)$, so that $k^2 = k_+ k_- - \mathbf{k}_T^2$, $P^+ \neq 0$ and $P^- = 0$.

Following the general scheme of LO PRA calculations, which is described in more detail in the Refs. [4, 5], we write down the modified MRK (mMRK) approximation for the partonic tensor of processes (2) and (3):

$$\hat{\omega}^{\mu\nu}_{q/g} = \frac{2g_s^2}{(-q_1^2)} \frac{P_{q(q/g)}(\tilde{z})}{\tilde{z}} \hat{\omega}^{\mu\nu}_{LO}, \tag{4}$$

where $q_1 = p_1 - k_2$, $g_s = \sqrt{4\pi\alpha_s}$ is the coupling contant of strong interaction, $\tilde{z} = q_1^+/p_1^+$, $P_{q(q/g)}(z)$ denotes the (non-regularized) DGLAP splitting functions: $P_{qq}(z) = C_F(1+z^2)/(1-z)$ or $P_{qg}(z) = T_R\left[z^2 + (1-z)^2\right]$ for process (2) and (3) respectively, $C_F = (N_c^2 - 1)/(2N_c)$, $T_R = 1/2$, $N_c = 3$, and $\hat{\omega}^{\mu\nu}_{LO}$ is the LO partonic tensor in PRA. The diagrammatic representation of Eq. (4) is given in the Fig. 1(a). The dashed lines (dashed lines with arrow) in the Fig. 1 denote Reggeized gluons(R) (quarks – Q). By definition of the mMRK-approximation, the q_1^- light-cone component of the momentum q_1 does not propagate to the γQq-scattering vertex, so that the LO partonic tensor is given by the hard subprocess $\gamma^\star(q) + Q(\tilde{q}_1) \to q(k_1)$, where $\tilde{q}_1^\mu = q_1^+ n_-^\mu/2 + q_{T1}^\mu$. The partonic tensor of this subprocess reads:

$$\hat{\omega}^{\mu\nu}_{LO} = \frac{e_q^2}{2} \mathrm{tr}\left[\hat{k}_1 \Gamma^\mu(\tilde{q}_1, q)\left(\frac{q_1^+}{2}\hat{n}_-\right)\Gamma^\nu(\tilde{q}_1, q)\right], \tag{5}$$

where $\Gamma^\mu(q_1, k) = \gamma^\mu + \hat{q}_1(n_-^\mu)/k^-$ is the gauge-invariant effective vertex of γQq-interaction (Fadin-Sherman vertex [6, 7]), and e_q is the electric charge of the quark in units of the electron charge.

Substituting the results (4) and (5) to the factorization formula of the CPM and projecting-out the F_2 structure function, one can rewrite the structure function in the k_T-*factorized form*:

$$F_{2q}^{(LO\ PRA)}(x_B, Q^2) = \int\limits_{x_B}^{1} \frac{dx_1}{x_1} \int\limits_0^\infty dt_1 \left[\tilde{\Phi}_q(x_1, t_1, Q^2) + \tilde{\Phi}_{\bar{q}}(x_1, t_1, Q^2)\right] \cdot C_2^{(0)}\left(\frac{x_B}{x_1}, \frac{t_1}{Q^2}\right), \tag{6}$$

where the HSC $C_2^{(0)}$ takes the form:

$$C_2^{(0)}\left(z, \frac{t_1}{Q^2}\right) = e_q^2 \cdot z\delta\left(\left(1 + \frac{t_1}{Q^2}\right)z - 1\right), \tag{7}$$

and the tree-level "unintegrated PDF" (unPDF) is defined as:

$$\tilde{\Phi}_i(x, t, \mu^2) = \frac{1}{t}\frac{\alpha_s}{2\pi} \sum_{j=q,\bar{q},g} \int\limits_x^1 d\tilde{z}\, P_{ij}(\tilde{z}) \cdot \frac{x}{\tilde{z}} f_j\left(\frac{x}{\tilde{z}}, \mu^2\right). \tag{8}$$

The F_L structure function is equal to zero in the LO of PRA, due to the properties of the Fadin-Sherman vertex.

Integration over t_1 in the Eq. (6) with "unPDF" (8) is logarithmically divergent at $t_1 \to 0$ and at $\tilde{z} \to 1$ (for the case of diagonal splitting fuctions). The latter divergence is regularized in PRA by the following cutoff on $\tilde{z}$-variable:

$$\tilde{z} < 1 - \Delta_{KMR}(t, \mu^2), \tag{9}$$

where the Kimber-Martin-Ryskin [8] cutoff function $\Delta_{KMR}(t, \mu^2) = \sqrt{t}/(\sqrt{\mu^2} + \sqrt{t})$ is given by the condition of ordering in rapidity between the particles produced in the hard process and the parton emitted on the last step of the ISR parton cascade in the kinematics of pp-collisions[2].

[2] In the case of DIS, the rapidity ordering leads to the weaker condition on $\tilde{z}$: $\tilde{z} < 1 - t_1/(Q^2 + 2t_1)$, but we will keep KMR condition for consistency.

Collinear divergence at $t_1 \to 0$ in (8) is regularized by the introduction of Sudakov formfactor, which resums the doubly-logarithmic corrections $\sim \log^2(t_1/\mu^2)$ in LLA, analogously to the treatment of this divergence in standard PS algorithms [9]. The exact expression of this formfactor is dictated by the DGLAP evolution for PDFs $f_i(x, \mu^2)$ and the following normalization condition for unPDFs:

$$\int_0^{\mu^2} dt \, \Phi_i(x, t, \mu^2) = x f_i(x, \mu^2), \tag{10}$$

which, as it will be explained in the Sec. 3, is equivelent to the normalization of single-scale observables on the corresponding LO CPM results up to NLO terms in α_s and power-supressed corrections. Condition (10) is obviously fulfilled by the following *derivative form* of unPDF:

$$\Phi_i(x, t, \mu^2) = \frac{\partial}{\partial t} \left[T_i(t, \mu^2, x) \cdot x f_i(x, t) \right], \tag{11}$$

where T_i is the Sudakov formfactor with the boundary conditions[3] $T_i(0, \mu^2, x) = 0$ and $T_i(\mu^2, \mu^2, x) = 1$. Eq. 11 can be shown to coincide exactly with the following *integral form* of unPDF:

$$\Phi_i(x, t, \mu^2) = \frac{T_i(t, \mu^2, x)}{t} \frac{\alpha_s(t)}{2\pi} \sum_{j=q,\bar{q},g} \int_x^1 dz \, P_{ij}(z) \cdot \frac{x}{z} f_j\left(\frac{x}{z}, t\right) \cdot \Theta(z, t, \mu^2), \tag{12}$$

where $\Theta(z, t, \mu^2) = \theta\left(1 - \Delta_{KMR}(t, \mu^2) - z\right)$, and the Sudakov formfactor is given by:

$$T_i(t, \mu^2, x) = \exp\left[-\int_t^{\mu^2} \frac{dt'}{t'} \frac{\alpha_s(t')}{2\pi} (\tau_i + \Delta\tau_i) \right], \quad \tau_i = \sum_j \int_0^1 dz \, \Theta(z, t', \mu^2) \cdot z P_{ji}(z),$$
$$\Delta\tau_i = \sum_j \int_0^1 dz \, \left(1 - \Theta(z, t', \mu^2)\right) \cdot \left[z P_{ji}(z) - \frac{\frac{x}{z} f_j(\frac{x}{z}, t')}{x f_i(x, t')} P_{ij}(z) \cdot \theta(z - x) \right]. \tag{13}$$

The integral form of unPDF (12) reproduces Eq. 8 when $t \sim \mu^2$, in this way it is connected with the derivation of factorization formula (6). Another important property of the Eqns. 12 – 13 is that they are equivalent to Eq. 11 in any order in α_s for the evolution kernels $P_{ij}(z)$ if the PDFs $f_i(x, \mu^2)$ are defined in the corresponding subtraction scheme, e. g. both in the $\overline{MS}$-scheme.

The LO framework of PRA has been formulated above, on the example of DIS process. Extension to the NLO would require the definition of subtraction scheme for double counting and *rapidity divergences* (see Ref. [1]) in the real and virtual NLO corrections to the HSC, as well as the definition of NLO unPDFs. As the main guidance to resolve all these issues we will employ the *physical normalization condition* (PNC), which simply states, that for a given set of collinear PDFs, defined in the $\overline{MS}$ scheme, the NLO PRA calculation of any single-scale observable with the scale Q^2 should reproduce the corresponding NLO CPM result up to corrections, which are formally of higher order in $\alpha_s(Q^2)$ or supressed by the negative power of Q^2. Validity of this condition in QCD is guaranteed by the standard collinear factorization theorems [3]. To formulate PNC, we coose the *structure-function of DIS on a specified quark flavor* – $F_{2q}(x_B, Q^2)$ as a benchmark single-scale gauge-invariant quantity,

[3]The first boundary condition is actually nonperturbative and introduced only to simplify the presentation of the analytic arguments. In our numerical calculations, the unPDF $t \cdot \Phi_i(x, t, \mu^2)$ is defined by quadratic polynomial for $t < Q_0^2 = 1$ GeV^2, which is adjusted to ensure continuity and the validity of Eq. 10. Other nonperturbative definitions are also possible.

which is well-defined in any order of QCD perturbation theory. In the present paper we will study the contributions to PNC, which are related to the NLO PRA subprocess:

$$\gamma^\star(q) + R(q_1) \to q(k_1) + \bar{q}(k_2), \tag{14}$$

and we will match them to the corresponding $O(\alpha_s)$ contribution of gluon PDF in the NLO of CPM for $F_{2q}(x_B, Q^2)$. The NLO CPM test function, which is to be compared with our NLO PRA results, reads:

$$\tilde{F}_{2q}^{(NLO\ CPM)}(x_B, Q^2) = (e_q^2 x_B)\left\{ f_q^{\overline{MS}}(x_B, Q^2) + f_{\bar{q}}^{\overline{MS}}(x_B, Q^2) + 2\left[C_{2g}^{\overline{MS}} \otimes f_g^{\overline{MS}} \right](x_B, Q^2) \right\}, \tag{15}$$

where the Mellin convolution is defined as: $[C \otimes f](x, Q^2) = \int\limits_x^1 \frac{dz}{z} C(x/z) f(z, Q^2)$ and the textbook [3] result for NLO HSC in $\overline{MS}$-scheme is: $C_{2g}^{\overline{MS}}(z) = T_R \frac{\alpha_s(Q^2)}{2\pi}\left[\left(z^2 + (1-z)^2\right)\log\left(\frac{1-z}{z}\right) - 8z^2 + 8z - 1 \right]$.

3 NLO contribution from k_T-factorization

In this section we will study the effects of nontrivial dependence of the LO HSC (7) on t_1. Substituting the derivative form of unPDF to the factorization formula, one obtains:

$$F_{2q}^{(LO\ PRA)}(x_B, Q^2) = \int\limits_{x_B}^{1} \frac{dx_1}{x_1} \int\limits_0^\infty dt_1 \frac{\partial}{\partial t_1}\left[T_q(t, Q^2, x_1)\cdot x_1 f_q(x_1, t) \right]\cdot C_2^{(0)}\left(\frac{x_B}{x_1}, \frac{t_1}{Q^2} \right) + \text{c.c.}, \tag{16}$$

where "c.c." denotes the antiquark term. Since the unPDFs decrease like a negative power of t_1 for $t_1 > Q^2$ and the HSC behaves as $1/t_1$ for $t_1 \gg Q^2$, one can conclude, that the contribution to the integral from $t_1 > Q^2$ is supressed by the negative power of Q^2, and therefore can be neglected, which is denoted by the ($\simeq$) sign in the following equations. With the help of integration by parts, the residual integral over the range $0 \le t_1 \le Q^2$ can be decomposed as follows:

$$F_{2q}^{(LO\ PRA)}(x_B, Q^2) \simeq e_q^2 x_B f_q(x_B, Q^2) + \Delta F_{2q}^{(f)}(x_B, Q^2) + \Delta F_{2q}^{(T)}(x_B, Q^2) + \text{c.c.}, \tag{17}$$

where the first term is the usual CPM expression, and the corrections, which originate from the dependence of $C_2^{(0)}$ on t_1, have the following form:

$$\Delta F_{2q}^{(f)}(x_B, Q^2) = \int\limits_0^1 dx_1 \int\limits_0^{Q^2} dt_1\, T_q(t_1, Q^2, x_1)\left[f_q(x_1, Q^2) - f_q(x_1, t_1) \right]\cdot \frac{\partial}{\partial t_1} C_2^{(0)}\left(\frac{x_B}{x_1}, \frac{t_1}{Q^2} \right), \tag{18}$$

$$\Delta F_{2q}^{(T)}(x_B, Q^2) = \int\limits_0^1 dx_1\, f_q(x_1, Q^2) \int\limits_0^{Q^2} dt_1 \left[1 - T_q(t_1, Q^2, x_1) \right]\cdot \frac{\partial}{\partial t_1} C_2^{(0)}\left(\frac{x_B}{x_1}, \frac{t_1}{Q^2} \right). \tag{19}$$

Expressions in the square brackets in Eqns. 18 and 19 are of $O(\alpha_s(Q^2))$, since both the running of PDFs with the scale and difference of Sudakov formfactor from unity are perturbative effects. Therefore both corrections $\Delta F_{2q}^{(f)}$ and $\Delta F_{2q}^{(T)}$ are of the NLO in α_s, and the condition (10) guarantees the physical normalization at LO, as it was stated in the Sec. 2.

To compute $\Delta F_{2q}^{(f)}$, one taylor-expands $f_q(x, Q^2) - f_q(x, t_1)$ in $\alpha_s(Q^2)\log(t_1/Q^2)$ using the DGLAP equations:

$$f_q(x, Q^2) - f_q(x, t_1) = -\frac{\alpha_s(Q^2)}{2\pi}\log\left(\frac{t_1}{Q^2} \right)\left\{ \left[P_{qq} \otimes f_q \right](x, Q^2) + \left[P_{qg} \otimes f_g \right](x, Q^2) \right\} + O(\alpha_s^2(Q^2)). \tag{20}$$

Since in the present paper we deal with the gluon-induced subprocess (14), we will be interested only in the contribution, which contains the gluon PDF. Contribution to $\Delta F_{2q}^{(f)}$ which contains quark PDFs, as well as the $\Delta F_{2q}^{(T)}$ should be considered together with the $\gamma^\star + Q \to q + g$ subprocess and loop correction to the $\gamma^\star + Q \to q$ subprocess respectively. To compute the $O(\alpha_s)$ contribution to $\Delta F_{2q}^{(f_g)}$, terms of higher order in α_s in (20) and the Sudakov formfactor in the Eq. 18 can be omitted altogether. Applying integration by parts to the integral over t_1 in (18), one obtains:

$$\Delta F_{2q}^{(f_g)} = (e_q^2 x_B)\frac{\alpha_s(Q^2)}{2\pi}\int\limits_0^1 \frac{dx_1}{x_1} f_g(x_1, Q^2)\left\{-P_{qg}\left(\frac{x_B}{x_1}\right)\log\left(\frac{Q^2}{\lambda^2}\right) + \int\limits_{\lambda^2}^{Q^2}\frac{dt_1}{t_1}\int\limits_{x_B}^{x_1} dz\, P_{qg}\left(\frac{z}{x_1}\right)\delta\left(z - x_B\frac{Q^2+t_1}{Q^2}\right)\right\}. \quad (21)$$

The spurious collinear divergence is regulated by the cutoff λ^2 in Eq. 21, and it cancels between two terms in curly brackets. Taking the integrals over z and t_1 one finds, that $\Delta F_{2q}^{(f_g)}(x_B, Q^2) = (e_q^2 x_B)\left[\Delta C_{qg}^{(k_T)} \otimes f_g\right](x_B, Q^2)$, where:

$$\Delta C_{qg}^{(k_T)}(z) = T_R \frac{\alpha_s(Q^2)}{2\pi}\left[\xi z\left((4+\xi)z - 2\right) + \left(z^2 + (1-z)^2\right)\log\xi\right], \quad (22)$$

and $\xi = \min(1, (1-z)/z)$. In such a way, the $O(\alpha_s)$ contributions, which arize from the t_1-dependence of LO HSC, can be calculated explicitly and taken into account in the PNC.

4 Gluon-induced NLO contribution

Contribution of the subprocess (14) to the HSC is depicted by the first term of diagrammatic expression in the Fig. 1(b) and explicitly depends on $\mathbf{q}_{T1}$. This contribution can be written down using Feynman Rules of Lipatov's theory, see e. g. Ref. [5] and references therein. It is convenient work in terms of kinematic variables:

$$z = \frac{x_B}{x_1},\ \tilde{z} = \frac{q_1^+ - k_2^+}{q_1^+}, \quad (23)$$

and express the transverse momentum $\mathbf{k}_{T1}$ in terms of $\mathbf{q}_{T1}$ and a new vector $\mathbf{k}_T$:

$$\mathbf{k}_{T1} = \mathbf{q}_{T1}\frac{\tilde{z} - z}{1 - z} + \mathbf{k}_T. \quad (24)$$

The square of $\mathbf{k}_T$ is fixed by the condition $k_1^2 = 0$:

$$\mathbf{k}_T^2 = \frac{(\tilde{z} - z)(1 - \tilde{z})}{1 - z}\left[\frac{Q^2}{z} - \frac{t_1}{1 - z}\right], \quad (25)$$

and we will integrate over azimuthal angle ψ between $\mathbf{k}_T$ and $\mathbf{q}_{T1}$. In this notation, contribution of the subprocess (14) to the F_{2q} structure function takes the form:

$$F_{2q}^{(1,g)}(x_B, Q^2) = (e_q^2 x_B)\int\limits_{x_B}^{1}\frac{dx_1}{x_1^2}\int\limits_0^{t_1^{\max}} dt_1\, \Phi_g(x_1, t_1, Q^2)\cdot C_2^{(1,g)}\left(\frac{x_B}{x_1}, \frac{t_1}{Q^2}\right), \quad (26)$$

where the upper limit $t_1^{\max} = Q^2(1-z)/z$ follows from the condition $(k_1 + k_2)^2 > 0$,

$$C_{2\ NS}^{(1,g)}\left(z, \frac{t_1}{Q^2}\right) = \frac{\alpha_s(Q^2)}{2\pi}\int\limits_z^1 \frac{d\tilde{z}}{(1 - z)}\int\limits_0^{2\pi}\frac{d\psi}{2\pi}\, C_2^{(1,g)}\left(z, \tilde{z}, t_1, \mathbf{k}_{T1}^2, Q^2\right), \quad (27)$$

where subscript NS denotes, that no double-counting subtractions has been performed yet, and

$$\begin{aligned} C_2^{(1,g)} &= \frac{T_R}{t_1\left(Q^2(\tilde{z}-1)(z-\tilde{z})+\mathbf{k}_{T1}^2 z^2\right)^2\left(Q^2\tilde{z}(z-\tilde{z})+\mathbf{k}_{T1}^2 z^2\right)^2} \\ &\times \left\{Q^{10}(\tilde{z}-1)^2(z-\tilde{z})^4+Q^8(\tilde{z}-1)(z-\tilde{z})^3(2\mathbf{k}_{T1}^2 z(z+3\tilde{z}-4)+t_1\tilde{z}(z-\tilde{z}))\right. \\ &+ Q^6\mathbf{k}_{T1}^2 z(z-\tilde{z})^2\left(\mathbf{k}_{T1}^2 z\left(z^2+2z(6\tilde{z}-7)+6(\tilde{z}-4)\tilde{z}+19\right)\right. \\ &+ t_1(z-\tilde{z})(z(2\tilde{z}-1)+2(\tilde{z}-1)\tilde{z})) \\ &+ Q^4\mathbf{k}_{T1}^4 z^2(z-\tilde{z})\left(6\mathbf{k}_{T1}^2(z-1)z(z+2\tilde{z}-3)+t_1(z-\tilde{z})\left(z^2+z(4\tilde{z}-2)+2(\tilde{z}-1)\tilde{z}\right)\right) \\ &+ \left.2Q^2\mathbf{k}_{T1}^6 z^4\left(3\mathbf{k}_{T1}^2(z-1)^2+t_1(z-\tilde{z})(z+2\tilde{z}-1)\right)+2\mathbf{k}_{T1}^8 t_1 z^6\right\}. \end{aligned}$$

Contribution of the subprocess (14) clearly contains double-counting with the contribution of the second diagram in the Fig. 1(a). To resolve this problem, we construct the corresponding mMRK double-counting subtraction (DCS) terms, which are depicted as the second and third terms of diagrammatic expression in the Fig. 1(b). In these DCS terms one should implement the same kinematic approximations, as in the LO diagrams of the Fig. 1(a), i. e. one have to neglect the q_t^- momentum component on the entrance to the γQq-scattering vertex. This operation can be consistently performed by the following substitution:

$$\begin{aligned} \delta(q_1^+ + q^+ - k_1^+ - k_2^+)\delta(q^- - k_1^- - k_2^-) &\to \int dq_t^+ dq_t^-\, \delta(q_1^+ - k_2^+ - q_t^+)\delta(k_2^- + q_t^-) \\ &\times\delta(q^+ + q_t^+ - k_1^+)\delta(q^- + \not{q_t^-} - k_1^-). \end{aligned}$$

In mMRK kinematics there is no need to introduce the vector $\mathbf{k}_T$, since the square of $\mathbf{k}_{T1}$ is expressed simply as:

$$\mathbf{k}_{\mathrm{T1(mMRK)}}^2 = Q^2\frac{\tilde{z}-z}{z}, \qquad (28)$$

and we will integrate over the angle ϕ between $\mathbf{k}_{T1}$ and $\mathbf{q}_{T1}$ instead of ψ. Also the factor $(1-z)$ in denominator of the Eq. 27 is replaced by $(1-\tilde{z})$.

Expression for the integrand of the t-channel DCS term reads:

$$\Delta C_2^{(t)} = \frac{P_{QR}(\tilde{z},\mathbf{q}_{T1},\mathbf{k}_{T1})}{\tilde{z}}\frac{Q^2}{(-t)}\cdot\Theta\left(\tilde{z},\mathbf{k}_{T1}^2,Q^2\right), \qquad (29)$$

where $t = (q_1-k_2)^2 = -\left[Q^2(\tilde{z}-z)+z\tilde{z}\cdot\mathbf{q}_{T1}(\mathbf{q}_{T1}+2\mathbf{k}_{T1})\right]/z(1-\tilde{z})$ and the Transverse-Momentum Dependent splitting function P_{QR} equals to:

$$P_{QR}(z,\mathbf{q}_{T1},\mathbf{q}_{T2}) = T_R\frac{z^2(\mathbf{q}_{T1}^2-2\mathbf{q}_{T1}\mathbf{q}_{T2})^2+2z(\mathbf{q}_{T1}\mathbf{q}_{T2})(\mathbf{q}_{T1}^2-2\mathbf{q}_{T1}\mathbf{q}_{T2})+\mathbf{q}_{T1}^2\mathbf{q}_{T2}^2}{\mathbf{q}_{T1}^2\left(\mathbf{q}_{T2}^2+z(\mathbf{q}_{T1}^2-2\mathbf{q}_{T1}\mathbf{q}_{T2})\right)}. \qquad (30)$$

Eq. 30 coincides with Eq. 13 of the Ref. [10] and reproduces the $P_{qg}(z)$ DGLAP splitting function in the collinear limit.

The u-channel DCS term can be obtained from the t-channel DCS term (29) via the substitution: $\tilde{z}\to 1-(\tilde{z}-z)$. Finally, subtracted HSC for the subprocess (14) takes the form:

$$C_2^{(1,g)}\left(z,\frac{t_1}{Q^2}\right) = \int\limits_z^1 d\tilde{z}\int\limits_0^{2\pi}\frac{d\psi}{2\pi}\left\{\frac{1}{(1-z)}C_2^{(1,g)} - \frac{1}{(1-\tilde{z})}\Delta C_2^{(t)}\Big|_{\phi=\psi}^{(\mathrm{mMRK})} - \frac{1}{(\tilde{z}-z)}\Delta C_2^{(u)}\Big|_{\phi=\psi}^{(\mathrm{mMRK})}\right\}. \qquad (31)$$

The analysis fully analogous to that of the Sec. 3 shows, that if the HSC (31) is finite in collinear limit $\mathbf{q}_{T1} \to 0$, then the $O(\alpha_s)$ contribution of the subprocess (14) is given by the convolution of the gluon PDF with the collinear limit of Eq. 31. Indeed, NLO HSC (31) is finite in this limit, despite the fact, that $C_2^{(1,g)}$ contains collinear singularities for $t \to 0$ and $u \to 0$ when $\mathbf{q}_{T1} = 0$. These singularities manifest themselves as $1/(\tilde{z}-z)$ and $1/(1-\tilde{z})$ poles in $C_2^{(1,g)}$, but $\Delta C_2^{(t)}$ and $\Delta C_2^{(u)}$ also contain this poles when $\mathbf{q}_{T1} = 0$, so that the integrand of (31) is free from any singularities even in the collinear limit.

It is convenient to study the collinear limit for Eq. 31, using dimensional regularization, since in this limit the unsubtracted NLO HSC in PRA coincides with the textbook result [3] for the unsubtracted NLO HSC of the subprocess (3):

$$C_{2\ NS}^{(1,g)}(z,0) = C_{2g\ NS}^{(NLO\ CPM)}(z) = 2\left[\frac{\bar{\alpha}_s(Q^2)}{2\pi}\left(-\frac{1}{\epsilon} - \log\frac{\mu^2}{Q^2}\right)P_{qg}(z) + C_{2g}^{\overline{MS}}(z) + O(\epsilon)\right], \tag{32}$$

where $\epsilon = (4-D)/2$, D is the dimension of space-time, $\bar{\alpha}_s = (4\pi/\mu^2)^\epsilon e^{-\epsilon\gamma_E}\alpha_s$ is the dimensionless coupling constant of QCD in the $\overline{MS}$-scheme and $\gamma_E \simeq 0.5772$ is the Euler-Mascheroni constant.

One can derive the collinear limit of t-channel DCS term in D dimensions, introducing the mMRK approximation for the partonic tensor into the standard calculation of the contribution of subprocess (3) in CPM. The result reads:

$$\Delta C_2^{(t)}(z,0) = \frac{\bar{\alpha}_s(Q^2)}{2\pi}\left(\frac{\mu^2 z}{Q^2}\right)^\epsilon \int\limits_z^1 \frac{d\tilde{z}}{(1-\epsilon)}\frac{P_{qg}(\tilde{z},\epsilon)}{(\tilde{z}-z)^{1+\epsilon}} \cdot \Theta\left(\tilde{z}, Q^2\frac{\tilde{z}-z}{z}, Q^2\right), \tag{33}$$

where the factor $(1-\epsilon)$ in the denomiantor stands for averaging over $D-2 = 2(1-\epsilon)$ polarizations of the on-shell gluon in the initial state, and $P_{qg}(z,\epsilon) = T_R\left[z^2 + (1-z)^2 - \epsilon\right]$ is the space-like DGLAP splitting function in D-dimensions, see e.g. Eq. 4.15 in Ref. [11]. The function Θ, which implements KMR kinematic constraint (9), leads to the upper bound on $\tilde{z} < \tilde{Z}(z)$, where $\tilde{Z}(z) < 1$ is the solution of a cubic equation:

$$\tilde{Z}(z) = \tfrac{1}{3}\left[d^{1/3} - 2z(3-2z)d^{-1/3} + 2z\right],\ d = \tfrac{z}{2}\left[4z(4z-9) + 3\left(9 + \sqrt{81 + 24z(2z-5)}\right)\right]. \tag{34}$$

Calculating integral (33) and expanding the result in ϵ one obtains

$$\Delta C_2^{(u)}(z,0) = \Delta C_2^{(t)}(z,0) = \frac{\bar{\alpha}_s(Q^2)}{2\pi}\left(-\frac{1}{\epsilon} - \log\frac{\mu^2}{Q^2}\right)P_{qg}(z) + \Delta C_{qg}(z) + O(\epsilon), \tag{35}$$

where

$$\Delta C_{qg}(z) = \frac{\alpha_s(Q^2)}{2\pi}\left\{\log\left(\frac{\tilde{Z}(z)-z}{z}\right)P_{qg}(z) + T_R\left[\tilde{Z}^2(z) - 2(1-z)\tilde{Z}(z) + z(4-5z)\right]\right\}. \tag{36}$$

Comparing Eqns. 35 and 32 one can see, that collinear divergences cancel explicitly, but the collinear limit of the subtracted HSC in PRA differs from the CPM result in $\overline{MS}$-scheme by the $\Delta C_{qg}(z)$ term:

$$C_2^{(1,g)}(z,0) = 2\left[C_{2g}^{\overline{MS}}(z) - \Delta C_{qg}(z)\right]. \tag{37}$$

5 Numerical results

The PRA analog of CPM test-function (15) reads:

$$\begin{aligned}\tilde{F}_{2q}^{(NLO\ PRA)}(x_B,Q^2) &= (x_B e_q^2)\Big\{f_q^{PRA}(x_B,Q^2) + f_{\bar{q}}^{PRA}(x_B,Q^2) + 2\left[\Delta C_{qg}^{(k_T)} \otimes f_g^{PRA}\right](x_B,Q^2) \\ &+ F_{2q}^{(1,g)}(x_B,Q^2)\Big\},\end{aligned} \tag{38}$$

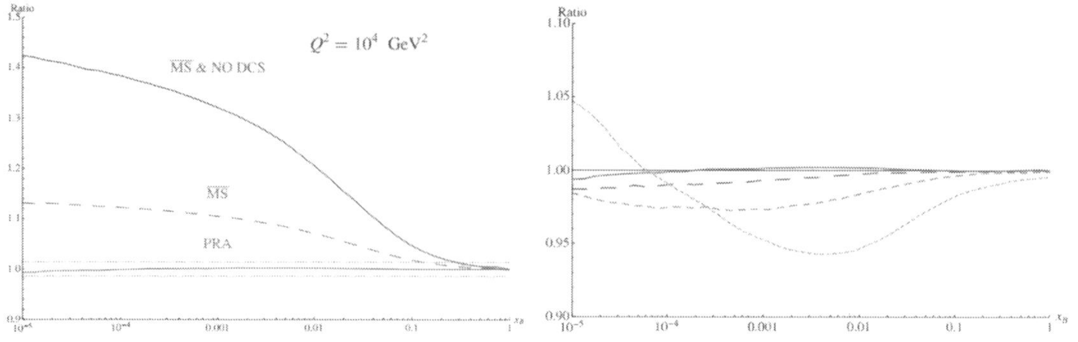

Figure 2. Numerical results for the ratio of $\tilde{F}_{2q}^{(NLO\ PRA)}(x_B, Q^2)/\tilde{F}_{2q}^{(NLO\ CPM)}(x_B, Q^2)$ as a function of x_B. **Left panel:** significance of different contributions to $\tilde{F}_{2q}^{(NLO\ PRA)}$ is illustrated. Curves from bottom to top: full PRA case; $\overline{MS} \to PRA$ scheme transformation is *turned off*; subtraction of double-counting is *turned off*. Dotted horizontal lines denote $1 \pm \alpha_s^2(Q^2)$ band. **Right panel:** scale-dependence of the ratio. Solid line – $Q^2 = 10^4$ GeV^2, dashed line – $Q^2 = 10^3$ GeV^2, short-dashed line – $Q^2 = 10^2$ GeV^2, dotted line – $Q^2 = 10$ GeV^2.

where f_i^{PRA} are PDFs in the *PRA scheme*, precise definition of which is given in Eq. 40 below. In the first line of Eq. 38, the $O(\alpha_s^0)$ and $O(\alpha_s)$ terms from the LO PRA contribution (6), relevant for our present analysis, are written. Term in the second line of Eq. 38 is given by Eq. 26, i. e. it takes into account the $\mathbf{q}_{T1}$-dependence of NLO HSC. Using the results of Sec. 4, one can collect all $O(\alpha_s)$ contributions in Eq. 38 together:

$$\begin{aligned}\tilde{F}_{2q}^{(NLO\ PRA)}(x_B, Q^2) &= (x_B e_q^2)\left\{f_q^{PRA}(x_B, Q^2) + f_{\bar{q}}^{PRA}(x_B, Q^2)\right. \\ &+ \left. 2\left[\left(C_{2g}^{\overline{MS}} - \Delta C_{qg} + \Delta C_{qg}^{(k_T)}\right) \otimes f_g^{PRA}\right](x_B, Q^2) + O(\alpha_s^2) + O\left((Q^2)^{-\gamma}\right)\right\}, \quad (39)\end{aligned}$$

where $\gamma > 0$. To ensure matching of the $O(\alpha_s^0)$ and $O(\alpha_s)$ terms in the Eq. 39 to the CPM result (15), PDFs in PRA scheme should be related with PDFs in $\overline{MS}$ scheme as follows:

$$f_g^{PRA}(x, \mu^2) = f_g^{\overline{MS}}(x, \mu^2),\ \ f_q^{PRA}(x, \mu^2) = f_q^{\overline{MS}}(x, \mu^2) + \left[\left(\Delta C_{qg} - \Delta C_{qg}^{(k_T)}\right) \otimes f_g^{\overline{MS}}\right](x, \mu^2). \quad (40)$$

In the complete NLO PRA calculation, all LO and NLO terms should be written in k_T-factorized form, *with the unPDFs normalized to the NLO PDFs in PRA scheme as prescribed by the Eq. 10.* To this end, the usual $\overline{MS}$ PDFs shold be transformed into PRA scheme by the transformation, similar to the transformation (40), and then the unPDFs can be constructed from them, using Eqns. 12 and 13. NLO splitting functions in this equations also recieve the $O(\alpha_s)$ corrections, due to the difference between $\overline{MS}$ and *PRA* schemes. These corrections are known, but the corresponding formulae are too lengthy to present them here. In contrast to the complete NLO calculation, in the present paper we have concentrated on the contribution of only one subprocess (14), and therefore, PDFs in PRA scheme appear explicitly in Eq. 38.

In the Fig. 2 the PRA test function (38) is compared numerically to the CPM test function (15). The NLO set of MSTW-2008 PDFs [12] has been used as the collinear input for this computation, togeher with the value of $\alpha_s(M_Z) = 0.1202$ from the PDF fit. As it is explained above, Eq. 38 should reproduce the CPM result up to $O(\alpha_s^2)$ terms and corrections supressed by the negative power of Q^2. Numerically, NLO PRA result (38) and NLO CPM result (15) indeed agree with the accuracy better than 1% for $Q^2 = 10^4$ GeV^2 (see the right panel of Fig. 2). Their agreement become less good with Q^2 decreasing, but even for the $Q^2 = 10$ GeV^2 the error does not exceed ±7%, which is compatible with the value of $\alpha_s^2(Q^2)$ at this scale.

In the left panel of Fig. 2, the role of different contributions in the PRA result is illustrated. Neglecting the scheme-transformation (40) one clearly misses some $O(\alpha_s) \sim O(10\%)$ correction, while turning off the DCS terms one overshoots the NLO CPM result by 40% in the small-x_B region.

In the present contribution, the problem of double-counting of real NLO corrections in PRA is adressed, and the procedure for the subtraction of this double counting, together with the definition of NLO unPDF in PRA is constructed, which resolves this problem and correctly takes into account the scheme-dependence of NLO PDFs in CPM. In such a way, significant part of the formalism of NLO calculations in PRA has been formulated.

Acknowledgements: This work was supported by the Ministry of Education and Science of Russia under Competitiveness Enhancement Program of Samara University for 2013-2020, project 3.5093.2017/8.9.

References

[1] M. Nefedov and V. Saleev, "Towards NLO calculations in the parton Reggeization approach," arXiv:1608.04201 [hep-ph].

[2] B. L. Ioffe, V. S. Fadin and L. N. Lipatov, *Quantum chromodynamics, perturbative and nonperturbative aspects* (Cambridge University Press, Cambridge, UK, 2011)

[3] J. C. Collins, *Foundations of preturbative QCD* (Cambridge University Press, Cambridge, UK, 2011)

[4] A. Karpishkov, M. Nefedov, V. Saleev and A. Shipilova, "Heavy quark production at the LHC in the Parton Reggeization Approach," arXiv:1702.05081 [hep-ph].

[5] A. Karpishkov, M. Nefedov and V. Saleev, "$B\bar{B}$ angular correlations at the LHC in parton Reggeization approach merged with higher-order matrix elements," arXiv:1707.04068 [hep-ph].

[6] V. S. Fadin and V. E. Sherman, JETP Lett. **23**, 599 (1976); JETP **45**, 861 (1977).

[7] A. V. Bogdan and V. S. Fadin, "A Proof of the reggeized form of amplitudes with quark exchanges," Nucl. Phys. B **740**, 36 (2006).

[8] M. A. Kimber, A. D. Martin and M. G. Ryskin, "Unintegrated parton distributions," Phys. Rev. D **63**, 114027 (2001); A. D. Martin, M. G. Ryskin and G. Watt, "NLO prescription for unintegrated parton distributions," Eur. Phys. J. C **66**, 163 (2010).

[9] A. Buckley *et al.*, "General-purpose event generators for LHC physics," Phys. Rept. **504**, 145 (2011).

[10] F. Hautmann, M. Hentschinski and H. Jung, "Forward Z-boson production and the unintegrated sea quark density," Nucl. Phys. B **865**, 54 (2012).

[11] S. Catani, S. Dittmaier, M. H. Seymour and Z. Trocsanyi, "The Dipole formalism for next-to-leading order QCD calculations with massive partons," Nucl. Phys. B **627**, 189 (2002).

[12] A. D. Martin, W. J. Stirling, R. S. Thorne and G. Watt, "Parton distributions for the LHC," Eur. Phys. J. C **63**, 189 (2009).

EPJ Web of Conferences **158**, 03012 (2017) DOI: 10.1051/epjconf/201715803012
QFTHEP 2017

Exotic meson decays in the environment with chiral imbalance

A. A. Andrianov[1,2,⋆] *V. A.* Andrianov[1,⋆⋆] *D.* Espriu[2,⋆⋆⋆] *A.V.* Iakubovich[1,⋆⋆⋆⋆] and *A.E.* Putilova[1,†]

[1] *Faculty of Physics, Saint Petersburg State University, Universitetskaya nab. 7/9, Saint Petersburg 199034, Russia, and*

[2] *Departament de Física Quàntica i Astrofísica and Institut de Ciéncies del Cosmos (ICCUB),Universitat de Barcelona, Martí i Franqués 1, 08028 Barcelona, Spain*

Abstract. An emergence of Local Parity Breaking (LPB) in central heavy-ion collisions (HIC) at high energies is discussed. LPB in the fireball can be produced by a difference between the number densities of right- and left-handed chiral fermions (Chiral Imbalance) which is implemented by a chiral (axial) chemical potential. The effective meson lagrangian induced by QCD is extended to the medium with Chiral Imbalance and the properties of light scalar and pseudoscalar mesons (π, a_0) are analyzed. It is shown that exotic decays of scalar mesons arise as a result of mixing of π and a_0 vacuum states in the presence of chiral imbalance. The pion electromagnetic formfactor obtains an unusual parity-odd supplement which generates a photon polarization asymmetry in pion polarizability. We hope that the above pointed indications of LPB can be identified in experiments on LHC, RHIC, CBM FAIR and NICA accelerators.

1 Topological charge, Chiral Imbalance and axial chemical potential

The behaviour of baryonic matter under extreme conditions has got recently a lot of interest [1, 2]. A medium generated in the heavy ion collisions may serve for detailed studies, both experimental and theoretical, of various phases of hadron matter. In this context new properties of QCD in the hot and dense environment are tested in current accelerator experiments on RHIC and LHC [3, 4].

In heavy ion collisions, in principle, there are two distinct experimental situations for peripheral and central collisions. In the first case the so-called Chiral Magnetic Effect (CME) can be detected, details see in [5] and also [6] for a review and additional references.

In the second case there are some experimental indications of an abnormal dilepton excess in the range of low invariant masses and rapidities and moderate values of the transverse momenta [7]–[11] (see the reviews in [12]), which can be thought of as a result of LPB in the medium (the details can be found in [13]). In particular, in heavy-ion collisions at high energies, with raising temperatures and baryon densities, metastable states can appear in the finite-volume fireball with a nontrivial topological

⋆e-mail: andrianov@icc.ub.edu
⋆⋆e-mail: v.andriano@rambler.ru
⋆⋆⋆e-mail: espriu@icc.ub.edu
⋆⋆⋆⋆e-mail: a.yakubovich@outlook.com
†e-mail: alyput@gmail.com

axial charge (due to fluctuations of gluonic fields) T_5, which is related to the gluon gauge field G_i,

$$T_5(t) = \frac{1}{8\pi^2}\int_{\mathrm{vol.}} d^3x\, \varepsilon_{jkl}\, \mathrm{Tr}\Big(G^j\partial^k G^l - i\frac{2}{3}G^jG^kG^l\Big), \quad j,k,l = 1,2,3, \tag{1}$$

where the integration is over the fireball volume. Its jump ΔT_5 can be associated with the space-time integral of the gauge-invariant Chern-Pontryagin density,

$$\begin{aligned}\Delta T_5 &= T_5(t_f) - T_5(0) = \frac{1}{16\pi^2}\int_0^{t_f} dt\int_{\mathrm{vol.}} d^3x\, \mathrm{Tr}(G^{\mu\nu}\widetilde{G}_{\mu\nu}) = \frac{1}{4\pi^2}\int_0^{t_f} dt\int_{\mathrm{vol.}} d^3x\, \partial^\mu K_\mu,\\ K_\mu &= \frac{1}{2}\epsilon_{\mu\nu\rho\sigma}\, \mathrm{Tr}\Big(G^\nu\partial^\rho G^\sigma - i\frac{2}{3}G^\nu G^\rho G^\sigma\Big).\end{aligned} \tag{2}$$

For the time being we adopt a static case and neglect a topological current flux through the fireball boundary during lifetime of the corresponding thermodynamic phase in a domain.

It is known that the divergence of isosinglet axial quark current $J_{5,\mu} = \bar{q}\gamma_\mu\gamma_5 q$ is locally constrained via the relation of partial conservation of axial current affected by the gluon anomaly,

$$\partial^\mu J_{5,\mu} - 2i\widehat{m}_q J_5 = \frac{N_f}{2\pi^2}\partial^\mu K_\mu; \quad J_5 = \bar{q}\gamma_5 q \tag{3}$$

This relation allows to find the relation of a nonzero topological charge with a non-trivial quark axial charge Q_5^q. Namely, integrating over a finite volume of fireball we come to the equality,

$$\begin{aligned}\frac{d}{dt}(Q_5^q - 2N_f T_5) &\simeq 2i\int_{\mathrm{vol.}} d^3x\, \widehat{m}_q\bar{q}\gamma_5 q,\\ Q_5^q &= \int_{\mathrm{vol.}} d^3x\, q^\dagger\gamma_5 q = \langle N_L - N_R\rangle,\end{aligned} \tag{4}$$

where $\langle N_L - N_R\rangle$ stands for the vacuum averaged difference between left and right chiral densities of baryon number. Therefrom it follows that in the chiral limit (when the masses of light quarks are taken zero) the axial quark charge is conserved in the presence of non-zero (metastable) topological charge. If for the lifetime of fireball and the size of hadron fireball of order $L = 5-10$ fm , the average topological charge is non-zero, $\langle\Delta T_5\rangle \neq 0$, then it may be associated with a topological chemical potential μ_T or an axial chemical potential μ_5 [14] for neglected masses of light u, d quarks. Thus we have,

$$\langle\Delta T_5\rangle \simeq \frac{1}{2N_f}\langle Q_5^q\rangle \iff \mu_5 \simeq \frac{1}{2N_f}\mu_T, \tag{5}$$

Thus adding to the QCD lagrangian the term $\Delta\mathcal{L}_{\mathrm{top}} = \mu_T\Delta T_5$ or $\Delta\mathcal{L}_q = \mu_5 Q_5^q$, we get the possibility of accounting for non-trivial fluctuations of topological charge (fluctons) in the nuclear (quark) fireball.

In the general, Lorentz covariant form the field dual to the fluctons is described by means of the classical pseudoscalar field $a(x)$, so that,

$$\Delta\mathcal{L}_a = \frac{N_f}{2\pi^2}K_\nu\partial^\nu a(x) \simeq \frac{1}{\pi^2}K_\nu b^\nu \iff b^\nu\bar{q}\gamma_\nu\gamma_5 q, \quad b_\nu \simeq \langle\partial_\nu a(x)\rangle \simeq \mathrm{const.} \tag{6}$$

Thus in a quasi-equilibrium situation the appearance of a nearly conserved chiral charge can be incorporated with the help of an axial (chiral) vector chemical potential b_ν. The appearance of a space vector part in b_ν can be associated with the non-equilibrium axial charge flow [15–17].

2 QCD-inspired effective meson Lagrangian for $SU_f(2)$ case

For the detection of Local Parity Breaking in the hadron fireball we implement the generalized sigma model with a background 4-vector of axial chemical potential [18], symmetric under $SU_L(N_f) \times SU_R(N_f)$, for u, d-quarks ($N_f = 2$),

$$L = \frac{1}{4}\mathrm{Tr}\,(D_\mu H (D^\mu H)^\dagger) + \frac{B}{2}\mathrm{Tr}\,[\,m(H + H^\dagger)] + \frac{M^2}{2}\mathrm{Tr}\,(HH^\dagger)$$
$$- \frac{\lambda_1}{2}\mathrm{Tr}\,\left[(HH^\dagger)^2\right] - \frac{\lambda_2}{4}[\,\mathrm{Tr}\,(HH^\dagger)]^2 + \frac{c}{2}(\det H + \det H^\dagger), \tag{7}$$

where $H = \xi\Sigma\xi$ is an operator for meson fields, m is an average mass of current quarks, M is a "tachyonic" mass generating the spontaneous breaking of chiral symmetry, $B, c, \lambda_1, \lambda_2$ are real constants.

The matrix Σ includes the singlet scalar meson σ, its vacuum average v and the isotriplet of scalar mesons a_0^0, a_0^-, a_0^+,

$$\Sigma = \begin{bmatrix} v + \sigma + a_0^0 & \sqrt{2}\,a_0^+ \\ \sqrt{2}\,a_0^- & v + \sigma - a_0^0 \end{bmatrix}. \tag{8}$$

The operator ξ realizes a nonlinear representation of the chiral group and is determined by the isotriplet π^0, π^-, π^+ of pseudoscalar mesons,

$$\xi = \exp\left(\frac{i\,\vec{\pi}\vec{\tau}}{2\,f_\pi}\right) \approx 1 + \frac{i\,\vec{\pi}\vec{\tau}}{2\,f_\pi} - \frac{(\vec{\pi}\vec{\tau})^2}{8\,f_\pi^2}, \tag{9}$$

$$\vec{\pi}\vec{\tau} = \begin{bmatrix} \pi^0 & \sqrt{2}\,\pi^+ \\ \sqrt{2}\,\pi^- & -\pi^0 \end{bmatrix}, \tag{10}$$

where $\vec{\tau}$ are Pauli matrix, f_π is a decay constant of π mesons.

The covariant derivative of H contains external gauge fields R_μ and L_μ,

$$D_\mu H = \partial_\mu H - iL_\mu H + iHR_\mu \tag{11}$$

These fields include the photon field A_μ and are supplemented also a background 4-vector of axial chemical potential $(b_\mu) = (b_0, \mathbf{b})$,

$$R_\mu = e\,Q_{em}A_\mu - b_\mu \cdot 1_{2\times2},$$
$$L_\mu = e\,Q_{em}A_\mu + b_\mu \cdot 1_{2\times2}, \tag{12}$$

where $Q_{em} = \frac{1}{2}\tau_3 + \frac{1}{6}1_{2\times2}$ is a matrix of electromagnetic charge.

The complete effective meson lagrangian has to include a P-odd part the Wess-Zumino-Witten effective action [19] which is modified in the chirally imbalanced medium. The relevant parts of WZW action read,

$$\Delta\mathcal{L}_{WZW} = -\frac{ie\,N_c b_\nu}{6\pi^2\,v^2}\,\epsilon^{\nu\sigma\lambda\rho}\,A_\rho(\partial_\sigma\pi^+)(\partial_\lambda\pi^-) - \frac{e^2 N_c}{24\,\pi^2 v}\,\epsilon^{\nu\sigma\lambda\rho}\,(\partial_\sigma A_\lambda)(\partial_\nu A_\rho)\pi^0 \tag{13}$$

3 Chiral (scalar) condensate depending on chiral chemical vector

The mass gap equation for the scalar condensate follows from (7),

$$-4(\lambda_1+\lambda_2)v^3+\left(2M^2+4b^2+2c\right)v+2Bm=0.$$

The general solution of this equation reads

$$\begin{aligned} v(b_\mu) &= \frac{1}{6^{\frac{2}{3}}(\lambda_1+\lambda_2)}\Bigg[9B\,m(\lambda_1+\lambda_2)^2 \\ &+ \sqrt{3(\lambda_1+\lambda_2)^3\left(27B^2m^2(\lambda_1+\lambda_2)-2\left(M^2+2b^2+c\right)^3\right)}\Bigg]^{1/3} \\ &+ \frac{M^2+2b^2+c}{6^{\frac{1}{3}}}\Bigg[9B\,m(\lambda_1+\lambda_2)^2 \\ &+ \sqrt{3(\lambda_1+\lambda_2)^3\left(27B^2m^2(\lambda_1+\lambda_2)-2\left(M^2+2b^2+c\right)^3\right)}\Bigg]^{-1/3}. \end{aligned}$$

There are different regions for chiral vector covariant under Lorentz transformations of fireball frame.

1. Chiral charge imbalance region when $b^2>0$, in the rest frame the chiral background $(b^\mu)=(\mu_5,0,0,0)$.
2. Chiral vector imbalance region with $b^2<0$, in the static frame the chiral background is taken along the beam axis $(b^\mu)=(0,0,0,b)$.
3. Transition region with $b^2=0$, in the light-cone background $(b^\mu)=(b,0,0,\pm b)$.

We stress that in a hot medium Lorentz invariance is broken by thermal bath and the physical effects depend on a particular set of components of (b^μ).

The plots for condensates display the enhancement of CSB and the restoration of chiral symmetry depending on the sign of $b_\mu b^\mu$.

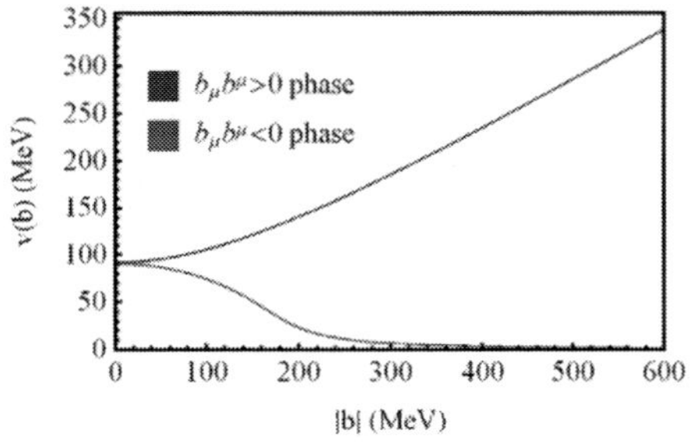

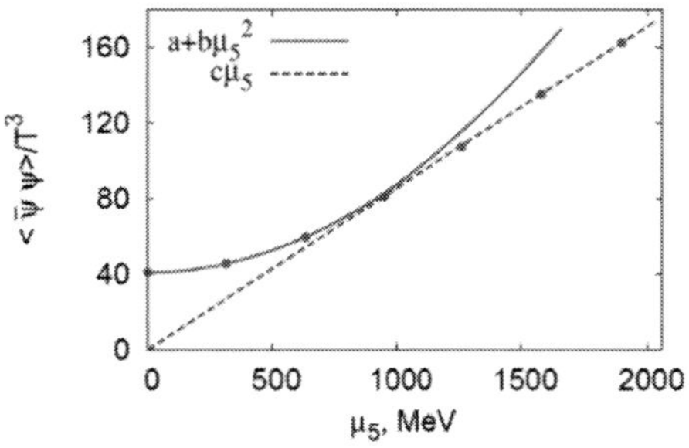

Figure 1. Left: the condensate in $b_\mu b^\mu>0$ and $b_\mu b^\mu<0$ region; right: the vacuum average $v(\mu_5)\sim\langle\bar{\psi}\psi\rangle/B$ from the lattice calculations [20].

Namely, in the chiral imbalance region with $b^2 > 0$ the increasing of chemical potentials causes the growth of chiral condensate, i.e. enhances Chiral Symmetry Breaking (CSB).
Instead, in the chiral vector imbalance region with $b^2 < 0$ the chiral condensate is decreasing with growing $|b^2|$ up to $|b^2| = \frac{1}{2}(M^2 + c)$. At this scale in the chiral limit $m \to 0$ the CSB parameter $v \to 0$ (see below) and spontaneous CSB is restored.
At last the CSB parameter v is insensitive to the light-cone b^μ.

For $b^2 > 0$ in the rest frame of vector $(b_\mu) = (\mu_5, 0, 0, 0)$ one can compare the predictions of our effective meson lagrangian (see Fig.1, the left plot) with the lattice estimations [20] (see Fig.1, the right plot) which clearly shows the enhancement of CSB.

However non-zero real space components of (b_μ) produce non-Hermitian purely imaginary vertices after euclidization of QCD which makes it difficult to compute on the lattice their contribution to the quark determinant.

4 Meson mass spectrum in different chiral imbalance regions

Introduce the definitions for meson state masses in the chiral imbalance environment. The mass matrix for scalar and pseudoscalar mesons on the diagonal takes the following values,

$$
\begin{aligned}
m_a^2 &= -2\left(M^2 - 2\,(3\lambda_1 + \lambda_2)\,v^2 - c + 2b^2\right),\\
m_\sigma^2 &= -2\left(M^2 - 6\,(\lambda_1 + \lambda_2)\,v^2 + c + 2b^2\right),\\
m_\pi^2 &= \frac{2B\,m}{v}.
\end{aligned}
\tag{14}
$$

After diagonalization we define distorted masses as m_{eff+} for the field $\tilde{a}$ and m_{eff-} for the field $\tilde{\pi}$

$$
m_{eff\pm}^2 = \frac{1}{2}\left(m_{\tilde{a}}^2 + m_{\tilde{\pi}}^2 \pm \sqrt{(m_{\tilde{a}}^2 - m_{\tilde{\pi}}^2)^2 + (8\,b^\mu\,k_\mu)^2}\right). \tag{15}
$$

4.1 Parameters of QCD-inspired generalized sigma model

Let us normalize the vacuum parameters of our model. We take $m_\pi = 139$ MeV, $m_a = 980$ MeV, $m_\sigma = 500$ MeV, $m = 5.5$ MeV, $\mu_5 = 0$, $M = 300$ MeV, $v = 92$ MeV. Then from the following Eqs.

$$
\begin{cases}
m_\sigma^2 = -2\left(M^2 - 6\,(\lambda_1 + \lambda_2)v^2 + c\right)\\
m_a^2 = -2\left(M^2 - 2\,(3\lambda_1 + \lambda_2)v^2 - c\right)\\
m_\pi^2 = \dfrac{2\,B\,m}{v}\\
v(\mu_5 = 0) = \sqrt{\dfrac{M^2 + c}{2(\lambda_1 + \lambda_2)}} + \dfrac{B\,m}{2(M^2 + c)}
\end{cases}, \tag{16}
$$

one can find λ_1, λ_2, c and b. We have for parameters $\lambda_1 = 1.64850 \times 10$, $\lambda_2 = -1.31313 \times 10$, $c = -4.46874 \times 10^4$ MeV^2, $B = 1.61594 \times 10^5$ MeV^2 .

 DOI: 10.1051/epjconf/201715803012

4.2 Diagonalization matrix

After diagonalization of mass matrix the states for a_0 and π mesons happen to be mixed. The eigenstates are defined as

$$\begin{aligned} a_0 &= C_{a\tilde{a}}\tilde{a} + C_{a\tilde{\pi}}\tilde{\pi}, \quad \pi = C_{\pi\tilde{a}}\tilde{a} + C_{\pi\tilde{\pi}}\tilde{\pi}, \\ C_{a\tilde{a}} &= iC_{\pi\tilde{\pi}} = C_+, \quad C_{a\tilde{\pi}} = -iC_{\pi\tilde{a}} = -C_-, \end{aligned} \tag{17}$$

with

$$C_{\pm} = \frac{1}{\sqrt{2}}\sqrt{1 \pm \frac{m_a^2 - m_\pi^2}{\sqrt{(m_a^2 - m_\pi^2)^2 + (8\, b^\mu k_\mu)^2}}}. \tag{18}$$

We use the notation $\tilde{a}, \tilde{\pi}$, indicating that these states tend to a_0, π when $b^\mu = 0$.

One can see (Fig.2, left plot) that for $b_\mu b^\mu > 0$ the growing chemical potential very quickly enforces the mixing between pions and isotriplet scalars so that distorted scalars happen to be involved into typical reactions of pion decay and pion formfactor. Yet with increasing of chemical potential the content of original pion states in distorted scalars and pseudoscalars diminishes.

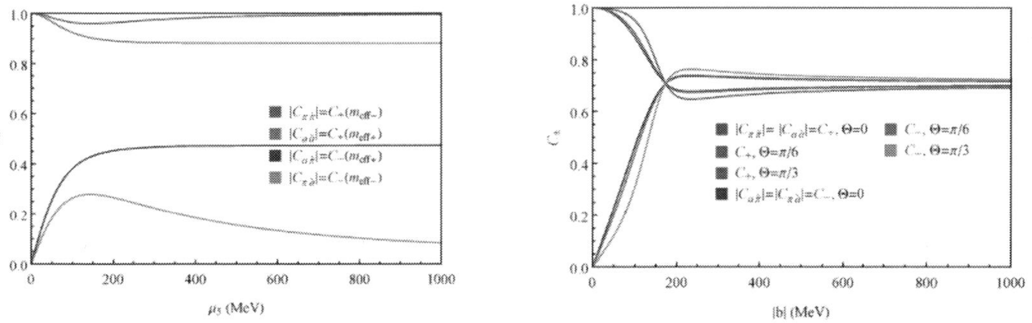

Figure 2. Left: mixing coefficients dependence on chemical potential μ_5 for $|\vec{k}| = 1000$ MeV in $b_\mu b^\mu > 0$ region; right: mixing coefficients dependence on chemical potential $|b|$ for $|\vec{k}| = 1000$ MeV in $b_\mu b^\mu < 0$ region .

In turn, for $b_\mu b^\mu < 0$ (Fig.2, right plot) the restoration of CS does not suppress the mixing of parity counterpartners and for increasing chemical potential tends to a nearly constant mixing coefficients with values in between 0 and 1.

4.3 Masses in CSB region with $b_\mu b^\mu > 0$

This is a region where CSB is enhancing and the $\tilde{a}_0$ and σ mesons become more heavy with growing chemical potential.

Meantime the $\tilde{\pi}$-meson effective mass is slowly decreasing at rest and decreasing faster in flight with $|\vec{k}| \neq 0$. One can see how the $\tilde{\pi}$-meson reaches the massless point and further on its mass squared becomes negative, "tachyonic" which however does not cause any causality problems. One can check that the group velocity of these states remains less than the light velocity.

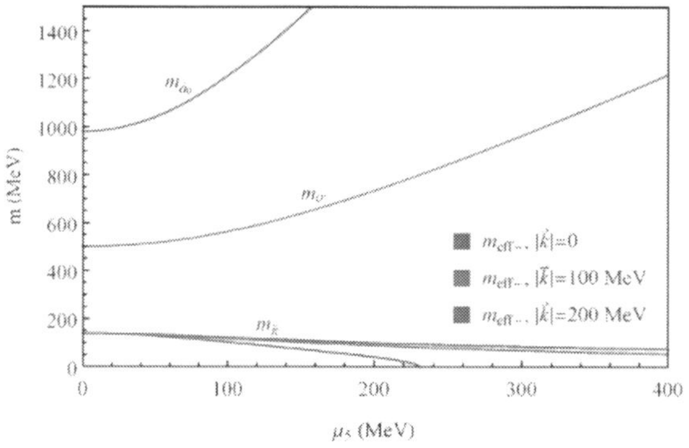

Figure 3. $\tilde{a}_0$-meson and $\tilde{\pi}$-meson effective mass dependence as well as σ-meson mass dependence on chemical potential $b = \mu_5$, for different values of $|\vec{k}|$ in $b_\mu b^\mu > 0$ region.

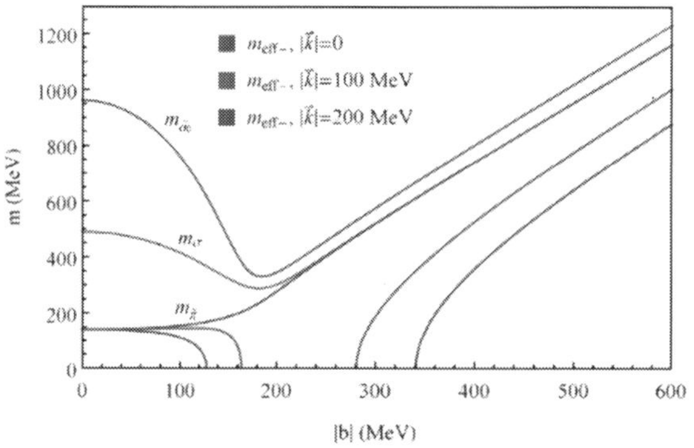

Figure 4. $\tilde{a}_0$-meson and σ-meson mass dependence on chemical potential b, $\tilde{\pi}$-meson effective mass dependence on b for different values of $|\vec{k}|$ in $b_\mu b^\mu < 0$ phase, the angle between $\vec{k}$ and $\vec{b}$ is $\theta = 0$

4.4 Masses in CSR region with $b_\mu b^\mu < 0$

In this region in the rest frame one can see clearly the CS restoration with merging masses of all scalars and pseudoscalars.

But in flight the behavior of pion masses is more peculiar. First pion masses vanish and then mass squared become negative. Next they reappear with positive mass squared and slowly approach to scalar masses in asymptotics. Thus the in-flight effect of chiral vector imbalance on pion spectrum deviates strongly from naive expectations.

5 The decays $\tilde{a}_0^\pm \to \tilde{\pi}^\pm\gamma$, $\tilde{\pi}^0 \to \gamma\gamma$, $\tilde{a}_0^0 \to \gamma\gamma$

After mixing $\pi, a_0^0 \to \tilde{\pi}, \tilde{a}_0^0$ the decay $\tilde{a}_0^\pm \to \tilde{\pi}^\pm\gamma$ arises which breaks space parity and therefore is forbidden in vacuum.

with $q = |\vec{q}|$ being space momenta of scalars.

As well in the decay process $\pi^0 \to \gamma\gamma$ an adjacent resonance decay $\tilde{a}_0^0 \to \gamma\gamma$ emerges after mixing. From the plots for mixing coefficients we conclude that these processes are comparable in decay widths.

DOI: 10.1051/epjconf/201715803012

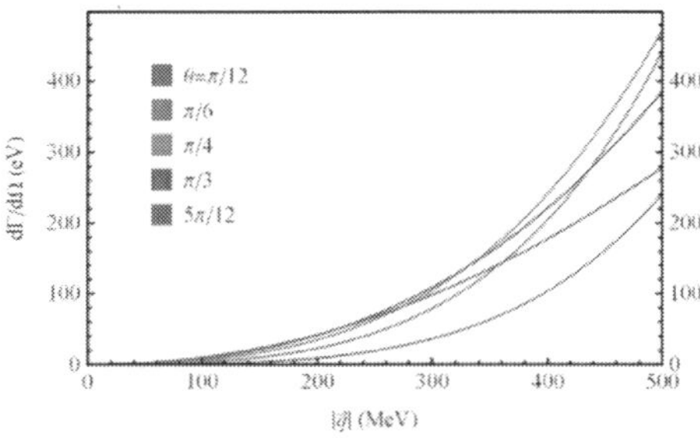

Figure 5. Decay width $a^{\pm} \to \pi^{\pm}\gamma$, $\mu_5 = 100$ MeV

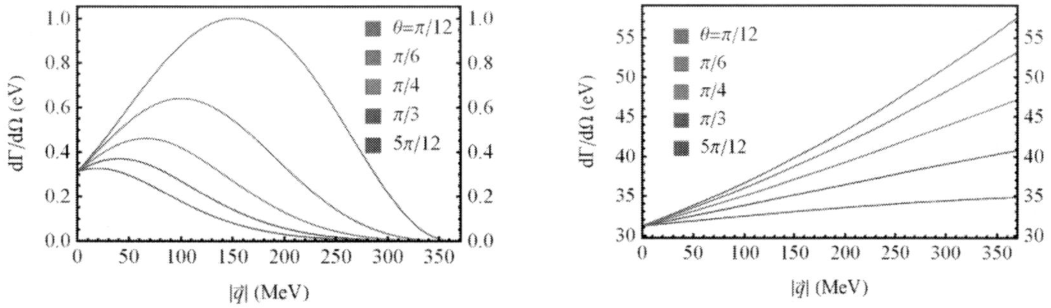

Figure 6. Decay widths; left: $\tilde{\pi}^0 \to \gamma\gamma$, right: $\tilde{a}_0^0 \to \gamma\gamma$, $\mu_5 = 100$ MeV

For pseudoscalars and scalars in flight the speeds of decays are considerably increasing. The effects are opposite to the Lorentz retardation.

6 Conclusions and outlook

In this work we described a possibility of local parity breaking (LPB) emerging in a dense hot baryon matter (hadron fireball) in heavy-ion collisions at high energies. The phenomenology of LPB in a fireball is based on introducing a topological (axial) charge and a topological (chiral) chemical potential. Topological charge fluctuations transmit their influence to hadronic physics via an axial chemical potential. We suggested QCD-motivated sigma model for the description of isotriplet pseudoscalar and isoscalar and isotriplet scalar mesons in the body of a fireball. We conclude:

- Strong CP violation is quite a challenging possibility to be revealed in heavy-ion collisions both at high energy densities (temperatures) and being triggered by large baryon densities.
- However the existing theoretical arguments for arising CP violation in FINITE volumes are not well sufficient to calculate the production rate of CP violating nuclear processes.
- There are two ways to improve the discovery potential: firstly, to elaborate the recipes for experimentalists to detect peculiar effects generated in CP-odd background, secondly, to measure the production of the mass states without a firm CP parity. In both cases the chiral chemical potential method helps a lot in predictions.
- In addition we already suggested [14] the vector meson dominance model with chiral imbalance: the spectrum of massive vector mesons splits into three components with different polarizations and with different effective masses that can be used to detect local parity breaking. The proposed

schemes for revealing local parity breaking helps to (partially) explain qualitatively and quantitatively the anomalous yield of dilepton pairs in the CERES, PHENIX, STAR, NA60, and ALICE experiments. Accordingly the identification of its physical origin might serve as a base for a deeper understanding of QCD properties in a medium under extreme conditions. Experimental collaborations should definitely check this possibility.

- Recently an interesting proposal was given in [21] to detect the LPB by measuring photon polarization asymmetry in the process $\pi^{\pm}\gamma \to \pi^{\pm}\gamma$. We extend this proposal indicating the resonance enhancement at energies comparable with the mass of $\tilde{a}_0^{\pm}$ scalars.

Acknowledgements

It is a pleasure to thank the organizers of the XXIII International Workshop on High Energy Physics and Quantum Field Theory, QFTHEP 2017 for a fruitful meeting and an excellent atmosphere. This work has been supported through grants FPA2013-46570, 2014-SGR-104 and Consolider CPAN. Funding was also partially provided by the Spanish MINECO under project MDM-2014-0369 of ICCUB (Unidad de Excelencia 'Maria de Maeztu'). A.A. and V.A. were supported by Grant RFBR projects 16-02-00348 and also got a financial support of SPbSU, in particular, by projects 11.41.417.2017, 11.41.415.2017, 11.42.697.2017 and 11.42.698.2017.

References

[1] P.Jacobs, D. Kharzeev, B. Muller, J. Nagle, K. Rajagopal and S. Vigdor, arXiv:0705.1930 [nucl-ex].

[2] J. -P. Blaizot, F. Gelis, J. -F. Liao, L. McLerran and R. Venugopalan, Nucl. Phys. A **873**, 68 (2012) [arXiv:1107.5296 [hep-ph]].

[3] A. Andronic, D. Blaschke, P. Braun-Munzinger, J. Cleymans, K. Fukushima, L. D. McLerran, H. Oeschler and R. D. Pisarski, Nucl. Phys. A **837**, 65 (2010) [arXiv:0911.4806 [hep-ph]].

[4] J. -P. Blaizot, Nucl. Phys. A **834**, 515 C (2010) [arXiv:0911.5059 [hep-ph]].

[5] D. Kharzeev, R. D. Pisarski and M. H. G. Tytgat, Phys. Rev. Lett. **81**, 512 (1998). D. E. Kharzeev, L. D. McLerran and H. J. Warringa, Nucl. Phys. A**803**, 227 (2008). K. Buckley, T. Fugleberg, A. Zhitnitsky, Phys. Rev. Lett., **84**,4814 (2000).

[6] N. Ajitanand, S. Esumi, R. Lacey, P. Collaboration, et al., P-and CP-odd effects in hot and dense matter, in: Proc. of the RBRC Workshops, Vol. **96**, 2010.

[7] P. Wurn et al. (CERES Collab.), Nucl. Phys. A, **5901-2**, 103-116 (1995); CERES Collaboration (Agakichiev, G. et al.) Phys. Rev. Lett. **75** , 1272 (1995); Phys. Lett. B **422**, 405(1998); Eur. Phys. J. C **41**, 475(2005).

[8] R. Arnaldi et al. (NA60 Collab.), Phys. Rev. Lett., **96** , 162302 (2006).

[9] A. Adare et al. (PHENIX Collab.), Phys. Rev. C, **81**, 034911 (2010).

[10] G. Agakichiev et al. (HADES Collab.), Phys. Rev. Lett., **98**, 052302 (2007). Phys. Lett. B, **663**, 43-48 (2008).

[11] K. O. Lapidus, V. M. Emel'yanov, Phys. Part. Nucl., **40**, 29 (2009).

[12] I. Tserruya, Electromagnetic Probes, arXiv: 0903.0415; G. E. Brown, M. Rho, Phys. Rev. Lett., **66**,2720-2723 (1991).

[13] A. A. Andrianov, V. A. Andrianov, D. Espriu, X. Planells, Theor.Math.Phys., **170**, 17 (2012); A. A. Andrianov, V. A. Andrianov,Theor.Math.Phys., **185**, 1370 (2015).

[14] A. A. Andrianov, V. A. Andrianov, D. Espriu and X. Planells, Phys. Lett. B **710** 230 (2012); Proc.Sci., QFTHEP, 025 (2013); Phys. Rev. D, **90**,034024 (2014).
[15] D. Kharzeev, Y. Kikuchi and R. Meyer, arXiv:1610.08986 [cond-mat.mes-hall].
[16] A. Sorin and O. Teryaev, Nucl. Part. Phys. Proc. **273-275**, 2587 (2016). doi:10.1016/j.nuclphysbps.2015.09.468 .
[17] K. Landsteiner, Acta Phys. Polon. B **47**, 2617 (2016) doi:10.5506/APhysPolB.47.2617 [arXiv:1610.04413 [hep-th]].
[18] A. A. Andrianov, D. Espriu and X. Planells, Eur. Phys. J. C **73**:2294 (2013).
[19] J. Wess and B. Zumino, Phys. Lett. **37**B, 95 (1971); E. Witten, Nucl. Phys. B**223**, 422 (1983); A. A. Andrianov, V. A. Andrianov, V. Yu. Novozhilov, and Yu. V. Novozhilov, Theor. Math. Phys.,**70**, 43 (1987).
[20] V. V. Braguta et al., Phys. Rev. D **93**, 034509 (2016); V. V. Braguta et al. AIP Conf. Proc. **1701**, 060002 (2016); V. V. Braguta, A. Yu. Kotov, Phys. Rev. D**93**, 105025 (2016).
[21] Kawaguchi M., Harada M., Matsuzaki S., Ouyang R.M., Phys. Rev. C **95**, 065204 (2017) .

EPJ Web of Conferences **158**, 03013 (2017) DOI: 10.1051/epjconf/201715803013
QFTHEP 2017

Observation and investigation of πK atoms

Valeriy Yazkov[1,*] on behalf of the DIRAC collaboration

[1] *D. V. Skobeltsyn Institute of Nuclear Physics, M. V. Lomonosov Moscow State University, Leninskie Gory 1, 119991 Moscow, Russia*

Abstract. There is accurate relation between lifetime of atoms, consisting of pion and kaon, and a difference of S-wave pion kaon scattering length with isospin 1/2 and 3/2: $|a_{1/2} - a_{3/2}|$. This difference is predicted by low energy QCD. Experiment DIRAC at CERN PS detects 349 ± 62 pairs from $K^+\pi^-$ and π^+K^- atoms and make observation of exotic atoms consist of pion and kaons and provides way for measurement of pion-kaon scattering length difference and check of theoretical predictions.

1 Introduction

In 2007, the DIRAC collaboration enlarged the scope of the dimesonic atom investigation by starting to search for the strange pion-kaon (πK) atom. In addition to the ongoing study of $\pi\pi$ atoms, the DIRAC experiment at the CERN proton synchrotron (CERN PS) also collected data, containing a kaon beside a pion in the final state. Using all the data since 2007 and optimizing data handling and analysis, the observation of the πK atom could be achieved for the first time with a significance of more than 5 standard deviations [1].

πK-atom ($A_{\pi K}$) is a hydrogen-like atom consisting of K^+ (K^-) and π^- (π^+) mesons with a Bohr radius of $a_B = 249$ fm, Bohr momentum of $p_B \simeq 0.8$ MeV/c and a ground state Coulomb binding energy of $E_B = 2.9$ keV .

The πK-atom lifetime (ground state 1S), $\tau = \frac{1}{\Gamma}$ is dominated by the annihilation process into $\pi^0 K^0$. There is a relation between the width of $A_{\pi K}$ decay and S-wave πK scattering lengths for isospin 1/2 and 3/2 [2]:

$$\Gamma_{1S,\pi^0 K^0} = \frac{1}{\tau_{1S}} = 8\alpha^3\mu^2 p^*(a_0^-)^2(1+\delta_K). \tag{1}$$

Here S-wave isospin-odd πK scattering length $a_0^- = \frac{1}{3}(a_{1/2} - a_{3/2})$ is defined in pure QCD for the quark masses $m_u = m_d$, α is the fine structure constant, μ is the reduced mass of the $\pi^\pm K^\mp$ system, p^* is the outgoing π^0 momentum in the πK atom system, and δ_K accounts for corrections, due to isospin breaking, at order α and quark mass difference $(m_u - m_d)$.

Chiral Perturbation Theory (ChPT) describes QCD processes at low energies. ChPT in 1-loop approximation predicts S-wave pion-kaon scattering lengths with isospin 1/2 ($a_0^{1/2}$) and 3/2 ($a_0^{3/2}$) to be [3, 4]:

$$a_0^{1/2} = 0.19 \pm 0.2,\ a_0^{3/2} = -0.05 \pm 0.02,\ a_0^{1/2} - a_0^{3/2} = 0.23 \pm 0.01\,. \tag{2}$$

*e-mail: valeri.yazkov@cern.ch

ChPT with $L^{(2)}$, $L^{(4)}$, $L^{(6)}$ in 2-loop approximation predicts S-wave scattering length difference [5]:

$$a_0^{1/2} - a_0^{3/2} = 0.267\,. \tag{3}$$

Scattering length difference also has been predicted, using Roy-Steiner equations [6]:

$$a_0^{1/2} - a_0^{3/2} = 0.269 \pm 0.015\,. \tag{4}$$

In the framework of lattice QCD, predictions for πK scattering length and their combination a_0^- have been obtained in [7] (Eq. 5), [8] (Eq. 6), [9] (Eq. 7), [10] (Eq. 7):

$$a_0^{1/2} = 0.1725^{+0.0026}_{-0.0157},\ a_0^{3/2} = -0.0574^{+0.0029}_{-0.0060}, \tag{5}$$

$$a_0^{1/2} = 0.183 \pm 0.039,\ a_0^{3/2} = -0.0602 \pm 0.0040, \tag{6}$$

$$a_0^- = \frac{1}{3}(a_0^{1/2} - a_0^{3/2}) = 0.0811 \pm 0.0143\,. \tag{7}$$

$$a_0^- = \frac{1}{3}(a_0^{1/2} - a_0^{3/2}) = 0.0745 \pm 0.0020\,. \tag{8}$$

Prediction of scattering length difference in Eq. (4) provides an estimation of lifetime of $A_{\pi K}$ in ground state:

$$\tau = (3.5 \pm 0.4) \times 10^{-15}\,. \tag{9}$$

There is differences in predictions of $a_0^{1/2} - a_0^{3/2}$ value, obtained in [5, 6] from one side and [7, 10] from another side. It is needed to test predictions experimentally and accuracy of experimental measurement is to be at the level 5%.

The measurement of the S-wave πK scattering lengths would test our understanding of the chiral $SU(3)_L \times SU(3)_R$ symmetry breaking of QCD (u, d and s quarks), while the measurement of $\pi\pi$ scattering lengths checks only the $SU(2)_L \times SU(2)_R$ symmetry breaking (u, d quarks). This is the principal difference between $\pi\pi$ and πK scattering.

Experimental data on the πK low-energy phases are absent. The only experimental pion-kaon scattering length measurement has been done with estimation of πK atom lifetime, using data collected in 2008-2010 with Nickel (Ni) target [11]:

$$|a_0^-|M_\pi = 0.107^{+0.093}_{-0.035}\,. \tag{10}$$

In the present publication experimental data for πK atom investigation, collected in 2008-2010 and data collected in 2007 with Platinum (Pt) target [12] have been analysed with improved description of the setup response.

2 Method of πK atom observation and investigation

A method of investigation for $\pi^+\pi^-$, πK and other atoms, consisted of two oppositely charged mesons, has been proposed in [13]. Pairs of K^+ (K^-) and π^- (π^+) mesons are producing in proton-target interactions. Pairs, which are generated from fragmentation and strong decays ("short-lived" sources), are affected by Coulomb interaction in the final state. Some of them form Coulomb bound states — atoms, other are generated as free pairs ("Coulomb pairs"). Number of produced atoms (N_A) is proportional to a number of "Coulomb pairs" (N_C) with low relative momentum Q in a pair C.M. system: $N_A = K \cdot N_C$. The coefficient K is calculated with an accuracy better than 1% [14].

DOI: 10.1051/epjconf/201715803013

If at least one meson is generated from long-lived sources (electromagnetically or weakly decaying mesons or baryons: $\eta, \eta', K_s^0, \ldots$), then such pairs ("non-Coulomb pairs") are not affected by interaction in the final states.

After production, $A_{\pi K}$ travel through the target and could to annihilate into $\pi^0 K^0$, or to be ionised due to interaction with the target matter, producing specific "atomic pairs". These pairs have small relative momentum ($Q < 3$ MeV/c) and a number of such pairs n_A could be measured experimentally. Ratio of "atomic pair" number to a number of atom produced is a breakup probability: $P_{br}(\tau) = n_A/N_A = n_A/(K \cdot N_C)$ [15, 16]. In Fig 1 dependence of $A_{\pi K}$ breakup probability is shown for two Ni target are used in experiment DIRAC for pair laboratory momentum range $5.1 \div 8.5$ GeV/c. Value is averaged, using experimentally measured spectrum of atoms.

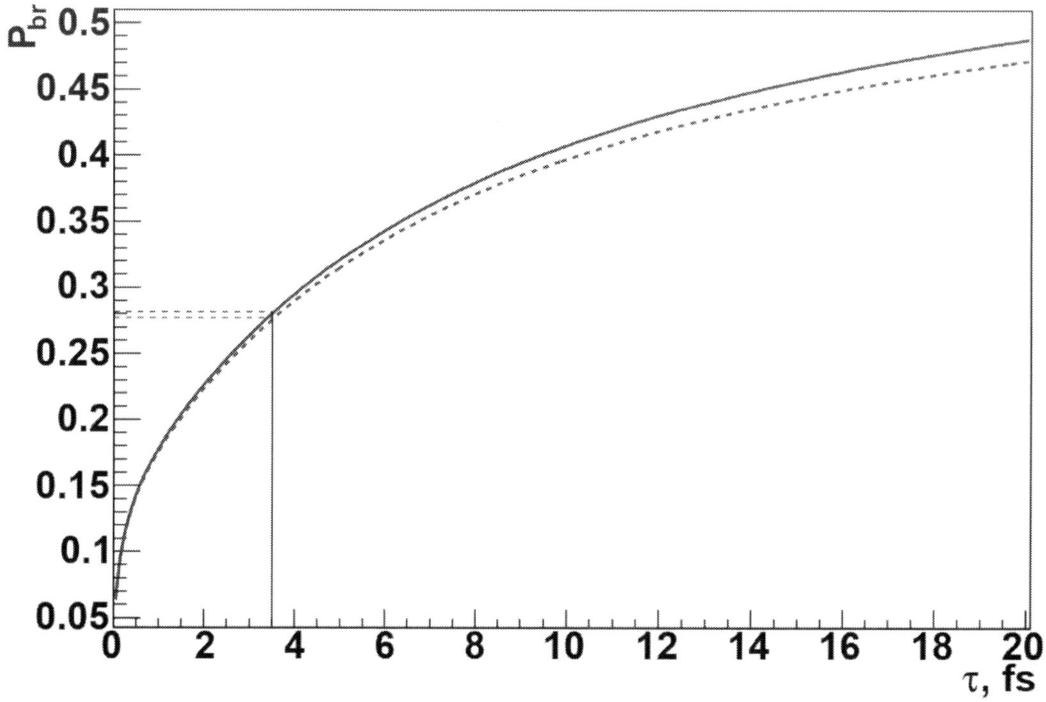

Figure 1. Dependence of the breakup probability P_{br} on $A_{\pi K}$ lifetime for 108μm (solid blue line) and 98μm (dashed red line) Ni targets, and an example how lifetime could be obtained from experimentally measured breakup probability

3 DIRAC setup

DIRAC setup was created to detect $\pi^+\pi^-$ with small relative momenta [17]. In 2004-2006 it has been modified in order to detect both $\pi^+\pi^-$ and πK pairs [18]. New detectors for particle identification have been added: Cherenkov detectors with heavy gas and aerogel for identification of K-mesons among background of pions and protons, correspondingly. Taking into account kinematic of πK "atomic pairs", new detectors cover only internal parts of each arm (see Fig. 2). Aerogel Cherenkov detector is mounted only in positive particle arm, because a flux of antiprotons is small relative to a flux of K^- mesons.

Coordinate resolution of the setup detector provide high resolution over components of relative momentum Q: transverse $\sigma_{Q_X} = \sigma_{Q_Y} = 0.5$ MeV/c and longitudinal $\sigma_{Q_L} = 0.9$ MeV/c (for πK pairs). Time resolution of vertical hodoscope $\sigma_T^{VH} = 100$ ps allows to use time-of-flight technique for suppression of wrongly identified particle pairs.

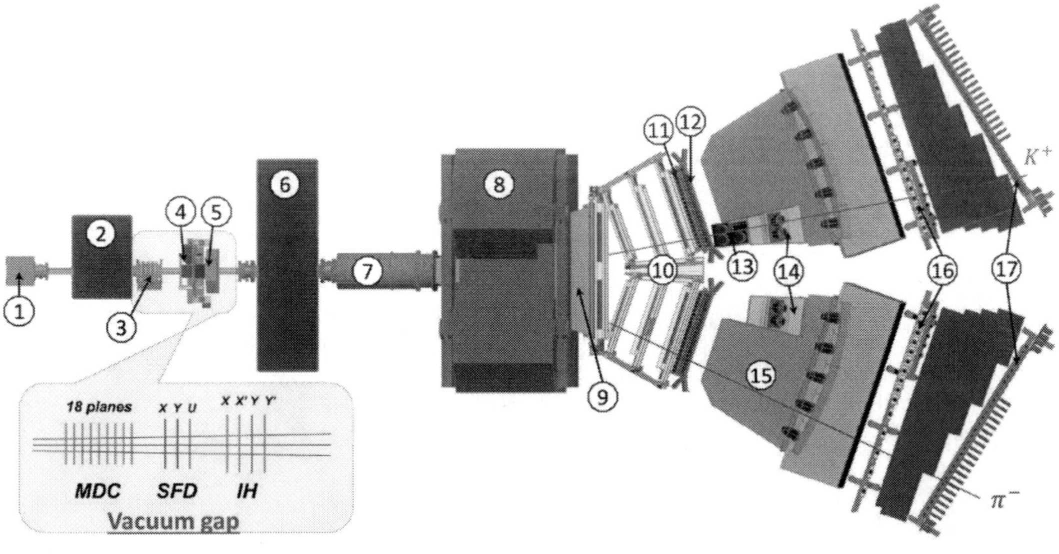

Figure 2. General view of upgraded DIRAC setup (1 – target station; 2 – first shielding; 3 – micro drift chambers (MDC); 4 – scintillating fiber detector (SFD); 5 – ionization hodoscope (IH); 6 – second shielding; 7 – vacuum tube; 8 – spectrometer magnet; 9 – vacuum chamber; 10 – drift chambers (DC); 11 – vertical hodoscope (VH); 12 – horizontal hodoscope (HH); 13 – aerogel cherenkov (ChA); 14 – heavy gas cherenkov (ChF); 15 – nitrogen cherenkov (ChN); 16 – preshower (PSh); 17 – muon detector (Mu).

4 Selection πK events

Analysis procedure selects events which have signals of detectors expected for π^+K^- and $K^+\pi^-$ pairs. Fig. 3(left) presents the distribution of selected events over the difference of the particle production times for K^+ mesons in the range (4.4–4.5) GeV/c. The distribution is fitted by the simulated distribution of $K^+\pi^-$, $\pi^+\pi^-$, pπ^- and accidental pairs. Fig. 3(right) shows the fit for K^+ in the range (5.4–5.5) GeV/c. The contribution of misidentified pairs was estimated and accordingly subtracted [19]. Fraction of non-suppressed background pairs is calculated as function of momentum and used for estimation of systematic uncertainty (see below).

Fig. 4a illustrates the Q_L distribution of potential $K^+\pi^-$ pairs requiring a ChF signal and $Q_T <$ 4 MeV/c. The dominant peak on the left side is due to pπ^- pairs from Λ decay. After requesting a ChA signal, the admixture of pπ^- pairs is decreased by a factor of 10 (Fig. 4b). By selecting compatible TOFs between target and VH, background pπ^- and $\pi^+\pi^-$ pairs can be substantially suppressed (Fig. 4c). In the final distribution, the well-defined $K^+\pi^-$ Coulomb peak at $Q_L = 0$ emerges beside the strongly reduced peak from Λ decays at $Q_L = -30$ MeV/c. The Q_L distribution of potential π^+K^- pairs shows a similar behaviour. Applying the ChF and TOF criteria provides a sufficient background

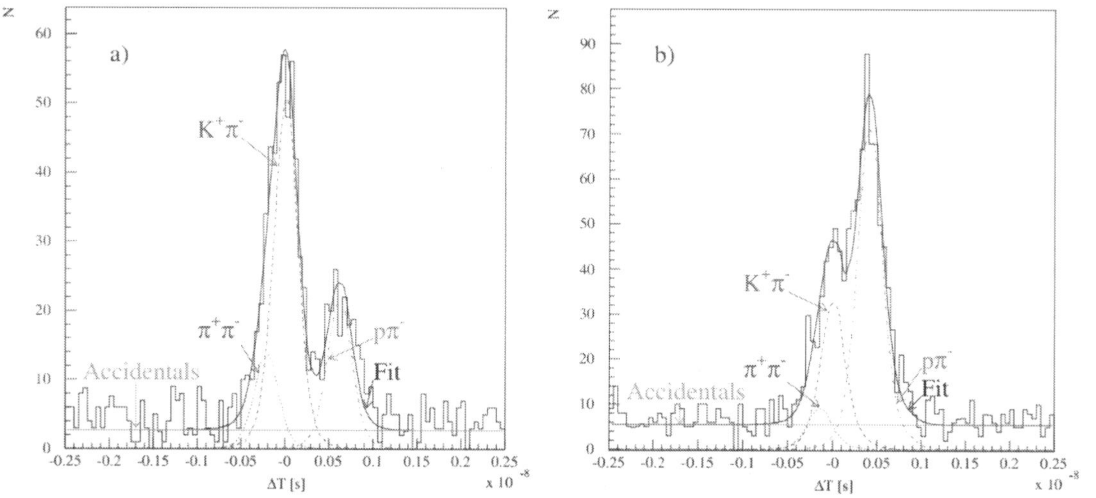

Figure 3. a) difference of particle generation times for events with positively charged particle momenta (4.4 ÷ 4.5) GeV/c. Experimental data (histogram) are fitted by the event sum (black, solid): $K^+\pi^-$ (red, dashed), $\pi^+\pi^-$ (blue, dotted), $p\pi^-$ (magenta, dotted-dashed) and accidentals (green, constant). b) similar distributions for events with positively charged particle momenta (5.4 ÷ 5.5) GeV/c.

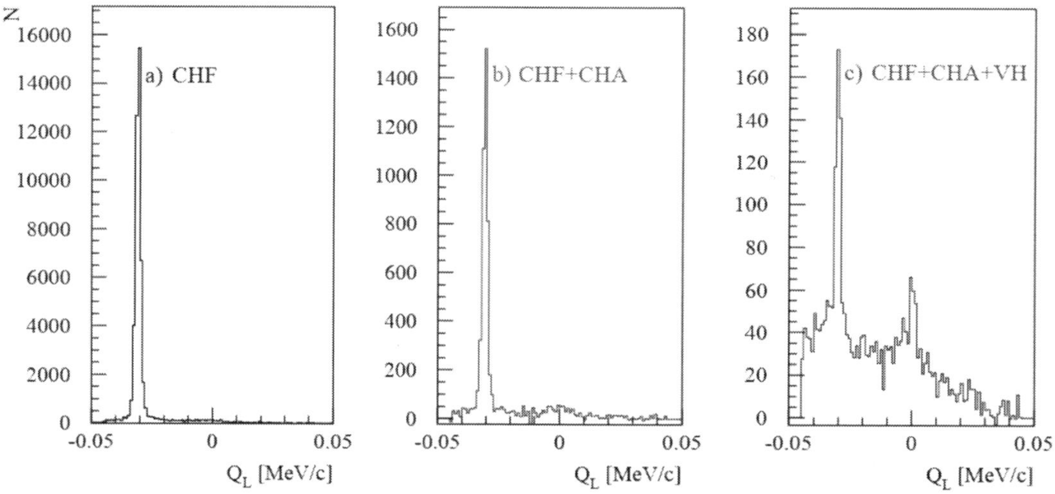

Figure 4. Q_L distribution of $K^+\pi^-$ pairs after applying different criteria (see text).

rejection. Fig. 5 presents the π^+K^- Coulomb peak at $Q_L = 0$ and a second peak from $\bar{\Lambda}$ decays at $Q_L = 30$ MeV/c.

For the final analysis, the DIRAC procedure selects events fulfilling the following criteria:

$$Q_T < 4\text{MeV}/c, |Q_L| < 20\text{MeV}/c\,.$$

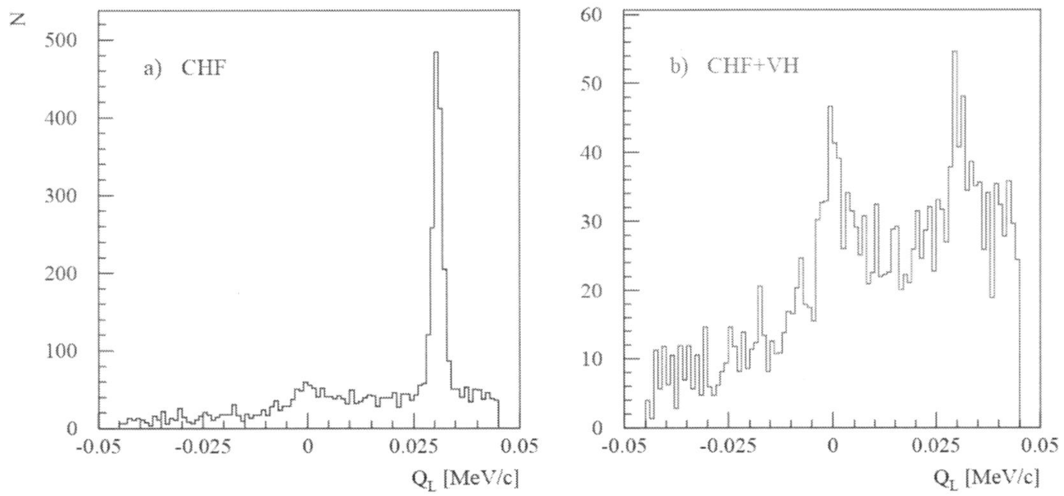

Figure 5. Q_L distribution of π^+K^- pairs after applying different criteria (see text).

Table 1. "Atomic pair" numbers n_A by analysing the 1-dimensional Q and $|Q_L|$ distributions and the 2-dimensional ($|Q_L|$, Q_T) distribution. Only statistical errors are given.

Analysis	$K^+\pi^-$	π^+K^-	$K^+\pi^-$ and π^+K^-
Q	243 ± 52 (4.7σ)	106 ± 32 (3.3σ)	349 ± 61 (5.7σ)
$\lvert Q_L\rvert$	164 ± 79 (2.1σ)	67 ± 47 (1.4σ)	230 ± 92 (2.5σ)
$\lvert Q_L\rvert$, Q_T	237 ± 50 (4.7σ)	78 ± 32 (2.5σ)	314 ± 59 (5.3σ)

5 Observation of π^+K^- and $K^+\pi^-$ atoms

Distributions of experimental data over relative momentum Q and its projections Q have been fitted by a sum of simulated distributions of "atomic", "Coulomb" and "non-Coulomb" pairs [20]. Contributions of simulated distributions are free parameters of fit. In order to reproduce distribution of experimental pairs over relative momentum Q and its projections, simulation procedure takes into account resolution and efficiency of the setup detectors, multiplicity of background particles and noise signals, multiple scattering in Pt (run 2007) and Ni (2008-2010) targets, detector planes and partitions.

Fig. 6a presents the experimental and simulated Q distributions of $K^+\pi^-$ pairs for the data obtained from the Pt target and Ni targets. Fig. 7a is for π^+K^- pairs. One observes an excess of events above the sum of "Coulomb" and "non-Coulomb" pairs in the low Q region, where atomic pairs are expected: these excess spectra are shown in Figs. 6b and 7b together with the simulated distribution of atomic pairs. The numbers of atomic pairs, found in the $K^+\pi^-$ and π^+K^- data, are $n_A(K^+\pi^-) = 243 \pm 51$ ($\chi^2/n = 36/37$, n = number of degrees of freedom) and $n_A(\pi^+K^-) = 106 \pm 32$ ($\chi^2/n = 42/37$). Comparing the experimental with the simulated distributions, demonstrates good agreement.

Experimental distribution for sum of π^+K^- and $K^+\pi^-$ data is presented in Fig. 8. Numbers of π^+K^- and $K^+\pi^-$ "atomic pairs" obtained with analysis of one-dimensional distributions over Q and $|Q_L|$ and two-dimensional ($|Q_L|$, Q_T) distribution are presented in Table 1 (Ni and Pt target together). The best statistical accuracy are achieved by analysis of Q and ($|Q_L|$, Q_T) distributions. Signal to error ratio is more than 5. The 1-dimensional $|Q_L|$ analysis for all πK data yields $n_A = 230 \pm 92$, which does not contradict the values, obtained in the other two statistically more precise analyses. Compared

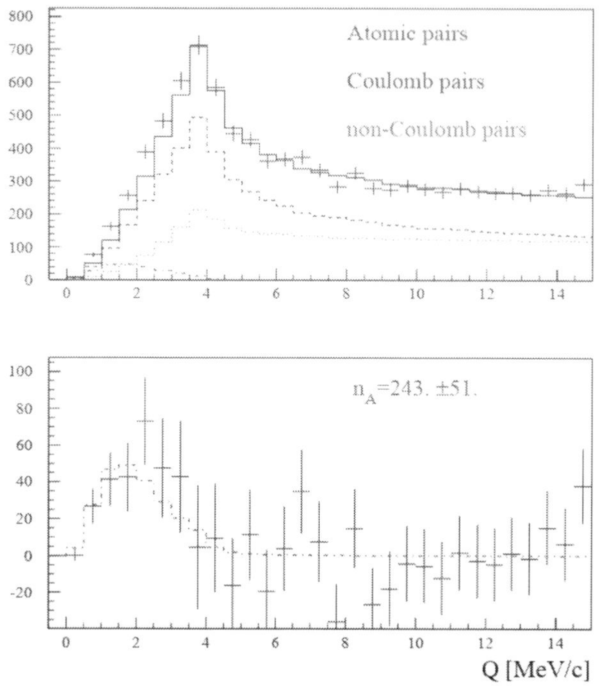

Figure 6. Experimental distribution of $K^+\pi^-$ pairs over Q (upper pictures), shown by points with error bars, is fitted by a sum of simulated distributions of "atomic", "Coulomb" and "non-Coulomb" pairs. The background distribution of free ("Coulomb" and "non-Coulomb") pairs is shown as solid black line. Differences of experimental and background distributions are shown in lower picture together with simulated distributions of "atomic pairs".

to the previous investigation [12], the Pt data was analysed including the upstream detectors. The consequence is a decrease of the statistics, but on the other hand an increase of the Q_T resolution. This better resolution improves the quality of data. Concerning the Ni target, the increase of n_A, compared to [11], is caused by optimizing the time-of-flight criteria, which decreases "atomic pair" losses for the same fraction of background in the final distributions.

The evaluation of the "atomic pair" number n_A is affected by several sources of systematic errors [19, 21]. These uncertainties lead to differences in the shapes of experimental and MC distributions for "atomic", "Coulomb" and to a much lesser extent for "non-Coulomb" pairs. The shape differences induce a bias in the value of the fit parameter n_A, corresponding to a systematic error of the "atomic pair" number. Sources of systematic error and estimation of error values are listed in Table 2.

Taking into account both statistical and systematic errors, the one-dimensional $\pi^\pm K^\mp$ analysis in Q yields $n_A = 349 \pm 61(stat) \pm 9(syst) = 349 \pm 62(tot)$ "atomic pairs" (5.6 σ) for both combinations of charge and two-dimensional analysis in $(|Q_L|, Q_T)$ yields $n_A = 314 \pm 59(stat) \pm 10(syst) = 314 \pm 60(tot)$ "atomic pairs" (5.2 σ). This is the first statistically significant observation of the strange dimesonic πK atom.

6 Measurement of S-wave isospin-odd πK scattering length

Experimentally measured numbers of "atomic pairs" n_A and produced atoms N_A allow (see section 2) to obtain breakup probability P_{br}.

Table 3 contains the P_{br} values obtained in the Q and $(|Q_L|, Q_T)$ analyses with statistical uncertainties.

Sources of systematic uncertainties of breakup probability are mainly the same like for number of "atomic pairs" (see Table 2). There only one new source - uncertainty in the $P_{\text{br}}(\tau)$ relation.

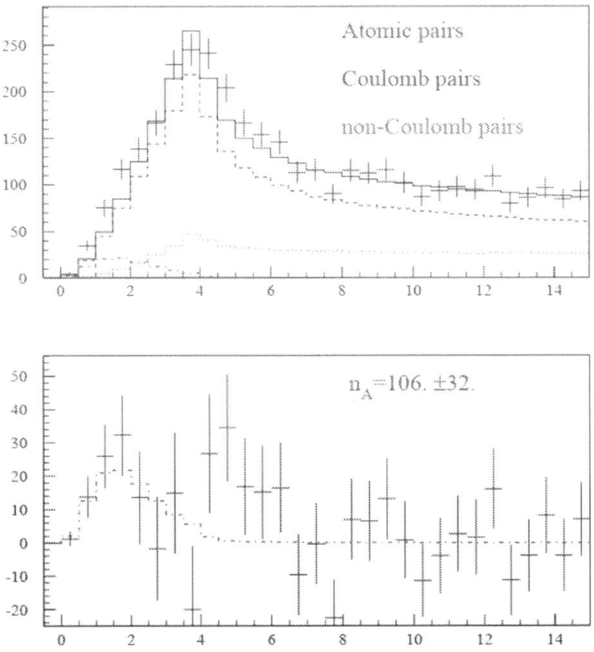

Figure 7. Experimental distribution of π^+K^- pairs analogous to Fig. 6.

Table 2. Estimations of systematic errors, which are induced by different sources, for analysis of data distribution over relative momentum Q, its longitudinal projection $|Q_L|$ and two dimensional distribution over $(|Q_L|, Q_T)$.

Sources of systematic errors	σ_Q^{syst}	$\sigma_{Q_L}^{syst}$	$\sigma_{\lvert Q_L\rvert,Q_T}^{syst}$
Uncertainty in Λ width correction	0.8	3.0	2.0
Uncertainty of multiple scattering in Ni (Pt) target	4.4	0.7	2.7
Accuracy of SFD simulation	0.2	0.0	0.1
Correction of Coulomb correlation function due to finite size production region	0.0	0.2	0.1
Uncertainty in πK pair laboratory momentum spectrum	3.3	5.4	7.8
Uncertainty in laboratory momentum spectrum of background pairs	6.6	1.6	5.4
Total	8.6	6.4	10.1

Estimations of systematic errors, induced by different sources, are presented in Table 4 for Pt data and Table 5 for Ni data. The total errors were calculated as the quadratic sum. The procedure of the πK atom lifetime estimation described below includes all systematic errors, although their contributions are insignificant compared to the statistical errors.

For estimating the lifetime of $A_{\pi K}$ in the ground state, the maximum likelihood method [22] is applied [23]:

$$L(\tau) = \exp\left(-U^T G^{-1} U/2\right), \tag{11}$$

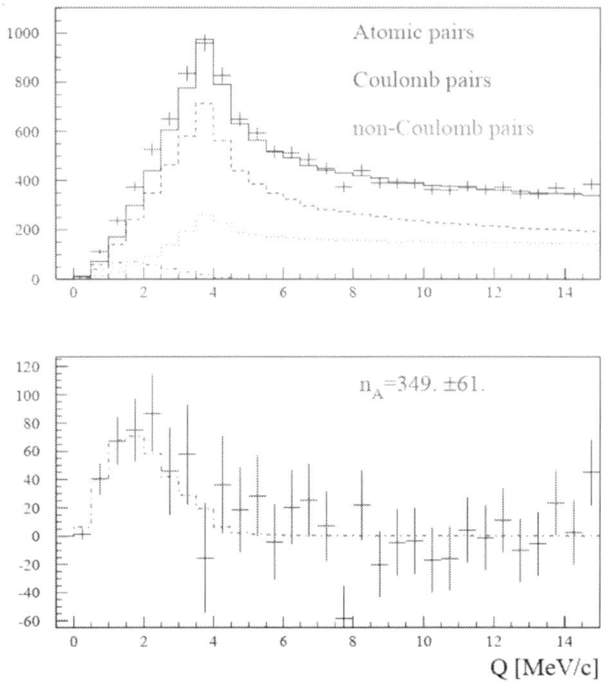

Figure 8. Distribution of π^+K^- and $K^+\pi^-$ pairs over Q (upper pictures), shown by points with error bars, is fitted by a sum of simulated distributions of "atomic" (red dotted-dashed), "Coulomb" (blue dashed) and "non-Coulomb" (magenta dotted) distributions ($\chi^2/n = 41/37$, n = number of degrees of freedom). A sum of background distributions ("Coulomb" and "non-Coulomb") is shown by a solid black line. Differences of experimental and background distributions are shown in lower picture together with simulated distributions of "atomic pairs". Quality of fit, performed in assuption that "atomic pairs" are absent, is worse: $\chi^2/n = 73/38$

Table 3. Experimental $P_{\rm br}$ from Q and ($|Q_L|$, Q_T) analyses. Only statistical uncertainties are cited.

Data	RUN	Target (μm)	P_{br}^{Q}	$P_{br}^{\lvert Q_L\rvert,Q_T}$
π^+K^-	1	Pt (25.7)	1.2 ± 1.3	0.27 ± 0.56
π^+K^-	2	Ni (98)	0.53 ± 0.39	0.42 ± 0.38
π^+K^-	3	Ni (108)	0.29 ± 0.20	0.33 ± 0.24
π^+K^-	4	Ni (108)	0.33 ± 0.22	0.21 ± 0.20
$K^+\pi^-$	1	Pt (25.7)	1.09 ± 0.52	1.44 ± 0.59
$K^+\pi^-$	2	Ni (98)	0.32 ± 0.20	0.44 ± 0.22
$K^+\pi^-$	3	Ni (108)	0.23 ± 0.16	0.16 ± 0.15
$K^+\pi^-$	4	Ni (108)	0.41 ± 0.17	0.34 ± 0.16
$\pi^+K^-\&K^+\pi^-$	1	Pt, 25.7	1.11 ± 0.48	0.83 ± 0.41

where $U_i = \Pi_i - P_{{\rm br},i}(\tau)$ is a vector of differences between measured Π_i ($P_{\rm br}$ in Table 3) and corresponding theoretical breakup probability $P_{{\rm br},i}(\tau)$ for a data sample i. The error matrix of U, named G, includes statistical (σ_i) as well as systematic uncertainties. Only the term corresponding to the uncertainty in the $P_{\rm br}(\tau)$ relation is considered as correlated between the Ni and Pt data, which is a conservative approach and overestimates this error. The other systematic uncertainties do not exhibit a correlation between the data samples from the Ni and Pt targets. On the other hand, systematic uncertainties of the Ni data samples are correlated. The likelihood functions of the ($|Q_L|$, Q_T) and Q analyses are shown in Fig. 9.

New estimation of $A_{\pi K}$ lifetime in ground state, based on two-dimensional analysis in ($|Q_L|$, Q_T):

$$\tau = (3.8^{+3.3}_{-2.0}|_{stat}\ ^{+1.0}_{-0.6}|_{syst})\ {\rm fs} = (3.8^{+3.5}_{-2.1}|_{tot})\ {\rm fs}\,, \qquad (12)$$

Table 4. Estimated systematic errors of $P_{\rm br}$ for Pt in Q and $(|Q_L|, Q_T)$ analyses.

Source	Q	$(\|Q_L\|, Q_T)$
Uncertainty in Λ width correction	0.011	0.073
Uncertainty of multiple scattering in the Pt target	0.0087	0.014
Accuracy of SFD simulation	0.	0.
Correction of the Coulomb correlation function on finite size production region	0.0001	0.0002
Uncertainty in πK pair lab. momentum spectrum	0.089	0.25
Uncertainty in the laboratory momentum spectrum of background pairs	0.22	0.21
Uncertainty in the $P_{\rm br}(\tau)$ relation	0.01	0.01
Total	0.24	0.34

Table 5. Estimated systematic errors of $P_{\rm br}$ for Ni in Q and $(|Q_L|, Q_T)$ analyses.

Source	Q	$(\|Q_L\|, Q_T)$
Uncertainty in Λ width correction	0.0006	0.0006
Uncertainty of multiple scattering in a Ni target	0.0051	0.0036
Accuracy of SFD simulation	0.0002	0.0003
Correction of the Coulomb correlation function on finite size production region	0.0001	0.0000
Uncertainty in πK pair lab. momentum spectrum	0.0052	0.0050
Uncertainty in the laboratory momentum spectrum of background pairs	0.0011	0.0011
Uncertainty in the $P_{\rm br}(\tau)$ relation	0.0055	0.0055
Total	0.0092	0.0084

which corresponds (see Fig. 10(left)) to isospin-odd πK scattering length estimation to be:

$$|a_0^-|M_\pi = 0.087^{+0.043}_{-0.024}|_{tot}\,. \tag{13}$$

Estimation, based on Q analysis:

$$\tau = (5.5^{+4.9}_{-2.8}|_{stat}\,{}^{+0.9}_{-0.5}|_{syst})\ \text{fs} = (5.5^{+5.0}_{-2.8}|_{tot})\ \text{fs}\,, \tag{14}$$

provides (see Fig. 10(right)) the following value of the πK scattering length a_0^-:

$$|a_0^-|M_\pi = 0.072^{+0.031}_{-0.020}|_{tot}\,. \tag{15}$$

All theoretical predictions are compatible with the measured value taking into account the experimental precision.

7 Possibility to improve accuracy with experiment at 450 GeV beam

Updated analysis allows to improve accuracy of $|a_0^-|$ measurement from 60% to 40%, but it is still unsufficient for testing predictions of CHPT and LQCD. As shown in [24], the number of πK atoms

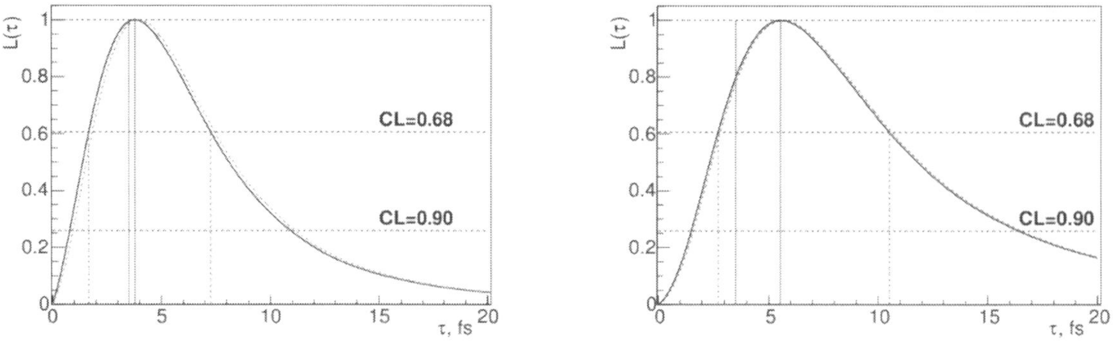

Figure 9. Likelihood functions $L(\tau)$ for $(|Q_L|, Q_T)$ (left) and Q (right) analyses with $Q_T < 4$ MeV/c. The likelihood functions on the basis of both statistical and systematic errors (dashed green line) and on the basis of only statistical error (solid blue line) are presented. The vertical blue lines indicate the best estimate for τ_{tot} and the corresponding confidence interval. The vertical red line is the theoretical prediction (9).

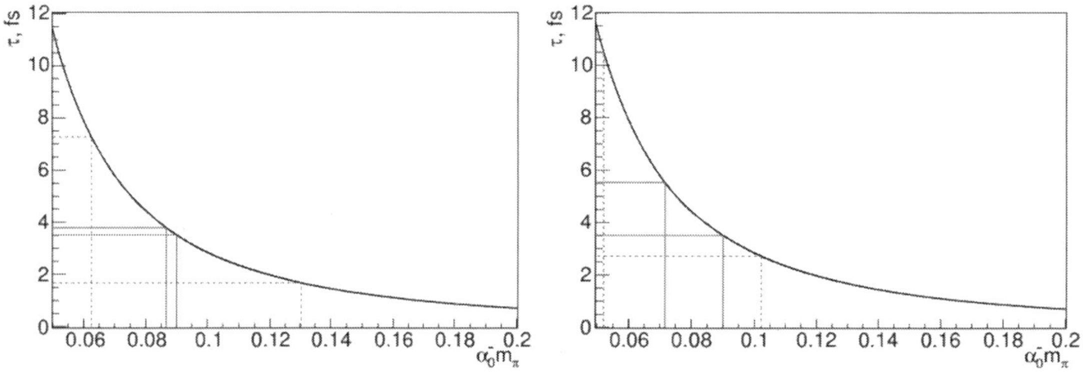

Figure 10. Ground state $A_{\pi K}$ lifetime τ_{1S} versus a_0^-. Experimental results (blue lines) are compared to the theoretical prediction (red lines). $(|Q_L|, Q_T)$ analysis (left) and Q analysis (right).

detected per time unit would be increased by a factor of 30 to 40, if the DIRAC experiment could exploit the CERN SPS 450 GeV/c proton beam. The simulation of atom and charged particle yields and the result of current experimental measurement have been used for estimation of time, which is needed for measurement a_0^- with accuracy 5% [25]. Table 6 presents expected beam time and run time are needed for achievement of this accuracy for present (Ni target only) and modified DIRAC setup. It is seen that experiment at SPS energy allows to measure S-wave pion-kaon scattering length difference with sufficient accuracy for checking theoretical predictions made by ChPT and LQCD.

Expected results of the experiment with 450 GeV beam:

- Expected statistic of πK "atomic pairs": $n_A \approx 13000$.
- Statistical accuracy of πK scattering length difference: $\sim 5\%$.
- Expected systematic error of πK scattering length difference: $\sim 2\%$.
- Expected statistic of $\pi^+\pi^-$ "atomic pairs": $n_A \approx 400000$.
- Statistical accuracy of pion-pion scattering length difference: $\sim 0.7\%$.

Table 6. Estimation of time needed for measurement a_0^- with statistical accuracy $\delta_{a_0^-}$ for present DIRAC setup and beam condition, and for versions Mod1 and Mod2, modified for proton beam energy E_p = 450 GeV and intensity I_B (proton per second). Angle between primary and secondary beams θ_{lab} and solid angle of secondary beam aperture are presented. It is assumed, that at 450 GeV beam the setup would obtain 3000 spills (4.5s) per day.

Setup	E_p	I_b	θ_{lab}	Solid angle	Beam time	Run time	$\delta_{a_0^-}$
	GeV	p/s		*sr*	s	months	%
Present	24	$2.7 \cdot 10^{11}$	5.7	$1.2 \cdot 10^{-3}$	$1.2 \cdot 10^{6}$	14.5	43.
Mod1	450	$1.0 \cdot 10^{11}$	4.0	$0.6 \cdot 10^{-3}$	$5.5 \cdot 10^{6}$	13.6	5.
Mod2	450	$1.0 \cdot 10^{12}$	4.0	$0.6 \cdot 10^{-3}$	$6.5 \cdot 10^{5}$	1.6	5.

- Expected systematic error of pion-pion scattering length difference: $\sim$ 2%.

8 Summary

In the DIRAC experiment at CERN, the dimesonic Coulomb bound states involving strangeness, the $K^+\pi^-$ and π^+K^- atoms, were observed for the first time with reliable statistics. The one-dimensional $\pi^\pm K^\mp$ analysis in Q yields 349 $\pm$ 62(*tot*) "atomic pairs" (5.6 σ) for both combinations of charge. Analogously, a two-dimensional analysis in ($|Q_L|$, Q_T) was performed with the result of 314 $\pm$ 60(*tot*) "atomic pairs" (5.2 σ).

The breakup probabilities for each atom type and each target are determined. By means of these probabilities, the lifetime of the πK atom in the ground state is evaluated to be:

$$\tau_{\rm tot} = (5.5^{+5.0}_{-2.8}\big|_{\rm tot}) \cdot 10^{-15}{\rm s}\,,$$

and the S-wave isospin-odd πK scattering length deduced:

$$\left|a_0^-\right| = \frac{1}{3}\left|a_{1/2} - a_{3/2}\right| = (0.072^{+0.031}_{-0.020}\big|_{\rm tot})M_\pi^{-1}\,.$$

DIRAC-like experiment at SPS energy provides possibility to check prediction of ChPT and LQCD for pion-kaon scattering lengths with accuracy at the level 5%.

We are grateful to CERN for support and the PS team for the excellent performance of the accelerator. This work was funded by CERN, INFN (Italy), INCITE and MICINN (Spain), IFIN-HH (Romania), the Ministry of Education and Science and RFBR grant 01-02-17756-a (Russia), the Grant-in-Aid from JSPS and Sentanken-grant from Kyoto Sangyo University (Japan).

References

[1] B. Adeva *et al.*, Phys. Rev. Lett. **117**, 112001 (2016)
[2] J. Schweizer, Phys. Lett. B **587**, 33 (2004)
[3] V. Bernard *et al.*, Phys. Rev. D **43**, 2757 (1991)
[4] A. Roessl, Nucl. Phys. B **555**, 507 (1999)
[5] J. Bijnens *et al.*, JHEP **0405**, 036 (2004)
[6] P. Buttiker *et al.*, Eur. Phys. J. C **33**, 409 (2004)
[7] S.R. Beane *et al.*, Phys. Rev. D **77**, 094507 (2008)
[8] C.B. Lang *et al.*, Phys. Rev. D **86**, 054508 (2012)

[9] K. Sasaki *et al.*, Phys. Rev. D **89**, 054502 (2014)
[10] T. Janowski et al., PoS **LATTICE2014** 080 (2015)
[11] B. Adeva *et al.*, Phys. Lett. B **735**, 288 (2014)
[12] B. Adeva *et al.*, Phys. Lett. B **674**, 11 (2009)
[13] L. Nemenov, Sov. J. Nucl. Phys. **41**, 629 (1985)
[14] L. Afanasyev and O. Voskresenskaya, Phys. Lett. B **453**, 302 (1999)
[15] L. Afanasyev and A. Tarasov, Phys. At. Nucl. **59**, 2130 (1996)
[16] M. Zhabitsky, Phys. At. Nucl. **71**, 1040 (2008)
[17] B. Adeva *et al.*, Nucl. Instrum. Methods A **515**, 467 (2003)
[18] B. Adeva *et al.*, Nucl. Instrum. Methods A **839**, 52 (2016)
[19] V. Yazkov and M. Zhabitsky, DN **2013-06** (DN = DIRAC Note), cds.cern.ch/record/1628544
[20] B. Adeva *et al.*, arXiv:1707.02184 [hep-ex]
[21] A. Benelli and V. Yazkov, DN **2016-01**, cds.cern.ch/record/2137645
[22] D. Drijard and M. Zhabitsky, DN **2008-07**, cds.cern.ch/record/1367888
[23] V. Yazkov and M. Zhabitsky, DN **2016-06**, cds.cern.ch/record/2252375
[24] O. Gorchakov and L. Nemenov, J. Phys. G: Nucl. Part. Phys. **43**, 095004 (2016)
[25] V. Yazkov, DN **2016-05**, cds.cern.ch/record/2207227

Highlights of top quark cross-section measurements at ATLAS

Peter Berta[1,⋆] *on behalf of the ATLAS Collaboration*

[1]*Johannes Gutenberg University Mainz, Institute of Physics, Staudingerweg 7, 55128 Mainz, Germany*

Abstract. The highlights of the measurements of top quark production in proton-proton collisions at the Large Hadron Collider with the ATLAS detector are presented. The inclusive measurements of the top-pair production cross section have reached high precision and are compared to the best available theoretical calculations. The differential cross section measurements, including results using boosted top quarks, probe our understanding of top-pair production in the TeV regime. The results are compared to Monte Carlo generators implementing LO and NLO matrix elements matched with parton showers. Measurements of the single top quark production cross section are presented in the t-channel and s-channel, and with associated production with a W boson. For the t-channel production, results on the ratio between top quark and antitop quark production cross sections and differential measurements are also included.

1 Introduction

Millions of top quarks have been produced in proton-proton collisions at the Large Hadron Collider (LHC) at three different center-of-mass energies: $\sqrt{s}$ = 7 TeV, $\sqrt{s}$ = 8 TeV, and $\sqrt{s}$ = 13 TeV allowing us to precisely measure its production cross sections. There are two main production modes at the LHC: top-pair ($t\bar{t}$) production and single top production. The single top production consists of three separate physics processes: t-channel production, s-channel production and production associated with a W boson (Wt). A pair of top quarks or a single top quark can be produced in association also with other objects (e.g. Z boson, Higgs boson), which is not covered in this overview.

There are several important reasons to measure precisely the top quark production cross sections. First of all, the measurements are precision tests of the Standard Model (SM). The $t\bar{t}$ production probes Quantum Chromodynamics (QCD) and the single top production probes the electroweak theory. The measurements are sensitive to many physics models beyond the SM. In particular, the measurements of the $t\bar{t}$ differential cross sections probe the TeV scale, and therefore a new particle with large coupling to the top quark can cause a significant increase in the $t\bar{t}$ invariant mass spectrum or can modify the top p_T spectrum. Furthermore, the measurements can constrain parton density function (PDF) fits and the parameters of the SM Monte Carlo (MC) generators. The measurement of the single top cross sections can be used to estimate the CKM matrix element V_{tb} and various differential cross section measurements can be used to extract the top quark pole mass.

In this overview, a selection of measurements from the ATLAS experiment [1] are presented. The top quark decays almost in ~100% of the cases to a W boson and a b-quark, and therefore the

⋆e-mail: peter.berta@cern.ch

signature of the top quark events in the detector depends on the decay mode of the W boson, which can be leptonic or hadronic.

2 $t\bar{t}$ production

2.1 $t\bar{t}$ inclusive cross sections

Precise measurements of the $t\bar{t}$ inclusive cross sections are performed in full and fiducial phase spaces at $\sqrt{s}$ = 7 TeV, $\sqrt{s}$ = 8 TeV [2], and $\sqrt{s}$ = 13 TeV [3] using events containing exactly one isolated electron (e) and one isolated muon (μ) of opposite charge. Further selection requirements are on the missing transverse energy and on the number of jets tagged by a track-based algorithm which identifies jets containing a b-quark (b-tagged jets). Figure 1 shows the number of b-tagged jets, in which the prediction is consistent with data. The number of events with one b-tagged jet and two b-tagged jets is used to determine the b-tagging efficiency in data which reduced the systematic uncertainty due to the difference of the b-tagging efficiencies between data and MC. The main systematic uncertainties arise from the luminosity and LHC beam energy measurements, as well as from the modeling of the $t\bar{t}$ events. The total uncertainty is ~4% for measurements at each of the three center-of-mass energies. The result of the full phase space measurement is compared to theoretical QCD calculations at next-to-next-to-leading order (NNLO), and it is found that all measurements are consistent with it.

Figure 2 summarizes the $t\bar{t}$ inclusive measurements in the full phase space as a function of the center-of-mass energy compared to the NNLO QCD prediction complemented with NNLL resummation (NNLO+NNLL QCD). The measurements from the two LHC experiments, ATLAS and CMS,

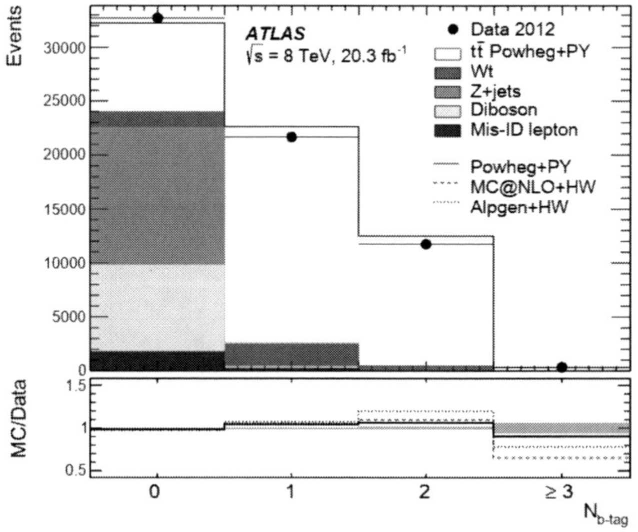

Figure 1. Distribution of the number of b-tagged jets in preselected opposite-charge $e\mu$ events in $\sqrt{s}$ = 8 TeV data [2]. The data are compared to the expectation from simulation, broken down into contributions from $t\bar{t}$, Wt single top, Z+jets, dibosons, and events with misidentified electrons or muons. The lower part of the figure shows the ratios of simulation to data, using various $t\bar{t}$ samples generated with Powheg + Pythia6 (PY), MC@NLO + Herwig (HW) and Alpgen + Herwig, and with the cyan band indicating the statistical uncertainty.

and also a combination of the Tevatron measurements from proton-antiproton collisions are shown. All the existing measurements of the $t\bar{t}$ inclusive cross section are consistent with the NNLO+NNLL QCD calculation.

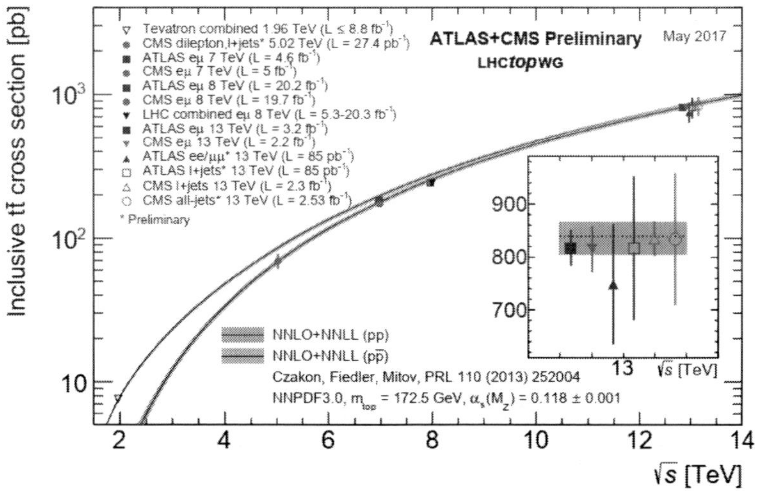

Figure 2. Summary of LHC and Tevatron measurements of the top-pair production cross-section as a function of the center-of-mass energy compared to the NNLO QCD calculation complemented with NNLL resummation (top++2.0) [4]. The theory band represents uncertainties due to the renormalisation and factorisation scale, parton density functions and the strong coupling. The measurements and the theory calculation assume m_{top} = 172.5 GeV. Measurements made at the same center-of-mass energy are slightly offset for clarity.

2.2 $t\bar{t}$ differential cross sections

The ATLAS experiment has an extensive program on the $t\bar{t}$ differential cross sections exploring them in various decay channels and phase spaces, and at all center-of-mass energies. Two measurements are described in the following.

The differential cross-section for $t\bar{t}$ events containing highly energetic top quarks is measured using all the recorded data (20.3 fb^{-1}) at $\sqrt{s}$ = 8 TeV [5]. The event selection focuses on the semileptonic $t\bar{t}$ decay in the boosted regime and consists of the following main criteria: one electron or one muon with $p_{\mathrm{T}} > 25$ GeV, $E_{\mathrm{T}}^{\mathrm{miss}} > 20$ GeV, at least one large radius jet with $p_{\mathrm{T}} > 300$ GeV (clustered with algorithm anti-k_t with distance parameter $R = 1$), and at least one b-tagged small radius jet ($R = 0.4$) with $p_{\mathrm{T}} > 25$ GeV. Jet substructure criteria are applied for the large radius jets to identify hadronically decaying top quarks contained in these large radius jets. The measurement is performed in the full phase space and unfolded to parton level, where the differential cross section is expressed as a function of the p_{T} of the top quark. Additionally, the unfolding was performed to particle level with the differential cross section expressed as a function of the p_{T} of the large radius jet in a fiducial phase space similar to the selection criteria. The parton level result is shown in Figure 3 and it is compared to the predictions of next-to-leading-order (NLO) and leading-order (LO) matrix-element MC generators interfaced with various parton shower (PS) MC generators. The main uncertainties arise from signal modeling and jet energy scale (JES) of large radius jets. The measurement probes

DOI: 10.1051/epjconf/201715804001

the top quarks at the TeV scale and it is consistent within uncertainties with all the tested QCD MC predictions.

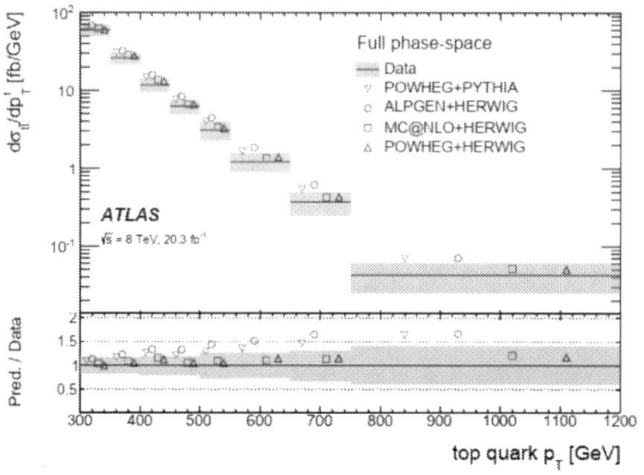

Figure 3. Parton-level differential cross-section as a function of the hadronically decaying top quark p_T [5]. Powheg+Pythia, Alpgen+Herwig, MC@NLO+Herwig, and Powheg+Herwig predictions are compared with the final results. MC samples are normalized to the NNLO+NNLL inclusive cross-section $\sigma^{t\bar{t}}$ = 253 pb. No electroweak corrections are applied to the predictions. The lower part of the figure shows the ratio of the MC prediction to the data. The shaded area includes the total statistical plus systematic uncertainties. The points of the various predictions are spaced along the horizontal axis for presentation only; they correspond to the same p_T range.

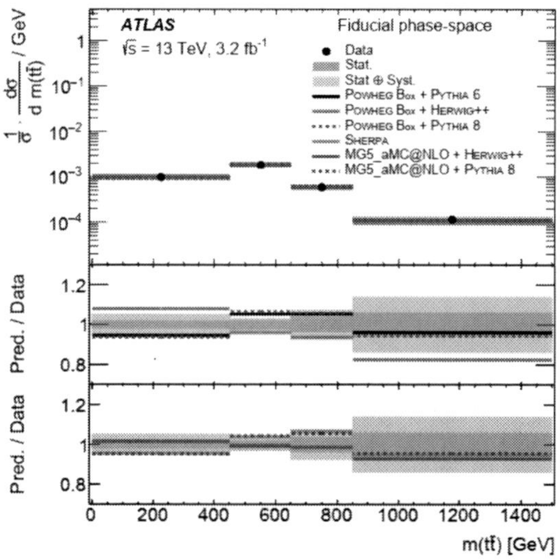

Figure 4. The measured normalized fiducial differential cross-section as a function of the particle-level invariant mass of the $t\bar{t}$ system [6]. The data are compared to the predictions from Powheg-Box (top ratio panel), Sherpa, and MG5_aMC@NLO (bottom ratio panel) interfaced to various PS generators.

The dilepton decay channel is exploited in a differential cross section measurement using data $\sqrt{s}$ = 13 TeV corresponding to an integrated luminosity of 3.2 fb^{-1} [6]. The differential cross section is measured in a fiducial phase space as a function of the p_T and absolute rapidity of the top quark, and of the p_T, absolute rapidity and invariant mass of the $t\bar{t}$ system, where the top quark is defined using particle-level objects. The event selection requires an electron and a muon with opposite charge, and at least two jets, from which at least one is b-tagged. Figure 4 shows the normalized spectrum of the invariant mass of the $t\bar{t}$ system compared to the predictions from various QCD MC generators. All the predictions are consistent with the measured data, except the one from Powheg-Box interfaced with Herwig++. A similar inconsistency between data and the Powheg-Box + Herwig++ prediction is observed also for the top quark p_T spectrum, while the other MC predictions are consistent with data.

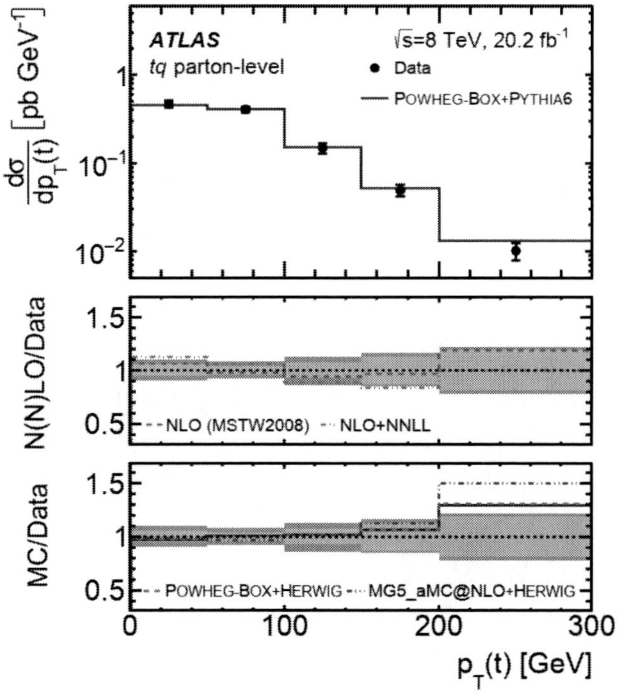

Figure 5. Absolute unfolded differential cross-section as a function of the p_T of the top quark (t) [7]. The unfolded distributions are compared to QCD NLO and NLO+NNLL calculations as well as various MC predictions. The vertical error bars on the data points denote the total uncertainty. The dashed (red) line in the central distribution shows the NLO prediction calculated using MCFM. The dash-dot (blue) line is the NLO+NNLL prediction. The bottom distribution compares the data with the MC predictions from Powheg-Box (orange dashed line) and MadGraph5_aMC@NLO (purple dash-dotted line). The inner (yellow) band in the bottom part of each figure represents the statistical uncertainty of the measurement, and the outer (green) band the total uncertainty.

3 Single top production - *t*-channel

Inclusive and differential cross sections for the t-channel single top production are measured in detail at $\sqrt{s} = 8$ TeV in ATLAS [7]. The leptonic decay of the top quark is selected by requiring one electron or muon, missing transverse energy, and two jets, of which one is b-tagged. Several kinematic variables are used to derive a neural-network discriminant. The charge of the lepton allows to distinguish between top quark particle and antiparticle (t and $\bar{t}$). The measured inclusive cross sections in the full phase space for t and $\bar{t}$ production have uncertainties of ~7% and ~9%, respectively, and are in agreement with the NNLO QCD predictions. The inclusive cross sections are measured as well in a fiducial phase space with lower uncertainties of ~6% and ~8% for t and $\bar{t}$ productions, respectively. The differential cross sections as a function of the p_{T} and rapidity of t and $\bar{t}$ are measured at parton and particle levels. Figure 5 shows the measured parton-level p_{T} spectrum of t compared to NLO QCD and NLO+NLL QCD calculations, and to various MC predictions. All the differential cross section measurements are in good agreement with the SM.

The ratio of the t and $\bar{t}$ production cross sections, R_t, is an interesting observable which can be determined more precisely than the individual cross sections due to the cancellation of several systematic uncertainties in the ratio. This observable is sensitive to the difference between PDFs for u and d quarks. The measured observable R_t is shown in Figure 6 as a function of the center-of-mass energy. It is compared to NLO QCD calculations using various PDF sets (CT14, NNPDF 3.0 and MMHT2014). All the measurements of R_t are consistent with all the predictions.

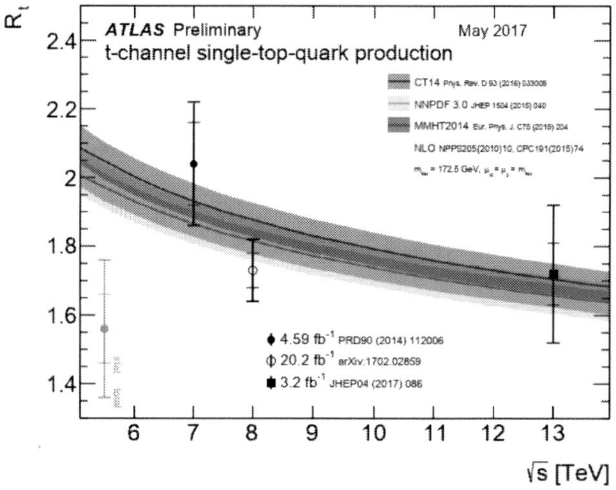

Figure 6. Summary of the ATLAS measurements of R_t, the ratio of the t-channel top-quark production cross-section to the t-channel antitop-quark production cross-section [4]. The data measurements are compared to NLO QCD calculations using the CT14, NNPDF 3.0 and MMHT2014 PDF sets. The coloured bands represent the uncertainties on the theoretical predictions (scale and PDF uncertainties).

4 Single top production - associated with a *W* boson

The inclusive cross section for Wt production at $\sqrt{s} = 8$ TeV is measured in the full phase space [8]. Two oppositely charged leptons (e or μ) and one b-tagged jet are required, but due to the overwhelming $t\bar{t}$ background, further discrimination is provided by boosted decision trees. The inclusive cross section in the full phase space has an uncertainty of ~20%, and it is consistent with the NLO+NNLL QCD prediction. Assuming both, Wt and $t\bar{t}$, processes as signal, the cross section is measured as well in a fiducial phase space, which is defined at particle level. The main requirements for the fiducial phase space are: two leptons with $p_T > 25$ GeV, one jet with $p_T > 20$ GeV, and $E_T^{\text{miss}} > 20$ GeV. The measured value of the fiducial cross section is shown in Figure 7 and it is compared to various MC NLO QCD predictions interfaced with PS generators. The predictions differ in the used PDF set, the scheme to handle the Wt and $t\bar{t}$ interference, overall normalization, and used MC generators. All the tested predictions are consistent with the measurement.

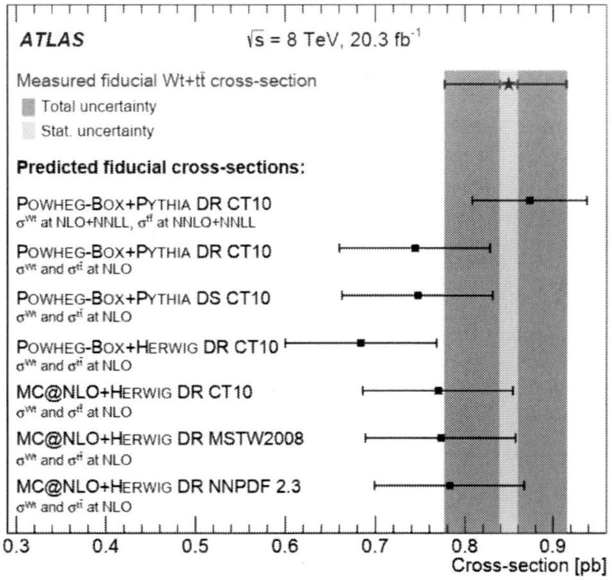

Figure 7. Comparison of the measured fiducial cross-section to theoretical predictions in a fiducial phase space requiring two leptons with $p_T > 25$ GeV and $|\eta| < 2.5$, one jet with $p_T > 20$ GeV and $|\eta| < 2.5$, and $E_T^{\text{miss}} > 20$ GeV [8]. The predictions are computed at NLO accuracy for the fiducial acceptance and the inclusive cross-section, except for the first one, for which the inclusive cross-sections for Wt and $t\bar{t}$ are computed at NLO+NNLL and NNLO+NNLL accuracy, respectively.

5 Single top production - *s*-channel

Evidence for the s-channel single top production at LHC was first published by ATLAS using $\sqrt{s} =$ 8 TeV data [9]. Events containing one electron or muon, large missing transverse momentum and two b-tagged jets are selected. The measurement is very challenging due to the low cross section and the large background contributions, which are mainly due to $t\bar{t}$ and W boson production in association with heavy-flavor jets. The matrix element (ME) method is optimized to derive a discriminant which allows to separate signal from background events. Figure 8 shows this discriminant for data and prediction. One can observe that there is reasonable agreement between data and prediction, and that the discriminant has good discrimination power between signal and background. The observed (expected) signal significance is 3.2 (3.9) standard deviations. The measured s-channel single top inclusive cross section of $\sigma = 4.8 \pm 0.8(\text{stat.})^{+1.6}_{-1.3}(\text{syst.})$ pb is consistent with the SM expectation.

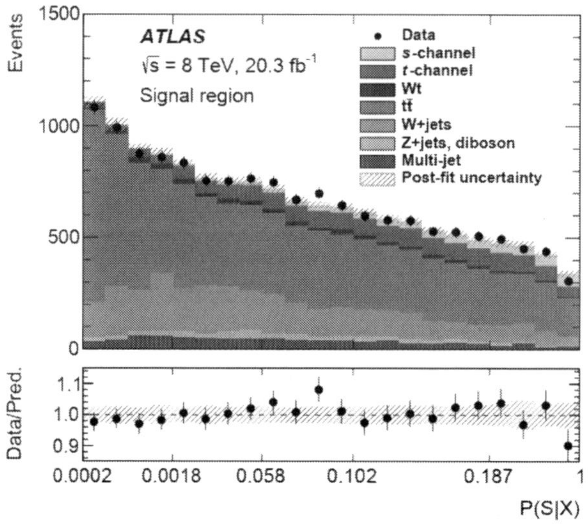

Figure 8. Post-fit distribution of the ME discriminant in the signal region [9]. All samples are scaled by the fit result utilizing all fit parameters. The hatched bands indicate the total uncertainty of the post-fit result including all correlations. The ME distributions are made using the optimized, non-equidistant binning which is also applied in the signal extraction fit.

6 Summary and outlook

The ATLAS experiment has an extensive program of inclusive and differential top cross section measurements. The uncertainties of the $t\bar{t}$ inclusive cross section measurements are only of a few percent. The $t\bar{t}$ differential cross section measurements probe the TeV scale and their main uncertainties are due to signal modeling and JES. There is evidence for s-channel single top production, while the t-channel and Wt productions have been explored in detail. The single top production measurements are dominated by data statistics, JES and MC modeling uncertainties. All the performed measurements are consistent with the SM.

The top quark production cross section measurements are further on-going at the ATLAS experiment while there are more data delivered by the LHC. More data allow to explore higher scales in the $t\bar{t}$ differential cross section measurements and they make the single top cross section measurements more precise.

References

[1] ATLAS Collaboration, JINST **3** (2008) S08003
[2] ATLAS Collaboration, Eur. Phys. J. C **74** (2014) 3109
[3] ATLAS Collaboration, Phys. Lett. B **761** (2016) 136–157
[4] ATLAS Collaboration, https://atlas.web.cern.ch/Atlas/GROUPS/PHYSICS/CombinedSummary Plots/TOP/ (May 2017)
[5] ATLAS Collaboration, Phys. Rev. D **93** (2016) 032009
[6] ATLAS Collaboration, Eur. Phys. J. C **77** (2017) 292
[7] ATLAS Collaboration, Eur. Phys. J. C **77** (2017) 531

[8] ATLAS Collaboration, JHEP **01** (2016) 064
[9] ATLAS Collaboration, Phys. Lett. B **756** (2016) 228-246

EPJ Web of Conferences **158**, 04002 (2017) DOI: 10.1051/epjconf/201715804002
QFTHEP 2017

Probing anomalous off-shell electroweak dipole moment of the top quark

Edward Boos[1,2], *Vyacheslav* Bunichev[1], *Lev* Dudko[1], and *Georgii* Vorotnikov[1]

[1]*Skobeltsyn Institute of Nuclear Physics M.V. Lomonosov Moscow State University, Moscow 119991, Russian Federation*

[2]*Faculty of Physics M.V.Lomonosov Moscow State University Leninskie Gory, Moscow 119991, Russian Federation*

Abstract. A possibility to search for anomalous Wtb interaction $\partial^{\mu} W^{-}_{\mu} \bar{b}(f^{L}_{W} P_{L} + f^{R}_{W} P_{R})t$ is considered for the first time. Despite such couplings do not contribute to both pair top quark production or three main electroweak production modes, they could provide non-zero contributions to associated single top quark production with W or Z bosons or in association with two forward light jets via electroweak fusion diagrams.

1 Introduction

The Standard Model (SM) is a theory, which is now tested and proved in a hundreds of different experiments. However, SM obliviously is not a final theory, but a very successful effective theory within the electroweak energy scale. One of the primary goal of the Large Hadron Collider is to search for physics beyond the Standard Model (BSM). A new physics may manifest itself through different ways. Many BSM theories predict new particles if the energy of collision is greater than the production threshold ($E_{\text{collision}} > E_{\text{threshold}}$). However, even in opposite scenario, when ($E_{\text{collision}} < E_{\text{threshold}}$) BSM physics could prove itself through anomalous couplings originating from higher dimensional gauge invariant operators or interference terms of new resonances with the SM particles, that leads to deviation from SM prediction in production cross sections and/or kinematic distributions. The top quark sector of SM owing to its unique properties such as large mass, small lifetime, spin correlations and very small mixing to the light quark generations, could be a window to BSM physics. The effective Lagrangian approach is common used in experimental physics for model-independent description possible effects of BSM physics. Thus, lowest-dimension Lagrangian [1] for the Wtb vertex usually written as:

$$\mathfrak{L} = \frac{g}{\sqrt{2}} \bar{\mathrm{b}} \gamma^{\mu} \left(f^{\mathrm{L}}_{\mathrm{V}} P_{\mathrm{L}} + f^{\mathrm{R}}_{\mathrm{V}} P_{\mathrm{R}} \right) \mathrm{t} \mathrm{W}^{-}_{\mu} - \frac{g}{\sqrt{2}} \bar{\mathrm{b}} \frac{\sigma^{\mu\nu} \partial_{\nu} \mathrm{W}^{-}_{\mu}}{M_{\mathrm{W}}} \left(f^{\mathrm{L}}_{\mathrm{T}} P_{\mathrm{L}} + f^{\mathrm{R}}_{\mathrm{T}} P_{\mathrm{R}} \right) \mathrm{t} + \mathrm{h.c.}, \quad (1)$$

where $P_{\mathrm{L,R}} = (1 \mp \gamma_5)/2$, $\sigma_{\mu\nu} = i(\gamma_{\mu}\gamma_{\nu} - \gamma_{\nu}\gamma_{\mu})/2$, g is the coupling constant of the weak interaction, the constant $f^{\mathrm{L}}_{\mathrm{V}}$ represents the left-handed vector coupling, presented in SM, $f^{\mathrm{R}}_{\mathrm{V}}$ is the anomalous right-handed vector coupling and $f^{\mathrm{L}}_{\mathrm{T}}$ ($f^{\mathrm{R}}_{\mathrm{T}}$) represents the anomalous left-handed (right-handed) tensor coupling. Such anomalous couplings nowadays are relatively well studied both theoretically [2]-[5] and experimentally [6]-[8]. In this paper we discuss additional anomalous couplings which were not considered yet from phenomenology point of view.

2 Relevant processes

As was mentioned in the paper [1] there is an additional contribution to the effective Lagrangian in case of off-shell W boson:

$$\partial^{\mu} W^{-}_{\mu} \bar{b} P_{\rm L} t.$$

Such an interaction which one can call "off-shell electroweak dipole moment" (osEDM) appears in the effective electroweak non-linear chiral Lagrangian approach [9]. Phenomenological study of this interaction term has significant difficulties. The interaction term is totally vanish for an on-shell W boson or in the case W couples to massless on-shell fermions. Therefore, the anomalous osEDM couplings do not contribute to the main top quark production processes, both pair or any of three electroweak production t-, s-, tW-channels.

Despite such couplings do not contribute to usual top quark production processes they could provide non-zero contributions to associated single top quark production with W or Z bosons or in association with two forward light jets via electroweak fusion diagrams. The representative Feynman diagrams with non-zero osEDM contribution are shown in Fig. 1. In this paper we consider more simple $2 \to 3$ processes of associated production top quark and electroweak gauge bosons.

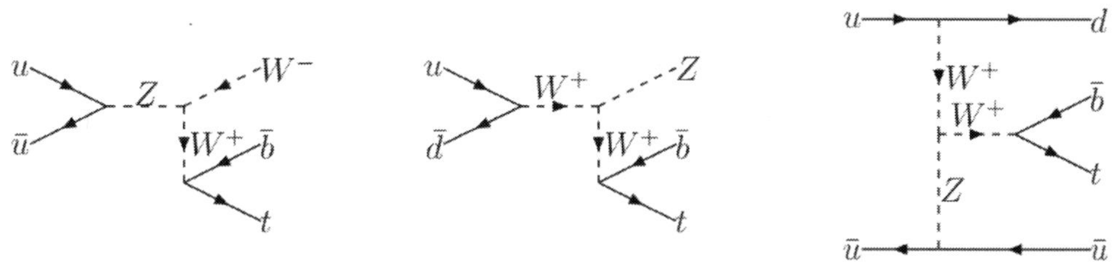

Figure 1. Feynman diagrams of the processes, which could be used to search for the anomalous off-shell W boson couplings.

The effective Lagrangian used in the analysis has the following form:

$$\mathcal{L}_{eff.} = \frac{g}{\sqrt{2}} \frac{v}{\Lambda^2} \partial^{\mu} W^{-}_{\mu} \bar{b} (f^{\rm L}_{\rm W} P_{\rm L} + f^{\rm R}_{\rm W} P_{\rm R}) t + {\rm h.c.}. \quad (2)$$

Here $\frac{v}{\Lambda^2}$ corresponds to natural dimensional factor of the anomalous operators, $v = 246$ GeV is vacuum expectation value, Λ is the scale of a new physics and constants $f^{\rm L}_{\rm W}$ and $f^{\rm R}_{\rm W}$ are the strengths of left and right anomalous couplings. In the purpose of CP conserving, we consider the case of a purely imaginary $f^{\rm L}_{\rm W}$ and $f^{\rm R}_{\rm W}$ values [9].

For studying processes with anomalous osEDM the CompHEP 4.5.2 package [10] was used to model kinematic distributions of the osEDM and SM contributions including their interference. The Feynman rules for the Wtb vertex were generated with the LanHEP package [11], and implemented into CompHEP.

The complete set of electroweak diagrams for the associated tbZ production is shown in Fig. 2. The first three diagrams give non-zero contribution with considered anomalous coupling while the last two contain only SM contributions. The second process tbW also contains non-zero osEDM contributions from the diagrams like the first diagram in Fig 1. However, in this case there are SM diagrams corresponding to top quark pair production and associated tW single top production with cross sections significantly larger than possible anomalous contributions. Moreover, in case of tWb final state there is contribution from gluon - gluon collision with huge SM rate and zero osEDM contribution.

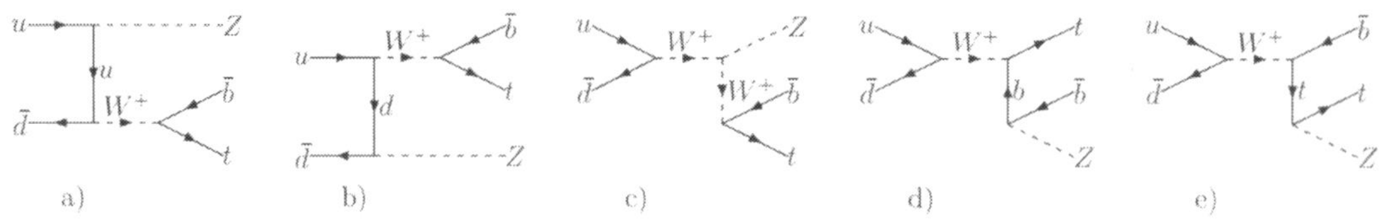

Figure 2. All relevant Feynman diagrams of the process of associated production of a top quark with Z boson.

3 Kinematic distribution and extracting potential

Since anomalous osEDM interactions are directly proportional to the momentum of the W boson, it gives approximately linear grow of sensitivity to the couplings with the energy of the colliding protons. Thus, for such projects as High Energy LHC or Future Circular Collider, efficiency of the searches for osEDM couplings will be increasing 5-10 times compare to current energies at the LHC.

The anomalous interactions could manifest itself not only in cross section rates, but also in different kinematics of the process. For example, on Fig. 3 one can see distribution of a transverse momentum of the Z boson in the process of its associated production with top quark. In the case of anomalous contributions, the momentum, in average, is much higher than for SM production, and is peaking at about 160 GeV instead of $\approx$50 GeV for SM. The same situation is observed for the transverse momenta of the t and b quarks.

The difference in transverse momentum distributions of the final state particles is provided mostly by their space bearings, as one can see on the Fig. 4. The asymmetry between the negative and positive parts of the plots is provided by choose of initial state quarks ($u\bar{d}$), which gives symmetric picture with a complementary $\bar{d}u$ initial state.

The angular distributions (Fig. 5) are also shown different behaviour of SM and osEDM production. The anomalous interactions are characterized by more smooth distribution both for Zt or tb angle.

However, one could see there are pure left and right osEDM interactions are totally indistinguishable on the Fig. 3-5. The way to see the difference between the couplings is in their interference with the SM part. The right anomalous interaction has a negative interference term with SM, while the left coupling has a small, but positive one (Fig. 6). Since there are no experimental limits on osEDM couplings we take such values for the couplings to get visible deviations from SM.

Another distinction between the coupling is the cross section rate for sea massive quarks in the initial state (Fig. 7). The difference is provided by diagrams a) and b) on the Fig 2, where quarks of different flavours are taken into account for propagators of the initial virtual quark exchange.

4 Conclusion

The possible anomalous osEDM interactions are phenomenologically studied for the first time. Such couplings do not contribute to both pair top quark production or three main electroweak production processes. However, there are several processes considered in this article which have non zero

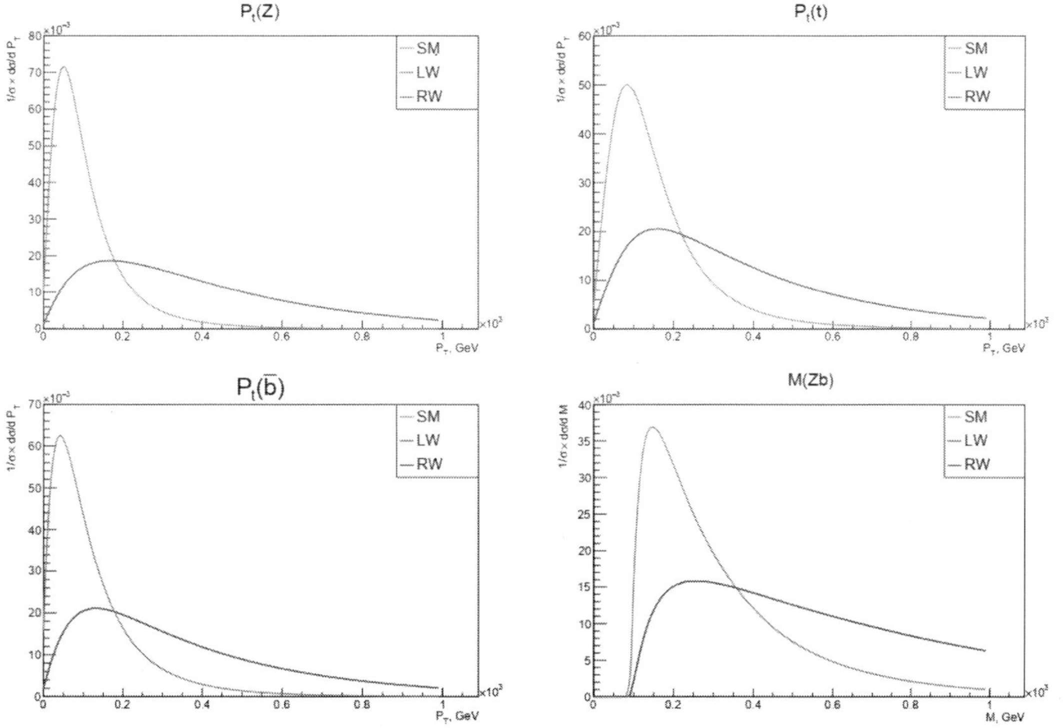

Figure 3. Transverse momentum distributions for the SM interaction (SM) and pure left (LW) and right (RW) osEDM in the process of associated production of a top quark with Z boson with the $u\bar{d}$ initial state. There are plots for Z boson (top left), top quark (top right) and bottom quark (bottom left), and invariant mass for Z and b (bottom right). The distributions of each coupling are normalized to the corresponding cross sections.

contribution from such anomalous couplings. Necessary calculations of these processes have been performed. The difference in the shape of the kinematic distributions owing to the contribution of the osEDM couplings are demonstrated in this article. The demonstrated difference in the kinematic distributions is promising for the search and achieving the first experimental limits on these couplings. More accurate estimation of the sensitivity to osEDM couplings requires simulation of the hadronization with detector response and careful estimation of the background contributions.

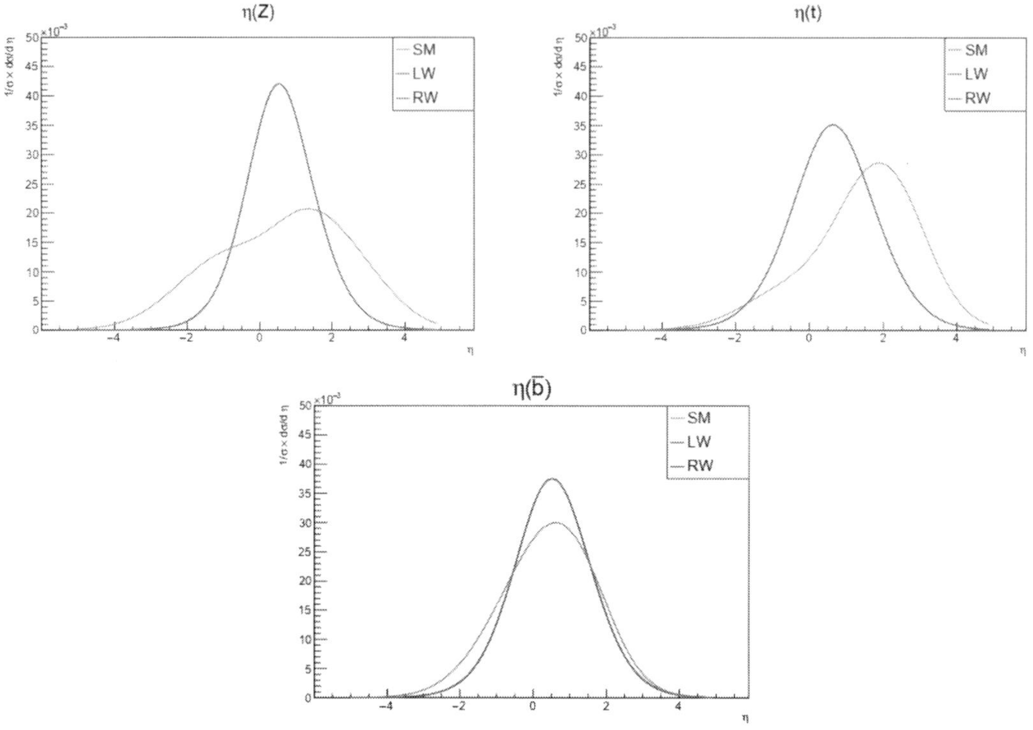

Figure 4. Pseudorapidity distributions for the SM interaction (SM) and pure left (LW) and right (RW) osEDM in the process of associated production of a top quark with Z boson with the $u\bar{d}$ initial state. There are plots for Z boson (top left), top quark (top right) and bottom quark (bottom). The distributions of each coupling are normalized to the corresponding cross sections.

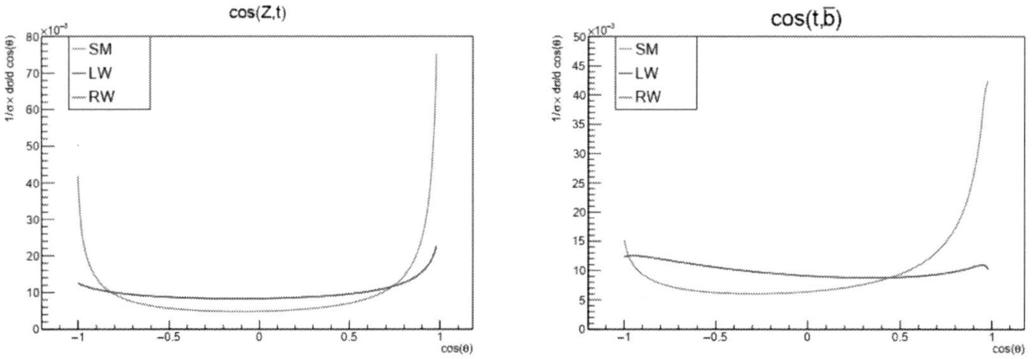

Figure 5. Angular distributions for the SM interaction (SM) and pure left (LW) and right (RW) osEDM in the process of associated production of a top quark with Z boson with the $u\bar{d}$ initial state. There are plots for cosine of angle between Z boson and top quark (left), and cosine of angle between top and bottom quarks (right). The distributions of each coupling are normalized to the corresponding cross sections.

DOI: 10.1051/epjconf/201715804002

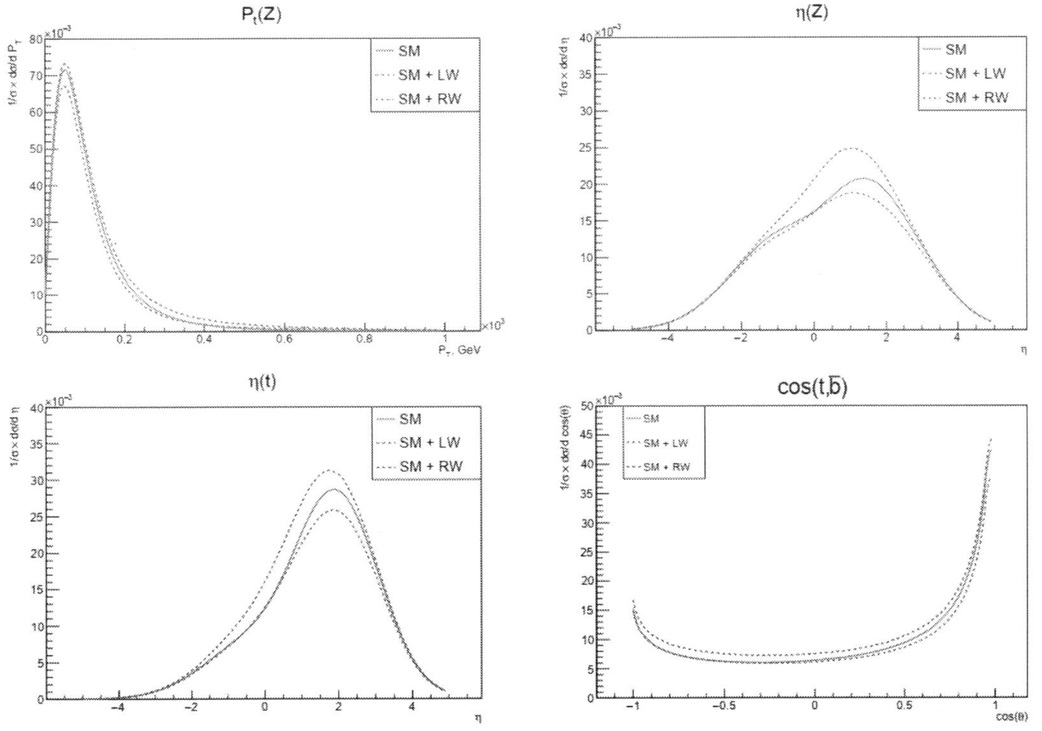

Figure 6. Distributions for the SM interaction (SM), SM and left osEDM (SM+LW), and SM and right anomalous coupling (SM+RW), including interference, in the process of associated production of a top quark with Z boson with the $u\bar{d}$ initial state. There are plots for Transverse momentum of Z boson (top left), pseudorapidity of Z boson (top right) and top quark (bottom left), and cosine of angle between top and bottom quarks (bottom right). The all distributions are normalized to the cross section of SM interaction.

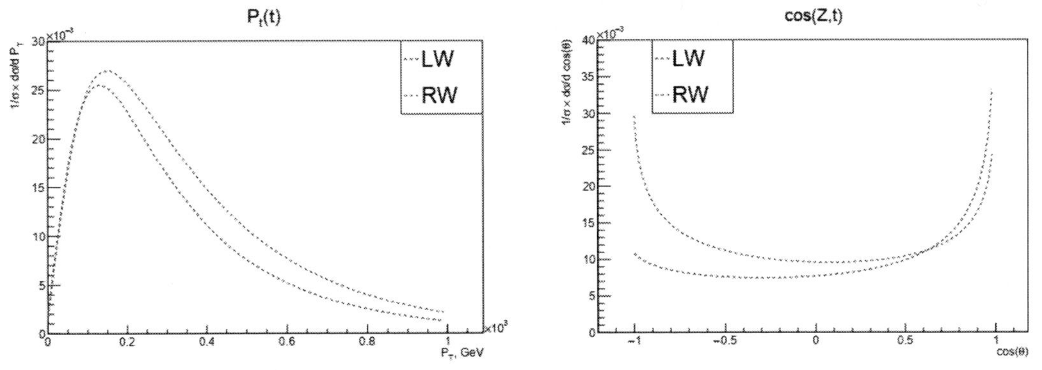

Figure 7. Distributions for the pure left (LW) and right (RW) anomalous couplings, in the process of associated production of a top quark with Z boson with the $c\bar{s}$ initial state. There are plots for Transverse momentum of top quark (left), and cosine of angle between Z boson and t quark (right). The all distributions are normalized to the cross section of left osEDM.

Acknowledgements

The work was supported by grant 16-12-10280 of Russian Science Foundation.

References

[1] G. L. Kane, G. A. Ladinsky and C. P. Yuan, Phys. Rev. D **45** (1992) 124.
[2] W. Buchmuller and D. Wyler, Nucl. Phys. B **268** (1986) 621.
[3] K. Whisnant, J. M. Yang, B. L. Young and X. Zhang, Phys. Rev. D **56** (1997) 467
[4] E. Boos, M. Dubinin, M. Sachwitz and H. J. Schreiber, Eur. Phys. J. C **16** (2000) 269
[5] J. A. Aguilar-Saavedra, J. Carvalho, N. F. Castro, F. Veloso and A. Onofre, Eur. Phys. J. C **50** (2007) 519
[6] V. M. Abazov *et al.* [D0 Collaboration], Phys. Lett. B **708** (2012) 21
[7] V. Khachatryan *et al.* [CMS Collaboration], JHEP **1501** (2015) 053
[8] V. Khachatryan *et al.* [CMS Collaboration], JHEP **1702** (2017) 028
[9] F. Larios and C. P. Yuan, Phys. Rev. D **55** (1997) 7218
[10] E. Boos *et al.* [CompHEP Collaboration], Nucl. Instrum. Meth. A **534** (2004) 250
[11] A. Semenov, Comput. Phys. Commun. **201** (2016) 167

EPJ Web of Conferences **158**, 04003 (2017) DOI: 10.1051/epjconf/201715804003
QFTHEP 2017

Modeling of associated tW single top production including the interference with $t\bar{t}$ diagrams

Edward Boos[1,2,⋆], *Viacheslav* Bunichev[1,⋆⋆], *Lev* Dudko[1,⋆⋆⋆], *Igor* Myagkov[1,⋆⋆⋆⋆], and *Maxim* Perfilov[1,†]

[1]*Lomonosov Moscow State University, Skobeltsyn Institute of Nuclear Physics (SINP MSU), 1(2), Leninskie gory, GSP-1, Moscow 119991, Russian Federation*
[2]*Faculty of Physics M.V.Lomonosov Moscow State University Leninskie Gory, Moscow 119991, Russian Federation*

Abstract. The associated tW single top production channel is one of the main mechanism of the electroweak production of top quarks at the LHC. Correct simulation of the SM prediction for this process is one of the important tasks in the SM measurements and BSM searches. One of the difficulties to simulate it correctly is a significant contribution of the interference terms with the pair top quark production. In this article authors compare most popular simulation schemes for tW production with simulation of the complete set of diagrams and propose new scheme to simulate interference terms without negative weights of the events.

1 Introduction

Electroweak top quark production is one of the most promising topic to search for New Physics. These are unique processes to measure some of the SM parameters. There are three different production channels. They are classified using the topology of diagrams which include W-boson [1] as t-channel, s-channel and associated tW production. The associative tW channel becomes important at the LHC and requires proper simulation. The correct simulation of the tW process is difficult due to the well known problem [2]-[8] that at NLO level tW diagrams interfere with pair top quark production diagrams. The top quark pair production has more than ten times larger cross section than the associated tW production and interference is an important part in case of the precise measurements of tW process. There are different approaches to simulate associated tW production with different advantages and disadvantages. It is also possible to simulate complete set of tree level diagrams including pair top quark production and interference terms. In the next sections we discuss different popular approaches, simulation of the complete set of Feynman diagrams and new scheme proposed

⋆e-mail: boos@theory.sinp.msu.ru
⋆⋆e-mail: bunichev@theory.sinp.msu.ru
⋆⋆⋆e-mail: dudko@sinp.msu.ru
⋆⋆⋆⋆e-mail: imyagkov@cern.ch
†e-mail: maxim.perfilov@cern.ch

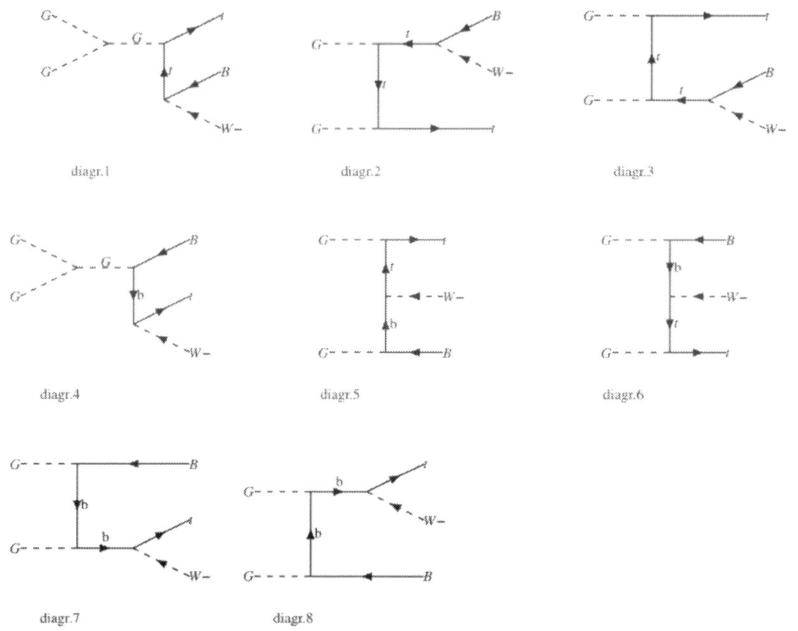

Figure 1: Diagrams for the process $gg \to t\bar{b}W^-$.

in this article which we call "contact interactions" (CI) scheme. The CI scheme solves number of simulation problems. These schemes were implemented in CompHEP [9, 10] package and the necessary simulations have been performed. The results are presented in the next sections.

2 Modeling approaches

The main LO contribution to the tW process comes from subprocesses with b-quark in the initial state. At NLO level this contribution yields from gluon collisions and leads to the process $gg \to tWb$ which includes 8 diagrams. The corresponding diagrams are presented in Fig. 1. Simulation of all these diagrams is the most precise approach to model this process. However, part of these diagrams correspond to the pair top quark production with significantly lager cross section than the other diagrams with single top quark production. It is possible to distinguish double resonant diagrams with pair top quark production and single resonant diagrams with single top quark production. In additional, there is significant interference between double and single resonant diagrams. Accurate simulation of the associated tW electroweak single top quark production have to take into account not only single resonant diagrams, but also the negative interference terms with pair top quark production. Different approximate schemes have been developed to properly model tW channel of single top quark production.

Diagram Removal. The most simple recipe is to remove all Feynman diagrams with pair top quark production, diagrams 1-3 in Fig 1. Such approach is called "Diagram Removal" (DR) scheme [2, 3, 7]. In this approach the interference of single resonant diagrams (diagrams 4-8 in Fig. 1) and double resonant diagrams (diagrams 1-3 in Fig. 1) are not taken into account.

Diagram Subtraction. More sophisticated approach called Diagram Subtraction [3, 7] takes into account interference terms. This method implies removing only squared double resonant diagrams

DOI: 10.1051/epjconf/201715804003

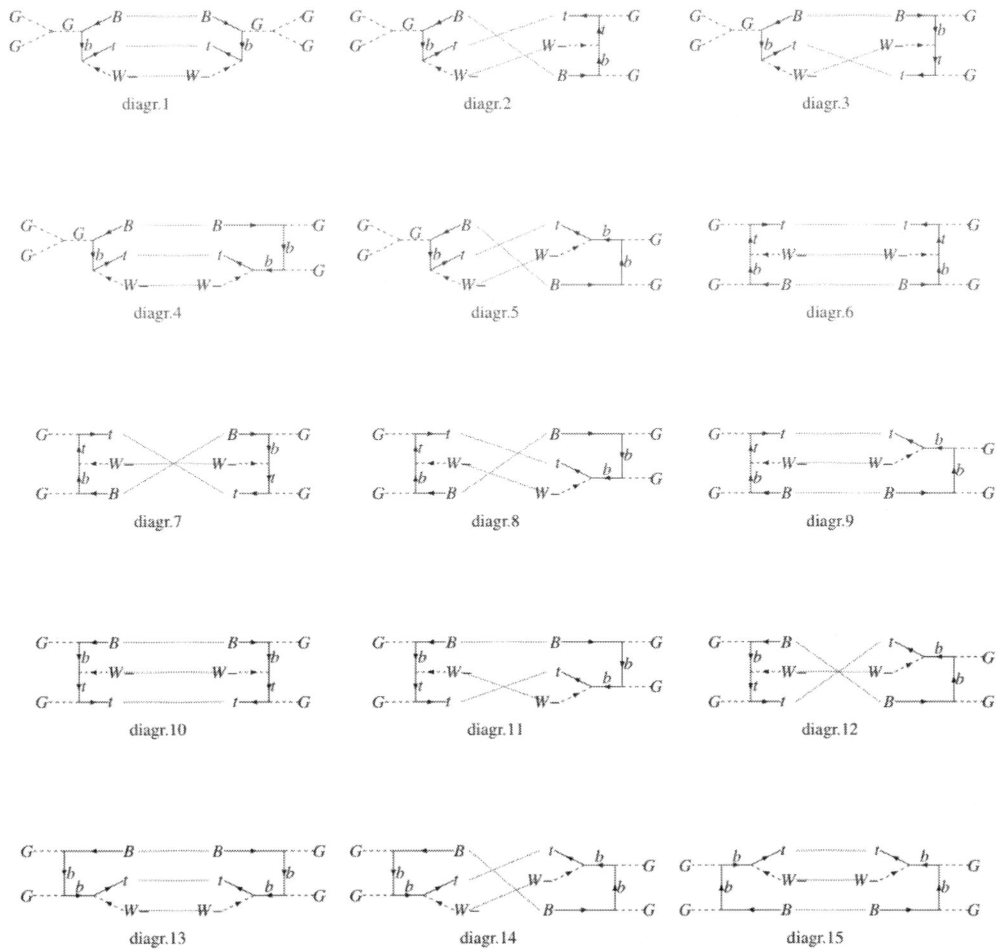

Figure 2: Squared single resonant diagrams for the process $gg \to t\bar{b}W^-$.

during the modeling. The remaining diagrams are shown in Fig. 2 for the single resonant contribution and in Fig. 3 for the interference between single and double resonant contributions. This approach allows to retain all of anticipated single top quark production properties in the Wtb vertex. On the other hand this subset of diagrams also introduces extra singularities from an uncoupled top quark decaying into W boson and b quark in either part of squared diagrams. This effect can be clearly observed in a distribution of invariant mass of system of W boson and b quark, and leads to negative weights of the significant part of the simulated events.

Contact interactions scheme. An improvement can be done to previous scheme in order to get rid of described singularities. These singularities can be excluded entirely by replacing top quark decaying into W boson and b quark by a single contact interaction in the double resonant part of the squared interference diagrams in Fig. 3. In this case the distribution of invariant mass of W and b turns up to be single-top-like while keeping all interference in place without negative contribution.

Full scheme. Modeling the complete set of diagrams is the most precise approach. For the SM measurements this is not very interesting approach since for the most of the SM measurements one

DOI: 10.1051/epjconf/201715804003

Figure 3: Squared interference diagrams of single and double resonant contributions for the process $gg \to t\bar{b}W^-$.

needs to separate top quark single electroweak production mode and pair production mode in strong interactions. In the full modeling scheme, which will be referenced as a "Full scheme", different terms contribute in total cross section. These are (as shown in Fig. 1) resonant paired production diagrams, resonant single top production diagrams, and their interfere terms. To retain all single top-sensitive interactions, interference terms need to be studied as well as resonant single top terms.

3 Comparison

In the previous section we discuss different schemes to simulate tW with different contributions. In this section we compare the common kinematic distributions achieved in different schemes. Figures 4-17 give a vision of different impacts of particular schemes. Since the cross sections are significantly different for double resonant and single resonant contributions each curve is normilised to its integral (total cross section for this contribution) for the Figs. 4-15. The curves for DR, DS, contact interactions approaches, full scheme and pair top quark production are shown on the plots. The curve for the interference term between paired and single production processes is shown on the plots as well. The transverse momenta and pseudorapidity of the t, W, b are shown in Figs. 4-6 and Figs. 8-10. The vector sum of W boson and b-quark transverse momenta and its pseudorapidity is shown in Fig. 7 and Fig. 11. The significant difference between simulation approaches is visible in the shapes of b-quark and *Wb* system distributions. The angular distributions are show in Figs. 12-15. The cosine of angle between W and t momenta in the laboratory rest frame and in Wb rest frame are shown in Figs. 12 and Figs. 13 correspondingly. The cosine of angle between t and b quarks momenta are shown in Figs. 14 and Figs. 15 correspondingly. Angular distributions demonstrate significant difference of DS and CI schemes from other simulation approaches. This is reasonable, since the DS and DI schemes takes into account the interference terms. The differential cross section of invariant mass of W boson and b-quark is shown in Fig. 16. The distribution of invariant mass of W boson and b-quark for the interference term is presented in Fig. 17.

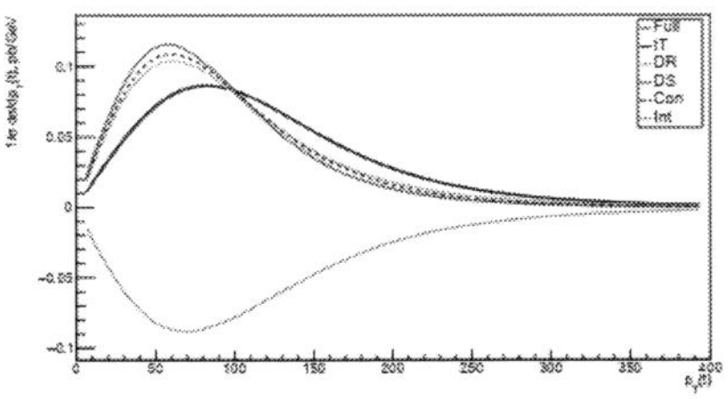

Figure 4: Transverse momentum of top quark

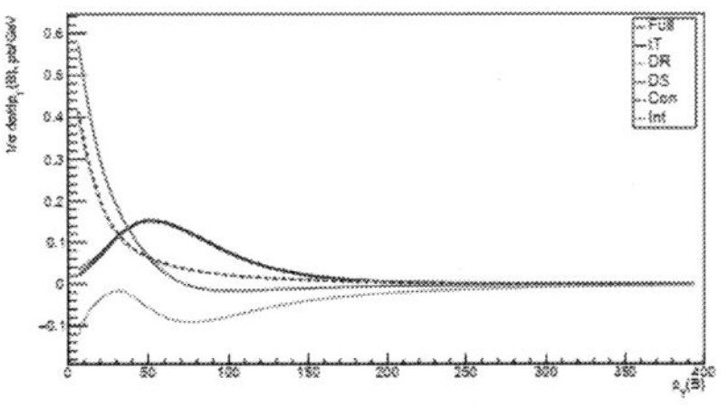

Figure 5: Transverse momentum of b quark.

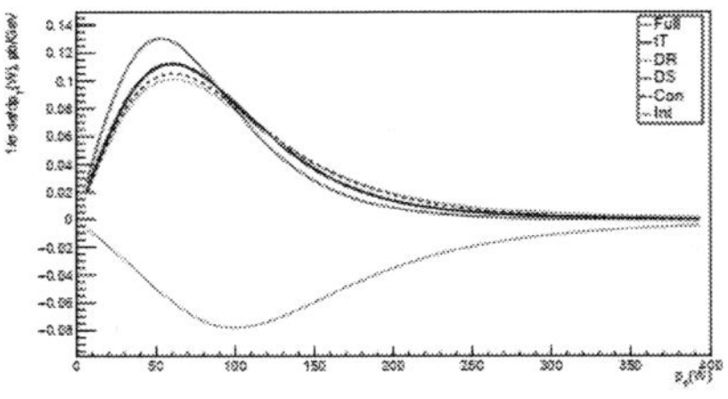

Figure 6: Transverse momentum of W boson

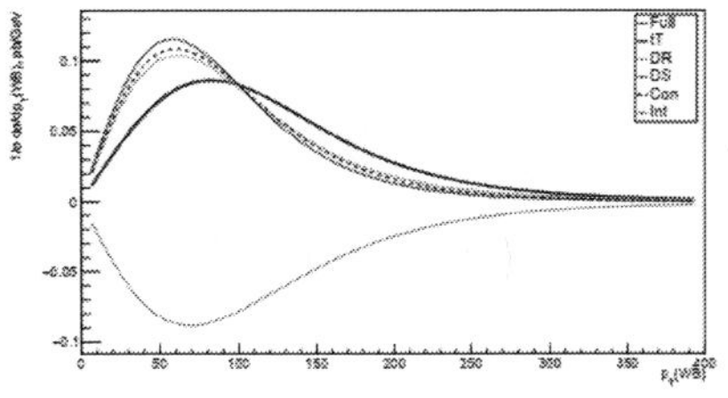

Figure 7: Transverse momentum of system of W boson and b quark.

4 Kinematic separation

Kinematic separation of double and single resonant contributions is also possible [2, 3]. Such approach takes into account interference terms, it is gauge invariant and avoids negative weights of the events. In this section we compare simulation in different approximate schemes with the simulation of complete set of diagrams. For the calculations we exclude phase space region $145 < M_{Wb} < 200$ GeV which corresponds to the second top quark pole. In this way the cross sections for all schemes are comparable. The correct simulation should correspond to single resonant diagrams with interference term, and off-shell contribution for the double resonant diagrams. Simulation of the complete set

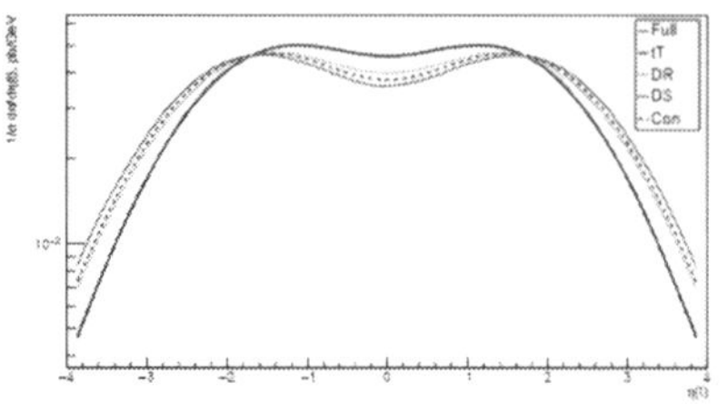

Figure 8: Pseudorapidity of top quark.

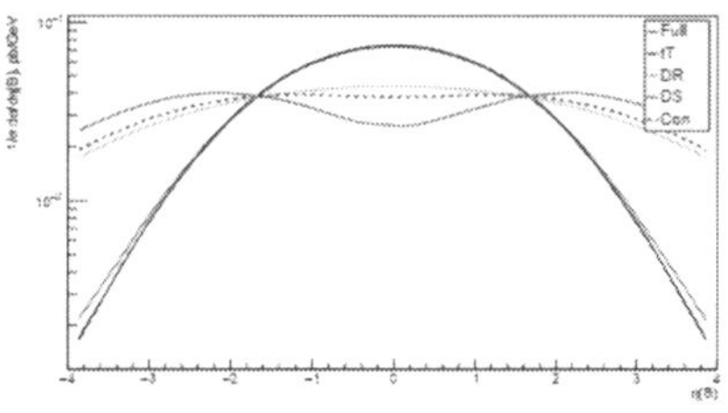

Figure 9: Pseudorapidity of b quark.

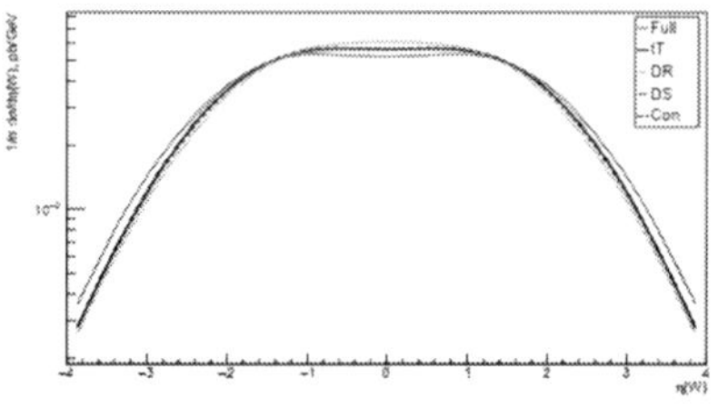

Figure 10: Pseudorapidity of W boson.

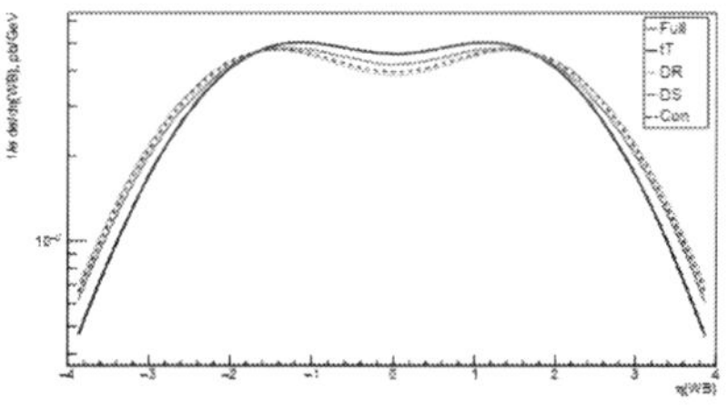

Figure 11: Pseudorapidity of system of W boson and b quark.

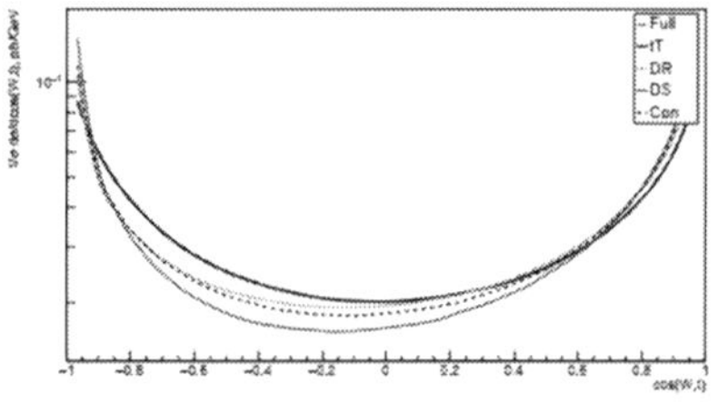

Figure 12: Cosine between W boson and top quark.

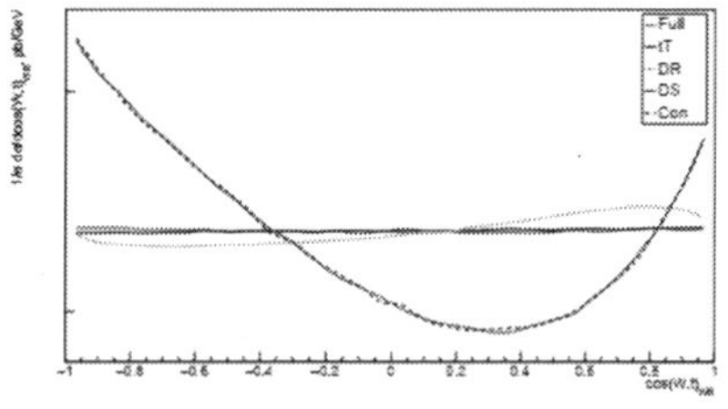

Figure 13: Cosine between W boson and top quark in rest frame of W and b.

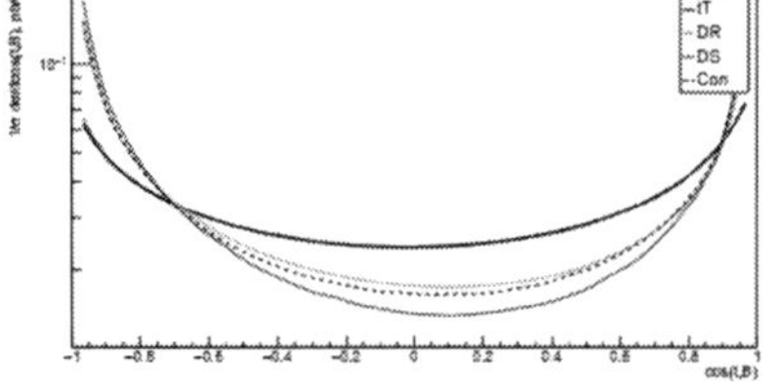

Figure 14: Cosine between top quark and b quark.

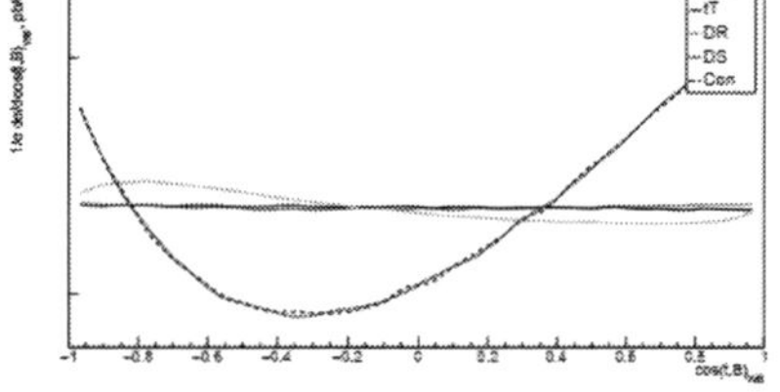

Figure 15: Cosine between top quark and b quark in rest frame of W and b.

DOI: 10.1051/epjconf/201715804003

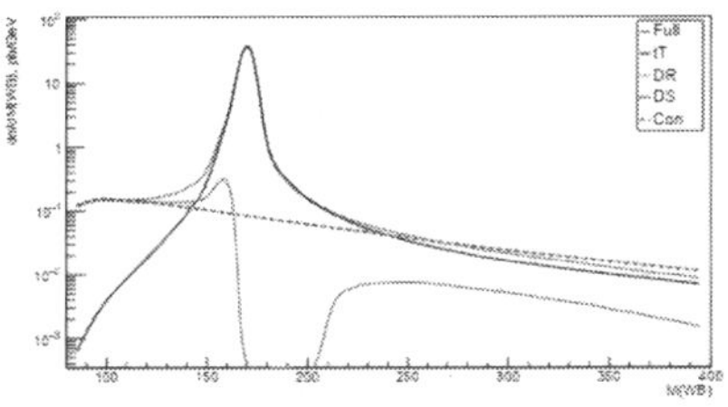

Figure 16: Invariant mass of W boson and b quark, no interference term.

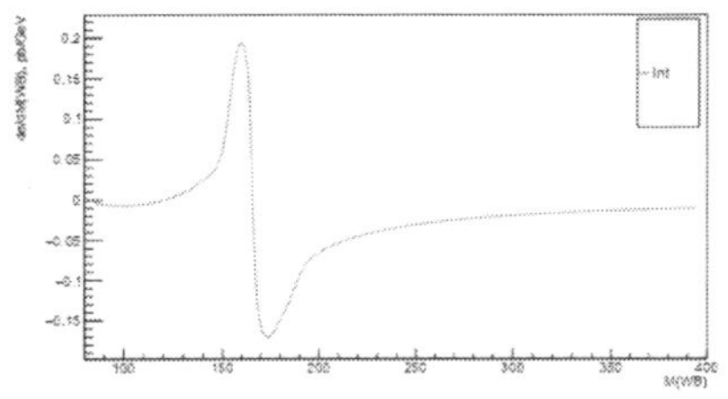

Figure 17: Invariant mass of W boson and b quark, only interference term.

of diagrams ("Full scheme") is most accurate way to model this process in this phase space region. Kinematic distributions in Figs. 18-27 demonstrate the difference between DR, DS, CI schemes and full scheme. The contribution of the negative interference is also shown as well as pure double resonant contribution. DR scheme should reproduce only single resonant contribution. DS scheme have to take into account single resonant contribution and negative interference. CI scheme is designed to reproduce kinematics of the single resonant contribution with interference term, but without negative events. Double resonant contribution reproduces only pair top quark production. Full scheme reproduces all contributions, single resonant, double resonant and their interference.

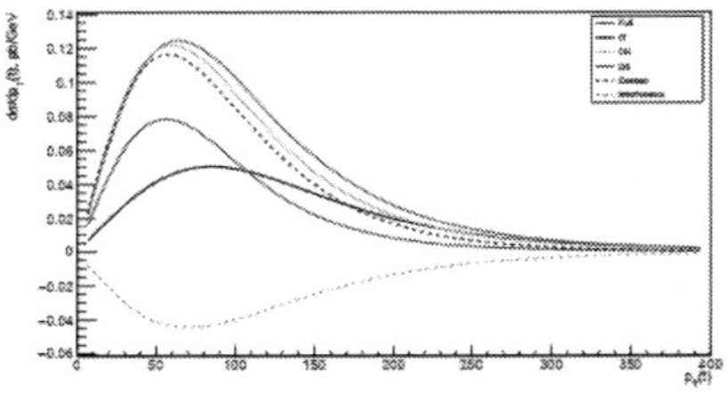

Figure 18: Transverse momentum of top quark with excluded region $145 < M_{Wb} < 200$ GeV

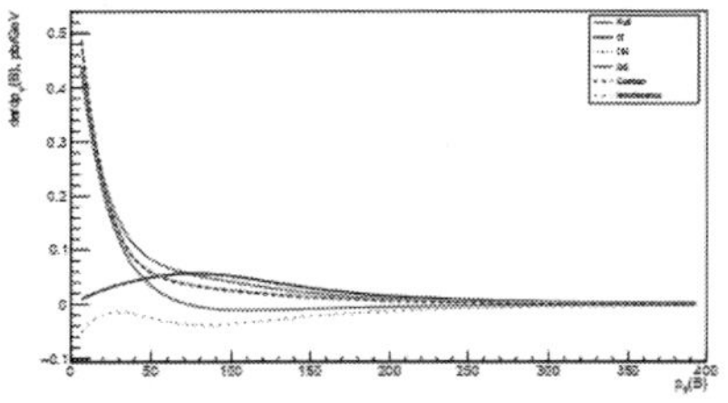

Figure 19: Transverse momentum of b quark with excluded region $145 < M_{Wb} < 200$ GeV.

5 Conclusion

Presented techniques allow different approaches to simulate tWb production and may be useful for different particular analyses. However, only modeling of the complete set of Feynman diagrams retains correct distributions of kinematic variables. The "Diagram Removal" scheme discards all interference terms which are significant part of single top quark production processes. The "Diagram Subtraction" scheme provides best differentiation in angular distributions and could be useful for separation of single top quark production, but introduces negative weights for the significant part of the events and barely can be used in the analysis. The improved "Contact interaction" scheme does not have events with negative weights and shows little difference from "Diagram subtraction" scheme for the angular distributions, thus could be used for effective simulation and separation of single top quark production process.

DOI: 10.1051/epjconf/201715804003

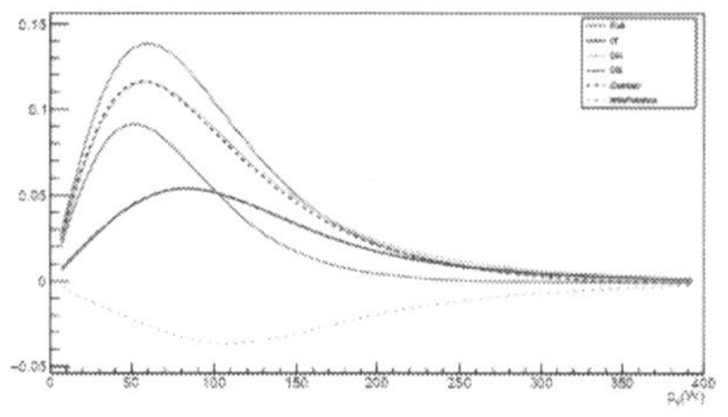

Figure 20: Transverse momentum of W boson with excluded region $145 < M_{Wb} < 200$ GeV

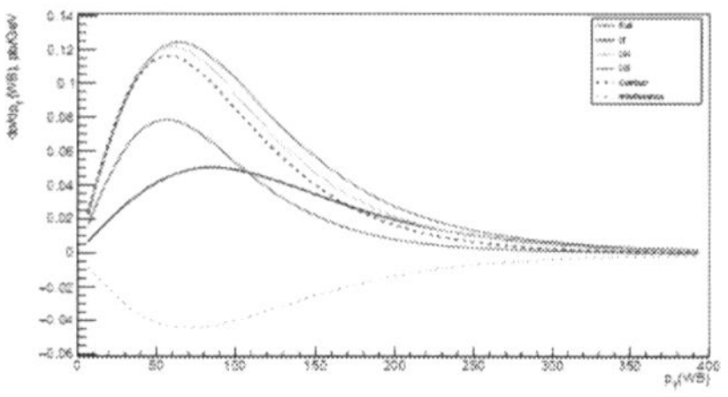

Figure 21: Transverse momentum of system of W boson and b quark with excluded region $145 < M_{Wb} < 200$ GeV.

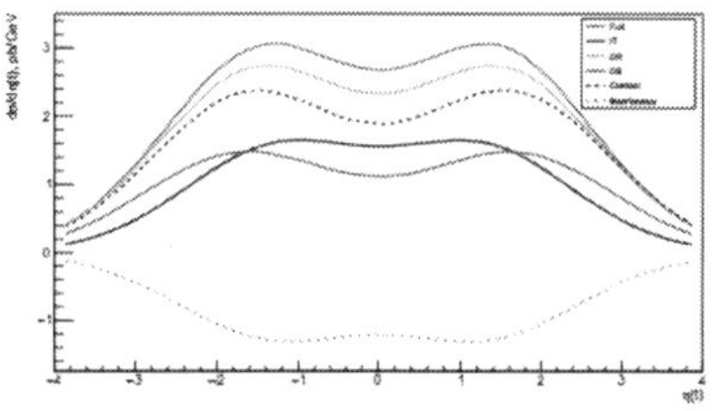

Figure 22: Pseudorapidity of top quark with excluded region $145 < M_{Wb} < 200$ GeV.

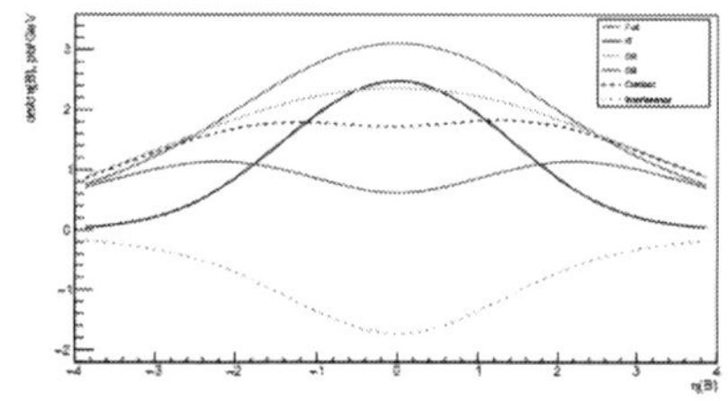

Figure 23: Pseudorapidity of b quark with excluded region $145 < M_{Wb} < 200$ GeV.

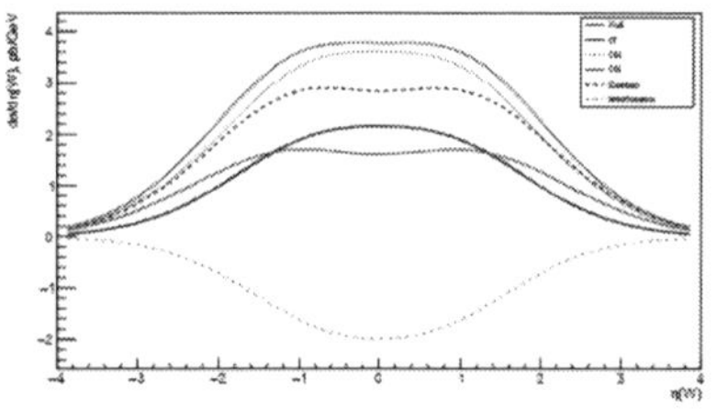

Figure 24: Pseudorapidity of W boson with excluded region $145 < M_{Wb} < 200$ GeV.

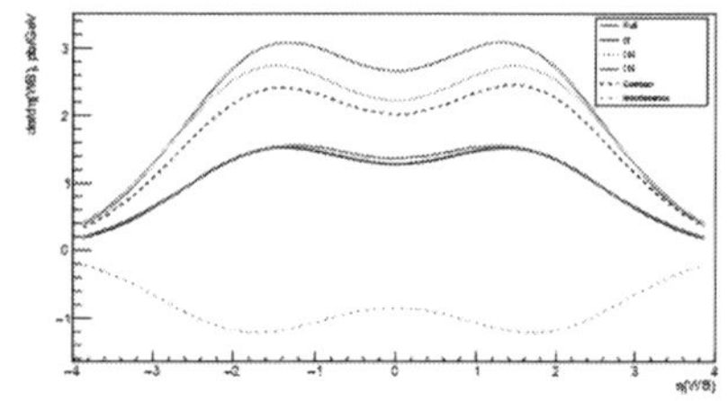

Figure 25: Pseudorapidity of system of W boson and b quark with excluded region $145 < M_{Wb} < 200$ GeV.

Kinematic cut based approach, as shown in Section 4 is generally less sensitive to electroweak properties since important part of phase space is discarded and the impact from paired top quark production is not reduced entirely. Nonetheless, such approach takes into account all necessary contributions and gives a more stable result because it uses purely gauge invariant kinematic reduction procedure. Kinematic separation does not have negative weights of the events and can be used in the searches for the deviations from the SM.

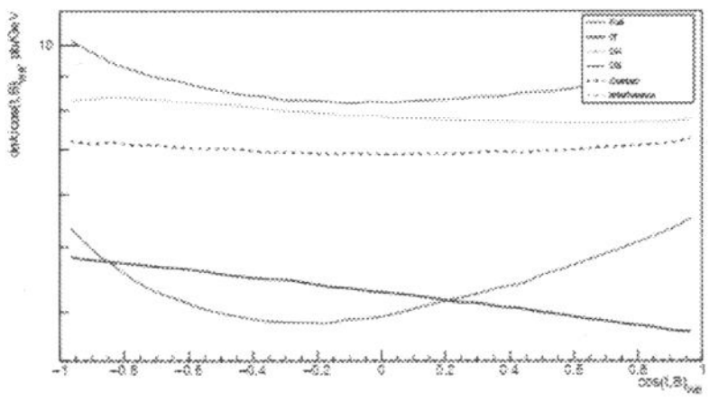

Figure 26: Cosine between top quark and b quark in rest frame of W and b with excluded region $145 < M_{Wb} < 200$ GeV.

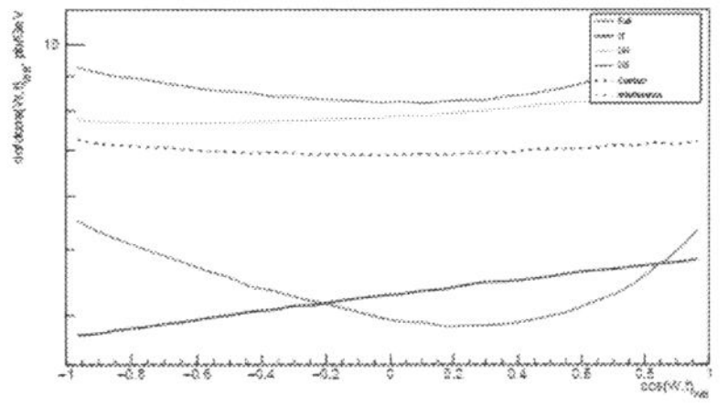

Figure 27: Cosine between W boson and top quark in rest frame of W and b with excluded region $145 < M_{Wb} < 200$ GeV.

Acknowledgments

The work was supported by grant 16-12-10280 of Russian Science Foundation.

References

[1] S. S. D. Willenbrock and D. A. Dicus, Phys. Rev. D **34**, 155 (1986). doi:10.1103/PhysRevD.34.155

[2] A. S. Belyaev, E. E. Boos and L. V. Dudko, Phys. Rev. D **59**, 075001 (1999) doi:10.1103/PhysRevD.59.075001 [hep-ph/9806332].

[3] T. M. P. Tait, Phys. Rev. D **61**, 034001 (1999) doi:10.1103/PhysRevD.61.034001 [hep-ph/9909352].

[4] S. Zhu, Phys. Lett. B **524**, 283 (2002) Erratum: [Phys. Lett. B **537**, 351 (2002)] doi:10.1016/S0370-2693(02)01952-4, 10.1016/S0370-2693(01)01404-6 [hep-ph/0109269].

[5] J. M. Campbell and F. Tramontano, Nucl. Phys. B **726**, 109 (2005) doi:10.1016/j.nuclphysb.2005.08.015 [hep-ph/0506289].

[6] B. P. Kersevan and I. Hinchliffe, JHEP **0609**, 033 (2006) doi:10.1088/1126-6708/2006/09/033 [hep-ph/0603068].

[7] S. Frixione, E. Laenen, P. Motylinski, B. R. Webber and C. D. White, JHEP **0807**, 029 (2008) doi:10.1088/1126-6708/2008/07/029 [arXiv:0805.3067 [hep-ph]].

[8] C. D. White, S. Frixione, E. Laenen and F. Maltoni, JHEP **0911**, 074 (2009) doi:10.1088/1126-6708/2009/11/074 [arXiv:0908.0631 [hep-ph]].

[9] E. Boos *et al.* [CompHEP Collaboration], Nucl. Instrum. Meth. A **534**, 250 (2004) doi:10.1016/j.nima.2004.07.096 [hep-ph/0403113].

[10] E. Boos *et al.*, PoS ACAT **08**, 008 (2008) [arXiv:0901.4757 [hep-ph]].

EPJ Web of Conferences **158**, 04004 (2017) DOI: 10.1051/epjconf/201715804004
QFTHEP 2017

Probing the Anomalous Wtb couplings in different schemes of tWb associated production modeling

Alexey Baskakov[1,*], *Eduard* Boos[1,2,**], *Viacheslav* Bunichev[1,***], *Georgi* Vorotnikov[1,****], *Lev* Dudko[1,†], *Igor* Myagkov[1,‡], and *Maxim* Perfilov[1,§]

[1] *Lomonosov Moscow State University, Skobeltsyn Institute of Nuclear Physics (SINP MSU), 1(2), Leninskie gory, GSP-1, Moscow 119991, Russian Federation*
[2] *Faculty of Physics, Lomonosov Moscow State University, Leninskie Gory, Moscow 119991, Russian Federation*

Abstract. The impact of anomalous Wtb couplings on tWb process of the single top quark production has been tested in different schemes of tWb modeling. The scenario with right-handed vector operator (RV) in Wtb vertex with and without the presence of Standard model (SM) left-handed vector operator in Wtb vertex has been considered. It is shown that for anomalous Wtb couplings searches in tWb process usage of the full gauge invariant set of diagrams, with the same final state for the pair top quark production as well as for the single top quark production, is more justified than the usage of the schemes with deletion of some part of diagrams.

Introduction

There are three well-known processes of a single top quark production at hadron colliders, they are called s-channel, t-channel and tW-associated single top production [1]. All of them, except the last one, have been observed at Tevatron collider [2],[3],[4]. Due to its tiny cross section on the Tevatron collider tW-associated production have been observed at the LHC collider only recently [5]. While considering the tW-associated production it is necessary to take into account next-to-leading order corrections, specifically for the gluon splitting $g \to b\bar{b}$ which leads to the additional b-quark appearance in the final state, that is to say tWb processes. The main difficulty here is the identical final state of the double top quark production and the single top production so the interference between single top quark production Feynman diagrams and double top quark production ones appears.

The full process $pp \to t\bar{b}W^-$ contains subprocesses with two quarks or two gluons in the initial state and the last ones are dominant. In Fig. 1 all the diagrams of $gg \to t\bar{b}W^-$ subprocess for the tWb final state are listed.

*e-mail: a.baskakov@cern.ch
**e-mail: Eduard.Boos@cern.ch
***e-mail: bunichev@theory.sinp.msu.ru
****e-mail: georgii.vorotnikov@cern.ch
†e-mail: lev.dudko@cern.ch
‡e-mail: igor.myagkov@cern.ch
§e-mail: maksim.perfilov@cern.ch

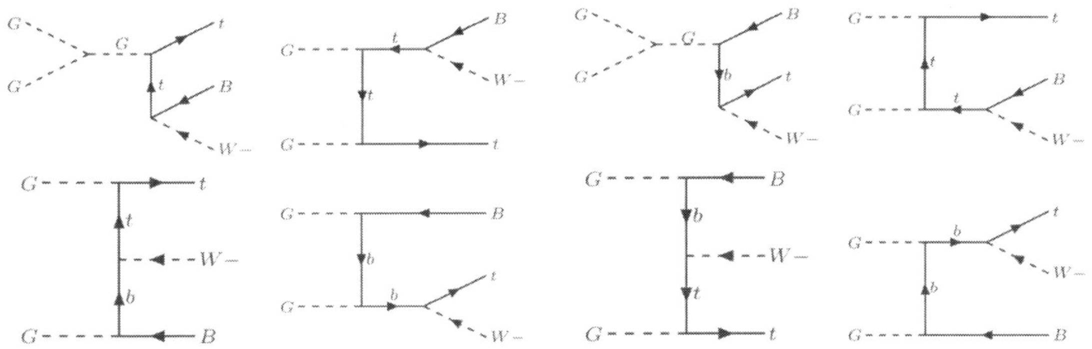

Figure 1: Diagrams for the process $gg \to t\bar{b}W^-$.

Let's consider the diagrams from Fig. 1 more circumstantially. The first, second and fourth diagrams from the top raw contain two top quarks in the final state with the subsequent decay of one of them - these are the diagrams of the double top production; other five diagrams correspond to the single top production. Different schemes of tWb modeling extract the single top production from the double top production in some or another way. E.g. in so-called *Diagram Removal* scheme [6],[7] all three double top production diagrams are removed and the other diagrams are considered. In this case the interference between double top and single top production diagrams has not taken into account. In a *Diagram Subtraction* scheme [7],[8] deletion of the double top production diagrams occurs after the squaring all the diagrams. In this case all the interference terms between double top and single top productions are kept, however this procedure leads to significant part of the simulated events with negative weights.

Among a very interesting physics tasks in the top quark sector, the searches for the anomalous operators in the Wtb vertex is one of the most promising ones due to direct impact of the anomalous operators to the cross sections and kinematic distributions of the single top quark production processes. There are direct searches for the anomalous operators in the t-channel single top production [9],[10],[11] however the same ones but in the tW-associated single top production have not been provided so far.

The aim of this paper is the study of the influence of the anomalous operators in the Wtb vertex on a different schemes of the tWb processes modeling. There is a short introduction to the anomalous Wtb operators formalism in Sec.1. In Sec.2 some distributions for different schemes of tWb processes modeling are shown for different contributions of the anomalous operators.

1 Anomalous structure of the Wtb vertex.

The most general, lowest dimension, CP conserving Lagrangian for the Wtb vertex has the following form [13, 14]:

$$\mathfrak{L} = -\frac{g}{\sqrt{2}}\bar{b}\gamma^{\mu}\left(f_{\mathrm{V}}^{\mathrm{L}}P_{\mathrm{L}} + f_{\mathrm{V}}^{\mathrm{R}}P_{\mathrm{R}}\right)tW_{\mu}^{-} - \frac{g}{\sqrt{2}}\bar{b}\frac{i\sigma^{\mu\nu}\partial_{\nu}W_{\mu}^{-}}{M_{\mathrm{W}}}\left(f_{\mathrm{T}}^{\mathrm{L}}P_{\mathrm{L}} + f_{\mathrm{T}}^{\mathrm{R}}P_{\mathrm{R}}\right)t + h.c. \tag{1}$$

where $P_{\mathrm{L,R}} = \frac{1\mp\gamma_5}{2}$, $\sigma_{\mu\nu} = \frac{i}{2}(\gamma_{\mu}\gamma_{\nu} - \gamma_{\nu}\gamma_{\mu})$, form factor $f_{\mathrm{V}}^{\mathrm{L}}$ ($f_{\mathrm{V}}^{\mathrm{R}}$) represents the left-handed (right-handed) vector coupling, $f_{\mathrm{T}}^{\mathrm{L}}$ ($f_{\mathrm{T}}^{\mathrm{R}}$) represents the left-handed (right-handed) tensor coupling. The SM has the following set of coupling values: $f_{\mathrm{V}}^{\mathrm{L}} = V_{\mathrm{tb}}$, $f_{\mathrm{V}}^{\mathrm{R}} = f_{\mathrm{T}}^{\mathrm{L}} = f_{\mathrm{T}}^{\mathrm{R}} = 0$.

DOI: 10.1051/epjconf/201715804004

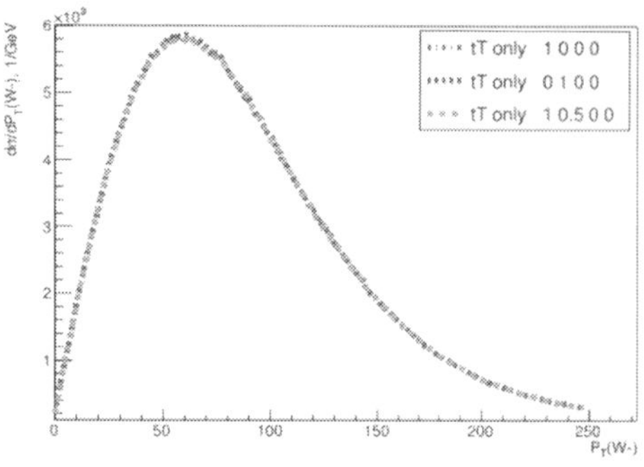

Figure 2: The distribution of the W-boson momenta for the top-pair production processes for different kinds of the *Wtb* vertex structure

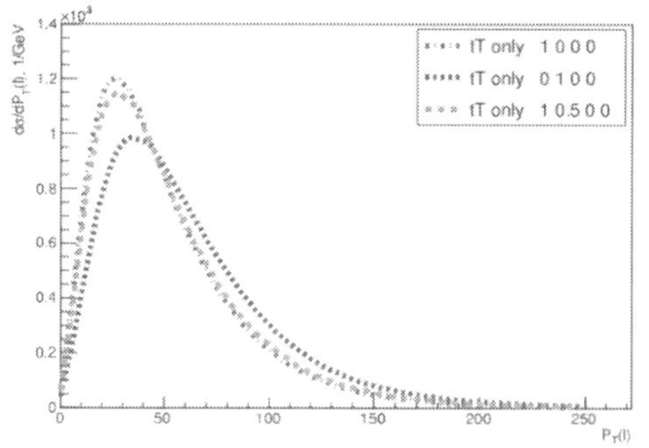

Figure 3: The distribution of the lepton from W-boson decay momenta for the top-pair production processes for different kinds of the *Wtb* vertex structure

At this paper the case with only vector operators are in the *Wtb* vertex (so called (f_V^L, f_V^R) scenario) is considered and all the notations have the form "f_V^L f_V^R 0 0".

2 The sensitivity of the different schemes of the tWb processes modeling to the presence of the anomalous right-handed vector operator in the Wtb vertex.

In this Sec. some kinematic distributions of the *tWb* process final state particles have been provided using the different subsets of the full diagram set of Fig. 1. The idea is to understand which scheme of the *tWb* processes modeling is the most correct to search for the anomalous couplings in the *Wtb* vertex. We tested the distribution of the transverse momenta of the W-boson of the process $gg \to t\bar{b}W^-$ (with 3 particles in the final state) and the transverse momenta of the lepton from the subsequent decay of the W-boson (4 particles in the final state, process $gg \to t\bar{b}l\bar{\nu}_l$).

The first, second and fourth diagram of the top raw from Fig. 1 correspond to the double top production. Of course the highlighting of these three diagrams with removing another ones is not the task of the *tWb* processes modeling, however in the pair top quark production processes all top quark decays through the *Wtb* vertex with 100% probability so they are more or less sensitive to the *Wtb* couplings. At the Fig 2 and Fig. 3 the transverse momentum of the W-boson from one of the top quarks decay and lepton from this W-boson decay are shown for the SM case (curve, labeled as "1 0 0 0"), the case with only right-handed vector operator is in the *Wtb* vertex only (curve, labeled as "0 1 0 0") and the case with the simultaneous presence of the SM part and the right-handed vector operator but with the value of its constant is equal to f_V^R = 0.5 (the curve "1 0.5 0 0") to show the situation with the SM and some contribution of new physics.

As it was mentioned earlier, pair top quark production has the top decay *Wtb* vertex only, that's the reason why the W-boson momentum distribution is (almost) not sensitive to the presence of the anomalous couplings in the *Wtb* vertex (all the curves completely coincide to each other). The presence of the additional part in the Lagrangian 1 leads to the enhancement of the production cross section. However, as it was mentioned in [12] the width of a top quark changes correspondingly. That's the reason why the shapes of all the curves from Fig 2 are similar to each other. Only the

DOI: 10.1051/epjconf/201715804004

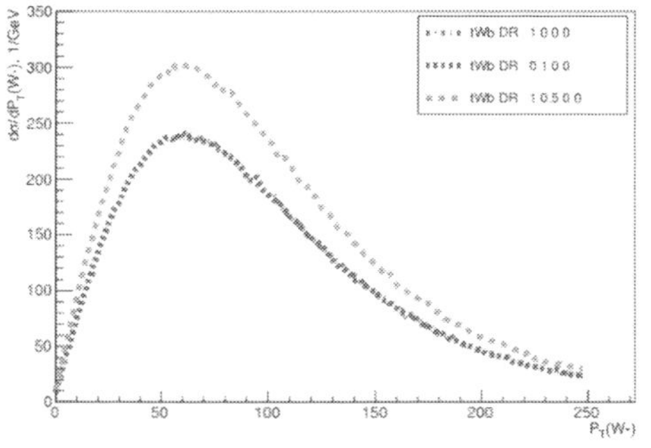

Figure 4: The distribution of the W-boson momenta for the Diagram Removal scheme of the *tWb* -processes modeling for different kinds of the *Wtb* vertex structure

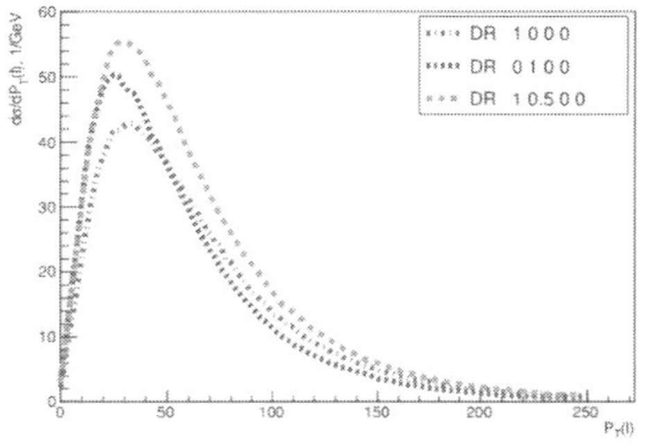

Figure 5: The distribution of the lepton from W-boson decay momenta for the Diagram Removal scheme of the *tWb* -processes modeling for different kinds of the *Wtb* vertex structure

lepton p_T distribution is sensitive to the anomalous right-handed vector operator due to the angular correlation of the top quark and its subsequent decay products.

One may expect the higher sensitivity to the anomalous couplings in the different schemes of *tWb* processes emphasizing due to the presence of two *Wtb* vertices in the single top quark production (and subsequent decays) in comparison to the double top quark production processes. At Fig. 4 and Fig. 5 the same distributions are shown but for the rest of the diagrams (3 diagrams from Fig.1, which correspond to the pair top quark production, have been deleted, other five are considered, this is *Diagram Removal* scheme of the *tWb* modeling).

W-boson p_T distribution for the "1 0 0 0" and "0 1 0 0" cases look exactly the similar to each other (due to the same cross section value in these cases; however the "1 0.5 0 0" curve differs from the another ones. It is explained by the cross section increment for $f_V^L = 1, f_V^R = 0.5$ values of couplings in comparison to the SM case ($f_V^L = 1$, other values of couplings are equal to zero). For the double top production the squared matrix element of the process has the top quark width in the denominator which leads to the same CS value for all three scenarios and the curve "1 0.5 0 0" coincides with the other two ones. It's not the case here, the squared matrix element of the diagrams for "Diagram Removal" scheme doesn't include the top quark width value, which leads to the cross section increment and the "1 0.5 0 0" curve places above the other two ones.

After squaring of all the diagrams from the Fig. 1 and deleting the six squared diagrams which are correspond to the double top production, the remaining squared diagrams will describe the single top production itself and the interference terms between double top and single top production, it is the *Diagram Subtraction* scheme of *tWb* processes modeling. One can expect almost the same sensitivity to the anomalous couplings in this case, that is shown in Figs. 6,7 where the chosen kinematic distributions are shown.

For the schemes considered above, the deletion of some of the diagrams from the complete set Fig.1 is needed. Specifically, using such kind of non-natural procedure leads to a violation of the gauge invariance (it results in the convergence of about 15% for the cross section of *tWb* process calculated in Feynman gauge beside the one calculated in unitary gauge for the *DiagramSubtraction* scheme.

The set of all diagrams from Fig. 1 is the full $tT + tWb$ process and represents itself the most gauge-invariant set of diagrams in comparison to the *Diagram Removal* and *Diagram Subtraction*

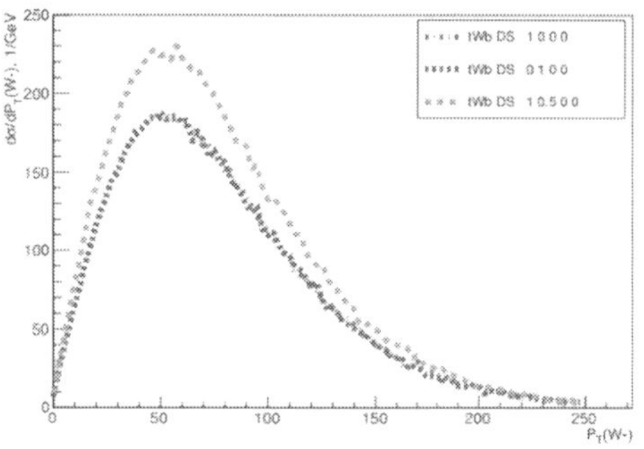

Figure 6: The distribution of the W-boson momenta for the Diagram Subtraction scheme of the *tWb* -processes modeling for different kinds of the *Wtb* vertex structure

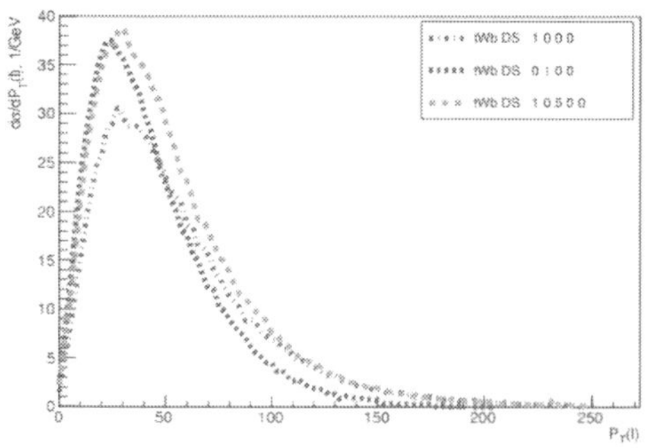

Figure 7: The distribution of the lepton from W-boson decay momenta for the Diagram Subtraction scheme of the *tWb* -processes modeling for different kinds of the *Wtb* vertex structure

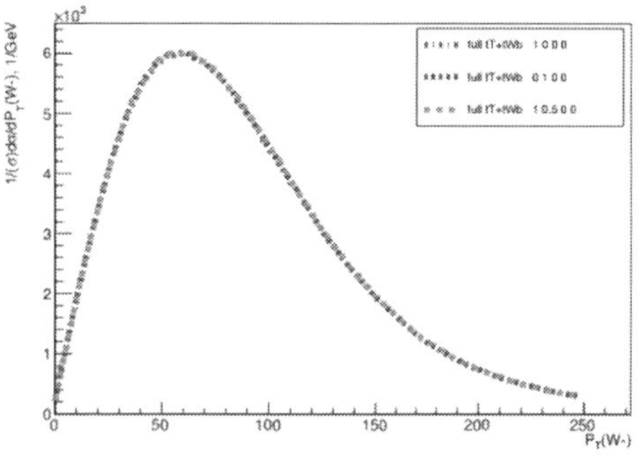

Figure 8: The distribution of the W-boson momenta for the full set of diagrams for different kinds of the *Wtb* vertex structure

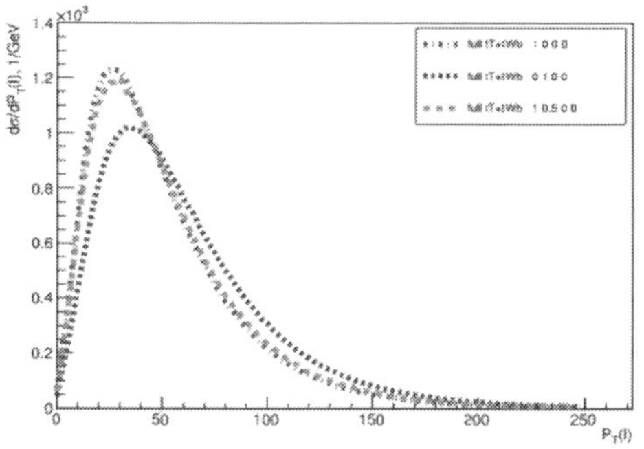

Figure 9: The distribution of the lepton from W-boson decay momenta for the full set of diagrams for different kinds of the *Wtb* vertex structure

schemes. The contribution of the pair top quark production diagrams is huge here, that's why the distributions from Fig. 8 and Fig. 9 look the same as for the double top production ones.

The sensitivity of the different subsets of the diagrams from the whole set of Fig. 1 is demonstrated on Figs. 10 and 11 for the case where right-handed operator only is presented in *Wtb* vertex with the value of its parameter is equal to $f_V^R = 1$ (other couplings are equal to 0). One can see here that different schemes have different sensitivity to anomalous operators.

Conclusion

The impact of the anomalous operators in *Wtb* vertex have been studied in different simulation schemes of *tWb* processes The special attention has been given to the correctness of the simulation and the sensitivity of the search for the anomalous right-handed vector operator in a *Wtb* vertex. As it was expected, manifestation of the anomalous contributions is more strong in the *Diagram Removal* and *Diagram Subtraction* simulation schemes in comparison to the full scheme which considers the

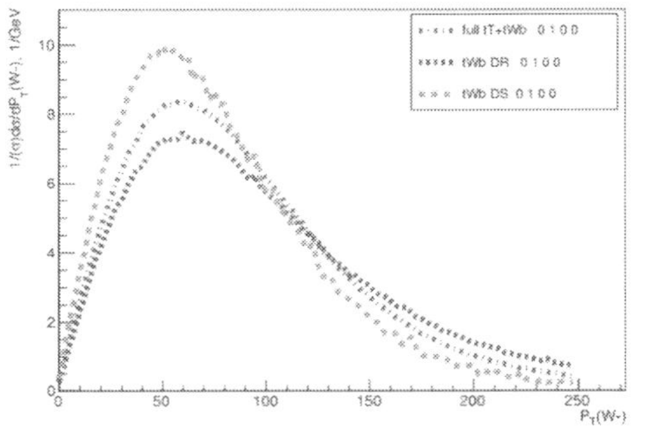

Figure 10: The distribution of the W-boson transverse momenta for the full set of diagrams, tWb Diagram Removal and tWb Diagram Subtraction schemes for the case when right-handed vector operator only is in the Wtb vertex. The curves are normalized to the corresponding cross sections.

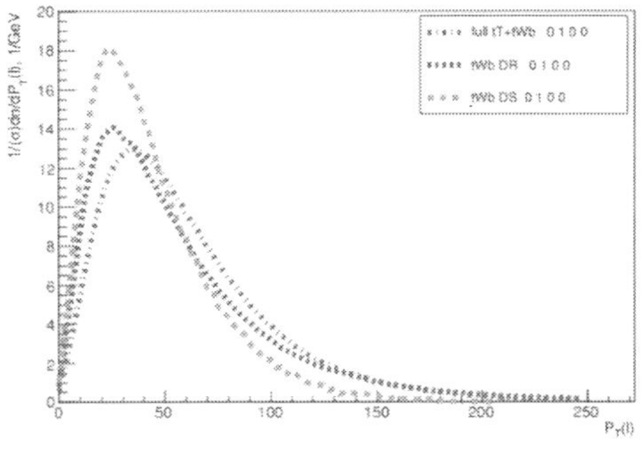

Figure 11: The distribution of the lepton from W-boson decay transverse momenta for the full set of diagrams, tWb Diagram Removal and tWb Diagram Subtraction schemes for the case when right-handed vector operator only is in the Wtb vertex. The curves are normalized to the corresponding cross sections.

complete set of the diagrams. Different behaviour is visible even for the distributions of the W-boson transverse momentum (without its decay) if the anomalous right-handed vector operator in the Wtb vertex turns on. However simulation of the complete set of tWb diagrams is preferable due to the usage of gauge invariant set of diagrams and correct simulation of all necessary properties. The similar sensitivity to the deviations in the couplings is demonstrated in the distributions of top quark decay products, like transverse momenta of a lepton which comes from W-boson decay and in angular variables.

References

[1] S. S. D. Willenbrock and D. A. Dicus, Phys. Rev. D **34**, 155 (1986). doi:10.1103/PhysRevD.34.155

[2] V. M. Abazov *et al.* [D0 Collaboration], Phys. Rev. Lett. **103**, 092001 (2009) doi:10.1103/PhysRevLett.103.092001 [arXiv:0903.0850 [hep-ex]].

[3] T. Aaltonen *et al.* [CDF Collaboration], Phys. Rev. Lett. **103**, 092002 (2009) doi:10.1103/PhysRevLett.103.092002 [arXiv:0903.0885 [hep-ex]].

[4] T. A. Aaltonen *et al.* [CDF and D0 Collaborations], Phys. Rev. Lett. **112**, 231803 (2014) doi:10.1103/PhysRevLett.112.231803 [arXiv:1402.5126 [hep-ex]].

[5] S. Chatrchyan *et al.* [CMS Collaboration], Phys. Rev. Lett. **112**, no. 23, 231802 (2014) doi:10.1103/PhysRevLett.112.231802 [arXiv:1401.2942 [hep-ex]].

[6] A. S. Belyaev, E. E. Boos and L. V. Dudko, Phys. Rev. D **59**, 075001 (1999) doi:10.1103/PhysRevD.59.075001 [hep-ph/9806332].

[7] T. M. P. Tait, Phys. Rev. D **61**, 034001 (1999) doi:10.1103/PhysRevD.61.034001 [hep-ph/9909352].

[8] S. Frixione, E. Laenen, P. Motylinski, B. R. Webber and C. D. White, JHEP **0807**, 029 (2008) doi:10.1088/1126-6708/2008/07/029 [arXiv:0805.3067 [hep-ph]].

[9] V. M. Abazov *et al.* [D0 Collaboration], Phys. Rev. Lett. **101**, 221801 (2008) doi:10.1103/PhysRevLett.101.221801 [arXiv:0807.1692 [hep-ex]].

[10] V. M. Abazov *et al.* [D0 Collaboration], Phys. Lett. B **708**, 21 (2012) doi:10.1016/j.physletb.2012.01.014 [arXiv:1110.4592 [hep-ex]].

[11] V. Khachatryan *et al.* [CMS Collaboration], JHEP **1702**, 028 (2017) doi:10.1007/JHEP02(2017)028 [arXiv:1610.03545 [hep-ex]].

[12] E. Boos, V. Bunichev, L. Dudko and M. Perfilov, Int. J. Mod. Phys. A **32**, no. 02n03, 1750008 (2016) doi:10.1142/S0217751X17500087 [arXiv:1607.00505 [hep-ph]].

[13] W. Buchmuller and D. Wyler, Nucl. Phys. B **268**, 621 (1986). doi:10.1016/0550-3213(86)90262-2

[14] G. L. Kane, G. A. Ladinsky and C. P. Yuan, Phys. Rev. D **45**, 124 (1992). doi:10.1103/PhysRevD.45.124

EPJ Web of Conferences **158**, 04005 (2017) DOI: 10.1051/epjconf/201715804005
QFTHEP 2017

On the Light Massive Flavor Dependence of the Top Quark Mass

André H. Hoang[1,2,⋆], *Christopher* Lepenik[1,⋆⋆], and *Moritz* Preisser[1,⋆⋆⋆]

[1] *University of Vienna, Faculty of Physics Boltzmanngasse 5, A-1090 Wien, Austria*
[2] *Erwin Schrödinger International Institute for Mathematical Physics, University of Vienna, Boltzmanngasse 9, A-1090 Wien, Austria*

Abstract. We provide a systematic renormalization group formalism for the mass effects in the relation of the pole mass and short-distance masses such as the $\overline{\text{MS}}$ mass of a heavy quark Q, coming from virtual loop insertions of massive quarks lighter than Q with the main focus on the top quark. The formalism reflects the constraints from heavy quark symmetry and entails a combined matching and evolution procedure that allows to disentangle and successively integrate out the corrections coming from the lighter massive quarks and the momentum regions between them and also to precisely control the large order asymptotic behavior. With the formalism we systematically sum logarithms of ratios of the lighter quark masses and heavy quark mass, predict the $O(\alpha_s^4)$ virtual quark mass corrections in the pole-$\overline{\text{MS}}$ mass relation and calculate the pole mass differences for the top, bottom and charm quarks with a precision of around 20 MeV.

1 Introduction

The masses of the heavy quarks, especially of the top, are among the most important parameters in the standard model, having strong impact on e.g. precise consistency tests of the standard model and the estimation of the electroweak vacuum stability. However, it must be kept in mind that the mass of a heavy quark Q, due to confinement, is not a physical observable but should be viewed as a formal theory parameter which depends on the renormalization scheme. This concept is incorporated most cleanly in the so-called $\overline{\text{MS}}$ mass $\overline{m}_Q(\mu)$, which is defined analogous to the $\overline{\text{MS}}$-renormalized strong coupling constant $\alpha_s(\mu)$. Like the strong coupling, the $\overline{\text{MS}}$ mass $\overline{m}_Q(\mu)$ depends on a renormalization scale μ. This scale can be interpreted as the scale above which short-distance information is contained in the mass and should be chosen around or higher than the mass scale. The $\overline{\text{MS}}$ mass has no kinematic meaning, but can be measured very precisely from experimental data. On the other hand, the so-called pole mass m_Q^{pole} is defined as the position of the single particle pole of correlation functions containing the heavy quark Q as an external on-shell state, which is identical to the statement that in the pole mass scheme all on-shell self-energy corrections are absorbed into the mass parameter, including contributions from energy regions below 1 GeV $\gtrsim \Lambda_{\text{QCD}}$. Since the $\overline{\text{MS}}$ mass does not have such contributions, it is called a "short-distance mass". The pole mass m_Q^{pole} can be interpreted as the kinematic mass of an isolated heavy quark Q in the context of perturbation theory. It is obvious

⋆e-mail: andre.hoang@univie.ac.at
⋆⋆Speaker, e-mail: christopher.lepenik@univie.ac.at
⋆⋆⋆e-mail: moritz.preisser@univie.ac.at

that the definition of the pole mass implies some conceptual and practical problems, since, due to confinement, external on-shell quarks do not exist and therefore they do not arise as single particle poles in physical correlation functions. Nevertheless it is unavoidable that the pole mass scheme appears in one way or another in higher order QCD calculations involving external massive quarks. For perturbative calculations involving the production of top quarks at hadron colliders, the pole mass scheme is the main top quark mass scheme used in the literature.

We now investigate the perturbative behavior of the pole mass for the top, bottom and charm quarks. For this purpose it is convenient to consider the perturbation series relating the pole and $\overline{\text{MS}}$ mass. In Refs. [1–8] this relation was computed to 4-loop accuracy, within the approximation that all quarks lighter then Q are massless. Using their results and the top, bottom and charm quark $\overline{\text{MS}}$ mass values $\overline{m}_t \equiv \overline{m}_t(\overline{m}_t) = 163$ GeV, $\overline{m}_b \equiv \overline{m}_b(\overline{m}_b) = 4.2$ GeV and $\overline{m}_c \equiv \overline{m}_c(\overline{m}_c) = 1.3$ GeV we obtain

$$m_t^{\text{pole}} = 163 + 7.5040 + 1.6005 + 0.4941 + (0.1944 \pm 0.0004)\,\text{GeV}\,, \tag{1}$$

$$m_b^{\text{pole}} = 4.2 + 0.3998 + 0.1986 + 0.1443 + (0.1349 \pm 0.0002)\,\text{GeV}\,, \tag{2}$$

$$m_c^{\text{pole}} = 1.3 + 0.2108 + 0.1984 + 0.2725 + (0.4843 \pm 0.0005)\,\text{GeV}\,, \tag{3}$$

where the terms show the series in powers of the strong coupling $\alpha_s(\overline{m}_Q)$ in the scheme that includes Q as a dynamical flavor. The fourth order coefficient displays the numerical uncertainties from [8], which are, however, much smaller than other types of uncertainties considered here.

The pole mass scheme is infrared-save and gauge-invariant [1, 9], but suffers from large perturbative corrections, as can be seen explicitly already at low orders from the series above. The bad perturbative behavior is related to the absorption of linearly infrared sensitive on-shell self-energy contributions into the pole mass. This infrared sensitivity of the on-shell diagrams grows with loop order. In the large-β_0 limit, the asymptotic large order behavior of the pole-$\overline{\text{MS}}$ relation has the form

$$m_Q^{\text{pole}} - \overline{m}_Q(\overline{m}_Q) \sim \mu \sum_{n=0}^{\infty} \frac{16}{3} \left(2\beta_0^{(n_\ell)}\right)^n n! \left(\frac{\alpha_s^{(n_\ell)}(\mu)}{4\pi}\right)^{n+1}, \tag{4}$$

where β_0 is defined via the QCD β-function describing the running of the strong coupling

$$\frac{\mathrm{d}\alpha_s^{(n_\ell)}(\mu)}{\mathrm{d}\log\mu} = \beta^{(n_\ell)}(\alpha_s(\mu)) = -2\,\alpha_s^{(n_\ell)}(\mu) \sum_{n=0}^{\infty} \beta_n^{(n_\ell)} \left(\frac{\alpha_s^{(n_\ell)}(\mu)}{4\pi}\right)^{n+1}. \tag{5}$$

The variable n_ℓ denotes the number of quarks lighter then Λ_{QCD}, which are taken to be massless throughout this work. The pattern of factorial divergence of the perturbation series and the linear dependence on the renormalization scale μ of the strong coupling are called the $\mathcal{O}(\Lambda_{\text{QCD}})$ renormalon of the pole mass [10, 11]. It is an important observation that, up to power corrections of order $\Lambda_{\text{QCD}}^2/m_Q$, the asymptotic large order behavior is independent of the heavy quark mass m_Q. This follows from Heavy Quark Symmetry (HQS) [12]: low momentum sensitivity grows with loop order and the high order terms are therefore governed almost exclusively by long-distance physics. Consequently, in this regime the heavy quarks are essentially static color sources whose precise mass values are irrelevant. On the other hand, it should be stressed that the renormalon is solely an artifact of the pole mass definition and not related to any physical effects. Another crucial point is that the high order asymptotics depends only on the number of massless quark flavors n_ℓ. This can again be understood from the HQS argument above. Due to the linear dependence of Eq. (4) to the renormalization scale μ the precise value which is chosen in practice is of high importance. E.g., to ensure the cancellation of an $\mathcal{O}(\Lambda_{\text{QCD}})$ renormalon between two series, the same μ-value must be chosen. We note that in the relation between the pole and other short-distance masses the same pattern can be seen: at asymptotic high orders one observes factorial growth and the series becomes linear in the renormalization scale of the stong coupling.

It is important to note that the renormalon behavior of the pole mass is not just an issue of pedagogical interest, but one that is relevant phenomenologically [13]. This is because for $\mu = \overline{m}_Q$ the known coefficients of the series in Eqs. (1)–(3) agree remarkably well with the corresponding large order asymptotic behavior already beyond the terms of $O(\alpha_s)$ (so that the terms of the series are known quite precisely to all orders) and because even for orders where the QCD corrections still decrease with order they can be very large numerically and make phenomenological applications difficult.

Although the perturbative series relating the pole and $\overline{\text{MS}}$ mass is diverging, the series is asymptotic and therefore not meaningless. It is known from the theory of asymptotic series that the best approximation of the LHS of series like in Eq. (4) is usually obtained when truncating the asymptotic series at the smallest correction term at order n_{min}. There is a region in the orders n around n_{min} of width Δn in which all series terms have a size close to the minimal term. At orders above $n_{\text{min}} + \Delta n/2$ the series diverges quickly and the series terms from these orders are useless even if they are known through an elaborate loop calculation. The uncertainty with which the pole mass can be determined in principle given the full information about the perturbative series is called the pole mass ambiguity. It is universal, independent of the choice of the renormalization scale μ and exists in equivalent size in any context without the possibility to be circumvented. Using the method of Borel summation the pole mass ambiguity can be estimated to be of order $\Lambda_{\text{QCD}}^{(n_\ell)}$, where the superscript (n_ℓ) stands for the dependence of the hadronization scale on the number of massless quark flavors.

For more details on the numerical values and derivations we refer to Ref. [14].

2 Light Massive Flavor Dependence

Due to the hierarchy in quark masses, in many applications of top quark physics the bottom and charm quarks may be taken as massless. However, when interested in the high order behavior of the pole mass, due to the increasing sensitivity to low scale physics, the effects of lighter massive quarks q with $m_q > \Lambda_{\text{QCD}}$ become relevant and actually have some impact. The mass of a virtual quark flavor in the on-shell self-energy diagram acts as an effective infrared cut-off at the mass scale and therefore this quark flavor is removed from the high order asymptotic behavior [15]. The removal of heavy lighter flavors from the high order asymptotics together with the n_ℓ-dependence of the QCD β-function lead to an increased infrared sensitivity of the heavy quark pole mass and a stronger divergence pattern of the series, as can be seen from Eq. (4). It is obvious that, since the masses of the lighter massive flavors alter the pattern of divergence of the renormalon series, their induced corrections are themselves plagued by a renormalon.

To study these corrections it is again convenient to consider the perturbative relation between the pole and $\overline{\text{MS}}$ mass. Including the bottom and charm quark as massive flavors, the pole-$\overline{\text{MS}}$ mass relation for the top quark can be written in the form [14]

$$m_t^{\text{pole}} - \overline{m}_t = \overline{m}_t \sum_{n=1}^{\infty} a_n(n_t+1) \left(\frac{\alpha_s^{(n_t+1)}(\overline{m}_t)}{4\pi} \right)^n + \overline{m}_t \left[\overline{\delta}_t^{(t,b,c)}(1, r_{bt}, r_{ct}) + \overline{\delta}_t^{(b,c)}(r_{bt}, r_{ct}) + \overline{\delta}_t^{(c)}(r_{ct}) \right], \quad (6)$$

where $\overline{m}_t \equiv \overline{m}_t(\overline{m}_t)$, $n_t = 5$ denotes the number of quark flavors lighter than the top, $a_n(n_t + 1)$ are the perturbative coefficients describing the QCD corrections from gluons and $n_t + 1$ massless virtual quarks, and $\alpha_s^{(n_t+1)}$ is the strong coupling that evolves with $n_t + 1$ active dynamical flavors, see Eq. (5). Results for the corrections from one lighter massive flavor up to $O(\alpha_s^3)$ were determined [2, 16]. The terms $\overline{\delta}_t$ contain the mass corrections coming from the top quark on-shell self-energy diagrams with insertions of virtual massive quark loops and can be written in a perturbative expansion in $\alpha_s^{(n_t+1)}$. The superscripts of the form $(q, q', \ldots)$ indicate that each diagram contains at least one insertion of the massive quark q and in addition all possible insertions of the (lighter) massive quarks $q', \ldots$ as well as of massless quark and gluonic loops. b and c relate to the bottom and charm quark respectively. From each diagram the corresponding diagram with all the quark loops in the massless limit is subtracted in the scheme compatible with the flavor number scheme for the strong coupling α_s. The fraction $r_{qq'} \equiv \overline{m}_q/\overline{m}_{q'}$ stands for the ratio of $\overline{\text{MS}}$ masses for quarks q and q'. By construction, the sum of all virtual quark mass corrections contained in the functions $\overline{\delta}_t$ are RG-invariant: they originate solely from the quark self-energy diagrams arising in the pole mass renormalization and, since the $\overline{\delta}_t$ denote the difference between massive and massless diagrams, the UV-divergences are canceled.

The aim now is to study the contributions of the light massive flavors in a coherent and systematic manner, which is done in a renormalization group framework. This framework allows to disentangle the contributing momentum modes and resum the logarithms of quark mass ratios which arise in the multi-scale problem. Consequently, it is possible to study the large order behavior and general structure of the mass corrections and the overall large order behavior of the pole mass, and use this knowledge in several applications.

3 Integrating Out Hard Modes from the Top Quark Pole Mass

3.1 MSR Mass

To set up the renormalization group framework and disentangle the different momentum regions below the top mass scale, we use the MSR mass $m_Q^{\text{MSR}}(R)$ which was introduced in Ref. [17, 18], however, without the inclusion of massive lighter quarks. In the presence of massive bottom and charm quarks the top quark MSR mass is defined through [14]

$$m_t^{\text{pole}} - m_t^{\text{MSR}}(R) = R \sum_{n=1}^{\infty} a_n(n_\ell) \left(\frac{\alpha_s^{(n_\ell)}(R)}{4\pi} \right)^n + \overline{m}_t \left[\delta_t^{(b,c)}(r_{bt}, r_{ct}) + \delta_t^{(c)}(r_{ct}) \right], \tag{7}$$

where the coefficients a_n are the same as in the pole-$\overline{\text{MS}}$ relation Eq. (6) and R is a momentum scale which is in principle arbitrary, but should be sufficiently larger than Λ_{QCD} to stay away from the Landau pole. The terms δ_t are derived from the respective $\overline{\delta}_t$ of Eq. (6) by replacing $\alpha_s^{(n_\ell+1)} \to \alpha_s^{(n_\ell)}$ in the perturbative expansions. The introduction of the MSR mass is necessary since the $\overline{\text{MS}}$ mass is not adequate to describe scales far below the heavy quark mass scale. The MSR mass fixes the two main problems in this context: first, the heavy quark is removed as a dynamical degree of freedom from the series (i.e. integrated out). Second, the MSR mass incorporates linear scale dependence by replacing $\overline{m}_t$ multiplying the coefficients a_n on the RHS of Eq. (6) with the arbitrary momentum scale R. Linear scale dependence is crucial in the low momentum region to describe the linear low momentum sensitivity of the pole mass. Indeed linear momentum dependence appears also in the pole-$\overline{\text{MS}}$ mass relation, but only at asymptotically high orders (see Eq. (4)). It should be also mentioned that the replacement of $\overline{m}_t$ and the removal of the heavy flavor in the definition of the MSR mass do not change the asymptotic high order behavior of the series, because the latter is independent of the heavy quark mass and the number of massive flavors. As argued at the end of Sec. 2, the mass corrections are RG-invariant and it is therefore natural to leave the mass correction term independent of R. So the MSR mass is defined in close analogy to the μ-dependent $\overline{\text{MS}}$ strong coupling whose renormalization group evolution only depends on the number of active dynamical quarks (which is typically the number of quarks lighter than μ) and where mass effects are implemented by threshold corrections when μ crosses a flavor threshold.

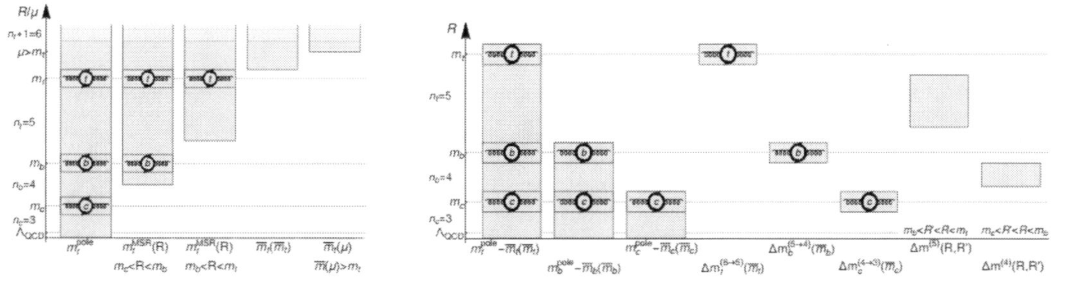

Figure 1. *Left:* Graphical illustration of the physical contributions contained in the pole, MSR and $\overline{\text{MS}}$ mass schemes coming from the different momentum scales for the case of the top quark. The quark loops stand for the contributions of the virtual massive quark loops contained in the masses. *Right:* Graphical illustration for pole-$\overline{\text{MS}}$ mass differences, the MSR-$\overline{\text{MS}}$ mass matching corrections and MSR mass differences for different R scales. They constitute the major contributions in the RG analysis of the heavy quark pole masses.

The resulting renormalization group equation in R is linear in R, as required by low scale physics, and is called the R-evolution equation. For an arbitrary heavy quark flavor Q, it takes the form

$$R\frac{\mathrm{d}}{\mathrm{d}R}m_Q^{\text{MSR}}(R) = -R\,\gamma^{R,(n_Q)}(\alpha_s^{(n_Q)}(R)) = -R\sum_{n=0}^{\infty}\gamma_n^{R,(n_Q)}\left(\frac{\alpha_s^{(n_Q)}(R)}{4\pi}\right)^{n+1}, \tag{8}$$

where n_Q denotes the number of quark flavors lighter than Q and the coefficients $\gamma_n^{R,(n_Q)}$ are known up to four loops and given in Refs. [17, 18]. It is easy to see that Eq. (8) is renormalon-free since the renormalon ambiguity of the series proportional to R is independent of R and therefore canceled when differentiated.

As mentioned below Eq. (5), it is crucial to choose the same renormalization scale when subtracting two series which contain an $O(\Lambda_{\text{QCD}})$ renormalon. This is e.g. the case for the difference of the MSR mass at two different R-scales in fixed-order perturbation theory using Eq. (7), which for widely separated scales leads inevitably to large logarithms. To avoid this problem one can use the solution of the R-evolution equation Eq. (8)

$$\Delta m^{(n_Q)}(R,R') = m_Q^{\text{MSR}}(R') - m_Q^{\text{MSR}}(R) = \sum_{n=0}^{\infty}\gamma_n^{R,(n_Q)}\int_{R'}^{R}\mathrm{d}R\left(\frac{\alpha_s^{(n_Q)}(R)}{4\pi}\right)^{n+1}, \tag{9}$$

to resum these logarithms in a renormalon free way.

One can interpret the MSR mass as the pole mass minus all self-energy contributions coming from scales below R and all virtual quark mass corrections from quarks lighter than the heavy quark, see the left plot of Fig. 1. This also illustrates that the MSR mass is a short-distance mass. $\Delta m^{(n_Q)}(R,R')$ represents the self-energy contributions to the mass in the presence of n_Q active dynamical flavors coming from the scales between R' and R. This is illustrated in the right plot of Fig. 1.

3.2 MSR-$\overline{\text{MS}}$ Mass Matching

The matching relation between the top MSR and $\overline{\text{MS}}$ mass at the common scale $\mu = R = \overline{m}_t$ can be obtained by eliminating the pole mass from Eqs. (6) and (7). The resulting matching coefficient then accounts for the virtual top quark contributions and can be written in the form

$$m_t^{\rm MSR}(\overline{m}_t) - \overline{m}_t = \Delta m_t^{(n_\ell+1\to n_\ell)}(\overline{m}_t) + \delta m_{t,b,c}^{(n_\ell+1\to n_\ell)}(\overline{m}_t)\,. \tag{10}$$

The term $\Delta m_t^{(n_\ell+1\to n_\ell)}(\overline{m}_t)$ contains the virtual top quark loop contributions in the approximation that all n_ℓ quarks lighter than the top are massless and can be expressed in a power series in the strong coupling in the $(n_\ell + 1)$ flavor scheme. The series only contains the hard corrections coming from the virtual heavy top and therefore does not have any $O(\Lambda_{\rm QCD})$ ambiguity, see the right plot of Fig. 1 for illustration. $\Delta m_t^{(n_\ell+1\to n_\ell)}(\overline{m}_t)$ can therefore be computed to high precision using the a_n coefficients which are known up to $O(\alpha_s^4)$ [1–8].

For the top quark the matching corrections amount to approximately 30 MeV and the analogous corrections for the bottom and charm amount to approximately 5 MeV, with an uncertainty at the level of 1 to 2 MeV.

The term $\delta m_{t,b,c}^{(n_\ell+1\to n_\ell)}(\overline{m}_t)$ represents the virtual top quark contributions arising from the finite masses of the lighter massive bottom and charm quarks. Since at $O(\alpha_s^2)$ only the loop of a top quark can be inserted, the series for $\delta m_{t,b,c}^{(n_\ell+1\to n_\ell)}(\overline{m}_t)$ starts at $O(\alpha_s^3)$, where only self-energy diagrams with one insertion of a top quark loop and one insertion of a loop of one of the lighter massive quarks can contribute. The expansions of $\delta m_{t,b,c}^{(n_\ell+1\to n_\ell)}(\overline{m}_t)$ in the mass ratios r_{bt} and r_{ct} start quadratically in these ratios indicating that the corrections are governed by the scale $\overline{m}_t$ just like the matching term $\Delta m_t^{(n_\ell+1\to n_\ell)}(\overline{m}_t)$ and do not have any linear sensitivity to small momenta and the lighter quark masses, in particular. This feature is realized at any order of perturbation theory. Because the finite mass corrections $\delta m_{t,b,c}^{n_\ell+1\to n_\ell}(\overline{m}_t)$ start at $O(\alpha_s^3)$ and are quadratic in the mass ratios they are extremely small and never exceed 0.01 MeV. This is also true for the finite mass corrections to the bottom quark due to the massive charm quark and we can expect that this is also exhibited at higher orders. Hence, this kind of correction can be neglected for all practical purposes and will not be considered and discussed any further here.

3.3 Top-Bottom and Bottom-Charm Mass Matching

Comparing the pole-MSR mass relation for the top quark of Eq. (7) to the pole-$\overline{\rm MS}$ mass relation for the next lighter massive quark, i.e. the bottom

$$m_b^{\rm pole} - \overline{m}_b = \overline{m}_b \sum_{n=1}^{\infty} a_n(n_b+1)\left(\frac{\alpha_s^{(n_b+1)}(\overline{m}_b)}{4\pi}\right)^n + \overline{m}_b\left[\overline{\delta}_t^{(b,c)}(1,r_{cb}) + \overline{\delta}_t^{(c)}(r_{cb})\right], \tag{11}$$

with $n_b = n_t - 1 = 4$, one immediately notices that for $R = \overline{m}_b$ the corrections are identical in the approximation that in the virtual quark loops all n_ℓ lighter quarks (including the bottom quark) are treated as massless. This identity is a consequence of heavy quark symmetry which states that the low-energy QCD corrections to the heavy quark masses coming from massless partons are flavor-independent. Obviously, this is also valid when comparing the bottom to the charm quark.

The resulting matching relation reads

$$\left[m_t^{\rm pole} - m_t^{\rm MSR}(\overline{m}_b)\right] - \left[m_b^{\rm pole} - \overline{m}_b\right] = \delta m_{b,c}^{(t\to b)}(\overline{m}_b, \overline{m}_c)\,, \tag{12}$$

where $\delta m_{b,c}^{(t\to b)}(\overline{m}_b, \overline{m}_c)$ encodes the heavy quark symmetry breaking corrections coming from the finite virtual charm and bottom quark masses. Their form can be extracted directly from Eqs. (6) and (7) and written as ($r_{qq'} = \overline{m}_q/\overline{m}_{q'}$)

$$\delta m_{b,c}^{(t\to b)}(\overline{m}_b, \overline{m}_c) = \overline{m}_t\left[\delta_t^{(b,c)}(r_{bt}, r_{ct}) + \delta_t^{(c)}(r_{ct})\right] - \overline{m}_b\left[\overline{\delta}_b^{(b,c)}(1, r_{cb}) + \overline{\delta}_b^{(c)}(r_{cb})\right], \tag{13}$$

DOI: 10.1051/epjconf/201715804005

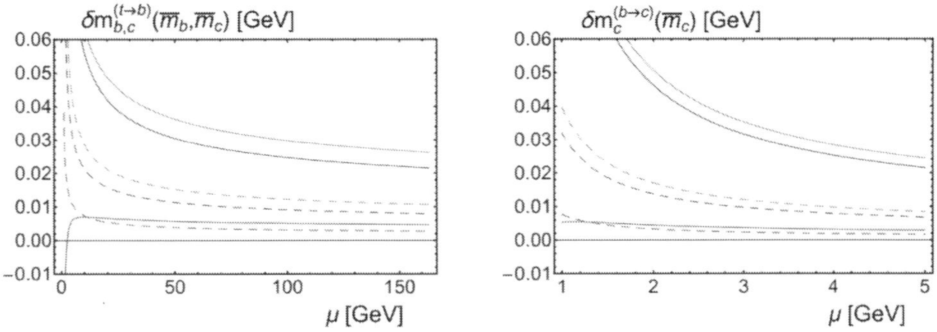

Figure 2. *Left:* Top-MSR bottom-$\overline{\text{MS}}$ mass matching correction $\delta m_{b,c}^{(t\to b)}(\overline{m}_b, \overline{m}_c)$ at $O(\alpha_s^2)$ (red dashed curve) and $O(\alpha_s^3)$ (red solid curve) over the renormalization scale μ. The virtual bottom and charm mass effects to the top quark self-energy of the first term in Eq. (13) (green curves) and the virtual bottom and charm mass effects to the bottom quark self-energy of the second term of Eq. (13) (blue curves) at $O(\alpha_s^2)$ (dashed) and $O(\alpha_s^3)$ (solid). For the masses of the top, bottom and charm quarks the values $(\overline{m}_t, \overline{m}_b, \overline{m}_c) = (163, 4.2, 1.3)$ GeV are used. *Right:* The bottom-MSR charm-$\overline{\text{MS}}$ mass matching correction $\delta m_c^{(b\to c)}(\overline{m}_c)$ at $O(\alpha_s^2)$ (red dashed curve) and $O(\alpha_s^3)$ (red solid curve) over the renormalization scale μ. The virtual charm mass effects to the bottom quark self-energy (green curves) and the virtual charm mass effects to the charm quark self-energy (blue curves) are shown at $O(\alpha_s^2)$ (dashed) and $O(\alpha_s^3)$ (solid).

where the first term on the RHS (multiplied by $\overline{m}_t$) represents the virtual bottom and charm mass effects in the top quark self-energy and the second term (multiplied by $\overline{m}_b$) represents the virtual bottom and charm mass effects in the bottom quark self-energy.

It is important that the quark mass corrections in (13) are expressed coherently in powers of α_s at the common scale μ because the individual δ_n terms carry infrared sensitive contributions and therefore each contain an $O(\Lambda_{\text{QCD}})$ renormalon ambiguity. In Eq. (13) these renormalon ambiguities cancel.

In the left plot of Fig. 2 the top-MSR bottom-$\overline{\text{MS}}$ mass matching correction $\delta m_{b,c}^{(t\to b)}(\overline{m}_b, \overline{m}_c)$ of Eq. (12) is displayed as a function of the renormalization scale μ at $O(\alpha_s^2)$ (red dashed line) and $O(\alpha_s^3)$ (red solid line) for $(\overline{m}_t, \overline{m}_b, \overline{m}_c) = (163, 4.2, 1.3)$ GeV. The matching correction at $O(\alpha_s^3)$ amounts to 6 MeV and has a scale variation of only 1 MeV for $\overline{m}_b \le \mu \le \overline{m}_t$. Compared to the $O(\alpha_s^2)$ result we see a strong reduction of the scale dependence at $O(\alpha_s^3)$. An analog matching contribution $\delta m_c^{(b\to c)}(\overline{m}_c)$ can be defined, describing the heavy quark symmetry breaking corrections for the bottom quark coming from the finite charm quark mass. Again, we observe excellent convergence and strongly reduced scale dependence, see the right plot in Fig. 2.

Through successive R-evolution with the appropriate flavor number (see Eq. (9)) and matching at the mass thresholds of the bottom and charm quark (see Eqs. (10) and (12) and their generalizations to the appropriate flavors) we can now decouple the different momentum regions in the pole-$\overline{\text{MS}}$ relation of Eq. (6). The resulting formula for the top quark pole mass reads

$$\begin{aligned} m_t^{\text{pole}} &= \overline{m}_t + \Delta m_t^{(6\to 5)}(\overline{m}_t) + \Delta m^{(5)}(\overline{m}_t, \overline{m}_b) + \delta m_{b,c}^{(t\to b)}(\overline{m}_b, \overline{m}_c) \\ &\quad + \Delta m_b^{(5\to 4)}(\overline{m}_b) + \Delta m^{(4)}(\overline{m}_b, \overline{m}_c) + \delta m_c^{(b\to c)}(\overline{m}_c) \\ &\quad + \Delta m_c^{(4\to 3)}(\overline{m}_c) + \Delta m^{(3)}(\overline{m}_c, R) + R \sum_{n=1}^{\infty} a_n(n_\ell = 3, 0) \left(\frac{\alpha_s^{(3)}(R)}{4\pi} \right)^n , \end{aligned} \tag{14}$$

where all logarithms $\log(\overline{m}_b/\overline{m}_t)$ and $\log(\overline{m}_b/\overline{m}_c)$ are systematically resummed. The sum of all terms in the first two lines is just $m_t^{\rm pole} - m_c^{\rm pole} + \overline{m}_c$, and the first term in the third line is the charm $\overline{\rm MS}$-MSR matching contribution. All quantities except for the last term are free from an $O(\Lambda_{\rm QCD})$ renormalon ambiguity and can be evaluated to high precision with uncertainties of the order of 20 MeV for the evolution contributions and 1 MeV for the matching contributions using the available 4-loop expressions for a_n [1–8] and 3-loop expressions for the mass corrections [2, 16]. The ambiguity of the top quark pole mass due to the $O(\Lambda_{\rm QCD})$ renormalon is contained solely in the R-dependent series which is just equal to $m_c^{\rm pole} - m_c^{\rm MSR}(R)$. This relation specifies the charm quark pole mass ambiguity, and it fully encodes the top and bottom quark pole mass ambiguities due to heavy quark symmetry. It should be noted that among all the terms shown in Eq. (14) the contributions from the MSR mass differences $\Delta m^{(5)}(\overline{m}_t, \overline{m}_b)$, $\Delta m^{(4)}(\overline{m}_b, \overline{m}_c)$ and $\Delta m^{(3)}(\overline{m}_c, R)$, determined with R-evolution, and the series proportional to R, which contains the $O(\Lambda_{\rm QCD})$ renormalon, constitute the numerically most important terms. Their sum is of the order of 10 GeV, exceeding the contributions from the matching corrections by far, which amount to only 50 MeV.

4 Some Applications

4.1 Light Virtual Quark Mass Corrections at $O(\alpha_s^4)$

The excellent perturbative convergence of the top-MSR bottom-$\overline{\rm MS}$ mass matching correction $\delta m_{b,c}^{(t\to b)}(\overline{m}_b, \overline{m}_c)$ discussed in the previous section illustrates that it is a short-distance quantity and free of an $O(\Lambda_{\rm QCD})$ renormalon ambiguity. This is also expected theoretically due to heavy quark symmetry. However, the facts that the overall size of the matching corrections only amounts to a few MeV, and that the $O(\alpha_s^3)$ corrections are only around 1 MeV as seen in Fig. 2 allows us to draw interesting conceptual implications for the large order asymptotic behavior of the virtual quark mass corrections in the mass relations of Eqs. (6) and (7). As a consequence we can predict the yet uncalculated virtual quark mass corrections at $O(\alpha_s^4)$ to within a few percent without an additional loop calculation.

Let's consider the matching correction $\delta m_q^{(Q\to q)}(\overline{m}_q)$ between the MSR mass of heavy quark Q and the $\overline{\rm MS}$ mass of the next lighter massive quark q assuming the massless approximation for all quarks lighter than quark q, i.e. $n_Q = n_q + 1 = n_\ell + 1$ and $n_\ell = n_q$ being the number of massless quarks. This situation applies to the matching relation for the top-MSR and the bottom $\overline{\rm MS}$ masses for a massless charm quark or to the matching relation between the bottom-MSR and the charm-$\overline{\rm MS}$ masses.

In the left plot of Fig. 2 we have displayed separately the virtual bottom and charm mass effects to the top quark self-energy (green curves) and the virtual bottom and charm mass effects to the bottom quark self-energy (blue curves), i.e. the first and second term of the RHS of Eq. (13), at $O(\alpha_s^2)$ (dashed) and $O(\alpha_s^3)$ (solid). We see that both types of contributions each are quite large and furthermore do not at all converge. The $O(\alpha_s^3)$ corrections are even bigger than the $O(\alpha_s^2)$ corrections, which indicates that the corresponding asymptotic large order behavior already dominates the $O(\alpha_s^2)$ and $O(\alpha_s^3)$ corrections.

The origin of this behavior has been already mentioned and is understood: the mass of the virtual quark q acts as an infrared cutoff and therefore modifies the infrared sensitivity of the self-energy diagrams (of quark Q and of quark q) with respect to the case where the virtual loops of quark q are evaluated in the massless approximation. As a consequence these corrections individually carry an $O(\Lambda_{\rm QCD})$ renormalon ambiguity. Moreover, at large orders in perturbation theory the sensitivity of the self-energy diagrams to infrared momenta increases due to high powers of logarithms from gluonic and massless quark loops. As a consequence, at large orders, the finite mass effects of the virtual loops of quark q in the self-energy diagrams of quark Q and the self-energy diagrams of quark q

 DOI: 10.1051/epjconf/201715804005

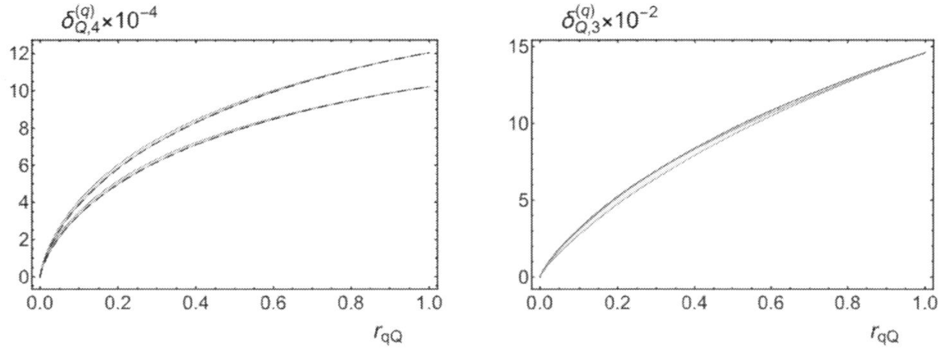

Figure 3. *Left:* Prediction for the $O(\alpha_s^4)$ virtual quark mass correction $\delta^{(q)}_{Q,4}(r_{qQ})$ for $\overline{m}_q \le \mu \le \overline{m}_Q$ (green bands) for $n_Q = n_\ell + 1 = 5$ (lower band) and $n_Q = n_\ell + 1 = 4$ (upper band). The black dashed lines show the prediction for $\mu = \overline{m}_Q$ which gives the simple approximation formula in Eq. (15). *Right*: The $O(\alpha_s^3)$ virtual quark mass correction $\delta^{(q)}_{Q,3}(r_{qQ})$ for $n_Q = n_\ell + 1 = 5$ (red curve). The green band is the prediction for $\delta^{(q)}_{Q,3}(r_{qQ})$ using the method used in the left plot showing excellent agreement to the exact result within errors.

become equivalent due to heavy quark symmetry. The strong cancellation in the sum of both types of corrections in $\delta m_q^{(Q\to q)}(\overline{m}_q)$ ($\sim$ 75% at $O(\alpha_s^2)$ and $\gtrsim$ 90% at $O(\alpha_s^3)$ for the cases displayed in Fig. 2) thus confirms that the known $O(\alpha_s^2)$ and $O(\alpha_s^3)$ self-energy corrections coming from virtual quark masses are already dominated by their large order asymptotic behavior.

The series for $\delta m_{b,c}^{(t\to b)}(\overline{m}_b, \overline{m}_c)$ and $\delta m_c^{(b\to c)}(\overline{m}_c)$ converge very well and their $O(\alpha_s^3)$ corrections amount to only about 1 MeV, thus we can expect that the two types of corrections that enter $\delta m_{b,c}^{(t\to b)}(\overline{m}_b, \overline{m}_c)$ agree even better than 1 MeV at $O(\alpha_s^4)$ and beyond. This allows to make an approximate prediction for the yet uncalculated $O(\alpha_s^4)$ finite mass corrections from virtual loops of a quark q in the pole-$\overline{\text{MS}}$ mass relations of a heavy quark Q by setting the $O(\alpha_s^4)$ correction in $\delta m_q^{(Q\to q)}(\overline{m}_q)$ to zero. The residual μ-dependence vanishes in the formal limit that the virtual quark q mass corrections are entirely dominated by their large order asymptotic behavior and can be used as an uncertainty estimate of our approximation. For $\mu = \overline{m}_Q$, we can provide the very simple closed analytic expression

$$\delta^{(q)}_{Q,4}(r_{qQ}) \approx r_{qQ}\Big[\,\delta^{(q)}_{q,4}(1) - \Big(6\beta_0^{(n_Q)}\delta^{(q)}_{q,3}(1) + 4\beta_1^{(n_Q)}\delta_2(1)\Big)\ln\big(r_{qQ}\big) + 12\,\delta_2(1)\big(\beta_0^{(n_Q)}\ln(r_{qQ})\big)^2\,\Big]. \qquad (15)$$

The coefficients $\delta^{(q)}_{q,n}(1)$ describe the corrections from virtual massive loops of the heavy quark q to the q self-energy and are known up to $O(\alpha_s^4)$ from the full a_n coefficients computed in Refs. [1–8].

In the left plot of Fig. 3 we show the prediction of $\delta^{(q)}_{Q,4}(r_{qQ})$ for the top ($n_Q = n_t = 5$, lower band) and bottom ($n_Q = n_b = 4$, upper band) with a scale variation of $\overline{m}_b \le \mu \le \overline{m}_t$ and $\overline{m}_c \le \mu \le \overline{m}_b$ respectively. The appropriate curves for Eq. (15) are shown as the black dashed lines. The uncertainty amounts to $\pm$3% (for $r_{qQ} \lesssim 0.1$) or smaller (for $r_{qQ} > 0.1$). The reliability of the prediction and uncertainty estimate was additionally tested by "predicting" the already exactly known $\delta^{(q)}_{Q,3}(r_{qQ})$ using the same method, see the right plot of Fig. 3. The prediction is fully compatible with the exact result and the uncertainty amounts to $\pm$10% (for $r_{qQ} \lesssim 0.1$) or smaller (for $r_{qQ} > 0.1$).

To conclude it should be mentioned that the prediction of Eq. (15) can be generalized to even higher orders in α_s by using the known asymptotic behavior of the coefficients a_n, as well as to the case of having a larger number of lighter massive quarks.

4.2 Pole Mass Differences

Another interesting application of the RG-setup introduced here is the determination of the difference of the pole masses of two massive quarks. Due to heavy quark symmetry, the differences of two heavy quark pole masses are free of $O(\Lambda_{\rm QCD})$ renormalon ambiguities and can therefore be determined to high precision. The matching corrections discussed above and the R-evolution of the MSR mass allow us to systematically sum logarithms of the mass ratios that would remain unsummed in a fixed-order calculation, and to achieve more precise perturbative predictions [18]. The resulting relations between the top, bottom and charm quark pole masses read

$$m_t^{\rm pole} - m_b^{\rm pole} = [\overline{m}_t - \overline{m}_b] + \Delta m_t^{(6\to5)}(\overline{m}_t) + \Delta m^{(5)}(\overline{m}_t, \overline{m}_b) + \delta m_{b,c}^{(t\to b)}(\overline{m}_b, \overline{m}_c)\,, \tag{16}$$

$$m_b^{\rm pole} - m_c^{\rm pole} = [\overline{m}_b - \overline{m}_c] + \Delta m_b^{(5\to4)}(\overline{m}_b) + \Delta m^{(4)}(\overline{m}_b, \overline{m}_c) + \delta m_c^{(b\to c)}(\overline{m}_c)\,, \tag{17}$$

$$\begin{aligned} m_t^{\rm pole} - m_c^{\rm pole} &= [\overline{m}_t - \overline{m}_c] + \Delta m_t^{(6\to5)}(\overline{m}_t) + \Delta m^{(5)}(\overline{m}_t, \overline{m}_b) + \delta m_{b,c}^{(t\to b)}(\overline{m}_b, \overline{m}_c) \\ &\quad + \Delta m_b^{(5\to4)}(\overline{m}_b) + \Delta m^{(4)}(\overline{m}_b, \overline{m}_c) + \delta m_c^{(b\to c)}(\overline{m}_c)\,. \end{aligned} \tag{18}$$

Each of the mass differences is the sum of universal matching and evolution building blocks which each can be computed to high precision as mentioned at the end of Sec. 3.3. The numerical evaluation using the values given in Ref. [14] for the case $(\overline{m}_t, \overline{m}_b, \overline{m}_c) = (163, 4.2, 1.3)$ GeV gives

$$m_t^{\rm pole} - m_b^{\rm pole} = 168.169 \pm 0.016~{\rm GeV}\,, \tag{19}$$

$$m_b^{\rm pole} - m_c^{\rm pole} = 3.331 \pm 0.017~{\rm GeV}\,, \tag{20}$$

$$m_t^{\rm pole} - m_c^{\rm pole} = 171.500 \pm 0.024~{\rm GeV}\,. \tag{21}$$

The uncertainties in the pole mass differences are between 16 and 24 MeV and should be considered as conservative estimates of the theoretical uncertainties due to missing higher order corrections.

5 Conclusions

We have provided a renormalization group framework which allows to study the mass effects of virtual massive quark loops in the relation between the pole mass $m_Q^{\rm pole}$ and short-distance masses such as the $\overline{\rm MS}$ mass $\overline{m}_Q(\mu)$ of a heavy quark Q, where we mean virtual loop insertions of quarks q with $\Lambda_{\rm QCD} < m_q < m_Q$. Special focus was put on the top, where these corrections are most important. In this context it is well-known that the virtual loops of a massive quark act as an infrared cut-off on the virtuality of the gluon exchange that eliminates the effects of that quark from the large order asymptotic behavior of the series. This effect arises from the $O(\Lambda_{\rm QCD})$ renormalon contained in the pole mass which means that the QCD corrections have a linear sensitivity to small momenta that increases with the order in the perturbative expansion.

It was examined (i) how the logarithms of mass ratios that arise in this multi-scale problem can be systematically summed to all orders, (ii) the large order asymptotic behavior and structure of the mass corrections themselves and (iii) the consequences of heavy quark symmetry. E.g. the framework allows to relate the QCD corrections in the top quark pole-$\overline{\rm MS}$ mass difference $m_t^{\rm pole} - \overline{m}_t(\overline{m}_t)$ coming from scales smaller than the bottom mass, to the bottom quark pole-$\overline{\rm MS}$ mass difference $m_b^{\rm pole} - \overline{m}_b(\overline{m}_b)$.

Within the RG framework, we find that the bulk of the lighter virtual quark mass corrections is determined by their large order asymptotic behavior already at $O(\alpha_s^3)$, which confirms earlier observations made in Refs. [19, 20] and [21]. We used this property to predict the previously unknown $O(\alpha_s^4)$ lighter virtual quark mass corrections to within a few percent from the available information on the

$O(\alpha_s^4)$ corrections for massless lighter quarks without an additional loop computation, see Eq. (15). Furthermore we calculated the differences of the top, bottom and charm quark pole masses with a precision of around 20 MeV.

Acknowledgments

We acknowledge partial support by the FWF Austrian Science Fund under the Doctoral Program No. W1252-N27 and the Project No. P28535-N27.

References

[1] R. Tarrach, Nucl. Phys. **B183**, 384 (1981)
[2] N. Gray, D.J. Broadhurst, W. Grafe, K. Schilcher, Z. Phys. **C48**, 673 (1990)
[3] K.G. Chetyrkin, M. Steinhauser, Phys. Rev. Lett. **83**, 4001 (1999), hep-ph/9907509
[4] K.G. Chetyrkin, M. Steinhauser, Nucl. Phys. **B573**, 617 (2000), hep-ph/9911434
[5] K. Melnikov, T.v. Ritbergen, Phys. Lett. **B482**, 99 (2000), hep-ph/9912391
[6] P. Marquard, L. Mihaila, J.H. Piclum, M. Steinhauser, Nucl. Phys. **B773**, 1 (2007), hep-ph/0702185
[7] P. Marquard, A.V. Smirnov, V.A. Smirnov, M. Steinhauser, Phys. Rev. Lett. **114**, 142002 (2015), 1502.01030
[8] P. Marquard, A.V. Smirnov, V.A. Smirnov, M. Steinhauser, D. Wellmann, Phys. Rev. **D94**, 074025 (2016), 1606.06754
[9] A.S. Kronfeld, Phys. Rev. **D58**, 051501 (1998), hep-ph/9805215
[10] I.I.Y. Bigi, M.A. Shifman, N.G. Uraltsev, A.I. Vainshtein, Phys. Rev. **D50**, 2234 (1994), hep-ph/9402360
[11] M. Beneke, V.M. Braun, Nucl. Phys. **B426**, 301 (1994), hep-ph/9402364
[12] N. Isgur, M.B. Wise, Phys. Lett. **B232**, 113 (1989)
[13] M. Beneke, Phys. Rept. **317**, 1 (1999), hep-ph/9807443
[14] A.H. Hoang, C. Lepenik, M. Preisser, JHEP **09**, 099 (2017), 1706.08526
[15] P. Ball, M. Beneke, V.M. Braun, Nucl. Phys. **B452**, 563 (1995), hep-ph/9502300
[16] S. Bekavac, A. Grozin, D. Seidel, M. Steinhauser, JHEP **10**, 006 (2007), 0708.1729
[17] A.H. Hoang, A. Jain, I. Scimemi, I.W. Stewart, Phys. Rev. Lett. **101**, 151602 (2008), 0803.4214
[18] A.H. Hoang, A. Jain, C. Lepenik, V. Mateu, M. Preisser, I. Scimemi, I.W. Stewart (2017), 1704.01580
[19] A.H. Hoang, A.V. Manohar, Phys. Lett. **B483**, 94 (2000), hep-ph/9911461
[20] A.H. Hoang (2000), hep-ph/0008102
[21] C. Ayala, G. Cvetič, A. Pineda, JHEP **09**, 045 (2014), 1407.2128

EPJ Web of Conferences **158**, 04006 (2017) DOI: 10.1051/epjconf/201715804006
QFTHEP 2017

The effect of the anomalous Wtb interactions on spin correlations in t-channel single top quark production with subsequent decay

Eduard Boos[1,2,⋆] and *Viacheslav* Bunichev[1,⋆⋆]

[1] *Skobeltsyn Institute of Nuclear Physics, Lomonosov Moscow State University Leninskie Gory, 119991, Moscow, Russia*
[2] *Faculty of Physics, Lomonosov Moscow State University, Leninskie Gory, 119991, Moscow, Russia*

Abstract. Spin correlations in the t-channel single top quark production and its subsequent decay are investigated for the case of contributions involving anomalous Wtb couplings. It is shown that shapes of certain multidimensional kinematic distributions of final state particles are significantly different for the contributions proportional to different combinations of the anomalous couplings. Observed properties are confirmed by two methods of computations either from the obtained symbolic expression for the differential width of polarized three-particle top quark decay or by means of the CompHEP program for the complete process involving the t-channel single top.

1 Introduction

With the Higgs boson discovery at the LHC the Standard Model (SM) is completed in a sense that all predicted particles are found and all the interactions between particles are fixed. However, most likely, the SM is a kind of effective field theory working (and working amazingly well) at the energy range determined by the electroweak (EW) energy scale. The top quark being the heaviest found fundamental particle with closest to the EW scale mass is a promising place to search for possible deviations from the SM (see, recent reviews on top quark [1–5]. In particular such deviations may be related to a presence of the top quark anomalous couplings which are usually parametrizes in terms of a number of the gauge invariant dimension 6 operators [6–10]

$$O_{\phi q}^{(3,3+3)} = \frac{i}{2}\left[\phi^{\dagger}(\tau^{I}D_{\mu} - \overleftarrow{D}_{\mu}\tau^{I})\phi\right](\bar{q}_{L3}\gamma^{\mu}\tau^{I}q_{L3}), \qquad O_{\phi\phi}^{33} = i(\tilde{\phi}^{\dagger}D_{\mu}\phi)(\bar{t}_{R}\gamma^{\mu}b_{R}),$$
$$O_{dW}^{33} = (\bar{q}_{L3}\sigma^{\mu\nu}\tau^{I}b_{R})\phi\, W_{\mu\nu}^{I}, \qquad O_{uW}^{33} = (\bar{q}_{L3}\sigma^{\mu\nu}\tau^{I}t_{R})\tilde{\phi}\, W_{\mu\nu}^{I}, \qquad (1)$$

These operators lead to the effective Lagrangian [11] allowed by the Lorentz invariance parametrizing the anomalous terms in Wtb vertex

⋆e-mail: boos@theory.sinp.msu.ru
⋆⋆e-mail: bunichev@theory.sinp.msu.ru

$$\mathcal{L} = \frac{g}{\sqrt{2}} \bar{b}\gamma^{\mu}(f_{LV}P_L + f_{RV}P_R)tW_{\mu}^{-} + \frac{g}{\sqrt{2}} \bar{b}\, \frac{\sigma^{\mu\nu}}{2M_W}(f_{LT}P_L + f_{RT}P_R)tW_{\mu\nu}^{-} + h.c., \tag{2}$$

where M_W is the W-boson mass, $P_{L,R} = (1 \mp \gamma_5)/2$ is the left(right)-handed projection operator, $W_{\mu\nu}^{-} = \partial_{\mu}W_{\nu}^{-} - \partial_{\nu}W_{\mu}^{-}$, $\sigma^{\mu\nu} = i/2[\gamma_{\mu}, \gamma_{\nu}]$, g is the weak isospin gauge coupling, parameters $f_{LV(T)}$ and $f_{RV(T)}$ are the dimensionless coefficients that parametrize strengths of the left-vector (tensor) and the right-vector (tensor) structures in the Lagrangian.

The couplings in the Lagrangian (2) are related in the following way to constants $C_{\phi q}^{(3,3+3)}, C_{\phi\phi}^{33}, C_{dW}^{33}, C_{uW}^{33}$ in front of the gauge-invariant dimension six operators : $f_{LV} = V_{tb} + C_{\phi q}^{(3,3+3)} \frac{v^2}{\Lambda^2}$, $f_{RV} = \frac{1}{2} C_{\phi\phi}^{33} \frac{v^2}{\Lambda^2}$, $f_{LT} = \sqrt{2} C_{dW}^{33} \frac{v^2}{\Lambda^2}$, $f_{RT} = \sqrt{2} C_{uW}^{33} \frac{v^2}{\Lambda^2}$.

If one assumes that naturally the constants in front of the operators are of the order of unity, the natural values for the anomalous couplings are of the order of $\frac{v^2}{\Lambda^2}$ and therefore rather small. This is confirmed by the recent most stringent experimental limits [12].

In the SM all fermions interact through the left-handed currents and all constants are equal to zero, except $f_{LV} = V_{tb}$ (CKM-matrix element). Note that the anomalous coupling parameters could be complex in the most general case. However here we consider only real anomalous couplings for simplicity.

In this short paper we discuss a simple idea allowing to improve further the limits on anomalous couplings. The idea is based on the top quark spin correlation properties in the t- and s-channel single top production with its subsequent decay. Spin correlations in presence of anomalous couplings have been studied previously in [9, 13–15]

As well known in the SM, the positively charged lepton from the top quark decay in its rest frame tends to follow the top quark spin direction [16, 17]. In the t- and s-channel single top production the direction of the top quark spin in the top rest frame is highly correlated with the d-quark momentum for the t-channel (outgoing light jet) and anti d-quark momentum (incoming parton corresponding to the beam axis) for the s-channel [18, 19]. One can understand this very simply by considering the single top production as a decay back in time [20]. The direction of preferred spin configuration in the single top production in presence of anomalous couplings is changed insignificantly comparing to SM due to smallness of the anomalous couplings. So one can chose the direction related to mentioned d-quark momentum as a top spin quantization axis and to make sure that this will be a preferred spin direction of top quark in its rest frame similar to SM. After production the highly polarized top quark decays (Fig. 1), and one can use properties of such a polarized decay.

2 Top decay

Symbolic computation gives the following expression for the differential decay width of the top quark in its rest frame. We integrate the matrix element squared of the polarized t-quark decay over the neutrino and b-quark momenta using the narrow width approximation for the W-boson decay and neglecting the b-quark mass in comparison to the top quark and W-boson masses:

$$\frac{d\Gamma_{t\to b\nu e^+}}{dE_e \cdot d\cos\theta \cdot d\phi} = \frac{g^4}{256 \cdot \pi^3 \cdot \Gamma_W \cdot m_W} \cdot [\tag{3}$$

$$+ \Big(f_{LV}^2 \cdot (E_{max} - E_e) + f_{LT}^2 \cdot (E_e - E_{min}) \Big) \cdot E_e \cdot (1 + \cos\theta)$$

$$+ \Big(f_{RT}^2 \cdot (E_{max} - E_e) + f_{RV}^2 \cdot (E_e - E_{min}) \Big) \times$$
$$\Big(E_{min} + E_{max} - E_e + m_W/E_e \cdot c_e \cdot \sin\theta\cos\phi + \big(m_W^2/(2E_e) + E_e - E_{min} - E_{max} \big) \cdot \cos\theta \Big)$$

$$- 2 \cdot (f_{LV} \cdot f_{RT} \cdot (E_{max} - E_e) + f_{LT} \cdot f_{RV} \cdot (E_e - E_{min})) \cdot (c_e \cdot \sin\theta\cos\phi + m_W/2 \cdot (1 + \cos\theta))]$$

where:

$$c_e = \sqrt{(E_{max} - E_e) \cdot (E_e - E_{min})}, \quad E_{max} = m_t/2, \quad E_{min} = m_W^2/(2m_t)$$

Here the angle θ is the angle between the charged lepton momentum and the direction of the top quark spin quantization axis, the angle ϕ is the angle in the plane perpendicular to the lepton momenta counted from the decay plain formed by the top quark decay products as shown in Fig. 2, E_e is the energy of the charged lepton.

The expression (3) with real anomalous couplings contains of 4 terms corresponding to different combinations the coupling products. Each coupling combination is multiplied by an individual polynomial kinematic function depending on θ, ϕ and E_e. Therefore such multidimensional distribution shape (multidimensional surface) can be used to separate better contributions from different anomalous coupling combinations.

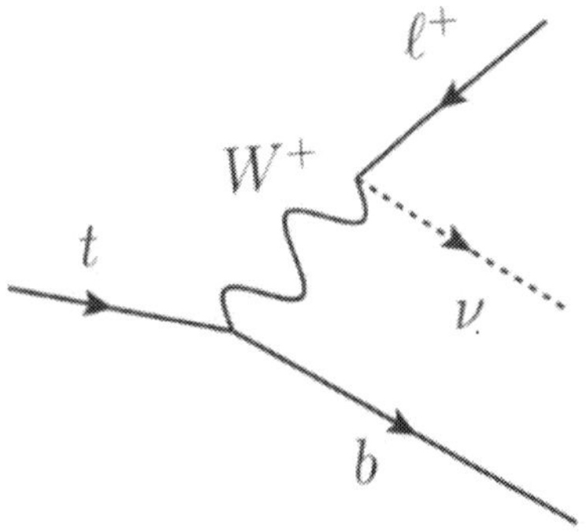

Figure 1. Top quark leptonic decay process.

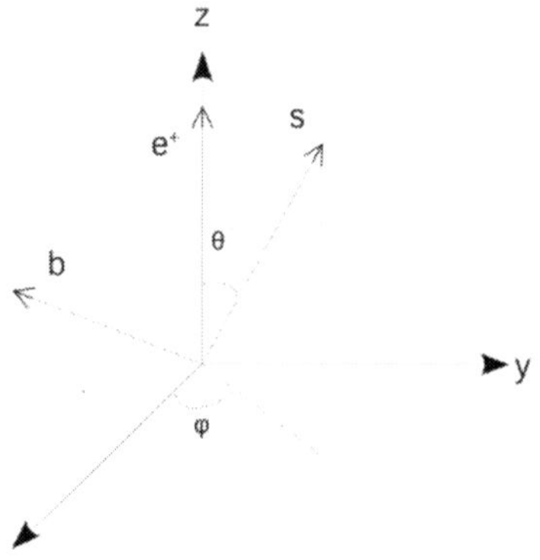

Figure 2. Top quark spin quantization axis.

The first term in the expression [16, 17] is the well-known SM like contribution corresponding to 100 % spin correlation behavior $(1 + \cos\theta)$ with the lepton energy distribution having a maximum at $m_t/4$. The interference term between the left-vector and the right-tensor terms coincides exactly with the expression presented in [9].

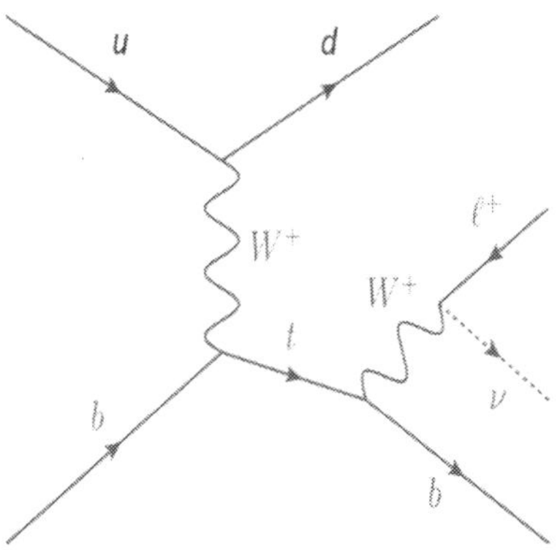

Figure 3. t-channel single top quark production and its leptonic decay processes.

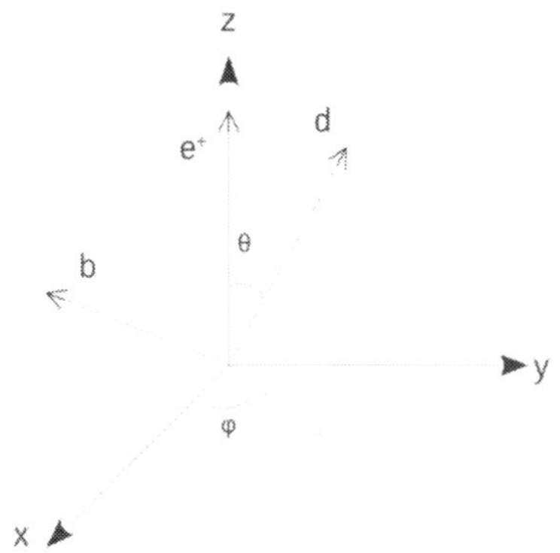

Figure 4. Top quark spin quantization axis in the top rest frame is chosen as the direction of d-quark momentum for the t-channel (outgoing light jet).

3 Numerical illustration

Numerical illustration we present for few cases performing computations by two different methods in each case. First, we take the differential width (3) and plot the corresponding surface. Second, we perform computations and Monte-Carlo event generation for the $2 \to 4$ process corresponding to the dominating t-channel single production with the top quark 3 body decay $ub \to l^+\nu_l bd$ (Fig. 4) using the CompHEP code version [21]. For the latter computation we show the same multidimensional distributions in the top rest frame for proton-proton collisions at 13 TeV energy taking all the angles with respect to the d-quark momentum. To make deviations from the SM more pronounce we subtract the SM contribution and we integrate in both ways of computations over the variable ϕ in the some range (for definiteness the range was taken from $-\pi/3$ to $+\pi/3$). We take values for the anomalous couplings from in correspondence with the recent experimental upper limits [12].

1. Non-zero left vector coupling f_{LV} ($|f_{LV} - V_{tb}| = 0.1$), other anomalous couplings are equal to zero. Contributing terms of the (3) are proportional to $(f_{LV} - V_{tb}) \cdot (f_{LV} - V_{tb})$ and $(f_{LV} - V_{tb}) \cdot V_{tb}$. Fig. 5 shows distribution $\frac{1}{\Gamma - \Gamma_{SM}} d(\Gamma - \Gamma_{SM})/(dE_e \cdot dcos\theta))$. Because of the dominating interference term of the left vector anomalous coupling $(f_{LV} - V_{tb})$ with the SM V_{tb} coupling the maximum of the distribution is located as expected at $cos\theta = 1$ and $E_e = m_t/4 \approx 43$ GeV. The distribution almost linearly drops to 0 with decreasing $cos\theta$. The distribution from the CompHEP simulation is shown in Fig. 6. Good agreement is observed.

2. Non-zero right vector coupling $f_{RV} = 0.16$, other anomalous couplings are equal to zero. The contributing term of the (3) is proportional to $|f_{RV}|^2$. Two dimensional distribution shown in Fig. 7 agrees well with the CompHEP computations given in Fig. 8. The maximum of is located now at the values $cos\theta \approx 0.3$ and $E_e \approx 60$. One can see that the shape of the distribution in this case is significantly different from that in case 1.

3. Non-zero left tensor coupling $f_{LT} = 0.057$, other anomalous couplings are equal to zero. The contributing term of the (3) is proportional to $|f_{LT}|^2$, and the distribution in Fig. 9 is very similar to that from the CompHEP computation (Fig. 10). The maximum is located at $cos\theta = 1$ and $E_e = E_{max} \approx$ 86.25 GeV. With decreasing $cos\theta$ and E_e the distribution quickly goes to 0.

4. Non-zero right tensor coupling $f_{RT} = 0.048$, other anomalous couplings are equal to zero. The contributing terms of the (3) are proportional to the squared $|f_{RT}|^2$ and to the interference $Ref_{RT} \cdot V_{tb}$

DOI: 10.1051/epjconf/201715804006

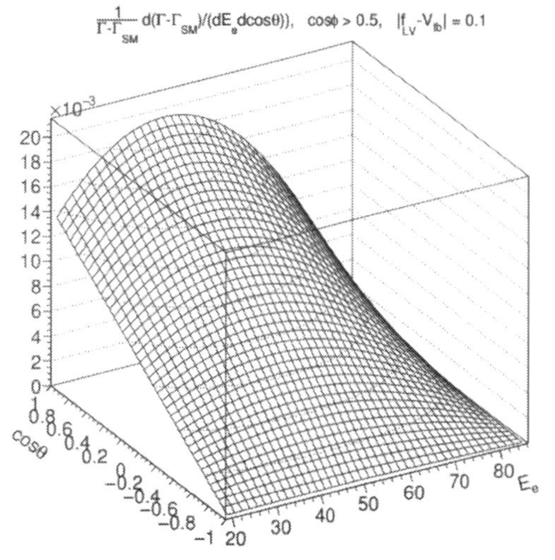

Figure 5. Top quark decay, $|f_{LV} - V_{tb}| = 0.1$.

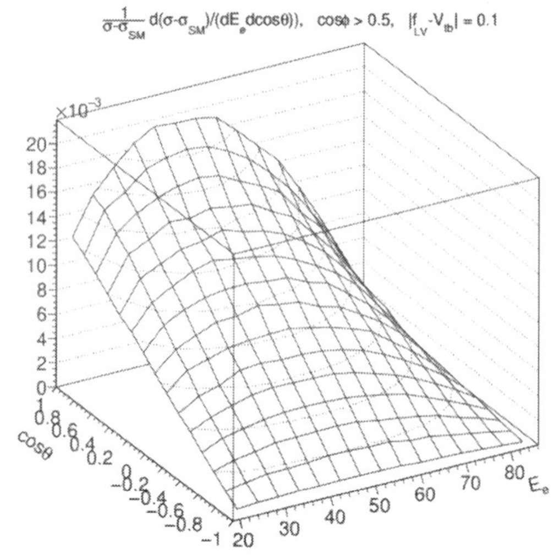

Figure 6. Top quark production and decay, $|f_{LV} - V_{tb}| = 0.1$.

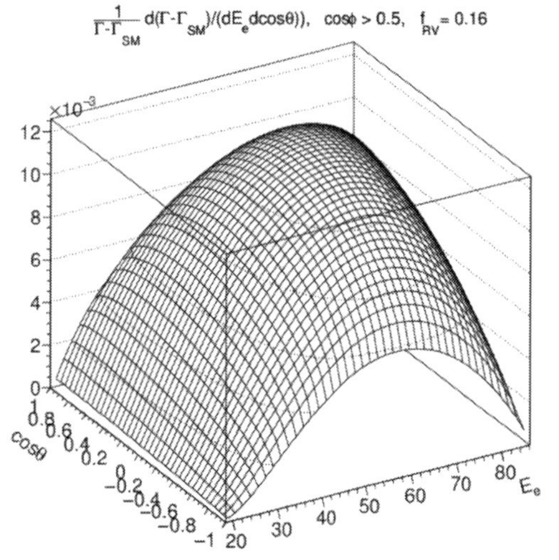

Figure 7. Top quark decay, $f_{RV} = 0.16$.

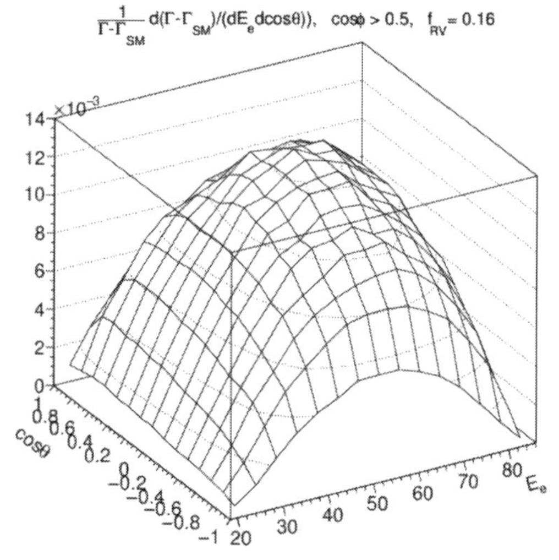

Figure 8. Top quark production and decay $f_{RV} = 0.16$.

parts. Once more the complete agreement between computations using the formula 3 for the width (Fig. 11) and by means of the CompHEP (Fig. 12). With decreasing $cos\theta$ and increasing E_e the distribution shape almost linearly drops to 0.

One can clearly see that the shapes of distributions proportional to different anomalous couplings are significantly different.

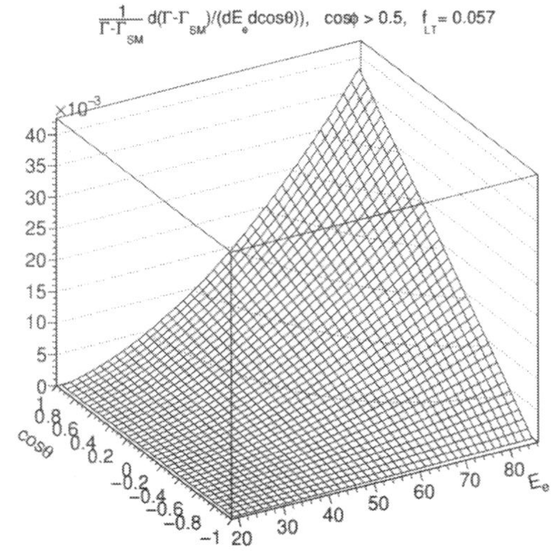

Figure 9. Top quark decay, $f_{LT} = 0.057$.

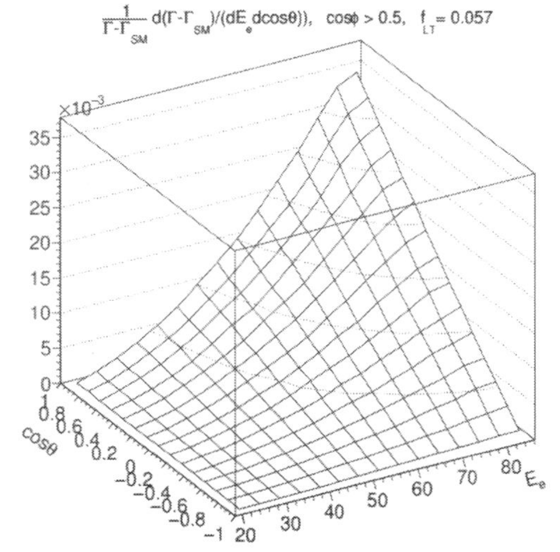

Figure 10. Top quark production and decay $f_{LT} = 0.057$.

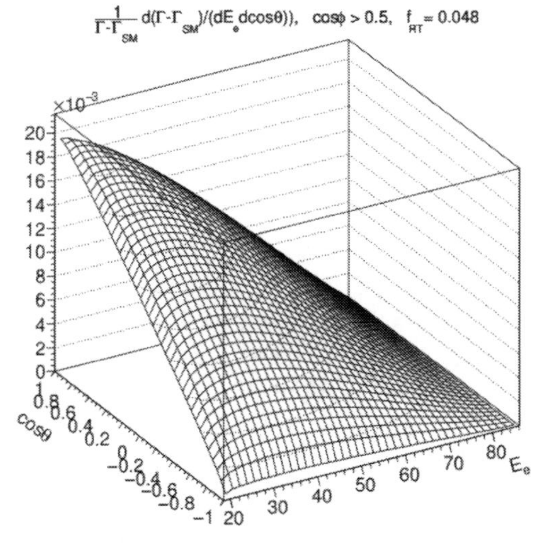

Figure 11. Top quark decay, $f_{RT} = 0.048$.

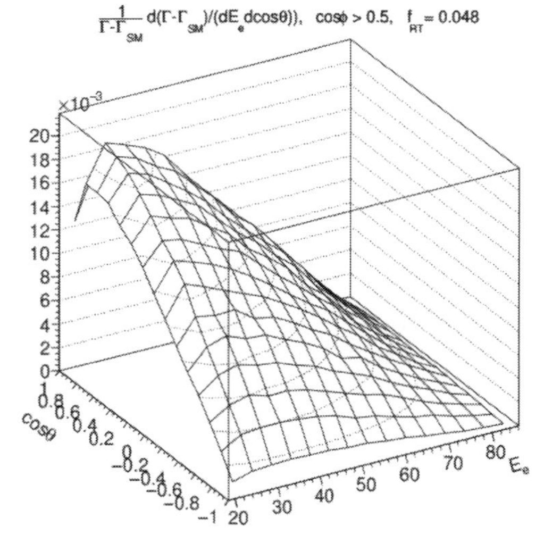

Figure 12. Top quark production and decay $f_{RT} = 0.048$.

4 Conclusions

We obtained the analytic expression for the differential width for the three-particle decay of polarized t-quark in the its rest frame as a function of orientation angles of the top quark spin quantization axis and charged lepton energy. The expression is presented for the case of real vector and tensor anomalous Wtb couplings. The parts of this expression containing the SM contribution and its interference with anomalous contributions are in complete agreement with published results.

Using this formula for the differential width we constructed two-dimensional distributions on the angle between the direction of the charged lepton momentum and the top quark spin quantization axis and on the p energy of the charged lepton. We plot such a distributions separately for each of the contribution proportional to anomalous couplings and shown significant differences in shapes of the distributions. These differences can be extracted form the final products of the t-channel single top production process with the top decay taking into account that due to smallness of the anomalous couplings the top is produced highly polarized in the direction of the d-quark produced giving correspondingly the best choice for top quark spin quantization axis.

We have confirmed the forms of the distributions by numerical computation of the complete process involving the t-channel top quark production and its decay and generation of the corresponding Monte-Carlo events using the Comp Hep package. From the generated events we constructed the same two-dimensional differential distributions for each anomalous contribution. It was shown that these distributions are in a very good agreement with the distributions predicted by the derived analytic formula. The differences in shapes of the distributions allows one to get limits on anomalous couplings in addition to the limits already extracted by other methods. Details of an application of such a method will be presented in a separate study.

Acknowledgments

The work was supported by grant 16-12-10280 of Russian Science Foundation. The authors are grateful to L. Dudko for useful discussions and critical remarks.

References

[1] U. Husemann, "Top-Quark Physics: Status and Prospects," Prog. Part. Nucl. Phys. **95** (2017) 48

[2] M. Cristinziani and M. Mulders, "Top-quark physics at the Large Hadron Collider," J. Phys. G **44** (2017) no.6, 063001

[3] E. Boos, O. Brandt, D. Denisov, S. Denisov and P. Grannis, "The top quark (20 years after its discovery)," Phys. Usp. **58**, no. 12, 1133 (2015) [Usp. Fiz. Nauk **185**, no. 12, 1241 (2015)]

[4] C. E. Gerber and C. Vellidis, "Review of Tevatron results: Top quark physics," Int. J. Mod. Phys. A **30**, no. 06, 1541005 (2015).

[5] F. Déliot, N. Hadley, S. Parke and T. Schwarz, "Properties of the Top Quark," Ann. Rev. Nucl. Part. Sci. **64**, 363 (2014).

[6] K. Whisnant, J. M. Yang, B. L. Young and X. Zhang, "Dimension-six CP conserving operators of the third family quarks and their effects on collider observables," Phys. Rev. D **56**, 467 (1997)

[7] E. Boos, M. Dubinin, M. Sachwitz and H. J. Schreiber, "Probe of the W t b coupling in t anti-t pair production at linear colliders," Eur. Phys. J. C **16** (2000) 269

[8] J. A. Aguilar-Saavedra, "Single top quark production at LHC with anomalous Wtb couplings," Nucl. Phys. B **804**, 160 (2008)

[9] C. Zhang and S. Willenbrock, "Effective-Field-Theory Approach to Top-Quark Production and Decay," Phys. Rev. D **83**, 034006 (2011)

[10] J. L. Birman, F. Déliot, M. C. N. Fiolhais, A. Onofre and C. M. Pease, "New limits on anomalous contributions to the *Wtb* vertex," Phys. Rev. D **93**, no. 11, 113021 (2016)

[11] G. L. Kane, G. A. Ladinsky and C. P. Yuan, "Using the Top Quark for Testing Standard Model Polarization and CP Predictions," Phys. Rev. D **45**, 124 (1992).

[12] V. Khachatryan *et al.* [CMS Collaboration], "Search for anomalous Wtb couplings and flavour-changing neutral currents in t-channel single top quark production in pp collisions at $\sqrt{s} = 7$ and 8 TeV," JHEP **1702**, 028 (2017)

[13] J. A. Aguilar-Saavedra and S. Amor Dos Santos, "New directions for top quark polarization in the t-channel process," Phys. Rev. D **89**, no. 11, 114009 (2014)

[14] J. A. Aguilar-Saavedra, "Missing Top Properties," arXiv:1411.5697 [hep-ph].

[15] J. A. Aguilar-Saavedra, C. Degrande and S. Khatibi, "Single top polarisation as a window to new physics," Phys. Lett. B **769** (2017) 498

[16] M. Jezabek and J. H. Kuhn, "V-A tests through leptons from polarized top quarks," Phys. Lett. B **329**, 317 (1994)

[17] M. Jezabek, "Top quark physics," Nucl. Phys. Proc. Suppl. **37B**, no. 2, 197 (1994)

[18] G. Mahlon and S. J. Parke, "Improved spin basis for angular correlation studies in single top quark production at the Tevatron," Phys. Rev. D **55**, 7249 (1997)

[19] G. Mahlon and S. J. Parke, "Single top quark production at the LHC: Understanding spin," Phys. Lett. B **476**, 323 (2000)

[20] E. E. Boos and A. V. Sherstnev, "Spin effects in processes of single top quark production at hadron colliders," Phys. Lett. B **534**, 97 (2002)

[21] E. Boos *et al.* [CompHEP Collaboration], "CompHEP 4.4: Automatic computations from Lagrangians to events," Nucl. Instrum. Meth. A **534**, 250 (2004)

DOI: 10.1051/epjconf/201715804007

On the top-quark width measurement using a combination of resonant and non resonant cross sections.

Alexey Baskakov[1,*], *Eduard* Boos[1,2,**], and *Lev* Dudko[1,***]

[1] *Skobeltsyn Institute of Nuclear Physics M.V. Lomonosov Moscow State University, Moscow 119991, Russian Federation*
[2] *Faculty of Physics M.V.Lomonosov Moscow State University Leninskie Gory, Moscow 119991, Russian Federation*

Abstract. Top quark width measurement is important and complex task. Most measurements are done in assumption of the SM top quark interactions. We would like to discuss model independent way of measuring the top quark width. Experimental accuracies of the top quark fiducial cross section measurements from 2% to 10% may allow to put constrains on the top quark width with 6% to 29% precision.

1 Introduction

The top quark was discovered more than 20 years ago [1, 2] and it still remains the heaviest known elementary particle. This fact probably makes it the most promising window to the Beyond Standard Model (BSM) physics. The top quark parameters measurement is critical for testing deviations from the Standard Model (SM) predictions. During LHC operation period mass and inclusive cross sections were measured at the percentage level [3]. Indirect top quark width measurements in assumption of only SM decay modes have reached ten percent accuracy [4, 5]. However direct measurements of the top quark width are far from that precision because of experimental resolution limitations of the top-quark mass spectrum [6]. Recent direct measurements presented by CMS and ATLAS collaborations are mostly limited by the jet energy scale and resolution [7, 8]. To improve accuracy of direct measurements usage of the b-charge asymmetry in the s- and t-channels of the single top production was recently proposed [9].
In our paper we present first steps of putting model independent and fully gauge invariant limits on the top quark width using combinations of resonant and non resonant cross sections.

2 Simulations

Model-independent constrains on the Higgs boson width can be done from comparison of the on-/off-shell regions. Importance of the off-shell region in Higgs production first time was mentioned in [10]. In paper [11], the method for deriving model-independent upper bound on the Higgs boson width

*e-mail: a.baskakov@cern.ch
**e-mail: boos@theory.sinp.msu.ru
***e-mail: lev.doudko@cern.ch

from $pp \to ZZ$, $pp \to H \to ZZ$ events was proposed. These ideas can not be directly applied to the top quark. First, the Higgs is match more narrow resonance than the top quark, in other words $\frac{\Gamma_H}{m_H} << \frac{\Gamma_t}{m_t}$. Second, in case of the Higgs, we can calculate separately amplitude for the pole, non pole and interference amplitude for $pp \to ZZ$ and $pp \to H \to ZZ$ contributions. In case of the top quark pair production with subsequent decay to Wb one can not separate contributions in gauge invariant way for the top pair and the single top production parts. Therefore we perform the computation of the complete gauge invariant set of diagrams and investigate sensitivity of different kinematic regions to deviations from the SM caused by the top quark width and related coupling. This approach allows to put model independent and fully gauge invariant constrains on the top quark width.

2.1 Modeling details

For simplicity we take into account only main signal events on the parton level for final states as the first step of our investigation. We model $gg \to tW^-\bar{b}$, when one top quark decays and another comes on its mass shell. Additional sub-processes with quarks in the initial state does not make significant impact on results since the gluon fusion gives up to 90% at LHC energies [13, 14]. Hadronization and fragmentation effects, as well as backgrounds impact are postponed to the next more realistic analysis. We use CompHEP generator [15] with MSTW2008 PDF [16] to generate events for the full gauge invariant diagram set Fig. 1 which include both single top quark and the top quark pair production. We take SM values of the top quark mass and width $M_t^{SM} = 172.5\,GeV$, $\Gamma_t^{SM} = 1.49\,GeV$. Process $gg \to tW^-\bar{b}$ at tree level includes eight diagrams Fig. 1. We do not take into account NLO corrections since

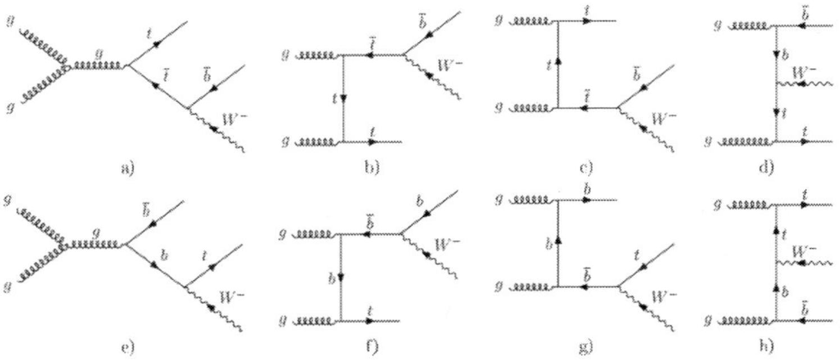

Figure 1. Full gauge invariant tree level diagram set for the $gg \to tW^-\bar{b}$ process.

their influence on the top quark width kinematic dependence is negligible. NLO QCD corrections to $W^+W^-b\bar{b}$ production at hadron colliders were calculated in [12]. We use notations shown in Eq. 1 and Eq. 2 to parametrize different kinematic regions of the $tW^-\bar{b}$ mass spectrum named as double pole (DP) and single pole (SP) respectively.

$$M_t^{SM} - n \cdot \Gamma_t^{SM} \leq M_{W\text{-}\bar{b}} \leq M_t^{SM} + n \cdot \Gamma_t^{SM} \tag{1}$$

$$M_{W\text{-}\bar{b}} \leq M_t^{SM} - k \cdot \Gamma_t^{SM} \quad or \quad M_t^{SM} + k \cdot \Gamma_t^{SM} \leq M_{W\text{-}\bar{b}} \tag{2}$$

Parameters n and k are taken to be integer numbers that allow to change kinematic regions. They should fill the following relation $n \leq k$ as regions may not include the hole mass spectrum but should not overlap. The DP region determined in Eq. 1 and shown in Fig. 2 by the dark blue includes the areas around to the top quark mass peaks, one of the top is on shell in our simple example. In the SP

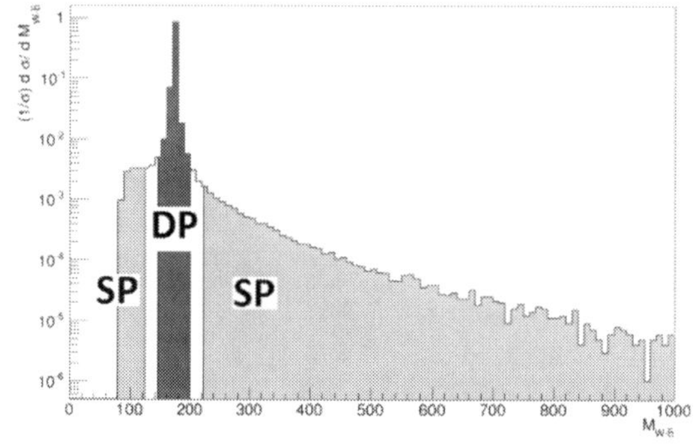

Figure 2. Differential mass spectrum of the $W^-\bar{b}$. Double pole(DP) corresponds to the $t\bar{t}$ kinematic region Eq. 1, single pole(SP) corresponds to the single top quark region Eq. 2.

regions (Eq. 2) shown by light blue in Fig. 2 one of Wb invariant mass is close to the top mass shell (on the mass shell in our example) while the second invariant mass is out of the top pole aria. Γ_t^{SM} is a constant, in its units it is shown how far from the top quark pole the Wb invariant mass is taken.

2.2 Top quark width parametrization

For the top quark width parametrization we use following notations

$$\Gamma_t = \xi^2 \cdot \Gamma_t^{SM} + \Delta, \tag{3}$$

reflecting the fact that the top quark width might be different from the SM width either by a modification of the Wtb coupling or by an existence of some non-SM additional decay modes [17–21]. In Eq. 3 the parameter ξ is used to reweight simultaneously the top quark width and the Wtb coupling, note that the production cross section times the branching ratio remains unchanged. It is useful to parametrize the deviation Δ also in terms of the SM width as $\Delta = \delta \cdot \Gamma_t^{SM}$. Details are given in Appendix 5. The invariant $W^-\bar{b}$ mass distribution for different values of ξ and δ is shown in Fig. 3. The

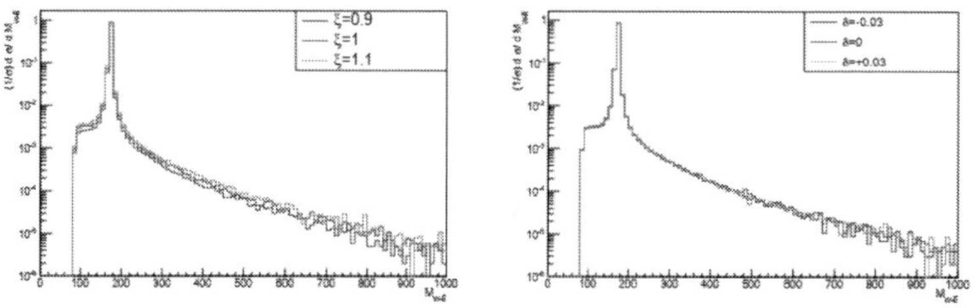

Figure 3. Differential mass spectrum of the $W^-\bar{b}$ for different ξ and δ.

influence of the n and k parameter variation on the cross section in DP and SP regions is demonstrated in Fig. 4. We see that DP region loose sensitivity to ξ variation for $n \geq 15$ Fig. 4 (top left) and influence of δ parameter keeps valuable for all n values Fig.4 (bottom left). In contrast for the SP region the sensitivity to ξ variation is visible for all physically meaningful k values, while δ modification dose not make any influence. This observation allows to put constrains on the ξ and δ using two dimension scan over the n and k parameters.

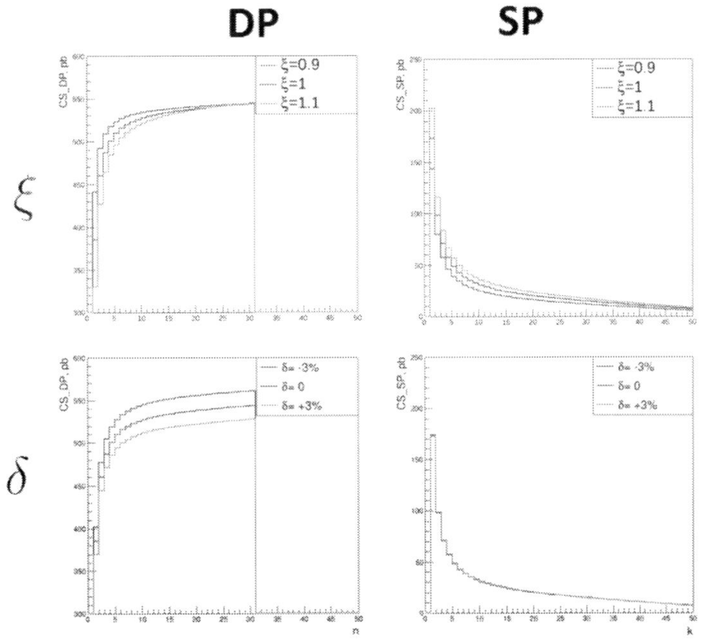

Figure 4. Influence of n and k parameters variation on the cross section in DP and SP regions.

2.3 Cross section parametrization

Now the cross sections in DP and SP regions can be parametrized in terms ξ and δ. Current experimental data [4–8] indicate that the deviations from the SM for the top width should be small. To keep deviations on both ξ and δ small let us make the following substitution

$$\epsilon = \xi^2 - 1 \tag{4}$$

and perform the linear expansion of the fiducial cross sections with respect to small parameters ϵ and δ as follows:

$$\sigma_{DP} = \sigma_{DP}^{SM} + A_{DP} \cdot \epsilon + B_{DP} \cdot \delta \tag{5}$$

$$\sigma_{SP} = \sigma_{SP}^{SM} + A_{SP} \cdot \epsilon + B_{SP} \cdot \delta \tag{6}$$

A, B are numerically calculated from DP and SP regions for different n, k.

Using the standard χ^2 method ($\chi^2(\sigma) = \left(\frac{\sigma^{SM}-\sigma}{\Delta\sigma}\right)^2$) and requiring that the SM and modified cross sections are within two standard deviations from each other, we derive an upper limits on ϵ and δ at the 95% confidence level. Figures 5, 6 show fits for different expected experimental measurements accuracies of 2, 5 or 10. As it was mentioned in 2.2, DP region has sensitivity mostly to δ parameter, while SP region is more sensitive to ϵ (or ξ see Eq. 4). Choice of the kinematic region boundary by varying n and k parameters make some shifts on the plots, but do not change the general dependence for each region. This fact allows to put much stronger combined limits on δ and ϵ Fig. 7, were two combinations $n = k = 15$ and $n = 15, k = 30$ are shown. We use Eq. 3 and take obtained parameters limits in quadratures to get predictions on the top quark width measurement precision Tab. 1.

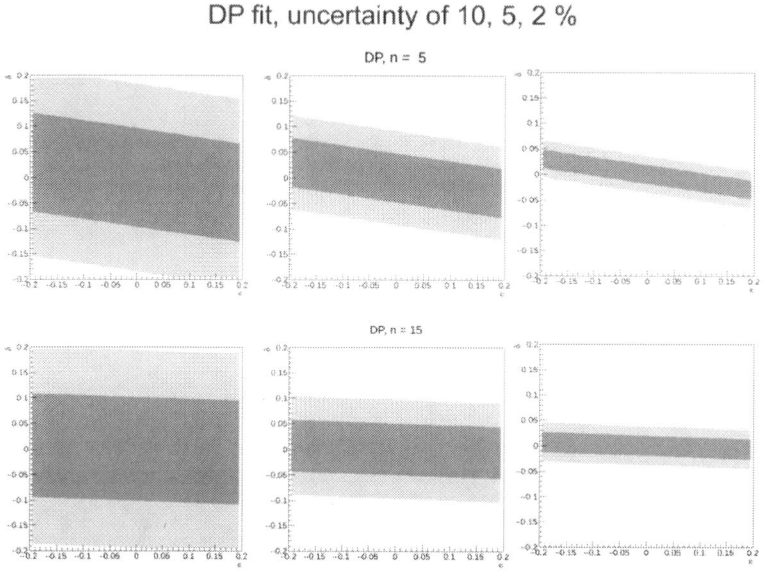

Figure 5. Constraints on the ϵ and δ parameters from DP region for boundary of ±5 (top row) and ±15 (bottom row) Γ_t^{SM} from the top quark mass value. Cross section experimental measurements accuracies are taken 10, 5 and 2% (from left to right). Green and yellow arias correspond to exclusion limits at 68% and 95% CL on ϵ and δ.

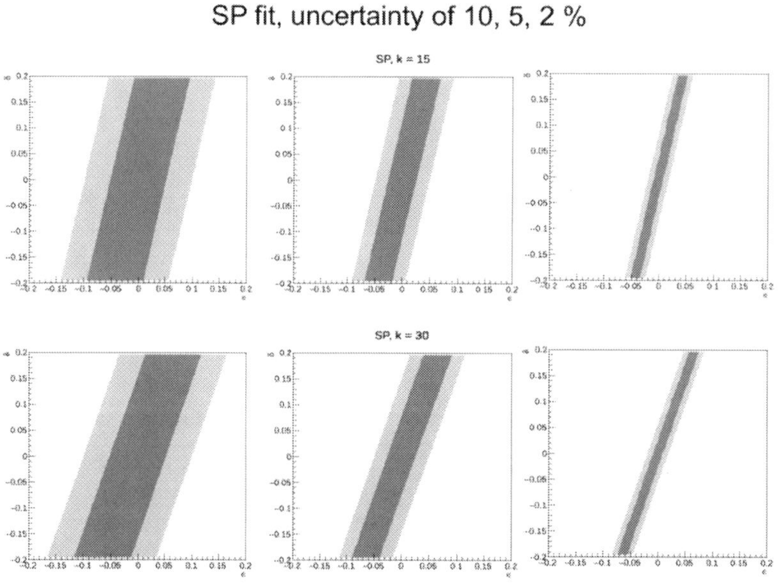

Figure 6. Constraints on the ϵ and δ parameters from SP region for $k = 15$ (top row) and $k = 30$ (bottom row). Cross section experimental measurements accuracies are taken 10, 5 and 2% (from left to right). Green and yellow arias correspond to exclusion limits at 68% and 95% CL on ϵ and δ.

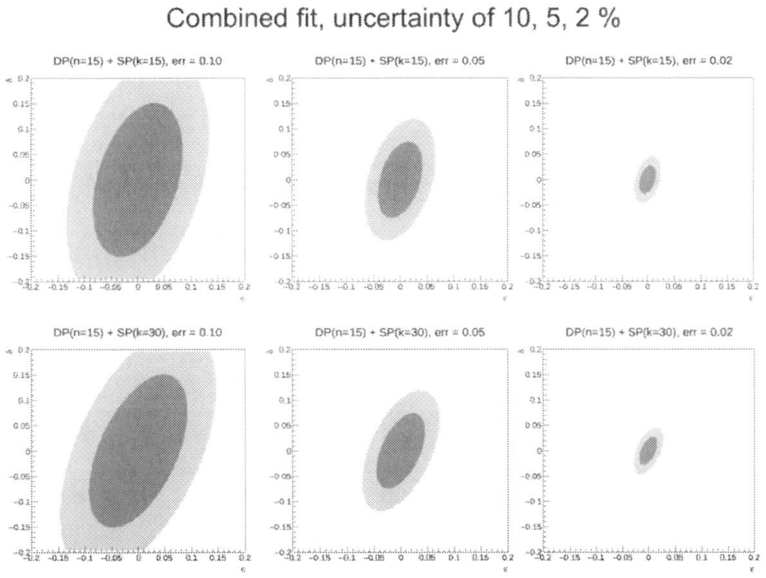

Figure 7. Limits on the ϵ and δ parameters obtained from DP and SP regions simultaneously. Region boundaries are taken as $n = k = 15$ (top row) and $n = 15, k = 30$ (bottom row). Green and yellow arias correspond to exclusion limits at 68% and 95% CL on ϵ and δ.

Experimental CS accuracy	Obtained limit on the top quark width
2%	$\geq 6\%$
5%	$\geq 14\%$
10%	$\geq 29\%$

Table 1. Estimations on achievable constrains on the top quark width for different experimental accuracies of the top quark fiducial cross section measurement.

3 Results

Gauge invariant estimation of deviations of the top quark width from its SM value is obtained in different kinematic regions. It is shown that top quark production cross section in the Double Pole(DP) region is more sensitive to δ, while in the Single Pole(SP) to ϵ (ξ) parameter. This fact allows to put combined limits on both parameters simultaneously and using these limits obtain constrains on the top quark width. Achievable constrains on the model independent top quark width measurement are estimated to be 6% to 29% for corresponding experimental accuracies from 2% to 10%. This results are achieved using simplified model, all effects as hadronization and fragmentation, detector response and background events will be taken into account in the following study.

Acknowledgments

The work was supported by grant 16-12-10280 of Russian Science Foundation.

References

[1] Abe, F. et al. (CDF Collaboration), Observation of top quark production in $\bar{p}p$ collisions, Phys. Rev. Lett. **74**, 2626 (1995).

[2] S. Abachi et al. (D0 Collaboration), Observation of the top quark, Phys. Rev. Lett. **74**, 2632 (1995).

[3] C. Patrignani et al. (Particle Data Group), Chin. Phys. **C40**, 100001 (2016).

[4] V. M. Abazov et al. (D0 Collaboration), Improved determination of the width of the top quark, Phys. Rev. **D85**, 091104 (2012).

[5] CMS collaboration, "Measurement of the ratio $\mathcal{B}(t \to Wb)/\mathcal{B}(t \to Wq)$ in pp collisions at $\sqrt{s}$ = 8 TeV", Phys. Lett. **B736**, 33-57 (2014).

[6] CDF Collaboration, "Direct Measurement of the Total Decay Width of the Top Quark", Phys. Rev. Lett. **111**, 202001, (2013).

[7] CMS Collaboration, "Bounding the top quark width using final states with two charged leptons and two jets at $\sqrt{s}$ = 13 TeV", (2016).

[8] ATLAS collaboration, "Direct top-quark decay width measurement in the $t\bar{t}$ lepton+jets channel at $\sqrt{s}$ = 8 TeV with the ATLAS experiment" ATLAS-CONF-2017-056, (2017).

[9] P.P. Giardino, C. Zhang, Phys. Rev. **D96**, 011901 (2017).

[10] N. Kauer and G. Passarino, JHEP **1208**, 116 (2012).

[11] F. Caola, K. Melnikov, Phys. Rev.**D88**, 054024 (2013).

[12] A. Denner, S. Dittmaier, S. Kallweit, S. Pozzorini, Phys. Rev. Lett.**106**, 052001 (2011).

[13] S. Moch and P. Uwer, Phys.Rev. D **78**, 034003 (2008).

[14] K. Olive et al., Particle Data Group Collaboration, Chin.Phys. C **38**,090001 (2014).

[15] E.E. Boos, V.E. Bunichev, L.V. Dudko, V.I. Savrin, A.V. Sherstnev, Phys.Atom.Nucl. **69**, 1317 (2006).

[16] A.D. Martin, W.J. Stirling, R.S. Thorne et al., Eur. Phys.J. C **63**, 189 (2009).

[17] W. Bernreuther, J. Phys. **G35**, 083001 (2008).

[18] F. Larios and R. Martinez and M. A. Perez, Int. J. Mod. Phys. **A21**, 3473-3494 (2006).

[19] T. Han, K. Whisnant, B.L. Young and X. Zhang, Phys. Rev. **D55**, 7241-7248 (1997).

[20] M.J. Herrero, 29th International Meeting on Fundamental Physics Barcelona, Spain (2011).

[21] A. Johan. R. Frederix, J. -M. Gerard, A. Giammanco, M. Herquet, S. Kalinin, E. Kou, V. Lemaitre and F. Maltoni, F. Eur. Phys. J. **C49**, 791-801 (2007).

[22] N. Kauer, D. Zeppenfeld, Phys. Rev. **D65**, 014021 (2001).

5 Appendix

The general answer for the Breit-Wigner type integral has the following form

$$I_{BW} = \int_{q^2_{max}}^{q^2_{min}} \frac{dq^2}{(q^2 - M^2)^2 + \Gamma^2 M^2} = \frac{1}{\Gamma M}\left[arctg(\frac{q^2_{max}}{M\Gamma} - \frac{M}{\Gamma}) - arctg(\frac{q^2_{min}}{M\Gamma} - \frac{M}{\Gamma})\right]. \quad (7)$$

If we define different kinematic regions as in Fig. 8, then I_1 can be interpreted as DP region Eq. 1 and $I_2 + I_3$ as SP region Eq. 2. In the limit $\Gamma << M$ and $\Delta M^{\pm} < M$

$$
\begin{aligned}
I_1 &= \frac{\pi}{M\Gamma} - \frac{1}{2M}\left(\frac{1}{\Delta M^+} - \frac{1}{\Delta M^-}\right) + ... \\
I_2 &= \frac{1}{2M}\left(\frac{1}{\Delta M^+} - \frac{1}{M} - \frac{2M}{M_{cut}^{+} - M^2} + ...\right) + ... \\
I_3 &= \frac{1}{2M}\left(\frac{1}{\Delta M^-} + \frac{1}{M} - \frac{2M}{M^2 - M_{cut}^{-}} + ...\right) + ...
\end{aligned}
\tag{8}
$$

The Equation 8 allows to understand qualitatively observed properties in Fig. 4 if one takes into

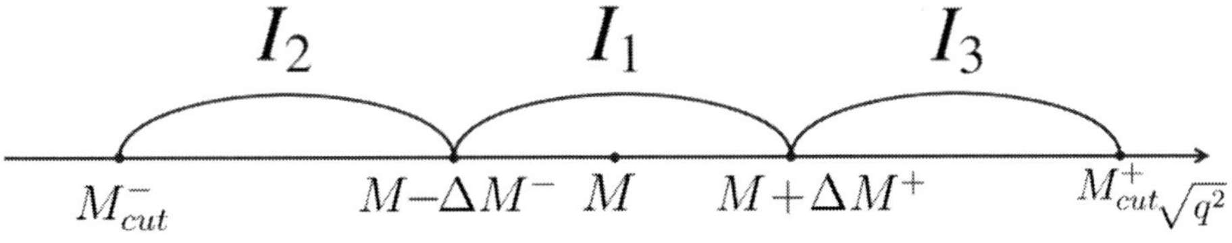

Figure 8. Definition of kinematic regions.

account the dependence of the top quark width on both parameters ϵ and δ and the fact that the numerator of the matrix element squared depends only on the parameter ϵ via the Wtb coupling. Furthermore, one can see that if $M_{cut}^- \rightarrow -\infty$ and $M_{cut}^+ \rightarrow +\infty$

$$I_{BW} = I_1 + I_2 + I_3 = \frac{\pi}{M\Gamma}. \tag{9}$$

At $\Delta M^+ = \Delta M^- = n \cdot \Gamma$ we get formula (19) [22]

$$I_1 = \frac{\pi}{M\Gamma}\left(1 - \frac{1}{n\Gamma} + ...\right). \tag{10}$$

For our consideration this formula (19) [22] is not useful since we want to study dependence on Γ, but ΔM^+ and ΔM^- are cuts and do not depend on Γ.

Domain Walls and Matter-Antimatter Domains in the Early Universe

A.D. Dolgov[1,2,⋆], *S.I.* Godunov[1,2,⋆⋆], and *A.S.* Rudenko[1,3,⋆⋆⋆]

[1] *Novosibirsk State University, Novosibirsk, 630090, Russia*
[2] *Institute for Theoretical and Experimental Physics, Moscow, 117218, Russia*
[3] *Budker Institute of Nuclear Physics, Novosibirsk, 630090, Russia*

Abstract. We suggest a scenario of spontaneous (or dynamical) C and CP violation according to which it is possible to generate domains of matter and antimatter separated by cosmologically large distances. Such $C(CP)$ violation existed only in the early universe and later it disappeared with the only trace of generated matter and antimatter domains. So this scenario does not suffer from the problem of domain walls.
According to this scenario the width of the domain wall should grow exponentially to prevent annihilation at the domain boundaries. Though there is a classical result obtained by Basu and Vilenkin that the width of the wall tends to the one of the stationary solution (constant physical width). That is why we considered thick domain walls in a de Sitter universe following paper by Basu and Vilenkin. However, we were interested not only in stationary solutions found therein, but also investigated the general case of domain wall evolution with time. When the wall thickness parameter, δ_0, is smaller than $H^{-1}/\sqrt{2}$, where H is the Hubble parameter in de Sitter space-time, then the stationary solutions exist, and initial field configurations tend with time to the stationary ones. However, there are no stationary solutions for $\delta_0 > H^{-1}/\sqrt{2}$. We have calculated numerically the rate of the wall expansion in this case and have found that the width of the wall grows exponentially fast for $\delta_0 \gg H^{-1}$. An explanation for the critical value $\delta_{0c} = H^{-1}/\sqrt{2}$ is also proposed.

1 Introduction

Our local cosmological neighborhood is made of baryons, while fraction of antimatter is vanishingly small. Therefore, it looks like the universe is 100% baryo-asymmetric (at least locally). The Standard Model (SM) cannot explain this asymmetry (BAU). Many quite different extensions of the SM and various scenarios for the BAU generation were suggested in the literature, for a review see e.g. Refs. [1–5]. Typically, consideration is restricted to the models where the universe is asymmetric globally. However, it is not excluded that the real universe may be globally symmetric. It may consist of domains of matter and antimatter, and if the domains are sufficiently large and far away, they may escape observational constraints on matter-antimatter annihilation at the domain boundaries. In the

⋆e-mail: dolgov@fe.infn.it
⋆⋆e-mail: sgodunov@itep.ru
⋆⋆⋆e-mail: a.s.rudenko@inp.nsk.su

DOI: 10.1051/epjconf/201715805001

simplest version of the scenario the distance to the nearest domain of antimatter should be close to the present day cosmological horizon [6].

Corresponding particle physics models, leading to the universe creation with abundant antimatter domains were suggested and developed in the past. While being more involved, the models of this type also suffer from the inherent problem — a domain wall problem [7]. However, there is a way out of this restriction: domain walls should exist only in the past and should disappear by now, though their effects in the form of matter and antimatter objects would survive to the present day.

One class of models where this can be achieved has been suggested in Refs. [8–11]. The main idea behind is a possibility of an unusual symmetry behavior at high temperatures. It is well known that a symmetry, which is broken in vacuum, at high temperatures tends to be restored. But in general, it is also possible for a symmetry to be broken only in a specific range of temperatures, for the particular models and details see [8–11]. This is just what is needed for a matter-antimatter domain generation without domain wall problem. However, if a model is based on the unusual symmetry behavior at high temperatures, then the size of domains will be too small from the cosmological point of view. We suggest the scenario in which this happens during inflationary stage. Domains with different sign of CP disappear by now, so the domain wall problem is absent. However, they appeared during inflation and survived at the baryogenesis epoch, therefore, cosmologically large[1] domains of matter and antimatter could be created.

The scenario described above inspired the consideration of the related problem: is it possible for the domain wall width to grow almost at exponential rate in expanding universe? The evolution of the domain walls in de Sitter space-time was considered by Basu and Vilenkin [12]. The authors argued that the width of the domain wall is determined by the ratio $C \equiv \lambda\eta^2/H^2 > 0$, where H is the Hubble parameter, which was assumed to be constant, η is the vacuum expectation value of the Higgs-like field which induced the spontaneous symmetry breaking, and λ is the coupling constant in the double-well potential, see (17). If $C \gg 2$, the width of the domain wall would be close to its flat space-time value, $\delta_0 = 1/(\sqrt{\lambda}\eta)$, which is microscopically small.

On the other hand, if $C < 2$, it is not excluded that the width of the domain wall may be astronomically large. We show that this is indeed the case. In ref. [12] only the stationary problem was considered, when the shape of the domain wall was a function of a single variable, the physical distance. In the paper [12] it was found numerically that the stationary solution exists only if $C > 2$.

In what follows we lift the assumption of the stationarity and consider the general plane solution being a function of both variables: the distance from the wall and time. It allows us to see how the solution approaches the stationary one and, in particular, what happens with initial configurations if $C \leq 2$, when the stationary solution does not exist. The related problems were considered in [13].

This talk is based on papers [14] (written in collaboration with I.I. Tkachev) and [15].

2 Generation of matter and animatter domains

2.1 Model

In the suggested model the difference between matter and antimatter is generated by pseudoscalar field χ which interacts with inflaton field Φ. We assume the following Lagrangian:

$$L = L_\Phi + L_\chi + L_{int}, \tag{1}$$

[1]The matter-antimatter domains should be separated by at least several megaparsec in terms of the present day scale to avoid excessive matter–antimatter annihilation. On the other hand, the distance should not be too large, otherwise the scenario would lead to too large angular fluctuations of CMB [6].

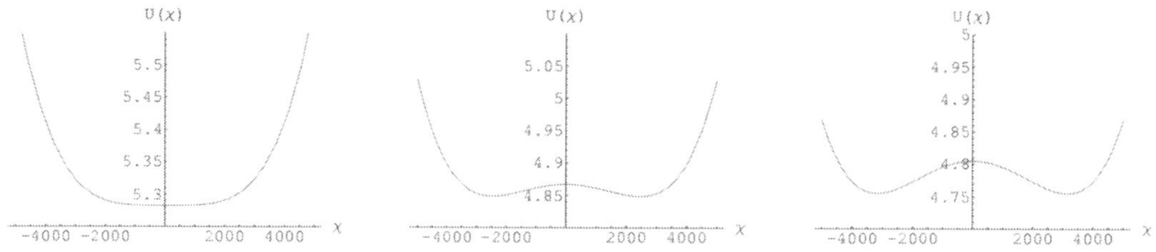

Figure 1. Potential $U(\Phi,\chi)$ at constant Φ. The left plot corresponds to the moment when the effective mass of χ becomes zero, i.e. when $m^2/2-\mu^2V(\Phi)=0$, or $\Phi=\Phi_0+2\Phi_1\sqrt{\ln(\sqrt{2}\mu/m)}$. The middle and the right plots show the moments when $\Phi=\Phi_0+\Phi_1$ and $\Phi=\Phi_0$, respectively. The parameters are $\Phi_0=3.1\,m_{Pl}$, $\Phi_1=0.02\,m_{Pl}$, $\mu=10^{-4}m_{Pl}$ and $m=10^{-10}m_{Pl}$. Field χ is measured in units of M and $U(\Phi,\chi)$ is measured in units of $10^{-12}\,m_{Pl}^4$.

where

$$L_\Phi=\frac{1}{2}(\partial\Phi)^2-\frac{1}{2}M^2\Phi^2,\qquad L_\chi=\frac{1}{2}(\partial\chi)^2-\frac{1}{2}m^2\chi^2-\frac{1}{4}\lambda_\chi\chi^4,\qquad L_{int}=\mu^2\chi^2V(\Phi).\tag{2}$$

Here the metric tensor enters into kinetic terms in the usual way, and M, m, λ_χ, μ are some constant parameters. The dimensionless function $V(\Phi)$ is chosen in the way that it is non-zero only when Φ is close to some constant value Φ_0. In this paper we choose it as Gaussian function

$$V(\Phi)=\exp\left[-\frac{(\Phi-\Phi_0)^2}{2\Phi_1^2}\right]\tag{3}$$

Parameters Φ_0 and Φ_1 have dimension of mass and indicate position of the "bell" center and characteristic width of the "bell", respectively.

The equations of motion have the form:

$$\ddot{\Phi}+3H\dot{\Phi}+M^2\Phi+\mu^2\chi^2\frac{\Phi-\Phi_0}{\Phi_1^2}V(\Phi)=0,\tag{4}$$

$$\ddot{\chi}+3H\dot{\chi}+m^2\chi+\lambda_\chi\chi^3-2\mu^2\chi V(\Phi)=0,\tag{5}$$

where $H=\dot{a}/a$ is the Hubble parameter, $a(t)$ is the cosmological scale factor which enters into the FLRW metric as

$$ds^2=dt^2-a^2(t)\,d\mathbf{x}^2.\tag{6}$$

It is assumed here that fields Φ and χ depend only on time, $\Phi=\Phi(t)$ and $\chi=\chi(t)$.

The Hubble parameter is expressed through the energy density ρ as

$$H=\sqrt{\frac{8\pi\rho}{3m_{Pl}^2}}=\sqrt{\frac{8\pi}{3m_{Pl}^2}\left(\frac{\dot{\Phi}^2}{2}+\frac{M^2\Phi^2}{2}+\frac{\dot{\chi}^2}{2}+\frac{m^2\chi^2}{2}+\frac{\lambda_\chi\chi^4}{4}-\mu^2\chi^2V(\Phi)\right)},\tag{7}$$

where $m_{Pl}\approx 1.2\cdot 10^{19}$ GeV is the Planck mass.

The interaction introduced above leads to the following scenario. During inflation the magnitude of the inflaton field Φ decreases and when it reaches vicinity of Φ_0 two minima appear in the potential

$$U(\Phi,\chi)=\left(\frac{1}{2}m^2-\mu^2V(\Phi)\right)\chi^2+\frac{1}{4}\lambda_\chi\chi^4+\frac{1}{2}M^2\Phi^2\tag{8}$$

at constant Φ, so the point $\chi = 0$ becomes local maximum (see Fig. 1).

Therefore, in the spatial regions where the field χ turns out to be positive (due to fluctuations) it rolls down to the positive minimum $\eta > 0$, and in the regions where χ is negative it goes to the negative minimum $-\eta < 0$. Thus one can substitute $\chi = \widetilde{\chi} + \eta$ and $\chi = \widetilde{\chi} - \eta$, respectively, where η and $-\eta$ have the meaning of the vacuum expectation values $\langle\chi\rangle$. Thus, due to initial fluctuations we should obtain domains of the field χ which will provide the source of CP violation, so in the domains with positive $\langle\chi\rangle$ we will have domain of matter while in the domain with negative $\langle\chi\rangle$ we will have domain of antimatter. After the interaction term $V(\Phi)$ is switched off there is again single minimum $\chi = 0$ so there is no domain walls problem at the present time.

We expect that the distances between domains, as well as domain sizes exponentially grow with the scale factor $a(t)$, so at the present time these domains are of cosmological size and they are separated by large distance which prevents the annihilation at their boundaries. According to [6], a very large piece of space between the domains, devoid of baryons, would lead to too large angular fluctuations of the CMB temperature. On the other hand, if the distance between domains is smaller than the baryon diffusion length, they would be able to meet successfully their counterparts and to annihilate. These arguments resulted in the conclusion that the nearest domains should be at the cosmologically large distance from us, about a few Gpc. On the other hand, the effects of the cosmological inhomogeneity, not yet studied, could inhibit the baryon diffusion and would relax the bound on the distance to the nearest antimatter domain.

Now let us consider what happens at different stages of the scenario, so we will be able to put the limits on initial conditions and parameters of the model.

For simplicity we assume, though it is not necessary, that the impact of field χ on cosmological expansion at inflationary stage is negligible. To this end we suppose that the energy density of Φ dominates over that of χ and the energy density of their interaction. In this approximation the second derivative term and the interaction term proportional to $V(\Phi)$ in the inflaton equation of motion (4) can be neglected. Therefore, we find

$$\dot{\Phi} = -\frac{M^2\Phi}{3H}, \qquad \Phi(t) = \Phi_{in} - \frac{M m_{Pl}\, t}{2\sqrt{3\pi}}, \tag{9}$$

where Φ_{in} is an initial value of Φ. In this scenario of inflation the Hubble parameter gradually decreases, $H \sim \Phi$, and to obtain a sufficiently long inflation it is sufficient to take $\Phi_{in} \gtrsim 3.3\, m_{Pl}$.

In order to avoid too large density perturbations one should choose the value of the inflaton mass M in the range $10^{-7} \lesssim M/m_{Pl} \lesssim 10^{-6}$; accordingly we take $M = 10^{-6}\, m_{Pl}$.

To arrange the desired scenario we should set the value of Φ_0 still at inflationary stage such that the distances between different domains and the domain sizes exponentially expanded up to cosmologically large scales. To ensure it we should take $\Phi_0 \gtrsim 3.1 \cdot m_{Pl}$ (see [14] for details).

Depending upon the value of Φ, there are three different regimes of the evolution of the field χ:

1. Initial stage: $\Phi > \Phi_0$ and $(\Phi - \Phi_0) \gg \Phi_1$.

   The mass of χ is supposed to be small in comparison with the expansion rate, $m < H$, so field χ sits near the minimum of its potential at $\chi = 0$ (see left plot in Fig. 1).

2. Second stage: $|\Phi - \Phi_0| \lesssim \Phi_1$.

   At this stage the squared effective mass of χ becomes negative. The minima of the potential (8) move from zero as $\eta(t)$ or $-\eta(t)$, where $\eta^2(t) = \frac{1}{\lambda_\chi}\left(2\mu^2 V(\Phi) - m^2\right)$. To have equal amplitudes of both positive and negative CP violation we need to impose the condition that χ reaches one or other minimum $\chi = \pm\eta(t)$ and stays there (approximately). This minimum moves

 DOI: 10.1051/epjconf/201715805001

exponentially fast and χ would follow it e.g. if $\mu \gg H$ (since in this case $\chi \propto \exp(\mu t)$). For $M = 10^{-6} m_{Pl}$, $\Phi_0 = 3.1\, m_{Pl}$ we get $\mu \gg H = \sqrt{4\pi/3}\, M/m_{Pl}\, \Phi \sim 6 \cdot 10^{-6} m_{Pl}$.

But even when χ grows exponentially there still should be enough time for χ to reach the minimum $\eta(t)$. If the inflaton Φ goes from $\Phi_0 + 2\Phi_1$ to $\Phi_0 - 2\Phi_1$ during the interval τ, we should require that

$$\mu\tau = \mu \frac{8\sqrt{3\pi}\Phi_1}{M m_{Pl}} \gtrsim \ln \frac{\eta_{max}}{\chi_{in}} = \frac{1}{2} \ln \frac{2\mu^2}{\lambda_\chi \chi_{in}^2} \Rightarrow \mu \gtrsim \frac{M m_{Pl}}{16\sqrt{3\pi}\Phi_1} \ln \frac{2\mu^2}{\lambda_\chi \chi_{in}^2}, \tag{10}$$

where $\eta_{max} \equiv 2\mu^2/\lambda_\chi$ and χ_{in} is initial value of χ.

According to this scenario $\chi^2 \sim \eta_{max}^2 \equiv 2\mu^2/\lambda_\chi$ is the largest value of χ^2. So to be sure that the inflaton field Φ always gives the main contribution to the energy density, we should impose the conditions:

$$M^2\Phi_0^2 \gg \mu^2 \chi^2\big|_{\Phi=\Phi_0} \sim \frac{\mu^4}{\lambda_\chi} \Rightarrow \mu^4 \ll M^2\Phi_0^2\lambda_\chi. \tag{11}$$

For $M = 10^{-6} m_{Pl}$, $\Phi_0 = 3.1\, m_{Pl}$ we get $\mu \ll 1.8 \cdot 10^{-3} m_{Pl} \sqrt[4]{\lambda_\chi}$.

After Φ passed Φ_0, $\eta(t)$ started to decrease exponentially and turned zero when $\Phi = \Phi_0 - 2\Phi_1 \sqrt{\ln(\sqrt{2}\mu/m)}$, so the potential (8) started to have only one minimum at $\chi = 0$ again. The field χ also began to decrease rapidly until $V(\Phi)$ became small in comparison with other terms in Eq. (5).

3. Final stage: $\Phi < \Phi_0$ and $(\Phi_0 - \Phi) \gg \Phi_1$.

   When the interaction term $\mu^2\chi^2 V(\Phi)$ vanishes, the evolution of field χ would be determined by the terms with λ_χ and m, see Eq. (5). If λ_χ is not small enough and χ is still large then the term $\lambda_\chi \chi^3$ dominates and the equation of motion turns into $3H\dot{\chi} + \lambda_\chi \chi^3 = 0$, which has the solution $\chi = \sqrt{\frac{3H}{2\lambda_\chi}\frac{1}{\sqrt{t-C}}}$, where C is a constant of integration. We see that in this regime χ decreases as a power of t.

   When χ becomes quite small, Eq. (5) turns into $\ddot{\chi} + 3H\dot{\chi} + m^2\chi = 0$, so the field χ slowly oscillates and decreases due to redshift related to cosmological expansion.

2.2 Numerical calculations

In order to check that the described scenario is indeed operative we performed the numerical calculations in the homogeneous case with the following set of parameters (all dimensional values are given in the Planck mass, m_{Pl}, units):

$$\Phi_{in} = 4,\ \Phi_0 = 3.1,\ \Phi_1 = 0.02,\ M = 10^{-6},\ \chi_{in} = 10^{-6},\ m = 10^{-10},\ \lambda_\chi = 2 \cdot 10^{-3},\ \mu = 10^{-4}. \tag{12}$$

In Figs. 2 and 3 the results of numerical simulation are shown. One can see that the interaction of the inflaton field Φ with the field χ does not break the standard inflation, $\Phi(t)$ only slightly deviates from straight line around $\Phi = \Phi_0 = 3.1\, m_{Pl}$.

In Fig. 3 one can see how the field χ rolls down to the minimum of the potential $\eta(t) > 0$ and oscillates there. Then it goes back to the old minimum at $\chi = 0$, but even at the end of inflation χ remains an order of magnitude larger than it was at the beginning ($\chi \sim 10\, M$ at $t \approx 25\, M^{-1}$, while $\chi_{in} = 1\, M$). At the moment $t \approx 35\, M^{-1}$ the field χ crosses zero and starts slow oscillations with very low frequency, which could be much lower than the characteristic rate of baryogenesis.

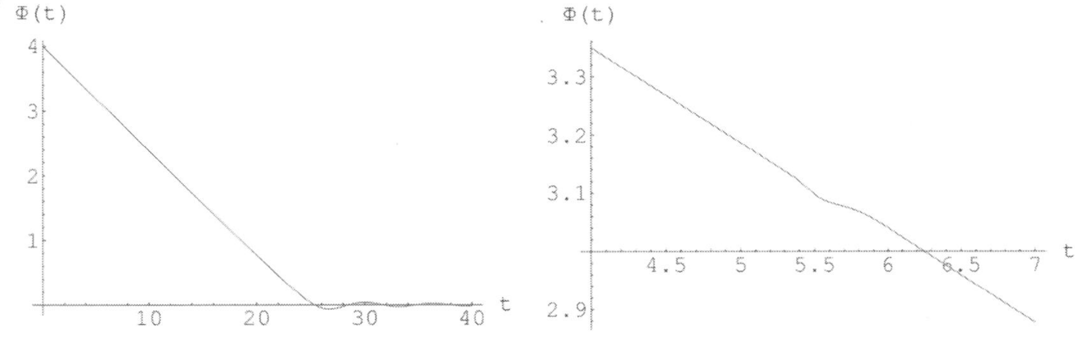

Figure 2. Time evolution of the inflaton field $\Phi(t)$ for the parameters (12). Time t is measured in units of M^{-1} and Φ is measured in units of m_{Pl}. The right plot shows the evolution of Φ in more detail near the point $\Phi = \Phi_0 = 3.1\, m_{Pl}$.

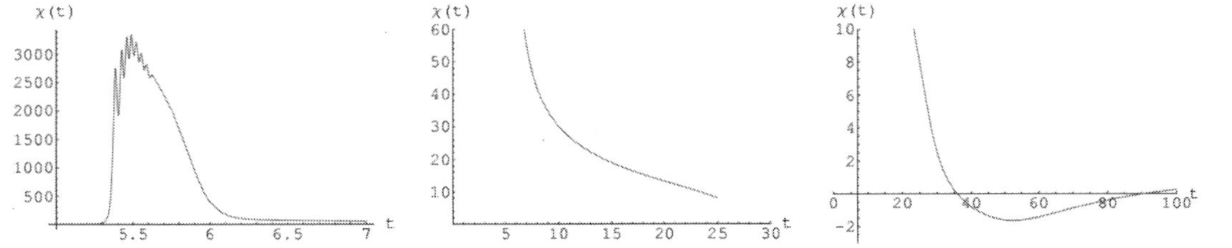

Figure 3. Time dependence of the field $\chi(t)$ for the parameters (12). Time t is measured in units of M^{-1} and χ is measured in units of M. All the three plots depict the evolution of χ during different time intervals. In the middle and right plots one can see that at the end of inflation, $t \approx 25\, M^{-1}$, the field χ remains quite large.

So we see that parameters (12) are chosen in a proper way. In this model the influence of χ on the process of inflation is insignificant, thus the domain size grows exponentially. The remained question is how the distance between domains grows. The related problem of domain wall width in expanding universe is considered in the Section 3.

2.3 Production of heavy particles after inflation and baryogenesis

As it is commonly known, the stage of inflation is followed by the stage of (re)heating, during which the very heavy X particles can be produced through decay of inflaton field (see e.g. book [16]) or from vacuum fluctuations in gravitational field [17].

Let us consider the situation when the inflaton field Φ interacts with X particles (for example, scalar bosons) through the coupling $g_X^2 \Phi^2 X^2$. As is shown in Refs. [18–20], the particle production can be strongly enhanced due to parametric resonance. For an unsuppressed production it is essential that the resonance is broad, which is true for the wide range of coupling constant, $g_X \gg 4 \cdot 10^{-6}$. Because of resonance, very heavy X-bosons with masses even of GUT scale $m_X \sim (M m_{Pl})^{1/2} \sim 10^{16}$ GeV can be produced under quite reasonable assumption $g_X \sim 1$ [16]. Moreover in this case of comparatively large coupling constant $g_X \sim 1$ the decay of the inflaton field occurs very rapidly, during only one or few oscillations.

Let us assume that the produced X-bosons in turn decay into fermions, for example, into quark-quark and antiquark-antilepton pairs, $X \to qq$ and $X \to \bar{q}\bar{l}$, respectively. If the corresponding coupling constants are large enough, the X-bosons decay very quickly. Therefore, the field $\chi(t)$ may remain non-zero yet to the moment when X-bosons have been completely decayed, indeed $\chi \sim 2\,M \sim 10^{13}$ GeV at $t = 30\,M^{-1}$ (see Fig. 3).

We want to provide the source of CP violation for this mechanism via the introduced field χ. It can be done if we suppose that χ interacts with the produced fermions as

$$L_{\chi\psi\psi} = g_{kl}\chi\bar{\psi}^k i\gamma_5\psi^l = g_{kl}\chi(\bar{\psi}^k_R i\gamma_5\psi^l_L + \bar{\psi}^k_L i\gamma_5\psi^l_R) = ig_{kl}\chi(\bar{\psi}^k_L\psi^l_R - \bar{\psi}^k_R\psi^l_L), \tag{13}$$

where k and l denote the fermion flavor (sum over repetitive indices is assumed), $\psi_R = ((1+\gamma_5)/2)\psi$, $\psi_L = ((1-\gamma_5)/2)\psi$. Since χ is supposed to be electrically neutral, it interacts with quarks with the same electric charge, so k and l either run over u, c, t or d, s, b and there are no cross terms.

The interaction of fermions with pseudoscalar field χ can be "rotated away" two unitary transformations $\psi_R \to \psi'_R = U_R\psi_R$ and $\psi_L \to \psi'_L = U_L\psi_L$. However, the interaction of fermions with vector (gauge) boson X remains the same under these transformations:

$$g_{Rkl}X_\mu\bar{\psi}^k_R\gamma^\mu\psi^l_R + g_{Lkl}X_\mu\bar{\psi}^k_L\gamma^\mu\psi^l_L \to g'_{Rab}X_\mu\bar{\psi}^a_R\gamma^\mu\psi^b_R + g'_{Lab}X_\mu\bar{\psi}^a_L\gamma^\mu\psi^b_L, \tag{14}$$

here $g'_R = U_R g_R U^\dagger_R$ and $g'_L = U_L g_L U^\dagger_L$ are matrices of coupling constants in mass eigenstate basis, so ψ^a describes a-th sort of fermion with definite mass. The constants g'_{ab} are complex in general case, and if there are at least three species of fermions, one cannot rotate away simultaneously all phases in complex matrices $g'_{R,L}$ [21]. The complexity of the coupling constants means that CP is violated in the X-boson decays [22]. The magnitude of this CP violation depends on the value of the field χ through the matrices $U_{R,L}$ and coupling constants $g'_{R,L}$. Since χ is essentially non-zero after the end of inflation and during baryogenesis, the CP-odd effects can be large enough.

We assume also that gauge interactions involve fermions with certain chirality, see (14), and thus these interactions break C-invariance.

C and CP violation is one of the necessary Sakharov conditions of generation of baryon asymmetry [23]. The another one is baryon number violation, so one needs to assume also that in the decays of X-bosons the baryon number is not conserved. Let δ be the baryon asymmetry generated in the decay of one X-boson. Then it can be easily demonstrated [16] that the ratio of the baryon number density to the entropy density is estimated as

$$\Delta_B = \frac{n_B - n_{\bar{B}}}{s} \sim \delta\frac{h}{g_X}\left(\frac{m_{th}}{m_{Pl}}\right)^{1/2}, \tag{15}$$

where h and g_X are typical coupling constants of X-boson with fermions and inflaton, respectively, m_{th} is mass scale of the theory. It is quite reasonable to believe that $h/g_X \sim 1$ and $m_{th} \sim M \sim 10^{-6}\,m_{Pl}$, so one has $\Delta_B \sim 10^{-3}\delta$. Thus, to get observed value $\Delta_B \simeq 0.86 \cdot 10^{-10}$ it is sufficient to have only $\delta \sim 10^{-7}$. Such small δ seems to be easily produced in the decay of X-boson. Therefore, the observed baryon asymmetry of the universe can be generated in the decays of X-bosons, without fine tuning of parameters of the theory.

3 Domain walls in expanding universe

3.1 Stationary solutions

In spatially flat section of de Sitter universe the expansion rate, $H = \dot{a}/a$, is constant and the scale factor evolves as $a(t) = \exp Ht$. The FLRW metric for such universe has the form:

$$ds^2 = dt^2 - e^{2Ht}\left(dx^2 + dy^2 + dz^2\right). \tag{16}$$

Let us consider a model of real scalar field φ with the Lagrangian

$$\mathcal{L} = \frac{1}{2} g^{\mu\nu} \partial_\mu \varphi \, \partial_\nu \varphi - \frac{\lambda}{2} \left(\varphi^2 - \eta^2 \right)^2 . \tag{17}$$

The corresponding equation of motion is

$$\frac{1}{\sqrt{-g}} \partial_\mu \left(\sqrt{-g} g^{\mu\nu} \partial_\nu \varphi \right) = -2\lambda \varphi \left(\varphi^2 - \eta^2 \right). \tag{18}$$

In flat space-time, $H = 0$, and in one-dimensional static case, $\varphi = \varphi(z)$, the equation takes the form $\frac{d^2\varphi}{dz^2} = 2\lambda\varphi\left(\varphi^2 - \eta^2\right)$ and has a kink-type solution, which describes a static infinite domain wall. Without loss of generality we can assume that the wall is situated at $z = 0$ in xy-plane:

$$\varphi(z) = \eta \tanh \frac{z}{\delta_0}, \tag{19}$$

where $\delta_0 = 1/(\sqrt{\lambda}\eta)$ has the meaning of the wall thickness (subscript 0 indicates that $H = 0$).

Now let us consider an expanding universe with constant $H > 0$. In this case, if one looks for stationary solution (see [12]), it is reasonable to suggest that the field φ depends only on $za(t) = z \exp Ht$, which is the proper distance from the wall. So, one can choose the following ansatz for φ:

$$\varphi = \eta \cdot f(u), \qquad \text{where} \quad u = Hze^{Ht}, \tag{20}$$

where u and f are dimensionless. Then the equation of motion takes the form:

$$\left(1 - u^2\right) f'' - 4uf' = -2Cf\left(1 - f^2\right). \tag{21}$$

Here prime means the derivative with respect to u. It is noteworthy that all parameters of the problem are combined into a single positive constant $C = 1/(H\delta_0)^2 = \lambda\eta^2/H^2 > 0$.

Since we are interested in kink-type solutions, the boundary conditions should be

$$f(0) = 0, \qquad f(\pm\infty) = \pm 1. \tag{22}$$

Corresponding numerical solutions for different values of parameter C are shown in Fig. 1 of [12, 15]. We see that the larger is C the closer is the solution to the flat space-time one ($H = 0$), as is naturally expected.

As it is noticed in ref. [12] the stationary solutions can be found only for $C > 2$, but no explanation of this observation is given therein. For the detailed explanation see [15].

3.2 Evolution of domain walls beyond the stationary limit

As we have seen in the previous section, Eq. (21) allows to find the field configurations, which describe stationary domain walls in expanding universe. However, it is also interesting to see how domain walls evolve from some initial states. Beyond the stationary approximation we can find not only solution for $C > 2$ but also for $C \leq 2$, for which the stationary approximation does not exist.

To this end one should solve the original equation of motion (18) in the case when the field φ is a function of two independent variables, z and t. It is convenient to introduce dimensionless variables $\tau = Ht$, $\zeta = Hz$ and function $f(\zeta, \tau) = \varphi(z, t)/\eta$. As a result one obtains the equation

$$\frac{\partial^2 f}{\partial \tau^2} + 3\frac{\partial f}{\partial \tau} - e^{-2\tau}\frac{\partial^2 f}{\partial \zeta^2} = 2Cf\left(1 - f^2\right), \tag{23}$$

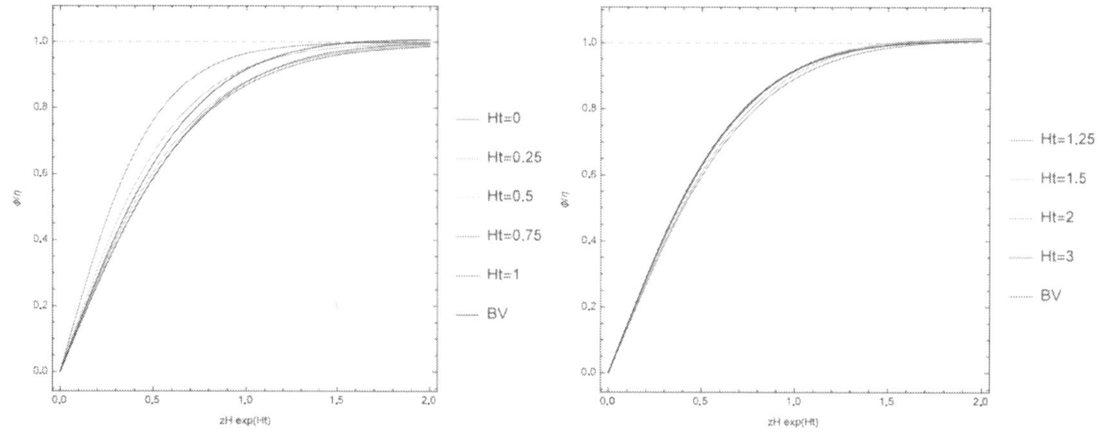

Figure 4. Evolution of domain wall for $C = 4$. Black curve corresponds to stationary solution.

where $C = \lambda\eta^2/H^2 = 1/(H\delta_0)^2 > 0$ as it was above.

The boundary conditions for the kink-type solution should be

$$f(0,\tau) = 0, \qquad f(\pm\infty,\tau) = \pm 1, \tag{24}$$

and we choose the initial configuration as the domain wall with "natural" thickness $1/\sqrt{C}$ (with respect to dimensionless coordinate ζ) and zero time derivative. Of course, this is a toy model with the artificial initial conditions. The realistic model should describe the evolution of domain walls from the very beginning, including the process of wall formation. This will be studied elsewhere.

The evolution of the domain wall for $C = 4$ is depicted in Fig. 4. The stationary solution is shown by black curve (it is denoted by "BV" because of Basu and Vilenkin who found it [12]). One sees that the domain wall evolves in somewhat non-trivial way. At the very beginning the wall starts to broaden in terms of the proper distance from the wall, $zH \exp Ht$, and at some moment it becomes wider than the stationary solution. However, afterwards the wall broadening changes to contraction. Finally, the wall comes to the stationary configuration after several damped oscillations around it.

When C is close to 2, the stationary solution is very broad and the solution of Eq. (23) converges to the stationary one very slowly. Unlike using plots like in Fig. 4, it is more convenient to look at the evolution of the wall width with time. At the initial moment in our model the wall is described by hyperbolic tangent. Although in the realistic course of the evolution the wall is not "pure" hyperbolic tangent, nevertheless one can use the same definition for the thickness $\delta(t)$ as the value of the coordinate z at which the field ϕ reaches the value $\phi/\eta = \tanh 1 \approx 0.76$.

In Fig. 5 the time dependence of the physical width of the wall for different values of C is presented. In the left plot ($C > 2$) one can see that all the curves indeed tend to constant values corresponding to stationary solutions (dashed lines). Oscillatory behavior mentioned above is also apparent. Right plot contains curves for $C < 2$. It is clear that for $C = 1$, i.e. for not very small values of C, the wall thickness increases slower than exponent. However, for smaller values of C, e.g. $C \lesssim 0.1$ the rate of the wall expansion is the exponential one with a good accuracy.

The Hubble constant, H, plays the role of friction in this problem and when it dominates over the potential term ($C \ll 1$), the field configuration as a function of coordinates becomes almost static. Therefore the width of the wall is growing nearly as the scale factor $a(t)$, i.e. almost exponentially. This is why there are no stable solutions for $C \ll 1$.

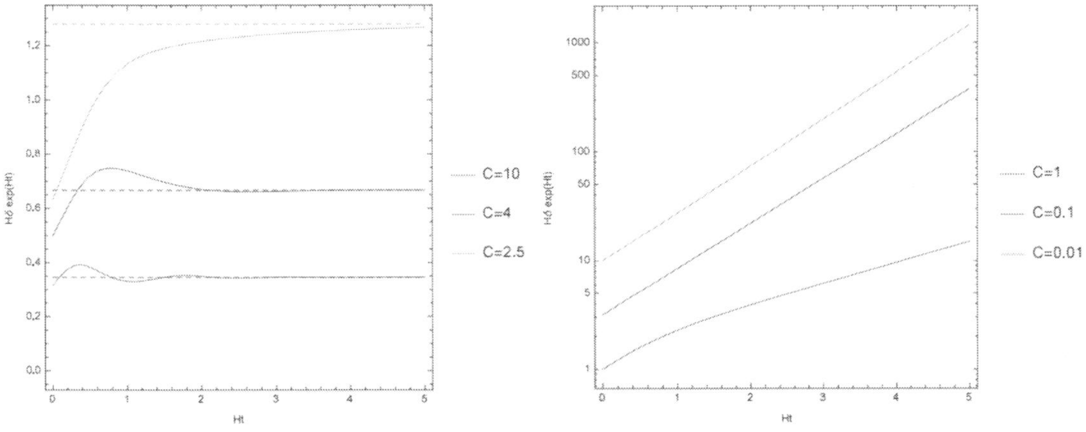

Figure 5. Time dependence of domain wall thickness for $C > 2$ (left plot) and $C < 2$ (right plot). Dashed lines describe the stationary solutions.

4 Conclusions

We have presented a model of baryogenesis which may lead to baryo-symmetric universe with cosmologically large domains of matter and antimatter, avoiding the domain wall problem. Breaking of charge symmetry is induced by a non-zero amplitude of a scalar field χ, which slowly relaxed down to equilibrium, much slower than the process of baryogenesis goes on. In classification of different types of CP violation which might be operative in cosmology this type is called dynamical one [24]. Later, after the baryon asymmetry was developed, χ evolved down to the equilibrium point $\chi = 0$ and thus domain walls disappeared rather early in the universe. Inflation is an essential ingredient of the scenario. A coupling to the inflaton field was introduced on purpose to generate a non-zero value of χ and to keep it non-zero during baryogenesis.

This scenario inspired us to consider time evolution of thick domain walls in a de Sitter universe. We have shown that for large values of parameter $C > 2$ the initial kink configuration in a de Sitter background tends to the stationary solution obtained by Basu and Vilenkin [12].

For $C < 2$ the stationary solution does not exist and the width of the wall infinitely grows with time. For $C \lesssim 0.1$ the rise is close to the exponential one within the precision of our numerical calculations. This result confirms the assertion made in ref. [14] that the transition region between matter and antimatter domain might be cosmologically large. This result is essential for application of spontaneous breaking of symmetry between particles and antiparticles to realistic cosmology.

We acknowledge support of the Grant of President of Russian Federation for the leading scientific Schools of Russian Federation, NSh-9022-2016.2. SG is also supported under the grants RFBR No. 16-32-60115, 16-32-00241, 16-02-00342. In addition, SG is grateful to Dynasty Foundation for support.

References

[1] A.D. Dolgov, Phys. Rept. **222**, 309 (1992)
[2] A.D. Dolgov, Surv. High Energ. Phys. **13**, 83 (1998)
[3] V.A. Rubakov and M.E. Shaposhnikov, Phys. Usp. **39**, 461 (1996)

[4] A. Riotto and M. Trodden, Ann. Rev. Nucl. Part. Sci. **49**, 35 (1999)
[5] M. Dine and A. Kusenko, Rev. Mod. Phys. **76**, 1 (2003)
[6] A.G. Cohen, A. De Rujula and S.L. Glashow, Astrophys. J. **495**, 539 (1998)
[7] Ya. B. Zeldovich, I.Yu. Kobzarev and L.B. Okun, Sov. Phys. JETP **40**, 1 (1974)
[8] V.A. Kuzmin, I.I. Tkachev and M.E. Shaposhnikov, Pisma Zh. Eksp. Teor. Fiz. **33**, 557 (1981)
[9] V.A. Kuzmin, M.E. Shaposhnikov and I.I. Tkachev, Phys. Lett. **B 105**, 159 (1981)
[10] V.A. Kuzmin, M.E. Shaposhnikov and I.I. Tkachev, Phys. Lett. **B 105**, 167 (1981)
[11] V.A. Kuzmin, M.E. Shaposhnikov and I.I. Tkachev, Nucl. Phys. **B 196**, 29 (1982) [Erratum ibid. **B 202**, 543 (1982)]
[12] R. Basu and A. Vilenkin, Phys. Rev. **D 50**, 7150 (1994)
[13] N.A. Voronov, A.L. Dyshko and N.B. Konyukhova, Phys. Atom. Nucl. **68**, 1218 (2005)
[14] A.D. Dolgov, S.I. Godunov, A.S. Rudenko and I.I. Tkachev, JCAP **10**, 027 (2015)
[15] A.D. Dolgov, S.I. Godunov, A.S. Rudenko, JCAP **10**, 026 (2016).
[16] D.S. Gorbunov, V.A. Rubakov, *Introduction to the theory of the early universe: Cosmological perturbations and inflationary theory*, World Scientific (2011).
[17] V.A. Kuzmin and I.I. Tkachev, Phys. Rev. **D 59**, 123006 (1999)
[18] Yu.V. Shtanov, J.H. Traschen and R.H. Brandenberger, Phys. Rev. **D 51**, 5438 (1995)
[19] L. Kofman, A.D. Linde and A.A. Starobinsky, Phys. Rev. Lett. **73**, 3195 (1994)
[20] L. Kofman, A.D. Linde and A.A. Starobinsky, Phys. Rev. **D 56**, 3258 (1997)
[21] M. Kobayashi and T. Maskawa, Prog. Theor. Phys. **49**, 652 (1973)
[22] D.V. Nanopoulos and S. Weinberg, Phys. Rev. **D 20**, 2484 (1979)
[23] A.D. Sakharov, JETP Lett. **5**, 24 (1967); Sov. Phys. Usp. **34** (1991) 392
[24] A.D. Dolgov, *CP violation in cosmology*, in the proceedings of 163rd Course of International School of Physics 'Enrico Fermi': CP Violation: From Quarks to Leptons, Varenna, Italy, July 19-29, 2005 [hep-ph/0511213].

DOI: 10.1051/epjconf/201715805002

Neutrino decay to electron and W boson in a superstrong magnetic field in the Early Universe

Alexander Kuznetsov[1,⋆], *Alexander* Okrugin[1,⋆⋆], and *Anastasiya* Shitova[1,⋆⋆⋆]

[1] *Division of Theoretical Physics, Department of Physics, Yaroslavl State P. G. Demidov University, Sovietskaya 14, 150000 Yaroslavl, Russian Federation*

Abstract. An intense electromagnetic field makes possible the process of the neutrino decay into electron and W boson that is kinematically forbidden in a vacuum. The exact formula for the probability of ultra high energy neutrino decay in a superstrong magnetic field via the representation for the propagators of the charged particles expanded as sums over Landau levels is derived. The obtained function clarifies previously known approximate formula. The neutrino mean free path is estimated. It is shown, that the resonance peaks present in the decay width for certain neutrino energies correspond to the electron and the W boson creation at specific Landau levels. Some interesting possible manifestations of the process in the conditions of the Early Universe are briefly discussed.

1 Introduction

In recent decades, the studies of electroweak processes in active media (such as plasma or strong magnetic fields) became noticeably popular. In many respects it is connected with rapidly developing concept of magnetars, i.e. neutron stars which evolution is driven largely by magnetic fields [1]. Other examples when a strong magnetic field could possibly manifest itself are the conditions in the Early Universe, and the experiments at modern colliders, e.g. non-central collisions of protons and heavy ions.

Extreme physical conditions actively influence the run of quantum processes, thus allowing or enhancing the transitions that are forbidden or strongly suppressed in vacuum. The external field effects should be taken into account on the basis of exact solutions of the field theory equations for a charged particle in an external electromagnetic field, but not on the basis of perturbation theory.

The strength of the fields we discuss is essentially greater than the so-called Schwinger limit $B_e = m_e^2/e \simeq 4.4 \times 10^{13}$ G, which is the naturally arising scale for the field strength as the quantizing field for an electron. Hereafter, we use the Planck units: $\hbar = 1, c = 1$, and the Minkowski metric with the signature $(+ - - -)$. Another limit that can be naturally introduced is defined by the mass of the gauge boson m_W and can serve as a boundary of applicability of Standard Model: $B_W = m_W^2/e \simeq 1.1\times10^{24}$ G. There arises a question whether the electroweak vacuum remains stable at $B \to B_W$. As it was shown in Ref. [2], the radiation corrections act to prevent the instability of the electroweak vacuum in strong fields.

⋆e-mail: avkuzn@uniyar.ac.ru
⋆⋆e-mail: okrugin@uniyar.ac.ru
⋆⋆⋆e-mail: pick@mail.ru

Magnetic fields exist at all the scales of the Universe structure: from planets to galaxy clusters. It is used to think that magnetic fields arised yet in the Early Universe, but until now it is not fully understood what is the preferential mechanism of formation of contemporary observable magnetic fields. There are plenty of ideas that can explain the evolution of some seed field to the modern scales [3, 4]. Primordial magnetic fields are worth to study not only for the reason to explain the observable fields but also because they open a possibility to study fundamentally new physics at the stage of the Early Universe.

A problem of studying possible influence of an active environment on the neutrino dispersion properties is quite important. As it was already mentioned, the intense magnetic field opens new possibilities for the processes that are kinematically forbidden in a vacuum. Probably, the most interesting processes of such kind are the transitions of a massless neutrino: the production of a lepton pair by the neutrino, $\nu \to \nu \ell^- \ell^+$, and the neutrino decay into W^+ boson and a charged lepton, $\nu_\ell \to \ell^- W^+$ ($\ell = e, \mu, \tau$) in a strong field. The external field influence on the propagation of charged particles is defined by it's specific charge, so, the lighter is a particle, the more sensitive it is. Thus, for the definiteness we consider an electron as the charged lepton, and the results can be generalized for the μ and the τ leptons straightforwardly. For the described conditions, only neutrinos with ultra high energy can play the role.

The calculation of the width of the process $\nu_e \to e^- W^+$ in an external magnetic field has a long history. Ref. [5] was probably the first publication on the problem. A detailed analysis of this process was also performed in Ref. [6]. Later on, another authors [7] remade the calculations and came to the contradictions with the previously obtained results. To the end, in Ref. [8] the result of Ref. [6] was confirmed. The most possible reason for the discrepancy between the results was as follows: the authors of Ref. [7] used the W boson propagator expanded over the field tensor $F^{\mu\nu}$ to the linear terms, while the quadratic terms were also essential.

Later, the process was also analysed in Ref. [9], where an expression for the decay width was obtained which was valid for an arbitrary relationship between the masses of the neutrino, charged lepton, and W boson, and in the case of a left-polarized standard neutrino, the result of Ref. [8] was reproduced exactly. In Ref. [10], on the basis of an instanton model, the author also generally confirmed the results of Ref. [8].

In the present paper, we discuss extremely strong magnetic fields $\sim B_W$ which could be quite important while studying the processes in the conditions of the Early Universe where the fields of that strength could probably exist [11–13].

2 Probability of the process $\nu_e \to e^- W^+$ in a strong external magnetic field

While solving series of various problems on charged particles interactions in external magnetic fields a great influence gained the method in which the contribution of the external field is considered not within the perturbation theory but on the basis of exact solutions of wave equations in external magnetic field. This method was firstly proposed by Sokolov and lately developed by Ternov while solving the problem of synchrotron radiation. The corresponding idea is now known as the Furry picture. For calculations of physical processes, the Feynman technique is used, the particles in the initial and final states are described by the wave equation for the particles in external field, and the internal lines are described by the propagators built on a base of those solutions. A more detailed description of this approach can be found e.g. in Refs. [14–19]. The method could be successfully applied in a case of strong magnetic fields when it becomes impossible to make calculations on the basis of perturbation theory. Taking into account a vacuum stability in superstrong magnetic fields, one may study processes for the values of B much greater than B_e.

The probability of the neutrino decay in an external magnetic field can be expressed via the neutrino self energy operator. The probability could be described in terms of the imaginary part of the invariant amplitude of the $\nu_e \to \nu_e$ transition via the following relation:

$$w\,(\nu_e \to e^- W^+) = \frac{1}{E}\,\mathrm{Im}\,\mathcal{M}(\nu_e \to \nu_e) = -\,\frac{1}{E}\,\mathrm{Im}\left\{\mathrm{Tr}\Big[\Sigma(p)\gamma_L(p\gamma)\Big]\right\}, \tag{1}$$

where we took into account the expression for the density matrix of left-handed neutrinos, $\rho(p) = \gamma_L(p\gamma)$; $p^\alpha = (E, \mathbf{p})$ is the neutrino 4-momentum, $\gamma_L = (1 - \gamma_5)/2$.

Neutrino self energy operator is defined by the relation:

$$\Sigma\,(p) = -\frac{\mathrm{i}g^2}{2}\left[\gamma^\alpha \gamma_L J^W_{\alpha\beta}(p)\gamma^\beta \gamma_L\right]. \tag{2}$$

The function $J^W_{\alpha\beta}(p)$ in Eq. (2) is an integral of the Fourier transforms of the translational-invariant parts of the electron and the W boson propagators, for details see e.g. Ref. [19]:

$$J^W_{\alpha\beta}\,(p) = \int \frac{\mathrm{d}^4 q}{(2\pi)^4}\, S(q)\, G^W_{\beta\alpha}(q - p). \tag{3}$$

The electron propagator expanded over Landau levels (see Ref. [20] and also Ref. [21]) takes the form:

$$S\,(q) = \sum_{n=0}^{\infty} \frac{\mathrm{i}}{q_\parallel^2 - m_e^2 - 2n\beta + \mathrm{i}\epsilon}\left\{[(q\gamma)_\parallel + m_e]\left[d_n(\alpha) - \frac{\mathrm{i}}{2}\,(\gamma\varphi\gamma)\, d_n'(\alpha)\right] - (q\gamma)_\perp\, 2n\,\frac{d_n(\alpha)}{\alpha}\right\}, \tag{4}$$

where $\beta = e\,B$, $\alpha = q_\perp^2/\beta$, the tensor indices for 4-vectors and tensors in parentheses are assumed to be contracted consecutively, for instance: $(\gamma\varphi\gamma) = \gamma^\alpha\varphi_{\alpha\beta}\gamma^\beta$, and the following notations are used: $(q\gamma)_\parallel = (q\,\tilde\varphi\tilde\varphi\,\gamma)$, $(q\gamma)_\perp = (q\,\varphi\varphi\,\gamma)$; the tensor $\varphi_{\alpha\beta} = F_{\alpha\beta}/B$ is the dimensionless electromagnetic tensor, $\tilde\varphi_{\alpha\beta} = \frac{1}{2}\,\varepsilon_{\alpha\beta\mu\nu}\,\varphi^{\mu\nu}$ is the dual dimensionless tensor ($\varepsilon^{0123} = -\varepsilon_{0123} = +1$). In the frame where the third space axis is directed along the vector of an external magnetic field $\mathbf{B}$, the 4-vectors with the indexes $\perp$ and $\parallel$ belong to the Euclidean $\{1, 2\}$-subspace and the pseudo-Euclidean $\{0, 3\}$-subspace respectively. Then, $(\varphi\varphi)_{\alpha\beta} = \mathrm{diag}(0, 1, 1, 0)$, $(\tilde\varphi\tilde\varphi)_{\alpha\beta} = \mathrm{diag}(1, 0, 0, -1)$, and $p_\perp = (0, p_1, p_2, 0)$, $p_\parallel = (p_0, 0, 0, p_3)$. The functions $d_n(\alpha)$ in Eq. (4) are expressed via the Laguerre polynomials:

$$d_n(\alpha) = (-1)^n \mathrm{e}^{-\alpha}[L_n(2\alpha) - L_{n-1}(2\alpha)], \tag{5}$$

with a completion of $L_{-1}(x) \equiv 0$. In particular, for $n = 0$, that is for the ground Landau level, one has:

$$S^{(0)}(q) = \mathrm{i}\,\frac{(q\gamma)_\parallel + m_e}{q_\parallel^2 - m_e^2 + \mathrm{i}\epsilon}\,\mathrm{e}^{-\alpha}\left(1 + \frac{\mathrm{i}}{2}(\gamma\varphi\gamma)\right). \tag{6}$$

For the W boson propagator, one can find the similar expression as the sum over the Landau levels [19, 22] (here, we use the 't Hooft–Feynman gauge):

$$\begin{aligned} G_{\mu\nu}(q - p) \;=\; & \sum_{k=0}^{\infty} \frac{\mathrm{i}\,(-1)^k\,\mathrm{e}^{-\alpha'}}{q_\parallel^2 - m_W^2 - \beta\,(2\,k - 1) + \mathrm{i}\epsilon}\Big\{2\,(\tilde\varphi\,\tilde\varphi)_{\mu\nu}\, L_{k-1}(2\alpha') \\ & + \;(\varphi\,\varphi)_{\mu\nu}\,[L_k(2\alpha') + L_{k-2}(2\alpha')] - \mathrm{i}\,\varphi_{\mu\nu}\,[L_k(2\alpha') - L_{k-2}(2\alpha')]\Big\}, \end{aligned} \tag{7}$$

where $\alpha' \equiv (q - p)_\perp{}^2/\beta$.

Unlike the electron propagator, the contribution from the ground level, $k = 0$, to the W boson propagator has a peculiarity:

$$G^{(0)}_{\mu\nu}(q) = \frac{\mathrm{i}}{q_{\|}^2 - m_W^2 + \beta + \mathrm{i}\epsilon}\, \mathrm{e}^{-q_\perp^2/\beta} \left[(\varphi\varphi)_{\mu\nu} - \mathrm{i}\,\varphi_{\mu\nu}\right], \tag{8}$$

and it is essential that it contains a pole at $q_{\|}^2 = m_W^2 - \beta$. So, if the magnetic field rises up to the critical value for the W boson, $B_W \simeq 1.1 \times 10^{24}$ G, then the already mentioned question of stability arises of the perturbation theory for the W boson vacuum [2, 12]. Gathering Eqs. (7), (4), (3) and (2) consistently, one can obtain the basic formula for the probability of the process $\nu_e \to e^- W^+$. Further integration over the 4-momentum q in Eq. (3) can be separated to independent integrals over the transverse momentum $q_\perp$ and the longitudinal momentum $q_{\|}$.

The integrals over the transverse momentum are of the scalar and the vector type:

$$\mathcal{I}_\perp = \int \frac{\mathrm{d}^2 q_\perp}{(2\pi)^2} \mathrm{e}^{-\alpha-\alpha'} L_n(2\alpha)\, L_k(2\alpha') = \frac{\beta}{8\pi}\, \mathrm{e}^{-\chi}\, L_n^{k-n}(\chi)\, L_k^{n-k}(\chi), \tag{9}$$

$$\begin{aligned} \mathcal{I}_\perp^\mu &= \int \frac{\mathrm{d}^2 q_\perp}{(2\pi)^2} \mathrm{e}^{-\alpha-\alpha'}\, \frac{4n}{\alpha}\, q_\perp^\mu \left(L_n(2\alpha) - L_{n-1}(2\alpha)\right) L_{k-1}(2\alpha') \\ &= -\frac{\beta}{2\pi}\, p_\perp^\mu\, \mathrm{e}^{-\chi}\, L_{n-1}^{k-n}(\chi)\, L_{k-1}^{n-k+1}(\chi), \end{aligned} \tag{10}$$

where $\chi \equiv p_\perp^2/(2\beta)$. Note that the associated Laguerre polynomials should be additionally determined so that $L_n^s(x) \equiv 0$ in the case when n is negative. The integral Eq. (9) appears also in the calculation of the fermion density matrix in a constant uniform field [23].

The integral over the longitudinal momentum $q_{\|}$ has the form:

$$\begin{aligned} \mathcal{I}_{\|}^\mu &= \mathrm{Im}\left\{\mathrm{i}\int \frac{\mathrm{d}^2 q_{\|}}{(2\pi)^2} \frac{q_{\|}^\mu}{\left(q_{\|}^2 - 2n\beta\right)\left((q-p)_{\|}^2 - m_W^2 - (2k-1)\beta + \mathrm{i}\epsilon\right)}\right\} \\ &= p_{\|}^\mu\, \frac{p_\perp^2 - z + 2n\beta}{4p_\perp^2\sqrt{4n\beta(n\beta - p_\perp^2 - z) + (p_\perp^2 - z)^2}}\, \theta\!\left[p_\perp^2 - \left(\sqrt{z} + \sqrt{2n\beta}\right)^2\right] \theta[z], \end{aligned} \tag{11}$$

where $z = m_W^2 + (2k-1)\beta$. Here, we have neglected the electron mass m_e as the smallest parameter, and the analysis has shown that this wouldn't cause any problems.

Combining the results of integration, one can obtain the resulting formula for the probability of the decay $\nu_e \to e^- W^+$:

$$\begin{aligned} w(\nu_e \to e^- W^+) &= \frac{G_\mathrm{F}\, m_W^2\, \beta}{\pi\sqrt{2}\,E} \sum_{k=0}^{k_{max}} \sum_{n=0}^{n_{max}(k)} \mathrm{e}^{-\chi} \sqrt{D^{-1}(\chi)}\, \Big\{(\chi - \mu - k + 1/2 + n) \\ &\times \left[\frac{k!}{n!}\, \chi^{n-k} \left(L_k^{n-k}(\chi)\right)^2 + \frac{(k-2)!}{(n-1)!}\, \chi^{n-k+1} \left(L_{k-2}^{n-k+1}(\chi)\right)^2\right] \\ &- 4(-1)^{k+n}\, \chi\, L_{k-1}^{n-k+1}(\chi)\, L_{n-1}^{k-n}(\chi)\Big\}\, \theta\!\left[\sqrt{\chi} - \sqrt{\mu + k - 1/2} - \sqrt{n}\right], \end{aligned} \tag{12}$$

where we use the following notations for the dimensionless variables: $\chi \equiv p_\perp^2/(2\beta)$, $\mu \equiv m_W^2/(2\beta)$, and

$$D(\chi) = (\chi - \mu - k + 1/2 + n)^2 - 4n\chi. \tag{13}$$

The theta-function in Eq. (12) allows to define the limits on summation over n and k. Thus, for fixed k, the value n should run from zero to its maximum, n_{max}:

$$n_{max}(k) = \text{IntegerPart}\left[\left(\sqrt{\chi} - \sqrt{\mu + k - 1/2}\right)^2\right], \tag{14}$$

while the maximum value for k is:

$$k_{max} = \text{IntegerPart}\left[\,\chi - \mu + 1/2\right]. \tag{15}$$

As a consequence there are kinematically allowed Landau states for creating W boson and electron originated from the threshold for neutrino energy.

Equation (12) should be compared with the asymptotic formula [8] obtained from the basic expressions for the electron and W boson propagators in the Fock–Schwinger proper time formalism [24, 25] in the strong field limit ($p_\perp^2 \gg m_W^2 \gg eB \gg m_e^2$):

$$w = \frac{G_F \beta^{3/2} p_\perp}{\pi \sqrt{2\pi} E} \Phi\left(\frac{4\beta p_\perp^2}{m_W^4}\right), \tag{16}$$

where

$$\Phi(x) = \frac{1}{x} \int_0^\infty \frac{dy}{y^{1/2}} \frac{(\sinh y)^2 - y \tanh y}{(y - \tanh y)^{3/2}} \frac{(\tanh y)^{1/2}}{(\sinh y)^2} \exp\left[-\frac{y \tanh y}{x(y - \tanh y)}\right]. \tag{17}$$

An independent evaluation of the process probability in a superstrong magnetic field was also derived within the instanton model [10].

The curves in Figs. 1, 2 display the probability $w(\chi)$ and the mean free path as functions of χ. The monotone curves correspond to the asymptotic expression (16), while the "saw-tooth" profiles refer to the exact solution (12). Numerical calculation confirms that taking into account the contribution from only the ground Landau level, $n = 0$, is insufficient in the context of the present task (see also Ref. [8]).

3 Discussion

The structure of the derived expression (12) indicates the existence of the abrupt increase of the probability (and decrease of the mean free path) at the energy values corresponding to the "turning on" the new Landau states. The exact expression affords to identify the quantum structure of the process. It is worth to note that the similar saw-tooth profile arises also in the problem of electron-positron pair creation by a photon in strong magnetic fields [26, 27]. As an example, in Ref. [26] the production of electron-positron pairs by single photons in magnetic fields of order $\geq 10^{12}$ G was investigated in detail. It was shown that the attenuation coefficient (which is similar to the decay probability) exhibits a sawtooth pattern as a function of the photon energy, and its value is significantly greater at some energies than that predicted by the asymptotic expression.

The analysis of Eq. (12) affords to identify the similar tendency: for some specific values of parameters, the neutrino mean free path is significantly lower than it is predicted by the asymptotic formula. That corresponds to generation of a great amount of electrons and W^+ bosons.

This fact could lead to several attractive conclusions. The subsequent W^+ boson decay can influence the picture of the baryosynthesis in the Early Universe.

The Early Universe is known to be a natural laboratory for the processes in the extreme physical conditions. There are a number of bright ideas connecting cosmology and high energy physics. One of them is the domain structure of the Universe.

The first papers which claimed the domain structure of vacuum as a consequence of CP-violation appeared in the 1970th. Thus, in Ref. [28] it was shown that while cooling the Universe the regions of two types should arise, divided by heavy domain walls that violate homogeneity. Domain structure became energeticaly not profitable so that every domain seeks to grow and absorb the others that leads to high velocities of domain walls [29]. The described processes could be extremely valuable for the generation of baryon asymmetry via anomalous electroweak processes (the so-called electroweak baryogenesis) in the models with thin or thick moving domain walls [30]. The important feature of such models is that the discussed energy region should be available at modern colliders.

In Ref. [31], the cosmological evolution of a universe filled with domains have been studied and it was concluded that at present the nearest wall should have gone beyond the horizon, leaving a trail in anisotropy in the black-body radiation [32].

Another attractive hypothesis claimed in Ref. [33] is that domain walls are ferromagnetic that allows to generate magnetic field strong enough to be a possible candidate of a primordial magnetic field leading to presently observable galactic fields. The idea was lately supported and discussed in Ref. [34–36], but received several contradict arguments [37, 38]. The authors of Ref. [39] claim that the submitted objections could only slightly decrease the effect but do not nullify it. It was suggested [35] that the effective surface tension of the domain walls can be made vanishingly small due to a peculiar magnetic condensation induced by fermion zero modes localized on the wall. As a consequence, the domain wall acquires a non-zero magnetic field perpendicular to the wall, and it becomes almost invisible as far as the gravitational effects are concerned.

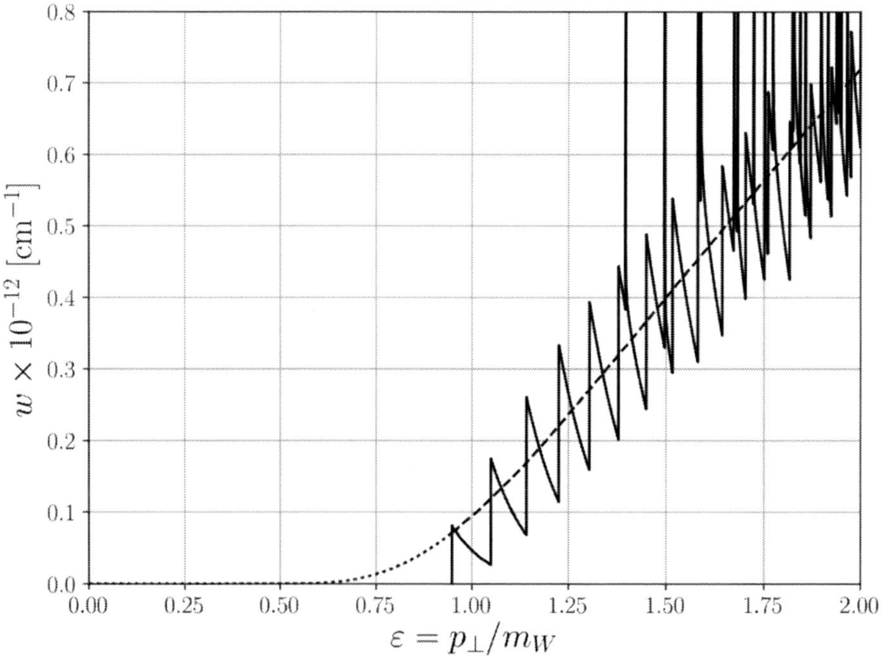

Figure 1. The decay $\nu_e \to e^- W^+$ width for $\beta = 0.1\, m_W^2$ (the "sawtooth" profile). Dashed line: approximate formula without considering the quantizing effect of the magnetic field.

It would be important that the process width (12) essentially depends on $p_\perp = E \sin\theta$, where θ is an angle between the neutrino momentum and the field direction. Suppose that the hypothesis of magnetic field generation in domain walls is correct. Then neutrinos moving almost perpendicular to the domain wall (i.e. almost parallel to the field direction) would fly it right through. On the other hand, neutrinos propagating almost parallel to the wall (in the way of the maximum magnetic field influence) with specific energies would predominantly decay, and would generate a great amount of W bosons in the thickness of the wall. The subsequent decay of the$W$ boson by dominant quark channels could lead to some overabundance of hadron matter and antimatter inside domain walls to outside.

If the lepton-antilepton asymmetry, induced by the CP violation in the lepton sector, has arisen before the electroweak phase transition, leading to overabundance of neutrinos over antineutrinos in the Universe, the considered mechanism would provide an overabundance of W^+ and e^- over W^- and e^+ inside domain walls. The W^+ bosons would finally transform to protons while the W^- bosons would dominantly transform to antiprotons. The subsequent formation of the baryon asymmetry would paint the final picture of the Universe.

Of course these interesting applications should be studied precisely in the future taking into account the influence of other effects such as gravity.

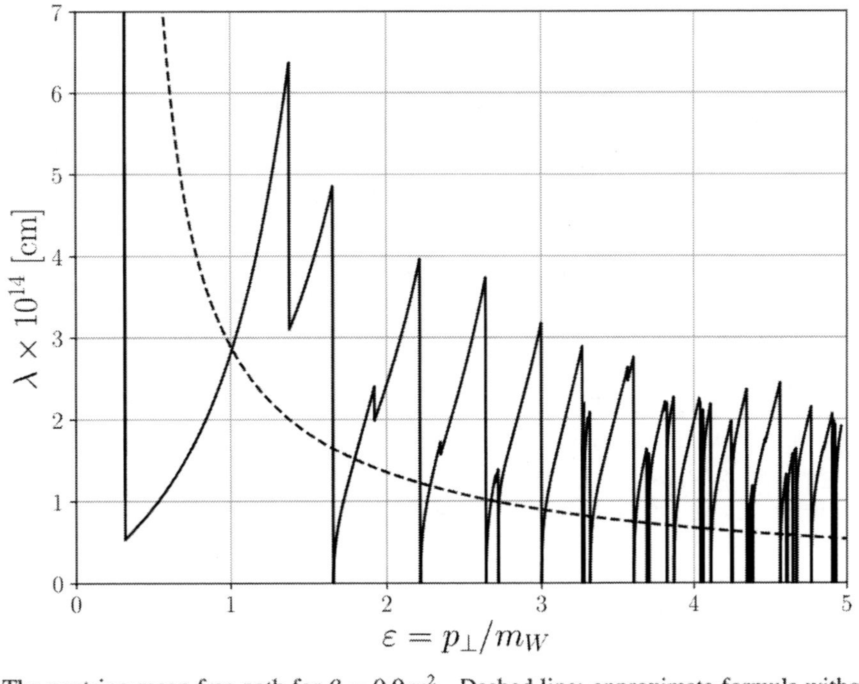

Figure 2. The neutrino mean free path for $\beta = 0.9\, m_W^2$. Dashed line: approximate formula without considering the quantizing effect of the magnetic field.

4 Conclusion

On the basis of the exact expressions of electron and W boson propagators expanded over Landau levels, a neutrino decay in a superstrong magnetic fields (which may be characteristic for the Early Universe) was analysed.

It was shown that resonance peaks, observed in the probability of the process for specific neutrino momentum $p_\perp = E \sin\theta$, lead to electron and W boson creation at specific Landau levels. It means that neutrino mean free path decreases sharply for certain values of $p_\perp$. The subsequent W boson decay into the dominant quark channels should probably influence the matter composition within the domain walls.

There are still some open questions. Could the neutrino decay in a superstrong magnetic field really influence the matter composition within the domain walls in Early Universe? Would a possible CP violation in the lepton sector (Pontecorvo–Maki–Nakagawa–Sakata phase) lead to overabundance of neutrinos to antineutrinos? Could it lead, together with the CP violating phase in the Cabibbo–Kobayashi–Maskawa matrix, to any consequences? These will constitute the subject of further investigations.

Acknowledgements

The authors wish to thank D.A. Rumyantsev for useful discussions.

The research is financed by the grant of the Russian Science Foundation (Project No. 15-12-10039).

References

[1] R. C. Duncan and C. Thompson, Astrophys. J. **392**, L9 (1992)
[2] V. V. Skalozub, Phys. At. Nucl. **77**, 901 (2014)
[3] L. M. Widrow et al., Space Sci. Rev. **166**, 37 (2012)
[4] K. Subramanian, Rep. Prog. Phys. **79**, 076901 (2016)
[5] A. V. Borisov, V. Ch. Zhukovskii, A. V. Kurilin and A. I. Ternov, Sov. J. Nucl. Phys. **41**, 473 (1985) [Original Russian text: Yad. Fiz. **41**, 743 (1985)]
[6] A. Erdas and M. Lissia, Phys. Rev. D **67**, 033001 (2003)
[7] K. Bhattacharya and S. Sahu, Eur. Phys. J. C **62**, 481 (2009)
[8] A. V. Kuznetsov, N. V. Mikheev and A. V. Serghienko, Phys. Lett. B **690**, 386 (2010)
[9] A. A. Dobrynina and N. V. Mikheev, JETP **118**, 54 (2014) [Original Russian text: Zh. Eksp. Teor. Fiz. **145**, 65 (2014)]
[10] P. Satunin, JETP Letters **101**, 657 (2014)
[11] T. Vachaspati, Phys. Lett. B **265**, 258 (1991)
[12] J. Ambjørn and P. Olesen, Electroweak magnetism, W-codensation and antiscreening, in: *Proc. of 4th Hellenic School on Elementary Particle Physics*, (Corfu, 1992)
[13] D. Grasso and H. R. Rubinstein, Phys. Rep. **348**, 163 (2001)
[14] V. I. Ritus, in *Quantum Electrodynamics of Phenomena in an Intense Field* (in Russian), Proc. P. N. Lebedev Physical Institute, **111**, 5 (1979)
[15] C. Itzykson and J.-B. Zuber, *Quantum Field Theory* (McGraw-Hill, New York, 1985)
[16] V. O. Papanian and V. I. Ritus, in *Issues in Intense-Field Quantum Electrodynamics*, ed. by V. L. Ginzburg (Nova Science Publishers, New York), 153 (1989)

[17] A. E. Shabad, in *Polarization of the Vacuum and a Quantum Relativistic Gas in an External Field*, ed. by V. L. Ginzburg (Nova Science Publishers, New York, 1992)

[18] I. M. Ternov, V. Ch. Zhukovskii and A. V. Borisov, *Quantum Processes in Strong External Field*, in Russian (Moscow State Univ., Moscow, 1989)

[19] A. V. Kuznetsov and N. V. Mikheev, *Electroweak Processes in External Active Media* (Berlin, Heidelberg, Springer-Verlag, 2013)

[20] A. Chodos, K. Everding and D. A. Owen, Phys. Rev. D **42**, 288 (1990)

[21] A. V. Kuznetsov and A. A. Okrugin, Int. J. Mod. Phys. A **26**, 2725 (2011)

[22] A. V. Kuznetsov, A. A. Okrugin and A. M. Shitova, Int. J. Mod. Phys. A **30**, 1550140 (2015)

[23] A. A. Gvozdev and E. V. Osokina, Theor. Math. Phys. **170**, 354 (2012) [Original Russian text: Teor. Mat. Fiz. **170**, 423 (2012)]

[24] V. A. Fock, Phyzik. Z. Sowjetunion **12**, 404 (1937)

[25] J. Schwinger, Phys. Rev. **82**, 664 (1951)

[26] J. K. Daugherty and A. K. Harding, Astrophys. J. **273**, 761 (1983)

[27] V. N. Baier and V. M. Katkov, Phys. Rev. D **75**, 073009 (2007)

[28] I. Yu. Kobsarev, L. B. Okun and Ya. B. Zeldovich, Phys. Lett. B **50**, 340 (1974)

[29] A. D. Dolgov and Ya. B. Zeldovich, Rev. Mod. Phys. **53**, 1 (1981)

[30] D. S. Gorbunov and V. A. Rubakov, *Introduction to the Theory of the Early Universe, Hot Big Bang Theory* (World Scientific Publishing Company, Singapore, 2011)

[31] V. A. Kuzmin, V. A. Rubakov and M. E. Shaposhnikov, Phys. Lett. B **155**, 36 (1985)

[32] L. B. Okun, *Vacua, vacuum: the physics of nothing, History of Original Ideas and Basic Discoveries in Particle Physics* (Plenum Press, New York, 1996)

[33] A. Iwazaki, Phys. Rev. D **56**, 2435 (1997)

[34] P. Cea and L. Tedesco, Phys. Lett. B **425**, 345 (1998)

[35] P. Cea and L. Tedesco, Phys. Lett. B **450**, 61 (1999)

[36] M. M. Forbes and A. R. Zhitnitsky, Phys. Rev. Lett. **85**, 5268 (2000)

[37] M. B. Voloshin, Phys. Lett. B **491**, 311 (2000)

[38] M. B. Voloshin, Phys. Rev. D **63**, 125012 (2001)

[39] L. Campanelli, P. Cea, G. L. Fogli and L. Tedesco, JCAP **0603**, 005 (2006)

Direct URCA-processes in neutron star quark core with strong magnetic field.

Vasily Belyaev[1,*] and *Alexander* Gvozdev[1,**]

[1] *Yaroslavl State University, Russia, Yaroslavl, Sovyetskaya st., 21*

Abstract. Neutrino emissivity in direct URCA processes in superdense strongly magnetized quark medium is calculated. In calculations, we use the covariant method of density matrixes for charge fermions in uniform constant magnetic field. One-gluon exchanges between quarks are also take into account.
In evaluations, the strength of magnetic field corresponds to the case, where the quarks of medium occupy a lot of Landau levels, while the electrons are in ground Landau level. The analytical dependence of neutrino emissivity on chemical potentials of quarks and electrons, temperature and magnetic field strength is obtained and briefly discussed. The result could be important in application to a massive strongly magnetized neutron star with quark core.

1 Introduction

Discovery of neutron stars in the $1.9 - 2.0 M_{sun}$ mass range [1] opened the wide discussion of the matter structure in the inner core of such a massive neutron star. In particular, model of neutron star core, consisting of strange quark matter, is discuss and develop at present time [2]. Neutrino cooling of neutron star core is very important, because of such a cooling is responsible for determination and calculation theoretically of neutron star cooling curve [3].

On the basis on observations made to date, there were discovered the group of neutron stars with strong magnetic fields, so called magnetars [4] with magnetic induction B up to $10^{15} G$ on the surface.

Hence, it is logically to investigate the neutrino cooling rate of a neutron star with strongly magnetized quark core. Without magnetic field, such a work was done and published in 1980 [5]. In this work cold electroneutral medium in beta-equilibrium consists of ideal gases of u-, d-, s-quarks and electrons was considered. It was shown, that it is necessary to include s-quark, because in that case the parameter "bound energy per quark" must be minimal [6].

To describe week interaction, we use Lagrangian of four-fermion interaction:

$$\mathcal{L} = \frac{G_F \cos\theta_c}{\sqrt{2}} [\bar{d}\gamma_\alpha (g_v + g_a \gamma^5) u][\bar{\nu}\gamma^\alpha (1 + \gamma^5) e], \tag{1}$$

where G_F is Fermi constant, θ_c is Cabbibo angle, g_v, g_a are vector and axial electroweak constants. We assume, that in the limit of cold quark matter the constants depend on chemical potentials of quarks

*e-mail: SCstalk@mail.ru
**e-mail: gvozdev@uniyar.ac.ru

only, and the following URCA-processes give leading contribution in neutrino emissivity:

$$u + e^- \to d + \nu_e, \quad \text{(crossing)} \tag{2}$$

$$d \to u + e^- + \bar{\nu}_e. \quad \text{(decay)} \tag{3}$$

One neglects contributions in neutrino emissivity of similar URCA-processes with s-quark, because they are suppressed by Cabbibo angle. If one takes into account one-gluon exchanges between quarks, then the neutrino emissivity is not negligible and proportional to $G_F^2 \alpha_s \mu_u \mu_d \mu_e T^6$ [5]. Here α_s is strong coupling constant, μ_u, μ_d, μ_e are chemical potentials of u-, d-quarks and electrons accordingly, T is temperature. This is well-known old result, cited in many reports and books on neutron stars.

In the present paper we consider the quark core of massive neutron star with strong magnetic field. We assume, that:

$$\mu_u^2, \mu_d^2 \gg m_q^2(\mu) > eB \gtrsim \mu_e^2 \gg T^2, \tag{4}$$

where

$$m_q^2(\mu) = \frac{2\alpha_s}{3\pi}\mu_q^2. \tag{5}$$

is well-known effective quark mass [7] in the one-gluon approximation of finite-temperature QCD in the limit $T \to 0$, B is the magnetic field strength. The first two inequations in (4) means, that magnetic field has small effect on QCD-dynamics, consequently, on effective quark masses and a lot of Landau levels are occupied by quarks. The third inequation mean, that electrons are occupied ground Landau level only. Substitute in inequation (4) reasonable values of chemical potentials quarks and electrons $\mu_{u,d} \approx 500$ MeV, $\mu_e \approx 30$ MeV, one obtain:

$$B \sim 10^{17}\text{G}. \tag{6}$$

We note, that this value of magnetic field is reasonable for neutron star core and is not effect substantially on equation of state of quark medium. We assume, that the quarks are ideal massive gas with mass $m(\mu)$, while electrons and neutrino are massless.

For evaluation neutrino emissivity we use matrix density technics for charged fermion in external magnetic field [8]. The conservation laws of 4-momentum for crossing and decay are:

$$\mathcal{P}_1 + p = \mathcal{P}_2 + k, \quad \mathcal{P}_2 = \mathcal{P}_1 + p + k, \tag{7}$$

where $\mathcal{P}_1, \mathcal{P}_2, p, k$ – 4-momenta of u-, d-quarks, electron and neutrino accordingly,

$$\mathcal{P}_1^\mu = (E_n, \vec{\mathcal{P}}_1), \quad \mathcal{P}_2^\mu = (E_{n'}, \vec{\mathcal{P}}_2), \quad p^\mu = (\epsilon_{n''}, \vec{p}), \quad k^\mu = (\omega, \vec{k}), \tag{8}$$

where $E_n = \sqrt{p_z^2 + 2qeBn + m^2}$ is the energy of charged particle in external magnetic field, q is module of particles charge, $\epsilon_{n''} = \epsilon_0 \equiv \epsilon$, because all electrons take ground Landau level.

Neutrino emissivities of decay and crossing are presented as follows:

$$Q^{(\text{dec})} = \frac{3}{V}\sum dn_d\, f_d \sum dn_u\,(1 - f_u) \sum dn_e (1 - f_e) \sum dn_\nu \omega \frac{|S_{if}^{(\text{dec})}|^2}{\tau}, \tag{9}$$

$$Q^{(\text{cr})} = \frac{3}{V}\sum dn_u\, f_u \sum dn_e\, f_e \sum dn_d\,(1 - f_d) \sum dn_\nu \omega \frac{|S_{if}^{(\text{cr})}|^2}{\tau}. \tag{10}$$

where sums mean summing under overall phase volumes of particles, f_i are the distributions functions of particles, ω is neutrino energy, V is normalized volume, τ is normalized time, $|S_{if}|^2$ is squared

S-matrix element. It is assumed in formulae, that emitting medium is transparent for neutrinos. The multiplier factors "3" take into account, because the quarks with all colours participate in all processes, and the colour conservates in processes.

We discuss below the crossing process, because the evaluations are similar for both processes. After simplifications, the neutrino emissivity of crossing $Q^{(cr)}$ could be present in the form:

$$Q^{(cr)} = \frac{3G_F^2\cos^2\theta_c g_v^2}{32(2\pi)^8}\sum_{n,n'} g_n g_{n'}\int\frac{d^3\mathcal{P}_1}{E_n}f_u\int\frac{d^3p}{\epsilon}f_e\cdot$$
$$\cdot\int\frac{d^3\mathcal{P}_2}{E_{n'}}(1-f_d)\int d^3k\,\delta^4(\mathcal{P}_1+p-\mathcal{P}_2-k)S\,p[\hat{\rho}_d\gamma_\alpha\cdot$$
$$\cdot(1+c\gamma^5)\hat{\rho}_u\gamma_\beta(1+c\gamma^5)]S\,p[\hat{\rho}_\nu\gamma^\alpha(1+\gamma^5)\hat{\rho}_e\gamma^\beta(1+\gamma^5)], \quad (11)$$

where $c = g_a/g_v$, $\hat{\rho}_u, \hat{\rho}_d, \hat{\rho}_e, \hat{\rho}_\nu$ - are density matrices of u-, d-quarks, electron, summed over spins, and neutrino accordingly, g_n is the factor of spins degeneration.

We take the density matrices for u, d, e from [8], and don't forget, that the quarks have fractional charges. One evaluates the traces by standard rules [8]. After long and tedious evaluations, we obtain:

$$Q^{(cr)} = \frac{6G_F^2\cos^2\theta_c g_v^2}{(2\pi)^8}\sum_{n,n'} g_n g_{n'}\int\frac{d^3\mathcal{P}_1}{E_n}f_u\int\frac{d^3p}{\epsilon}f_e\cdot$$
$$\cdot\int\frac{d^3\mathcal{P}_2}{E_{n'}}(1-f_d)\int d^3k\,\delta^4(\mathcal{P}_1+p-\mathcal{P}_2-k)(-1)^{n+n'}exp\Big(-\frac{u_u+u_d+u_e}{2}\Big)\cdot$$
$$\cdot\Big([(1+c)^2L_{n'-1}(u_d)L_n(u_u)+(1-c)^2L_{n-1}(u_u)L_{n'}(u_d)]\Phi_{\|}+8(1+c)^2L^1_{n'-1}(u_d)L_n(u_u)(\mathcal{P}_2\Lambda k)(\mathcal{P}_1(\tilde{\Lambda}+\tilde{\phi})p)+$$
$$+8(1-c)^2L^1_{n-1}(u_u)L_{n'}(u_d)(\mathcal{P}_1\Lambda k)(\mathcal{P}_2(\tilde{\Lambda}+\tilde{\phi})p)+$$
$$+4L_{n'}(u_d)L_n(u_u)(k(\tilde{\Lambda}+\tilde{\phi})p)[(1+c^2)(\mathcal{P}_2\tilde{\Lambda}\mathcal{P}_1)-(1-c^2)m_u m_d+2c(\mathcal{P}_1\tilde{\phi}\mathcal{P}_2)]\Big), \quad (12)$$

where

$$\Phi_{\|} \equiv 2(\mathcal{P}_2\tilde{\Lambda}\mathcal{P}_1)(k\tilde{\Lambda}p)-2(\mathcal{P}_2\tilde{\Lambda}p)(\mathcal{P}_1\tilde{\Lambda}k)-2(\mathcal{P}_2\tilde{\Lambda}k)(\mathcal{P}_1\tilde{\Lambda}p)-$$
$$-(\mathcal{P}_2\tilde{\Lambda}k)(\mathcal{P}_1\tilde{\phi}p)-(\mathcal{P}_1\tilde{\Lambda}k)(\mathcal{P}_2\tilde{\phi}p)-(\mathcal{P}_1\tilde{\Lambda}p)(\mathcal{P}_2\tilde{\phi}k)-(\mathcal{P}_2\tilde{\Lambda}p)(\mathcal{P}_1\tilde{\phi}k). \quad (13)$$

Here $\phi_{\mu\nu} = F_{\mu\nu}/B$, $\tilde{\phi}_{\mu\nu} = \tilde{F}_{\mu\nu}/B$ are dimensionless tensor, dual tensor of magnetic field, $\Lambda_{\mu\nu} = (\phi\phi)_{\mu\nu}$, $\tilde{\Lambda}_{\mu\nu} = (\tilde{\phi}\tilde{\phi})_{\mu\nu}$, L_n^k are Laguerre polynomials, $u_d = 6\mathcal{P}_{2\perp}^2/eB$, $u_u = 3\mathcal{P}_{1\perp}^2/eB$, $u_e = 2p_\perp^2/eB$, $p_\perp^2$ is denoted the transversal to magnetic field strength components of charge particles momenta, for example, electron squared.

2 Integrals on transversal to magnetic field strength components of charge particles momenta

To calculate the neutrino emissivity (12), it is necessary to evaluate the following scalar and vector integrals:

$$\mathcal{J}^{(n,m)} = \int d\mathcal{P}_{1x}d\mathcal{P}_{1y}\int d\mathcal{P}_{2x}d\mathcal{P}_{2y}\int dp_x dp_y L_n(u_u)L_m(u_d)\cdot$$
$$exp[-\frac{u_u+u_d+u_e}{2}]\delta_\perp(\mathcal{P}_1+p-\mathcal{P}_2-k), \quad (14)$$

$$\mathcal{J}^{(n,m)\alpha} = \int d\mathcal{P}_{1x} d\mathcal{P}_{1y} \int d\mathcal{P}_{2x} d\mathcal{P}_{2y} \int dp_x dp_y \mathcal{P}_1^\alpha L^1_{n-1}(u_u) L_m(u_d) \cdot \\ exp[-\frac{u_u + u_d + u_e}{2}] \delta_\perp(\mathcal{P}_1 + p - \mathcal{P}_2 - k), \quad (15)$$

$$\tilde{\mathcal{J}}^{(n,m)\alpha} = \int d\mathcal{P}_{1x} d\mathcal{P}_{1y} \int d\mathcal{P}_{2x} d\mathcal{P}_{2y} \int dp_x dp_y \mathcal{P}_2^\alpha L_n(u_u) L^1_{m-1}(u_d) \cdot \\ exp[-\frac{u_u + u_d + u_e}{2}] \delta_\perp(\mathcal{P}_1 + p - \mathcal{P}_2 - k), \quad (16)$$

where $\delta_\perp(\mathcal{P}_1 + p - \mathcal{P}_2 - k) = \delta(\mathcal{P}_{1x} + p_x - \mathcal{P}_{2x} - k_x)\delta(\mathcal{P}_{1y} + p_y - \mathcal{P}_{2y} - k_y)$. We have calculated these integrals similarly to [8], and obtained:

$$\mathcal{J}^{(n,m)}(v) = \frac{\pi^2}{3}(eB)^2(-1)^{n+m} I_0^{(n,m)}(v), \quad (17)$$

$$\mathcal{J}^{(n,m)\alpha}(v) = k_\perp^\alpha \frac{\pi^2}{6}(eB)^2(-1)^{n+m+1} n\Big(I_1^{(n-1,m)}(v) - I_1^{(n,m)}(v)\Big), \quad (18)$$

$$\tilde{\mathcal{J}}^{(n,m)\alpha}(v) = k_\perp^\alpha \frac{\pi^2}{6}(eB)^2(-1)^{n+m} m\Big(I_1^{(n,m-1)}(v) - I_1^{(n,m)}(v)\Big). \quad (19)$$

Here $v^2 = k_\perp^2/eB$, $k_\perp^2 = (k_x)^2 + (k_y)^2$ is the neutrino transverse momenta squared and integrals $I_i^{n,m}$ are determined, as follows:

$$I_0^{(n,m)}(v) \equiv \int_0^\infty dx\, e^{-3x}\, L_n(2x) L_m(x) J_0(\sqrt{6} v \sqrt{x}), \quad (20)$$

$$I_1^{(n,m)}(v) \equiv \sqrt{\frac{2}{3}} \frac{1}{v} \int_0^\infty dx \frac{e^{-3x}}{\sqrt{x}}\, L_n(2x) L_m(x) J_1(\sqrt{6} v \sqrt{x}), \quad (21)$$

where J_0, J_1 - are Bessel functions. The functions $I_0^{(n,m)}(v), I_1^{(n,m)}(v)$ at fixed values of v have asymmetric delta-like "burst" behavior at $\mathbf{m = 2n - 1}$. If values n, m turn from tens to thousands, then the "burst" diffuses fast.

From this point we use various approximations. First, we assume, that the parameter $v^2 = k_\perp^2/eB$ is small ($v \ll 1$). Hence, one evaluate approximately functions I_0, I_1. One approximate $J_1(z) \approx z/2$, $J_0(z) \approx 1$, after that the leading terms in expansion series on v are:

$$I_1^{(n,m)}(v) = I_0^{(n,m)}(v) \approx \int_0^\infty dx e^{-3x}\, L_n(2x) L_m(x) \equiv C_0^{(n,m)}. \quad (22)$$

One can evaluate the function $C_0^{(n,m)}$ exactly [9]

$$C_0^{(n,m)} = \frac{\Gamma(n+m+1)}{n!m!} \frac{2^m}{3^{n+m+1}}. \quad (23)$$

Second, we assume, that the leading contribution in neutrino emissivity gives the region $n, n' \gg 1$. Hence, we replace

$$g_n = g_{n'} = 2. \quad (24)$$

Moreover, where it does not cause confusion, we replace

$$(n-1) \to n, \quad (n'-1) \to n'. \quad (25)$$

DOI: 10.1051/epjconf/201715805003

In expressions like $(I_1^{(n,n'-1)} - I_1^{(n,n')})$ this replacement is not valid. Therefore, we calculate approximately by using (22), (23) the ratios:

$$\frac{I_1^{(n,n'-1)}}{I_1^{(n,n')}} \approx \frac{3n'}{2(n+n')}, \quad \frac{I_1^{(n-1,n')}}{I_1^{(n,n')}} \approx \frac{3n}{n+n'}. \tag{26}$$

Third, we assume, that the leading contribution in neutrino emissivity gives at fixed n by small vicinity of $m = 2n$, because functions $I_0^{(n,m)}(v), I_1^{(n,m)}(v)$ have delta-like burst in that domain. Hence, we replace $n' = 2n$, where it does not cause confusion:

$$n'\frac{n'-2n}{2(n+n')} \to \frac{n'-2n}{3}, \quad n\frac{2n-n'}{n+n'} \to \frac{2n-n'}{3}. \tag{27}$$

After integrations on transverse components and simplifications, the neutrino emissivity (12) could be presented in the forms:

$$Q^{(\mathrm{cr})} = \frac{G_F^2\cos^2\theta_c g_v^2}{16\pi^6}(eB)^2\sum_{n,n'}\int d^3k\int\frac{dp_z}{\epsilon}f_e\int\frac{d\mathcal{P}_{1z}}{E_n}f_u\int\frac{d\mathcal{P}_{2z}}{E_{n'}}(1-f_d)\cdot$$
$$\cdot\,\delta_{\parallel}(\mathcal{P}_1+p-\mathcal{P}_2-k)\Big(I_0^{(n,n')}\Big(-(1+c)^2\Phi_{\parallel}+2[k(\tilde{\Lambda}+\tilde{\phi})p][(1+c^2)(\mathcal{P}_1\tilde{\Lambda}\mathcal{P}_2)-$$
$$-(1-c^2)m_u m_d+2c(\mathcal{P}_1\tilde{\phi}\mathcal{P}_2)]\Big)+$$
$$+\,v^2 I_1^{(n,n')}\,{}_{\Delta}\mathcal{P}_{\perp}^2\Big((1+c)^2[\mathcal{P}_1(\tilde{\Lambda}+\tilde{\phi})p]+(1-c)^2[\mathcal{P}_2(\tilde{\Lambda}+\tilde{\phi})p]\Big)\Big), \tag{28}$$

where ${}_{\Delta}\mathcal{P}_{\perp}^2 = 2eB(n'-2n)/3$, $\delta_{\parallel}(\mathcal{P}_1+p-\mathcal{P}_2-k) = \delta(E_n+\epsilon-E_{n'}-\omega)\delta(\mathcal{P}_{1z}+p_z-\mathcal{P}_{2z}-k_z)$.

3 Integrals on parallel components

Firstly, for unification of calculations both processes, we rework artificially by using the conservation of energy and parallel component of momentum in the crossing process:

$$f_u f_e(1-f_d) = f_d(1-f_u)(1-f_e)exp(-\frac{\omega}{T})exp(\frac{\mu_e - {}_{\Delta}\mu}{T}), \tag{29}$$

where ${}_{\Delta}\mu \equiv \mu_d - \mu_u$. As one can see, the expression (28) for neutrino emissivity contain manifestly covariant, from the viewpoint of Lorenz transformations along the magnetic field strength, integrals on parallel components of quarks:

$$I_s \equiv \int_{-\infty}^{\infty}\frac{d\mathcal{P}_{1z}}{E_n}\int_{-\infty}^{\infty}\frac{d\mathcal{P}_{2z}}{E_{n'}}f_d(1-f_u)\delta_{\parallel}(\mathcal{P}_2-\mathcal{P}_1-q), \tag{30}$$

$$I_V^{(1)\mu} \equiv \int_{-\infty}^{\infty}\frac{d\mathcal{P}_{1z}}{E_n}\int_{-\infty}^{\infty}\frac{d\mathcal{P}_{2z}}{E_{n'}}f_d(1-f_u)\delta_{\parallel}(\mathcal{P}_2-\mathcal{P}_1-q)\mathcal{P}_{1\parallel}^{\mu}, \tag{31}$$

$$I_V^{(2)\mu} \equiv \int_{-\infty}^{\infty}\frac{d\mathcal{P}_{1z}}{E_n}\int_{-\infty}^{\infty}\frac{d\mathcal{P}_{2z}}{E_{n'}}f_d(1-f_u)\delta_{\parallel}(\mathcal{P}_2-\mathcal{P}_1-q)\mathcal{P}_{2\parallel}^{\mu}, \tag{32}$$

$$\mathcal{I}_T^{\alpha\beta} \equiv \int_{-\infty}^{\infty} \frac{d\mathcal{P}_{1z}}{E_n} \int_{-\infty}^{\infty} \frac{d\mathcal{P}_{2z}}{E_{n'}} f_d(1-f_u)\delta_{\|}(\mathcal{P}_2 - \mathcal{P}_1 - q)\mathcal{P}_{1\|}^{\alpha}\mathcal{P}_{2\|}^{\beta}, \tag{33}$$

$$\mathcal{I}_\phi \equiv \int_{-\infty}^{\infty} \frac{d\mathcal{P}_{1z}}{E_n} \int_{-\infty}^{\infty} \frac{d\mathcal{P}_{2z}}{E_{n'}} f_d(1-f_u)\delta_{\|}(\mathcal{P}_2 - \mathcal{P}_1 - q)(\mathcal{P}_1\tilde{\phi}\mathcal{P}_2) = 0, \tag{34}$$

where $\mathcal{P}_{1\|}^{\alpha} = (\mathcal{P}_1\tilde{\Lambda})^{\alpha}$, indices μ, α, β runing values of $(0,3)$, $q \equiv p + k$ - for decay and $q \equiv p - k$ - for crossing. As kinematics analysis shows, $q_0 = \epsilon + \omega > 0$ $(n' - 2n > 0)$ for decay and $q_0 = \epsilon - \omega < 0$ $(n' - 2n < 0)$ for crossing. We note also, that similar integrals over structures like $(p_1\tilde{\phi}p_2)$ are zero, while integrals $\mathcal{I}_V^{(1)\mu}, \mathcal{I}_T^{\alpha\beta}$ can be expressed in terms of $\mathcal{I}_s$:

$$\mathcal{I}_v^{(1)\mu} = \frac{\Delta\mathcal{P}_\perp^2 - q_\|^2}{2q_\|^2}\mathcal{I}_s q_\|^\mu, \quad \mathcal{I}_v^{(2)\mu} = \frac{\Delta\mathcal{P}_\perp^2 + q_\|^2}{2q_\|^2}\mathcal{I}_s q_\|^\mu, \tag{35}$$

$$\mathcal{I}_T^{\alpha\beta} = \left(a\tilde{\Lambda}^{\alpha\beta} + b\frac{q_\|^\alpha q_\|^\beta}{q_\|^2}\right)\mathcal{I}_s, \tag{36}$$

$$a = -\frac{\Delta\mathcal{P}_\perp^2 + q_\|^2}{2} + \frac{2}{3}eBn' + m_d^2 - \frac{(\Delta\mathcal{P}_\perp^2)^2 - (q_\|^2)^2}{4q_\|^2}, \tag{37}$$

$$b = -\frac{(\Delta\mathcal{P}_\perp^2)^2 - (q_\|^2)^2}{2q_\|^2} + \frac{\Delta\mathcal{P}_\perp^2 + q_\|^2}{2} - \frac{2}{3}eBn' - m_d^2, \tag{38}$$

where $q_\|^2 = (q\tilde{\Lambda}q)$.

One can evaluate the scalar integral $\mathcal{I}_s$ by using the approximation [10]:

$$f_d(1-f_u) \approx S \cdot \delta(E_{n'} - \mu_d), \tag{39}$$

where the area of overlap of strongly degenerate fermi-distributed functions is:

$$S = \int_{M_1}^{\infty} dE_n \int_{M_2}^{\infty} dE_{n'} f_d(1-f_u)\delta(E_{n'} - E_n - q_0) \approx \frac{\Delta\mu - q_0}{1 - exp(-\frac{\Delta\mu - q_0}{T})}, \tag{40}$$

where $M_1 = \sqrt{4eBn/3 + m_u^2}$, $M_2 = \sqrt{2eBn'/3 + m_d^2}$. By using (39), (40), assuming $\mu_d \gg \epsilon, \omega$ and neglecting $m_{u,d}^2$ in comparison with μ_d^2, eBn', one obtains roughly:

$$\mathcal{I}_s \approx \frac{2\delta(\cos\theta - \cos\theta_1^*) + \delta(\cos\theta - \cos\theta_2^*)}{\mu_d^2 v_F^2 \omega}, \tag{41}$$

where $v_F^2 \equiv 1 - 2eBn'/3\mu_d^2$ is Fermi velocity squared for d-quark, θ is the angle between neutrino momentum $\vec{k}$ and magnetic field strength $\vec{B}$. The explicit form for $\cos\theta_1^*$ $(\cos\theta_2^* = -\cos\theta_1^*)$ is not required, because, after substitution (29), (35), (36), (41) in neutrino emissivity (28), the integrand does not depend on it. Thus, the integration over $d^3k = 2\pi\omega^2 d\omega d(-\cos\theta)$ gives effectively $4\pi\omega^2 d\omega$.

We note, that approximation (39) take place in the rest frame of the quark matter. If the matter moves along the magnetic field, $\delta(E_{n'} - \mu_d)$ will be transformed to:

$$\delta(E_{n'} - \mu_d) \rightarrow \frac{\sqrt{q_\|^2}}{\mu_d}\delta(\sqrt{q_\|^2} - \Delta), \tag{42}$$

where $\Delta = \sigma\frac{eB}{3\mu_d}(n' - 2n)$, and σ - is sign function of q_0, or $(n' - 2n)$ ($\sigma > 0$ in decay and $\sigma < 0$ in crossing). Thus, we have the the constraint on Lorenz-invariant $\sqrt{q_{\parallel}^2} \approx \Delta$. Together with conservation laws of energy and parallel momentum, that constraint determine the region of integration on ϵ, ω - variables. We note also, that $\sqrt{q_{\parallel}^2}/\mu_d$ is a small number ($\sqrt{q_{\parallel}^2}/\mu_d \ll 1$).

By assuming, that:

$$\mu_d^2, eBn' \gg \Delta^2, m_{u,d}^2, \sigma eB(n' - 2n), \tag{43}$$

we take the leading contributions in neutrino emissivities:

$$Q^{(cr)} = \frac{G_F^2 \cos^2\theta_c(g_v^2 + g_a^2)}{\pi^5}(eB)^2 e^{\frac{\mu_e - {}_\Delta\mu}{T}} \sum_{n,n'} \frac{1}{v_F^2} \int_{\mu_e}^{\infty} d\epsilon \int_{\epsilon+\Delta}^{\infty} d\omega\omega e^{-\frac{\omega}{T}} \frac{{}_\Delta\mu - \epsilon + \omega}{1 - e^{-\frac{{}_\Delta\mu - \epsilon + \omega}{T}}} \cdot$$
$$\cdot (\omega - \epsilon - \sqrt{(\epsilon - \omega)^2 - \Delta^2}) I_0^{(n,n')} \Big(2 + [1 + v_F^2] \frac{2\epsilon(\omega - \epsilon - \sqrt{(\epsilon - \omega)^2 - \Delta^2})}{\Delta^2}\Big), \tag{44}$$

$$Q^{(dec)} = \frac{G_F^2 \cos^2\theta_c(g_v^2 + g_a^2)}{\pi^5}(eB)^2 \sum_{n,n'} \frac{1}{v_F^2} \int_{\mu_e}^{\infty} d\epsilon \int_{\Delta^2/(2\epsilon)}^{\infty} d\omega\omega \frac{(\epsilon + \omega - {}_\Delta\mu)}{e^{\frac{\epsilon + \omega - {}_\Delta\mu}{T}} - 1} \cdot$$
$$\cdot (\omega + \epsilon - \sqrt{(\epsilon + \omega)^2 - \Delta^2}) I_0^{(n,n')} \Big(2 + [1 + v_F^2] \frac{2\epsilon(\omega + \epsilon - \sqrt{(\epsilon + \omega)^2 - \Delta^2})}{\Delta^2}\Big). \tag{45}$$

In evaluating the rest of the integrals (44), (45) over (ϵ, ω) - variables we assume, that the deviation from beta-equilibrium $\mu_e - {}_\Delta\mu$ is positive and small enough number ($\mu_e - {}_\Delta\mu > 0$, $\mu_e - {}_\Delta\mu \ll \mu_e$), while $(\mu_e - {}_\Delta\mu)/T \gg 1$. As a consequence of exponentially suppressed integrands, the leading contributions in integrals gives the regions $\epsilon - \mu_e \sim T$, $\omega - \mu_e - \Delta \sim T$ (crossing) and $\epsilon - \mu_e \sim T$, $\omega - \Delta^2/(2\mu_e) \sim T$ (decay). Thus, the leading contributions in neutrino emissivities could be presented in the form:

$$Q^{(cr)} = \frac{2G_F^2 \cos^2\theta_c(g_v^2 + g_a^2)}{\pi^5}(eB)^2 \, {}_\Delta\mu \mu_e^2 T^2 e^{-\frac{{}_\Delta\mu}{T}} \sum_{n,n'} \frac{1 + v_F^2}{v_F^2} e^{-\frac{\Delta}{T}} I_0^{(n,n')}, \tag{46}$$

$$Q^{(dec)} = \frac{G_F^2 \cos^2\theta_c(g_v^2 + g_a^2)}{4\pi^5}(eB)^2 \frac{(\mu_e - {}_\Delta\mu)}{\mu_e^2} T^2 e^{-\frac{\mu_e - {}_\Delta\mu}{T}} \sum_{n,n'} \frac{1 - v_F^2}{v_F^2} \Delta^4 e^{-\frac{\Delta^2}{2\mu_e T}} I_0^{(n,n')}. \tag{47}$$

4 Summations over n, n'

We introduce new variables $k = 2n - n'$, $k = n' - 2n$ for crossing and decay accordingly, and replace sums:

$$\sum_{n,n'} \to \sum_{n=1}^{N_1} \sum_{k=1}^{2n}, \qquad \sum_{n,n'} \to \sum_{n'=0}^{N_2} \sum_{k=1}^{n'}. \tag{48}$$

where $N_1 = 3\mu_u^2/4eB$, $N_2 = 3\mu_d^2/2eB$.

Next, we express for crossing $n' = 2n - k$ in I_0 and v_F^2. We take only the first component of sum of k because of suppressing exponent, and obtain:

$$Q^{(cr)} \approx \frac{2G_F^2 \cos^2\theta_c(g_v^2 + g_a^2)}{(\pi)^5}(eB)^2 \mu_e^2 \, {}_\Delta\mu T^2 exp(-\frac{{}_\Delta\mu}{T}) exp(-\frac{eB}{3\mu_d T}) \sum_n I_0^{(n,2n-1)} \frac{(1 + v_F^2)}{v_F^2}, \tag{49}$$

where $v_F^2 = 1 - 2eB(2n-1)/3\mu_d^2$.

We express in decay process $n = (n' - k)/2$ and neglect in $I_0^{(n,n')}$ dependence of k. This action is giving too high values of $I_0^{(n,n')}$ by summing over n'. Hence, we obtain the upper bound on neutrino emissivity of decay. We use roughly the Euler-Macloren formula to sum the rest terms on k. In that case, the upper bound on neutrino emissivity of decay process is:

$$Q^{(\mathrm{dec})} \leqslant \frac{9\sqrt{\pi}G_F^2\cos^2\theta_c(g_v^2+g_a^2)}{4\sqrt{2}(\pi)^5}(eB)\mu_d\sqrt{\mu_e}(\mu_e - {}_\Delta\mu + \frac{5}{2}T)T^{\frac{9}{2}}\cdot$$

$$\cdot\, exp(-\frac{\mu_e - {}_\Delta\mu}{T})\sum_{n'} I_0^{(\frac{n'}{2},n')}\frac{1-v_F^2}{v_F^2}, \quad (50)$$

where $v_F^2 = 1 - 2eBn'/3\mu_d^2$.

5 Conclusions

We research the covariant method of evaluation of neutrino emissivity in direct URCA-processes. In our model the superdense strong magnetized quark medium emits neutrinos. In evaluations, the values of magnetic field corresponds to the case, where the quarks take a lot of Landau-levels, and the electrons take ground Landau-level. In this case the analytically dependence neutrino emissivity of task parameters (chemical potentials quarks and electrons, temperature, magnetic field) is obtained for the first time. In using approximations

$$\mu_u^2, \mu_d^2 \gg m_q^2(\mu) > eB \gtrsim \mu_e^2 \gg T^2,$$

$$\mu_d^2, eBn' \gg \Delta^2, m_{u,d}^2, \sigma eB(n'-2n),$$

the main result is an exponentially suppression of both processes emissivities. As one can see from formulae (49), (50), in beta-equilibrium the decay goes on and the crossing is suppressed. Without beta-equilibrium both processes are suppressed. Thus, for our medium the stable state is the state with small deflection from beta-equilibrium and forbidden direct URCA-processes.

The decay suppression is a consequence of very small area S (40) of strongly degenerate fermi-distributed functions $f_d(1-f_u)$ overlap.

The crossing suppression is a consequence of one-dimensional kinematics, which gives the restriction $\omega > \epsilon$, while $\epsilon \geqslant \mu_e$.

The main question for further investigation is how do URCA-processes go on at gradual decrease of magnetic field?

The paper is done within "YSU Initiative Scientific Researches" (Project AAAA-A16-116070610023-3).

References

[1] Ozel F., Freire P. Masses, radii, and equation of state of neutron star. Ann. Rev. Astron. Astrophys., **54**, 401-440, (2016).

[2] Weber F., Contera G. A., et al. Properties of high-density matter in neutron stars. Mod. Phys Lett. A, **29**, 1430022, (2014).

[3] Potekhin A. Y., Pons J. A., Page D. Neutron stars - cooling and transport. Space. Sci. Rev. **191**, 239, (2015).

[4] Turolla R., Zane S., Watts A. Magnetars: the physics behind observations. Rep. Prog. Phys. **78**, 116901, (2015).

[5] Iwamoto N. Quark Beta Decay and the Cooling of Neutron Stars. Phys.Rev.Lett., **44**, 1637, (1980).

[6] Schmitt A. Dense matter in compact stars. e-Print Arxiv: astro-ph.SR 1001.3294v2., (2010).

[7] Le Bellac M. *Thermal Field Theory.* (Cambridge University Press, Cambridge, 1996) 269.

[8] Gvozdev A. A., Osokina E. V. Neutrino processes in an external magnetic field in the density matrix formalism, TMP, **170**:3, pp 354 - 375, (2012).

[9] Gradshteyn I. S., Ryzhik I. M. *Table of Integrals, Series, and Products (7ed).* (Academic Press, Oxford, 2007) 1161.

[10] Mikheev N. V., Narynskaya E. N. Neutrino-electron processes in a dense magnetized plasma. Modern Physics Letters A., **15**:25, 1551-1556, (2000).

EPJ Web of Conferences **158**, 05004 (2017) DOI: 10.1051/epjconf/201715805004
QFTHEP 2017

Direct URCA Processes in Supernovae and Accretion Disks with Arbitrary Magnetic Field

Igor Ognev[1,⋆]

[1] *P. G. Demidov Yaroslavl State University, Sovietskaya 14, 150003 Yaroslavl, Russia*

Abstract. An effect of a magnetic field of an arbitrary strength on the beta-decay and reactions related with it by the crossing symmetry (the beta-processes) in supernovae and accretion disks around black holes is analyzed. Rates of the beta-processes and the energy and momentum transfered through them to an optically transparent matter are calculated. It is shown that the macroscopic momentum transferred to the medium increases linearly with the magnetic field strength and can substantially affect the dynamics of supernovae and accretion disks especially when a matter inside is degenerate. It is also demonstrated that the rates of the beta-processes and the energy deposition in these reactions for the magnetic field strength $B \lesssim 10^{15}$ G, which is assumed to be typical in supernovae and accretion disks, are lower than in the absence of the field. This suppression is more pronounced for reactions with neutrinos.

1 Introduction

Neutrinos can strongly affect the processes occurring in many astrophysical objects, especially explosive processes accompanied by the release of a large amount of energy and heating of the surrounding medium. Such processes include the core collapse supernova which is the final stage in an evolution of massive stars as well as a merger of close binary systems. Despite the fact that both these phenomena are fundamentally different, their common features are the powerful neutrino radiation and possible connection with cosmological gamma-ray bursts [1] which remain an unsolved problem in theoretical astrophysics for many years. In a supernova explosion, a source of the neutrino emission is the proto-neutron star that forms as the result of the core collapse of a massive star at the final stage of its evolution. It should be noted that neutrinos are the main engine leading to a successive explosion of supernovae. A merger of stars in a close binary system leads to the formation of a central object (as a rule, it is a black hole or, very rarely, a massive neutron star) and a hot accretion disk around it. In this case, the disk is a source of powerful fluxes of neutrinos. Note that an analogous system consisting of a central black hole and accretion disk can be formed during a failed supernova explosion called the collapsar model [2]. In this model, a black hole is appeared as the result of the core collapse of a massive star, rather than a proto-neutron star. The absence of a powerful neutrino radiation from the central part leads to the accretion of the remaining matter to the black hole and formation of a hot disk around the hole. General interest in such systems is connected with the fact that a large amount of energetic neutrino is emitted from the hot disk and produces a relativistic electron-positron plasma

⋆e-mail: ognev@uniyar.ac.ru

which could be a source of gamma-ray bursts [1]. However, the merger of compact objects (neutron stars or black holes) apparently leads to the formation of short-duration gamma-ray bursts [3]. Long-duration gamma-ray bursts are usually associated with the supernova explosions with a large energy release and probably with collapsars [1]. It should be noted that the merger of a compact object with the massive Wolf-Rayet star can also be the source of the long-duration gamma-ray bursts [4].

The dense and hot medium that forms after a supernova explosion, as well as the merger of stars in a close binary system, is not transparent to photons, and emission of neutrinos is the main channel for its cooling. Due to high temperatures, the medium predominantly consists of free nucleons, photons, and relativistic electrons and positrons being in the local thermodynamic equilibrium. Depending on the specific conditions, neutrinos in such a medium can either be in equilibrium, like in supernova cores, or be freely emitted, like in accretion disks or envelopes of supernovae. The main neutrino reactions in such media are the beta-processes (called often the direct URCA-processes) [5]:

$$p + e^- \rightleftarrows n + \nu_e, \tag{1}$$

$$n + e^+ \rightleftarrows p + \bar{\nu}_e, \tag{2}$$

$$n \rightleftarrows p + e^- + \bar{\nu}_e, \tag{3}$$

which are important primarily as the main channel of the energy exchange between the medium and neutrinos in the case of supernovae and the energy source of gamma-ray bursts in the merger of stars in the close binary system. An important feature in such astrophysical cataclysms is that a strong magnetic field can be generated. Simulations show that the magnetic field strength in supernovae [6, 7] and in accretion disks [4, 8] can reach the values $B \sim 10^{15}$ G. Such a strong magnetic field not only significantly affects the dynamics of supernovae and accretion disks but also modifies the interaction of neutrinos with their matter (see, for example, [9–11]).

In this paper, the modifications of the beta-processes taking place in supernovae and accretion disks, by the magnetic field of an arbitrary strength are discussed. The results of calculations of the neutrino- and antineutrino-production rates and the energy-momentum deposition to optically transparent medium are presented. A dependence of these quantities on the magnetic field strength and parameters of the medium is demonstrated. Other details of the processes presented here can be found in the recent paper [12].

In this paper we use the system of units in which $c = \hbar = k = 1$, where c is the velocity of light, $\hbar$ is the Planck constant, and k is the Boltzmann constant.

2 General formalism

The macroscopic quantities, being the most interesting for astrophysical applications, are the reaction rate (the number of processes occurring in a unit volume per unit time):

$$\Gamma = \frac{1}{V} \int \sum_{i,f} \frac{|S_{if}|^2}{\mathcal{T}} \prod_{i,f} f_i \left(1 - f_f\right) dn_i \, dn_f, \tag{4}$$

and the energy and momentum transferred from neutrinos to a unit volume of the medium per unit time:

$$Q^\mu = \left(Q, \vec{\mathcal{F}}\right) = \pm \frac{1}{V} \int \sum_{i,f} q^\mu \frac{|S_{if}|^2}{\mathcal{T}} \prod_{i,f} f_i \left(1 - f_f\right) dn_i \, dn_f, \tag{5}$$

where V is the normalization volume, $|S_{if}|^2/\mathcal{T}$ is the transition probability from the initial (i) state to final (f) one per unit time, $q^\mu = (\omega, \vec{q})$ is the neutrino four-momentum, and the sign "±" in (5) specifies

DOI: 10.1051/epjconf/201715805004

either absorption or emission neutrino process is considered. The summation and integration is carried out over the phase space $dn_{i,f}$ of all particles taking into account their distribution functions $f_{i,f}$ (it is implicitly assumed that all final particles are fermions). It is convenient to separate the integration over the neutrino momentum from all other. One can define the quantity called the emissivity [12]:

$$\mathcal{K}(q) = \sum_{i,f\neq\nu} \int \frac{|S_{if}|^2}{\mathcal{T}} \prod_{i,f\neq\nu} f_i\left(1-f_f\right) dn_i\, dn_f, \tag{6}$$

where the reaction with neutrino in the final state is assumed. It completely determines the property of the matter to emit or absorb the neutrino with the four-momentum q in the transition $i \leftrightarrow f$.

Under conditions of supernovae and accretion disks, the total emissivity for neutrino and antineutrino in beta-processes can be presented in the simple analytical form [12]:

$$\begin{aligned}
\mathcal{K}_\nu(q) &= K_0\, \Phi(b, x_\nu, \beta_\nu)\, \frac{e^{(\delta\chi+\Delta)/t-a}}{e^{x_\nu/t-a}+1}, \\
\mathcal{K}_{\bar\nu}(q) &= K_0\, \Phi(b, x_{\bar\nu}, \beta_{\bar\nu})\, \frac{1}{e^{x_{\bar\nu}/t+a}+1}, \\
K_0 &= \frac{\left(g_v^2+3g_a^2\right)\cos^2\theta_c}{2\pi}\, G_F^2\, N_n\, m_e^2, \\
\Phi(x,b,\beta) &= 2\,x\,\sqrt{x^2-1}\,\theta(x-\sqrt{1+2b}) + \\
&+ \frac{b\,x}{\sqrt{x^2-1}}\left[\theta(x-1)-\theta(x-\sqrt{1+2b})\right] - g_{va}\,\frac{b\,x\cos\beta}{\sqrt{x^2-1}}\,\theta(x-1),
\end{aligned} \tag{7}$$

where m_n, m_p, m_e and μ_n, μ_p, μ_e are the masses and local chemical potentials of neutrons, protons and electrons, $\Delta = (m_n - m_p)/m_e$, $\delta\chi = (\mu_e + \mu_p - \mu_n)/m_e$, T is the local temperature of the medium, $t = T/m_e$, $a = \mu_e/T$, N_n is the neutrons number density, $x_{\nu,\bar\nu} = \omega_{\nu,\bar\nu}/m_e \pm \Delta$ are the scaled neutrino and antineutrino energy, $b = eB/m_e^2 = B/B_0$ is the magnetic field strength measured in units of the Schwinger value $B_0 \approx 4.41 \times 10^{13}$ G, $e > 0$ is the elementary charge, $\beta_{\nu,\bar\nu}$ is the angle between the (anti)neutrino momentum and magnetic field direction. Here, the following constants are used [13]: $g_v \approx 1$, $g_a \approx -1.27$ are the vector and axial constants of the nucleon charged current, $g_{va} = (g_a^2 - g_v^2)/(3g_a^2 + g_v^2) \approx 0.11$, θ_c is the Cabibbo angle, $\cos\theta_c \approx 0.97$, $G_F \approx 1.17 \times 10^{-11}$ MeV^{-2} is the Fermi constant. The dependence on the magnetic-field strength is entering in $\mathcal{K}_\nu(q)$ and $\mathcal{K}_{\bar\nu}(q)$ through the function $\Phi(x,b,\beta)$ only. The first term in this function coincides with the one appeared in the absence of the field and describes the magnetized matter ability to emit and absorb neutrinos with high energies ($x \geqslant \sqrt{1+2b}$). The second term in $\Phi(x,b,\beta)$ dominates in the limit of the strong magnetic field and responsible for the low-energy neutrino spectrum ($x \leqslant \sqrt{1+2b}$). Note that this term is the result of exact calculations [12] of neutrino and antineutrino emissivities for the beta-processes under the supernova and accretion disk conductions. The last term in $\Phi(x,b,\beta)$ is linear in $\cos\beta$ and determines the asymmetric part of the emissivity. It existence leads to an uncompensated momentum transfered to the medium along the magnetic field direction.

3 Medium transparent for neutrino

The case of a medium transparent for neutrinos is the most simple for calculations. However, it is quite applicable in astrophysics. In particular, the matter of accretion disks and an external part of supernova envelopes is transparent for neutrinos and antineutrinos and their distribution functions can be well approached by unit, $f_{\nu,\bar{\nu}} \ll 1$, in (4) and (5). In this case, the neutrino- and antineutrino-production number and the energy emitted from the unit volume of the medium per unit time in the beta-processes can be written as follows:

$$\begin{aligned} \Gamma_{\nu,\bar{\nu}} &= \frac{1}{(2\pi)^3} \int \mathcal{K}_{\nu,\bar{\nu}}(q)\, d^3 q, \\ Q_{\nu,\bar{\nu}} &= \frac{1}{(2\pi)^3} \int \omega\, \mathcal{K}_{\nu,\bar{\nu}}(q)\, d^3 q. \end{aligned} \tag{8}$$

As noted above, the existence of the asymmetric part in the emissivity (the third term in $\Phi(x, b, \beta)$ in Eq. (7)) leads to a macroscopic momentum transferred to the medium. In the case of the transparent for (anti)neutrino matter only the component along the magnetic field direction is non zero:

$$\mathcal{F}_{\nu,\bar{\nu}\,\|} = -\frac{1}{(2\pi)^3} \int \omega\, \cos\beta\, \mathcal{K}_{\nu,\bar{\nu}}(q)\, d^3 q. \tag{9}$$

This quantity becomes quite simple in the case of the ultrarelativistic electron-positron plasma which exists in supernovae and accretion disks. In this limit, the force density along the magnetic field arising from neutrino and antineutrino emission has the form [12]:

$$\begin{aligned} &\mathcal{F}_{\nu\|} = \mathcal{F}_0\, I_3(a)\, Y\,/\,(1+Y), \quad \mathcal{F}_{\bar{\nu}\|} = \mathcal{F}_0\, I_3(-a)\,/\,(1+Y), \\ &\mathcal{F}_0 = \frac{\left(g_a^2 - g_v^2\right)\cos^2\theta_c}{12\pi^3}\, G_F^2\, eB\, T^4 \frac{\rho}{m_N}, \quad I_3(a) = \int\limits_0^\infty \frac{y^3\, dy}{e^{y-a}+1}, \end{aligned} \tag{10}$$

where $Y = N_p/N_n$ is the proton-to-neutron number density ratio, ρ is the density of the matter, $m_N \approx 940$ MeV is the nucleon mass [13]. Here, $\mathcal{F}_0 \approx 1.85 \times 10^{14}\, b\, t^4 \rho_{12}$ dyn/cm^3 gives the typical value for the force density in the nondegenerate electron-positron plasma with $\rho_{12} = \rho/10^{12}$ g/cm^3. In the limit of large values of a, the integral $I_3(a)$ in (10) can be approximated by the following expressions [12]:

$$I_3(a) \approx \frac{a^4}{4} + \frac{\pi^2}{2} a^2 + 3.2\, a + \frac{7\pi^4}{120}, \qquad I_3(-a) \approx 6\, e^{-a}. \tag{11}$$

So, for large values of the electron degeneracy parameter $a = \mu_e/T$ the momentum transferred to the medium increases by the factor $\sim a^4$ for the processes with the neutrino participation and exponentially suppressed for reactions with antineutrino.

In the case of an ultrarelativistic electron-positron plasma, the neutrino- and antineutrino-production rates and the energy emitted in these processes depend not on the magnetic field strength itself but on dimensionless parameter $y_b = \sqrt{2b}/t = \sqrt{2eB}/T$ which contains the plasma temperature [12]:

$$\begin{aligned} &\Gamma_\nu = \Gamma_0\, J_2(y_b, a)\, Y\,/\,(1+Y), \quad Q_\nu = Q_0\, J_3(y_b, a)\, Y\,/\,(1+Y), \\ &\Gamma_{\bar{\nu}} = \Gamma_0\, J_2(y_b, -a)\,/\,(1+Y), \quad Q_{\bar{\nu}} = Q_0\, J_3(y_b, -a)\,/\,(1+Y), \\ &\Gamma_0 = \frac{\left(3g_a^2 + g_v^2\right)\cos^2\theta_c}{8\pi^3}\, G_F^2\, T^5 \frac{\rho}{m_N}, \quad Q_0 = T\,\Gamma_0, \\ &J_n(y_b, a) = y_b^2 \int\limits_0^{y_b} \frac{y^n dy}{\exp(y-a)+1} + 4 \int\limits_{y_b}^\infty \frac{y^{n+2} dy}{\exp(y-a)+1}. \end{aligned} \tag{12}$$

Here, the values $\Gamma_0 \approx 9.64 \times 10^{31}\, t^5 \rho_{12}$ cm^{-3} s^{-1} and $Q_0 \approx 7.88 \times 10^{25}\, t^6 \rho_{12}$ erg cm^{-3} s^{-1} give the typical values for the neutrino- and antineutrino-production rates and the energy flux emitted in the beta-processes under conditions of the nondegenerate electron-positron plasma. Obviously, the function $J_n(y_b, a)$ introduced in (12) is quadratic in y_b in the limit of the strong magnetic field ($y_b \gg a$). So, $\Gamma_{\nu,\bar{\nu}}$ and $Q_{\nu,\bar{\nu}}$ must increase linearly with the growth of the magnetic field in this limit. However, in the case of a weak magnetic field, the rate of increase in the first term in $J_n(y_b, a)$ is lower than the rate of the decrease in the second term. Therefore, with the magnetic field increased, the beta-process rates and the energy emitted in these processes first decrease, come through their minimum values, and only then their linear increase begin as shown in Fig. 1 for the scaled reaction rates, $\Gamma_{\nu,\bar{\nu}}(b)/\Gamma_{\nu,\bar{\nu}}(0)$, and emitted energy, $Q_{\nu,\bar{\nu}}(b)/Q_{\nu,\bar{\nu}}(0)$. It can be seen that the minimum for the antineutrino-production rate

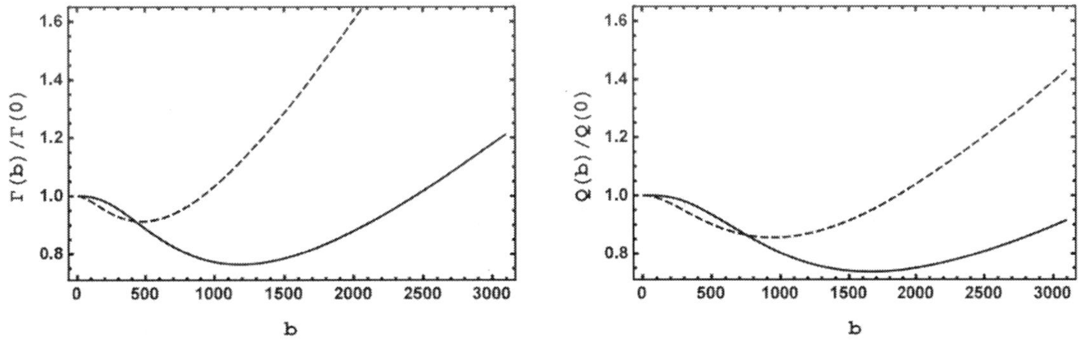

Figure 1. The beta-process rates (the left panel) and energy emitted through (anti)neutrino in these processes (the right panel) are plotted for the dimensionless temperature $t = 8$ and electron degeneracy parameters $a = 3$ as functions of the magnetic field $b = B/B_0$. The processes with the neutrino and antineutrino participation are shown by the solid and dashed curves, respectively.

and energy emitted in such reactions is not as deep as in the processes with neutrinos and is reached at lower values of the magnetic field strength.

Another important fact following from Fig. 1 should also be mentioned. A transition to an asymptotic linear increase for the beta-processes with antineutrinos occurs in lower magnetic fields in comparison to the reactions with neutrino. Nevertheless, the transition to asymptotic behavior even for these processes corresponds to the magnetic field with the strength $B \gtrsim 10^{16}$ G which exceeds substantially the values typical for supernovae and accretion disks. Consequently, the increase in the magnetic field in these objects leads not to the intensification but conversely to the beta-process suppression which is manifested more strongly for the reactions with neutrinos. In addition, the suppression of the energy emitted is stronger than for the rates of the beta-processes.

Let us consider separately the effect of other parameters on the rates of the beta-processes and energy emitted in them. Numerical calculations show [12] that the dependence of these quantities on the temperature of the medium is similar for both reactions which is demonstrated in Fig. 2 for the relative energy $Q_{\nu,\bar{\nu}}(b)/Q_{\nu,\bar{\nu}}(0)$ emitted in neutrino and antineutrino. It can be seen that with the temperature increased, the position of the minimum of the relative energy emitted is shifted towards stronger magnetic fields. Its depth for each reaction separately remains unchanged but the depth of the minimum for the beta-processes with the neutrino participation is larger. It should be noted that the dependence of the reaction rates on the temperature of the medium is analogous to that shown in Fig. 2 but the depth of the relative minimum is slightly smaller for these quantities.

A different situation is observed in the dependence of the rates of the beta-processes and the energy emitted in them on the electron degeneracy parameter a. Numerical analysis has shown [12] that such

 DOI: 10.1051/epjconf/201715805004

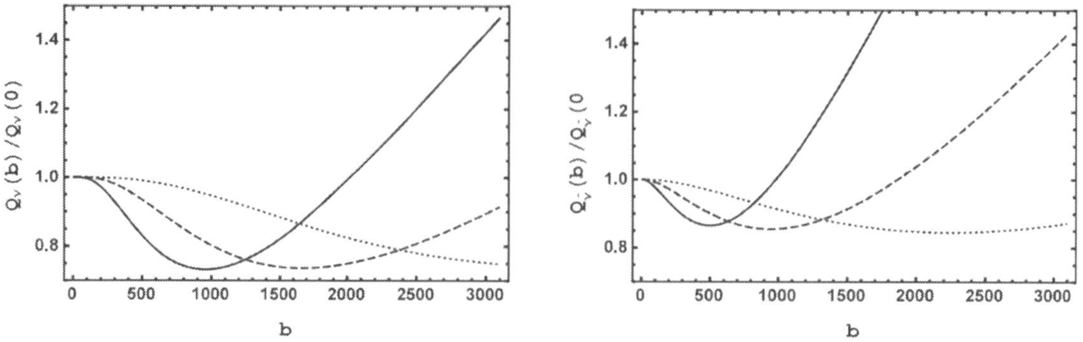

Figure 2. The relative energy emitted in neutrinos (the left panel) and antineutrinos (the right panel), corresponding to the electron degeneracy parameter $a=3$ and scaled temperatures $t=6$ (solid curves), $t=8$ (dashed curves), and $t=12$ (dotted curves) as functions of the magnetic field strength $b=B/B_0$.

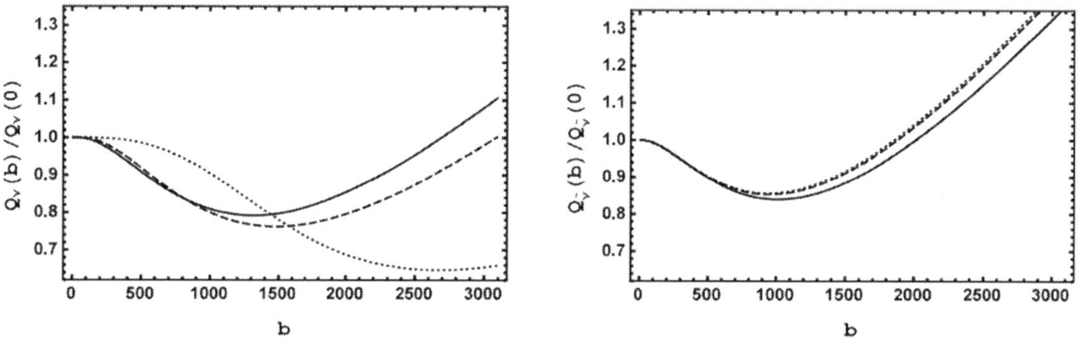

Figure 3. The relative energy emitted in neutrinos (the left panel) and antineutrinos (the right panel), corresponding to the scaled temperature $t = 8$ and the electron degeneracy parameters $a = 0$ (solid curves), $a = 2$ (dashed curves), and $a=6$ (dotted curve) as functions of the magnetic field strength $b=B/B_0$.

dependences for the reactions with neutrino and antineutrino differ substantially as shown in Fig. 3 for the relative energy emitted, $Q_{\nu,\bar{\nu}}(b)/Q_{\nu,\bar{\nu}}(0)$. It can be seen from Fig. 3 that the maximum suppression of the reaction with the antineutrino participation is observed for the nondegenerate plasma ($a = 0$). However, the dependence of the relative energy emitted in these processes on the electron degeneracy parameter a at a fixed temperature is very weak. The relative rate of the antineutrino generation demonstrates analogous behavior. It should be noted that this holds only for the relative quantities while the absolute values considerably depend on this parameter in terms of $\Gamma_{\bar{\nu}}(0)$ and $Q_{\bar{\nu}}(0)$. For the processes with neutrino, the situation is radically different because even relative values of the quantities strongly depend on the electron degeneracy parameter a. As seen in Fig. 3, the increase in this parameter leads to a displacement of the minima in the relative energy emitted in neutrino towards larger values of the field, and the minimum itself becomes deeper. The relative rate of the neutrino production behaves analogously but the depth of the minimum of this quantity is slightly smaller. Thus, the magnetic field suppresses the beta-processes with the neutrino participation the most strongly, and this suppression is the highest in a degenerate medium.

In the case of the strong degeneracy of the electron-positron plasma, the following simple analytic expressions can be obtained for the integrals $J_2(y_b, a_e)$ and $J_3(y_b, a_e)$ [12]:

$$\begin{aligned} J_2(y_b, a_e) &\approx 4/5\, a_e^5 - 7/15\, y_b^5, \quad J_3(y_b, a_e) \approx 2/3\, a_e^6 - 5/12\, y_b^6, \quad y_b \leqslant a_e, \\ J_2(y_b, a_e) &\approx 1/3\, y_b^2 a_e^3, \quad J_3(y_b, a_e) \approx 1/4\, y_b^2 a_e^4, \quad y_b \geqslant a_e. \end{aligned} \tag{13}$$

It can easily be seen that the relative neutrino production rate, $\Gamma_\nu(z_b)/\Gamma_\nu(0)$, and energy emitted, $Q_\nu(z_b)/Q_\nu(0)$, depend only on one variable $z_b = y_b/a_e = \sqrt{2b}\, m_e/\mu_e$, and this dependence for the degenerate plasma is universal as regards μ_e. The transition to such an asymptotic behavior is demonstrated in Fig. 4. The expressions obtained for $J_n(y_b, a_e)$ in the strongly degenerate plasma also show

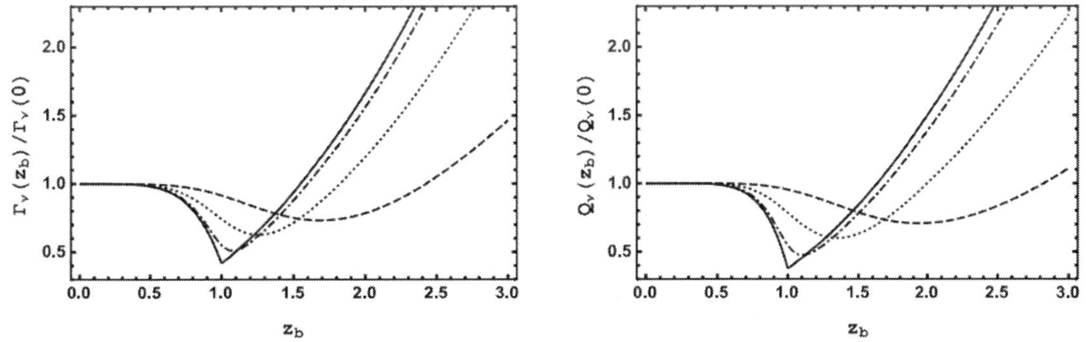

Figure 4. The relative rate of the neutrino production (the left panel) and energy emitted in this process (the right panel), corresponding to the scaled temperature $t = 8$ and degeneracy parameters $a = 4$ (dashed curves), $a = 8$ (dotted curves), and $a = 20$ (dash-dotted curves) as functions of the dimensionless parameter $z_b = \sqrt{2eB}/\mu_e$. Solid curves correspond to the asymptotic form of the degenerate plasma.

that the largest suppression of the beta-processes with neutrino by the magnetic field occurs for the strength $B_\nu \approx \mu_e^2/(2m_e^2)\, B_0$, and the magnitude of the suppression relative to the field-free value is $\Gamma_\nu(B_\nu)/\Gamma_\nu(0) = 5/12 \approx 0.42$ for the neutrino-production rate and $Q_\nu(B_\nu)/Q_\nu(0) = 3/8 \approx 0.38$ for the energy emitted.

For the beta-processes with the antineutrino participation, the convenient variable is the dimensionless parameter $y_b = \sqrt{2b}/t$. Numerical calculations show [12] that the relative rate of the antineutrino production, $\Gamma_{\bar\nu}(y_b)/\Gamma_{\bar\nu}(0)$, and energy, $Q_{\bar\nu}(y_b)/Q_{\bar\nu}(0)$, emitted in such processes as functions of this variable weakly depend on the temperature of the medium and the plasma degeneracy parameter as shown in Fig. (5). The same figure depicts the asymptotic form of the nondegenerate ($a_e = 0$) ultra-relativistic plasma in which, as noted above, the suppression this process by the magnetic field is the strongest. It follows from the results of numerical calculations [12] that the maximum suppression of the rate of the antineutrino production occurs in the magnetic field with the strength $B_{\Gamma_{\bar\nu}} \approx 10\, t^2\, B_0$, and the magnitude of the suppression in this case is $\Gamma_{\bar\nu}(B_{\Gamma_{\bar\nu}})/\Gamma_{\bar\nu}(0) \simeq 0.86$. The maximum suppression of the energy emitted in antineutrinos is shifted towards stronger fields, $B_{Q_{\bar\nu}} \approx 18\, t^2\, B_0$, and the suppression itself is slightly stronger, $Q_{\bar\nu}(B_{Q_{\bar\nu}})/Q_{\bar\nu}(0) \simeq 0.82$. Comparison with analogous quantities obtained for processes with the neutrino participation shows that the maximum suppression of these reactions with the neutrino participation by the magnetic field occurs in the nondegenerate medium, and the suppression itself is approximately half as strong.

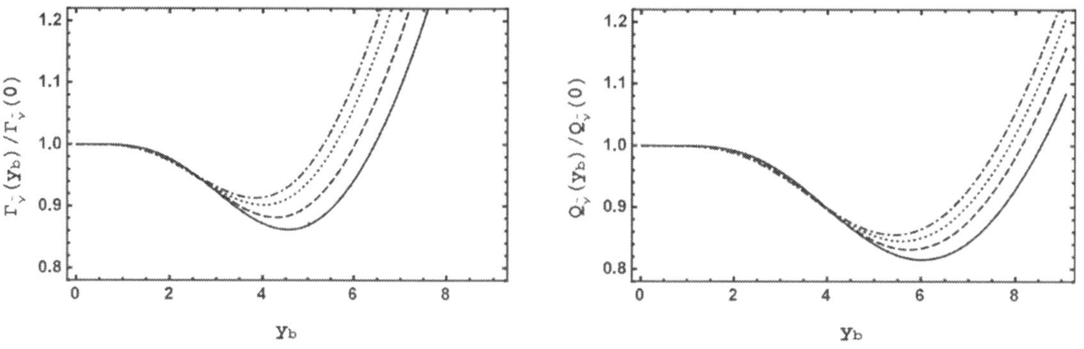

Figure 5. The relative rate of the antineutrino production (the left panel) and energy emitted in this process (the right panel) as functions of the dimensionless parameter $y_b = \sqrt{2b}/t$. Solid curves correspond to the asymptotic form of an ultrarelativistic nondegenerate plasma ($a=0$) in which process (2) has the most strong suppression by the magnetic field. The remaining curves correspond to the following dimensionless parameters: $t=12$ and $a=0$ (dashed curves), $t=12$ and $a=3$ (dotted curves), and $t=8$ and $a=3$ (dash-dotted curves).

4 Conclusions

In this paper, we have analyzed the beta-processes (1)–(3) in the medium of supernovae and accretion disks with an arbitrary magnitude of the magnetic field. The results obtained for the transparent medium were used to calculate the neutrino and antineutrino production rates, the energy emitted in these processes, and the macroscopic momentum transferred to the medium.

The simplest dependence on the magnetic field was obtained for the macroscopic momentum transferred to the matter. It was shown that in the case of an transparent medium, it increases linearly with the magnetic field and is directed along the field-strength lines. It was demonstrated that this effect is most significant in the degenerate electron-positron plasma. Unlike the momentum transferred, the rates of the beta-processes and energy emitted in them exhibit a more complex dependence on the magnetic field. In the limit of the strong magnetic field, these quantities, as well as the momentum transferred, increase linearly with the growth of the magnetic field. It was shown, however, that for the values $B \lesssim 10^{15}$ G which are typical for supernovae and accretion disks, the values of these quantities are suppressed as compared to the field-free case. The maximum suppression of the reactions with the neutrino participation occurs in the degenerate medium in the field with the strength $B_\nu \sim \mu_e^2/(2m_e^2)\, B_0$ which falls into the range of values typical of the objects studied. The suppression can reach 2.5 times as compared to the field-free case. The effect of the magnetic field on the antineutrino processes is weaker and its maximum suppression is 1.2 times as compared to the field-free case. However, such a suppression occurs in the nondegenerate plasma (i. e., in regions of supernovae and accretion disks with a moderate density). The magnetic field corresponding to the maximum suppression of these processes is $B_{\tilde{\nu}} \sim 10\, t^2\, B_0$ which also falls in the range of values typical of the objects considered at moderate temperatures. Thus, the suppression of the processes with the neutrino and antineutrino participation, by the magnetic field occurs in different parts of supernovae and accretion disks, and the effect itself can be significant.

It should be noted that the form of the dependence of the rates of the beta-processes and the energy-momentum transferred in these processes on the magnetic field and parameters of the matter presented here does not change significantly in the more general case of a medium partly transparent to neutrinos and antineutrinos.

I'd like thank to Thomas Janka for numerous discussions on this topic and possible extensions of this research. It is a pleasure to thank G. S. Bisnovatyi-Kogan, Alexander Parkhomenko and Alexandra Dobrynina for stimulating discussions and critical comments. This study was partially supported by the Russian Foundation for Basic Research (project no. 15-02-06033-a) and by the Michael Lomonosov Program (project no. 1.721.2016/2.2).

References

[1] P. Kumar, B. Zhang, Phys. Rept. **561**, 1 (2014), 1410.0679
[2] A.I. MacFadyen, S.E. Woosley, Astrophys. J. **524**, 262 (1999), astro-ph/9810274
[3] S. Rosswog, Int. J. Mod. Phys. **D24**, 1530012 (2015), 1501.02081
[4] M.V. Barkov, S.S. Komissarov, Mon. Not. Roy. Astron. Soc. **401**, 1644 (2010), 0908.0695
[5] S.W. Bruenn, Astrophys. J. Suppl. **58**, 771 (1985)
[6] S.G. Moiseenko, G.S. Bisnovatyi-Kogan, Astron. Rep. **59**, 573 (2015)
[7] H. Sawai, S. Yamada, Astrophys. J. **817**, 153 (2016), 1504.03035
[8] I. Zalamea, A.M. Beloborodov, Mon. Not. Roy. Astron. Soc. **410**, 2302 (2011), 1003.0710
[9] A.A. Gvozdev, I.S. Ognev, JETP Lett. **74**, 298 (2001), [Pisma Zh. Eksp. Teor. Fiz. 74, 330 (2001)], astro-ph/0201346
[10] A.A. Gvozdev, I.S. Ognev, J. Exp. Theor. Phys. **94**, 1043 (2002), [Zh. Eksp. Teor. Fiz. 121, 1219 (2002)], astro-ph/0403011
[11] A.A. Gvozdev, I.S. Ognev, Astron. Lett. **31**, 442 (2005)
[12] I.S. Ognev, J. Exp. Theor. Phys. **123**, 643 (2016), [Zh. Eksp. Teor. Fiz. 150, 744 (2016)]
[13] C. Patrignani et al. (Particle Data Group), Chin. Phys. **C40**, 100001 (2016)

 DOI: 10.1051/epjconf/201715805005

Radiative decay of keV-mass sterile neutrino in magnetized electron plasma

Alexandra Dobrynina[1,⋆], *Nicolay* Mikheev[1], and *Georg* Raffelt[2]

[1] *P. G. Demidov Yaroslavl State University, Sovietskaya 14, 150003 Yaroslavl, Russia*
[2] *Max-Planck-Institut für Physik (Werner-Heisenberg-Institut), Föhringer Ring 6, 80805 München, Germany*

Abstract. The radiative decay of sterile neutrinos with typical masses of 10 keV is investigated in the presence of an external magnetic field and degenerate electron plasma. Full account is taken of the modified photon dispersion relation relative to vacuum. The limiting cases of relativistic and nonrelativistic plasma are analyzed. The decay rate calculated in a strongly magnetized plasma, as a function of the electron number density, is compared with the unmagnetized plasma limit. It is found that the presence of the strong magnetic field in the electron plasma suppresses the catalyzing influence of the plasma by itself on the sterile-neutrino decay rate.

1 Introduction

It is not a secret that stars as laboratories for fundamental physics complement an effort from the existing accelerator facilities on particle physics. The properties and dynamics of astrophysical objects require detailed understanding of quantum processes under an influence of an external active medium. A dense electron-positron and nucleon plasma as well as an external electromagnetic field can play this role both in the interior and envelope of compact astrophysical objects.

An influence of an external magnetic field on properties of leptons and their interactions becomes especially important when the field strength reaches the critical Schwinger value of $B_e = m_e^2/e = 4.41 \times 10^{13}$ Gauss, where m_e is the electron mass and e is the elementary charge. Such fields are not accessible in accelerator experiments but can exist in strongly magnetized neutron stars called "magnetars" [1–3]. In astrophysics, "magnetars" are usually identified with objects like Soft Gamma-ray Repeaters (SGR) and Anomalous X-ray Pulsars (AXP) [4]. Recent simulations of core-collapse supernovae [5–7] and the formation of accretion discs at the merger of compact objects in a close binary system [8, 9] demonstrate that the magnetic field strength could reach the value of $B \sim 10^{15}$ Gauss.

Although all these phenomena are different, they are connected with a powerful neutrino emission. In a supernova explosion, the neutrino radiation source is the protoneutron star that is formed as a result of the core collapse of a massive star at the final stage of its evolution. A merger of compact objects in a close binary system leads to the formation of a central object and the hot accretion disk around it. In this case, the disk is a source of a powerful neutrino radiation. The emission of neutrinos is a basic channel of magnetar cooling.

⋆e-mail: dobrynina@uniyar.ac.ru

The plasma and magnetic field manifest themselves as optically active medium and can change dispersion relations of particles [10, 11]. As a result, the processes which are kinematically forbidden or suppressed in vacuum, could be opened or significantly enhanced in plasma and magnetic field. In particular, they can significantly influence the photon-neutrino interaction that arises in vacuum at the loop level and turns out to be extremely weak. Good example is the effect of neutrino luminosity by a plasma in the $\gamma \to \nu\bar{\nu}$ decay which is observed experimentally.

Studies of the radiative decay of a neutrino discussed in this paper have a long history. This process has been considered by several groups in the different cases: vacuum [12], electron plasma [13–18], magnetic fields of different configurations [19–23] and magnetized plasma [24–26]. The main point of the work we have done [26] was to extend the previous analysis by including the modified photon dispersion relation. The results obtained in [26] are briefly presented in this contribution.

2 Strongly magnetized plasma

In a strongly magnetized plasma the neutrino-photon interaction is mainly determined by electrons occupying the lowest Landau level. We assume that the magnetic field is directed along the third axis, $\vec{B} = (0, 0, B)$, and suppose the following hierarchy $2eB > \mu_e^2 - m_e^2 \gg T^2$ of plasma parameters, where m_e is the electron mass, e is the elementary charge, μ_e is the electron chemical potential, and T is the plasma temperature. In the presence of the strong magnetic field and degenerate plasma the photon is a particle with an effective mass [27]:

$$\Omega_0^2 = \frac{2\alpha}{\pi} \, \frac{eB\, p_F}{\sqrt{p_F^2 + m_e^2}} \, . \tag{1}$$

Here, α is the fine-structure constant and p_F is the electron Fermi momentum. The effective mass Ω_0 is often called the plasma frequency. The electron number density in the strong magnetic field is defined by the expression:

$$n_e = \frac{eB\, p_F}{2\pi^2} \, . \tag{2}$$

This relation allows us to express the plasma frequency in terms of the electron number density and magnetic field as follows [26]:

$$\Omega_0 \simeq 37.1 \text{ keV} \left[\frac{n_{30}^2 b^2}{b^2 + 1.3\, n_{30}^2} \right]^{1/4} , \tag{3}$$

where $b = B/B_e$ and $n_{30} = n_e / \left(10^{30} \text{ cm}^{-3}\right)$. The benchmark number density corresponds to the value of the mass density, where degenerate electrons would still be nonrelativistic. We can see that the typical mass scale of plasma frequency is in the keV region.

Standard neutrinos have sub-eV masses [28] and their radiative decays are forbidden. It is clear that radiative decays would be of interest only for sterile neutrinos with keV masses and above. The sterile neutrinos with keV-masses are very popular as candidates for a dark matter in the Universe [29]. Under minimal assumptions, the mass of DM sterile neutrinos should be in the range of 0.4–50 keV [30–32].

Sterile neutrinos can mix with massive standard ones and in this way interact with a matter. Let us consider one sterile neutrino which is mixing with one active neutrino of a definite flavor. In this case, the neutrino mass states are the mixture of the active and sterile ones:

$$|\nu_1\rangle = \cos\theta_s \, |\nu_a\rangle - \sin\theta_s \, |\nu_s\rangle, \qquad |\nu_2\rangle = \sin\theta_s \, |\nu_a\rangle + \cos\theta_s \, |\nu_s\rangle, \tag{4}$$

where θ_s is the mixing angle. When the mixing angle is negligibly small, the first mass state practically coincides with the active neutrino, $|\nu_1\rangle \simeq |\nu_a\rangle$, and the second mass state is close to sterile neutrino, $|\nu_2\rangle \simeq |\nu_s\rangle$. The smallness of the mixing angle is confirmed by the existing bound [31]:

$$\theta_s^2 \lesssim 1.8 \times 10^{-5} \left(\frac{1\ \text{keV}}{m_s}\right)^5 , \tag{5}$$

which is obtained from analysis of the radiative decay $\nu_s \to \nu_a \gamma$ of sterile neutrino in cosmology. Note that this bound depends substantially on the sterile neutrino mass m_s.

In presence of an electron plasma and magnetic field, the neutrino-photon interaction is induced by real electrons from the magnetized plasma. The neutrino-electron interaction can be described by the effective local Lagrangian:

$$\mathcal{L}_{\text{eff}} = -\frac{G_F}{\sqrt{2}} \left[\bar{\Psi}_e \gamma^\alpha (C_V - C_A \gamma_5) \Psi_e\right] j_\alpha . \tag{6}$$

Here, G_F is the Fermi constant, $C_V = 1/2 + 2\sin^2\theta_W$ and $C_A = 1/2$ are the vector and axial-vector constants of the electron, θ_W is the Weinberg angle, and Ψ_e is the electron quantum field. The neutrino current j_α of our interest is responsible for the transition of a sterile neutrino to an active one [26]:

$$j_\alpha = \cos\theta_s \sin\theta_s \left[\bar{\nu}_a \gamma_\alpha (1 - \gamma_5) \nu_s\right] , \tag{7}$$

where the dependence on the mixing angle is organized in the product $\sin\theta_s \cos\theta_s$. Further, we fix the flavor of the active neutrino to be the electron one. In addition, we assume that this neutrino is massless and neglect the modification of its dispersion relation by the magnetized plasma.

In the strong magnetic field limit, we may transform the axial-vector current of electrons in the Lagrangian (6) to the vector current [26]:

$$\bar{\Psi}_e \gamma_\alpha \gamma_5 \Psi_e = \bar{\Psi}_e \Pi_- \gamma_\alpha \gamma_5 \Pi_- \Psi_e = \bar{\Psi}_e (\tilde{\varphi}\gamma)_\alpha \Psi_e , \tag{8}$$

where $\Pi_- = (1 - i\gamma^1\gamma^2)/2$ is the projection operator [10, 11], $\varphi_{\alpha\beta} = F_{\alpha\beta}/B$ and $\tilde{\varphi}_{\alpha\beta} = \varepsilon_{\alpha\beta\rho\sigma}\varphi^{\rho\sigma}/2$ are dimensionless tensor of the external magnetic field and its dual, and $(\tilde{\varphi}\gamma)_\alpha = \tilde{\varphi}_{\alpha\beta}\gamma^\beta$. Taking into account Eq. (8), the effective Lagrangian (6) becomes:

$$\mathcal{L}_{\text{eff}} = e\left(\bar{\Psi}_e \gamma^\alpha \Psi_e\right) V_\alpha , \tag{9}$$

where the following local operator (the effective neutrino current) is introduced [26]:

$$V_\alpha = -\frac{G_F}{e\sqrt{2}} \left[C_V (\tilde{\varphi}\tilde{\varphi} j)_\alpha - C_A (\tilde{\varphi} j)_\alpha\right] . \tag{10}$$

The effective Lagrangian (9) has the same form as the usual Lagrangian of the electromagnetic interaction of electrons (the QED Lagrangian) [33, 34]:

$$\mathcal{L}_{\text{EM}} = e\left(\bar{\Psi}_e \gamma^\alpha \Psi_e\right) A_\alpha . \tag{11}$$

Therefore, the decay amplitude of the sterile neutrino in the magnetized plasma is described by two Feynman diagrams shown in Fig. 1. These diagrams are identical to the ones presented in Fig. 2, after one of the photon lines is replaced by the neutrino current V_α (10). Note that the diagrams in Fig. 2 determine the amplitude of the Compton scattering of photons on plasma electrons.

DOI: 10.1051/epjconf/201715805005

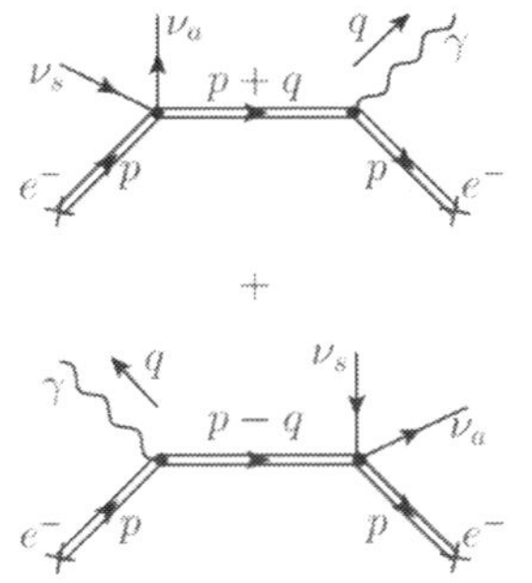

Figure 1. The Feynman graphs of the $\nu_s \to \nu_a + \gamma$ decay in a strongly magnetized electron plasma. The crosses at the ends of the electron lines specify that electrons pertain to plasma. Double lines indicate that the effect of an external magnetic field on the electron fields is taken into account exactly.

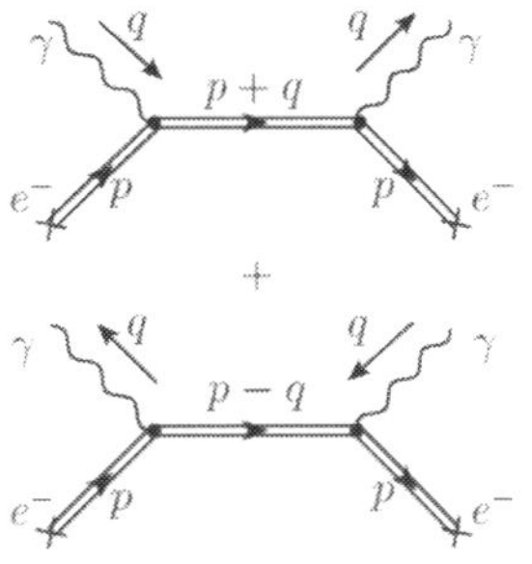

Figure 2. The diagrams of the photon forward scattering on electrons of the strongly magnetized plasma. The crosses at the ends of the electron lines are shown to indicate that electrons pertain to plasma. Double lines specify that the influence of an external magnetic field on electron fields is taken into account exactly.

It is well known [10, 11, 35] that the amplitude of the Compton scattering of photons on plasma electrons, $\gamma \to \gamma$, shown in Fig. 2 contributes to the photon polarization operator $\Pi_{\alpha\beta}$:

$$M_{\gamma\to\gamma} = -\mathcal{E}^*_\alpha \, \Pi^{\alpha\beta} \, \mathcal{E}_\beta \,. \tag{12}$$

In strongly magnetized plasma, the correspondence of the diagrams presented in Figs. 1 and 2 allows to express the sterile-neutrino decay amplitude in terms of the photon polarization operator $\Pi_{\alpha\beta}$:

$$M_{\text{pl+f}} = -\mathcal{E}^*_\alpha \, \Pi^{\alpha\beta} \, V_\beta \,. \tag{13}$$

In such plasma, there are only two physical photon states with the polarization vectors [36]:

$$\mathcal{E}^{(1)}_\alpha \approx \frac{(q\varphi)_\alpha}{\sqrt{q^2_\perp}}\,, \qquad \mathcal{E}^{(2)}_\alpha \approx \frac{(q\tilde{\varphi})_\alpha}{\sqrt{q^2_\parallel}}\,, \tag{14}$$

where $q^2_\parallel = (q\tilde{\varphi}\tilde{\varphi}q)$ and $q^2_\perp = (q\varphi\varphi q)$, which practically coincide with the eigenvectors of the photon polarization operator in the pure magnetic field [10, 11]. It is also possible to determine the set of eigenvalues Π_λ ($\lambda = 1, 2$) from the equation:

$$\Pi_{\alpha\beta} \, \mathcal{E}^{(\lambda)}_\beta = \Pi_\lambda \, \mathcal{E}^{(\lambda)}_\alpha. \tag{15}$$

In the strong magnetic field limit, the eigenvalues Π_1 and Π_2 under the kinematical conditions $\omega \lesssim m_s \ll m_e$ can be written in the form [36]:

$$\Pi_1 \approx -\frac{2\alpha}{\pi} \frac{\omega V_F \mu_e \sqrt{q^2}}{\sqrt{q^2_\parallel}}\,, \quad \Pi_2 \approx \frac{2eB\alpha}{\pi} \frac{q^2_\parallel V_F}{\omega^2 - V_F^2 k_3^2}, \tag{16}$$

where $\alpha = e^2/(4\pi)$ is the fine-structure constant, μ_e is the electron chemical potential, and V_F is the Fermi velocity. From the comparison of these eigenvalues with each other under the conditions on the plasma parameters, $2eB > \mu_e^2 - m_e^2 \gg T^2$, one can easily see that the contribution of Π_2 to the decay amplitude dominates [26].

With account of the discussion presented above, the sterile-neutrino decay amplitude in the conditions of the strongly magnetized plasma can be written as follows [26]:

$$M_{\text{pl+f}} = \frac{G_F \Omega_0^2}{e\sqrt{2}} \sqrt{q_{\parallel}^2} \, \frac{C_V (q\tilde{\varphi} j) + C_A (q\tilde{\varphi}\tilde{\varphi} j)}{\omega^2 - V_F^2 k_3^2}, \tag{17}$$

where j^μ is the neutrino current in the momentum space. The decay width is convenient to calculate in the rest frame of the sterile neutrino, $p_s^\mu = (m_s, \vec{0})$. As all the particles participating in the decay are electrically neutral, the standard procedure — an integration over the phase space of the final-state particles, including their appropriate dispersion properties — can be used for the width evaluation:

$$W_{\text{pl+f}} = \frac{1}{32\pi^2 m_s} \int \frac{d^3 p_a}{E_a} \frac{d^3 k}{\omega} \delta^{(4)}(p_s - p_a - q) \left| M_{\text{pl+f}} \right|^2, \tag{18}$$

where $p_a^\mu = (E_a, \vec{p}_a)$ and $q^\mu = (\omega, \vec{k})$ are the four-momenta of the active neutrino and photon, respectively. In addition, the active neutrino is assumed to be massless, $E_a = |\vec{p}_a|$.

As it was mentioned earlier, the presence of the strongly magnetized plasma is changing the photon dispersion properties. To get the modified dispersion for the photon of the second mode $\mathcal{E}_\alpha^{(2)}$ (14) one needs to resolve the following dispersion relation:

$$q^2 = \Pi_2, \tag{19}$$

where the explicit value of Π_2 is presented in Eq. (16), or the equivalent relation written in terms of the photon energy and momentum components:

$$\omega^2 = k_3^2 + k_\perp^2 + \Omega_0^2 \, \frac{\omega^2 - k_3^2}{\omega^2 - V_F^2 k_3^2} \,. \tag{20}$$

Because the photon dispersion is non-trivial, the integrand in the decay width (18) is complicated and becomes a relatively simple function only in limiting cases, for example, in the nonrelativistic and relativistic limits where the integral in the r.h.s. of (18) can be taken analytically.

We adopt the following parameter ranges $m_s = 2 - 20$ keV and $B = 1 - 100\,B_e$ in numerical analysis. In the nonrelativistic limit the decay width can be presented in the form [26]:

$$\begin{aligned} W_{\text{pl+f}}^{\text{n-rel}} = W_{\text{vac}} \, \frac{32\,\pi^2}{2835\,\alpha^2} \left(C_V^2 + C_A^2\right) \Big[& \theta(1 - 2x_0)\, 4x_0^4 \left(21 + 6x_0^2 - 8x_0^4\right) \\ & + \theta(2x_0 - 1) \left(-\frac{11}{x_0} + 129x_0 - 210x_0^2 + 168x_0^3 - 84x_0^4 - 24x_0^6 + 32x_0^8 \right) \Big], \end{aligned} \tag{21}$$

where $x_0 = \Omega_0/m_s$ is the dimensionless plasma frequency and, as the normalization scale, the vacuum rate of the sterile neutrino is accepted:

$$W_{\text{vac}} = \frac{9\alpha G_F^2}{2048\pi^4} \, m_s^5 \sin^2(2\theta_s) \,. \tag{22}$$

Note that at $C_V = C_A = 1/2$ and in the limit $x_0 \ll 1$ we reproduce the result of Ref. [25].

Under the conditions of relativistic strongly-magnetized plasma, the plasma frequency can be written as [26]:

$$\Omega_0 \simeq 34.7 \text{ keV} \sqrt{B/B_e} \,. \tag{23}$$

One can relatively easy obtain the analytical results for the sterile-neutrino decay width in two limiting cases [26]: the plasma frequency is small ($x_0 \ll m_e/\mu_e$):

$$W^{\rm rel}_{\rm pl+f} \simeq \frac{(G_F\Omega_0^2)^2}{64\pi^2\alpha}\, m_s \sin^2(2\theta_s)\left(C_V^2 + C_A^2\right)\frac{\ln(2\mu_e/m_e) - 5/4}{1 - e^{-m_s/(2T)}}\,, \tag{24}$$

and in the opposite limit — of large x_0 ($m_e/\mu_e \ll x_0 < 1$):

$$W^{\rm rel}_{\rm pl+f} \simeq \frac{(G_F m_s^2)^2}{64\pi^2\alpha}\, m_s \sin^2(2\theta_s)\frac{\left(C_V^2 + C_A^2\right)x_0^4}{1 - e^{-m_s(1+x_0^2)/(2T)}}\left[\left(1 + x_0^2\right)\ln\frac{1}{x_0} - \frac{1}{8}\left(1 - x_0^2\right)\left(3 + x_0^2\right)\right]. \tag{25}$$

It is interesting to compare the decay width in the strongly magnetized electron plasma with a width in the unmagnetized plasma, which is presented in the next section.

3 Unmagnetized electron plasma

In the unmagnetized case, the neutrino-electron interaction is defined by the same effective Lagrangian (6). The vector current of electrons in this Lagrangian is similar to the electromagnetic one in QED Lagrangian (11). Let us also apply the procedure explained in Sec. 2 that was successful in the case of the strongly magnetized plasma. After the replacement of the photon polarization vector by the neutrino current, one can express the vector part of the sterile-neutrino decay amplitude through the photon polarization operator [26]:

$$M^V_{\rm pl} = \frac{C_V\, G_F}{e\sqrt{2}}\,(j\Pi\varepsilon^*)\,. \tag{26}$$

The corresponding axial-vector contribution to the decay amplitude is much smaller [26]. In a nonrelativistic plasma, one finds explicitly that it is suppressed by the factor $(C_A/C_V)(m_s/m_e) \ll 1$. In a relativistic plasma, the electron mass should be replaced by the chemical potential μ_e and the condition above remains correct again.

As mentioned earlier, photons in plasma have an effective mass called the plasma frequency. In the nonrelativistic limit, the plasma frequency can be expressed in terms of an electron number density [35]:

$$\omega_0^2 = \frac{4\pi\alpha n_e}{m_e}\,, \tag{27}$$

where the number density of degenerate electrons is defined as $n_e = p_F^3/(3\pi^2)$. Photons in plasma have three polarization states, one longitudinal ε^ℓ_α and two transverse ε^t_α, where $t = 1, 2$. Eigenvalues of the polarization operator for longitudinal and transverse modes are different. For a nonrelativistic plasma, the eigenvalues are well known [35]:

$$\Pi_t \approx \omega_0^2\,, \qquad \Pi_\ell \approx \omega_0^2\left(1 - \frac{k^2}{\omega^2}\right). \tag{28}$$

These eigenvalues determine dispersion relations of photon modes through the equation:

$$q^2 = \omega^2 - k^2 = \Pi_\lambda\,, \tag{29}$$

which should be included in the integration over the photon momentum.

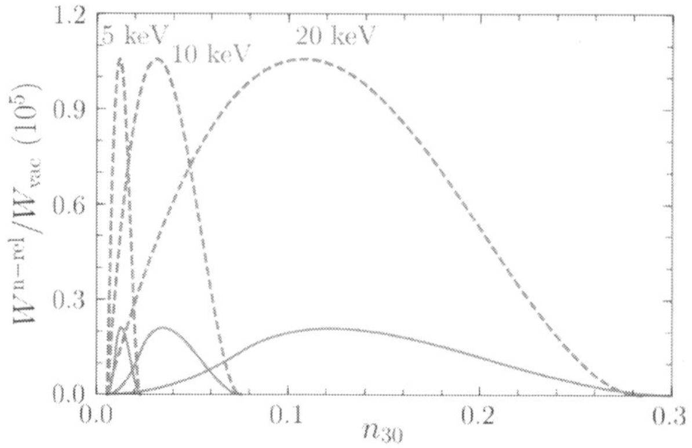

Figure 3. The decay probability $W^{\text{n-rel}}$ (in units of $10^5\ W_{\text{vac}}$) in the strongly magnetized (solid lines) and unmagnetized (dashed lines) non-relativistic plasma as a function of the electron number density $n_{30} = n_e/(10^{30}\ \text{cm}^{-3})$. The solid curves correspond to the magnetic field strength $B = B_e = 4.41\times10^{13}$ Gauss. Three values of the sterile-neutrino mass, $m_s = 5,\ 10,\ 20$ keV, (explicitly specified) are used to demonstrate the corresponding dependence of the decay width.

The decay width for longitudinal and transverse modes are known analytically [26]:

$$W^t_{\text{pl}} = W_{\text{vac}}\,\frac{16\pi^2}{9\alpha^2}\,C_V^2\,x_0^4(1-x_0^2)^2\,, \qquad W^\ell_{\text{pl}} = W_{\text{vac}}\,\frac{32\pi^2}{9\alpha^2}\,C_V^2\,x_0(1-x_0)^2\,, \tag{30}$$

where $x_0 = \omega_0/m_s$ is the scaled plasma frequency. The decay width W^t_{pl} and W^ℓ_{pl} are expressed in terms of the vacuum rate (22). The lifetime of the sterile neutrino with the mass of 10 keV is equal to 1.2×10^{26} years under the assumption that the mixing angle squared is $\theta_s^2 \simeq 10^{-11}$. Comparing this result with the age of the our Universe, $\tau_{\text{Univ}} = 1.37\times10^{10}$ years [28], we see that the sterile neutrino is stable to this decay even if we take into account the influence of the electron plasma or external magnetic field.

In Fig. 3 we compare the decay rate in the unmagnetized (dashed lines) and strongly magnetized (solid lines) plasma as a function of the electron number density for several fixed sterile-neutrino mass values. The strong catalyzing effect of the plasma is clearly seen with an enhancement of up to 5 orders of magnitude compared with vacuum. When the neutrino mass increases, the distribution over the electron number density becomes wider and maximum of this function moves to larger value of n_e. For the chosen field strength, $B = B_e = 4.41\times10^{13}$ Gauss, the decay rate is strongly suppressed as compared with the case of unmagnetized plasma, but it is still much larger than in vacuum. It is clear from this figure that the influence of plasma plays an important role in the relatively narrow range of plasma parameters, in particular, the electron number density. This could be explained by the fact that we have two opposite effects of the plasma influence on the decay width of sterile neutrino. On one hand, when the value of n_e increases, the amplitude and width of the decay in both unmagnetized and strongly magnetized plasma grow up. On the other hand, when the value of n_e increases, the plasma frequency grows up and can reach the sterile neutrino mass or becomes even larger. In this case the radiative decay of sterile neutrino will be kinematically suppressed or even forbidden.

4 Conclusions

We study the radiative decay of sterile neutrinos with keV-masses in presence of a strongly magnetized and unmagnetized electron plasma. The modified photon dispersion relation is included in the analysis. The limiting cases of relativistic and nonrelativistic magnetized plasma are considered. The decay rate in a strongly magnetized plasma as a function of the electron number density is compared with the one in the unmagnetized plasma. We found that the strong magnetic field suppresses by the order of magnitude the catalyzing influence of the plasma by itself on the decay rate.

Acknowledgments

A.D. acknowledges financial support by the *Russian Foundation for Basic Research* (Project No. 16-32-00066 мол-а and 15-02-06033-a), and by the *Dynasty Foundation*. Partial support by the Deutsche Forschungsgemeinschaft (DFG) under Grant No. EXC-153 (Cluster of Excellence "Origin and Structure of the Universe") and by the European Union under Grant No. PITN-GA-2011-289442 (FP7 Initial Training Network "Invisibles") is acknowledged.

References

[1] R.C. Duncan, C. Thompson, Astrophys. J. **392**, L9 (1992)
[2] C. Thompson, R.C. Duncan, Mon. Not. Roy. Astron. Soc. **275**, 255 (1995)
[3] C. Thompson, R.C. Duncan, Astrophys. J. **473**, 322 (1996)
[4] S.A. Olausen, V.M. Kaspi, Astrophys. J. Suppl. **212**, 6 (2014), 1309.4167
[5] A. Heger, S.E. Woosley, H.C. Spruit, Astrophys. J. **626**, 350 (2005), astro-ph/0409422
[6] H. Sawai, S. Yamada, Astrophys. J. **817**, 153 (2016), 1504.03035
[7] G.S. Bisnovatyi-Kogan, S.G. Moiseenko, N.V. Ardeljan, Acta Polytech. Proc. **1**, 181 (2014), 1408.2395
[8] M.V. Barkov, S.S. Komissarov, Mon. Not. Roy. Astron. Soc. **401**, 1644 (2010), 0908.0695
[9] I. Zalamea, A.M. Beloborodov, Mon. Not. Roy. Astron. Soc. **410**, 2302 (2011), 1003.0710
[10] A.V. Kuznetsov, N.V. Mikheev, *Electroweak Processes in External Electromagnetic Fields*, Vol. 197 of *Springer Tracts in Modern Physics* (Springer-Verlag, New York, 2004)
[11] A.V. Kuznetsov, N.V. Mikheev, *Electroweak Processes in External Active Media*, Vol. 252 of *Springer Tracts in Modern Physics* (Springer-Verlag, New York, 2013)
[12] P.B. Pal, L. Wolfenstein, Phys. Rev. **D25**, 766 (1982)
[13] V.N. Tsytovich, Sov. Phys. JETP **18**, 816 (1964)
[14] V.N. Oraevsky, V.B. Semikoz, Y.A. Smorodinsky, JETP Lett. **43**, 709 (1986)
[15] J.C. D'Olivo, J.F. Nieves, P.B. Pal, Phys. Rev. Lett. **64**, 1088 (1990)
[16] R.F. Sawyer, Phys. Rev. **D46**, 1180 (1992)
[17] J.C. D'Olivo, J.F. Nieves, P.B. Pal, Phys. Lett. **B365**, 178 (1996), hep-ph/9509415
[18] S.J. Hardy, D.B. Melrose, Phys. Rev. **D54**, 6491 (1996)
[19] D.V. Gal'tsov, N.S. Nikitina, Sov. Phys. JETP **35**, 1047 (1972)
[20] V.V. Skobelev, Sov. Phys. JETP **44**, 660 (1976)
[21] A.N. Ioannisian, G.G. Raffelt, Phys. Rev. **D55**, 7038 (1997), hep-ph/9612285
[22] A.A. Gvozdev, N.V. Mikheev, L.A. Vasilevskaya, Phys. Rev. **D54**, 5674 (1996), hep-ph/9610219

[23] A.A. Gvozdev, N.V. Mikheev, L.A. Vasilevskaya, Phys. Lett. **B410**, 211 (1997), hep-ph/9702285
[24] M.V. Chistyakov, N.V. Mikheev, Phys. Lett. **B467**, 232 (1999), hep-ph/9907345
[25] A.I. Ternov, P.A. Eminov, Phys. Rev. **D87**, 113001 (2013)
[26] A.A. Dobrynina, N.V. Mikheev, G.G. Raffelt, Phys. Rev. **D90**, 113015 (2014), 1410.7915
[27] H. Perez Rojas, A.E. Shabad, Annals Phys. **121**, 432 (1979)
[28] C. Patrignani et al. (Particle Data Group), Chin. Phys. **C40**, 100001 (2016)
[29] M. Drewes et al., JCAP **1701**, 025 (2017), 1602.04816
[30] S. Tremaine, J.E. Gunn, Phys. Rev. Lett. **42**, 407 (1979)
[31] A. Boyarsky, O. Ruchayskiy, M. Shaposhnikov, Ann. Rev. Nucl. Part. Sci. **59**, 191 (2009), 0901.0011
[32] M. Laine, M. Shaposhnikov, JCAP **0806**, 031 (2008), 0804.4543
[33] V.B. Berestetsky, E.M. Lifshitz, L.P. Pitaevsky, *Course of Theoretical Physics. Vol. 4. Quantum Electrodynamics* (Pergamon, Oxford, 1982)
[34] M.E. Peskin, D.V. Schroeder, *An Introduction to Quantum Field Theory* (Addison-Wesley, Reading, Massachusetts, 1995)
[35] G.G. Raffelt, *Stars as Laboratories for Fundamental Physics* (University of Chicago Press, Chicago & London, 1996)
[36] M.V. Chistyakov, D.A. Rumyantsev, J. Exp. Theor. Phys. **107**, 533 (2008), 0811.4526

EPJ Web of Conferences **158**, 05006 (2017) DOI: 10.1051/epjconf/201715805006
QFTHEP 2017

Sterile neutrino dark matter production

Dmitry Gorbunov[1,⋆]

[1] *Institute for Nuclear Research of the Russian Academy of Sciences, 60th October Anniversary prospect 7a, Moscow 117312, Russia*

Abstract. Sterile neutrinos provide active neutrinos with masses and mixing, and hence is one of the well-motivated candidate for dark matter. We discuss the sterile neutrino production mechanisms operating in the early Universe and show that additional scalar coupled to sterile neutrino can significantly change the situation, making moderate sterile-neutrino mixing and small sterile neutrino masses consistent with current cosmological and astrophysical bounds. Further searches for a narrow line in galactic X-rays and even direct searches for keV-scale sterile neutrinos in particle physics experiments can probe the suggested setup.

1 Introduction

Charged fermion sector of the Standard Model of particle physics (SM) exhibits perfect symmetry between left-handed and right-handed particles. The rest of the fermions, neutrinos, drastically violate this symmetry. All three neutrinos, electron ν_e, muon ν_μ and τ-neutrino ν_τ, are left-handed fermions, while their antiparticles, $\bar{\nu}_e$, $\bar{\nu}_\mu$ and $\bar{\nu}_\tau$, are right-handed. The absence of right-handed neutrino counterparts is inherent characteristic of the SM, where neutrinos are the only massless fermions. Thus, neutrino sector in the SM is exceptional from the theoretical point of view.

In the SM the neutrino sector is exceptional also from the phenomenological point of view. Indeed, this is the only place where the established theory, SM, fails to explain results of particle physics experiments. The phenomenon is called *neutrino oscillations* and consists in changing the neutrino type, called *flavor*, while the particles propagate between the source (neutrino emitter) and detector (neutrino absorber).

One can describe the oscillations assuming that neutrinos are actually massive particles, with masses m_i, $i = 1, 2, 3$, and the flavor basis, $|\nu_\alpha\rangle$, $\alpha = e,\ \mu,\ \tau$, and mass basis, $|\nu_i\rangle$, are not aligned. The two bases are related through the unitary 3×3 mixing matrix, dubbed Pontecorvo–Maki–Nakagawa–Sakata (PMNS) matrix $\hat{U}$ as follows,

$$|\nu_i\rangle = U_{\alpha i}|\nu_\alpha\rangle \ .$$

Then, in the flavor basis the elements of neutrino mass matrix $\hat{M}$ are defined as

$$M_{\alpha\beta} = \langle\nu_\alpha|\hat{M}|\nu_\beta\rangle = (\hat{U}\hat{M}^{(m)}U^\dagger)_{\alpha\beta} \ , \quad \text{where} \quad M^{(m)}_{ij} = m_i\delta_{ij} \ ,$$

that is how the neutrino mass matrix looks in the mass basis: it is diagonal by definition of the basis.

⋆e-mail: gorby@ms2.inr.ac.ru

In its rest-frame a neutrino mass eigenstate evolves in time (from zero to t) as any other mass state does,

$$|\nu_j(t)\rangle = \mathrm{e}^{-im_jt}|\nu_j(0)\rangle\,.$$

Likewise, in the laboratory frame, neutrino evolves in time and propagates in space (covering the distance L) according to

$$|\nu_j(t,L)\rangle = \mathrm{e}^{-i(E_jt-p_jL)}|\nu_j(0)\rangle\,,$$

where E and p are energy and 3-momentum, respectively. Remarkably, in all experiments with neutrinos, they are produced being ultrarelativistic, so one can expand $p = \sqrt{E^2 - m_j^2} = E - m_j^2/2E + \ldots$ and hence obtain for the neutrino mass eigenstate at distance L from the source

$$|\nu_j(L)\rangle = \mathrm{e}^{-i\frac{m_j^2}{2E}L}|\nu_j(0)\rangle\,.$$

One concludes, that the quantity

$$H_{eff} = \frac{M^2}{2E} \tag{1}$$

plays the role of effective Hamiltonian for the neutrino mass state.

While neutrino propagates in the mass state, it is produced and absorbed in the flavor state: only this state is observable. The neutrino emerges in a weak processes as a particular flavor state, then recasts into the mass eigenstates which propagate for some distance and must be recast back into the flavor eigenstates to describe the neutrino detection by weak interactions. This way neutrino changes its flavor. The transition amplitude of neutrino ν_α to neutrino ν_β is given by

$$A(\alpha\to\beta) = \sum_j \langle\nu_\beta|\nu_j(L)\rangle\langle\nu_j(0)|\nu_\alpha\rangle = \sum_j \langle\nu_\beta|\nu_j\rangle\,\mathrm{e}^{-i\frac{m_j^2}{2E}L}\langle\nu_j|\nu_\alpha\rangle = \sum_j U_{\beta j}\,\mathrm{e}^{-i\frac{m_j^2}{2E}L}U^*_{\alpha j}\,.$$

Consequently, the transition probability reads

$$\begin{aligned} P(\nu_\alpha\to\nu_\beta) &= |A(\alpha\to\beta)|^2 \\ &= \delta_{\alpha\beta} - 4\sum_{j>i}\mathrm{Re}[U^*_{\alpha j}U_{\beta j}U_{\alpha i}U^*_{\beta i}]\sin^2\left(\frac{\Delta m^2_{ji}}{4E}L\right) + 2\sum_{j>i}\mathrm{Im}[U^*_{\alpha j}U_{\beta j}U_{\alpha i}U^*_{\beta i}]\sin\left(\frac{\Delta m^2_{ji}}{2E}L\right)\,, \end{aligned} \tag{2}$$

where we introduced the squared mass differences $\Delta m^2_{ji} \equiv m_j^2 - m_i^2$. They, but not the masses themselves, determine the transition.

In many cases the transition involves not all the three but only two states. Then equation (2) gets simplified. In $2 \leftrightarrow 2$ approximation to neutrino oscillations the *transition probability* (probability to change the neutrino flavor) becomes

$$P(\nu_\alpha\to\nu_{\beta\neq\alpha}) = \sin^2 2\theta\cdot\sin^2\left(\frac{\Delta m^2}{4E}L\right)\,,$$

while the *survival probability* reduces to

$$P(\nu_\alpha\to\nu_\alpha) = 1 - \sin^2 2\theta\cdot\sin^2\left(\frac{\Delta m^2}{4E}L\right)\,.$$

DOI: 10.1051/epjconf/201715805006

The typical distance characterizing the flavor change is *oscillation length*,

$$L_{osc} = \frac{4\pi E}{\Delta m^2} = (2.5\text{ km}) \cdot \frac{E}{\text{GeV}} \frac{\text{eV}^2}{\Delta m^2} \,.$$

There is also so-called *matter effect*, which may distort the introduced above *vacuum oscillations*, when neutrino propagates in a medium, filled with particles participating in the weak interactions. The classical example is electron neutrino produced by nuclei fusion in the Sun. These *solar neutrinos* propagate in the medium highly populated by electrons. The Fermi interaction

$$\mathcal{L} = -2\sqrt{2} G_F \bar{\nu}_e \gamma^\mu e \cdot \bar{e} \gamma_\mu \nu_e \,,$$

without macroscopic currents, $\langle\langle \bar{e}_k \gamma^i_{kl} e_l \rangle\rangle = 0$, but with high density of electrons, $\langle\langle \bar{e}_k \gamma^0_{kl} e_l \rangle\rangle = \langle\langle e^\dagger e \rangle\rangle = n_e$, yields after averaging $\langle\langle e_k \bar{e}_l \rangle\rangle = -\frac{1}{4}\gamma^0_{kl} \cdot n_e$ the neutrino-medium interaction term,

$$\mathcal{L}_{eff} = -\sqrt{2} G_F n_e \bar{\nu}_e \gamma^0 \nu_e \,,$$

It gives the effective Hamiltonian in neutrino sector,

$$V = \sqrt{2} G_F n_e \tag{3}$$

to compete with the vacuum Hamiltonian (1). At sufficiently small energies the latter dominates, which gives rise to the vacuum oscillations, while the weak interactions are feeble. At larger energies the Hamiltonian (1) decreases, so that (3) governs the neutrino propagation, because the weak interaction grows with energy. In fact, electron neutrinos of high energies do not oscillate inside the Sun: the weak interactions are sufficiently fast to prevent this processes by collapsing the neutrino wave function back into the flavor basis. The process is known as Mikheev–Smirnov–Wolfenstein effect. The corresponding change in electron neutrino survival probability with energy has been observed by experiment BOREXINO [1].

To explain the neutrino oscillations, one must make neutrino massive, which implies departure from the SM. So far neutrino oscillations give the only *direct evidence* for the need of the physics beyond the SM.

2 Sterile neutrinos

One of the possible ways to explain the neutrino oscillations within the renormalizable extension of the SM is introducing *sterile neutrinos*. These are fermions singlet with respect to the SM gauge group (and hence *sterile*), which mix with the SM (or *active*) neutrinos. The mixing results in active neutrino masses and mixing responsible for the neutrino oscillation phenomena. Sterile neutrinos form right-handed neutrino components making the SM matter structure fully symmetric with respect to left-right transformations.

With sterile neutrinos one can explain the smallness of the active neutrino masses: why the neutrino scale is much lower than scales of all other electrically charged fermions of the SM, which masses range from 0.5 MeV for electron to 173 GeV for top-quark. The corresponding scheme is called seesaw type I mechanism [2–6] and works as follows.

Consider one sterile neutrino N and one neutrino flavor, let it be electron. To the SM lagrangian one adds the new terms

$$\mathcal{L}_N = \overline{N} i \partial\!\!\!/ N - f \overline{L}^c_e \tilde{H} N - \frac{M_N}{2} \overline{N}^c N + \text{h.c.}, \tag{4}$$

and we assume that the sterile neutrino mass $M_N \gg 1$ eV, which is much bigger than the active neutrino masses, constrained from above by particle physics experiments and cosmological observations. Sterile neutrino mass is of Majorana type, so we introduce only two degrees of freedom. When the SM Higgs doublet H acquires non-zero vacuum expectation value $v/\sqrt{2}$, the Yukawa term in (4), which includes SM lepton doublet L_e with left electron and neutrino and $\tilde{H}_i \equiv \epsilon_{ij} H_j^*$ with ϵ_{ij}, $i, j = 1, 2$ being fully antisymmetric 2×2 matrix, yields mass mixing in the neutral fermion sector,

$$\mathcal{V}_N = \frac{1}{2}\left(\bar{\nu}_e\,,\,\overline{N}^c\right)\begin{pmatrix} 0 & v\frac{f}{\sqrt{2}} \\ v\frac{f}{\sqrt{2}} & M_N \end{pmatrix}\begin{pmatrix} \nu_e \\ N \end{pmatrix} + \text{h.c.}$$

For the chosen hierarchy $M_N \gg M^D = v\frac{f}{\sqrt{2}}$ we have the flavor state

$$\nu_e = U\,\nu_1 + \theta N \tag{5}$$

with $U \approx 1$ and active-sterile mixing

$$\theta = \frac{M^D}{M_N} = \frac{v f}{2M_N} \ll 1$$

The mass eigenvalues are

$$\approx M_N \quad \text{and} \quad -m_a = \theta^2 M_N \lll M_N$$

which naturally explains the smallness of neutrino masses by doubling the small parameter θ. Note, that neither its value, nor the value of the sterile neutrino mass are fixed by the seesaw requirement to explain the active neutrino masses. In terms of the parameters of largangian (4) it means, that one and the same value of the active neutrino mass we obtain with larger Yukawas and heavier sterile neutrinos or with smaller Yukawas and lighter sterile neutrinos. In other words, *the mass scale of sterile neutrinos is not fixed by the success of explaining the active neutrino masses within the framework of the seesaw mechanism.*

The scheme above is straightforwardly extended to the case of several sterile neutrinos. To explain the presently confirmed results of neutrino oscillation experiments at least two non-zero active neutrino masses are needed, which asks for at least two sterile neutrinos. It looks weird to have the third active neutrino massless. If all the three are massive, one must introduce three or more sterile neutrinos, which allows us to address the issue of dark matter (see Ref. [7] for recent review on sterile neutrino dark matter).

3 Sterile neutrino as dark matter

Take sterile neutrino involved into the seesaw mechanism. Then its Majorana mass M_N and Yukawa coupling f in (4) are related to active neutrino mass m_a as

$$m_a \sim \frac{f^2 v^2}{M_N^2} M_N \sim \theta^2 M_N\,. \tag{6}$$

Sterile neutrino mixes with active neutrinos, and hence, being (much) heavier, is always unstable: its decay into three active neutrinos, $N \to 3\nu$, is kinematically allowed. To form dark matter, sterile neutrino lifetime must exceeds the age of the present Universe, 14 Gyr, that places constraint on the model parameters[1],

$$\theta^2 < 3 \times 10^{-3}\left(\frac{10\,\text{keV}}{M_N}\right)^5\,. \tag{7}$$

[1]Hereafter we consider the sterile neutrino lighter than the weak bosons, $M_N \ll 100$ GeV, and adopt the Fermi effective theory for the weak processes.

Plugging seesaw relation (6) into this inequality (7) we obtain for the sterile neutrino lifetime with respect to decay into three active neutrinos,

$$\tau_{N\to 3\nu} \sim 1/\left(G_F^2 M_N^5 \theta^2\right) \sim 1/\left(G_F^2 M_N^4 m_\nu\right) \sim 10^{11}\,\text{yr}\,(10\,\text{keV}/M_N)^4\,.$$

The obtained results severely constrain the mass scale of sterile neutrino candidate for dark matter: only keV region turns to be suitable. Note that lighter sterile neutrinos do not fit to the dark matter phase space density, see e.g. [8].

Even if very long-lived, such dark matter is unstable, that provides a unique opportunity to search for its decay products. The golden mode is one-loop radiative decay $N \to \gamma\nu$, which goes through the loop of charged lepton and W-boson (sterile neutrino mass state emits them because of tiny, suppressed by $\theta \ll 1$, admixture of active neutrino flavor state (5)) with photon outgoing from either of the virtual particles. This is two-body decay into massless states, so in the neutrino rest frame the photon energy is exactly half of the neutrino mass, $E_\gamma = M_N/2$. For sterile neutrino of Galactic dark matter the produced photons would form a narrow line, which relative width $\sim 10^{-3}$ is determined by the Doppler effect due to dark matter velocity dispersion in the Galaxy. The signature is very pronounced, and extensive investigations of Galactic X-ray spectrum and spectra of nearby dwarf and normal galaxies, galaxy clusters, put very strong limit on the sterile neutrino mixing angle (see e.g. [7]),

$$\frac{\theta^2}{10^{-11}} \lesssim \left(\frac{10\,\text{keV}}{M_N}\right)^4$$

surpassing that from the Universe age (7) by almost eight orders of magnitude. This limit is obtained assuming the sterile neutrino form all the dark matter component we needed in cosmology. If only a fraction of dark matter is due to sterile neutrinos, the limit becomes weaker proportionally to the fraction.

About three years ago two groups analyzing the data of Chandra telescope announced a feature consistent with the expected signal from decays of sterile neutrino dark matter [9, 10]. The feature was found in stacking signal from many galaxies in Perseus cluster and in nearby Andromeda galaxy. Though the signal was recognized only at about $3\,\sigma$ level, and at that level was missed in stacking signal from some other sets of galaxies, it nevertheless excited community which intensified hunting the sterile neutrino dark matter. The reason was that the model parameters, sterile neutrino mass and sterile-active mixing angle, were consistent with predictions of one of the realistic mechanisms of sterile neutrino production in the early Universe due to oscillations in primordial plasma.

To calculate the sterile neutrino production one can start with the Boltzmann equation for the sterile neutrino distribution function $f_S(t, \mathbf{p})$. In the Universe expanding at the Hubble rate $\mathcal{H}$ the Boltzmann equation reads (see e.g. [11])

$$\frac{\partial}{\partial t} f_s(t,\mathbf{p}) - \mathcal{H}\mathbf{p}\,\frac{\partial}{\partial \mathbf{p}} f_s(t,\mathbf{p}) = \Gamma_\alpha\, P(\nu_\alpha \to \nu_s)\, f_\alpha(t,\mathbf{p})\,,$$

where $\Gamma_\alpha \sim G_F^2 T^4 E$ is the weak interaction rate in plasma. This interaction collapses neutrino wave function and thus fixes the produced sterile neutrino state. The transition probability in the plasma

 DOI: 10.1051/epjconf/201715805006

can be written as

$$P(\nu_\alpha \to \nu_s) = \sin^2 2\theta_\alpha^{\rm mat} \cdot \sin^2\left(\frac{t}{2t_\alpha^{\rm mat}}\right),$$

$$t_\alpha^{\rm mat} = \frac{t_\alpha^{vac}}{\sqrt{\sin^2 2\theta_\alpha + (\cos 2\theta_\alpha - V_{\alpha\alpha} \cdot t_\alpha^{vac})^2}},$$

$$\sin 2\theta_\alpha^{\rm mat} = \frac{t_\alpha^{\rm mat}}{t_\alpha^{vac}} \cdot \sin 2\theta_\alpha, \qquad t_\alpha^{vac} = \frac{2E}{M_N^2}$$

with effective plasma potential for active neutrinos consisting of two terms (below we omit flavor and temperature dependent numerical coefficients in front of the terms),

$$V_{\alpha\alpha} \sim -\# G_F^2 T^4 E + \# G_F T^2 \mu_{L_\alpha}, \tag{8}$$

lepton-symmetric and lepton-asymmetric, respectively. The latter is essentially of the same nature as the term (3) governing the propagation of energetic electron neutrinos in the Sun, where only electrons, but not positrons, are present in the plasma. In (8) μ_{L_α} is the chemical potential with respect to leptonic charge L_α.

The first term in (8) refers to the standard cosmology, when the plasma consists of all particles and antiparticles relativistic at a given temperature. It contributes to effective neutrino mass in plasma and suppresses the oscillations. The production rate is low, and sterile neutrino never came to equilibrium in the primordial plasma. The mechanism is called Dodelson–Widrow or non-relativistic production [12]. The amount of produced sterile neutrinosis proportional to mixing squared angle. For a given sterile neutrino mass the requirement to form the dominant component of the dark matter fixes the value of mixing angle. This is a line on (θ^2, M_N) plane, and at present it is entirely forbidden by negative results of searches for a single line in X-rays and studies of matter perturbation evolution at dwarf galaxy scales, see Fig. 1.

The second term in (8) emerges in the primordial plasma with lepton asymmetry. If the letter is generated before the epoch of electroweak transition, the asymmetry is redistributed between baryons and leptons, so that both asymmetries are of the same order. In this case the second term in (8) is too small to play any role in sterile neutrino production. In specific extensions of the SM (e.g. νMSM [15, 16]) the asymmetry can be produced later, and then it can be much larger than the asymmetry in the baryon sector, reaching the values as large as 10^{-3}, being constrained from the primordial Big Bang Nucleosynthesis and anisotropy of the Cosmic Microwave Background radiation. The second term in (8) is linear in the chemical potential and hence, depending on the sign of asymmetry, can cancel the impact of the first term exhibiting the resonance in oscillations by significant increase of the effective mixing angle in plasma. Thus, the same amount of sterile neutrinos can be generated *with smaller vacuum mixing angle* θ [17]. In this setup the production depends both on vacuum mixing angle and the amount of lepton asymmetry in the primordial plasma. Then the region in the model parameter space, where sterile neutrino can form the entire dark matter, is not a line, but rather a wide band placed below the line of non-resonant production, see Fig. 1. The extensive studies of X-ray satellites data significantly shrank the viable region of parameter space asking for new mechanism of sterile neutrino production in the early Universe.

4 Sterile neutrino production in models with additional scalars

The prediction of sterile neutrino production can be drastically changed by introducing a scalar field ϕ coupled to the sterile neutrino,

$$\mathcal{L} \propto \phi \bar{N}^c N + h.c.\,. \tag{9}$$

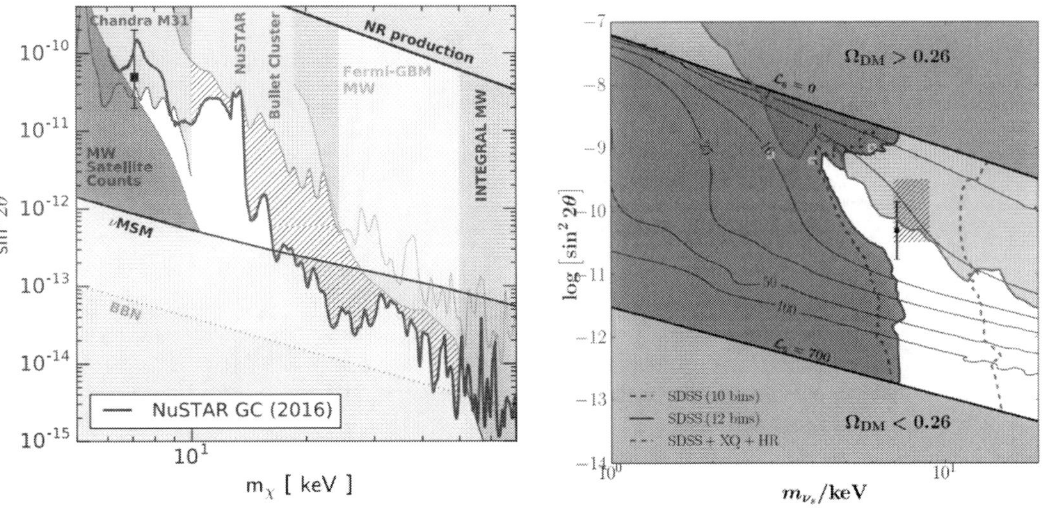

Figure 1. Present limits [13, 14] on the model parameters of sterile neutrino dark matter and predictions of dark matter production by non-resonant (lines 'NR production' and '$\mathcal{L}_6 = 0$') and resonant oscillations in primordial plasma (with lepton asymmetry $\mathcal{L}_6 \equiv L_e/10^6$).

This coupling is well-motivated since it violates lepton symmetry being of the same nature as the Majorana mass term in largangian (4). Depending on the behavior of the scalar field in the early Universe, there are several possible impacts on the time-evolution of the sterile neutrino sector.

One of the options considered in Refs. [18, 19] is ϕ being the inflaton. Sterile neutrinos can indeed be produced by inflaton decays and can form dark matter. However, sterile-active mixing plays no role in this production mechanism, and all searches for sterile neutrinos aimed at measuring the mixing angle are irrelevant in this case.

Recently, two other options have been put forward [20], which allow for significant suppression of sterile neutrino production in the early Universe which still goes mostly through the active-sterile mixing. The first option exploits the very light free scalar field, which is frozen at high value in the early Universe. Then coupling (9) makes sterile neutrino superheavy in the early Universe, which kinematically suppresses all possible oscillations. The scalar remains frozen until the Hubble parameter drops to the value of the scalar mass. Then the field starts to oscillate with decreasing amplitude and by the present epoch the scalar contribution to the sterile neutrino mass (9) can be neglected. Some amount of sterile neutrinos can be produced in neutrino oscillations while the scalar amplitude drops. However, by choosing the scalar parameters one can make this amount as small as possible.

Another option is related to the scalar field of the hidden sector, which dynamics initiates the phase transition in the late Universe, so that ϕ gains non-zero vacuum expectation value. Then, with zero sterile neutrino mass term in lagrangian (4) one has massless sterile neutrino in the early Universe, and hence no sterile-active oscillations at all. Later, after the phase transition, sterile neutrino becomes massive and hence participating in neutrino oscillations. Since the non-resonant production mechanism is mostly efficient at temperature [12]

$$T_{max} \approx 133\,\text{MeV}\left(\frac{1\,\text{keV}}{M_N}\right)^{1/3}$$

the phase transition must occur later, at lower temperature, which guarantees the suppression of sterile neutrino production in oscillations as compared to the non-resonant mechanism. The results are illustrated in Fig. 2.

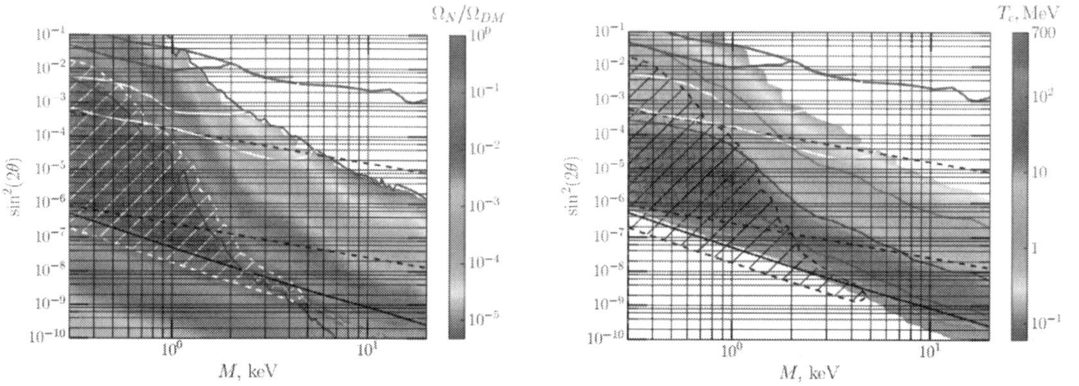

Figure 2. Dark matter sterile neutrino production in the extension with scalar from the hidden sector exhibiting the phase transition at temperature T_c, which makes sterile neutrino massive [8]. Green lines refer to present direct limits on sterile-neutrino mixing, while white lines indicate the prospect of future direct searches. The dotted black lines show the regions where the sterile neutrino can truly participate in seesaw mechanism providing active neutrinos with masses of viable scale. The limits from X-rays are avoided at relatively large and moderate mixings because of smaller sterile neutrino fraction in dark matter, $\Omega_N/\Omega_{DM} < 1$. See Ref. [8] for more details.

While the suppression of sterile neutrinos in the early Universe makes models with moderate and even relatively large mixing angles consistent with cosmology, the price to be paid is the unsolved dark matter problem. The limits from X-rays are rather severe, and constrain the models parameters for any dark matter production in the early Universe. They can be avoided only assuming that the sterile neutrinos form only a (small) fraction of dark matter, $\Omega_N < \Omega_{DM}$. Then the X-ray bounds rescale as

$$\theta^2_{X-ray} = \theta^2_{\alpha I}\,\frac{\Omega_N}{\Omega_{DM}}$$

opening a possibility for the keV-scale sterile neutrino to fully participate in seesaw mechanism and even being tested in future experiments on direct searches for light sterile neutrinos.

Note in passing that red region on left panel of Fig. 2, consistent with all bounds from astrophysics and direct searches, implies $\Omega_N \sim \Omega_{DM}$, so the sterile neutrinos form dominant dark matter component. However, sterile neutrinos are light here, and being produced in oscillations (hence with momenta $\sim 3T$) happen to be too fast at the onset of structure formation in the early Universe. They would suppress the formation of protohalo for future dwarf galaxies, which be observe today, though at a bit smaller amount than one expects for the cold dark matter particles. The problem is solved if sterile neutrino contribution to dark matter is only sub-dominant here. To make this region cosmologically viable with sterile neutrino forming entire dark matter component, one can involve another production mechanism, where sterile neutrino would be cold. One of the possibility is to use the oscillating scalar field with coupling (9) as the neutrino source exploiting the first aforementioned option with initially frozen and then oscillating scalar ϕ.

5 Conclusions

Sterile neutrinos is one of the optional physics beyond the SM, which may be (partly) responsible for the dark matter phenomenon. Introduced to explain neutrino oscillations, light sterile neutrinos can be produced in the early Universe and are long-lived enough to survive until the present days and populate galaxy and galaxy clusters. While the minimal model was extensively studied and the viable parameter space shrank significantly, we show that natural modification of the model with additional scalar field (possibly playing a role of Majoron) opens up new room available even for direct searches and promising interesting cosmological and phenomenological consequences.

The work is supported by the RSF grant 17-22-01547.

References

[1] M. Agostini et al. (Borexino) (2017), 1709.00756

[2] P. Minkowski, Phys. Lett. **B67**, 421 (1977)

[3] M. Gell-Mann, P. Ramond, R. Slansky, Conf. Proc. **C790927**, 315 (1979), 1306.4669

[4] R.N. Mohapatra, G. Senjanovic, Phys. Rev. Lett. **44**, 912 (1980)

[5] T. Yanagida, Prog. Theor. Phys. **64**, 1103 (1980)

[6] J. Schechter, J.W.F. Valle, Phys. Rev. **D22**, 2227 (1980)

[7] M. Drewes et al., JCAP **1701**, 025 (2017), 1602.04816

[8] D. Gorbunov, A. Khmelnitsky, V. Rubakov, JCAP **0810**, 041 (2008), 0808.3910

[9] E. Bulbul, M. Markevitch, A. Foster, R.K. Smith, M. Loewenstein, S.W. Randall, Astrophys. J. **789**, 13 (2014), 1402.2301

[10] A. Boyarsky, O. Ruchayskiy, D. Iakubovskyi, J. Franse, Phys. Rev. Lett. **113**, 251301 (2014), 1402.4119

[11] D.S. Gorbunov, V.A. Rubakov, *Introduction to the theory of the early universe: Hot big bang theory* (World Scientific, Hackensack, 2011), http://www.DESY.eblib.com/patron/FullRecord.aspx?p=737614

[12] S. Dodelson, L.M. Widrow, Phys. Rev. Lett. **72**, 17 (1994), hep-ph/9303287

[13] K. Perez, K.C.Y. Ng, J.F. Beacom, C. Hersh, S. Horiuchi, R. Krivonos, Phys. Rev. **D95**, 123002 (2017), 1609.00667

[14] J. Baur, N. Palanque-Delabrouille, C.Y. Eche, A. Boyarsky, O. Ruchayskiy, E. Armengaud, J. Lesgourgues (2017), 1706.03118

[15] T. Asaka, S. Blanchet, M. Shaposhnikov, Phys. Lett. **B631**, 151 (2005), hep-ph/0503065

[16] T. Asaka, M. Shaposhnikov, Phys. Lett. **B620**, 17 (2005), hep-ph/0505013

[17] X.D. Shi, G.M. Fuller, Phys. Rev. Lett. **82**, 2832 (1999), astro-ph/9810076

[18] M. Shaposhnikov, I. Tkachev, Phys. Lett. **B639**, 414 (2006), hep-ph/0604236

[19] F. Bezrukov, D. Gorbunov, JHEP **05**, 010 (2010), 0912.0390

[20] F. Bezrukov, A. Chudaykin, D. Gorbunov, JCAP **1706**, 051 (2017), 1705.02184

Neutrino-electron scattering in a dense strongly magnetized plasma

Alexander Kuznetsov[1,★] and *Vasiliy* Savin[1,★★]

[1] *Division of Theoretical Physics, Department of Physics, Yaroslavl State P. G. Demidov University, Sovietskaya 14, 150000 Yaroslavl, Russian Federation*

Abstract. We investigate the process of neutrino-electron scattering in a dense plasma and magnetic field of arbitrary strength, where electrons can occupy the states corresponding to excited Landau levels. We calculate the total probability of this process, summarized over all initial states of the plasma electrons which is only physically meaningful. Possible astrophysical manifestations of the process are briefly discussed.

1 Introduction

Astroparticle physics has manifested itself in recent decades as a vigorously growing and prospective line of investigation at the junction of particle physics, astrophysics, and cosmology; see e.g. [1–3]. An important stimulus of its development is an understanding of the essential role of quantum processes in the dynamics of astrophysical objects and of the early Universe. A decisive role in astrophysical cataclysms such as supernova explosions and coalescence of neutron stars, as well as in the early Universe, belongs to neutrino physics. Consequently, studies of neutrino interactions and in particular neutrino-electron processes in an external active medium are of considerable interest. At the same time, an investigation of neutrino processes under such extreme physical conditions is interesting from the conceptual viewpoint since it affects fundamental problems of quantum field theory.

For self-consistent analysis of the neutrino propagation process in a hot dense plasma in the presence of a strong magnetic field, consideration is required of a complete set of neutrino-electron processes. Then only the probability of the process summarized over all initial states of the plasma electrons is physically meaningful. In Ref. [4], numerical calculations of the differential cross-section of the neutrino-electron scattering in dense magnetized plasma were performed in the limit of rather weak magnetic field B, $eB < \mu E$, where μ is the plasma chemical potential, E is the typical neutrino energy (we use natural units in which $c = \hbar = 1$, $e > 0$ is the elementary charge). In Refs. [5, 6], the probability was evaluated of the total sum of νe processes ($\nu \to \nu e^- e^+$, $\nu e^\mp \to \nu e^\mp$, $\nu e^- e^+ \to \nu$) and the volume density of the neutrino energy and momentum losses, integrated over the momenta of the plasma electrons, in a strong magnetic field, $eB \gg (\mu^2, T^2, E^2) \gg m_e^2$, ($T$ is the plasma temperature); in such conditions electrons and positrons occupied only the ground Landau levels. In Refs. [7, 8], the probability of the $\nu e \to \nu e$ process and the volume density of the neutrino energy and momentum losses, summarized over all initial states of the plasma electrons, were evaluated under the physical

★e-mail: avkuzn@uniyar.ac.ru
★★e-mail: vs_post07@mail.ru

situation where the magnetic field is moderate, while the density of plasma is large, so the conditions are satisfied: $\mu^2 > eB \gg (T^2, E^2) \gg m_e^2$, where T is the plasma temperature, and $eB \gg \mu E$. Calculations were performed for the case when both initial and final electrons occupy the same Landau levels, because, as it was concluded in Refs. [7, 8], such transitions were dominating. The purpose of the present research is to calculate analytically the probability of this process for a more general case when the initial and final electrons could occupy any physically allowed Landau levels. Some details of the calculation technique can be found e.g. in Refs. [9–12].

2 Solutions of the Dirac equation for an electron in a magnetic field

There exist several descriptions of the procedure of obtaining the electron wave functions in the presence of an external magnetic field by solving the Dirac equation, see e.g. Refs. [13–19] and also Refs. [9, 10]. In the most cases, the solutions are presented in the form with the upper two components of the bispinor corresponding to the electron states with the spin projections 1/2 and -1/2 on the magnetic field direction. As the analysis has shown, it is optimal to use the electron wave functions as the eigenstates of the covariant operator of a magnetic polarization $\hat{\mu}_z$ [15, 16]:

$$\hat{\mu}_z = m_e \Sigma_z - \mathrm{i}\gamma_0\gamma_5[\boldsymbol{\Sigma} \times \boldsymbol{P}]_z \,, \tag{1}$$

where $\boldsymbol{P} = -\mathrm{i}\,\boldsymbol{\nabla} + e\boldsymbol{A}$. We take the frame where the field is directed along the z axis, and the Landau gauge where the four-potential is: $A^\lambda = (0, 0, xB, 0)$. In this approach, the electron wave functions have the form

$$\Psi^s_{p,n}(X) = \frac{1}{\sqrt{4\varepsilon_n M_n(\varepsilon_n + M_n)(M_n + m_e)L_y L_z}}\, \mathrm{e}^{-\mathrm{i}(\varepsilon_n t - p_y y - p_z z)}\, U^s_n(\xi)\,, \tag{2}$$

where $\varepsilon_n = \sqrt{M_n^2 + p_z^2}\,, \quad M_n = \sqrt{m_e^2 + 2\beta n}\,, \quad \beta = eB\,, \quad \xi = \sqrt{\beta}\left(x + p_y/\beta\right)$. The functions $\Psi^s_{p,n}(X)$ satisfy the equation: $\hat{\mu}_z\, \Psi^s_{p,n}(X) = s\, M_n\, \Psi^s_{p,n}(X)\, (s = \pm 1)$. The bispinors $U^s_n(\xi)$ in Eq. (2) take the form:

$$U^-_n(\xi) = \begin{pmatrix} -\mathrm{i}\sqrt{2\beta n}\, p_z V_{n-1}(\xi) \\ (\varepsilon_n + M_n)(M_n + m_e)V_n(\xi) \\ -\mathrm{i}\sqrt{2\beta n}(\varepsilon_n + M_n)V_{n-1}(\xi) \\ -p_z(M_n + m_e)V_n(\xi) \end{pmatrix}, \quad U^+_n(\xi) = \begin{pmatrix} (\varepsilon_n + M_n)(M_n + m_e)V_{n-1}(\xi) \\ -\mathrm{i}\sqrt{2\beta n}\, p_z V_n(\xi) \\ p_z(M_n + m_e)V_{n-1}(\xi) \\ \mathrm{i}\sqrt{2\beta n}(\varepsilon_n + M_n)V_n(\xi) \end{pmatrix}. \tag{3}$$

Here, $V_n(\xi)\, (n = 0, 1, 2, \dots)$ are the well-known normalized harmonic oscillator functions, which are expressed in terms of the Hermite polynomials $H_n(\xi)$:

$$V_n(\xi) = \frac{\beta^{1/4}\mathrm{e}^{-\xi^2/2}}{\sqrt{2^n n!\,\sqrt{\pi}}}\, H_n(\xi)\,, \tag{4}$$

With using the solutions of the Dirac equation for an electron in a magnetic field in the form (2), (3), the amplitude of any process will have an explicit Lorentz invariant structure. It is in contrast to the case when the solutions are presented in the form with the upper two components of the bispinor corresponding to the electron states with the spin projections 1/2 and -1/2 on the magnetic field direction.

 DOI: 10.1051/epjconf/201715805007

3 The process of the $\nu e \to \nu e$ scattering

The effective local Lagrangian of the neutrino-electron interaction is:

$$\mathcal{L} = -\frac{G_{\rm F}}{\sqrt{2}}[\bar{e}\gamma_\alpha(C_V - C_A\gamma_5)e]\,[\bar{\nu}\gamma^\alpha(1-\gamma_5)\nu], \quad C_V = \pm\frac{1}{2} + 2\sin^2\theta_{\rm W}\,,\ C_A = \pm\frac{1}{2}\,, \tag{5}$$

where the upper signs correspond to ν_e and the lower signs correspond to $\nu_{\mu,\tau}$, $\theta_{\rm W}$ is the Weinberg angle.

The S matrix element of the subprocess $\nu e^-_{(\ell)} \to \nu e^-_{(n)}$ takes the form

$$S = {\rm i}\,\frac{G_{\rm F}}{\sqrt{2}}\,\frac{(2\pi)^3\,\delta(\varepsilon'_n - \varepsilon_\ell - q_0)\,\delta(p'_y - p_y - q_y)\,\delta(p'_z - p_z - q_z)}{\sqrt{2EV\,2E'V\,2\varepsilon_\ell L_y L_z\,2\varepsilon'_n L_y L_z}}\,{\rm e}^{-q_\perp^2/4eB - {\rm i}q_x(p_y+p'_y)/2eB}$$
$$\times\ [\bar{u}(p')\,(j\gamma)\,(C_V - C_A\gamma_5)\,u(p)]\,, \tag{6}$$

where $q = P - P' = p' - p$ is the 4-momentum transferred, ε_ℓ and ε'_n are the energies of the initial and final electrons, $q_\perp$ is the projection of the vector $\boldsymbol{q}$ on the plane perpendicular to the vector $\boldsymbol{B}$, $q_\perp^2 = q_x^2 + q_y^2$, and $j_\alpha = \bar{\nu}(P')\gamma_\alpha(1-\gamma_5)\nu(P)$ is the Fourier transform of the current of the left-handed neutrinos.

The total process probability per unit time can be presented in the form

$$W(\nu e^- \to \nu e^-) = \frac{1}{\mathcal{T}}\sum_\ell\sum_n\sum_{s,s'}\int |\,S\,|^2\,{\rm d}n_{e^-}\,{\rm d}n'_{e^-}\,\frac{{\rm d}^3P'\,V}{(2\pi)^3}\,(1 - f(E'))\,, \tag{7}$$

where ${\rm d}n_{e^-} = f(\varepsilon_\ell)\,{\rm d}p_y\,{\rm d}p_z\,L_y\,L_z/(2\pi)^2$, ${\rm d}n'_{e^-} = (1 - f(\varepsilon'_n))\,{\rm d}p'_y\,{\rm d}p'_z\,L_y\,L_z/(2\pi)^2$, $f(\varepsilon_\ell)$ is the fermion distribution function, e.g. for initial electrons it is: $f(\varepsilon_\ell) = [{\rm e}^{(\varepsilon_\ell-\mu)/T} + 1]^{-1}$; $\mathcal{T}$ is the total time of interaction, $V = L_x\,L_y\,L_z$ is the total volume of the interaction region.

For the probability per unit time of the subprocess $\nu e^-_{(\ell)} \to \nu e^-_{(n)}$ one obtains

$$W_{\ell n} = \frac{\beta}{(2\pi)^4 16E}\int\frac{{\rm d}^3P'}{E'}\,(1 - f(E'))$$
$$\times\ \int\frac{{\rm d}p_z}{\varepsilon'_n\,\varepsilon_\ell}\,\delta(\varepsilon'_n - \varepsilon_\ell - q_0)\,f(\varepsilon_\ell)\,(1 - f(\varepsilon'_n))\sum_{s,s'}|\mathcal{M}^{ss'}_{\ell n}|^2\,. \tag{8}$$

Taking into account possible polarisation states of the initial and final electrons, there exist four invariant polarization amplitudes $\mathcal{M}^{ss'}_{\ell n}$, where $s, s' = \pm 1$, which can be presented in the form:

$$\mathcal{M}^{ss'}_{\ell n} = \eta^{ss'}\,\frac{G_{\rm F}}{2\sqrt{2}}\left\{A^{ss'}_{\ell n}\frac{(j\tilde{\varphi}q)}{q_\parallel^2} + B^{ss'}_{\ell n}\frac{(j\widetilde{\Lambda}q)}{q_\parallel^2} + C^{ss'}_{\ell n}\frac{(j\Lambda q)}{q_\perp^2} + D^{ss'}_{\ell n}\frac{{\rm i}\,(j\varphi q)}{q_\perp^2}\right\}, \tag{9}$$

where $\eta^{ss'}$ is an inessential phase factor, $\varphi_{\alpha\beta} = F_{\alpha\beta}/B$ is the dimensionless tensor of the external magnetic field, $\tilde{\varphi}_{\alpha\beta} = \frac{1}{2}\,\varepsilon_{\alpha\beta\mu\nu}\varphi^{\mu\nu}$ is the dual dimensionless tensor; the four-vectors with the indices $\perp$ and $\parallel$ belong to the Euclidean $\{1,2\}$-subspace and the pseudo-Euclidean $\{0,3\}$-subspace, correspondingly, and $\Lambda_{\alpha\beta} = (\varphi\varphi)_{\alpha\beta} = {\rm diag}(0,1,1,0)$, $\widetilde{\Lambda}_{\alpha\beta} = (\tilde{\varphi}\tilde{\varphi})_{\alpha\beta} = {\rm diag}(1,0,0,-1)$. In Eq. (9), auxiliary functions are introduced: A, B, C, D. We give here their explicit forms for all possible sets of polarization.

For the case $s = s' = -1$, one obtains:

$$A_{\ell n}^{--} = \sqrt{\left(1+\frac{m}{M_\ell}\right)\left(1+\frac{m}{M_n}\right)}\,\mathcal{I}_{\ell,n}\,(C_V\,\tau + C_A\,\kappa) + \sqrt{\left(1-\frac{m}{M_\ell}\right)\left(1-\frac{m}{M_n}\right)}\,\mathcal{I}_{\ell-1,n-1}(C_V\,\tau - C_A\,\kappa)\,, \tag{10}$$

$$B_{\ell n}^{--} = \sqrt{\left(1+\frac{m}{M_\ell}\right)\left(1+\frac{m}{M_n}\right)}\,\mathcal{I}_{\ell,n}\,(C_V\,\kappa + C_A\,\tau) + \sqrt{\left(1-\frac{m}{M_\ell}\right)\left(1-\frac{m}{M_n}\right)}\,\mathcal{I}_{\ell-1,n-1}(C_V\,\kappa - C_A\,\tau)\,, \tag{11}$$

$$C_{\ell n}^{--} = \sqrt{q_\perp^2}\left[-\sqrt{\left(1+\frac{m}{M_\ell}\right)\left(1-\frac{m}{M_n}\right)}\,\mathcal{I}_{\ell,n-1}\,(C_V\,u + C_A\,v) - \sqrt{\left(1-\frac{m}{M_\ell}\right)\left(1+\frac{m}{M_n}\right)}\,\mathcal{I}_{\ell-1,n}\,(C_V\,u - C_A\,v)\right], \tag{12}$$

$$D_{\ell n}^{--} = \sqrt{q_\perp^2}\left[\sqrt{\left(1+\frac{m}{M_\ell}\right)\left(1-\frac{m}{M_n}\right)}\,\mathcal{I}_{\ell,n-1}\,(C_V\,u + C_A\,v) - \sqrt{\left(1-\frac{m}{M_\ell}\right)\left(1+\frac{m}{M_n}\right)}\,\mathcal{I}_{\ell-1,n}\,(C_V\,u - C_A\,v)\right], \tag{13}$$

where, for $n \geqslant \ell$

$$\mathcal{I}_{n,\ell}(x) = \sqrt{\frac{\ell!}{n!}}\,\mathrm{e}^{-x/2}x^{(n-\ell)/2}L_\ell^{n-\ell}(x)\,, \quad \mathcal{I}_{\ell,n}(x) = (-1)^{n-\ell}\mathcal{I}_{n,\ell}(x)\,, \tag{14}$$

$L_n^k(x)$ are the generalized Laguerre polynomials, and also:

$$\kappa = u\,(M_n - M_\ell)\,,\ \tau = v\,(M_n + M_\ell)\,,\ u = \sqrt{(M_n + M_\ell)^2 - q_\parallel^2}\,,\ v = \alpha\,\zeta\,\sqrt{(M_n - M_\ell)^2 - q_\parallel^2}\,.$$

Here, $\alpha = q_0/|q_0|$ is the sign of q_0, $\zeta = \pm 1$ is the sign factor associated with the two roots of the equation, corresponding to the zeros of the δ function argument in Eq. (8). In the frame where $q_z = 0$, ζ is the sign of the p_z component, which is not fixed by the equation.

For the case $s = s' = +1$, one obtains:

$$A_{\ell n}^{++} = \sqrt{\left(1-\frac{m}{M_\ell}\right)\left(1-\frac{m}{M_n}\right)}\,\mathcal{I}_{\ell,n}\,(C_V\,\tau + C_A\,\kappa) + \sqrt{\left(1+\frac{m}{M_\ell}\right)\left(1+\frac{m}{M_n}\right)}\,\mathcal{I}_{\ell-1,n-1}(C_V\,\tau - C_A\,\kappa)\,, \tag{15}$$

 DOI: 10.1051/epjconf/201715805007

$$B_{\ell n}^{++} = \sqrt{\left(1-\frac{m}{M_\ell}\right)\left(1-\frac{m}{M_n}\right)}\mathcal{I}_{\ell,n}\,(C_V\,\kappa + C_A\,\tau) + \sqrt{\left(1+\frac{m}{M_\ell}\right)\left(1+\frac{m}{M_n}\right)}\mathcal{I}_{\ell-1,n-1}(C_V\,\kappa - C_A\,\tau)\,, \tag{16}$$

$$C_{\ell n}^{++} = \sqrt{q_\perp^2}\left[-\sqrt{\left(1-\frac{m}{M_\ell}\right)\left(1+\frac{m}{M_n}\right)}\mathcal{I}_{\ell,n-1}\,(C_V\,u + C_A\,v) - \sqrt{\left(1+\frac{m}{M_\ell}\right)\left(1-\frac{m}{M_n}\right)}\mathcal{I}_{\ell-1,n}\,(C_V\,u - C_A\,v)\right], \tag{17}$$

$$D_{\ell n}^{++} = \sqrt{q_\perp^2}\left[\sqrt{\left(1-\frac{m}{M_\ell}\right)\left(1+\frac{m}{M_n}\right)}\mathcal{I}_{\ell,n-1}\,(C_V\,u + C_A\,v) - \sqrt{\left(1+\frac{m}{M_\ell}\right)\left(1-\frac{m}{M_n}\right)}\mathcal{I}_{\ell-1,n}\,(C_V\,u - C_A\,v)\right]. \tag{18}$$

For the case $s = +1$, $s' = -1$, one obtains:

$$A_{\ell n}^{+-} = \sqrt{\left(1-\frac{m}{M_\ell}\right)\left(1+\frac{m}{M_n}\right)}\mathcal{I}_{\ell,n}\,(C_V\,\kappa + C_A\,\tau) - \sqrt{\left(1+\frac{m}{M_\ell}\right)\left(1-\frac{m}{M_n}\right)}\mathcal{I}_{\ell-1,n-1}(C_V\,\kappa - C_A\,\tau)\,, \tag{19}$$

$$B_{\ell n}^{+-} = \sqrt{\left(1-\frac{m}{M_\ell}\right)\left(1+\frac{m}{M_n}\right)}\mathcal{I}_{\ell,n}\,(C_V\,\tau + C_A\,\kappa) - \sqrt{\left(1+\frac{m}{M_\ell}\right)\left(1-\frac{m}{M_n}\right)}\mathcal{I}_{\ell-1,n-1}(C_V\,\tau - C_A\,\kappa)\,, \tag{20}$$

$$C_{\ell n}^{+-} = \sqrt{q_\perp^2}\left[\sqrt{\left(1-\frac{m}{M_\ell}\right)\left(1-\frac{m}{M_n}\right)}\mathcal{I}_{\ell,n-1}\,(C_V\,v + C_A\,u) - \sqrt{\left(1+\frac{m}{M_\ell}\right)\left(1+\frac{m}{M_n}\right)}\mathcal{I}_{\ell-1,n}\,(C_V\,v - C_A\,u)\right], \tag{21}$$

$$D^{+-}_{\ell n} = \sqrt{q_\perp^2}\Big[-\sqrt{\left(1-\frac{m}{M_\ell}\right)\left(1-\frac{m}{M_n}\right)}\,\mathcal{I}_{\ell,n-1}\,(C_V\, \upsilon + C_A\, u)$$

$$- \sqrt{\left(1+\frac{m}{M_\ell}\right)\left(1+\frac{m}{M_n}\right)}\,\mathcal{I}_{\ell-1,n}\,(C_V\, \upsilon - C_A\, u)\Big]. \qquad (22)$$

For the case $s = -1$, $s' = +1$, one obtains:

$$A^{-+}_{\ell n} = \sqrt{\left(1+\frac{m}{M_\ell}\right)\left(1-\frac{m}{M_n}\right)}\,\mathcal{I}_{\ell,n}\,(C_V\, \kappa + C_A\, \tau)$$

$$- \sqrt{\left(1-\frac{m}{M_\ell}\right)\left(1+\frac{m}{M_n}\right)}\,\mathcal{I}_{\ell-1,n-1}(C_V\, \kappa - C_A\, \tau), \qquad (23)$$

$$B^{-+}_{\ell n} = \sqrt{\left(1+\frac{m}{M_\ell}\right)\left(1-\frac{m}{M_n}\right)}\,\mathcal{I}_{\ell,n}\,(C_V\, \tau + C_A\, \kappa)$$

$$- \sqrt{\left(1-\frac{m}{M_\ell}\right)\left(1+\frac{m}{M_n}\right)}\,\mathcal{I}_{\ell-1,n-1}(C_V\, \tau - C_A\, \kappa), \qquad (24)$$

$$C^{-+}_{\ell n} = \sqrt{q_\perp^2}\Big[\sqrt{\left(1+\frac{m}{M_\ell}\right)\left(1+\frac{m}{M_n}\right)}\,\mathcal{I}_{\ell,n-1}\,(C_V\, \upsilon + C_A\, u)$$

$$- \sqrt{\left(1-\frac{m}{M_\ell}\right)\left(1-\frac{m}{M_n}\right)}\,\mathcal{I}_{\ell-1,n}\,(C_V\, \upsilon - C_A\, u)\Big], \qquad (25)$$

$$D^{-+}_{\ell n} = -\sqrt{q_\perp^2}\Big[\sqrt{\left(1+\frac{m}{M_\ell}\right)\left(1+\frac{m}{M_n}\right)}\,\mathcal{I}_{\ell,n-1}\,(C_V\, \upsilon + C_A\, u)$$

$$+ \sqrt{\left(1-\frac{m}{M_\ell}\right)\left(1-\frac{m}{M_n}\right)}\,\mathcal{I}_{\ell-1,n}\,(C_V\, \upsilon - C_A\, u)\Big]. \qquad (26)$$

Calculations of the process probability per unit time (7) with Eqs. (8)–(26) should be performed numerically for different values of the physical parameters B, T, μ, E etc., and these calculations now are in progress. It should be noted, that, in contrast to calculations of Refs. [11, 12], in the present case no upper limits arise on the Landau level numbers ℓ, n from kinematics, and the suppression of the large number contributions is provided by the distribution functions of initial and final electrons.

To illustrate, we present in the figure 1 the results of numerical calculations of the subprocess $\nu e^-_{(\ell)} \to \nu e^-_{(n)}$ probability per unit time (8) as the function of Landau level numbers of the initial and final electrons, for fixed values of the physical parameters B, T, μ, E.

As the analysis shows, the assumption made in Refs. [7, 8], that the subprocesses were dominating where both initial and final electrons occupied the same Landau levels, is not confirmed. This means that the results for the probability of the $\nu e \to \nu e$ process obtained in Refs. [7, 8], were underestimated.

 DOI: 10.1051/epjconf/201715805007

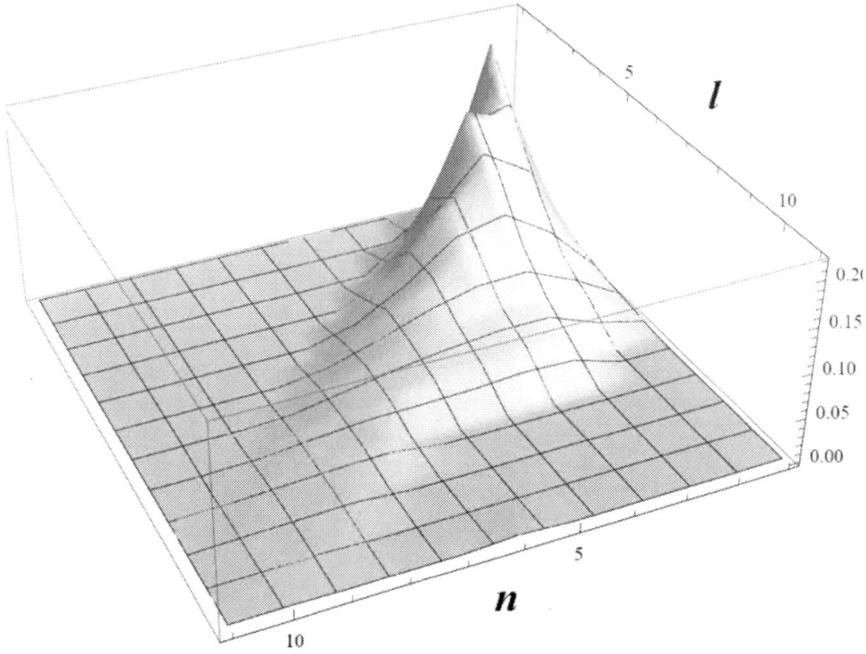

Figure 1. The dependence of the subprocess $\nu e^-_{(\ell)} \to \nu e^-_{(n)}$ probability per unit time on the Landau level numbers of the initial and final electrons

4 Conclusions

The probability of the $\nu e \to \nu e$ process in a dense magnetized plasma is calculated analytically, for a general case when the initial and final electrons could occupy any physically allowed Landau levels. The analysis shows, that the assumption made in previous calculations, that the subprocesses were dominating where both initial and final electrons occupied the same Landau levels, is not confirmed.

In astrophysical applications, the mean values of the neutrino energy and momentum losses could be more interesting:

$$Q^\alpha = E \int (P - P')^\alpha \, \mathrm{d}W = -E(\mathcal{I}, \boldsymbol{F}), \tag{27}$$

where $\mathrm{d}W$ is the total differential probability of the process which can be extracted from Eq. (7). The zeroth component of Q^α is connected with the mean energy lost by a neutrino per unit time due to the process considered, $\mathcal{I} = \mathrm{d}E/\mathrm{d}t$. The space components of the four-vector Q^α are similarly connected with the mean neutrino momentum loss per unit time, $\boldsymbol{F} = \mathrm{d}\boldsymbol{P}/\mathrm{d}t$. An analysis of the four-vector Q^α in a general case for the magnetic field of arbitrary strength, where electrons can occupy the states corresponding to excited Landau levels, now is in progress. The force density $\boldsymbol{F}$ could lead to a very interesting consequences if a strong toroidal magnetic field is generated in the supernova envelope, providing an asymmetry of the supernova explosion and, in particular, it can explain the phenomenon of high pulsar kick-velocities, for details see Ref. [10].

Acknowledgement

The paper is done within "YSU Initiative Scientific Researches" (Project No. AAAA-A16-116070610023-3).

References

[1] G. G. Raffelt, *Stars as Laboratories for Fundamental Physics* (Univ. Chicago Press, Chicago, 1996)
[2] M. Yu. Khlopov, *Cosmoparticle Physics* (World Scientific, Singapore, 1999)
[3] H. V. Klapdor-Kleingrothaus and K. Zuber, *Particle Astrophysics* (Inst. of Phys., Bristol, 2000)
[4] V. G. Bezchastnov and P. Haensel, Phys. Rev. D **54**, 3706 (1996)
[5] A. V. Kuznetsov and N. V. Mikheev, Mod. Phys. Lett. A **14**, 2531 (1999)
[6] A. V. Kuznetsov and N. V. Mikheev, Zh. Eksp. Teor. Fiz. **118**, 863 (2000) [JETP **91**, 748 (2000)]
[7] N. V. Mikheev and E. N. Narynskaya, Mod. Phys. Lett. A **15**, 1551 (2000)
[8] N. V. Mikheev and E. N. Narynskaya, Centr. Eur. J. Phys. **1**, 145 (2003)
[9] A. V. Kuznetsov and N. V. Mikheev, *Electroweak Processes in External Electromagnetic Fields* (Springer-Verlag, New York, 2003)
[10] A. V. Kuznetsov and N. V. Mikheev, *Electroweak Processes in External Active Media* (Springer-Verlag, Berlin, Heidelberg, 2013)
[11] A. V. Kuznetsov, D. A. Rumyantsev, and V. N. Savin, Int. J. Mod. Phys. A **29**, 1450136 (2014)
[12] A. V. Kuznetsov, D. A. Rumyantsev, and V. N. Savin, J. Phys.: Conf. Ser. **675**, 032019 (2016)
[13] M. H. Johnson and B. A. Lippmann, Phys. Rev. **76**, 828 (1949)
[14] A. I. Akhiezer and V. B. Berestetskii, *Quantum Electrodynamics* (Wiley, New York 1965)
[15] A. A. Sokolov and I. M. Ternov, *Synchrotron Radiation* (Pergamon, Oxford, 1968)
[16] D. B. Melrose and A. J. Parle, Aust. J. Phys. **36**, 755 (1983)
[17] A. A. Sokolov and I. M. Ternov, *Radiation from Relativistic Electrons* (American Institute of Physics, New York 1986)
[18] K. Bhattacharya and P. B. Pal, Pramana J. Phys. **62**, 1041 (2004)
[19] I. A. Balantsev, Yu. V. Popov, and A. I. Studenikin, J. Phys. A **44**, 255301 (2011)

Neutrino photoproduction on the electron in dense magnetized medium

Alexander Kuznetsov[1,⋆], *Dmitriy* Rumyantsev[1,⋆⋆], and *Denis* Shlenev[1,⋆⋆⋆]

[1] *Yaroslavl State University*

Abstract. The process of neutrino photoproduction on an electron, $e\gamma \to e\nu\bar{\nu}$, in a strongly magnetized cold plasma in resonant case has been considered. The contribution of this process to the neutrino emissivity has been calculated. It has been shown that under such conditions neutrino emissivity due to process $e\gamma \to e\nu\bar{\nu}$ could be expressed in terms of emissivity of it's subprocess, $e \to e\nu\bar{\nu}$.

1 Introduction

Nowadays, there exists rather keen interest to astrophysical objects with the scale of the magnetic field strength near the critical value of $B_e = m^2/e \simeq 4.41 \times 10^{13}$ G (we use natural units $c = \hbar = k_B = 1$, m is the electron mass, and e is the elementary charge). This group of objects includes the radio pulsars and the so-called magnetars, which are the neutron stars featuring the magnetic field strengths from 10^{12} G (radio pulsars) to 4×10^{14} G (magnetars), see Ref. [1] and the papers cited therein. The spectra analysis of these objects also provides an evidence for the presence of electron-positron plasma in the radio pulsar and magnetar environment, with the minimum magnetospheric plasma density being of the order of the Goldreich-Julian density [2]:

$$n_{GJ} \simeq 3 \times 10^{13}\,\text{cm}^{-3}\left(\frac{B}{100B_e}\right)\left(\frac{10\,\text{s}}{P}\right), \tag{1}$$

where P is the rotational period. It is well-known that strong magnetic field and/or plasma could have an essential influence on various quantum processes [3–6], because the external active medium catalyses the processes, by changing their kinematics and inducing new interactions. Therefore, the effects of magnetized plasma on microscopic physics should be incorporated in the magnetosphere models of strongly magnetized neutron stars.

All known theoretical models of internal structure of neutron stars give the parameters of the medium (density and temperature) at which the magnetized plasma is transparent to neutrinos. In this case, reactions with a neutrino-antineutrino pair in the final state are decisive for neutrino cooling. In this connection, the neutrino photoproduction (the so called photoneutrino process), $e\gamma \to e\nu\bar{\nu}$, was studied by multiple authors. In papers [7–9] the formulas for neutrino emissivity, i.e. energy carried away by the neutrino pair from the unit volume of a star by unit time, were calculated for

⋆e-mail: avkuzn@uniyar.ac.ru
⋆⋆e-mail: rda@uniyar.ac.ru
⋆⋆⋆e-mail: allen_caleb@rambler.ru

DOI: 10.1051/epjconf/201715805008

both nonrelativistic and relativistic plasmas. However, in those papers the anisotropy in the dispertion of photons was disregarded, which can change the results essentially. In recent paper [10] neutrino emisivity was calculated with taking into account the dispertion properties of photon in plasma in non-resonant case.

Into the boundary between inner and outer crust of magnetar, where plasma density may exceed value $\rho_9 = 10^9$ g/sm^3, the higher Landau levels of virtual electron are begin to be excited and the resonance on virtual electron may occur.

In this work we study in detail the process of photoproduction of neutrino-antineutrino pair, $e\gamma \to e\nu\bar{\nu}$, and the neutrino emissivity owing to this process in dense magnetized medium in resonance case.

2 Neutrino emissivity

A general expression for neutrino emissivity can be defined as follows:

$$Q = \frac{1}{V} \int \prod_i \mathrm{d}\Gamma_i f_i \prod_f \mathrm{d}\Gamma_f (1 \pm f_f)\, q_0' \,\frac{|S_{if}|^2}{\tau}\,, \qquad (2)$$

where $\mathrm{d}\Gamma_i$ ($\mathrm{d}\Gamma_f$) are the number of states of initial (final) particles; f_i (f_f) are the corresponding distribution functions, the sign + (−) corresponds to final bosons (fermions); q_0' is the neutrino pair energy; V is the plasma volume, τ is the interaction time, S_{if} is the S-matrix element.

To calculate S-matrix element of photoneutrino process we will utilize Lagrangian for electroweak interaction in the case of relatively small momentum transfers compared with W boson mass, $|q'^2| \ll m_W^2$, where q' is the four-vector of neutrino pair. Under this condition, the effective Lagrangian can be written as follows:

$$\mathcal{L} = \frac{G_F}{\sqrt{2}} \left[\bar{\Psi}\gamma_\alpha(C_V - C_A\gamma_5)\Psi\right] j_\alpha + e(\bar{\Psi}\gamma_\alpha\Psi)A_\alpha\,, \qquad (3)$$

where $j_\alpha = \bar{\nu}\gamma_\alpha(1-\gamma_5)\nu$ is the neutrino current, A_α is the four-potential of the photon field, $C_V = \pm 1/2 + 2\sin^2\theta_W$, $C_A = \pm 1/2$, and θ_W is the Weinberg angle. Here, the upper sign corresponds to electron neutrinos ($\nu = \nu_e$), when there is exchange reaction both of W and Z bozons. The lower sign corresponds to the μ and τ neutrinos, when there is only Z bozon exchange.

The S-matrix element in the tree-level approximation is described by Feynman diagrams shown in Fig. 1. It should be noted that the resonance on the virtual electron occurs in the s-channel (the first diagram in Fig.1) only. In this case we can present the S-matrix element in the next form:

$$S^{s's}_{e\gamma\to e\nu\bar{\nu}} = \frac{eG_f}{\sqrt{2}} \int \mathrm{d}^4X\, \mathrm{d}^4Y\, j^*_\alpha(Y) A_\beta(X) [\bar{\Psi}^{s'}_{p'l'}(Y)\gamma_\alpha(C_V - C_A\gamma_5)\hat{S}(Y,X)\gamma_\beta \Psi^s_{pl}(X)]\,. \qquad (4)$$

The currents in Eq. (4) can be expressed through the amplitudes in the momentum space:

$$A_\beta(X) = \frac{e^{-\mathrm{i}(qX)}}{\sqrt{2q_0V}} A_\beta(q)\,, \qquad (5)$$

$$j_\alpha(Y) = \frac{e^{-\mathrm{i}(q'Y)}}{\sqrt{2q_0'V}} j_\alpha(q')\,. \qquad (6)$$

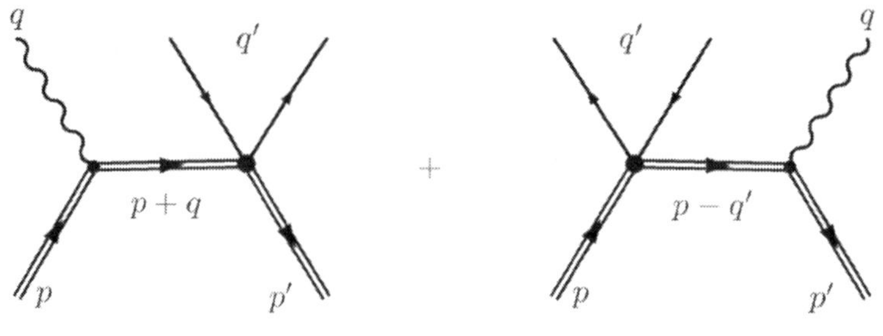

Figure 1. The Feynman diagrams for the reaction $\gamma e \to e\nu\bar{\nu}$. Double lines mean that the effects of external field on the initial and the final electron states and on the electron propagator are exactly taken into account.

There exist several descriptions of the procedure of obtaining the electron wave functions in the presence of an external magnetic field by solving the Dirac equation, see e.g. Refs. [14]–[20] and also Refs. [5, 6]. In the most cases, the solutions are presented in the form with the upper two components of the bispinor corresponding to the electron states with the spin projections 1/2 and -1/2 on the magnetic field direction. Here, we have found it more convenient to use another representation of the electron wave functions, being the eigenstates of the magnetic moment operator μ_z [16, 17]

$$\mu_z = m\Sigma_z - \mathrm{i}\gamma_0\gamma_5[\mathbf{\Sigma} \times \boldsymbol{P}]_z \,, \tag{7}$$

where $\boldsymbol{P} = -\mathrm{i}\boldsymbol{\nabla} + e\boldsymbol{A}$. We take the frame where the field is directed along the z axis, and the Landau gauge where the four-potential is: $A^\lambda = (0, 0, xB, 0)$. It is convenient to use the notation $\beta = eB$, and to introduce the sign of the fermion charge, η.

Our choice of the Dirac equation solutions as the eigenfunctions of the operator $\hat{\mu}_z$ is caused by the following arguments. Calculations of the process widths with two or more vertices in an external magnetic field by the standard method, including the squaring the amplitude with all the Feynman diagrams and with summation or averaging over the electron polarization states, contain significant computational difficulties. In this case, it is convenient to calculate partial contributions to the amplitude from the channels with different electron polarization states and for each diagram separately, by direct multiplication of the bispinors and the Dirac matrices. The result, up to a total for both diagrams non-invariant phase, will have an explicit Lorentz invariant structure. On the contrary, the amplitudes obtained with using the solutions for a fixed direction of the spin, do not have Lorentz invariant structure. Only the amplitude squared, summed over the electron polarization states, is manifestly Lorentz-invariant with respect to a boost along the magnetic field direction.

The electron wave functions having the form

$$\Psi^s_{p,n}(X) = \frac{e^{-\mathrm{i}(E_n X_0 - p_y X_2 - p_z X_3)}\; U^s_n(\xi)}{\sqrt{4E_n M_n (E_n + M_n)(M_n + m) L_y L_z}} \,, \tag{8}$$

where

$$E_n = \sqrt{M_n^2 + p_z^2}\,, \quad M_n = \sqrt{m^2 + 2\beta n}\,, \tag{9}$$

are the solutions of the equation

$$\hat{\mu}_z \Psi^s_{p,n}(X) = s\, M_n\, \Psi^s_{p,n}(X)\,, \quad s = \pm 1\,. \tag{10}$$

DOI: 10.1051/epjconf/201715805008

It is convenient to present the bispinors $U_n^s(\xi)$ in the form of decomposition over the solutions for negative and positive fermion charge, $U_{n,\eta}^s(\xi)$:

$$U_n^s(\xi) = \frac{1-\eta}{2}\, U_{n,-}^s(\xi) + \frac{1+\eta}{2}\, U_{n,+}^s(\xi)\,, \tag{11}$$

where

$$U_{n,-}^-(\xi) = \begin{pmatrix} -\mathrm{i}\,\sqrt{2\beta n}\, p_z V_{n-1}(\xi) \\ (E_n + M_n)(M_n + m)V_n(\xi) \\ -\mathrm{i}\,\sqrt{2\beta n}(E_n + M_n)V_{n-1}(\xi) \\ -p_z(M_n + m)V_n(\xi) \end{pmatrix}, \tag{12}$$

$$U_{n,-}^+(\xi) = \begin{pmatrix} (E_n + M_n)(M_n + m)V_{n-1}(\xi) \\ -\mathrm{i}\,\sqrt{2\beta n}\, p_z V_n(\xi) \\ p_z(M_n + m)V_{n-1}(\xi) \\ \mathrm{i}\,\sqrt{2\beta n}(E_n + M_n)V_n(\xi) \end{pmatrix}, \tag{13}$$

$$U_{n,+}^-(\xi) = \begin{pmatrix} \mathrm{i}\,\sqrt{2\beta n}\, p_z V_n(\xi) \\ (E_n + M_n)(M_n + m)V_{n-1}(\xi) \\ \mathrm{i}\,\sqrt{2\beta n}(E_n + M_n)V_n(\xi) \\ -p_z(M_n + m)V_{n-1}(\xi) \end{pmatrix} \tag{14}$$

$$U_{n,+}^+(\xi) = \begin{pmatrix} (E_n + M_n)(M_n + m)V_n(\xi) \\ \mathrm{i}\,\sqrt{2\beta n}\, p_z V_{n-1}(\xi) \\ p_z(M_n + m)V_n(\xi) \\ -\mathrm{i}\,\sqrt{2\beta n}(E_n + M_n)V_{n-1}(\xi) \end{pmatrix}, \tag{15}$$

$V_n(\xi)$ $(n = 0, 1, 2, \dots)$ are the normalized harmonic oscillator functions, which are expressed in terms of the Hermite polynomials $H_n(\xi)$ [21]:

$$V_n(\xi) = \frac{\beta^{1/4} \mathrm{e}^{-\xi^2/2}}{\sqrt{2^n n!\, \sqrt{\pi}}}\, H_n(\xi)\,, \tag{16}$$

$$\xi = \sqrt{\beta}\left(X_1 - \eta \frac{p_y}{\beta}\right). \tag{17}$$

Let us define functions $\phi_{p\ell}^s(X_1)$ (index p corresponds to electron momentum), based on bispinors $U_\ell^s(\xi)$:

$$\phi_{p\ell}^s(X_1) = \frac{U_\ell^s[\xi(X_1)]}{\sqrt{2M_\ell(E_\ell + M_\ell)(M_\ell + m)}}\,. \tag{18}$$

 DOI: 10.1051/epjconf/201715805008

We use the electron propagator in the form of the sum over the Landau levels [6, 22]:

$$\hat{S}(X,X') = \sum_{n=0}^{\infty} \hat{S}_n(X,X'), \tag{19}$$

$$\hat{S}_n(X,X') = \sum_{s=\pm 1} \hat{S}_n^s(X,X'), \tag{20}$$

$$\hat{S}_n^s(X,X') = \int \frac{\mathrm{d}p_0 \mathrm{d}p_y \mathrm{d}p_z}{(2\pi)^3} \frac{\mathrm{e}^{-\mathrm{i}(p(X-X'))_\parallel + \mathrm{i}p_y(X_2-X_2')}}{p_\parallel^2 - M_n^2 - \mathcal{R}_\Sigma^s(p) + \mathrm{i}\,\mathcal{I}_\Sigma^s(p)} \phi_{pn}^s(X_1)\bar{\phi}_{pn}^s(X_1'), \tag{21}$$

where ξ is defined by Eq. (17) and ξ' is obtained from ξ by substituting $X_1 \to X_1'$.

Hereafter we use the following notations: four-vectors with the indices $\perp$ and $\parallel$ belong to the Euclidean {1, 2} subspace and the Minkowski {0, 3} subspace correspondingly. Then for arbitrary 4-vectors A_μ, B_μ one has

$$A_\perp^\mu = (0, A_1, A_2, 0), \quad A_\parallel^\mu = (A_0, 0, 0, A_3),$$
$$(AB)_\perp = (A\Lambda B) = A_1 B_1 + A_2 B_2,$$
$$(AB)_\parallel = (A\widetilde{\Lambda} B) = A_0 B_0 - A_3 B_3,$$

where the matrices $\Lambda_{\mu\nu} = (\varphi\varphi)_{\mu\nu}$, $\widetilde{\Lambda}_{\mu\nu} = (\tilde{\varphi}\tilde{\varphi})_{\mu\nu}$ are constructed with the dimensionless tensor of the external magnetic field, $\varphi_{\mu\nu} = F_{\mu\nu}/B$, and the dual tensor, $\tilde{\varphi}_{\mu\nu} = \frac{1}{2}\varepsilon_{\mu\nu\rho\sigma}\varphi^{\rho\sigma}$. The matrices $\Lambda_{\mu\nu}$ and $\widetilde{\Lambda}_{\mu\nu}$ are connected by the relation $\widetilde{\Lambda}^{\mu\nu} - \Lambda^{\mu\nu} = g^{\mu\nu} = \mathrm{diag}(1,-1,-1,-1)$, and play the roles of the metric tensors in the perpendicular ($\perp$) and the parallel ($\parallel$) subspaces respectively.

Real part $\mathcal{R}_\Sigma^{s''}(P)$ and imaginary part $\mathcal{I}_\Sigma^{s''}(P)$ of eigenvalue of electron mass operator are the result of calculation of radiative corrections to the electron mass, caused by combined action of magnetic field and plasma. The imaginary part $\mathcal{I}_\Sigma^{s''}(P)$ can be obtained from the results of [11, 12] and has the following form:

$$\mathcal{I}_\Sigma^{s''}(P) = -\frac{1}{2} P_0 \Gamma_n^{s''} \tag{22}$$

where $\Gamma_n^{s''}$ is the total width of the change of electron state. This width can be expressed in terms of the width of electron creation [23]:

$$\Gamma_n^{s''} = \Gamma_n^{(abs)\,s''} + \Gamma_n^{(cr)\,s''} \simeq \Gamma_{e_n \to e_\ell \gamma}^{(cr)\,s''} \left[1 + \mathrm{e}^{(E_n''-\mu)/T}\right] \tag{23}$$

$$\Gamma_n^{(cr)\,s''} = \sum_{\ell=0}^{n-1} \sum_{s''} \frac{1}{2E_n''} \int \frac{d^3k}{2q_0(2\pi)^3} f_\gamma(q_0) \frac{d^2p}{2E_\ell} f_e(E_\ell)(2\pi)^3 \delta^{(3)}(P - p'')|M_{e_\ell\gamma\to e_n}|^2 \tag{24}$$

where $f_\gamma(q_0)$ is the equillibrium distribution fuction of initial photon, $f_e(E_\ell)$ is the equillibrium distribution fuction of initial electron in the plasma rest frame.

Real part of mass operator, $\mathcal{R}_\Sigma^{s''}(P)$, determines the change of the electron dispertion properties in presence of magnetized plazma. It was shown in [5, 18] that mass corrections are negligibly small under conditions $B \lesssim 10^{16}$ G.

After integration in Eq. (4) over d^4X and d^4Y we obtain:

$$S^{s's}_{e\gamma\to e\nu\bar{\nu}} = \frac{i(2\pi)^3\delta^{(3)}(P-p'-q')}{\sqrt{2q_0V2q_0'V2E_\ell L_yL_z2E'_{\ell'}L_yL_z}}\mathcal{M}^{s's}_{e\gamma\to e\nu\bar{\nu}}, \tag{25}$$

where $\delta^3(P-p'-q') = \delta(P_0-E'_{\ell'}-q_0')\delta(P_y-p_y'-q_y')\delta(P_z-p_z'-q_z')$, $P_\alpha = (p+q)_\alpha$, $\alpha = 0,2,3$, and the amplitude $\mathcal{M}^{s's}_{e\gamma\to e\nu\bar{\nu}}$ can be presented in the following form:

$$\mathcal{M}^{s's}_{e\gamma\to e\nu\bar{\nu}} = i\frac{eG_F}{\sqrt{2}}\sum_{n=0}^{\infty}\sum_{s''}j^*_\alpha(q')A_\beta(q)\times \tag{26}$$

$$\times\int dX_1dY_1e^{-iX_1q_x+iY_1q'_x}\frac{\bar{\phi}^{s'}_{p'\ell'}(Y_1)\gamma_\alpha(C_V-C_A\gamma_5)\phi^{s''}_{Pn}(Y_1)\phi^{s''}_{Pn}(X_1)\gamma_\beta\phi^{s}_{p\ell}(X_1)}{P_\parallel^2-M_n^2+i\mathcal{I}^{s''}_\Sigma(P)(X_1)}$$

Assuming that the main contribution to amplitude is given by resonance region, we can approximately replace the part of the amplitude by the δ-function [13]:

$$|\mathcal{M}^{s's}_{e\gamma\to e\nu\bar{\nu}}|^2 \simeq \frac{(eG_F)^2}{2}\sum_{n=0}^{\infty}\sum_{s''}\frac{\pi}{P_0\Gamma^{s''}_n}\delta(P_\parallel^2-M_n^2)\big|j_\alpha(q')A_\beta(q)\times \tag{27}$$

$$\times\int dX_1dY_1\bar{\phi}^{s'}_{p'\ell'}(Y_1)\gamma_\alpha(C_V-C_A\gamma_5)\phi^{s''}_{Pn}(Y_1)\phi^{s''}_{Pn}(X_1)\gamma_\beta\phi^{s}_{p\ell}(X_1)\big|^2.$$

After applying the delta-finction properties:

$$\delta(P_\parallel^2-M_n^2) = \frac{1}{2E''_n}\delta(P_0-E''_n), \tag{28}$$

where $E''_n = \sqrt{p_z''^2+M_n^2}$, it is straightforward to show that amplitude squared, averaged over polarizations of initial photon can be factored by amplitudes of sub-processes:

$$|\mathcal{M}_{\gamma e\to e\nu\bar{\nu}}|^2 \simeq \sum_{n=1}^{\infty}\frac{\pi}{P_0\Gamma^{s''}_n}\delta(P_\parallel^2-M_n^2)\,|\mathcal{M}_{e_\ell\gamma\to e_n}|^2\,|\mathcal{M}_{e_n\to e_{\ell'}\nu\bar{\nu}}|^2 \tag{29}$$

Substituing Eq. (29)into the expression for the the luminocity (2), we obtain:

$$Q_{\gamma e\to e\nu\bar{\nu}} = \sum_{n=1}^{\infty}\sum_{\ell'=0}^{n-1}Q_{e_n\to e_{\ell'}\nu\bar{\nu}}, \tag{30}$$

where

$$Q_{e_n\to e_{\ell'}\nu\bar{\nu}} = \frac{1}{L_x}\int\frac{dp_y''dp_z''}{(2\pi)^2\,2E''_n}f_e(E''_n)\frac{dp_y'dp_z'}{(2\pi)^2\,2E'_{\ell'}}\left[1-f_e(E'_{\ell'})\right]\times$$

$$\times\frac{d^3p_1}{(2\pi)^3\,2E_1}\frac{d^3p_2}{(2\pi)^3\,2E_2}q_0'(2\pi)^3\,\delta^3_{0,y,z}(p''-p'-q')|\mathcal{M}_{e_n\to e_{\ell'}\nu\bar{\nu}}|^2 \tag{31}$$

is the neutrino luminocity due to process $e_n \to e_{\ell'} \nu\bar{\nu}$ [24]. Squared amplitude of the process $e_n \to e_{\ell'} \nu\bar{\nu}$ has the form:

$$
\begin{aligned}
|\mathcal{M}_{e_n \to e_{\ell'} \nu\bar{\nu}}|^2 = {} & G_{\mathrm{F}}^2 \left(\overline{C_V^2} + \overline{C_A^2}\right) \Big\{ \left[2q^2(\beta(n+\ell) + m^2) + m^2 q_\perp^2 \right] \times \\
& \times \left(\mathcal{I}_{n-1,\ell}^2 + \mathcal{I}_{n,\ell-1}^2 - \mathcal{I}_{n,\ell}^2 - \mathcal{I}_{n-1,\ell-1}^2 \right) - q^4 \left(\mathcal{I}_{n-1,\ell}^2 + \mathcal{I}_{n,\ell-1}^2 \right) + \\
& + m^2 q^2 \left(\mathcal{I}_{n,\ell}^2 - \mathcal{I}_{n-1,\ell-1}^2 \right) \Big\} - G_{\mathrm{F}}^2 \left(\overline{C_V^2} - \overline{C_A^2}\right) m^2 \Big\{ (2q_\parallel^2 - q_\perp^2) \times \\
& \times \left(\mathcal{I}_{n-1,\ell}^2 + \mathcal{I}_{n,\ell-1}^2 - \mathcal{I}_{n,\ell}^2 - \mathcal{I}_{n-1,\ell-1}^2 \right) + 3q^2 \left(\mathcal{I}_{n-1,\ell}^2 + \mathcal{I}_{n,\ell-1}^2 \right) \Big\} .
\end{aligned}
\tag{32}
$$

Here $I_{n,\ell} \equiv I_{n,\ell}\left(q_\perp^2/(2\beta)\right)$, $I_{n,\ell}(x) = \sqrt{\frac{\ell!}{n!}}\, \mathrm{e}^{-x/2} x^{(n-\ell)/2} L_\ell^{n-\ell}(x)$. and $L_n^k(x)$ – generalized Laguerre polinomials [21], The constants $\overline{C_V^2} = 0.93$ and $\overline{C_A^2} = 0.75$ are the results of summation over all channels of the neutrino production of the types ν_e, ν_μ, ν_τ.

The obtained luminocity (31) coincides up to notation with results of [24]. It also generalizes obtained in [13] expression to the case of arbitrary initial and final electron Landau level.

3 Conclusion

We have considered the neutrino photoproduction on an electron, $e\gamma \to e\nu\bar{\nu}$, in dense magnetized medium in resonant case.

It has been shown that in the case of resonance on the virtual electron, the neutrino emissivity due to the process $\gamma e_\ell \to e_{\ell'} \nu\bar{\nu}$ can be expressed in terms of the neutrino emissivity due to the process $e_n \to e_{\ell'} \nu\bar{\nu}$. Obtained results coincide up to notation with obtained earlier in [24]. They also generalize the results of recent studies [13] to the case, when initial and final electron occupy arbitrary Landau level.

References

[1] S. A. Olausen & V. M. Kaspi, Astrophys. J. Suppl. **212**, 6 (2014).

[2] P. Goldreich & W. H. Julian, Astrophys. J. **157**, 869 (1969).

[3] D. Lai, Rev. Mod. Phys. **73**, 629 (2001).

[4] A. K. Harding & D. Lai, Rept. Prog. Phys. **69**, 2631 (2006).

[5] A. V. Kuznetsov & N. V. Mikheev, *Electroweak Processes in External Electromagnetic Fields* (Springer-Verlag, New York, 2003).

[6] A. V. Kuznetsov & N. V. Mikheev, *Electroweak Processes in External Active Media* (Springer-Verlag, Berlin, Heidelberg, 2013).

[7] D.A. Rumyantsev and M.V. Chistyakov J. Exp. Theor. Phys. **107**, 533 (2008)

[8] A.V. Borisov, B.K. Kerimov and P. E. Sizin, in: Particle Physics at the Tercentenary of Mikhail Lomonosov, Proc. 15th Lomonosov Conf. on Elementary Particle Physics, Moscow, 2011. Ed. A.I. Studenikin. World Sci., Singapore, 2013, p. 204-205

[9] A.V. Borisov, B.K. Kerimov and P. E. Sizin Phys. At.Nucl. **75** (11), 1305 (2012)

[10] N.V. Mikheev, D.A. Rumyantsev, M.V. Chistyakov J. Exp. Theor. Phys. **146**, 289 (2014)

[11] A. V. Borisov, A. S. Vshivtsev, V. C. Zhukovsky, and P. A. Eminov, Phys. Usp. **40**, 229 (1997) A. V. Kuznetsov, D. A. Rumyantsev, and D. M. Shlenev, Int. J. Mod. Phys. A **30**, 1550049 (2015)
[12] V. C. Zhukovsky, P. G. Midodashvili, and P. A. Eminov, JETP **79**, 509 (1994)
[13] A. V. Kuznetsov, D. A. Rumyantsev, and D. M. Shlenev, Int. J. Mod. Phys. A **30**, 1550049 (2015)
[14] M. H. Johnson & B. A. Lippmann, Phys. Rev. **76**, 828 (1949).
[15] A. I. Akhiezer & V. B. Berestetskii, *Quantum Electrodynamics*. 2nd ed. (Wiley, New York 1965).
[16] A. A. Sokolov & I. M. Ternov, *Synchrotron Radiation* (Pergamon, Oxford, 1968).
[17] D. B. Melrose & A. J. Parle, Aust. J. Phys. **36**, 755 (1983).
[18] A. A. Sokolov & I. M. Ternov, *Radiation from Relativistic Electrons* (American Institute of Physics, New York 1986).
[19] K. Bhattacharya & P. B. Pal, Pramana J. Phys. **62**, 1041 (2004).
[20] I. A. Balantsev, Yu. V. Popov, & A. I. Studenikin, J. Phys. A **44**, 255301 (2011).
[21] I. S. Gradshteyn & I. M. Ryzhik, *Table of Integrals, Series, and Products* (Academic, New York, 1980).
[22] A. V. Kuznetsov & A. A. Okrugin, Int. J. Mod. Phys. A **26**, 2725 (2011).
[23] H. A. Weldon, Phys. Rev. D **28**, 2007 (1983).
[24] D.G. Yakovlev, A.D. Kaminker, O.Y. Gnedin and P. Haensel, Phys. Rep. **354**, 1

Astronomical Tests for Extended Gravity: Possible Constraints on f(R) with Vanishing Cosmological Constant

Stanislav Alexeyev[1,*], *Boris* Latosh[2,3], and *Vsevolod* Echeistov[4]

[1]*Sternberg Astronomical Institute, Lomonosov Moscow State University, Universitetsky Prospekt, 13, Moscow 119234, Russia*
[2]*Department of Theoretical Physics, Faculty of Natural and Engineering Science, Dubna International University, Universitetskaya Str., 19, Dubna, Russia*
[3]*Bogoliubov Laboratory of Theoretical Physics, Joint Institute for Nuclear Research, Joliot-Curie 6, 141980 Dubna, Russia*
[4]*Department of Astrophysics and Stellar Astronomy, Physics Faculty, Lomonosov Moscow State University, Vorobievi Gori, 2/1, Moscow 119234, Russia*

Abstract. We discuss the predictions of Starobinsky model of $f(R)$-gravity with vanishing cosmological constant at the scales, typical for galaxies and galaxy clusters. We show that at current observational accuracy there is no significant difference in cut-off radius values for Starobinsky model and general relativity. This occurs from the masses of galaxies (from $10^9 M_{Sun}$), till galaxy clusters ones (till $10^{18} M_{Sun}$). In addition we demonstrate that cut-off radius value dependence against Starobinsky model parameter n increases during increasing n for all mass range. Despite the fact that some models provide the cut-off radius value less than in general relativity no any model comes outside the existing observational accuracy frames.

1 Introduction

The general relativity (GR) is widely treated as main theory of gravitation. It successfully describes all observable gravitational phenomena on the scales of the solar system [1]. The recent discovery of gravitational waves [2] confirmed the applicability of GR to describe the gravitational interaction [3] over a wide range of distances. Nevertheless, there is reason to believe that GR is not able to provide a fully correct description of gravity on large scales and in the regime of huge energies. Indication of the incomplete astrophysical applicability of GR is the Universe accelerated expansion [4]. This phenomenon is described by a cosmological constant most accurately. As a consequence, the introduction of an additional constant term to the action causes difficulties in matching with the quantum theory [5, 6]. The incomplete applicability of GR in the high-energy regime is indicated by several phenomena. First, GR is not renormalizable [7], therefore, in the quantum regime, it could be treated only as an effective field theory [8]. Further, the construction of self-consistent cosmological models requires the introduction of an inflationary stage, which cannot be obtained directly from GR [9]. To try to solve this problem the gravity models were extended by additional

*e-mail: alexeyev@sai.msu.ru

fields and curvature expansions [10–12]. Here we discuss of $f(R)$-gravity (the Lagrangian is a function f of scalar curvature R [13]). The choice of function f is not arbitrary, otherwise instabilities may appear [14, 15]. Lagrangians of such type arise during taking into account the radiative corrections of quantum field theory in curved space-time [16], and also in string theories [17].

Further, the Newtonian theory of gravitation can be treated as asymptotes of a more general theory of gravity (GR) in weak gravitational fields. Following this logic, we can assume that GR itself is an approximation on the scales of the solar system of a more general theory. Such approach allows considering the extended theories of gravity of various kinds necessary for the description of dark matter and dark energy. For an explanation of dark matter and dark energy problems are offered, on the one hand, the undiscovered population of objects of known types (black holes and other objects) and new types of particles and fields (" new physics "), on the other is extended theory of gravity [12, 18] (" new gravity "). The combination of these approaches makes possible to use the strengths of each of them. For example, in the framework of the scalar-tensor gravity [10].

The problem of dark energy and, consequently, the cosmological constant, is a strong stimulus for the modification of GR. Some models treat dark energy as the effect of the space-time geometry: massive gravity, where the gravitational interaction carrier (graviton) itself has a nonzero mass, theories with violation of Lorentz invariance (under extreme conditions), or, theories with additional fields or space-time dimensions (which can also be represented as additional fields) [10, 19]. Taking into account the relationship between $f(R)$ gravity and scalar-tensor models, their mutual transformations [20], as a first step we discuss the model proposed in [15]. It is constructed to satisfy two requirements. The function $f(R)$ must be regular and becomes equal to zero for $R = 0$. Models in which the function f diverges in the region of small curvature are indistinguishable from GR or do not provide an adequate description of the existing observational data (for a detailed discussion, see [15]). In addition, the scale of the curvature Λ is introduced to the model and is interpreted as the magnitude of the effective cosmological constant. For $R \gg \Lambda$, the function f tends to $R - 2\Lambda$. Here the effective cosmological constant disappear in the low curvature regime, therefore models of such class are called " models with a vanishing cosmological constant". In the framework of this paper, for the sake of simplicity, we refer to the model proposed in [15] as "Starobinsky model".

In [15] the following form of the function f is suggested:

$$f(R) = R + \lambda R_0 \left(\left(1 + \frac{R^2}{R_0^2} \right)^{-n} - 1 \right), \tag{1}$$

where R is the scalar curvature, λ, R_0 and n are the model parameters. It is assumed that $\lambda > 0$, and R_0 coincides with the cosmological constant in order of magnitude. As it was shown in [15] for large values of curvature (effective) the cosmological constant is expressed in terms of the parameters as:

$$\Lambda = \frac{\lambda}{2} R_0 \, . \tag{2}$$

Therefore for convenience we suppose that $\lambda = 2$ and $R_0 = \Lambda$.

According to [15], in the intermediate curvature mode, when $0 < R < \Lambda$, the Starobinsky model differs from GR and can describe the effects of dark matter. Laboratory tests and observations in the solar system provide the following limitation: $n \geqslant 2$. The model should describe gravitational phenomena on all scales; therefore we study its manifestations on the scales of galaxies and clusters of galaxies.

In order to verify the model, it is necessary to use values associated with real observational data. It is useful to address the effects in gravitationally coupled systems. At this level, the Universe expansion manifests itself as a force acting equally on all bodies and preventing their gravitational connection.

The balance of the force of gravitational attraction and the force associated with the accelerated expansion defines a specific scale. This manifold is called as "the turning radius". It determines the maximum size of the gravitational bound system [21, 22].

Thus, it is possible to compare the turning radii predicted by GR and Starobinsky model and with the observed spatial scales of galaxies and galaxy clusters. Such an activity could make possible to establish additional restrictions on the parameters of the Starobinsky model on these scales.

2 Turning radius calculation

The spheroidal and some elliptical galaxies, as well as the so-called regular clusters of galaxies, have a spherical or close to it shape, small rotational velocities (in comparison with the dispersion of velocities), and a pronounced central part of [23–25]. Therefore, to describe the gravitational field of the galaxy, one can use a spherically symmetric metric, assuming that the most part of the mass is concentrated in the center. So, we can use the Schwarzschild type of the metric, which we represent in the form:

$$ds^2 = e^A dt^2 - e^{-A} dr^2 - r^2 d\Omega, \tag{3}$$

where $A = A(r)$ is unknown function.

Field equations in $f(R)$-gravity have the following form [13, 15]:

$$f'(R)R_{\mu\nu} - \frac{1}{2} f(R)g_{\mu\nu} - \left(\nabla_\mu \nabla_\nu - g_{\mu\nu}\Box\right) f'(R) = 0, \tag{4}$$

where $f'(R) = df(R)/dR$, ∇_μ is covariant divergence, $\Box = \nabla_\mu \nabla^\mu$ is d'Alambert operator. Eqs ((4)) represent nonlinear system of second order, which could be solved numerically. To simplify we descrease the equations order by the introducing a new variable a:

$$a = \frac{dA}{dr}. \tag{5}$$

Similarly, for simplicity, we assume the scalar curvature as the independent variable. As an equation connecting it with A and a we use its explicit form:

$$R = -\frac{2}{r^2} + e^A \left(\frac{da}{dr} + a^2 + \frac{4a}{r} + \frac{2}{r^2}\right). \tag{6}$$

From $(1, 1)$-component of eqs ((4)) it is possible to rewrite $d^2 f/dr^2$ and using $(0, 0)$-component of ((4)), to show that:

$$\frac{dR}{dr} = \frac{f'(R) \cdot R_{00}/g_{00} - f(R)/2}{(e^A(a + 2/r) + a/2)f''(R)}. \tag{7}$$

So, we consider the equations ((5)), ((6)) and (7) together:

$$\begin{cases} \dfrac{dA}{dr} = a\,, \\ \dfrac{da}{dr} = e^{-A}\left(R + \dfrac{2}{r^2}\right) - a^2 - \dfrac{4a}{r} - \dfrac{2}{r^2}\,, \\ \dfrac{dR}{dr} = \dfrac{f'(R) \cdot R_{00}/g_{00} - f(R)/2}{(e^A(a + 2/r) + a/2)f''(R)}\,, \end{cases} \tag{8}$$

where f is defined in ((1)) and Ricci tensor components are calculated using the metrics ((3)).

 DOI: 10.1051/epjconf/201715805009

The system ((8)) is solved with the modified Runge-Kutta method (TR-BDF2, [26]). The mass spectrum is chosen from $10^9 M_{Sun}$ to $10^{18} M_{Sun}$ because these values that are typical for individual galaxies (Mass of the Milky Way Galaxy — 4 $times 10^{11} M_{Sun}$ [27]) and for clusters (the characteristic masses of the clusters are $\sim 10^{15} M_{Sun}$ [28–30]) respectively. We checked the values of the parameter n from 1 to 9 using the following definition of the gravitational potential:

$$\phi = \frac{c^2}{2}(g_{00} - 1) = \frac{c^2}{2}(e^A - 1)\,, \tag{9}$$

as well as its dimensionless version:

$$\phi/c^2 = \frac{e^A - 1}{2}. \tag{10}$$

The turning radius, as indicated above, is located at the point where the force of gravitational attraction is balanced by the one associated with the Universe accelerated expansion. At this point the gravitational potential must vanish:

$$\frac{d\phi}{dr} = 0\,. \tag{11}$$

Since a is an independent variable, one can look for r_1 and r_2 for which $a(r_1) > 0$ and $a(r_2) < 0$, respectively. Further it is easy to calculate the desired value of the turning radius using the linear interpolation. The dependence of the radius against the mass of the object (a galaxy or a cluster of galaxies) for Starobinsky's models with various n values and GR is shown in Fig. 1. The dependencies are very close to the power law, so, on a logarithmic scale, they look linear. We approximate the dependence of the radius R_{TAR} of the turning one by the following form:

$$R_{TAR} = M^{\alpha} \times R_{TAR}(M_{Sol}), \tag{12}$$

Where R_{TAR} is expressed in kiloparsecs (kpc), and the mass of the object M is expressed in the masses of the Sun (M_{Sun}). In this logarithmic form:

$$log_{10}(R_{TAR}) = \alpha log_{10}(M) + \beta, \tag{13}$$

We illustrate the values of the ratio of the radii of the rotation of the Starobinsky model and GR against the mass graphically in Fig. 2. In the whole range of considered masses, the Starobinsky models with $n = 1$ and $n = 2$ have the least radii of zero mass, and therefore they should be compared with observations first of all.

3 Comparing with observational data

Direct measurements of the turning radius at this stage of tool development are unreal. Really it is necessary to observe directly the motion of a test body, for example, a star in a distant galaxy. Such observations are still beyond modern possibilities. But, taking the density profile obtained from gravitational lensing one can use it as a down limit of the size of the object. The turning radius would limit the maximum size of the object. It cannot be less than the size of the density profile. Similar profiles were obtained from the analysis of CLASH [28, 29] and other clusters of galaxies [30]. All these data are comparable with $n = 1$ and $n = 2$ models of Starobinsky, which have the smallest value of the turning radius, and also with GR.

Despite significant differences, none of the models discussed (even with $n = 1$ that does not pass the solar system tests [15]) does not deviate further than by 20% from the radius determined by GR and fixed by the observational data. Thus, the current level of accuracy of astronomical observations does not allow strengthening or confirming the earlier limitations on the Starobinsky model.

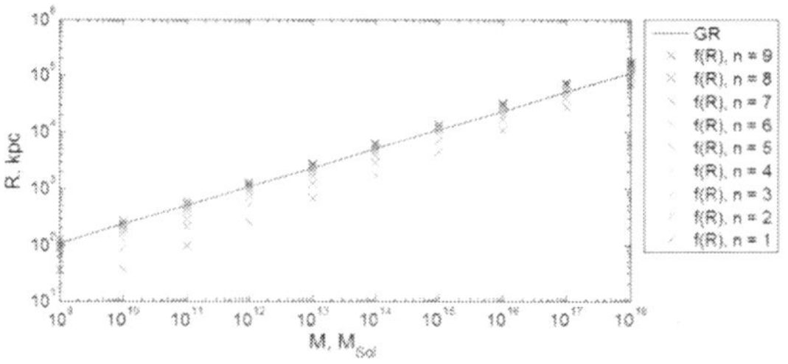

Figure 1. Dependence of the radius against the mass for the Starobinsky models with $n = 1..9$ (from the bottom to the top), as well as GR (direct line) in the logarithmic scale

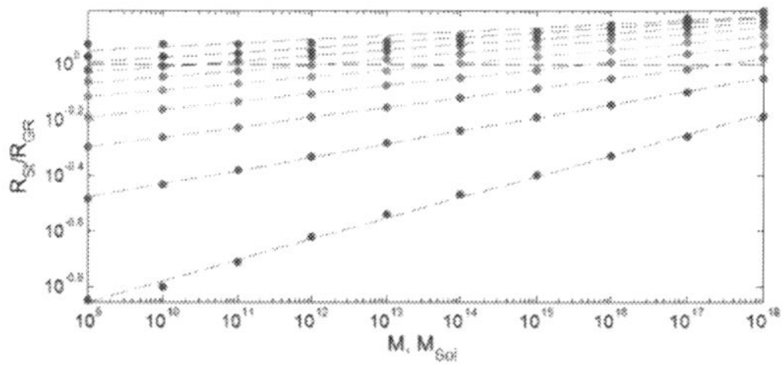

Figure 2. The ratio of the rotation radius of the Starobinsky model to the same value in GR against the mass. Here the points represent the results of numerical calculations; the dashed lines are approximations by linear dependences. For large n, nonlinear effects begin to appear.

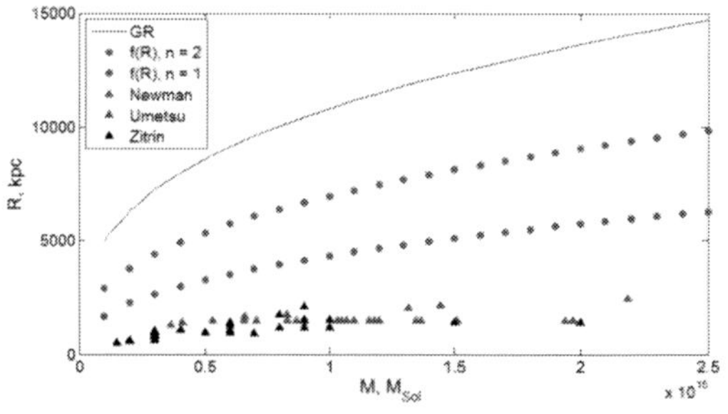

Figure 3. Comparison of the theoretical values of the turning radius with those obtained from observations of masses of clusters. The line denotes the values given by GR. The values for the Starobinsky models with $n = 1$ and $n = 2$ are denoted by round points, the values for $n = 2$ are above. Real data values are indicated by triangles.

4 Discussion and Conclusions

We considered the model of $f(R)$ - gravity with vanishing cosmological constant (the Starobinsky model) on scales of galaxies and their clusters. Turning radius values for masses of galaxy clusters are obtained, as a strict upper bound on the size of the clusters. As a result, it was found that the dependence of the turning radius against the mass of the object for the Starobinsky model differs from that obtained in GR very little. It is established that the turning radius increases with increasing parameter n in all mass range considered. Despite the fact that some models give a much smaller turning radius than GR, no model exceeded the limits set by observational data: 20% of the radius given by GR.

We can only conclude that the achieved level of accuracy does not allow obtaining additional restrictions on the Starobinsky model with a vanishing cosmological constant on the scales of clusters of galaxies. Thus, at the existing level of accuracy, the Starobinsky model of $f(R)$ gravity, and, consequently, the scalar-tensor models associated with this one by the transformations do not contradict the astronomical data on the scales of galaxies and their clusters. We also want to emphasize the perspectives of the proposed method for testing other extended gravity models.

Acknowledgements

The work was supported by the RFBR grant 16-02-00682. The authors are grateful to Maxim Pshirkov for interesting discussions and critical remarks.

References

[1] C.M. Will, Living Rev. Rel. **17**, 4 (2014), 1403.7377
[2] B.P. Abbott et al. (Virgo, LIGO Scientific), Phys. Rev. Lett. **116**, 061102 (2016), 1602.03837
[3] B.P. Abbott et al. (Virgo, LIGO Scientific), Phys. Rev. Lett. **116**, 221101 (2016), 1602.03841
[4] P.A.R. Ade et al. (Planck), Astron. Astrophys. **594**, A13 (2016), 1502.01589
[5] Ya.B. Zel'dovich, A. Krasinski, Ya.B. Zeldovich, Sov. Phys. Usp. **11**, 381 (1968), [Usp. Fiz. Nauk95,209(1968)]
[6] S. Weinberg, Rev. Mod. Phys. **61**, 1 (1989)
[7] M.H. Goroff, A. Sagnotti, Nucl. Phys. **B266**, 709 (1986)
[8] C.P. Burgess, Living Rev. Rel. **7**, 5 (2004), gr-qc/0311082
[9] A.D. Linde, Lect. Notes Phys. **738**, 1 (2008), 0705.0164
[10] S. Capozziello, M. De Laurentis, Phys. Rept. **509**, 167 (2011), 1108.6266
[11] T. Clifton, P.G. Ferreira, A. Padilla, C. Skordis, Phys. Rept. **513**, 1 (2012), 1106.2476
[12] E. Berti et al., Class. Quant. Grav. **32**, 243001 (2015), 1501.07274
[13] A. De Felice, S. Tsujikawa, Living Rev. Rel. **13**, 3 (2010), 1002.4928
[14] V. Faraoni, Phys. Rev. **D72**, 124005 (2005), gr-qc/0511094
[15] A.A. Starobinsky, JETP Lett. **86**, 157 (2007), 0706.2041
[16] A.A. Starobinsky, Phys. Lett. **B91**, 99 (1980)
[17] S. Nojiri, S.D. Odintsov, Phys. Lett. **B576**, 5 (2003), hep-th/0307071
[18] S. Capozziello, M. Francaviglia, Gen. Rel. Grav. **40**, 357 (2008), 0706.1146
[19] C. Charmousis, Lect. Notes Phys. **892**, 25 (2015), 1405.1612
[20] J. Ntahompagaze, A. Abebe, M. Mbonye, Int. J. Geom. Meth. Mod. Phys. **14**, 1750107 (2017), 1706.07722

[21] R. Nandra, A.N. Lasenby, M.P. Hobson, Mon. Not. Roy. Astron. Soc. **422**, 2931 (2012), 1104.4447
[22] V. Faraoni, Phys. Dark Univ. **11**, 11 (2016), 1508.00475
[23] J. Kormendy, D.B. Fisher, M.E. Cornell, R. Bender, Astrophys. J. Suppl. **182**, 216 (2009), 0810.1681
[24] B. Binggeli, Astronomy and Astrophysics **107**, 338 (1982)
[25] A.I. Zabludoff, M.J. Geller, Astron. J. **107**, 1929 (1994)
[26] R.B. et al., IEEE Transactions on Electron Devices **39**, 1031 (1983)
[27] M. Fich, S. Tremaine, Ann. Rev. Astron. Astrophys. **29**, 409 (1991)
[28] A. Zitrin et al., Astrophys. J. **801**, 44 (2015), 1411.1414
[29] K. Umetsu et al., Astrophys. J. **795**, 163 (2014), 1404.1375
[30] A.B. Newman, T. Treu, R.S. Ellis, D.J. Sand, C. Nipoti, J. Richard, E. Jullo, Astrophys. J. **765**, 24 (2013), 1209.1391

DOI: 10.1051/epjconf/201715805010

Inter-brane distance stabilization by bulk Higgs field in RS model

Vadim Egorov[1,2,★] and *Igor* Volobuev[1,★★]

[1] *Skobeltsyn Institute of Nuclear Physics, Lomonosov Moscow State University, 119991 Moscow, Russia*
[2] *Faculty of Physics, Lomonosov Moscow State University, 119991 Moscow, Russia*

Abstract. An extension of the Standard Model based on the Randall-Sundrum model with two branes is considered. The size of the extra dimension is stabilized by a five-dimensional two-component complex scalar field, the bulk Higgs field. All the matter fields, which are supposed to be localized on the brane with negative tension, acquire their masses due to the interaction with the boundary value of this field. The gauge invariance of the theory demands that the electroweak gauge fields also propagate in the bulk. The equations of motion for the background fields and their fluctuations against a background solution are found. The second variation Lagrangian is derived. The interactions of the bulk Higgs field with the multidimensional gauge and fermion fields are studied and a possible background solution is obtained.

1 Introduction

Models with extra space-time dimensions are widely discussed in theoretical physics in these latter days. One of the most interesting models of such type is the Randall-Sundrum model. Within its framework one considers two four-dimensional branes with tension interacting with gravity in a five-dimensional space-time [1]. The extra dimension forms the orbifold S^1/Z_2 and the branes are located at its fixed points. It is assumed that our world is located on one of the branes. The background metric is not flat, it has an exponential warp factor, which provides a solution to the hierarchy problem of the gravitational interaction.

In case the inter-brane distance is not stabilized, i.e. it is arbitrary, the model predicts the existence of a massless scalar mode, the radion, which couples to matter much stronger than gravity. It is not **in agreement** already with the classical gravity. In order to make the RS model phenomenologically acceptable, the Goldberger-Wise field is introduced into it [2, 3]. It is a 5D scalar field with a bulk potential and additional potentials on the branes. This allows one to stabilize the extra dimension size, giving a mass to the radion. The stabilization is achieved due to boundary conditions for the Goldberger-Wise field on the branes.

The idea of this report is to stabilize the size of extra dimension by a two-component complex scalar field propagating in the bulk and carrying the same representation of the gauge group $SU(2) \times U(1)$ as the usual Higgs field. It will act as the Higgs field on the brane, where our world is supposed to be located, providing spontaneous symmetry breaking. All the Standard Model fermion fields,

★e-mail: egorov@theory.sinp.msu.ru
★★e-mail: volobuev@theory.sinp.msu.ru

localized on the brane, will acquire their masses through the interaction with the boundary value of this field. Thus, we introduce an object, which can be called the bulk Higgs field.

The stabilization of the extra dimension size by the 5D Higgs field was discussed earlier in papers by L. Vecchi [4] and M. Geller et al. In the latter paper a perturbative background solution for gravity and the bulk scalar field was considered. Furthermore, the fact that the 5D theory is gauge invariant implies that there are corresponding gauge fields in the bulk. These fields have not been taken into consideration in the mentioned papers. Here we try to find an exact vacuum solution for gravity and the bulk Higgs field, while taking the 5D gauge fields into account in the linearized theory.

2 Equations of motion for background fields

Let us consider gravity interacting with two branes, a two-component scalar field $\phi = \begin{pmatrix} \phi_1 \\ \phi_2 \end{pmatrix}$ and bulk gauge fields A_M and B_M in a five-dimensional space-time $E = M_4 \times S^1/Z_2$. The action of the model can be written as follows:

$$S = S_g + S_\phi + S_{gauge} + S_{brane+SM}, \tag{1}$$

where the gravitational action S_g is

$$S_g = 2M^3 \int d^4x \int_{-L}^{L} dy R \sqrt{g}, \tag{2}$$

S_ϕ stands for the action of the two-component complex scalar field,

$$S_\phi = M \int d^4x \int_{-L}^{L} dy \left[(D_M\phi)^+ D^M\phi - V(\phi^+\phi) \right] \sqrt{g}, \tag{3}$$

the action S_{gauge} of the gauge fields is given by

$$S_{gauge} = -\int d^4x \int_{-L}^{L} dy \left[\frac{1}{4p^2} A^a_{MN} A^{a,MN} + \frac{1}{4q^2} B_{MN} B^{MN} \right] \sqrt{g}, \tag{4}$$

and $S_{brane+SM}$ is the action of the branes and the Standard Model,

$$S_{brane+SM} = -\int_{y=0} d^4x \lambda_1 (\phi^+\phi) \sqrt{-\tilde{g}} + \int_{y=L} d^4x \left[-\lambda_2 (\phi^+\phi) + L_{SM-HP}(\phi, \phi^+) \right] \sqrt{-\tilde{g}}. \tag{5}$$

Here D_M is the covariant derivative, $M = \frac{1}{2}(4\pi\hat{G})^{-\frac{1}{3}}$ stands for the fundamental five-dimensional energy scale of the theory, where $\hat{G}$ is the five-dimensional gravitational constant; p and q denote the gauge coupling constants of the groups $SU(2)$ and $U(1)$, respectively. L_{SM-HP} is the Lagrangian of the Standard model without the Higgs potential and the electroweak gauge field Lagrangian. The scalar field Lagrangian is chosen in the way that the field ϕ has the dimension of mass, i.e. the same dimension as a four-dimensional scalar field.

A background solution is sought in the standard form, which preserves the Poincaré invariance in any four-dimensional subspace $y = const$:

$$ds^2 = \gamma_{MN} dx^M dx^N = e^{-2A(y)} \eta_{\mu\nu} dx^\mu dx^\nu - dy^2, \tag{6}$$

$$\phi(x, y) = \phi(y)$$

and

$$A_\mu(x, y) = 0, \qquad A_4(x, y) = A_4(y), \qquad B_\mu(x, y) = 0, \qquad B_4(x, y) = B_4(y).$$

Variation of the action gives the equations of motion for the background fields:

$$\frac{1}{2}\left(\phi'^+\phi' + V + \frac{\lambda_1}{M}\delta(y) + \frac{\lambda_2}{M}\delta(y - L)\right) = 2M^2\left(3A'' - 6(A')^2\right), \tag{7}$$

$$12M^2(A')^2 + \frac{1}{2}(V - \phi'^+\phi') = 0, \tag{8}$$

$$\frac{dV}{d\phi} + \frac{1}{M}\frac{d\lambda_1}{d\phi}\delta(y) + \frac{1}{M}\frac{d\lambda_2}{d\phi}\delta(y - L) = \phi''^+ - 4A'\phi'^+, \tag{9}$$

$$\frac{dV}{d\phi^+} + \frac{1}{M}\frac{d\lambda_1}{d\phi^+}\delta(y) + \frac{1}{M}\frac{d\lambda_2}{d\phi^+}\delta(y - L) = \phi'' - 4A'\phi', \tag{10}$$

$$A_M = 0, \tag{11}$$

$$B_M = 0. \tag{12}$$

The requirement of the Poincare invariance leads to the vanishing of the background gauge field configurations. Thus, they do not affect the vacuum solution for the metric and the Higgs field.

3 Equations for the field fluctuations

Suppose we have a solution to these equations. In order to construct the linearized theory we represent the metric, the scalar field and the vector fields in the form "background solution + fluctuation":

$$g_{MN}(x, y) = \gamma_{MN}(y) + \frac{1}{\sqrt{2M^3}} h_{MN}(x, y), \tag{13}$$

$$\phi(x, y) = \phi_0(y) + f(x, y).$$

The gauge fields can be treated as fluctuations, because their vacuum values are zero.

Substituting this representation into the Lagrangian of the model and retaining only the terms of the second order in the deviations, we derive the so-called the second variation Lagrangian for the field fluctuations against a vacuum solution:

$$
\begin{aligned}
\frac{L^{(2)}}{\sqrt{\gamma}} &= \frac{1}{4}\left(\nabla_S h_{MN}\nabla^S h^{MN} + 2\nabla_M h^{MN}\nabla_N h - 2\nabla_M h^{MN}\nabla^S h_{SN} - \nabla_S h \nabla^S h\right) - \\
&- A''\left(h_{MN}h^{MN} - \frac{1}{2}h\tilde{h} + \frac{1}{2}h_{M\nu}h^{M\nu}\right) + (A')^2\left(\frac{7}{2}h_{MN}h^{MN} - hh\right) + \\
&+ \frac{1}{4M^2}\Bigg[\frac{V}{2}\left(h_{MN}h^{MN} - \frac{1}{2}hh\right) + \frac{1}{2M}\left[\lambda_1\delta(y) + \lambda_2\delta(y-L)\right]\left(h_{\mu\nu}h^{\mu\nu} - \frac{1}{2}\tilde{h}\tilde{h}\right) + \\
&\qquad + \phi_0'^{+}\phi_0'\left(-\frac{1}{4}hh + \frac{1}{2}h_{MN}h^{MN} - hh_{44} + 2h_{M4}h^{M4}\right)\Bigg] + \\
&+ M\Bigg[\partial_M f^+ \partial^M f + \phi_0^+\left(\frac{1}{4}A^a_M A^{a,M} + B_M B^M + A^a_M b^M \tau^a\right)\phi_0 + \\
&\qquad + iA^a_M\left(\phi_0^+\frac{\tau^a}{2}\partial^M f - \partial^M f^+\frac{\tau^a}{2}\phi_0\right) - iA^a_4\left(f^+\frac{\tau^a}{2}\phi_0' - \phi_0'^{+}\frac{\tau^a}{2}f\right) + \\
&\qquad + iB_M\left(\phi_0^+\partial^M f - \partial^M f^+\phi_0\right) - iB_4\left(f^+\phi_0' - \phi_0'^{+}f\right) - \\
&\qquad - \frac{1}{2}\left(\frac{d^2V}{(d\phi)^2}ff + 2f^+\frac{d^2V}{d\phi^+ d\phi}f + f^+f^+\frac{d^2V}{(d\phi^+)^2}\right)\Bigg] - \\
&- \frac{1}{2}\left[\frac{1}{p^2}\left(\partial_M A^a_N - \partial_N A^a_M\right)\partial^M A^a_K + \frac{1}{q^2}\left(\partial_M B_N - \partial_N B_M\right)\partial^M B_K\right]\gamma^{NK} - \\
&- \frac{1}{2}\Bigg[\left(\frac{d^2\lambda_1}{(d\phi)^2}ff + 2f^+\frac{d^2\lambda_1}{d\phi^+ d\phi}f + f^+f^+\frac{d^2\lambda_1}{(d\phi^+)^2}\right)\delta(y) + \\
&\qquad + \left(\frac{d^2\lambda_2}{(d\phi)^2}ff + 2f^+\frac{d^2\lambda_2}{d\phi^+ d\phi}f + f^+f^+\frac{d^2\lambda_2}{(d\phi^+)^2}\right)\delta(y-L)\Bigg] + \\
&+ \frac{1}{\sqrt{2M}}\left[\left(\phi_0'^{+}\partial^M f + \partial^M f^+\phi_0'\right)h_{M4} - \frac{1}{2}\left(\phi_0'^{+}f' + f'^{+}\phi_0' + \frac{dV}{d\phi}f + f^+\frac{dV}{d\phi^+}\right)h\right] - \\
&- \frac{1}{2\sqrt{2M^3}}\left[\left(\frac{d\lambda_1}{d\phi}f + f^+\frac{d\lambda_1}{d\phi^+}\right)\delta(y) + \left(\frac{d\lambda_2}{d\phi}f + f^+\frac{d\lambda_2}{d\phi^+}\right)\delta(y-L)\right]\tilde{h}. \qquad (14)
\end{aligned}
$$

Here $\gamma = \det\gamma_{MN}$, $h = h_{MN}\gamma^{MN}$, $\tilde{h} = h_{\mu\nu}\gamma^{\mu\nu}$ and ∇_M denotes the covariant derivative with respect to the background metric γ_{MN}. The indices are raised using the metric γ^{MN}.

Varying the action built with this Lagrangian one obtains the equations of motion for the fluctuations. Choosing a vacuum solution $\phi_0(y)$ having only a real lower component $v(y)/\sqrt{2}$, which breaks the gauge group $SU(2)\times U(1)$ to $U(1)_{em}$, and passing to the gauge, where the scalar field $f(x,y)$ also has only a real lower component, we get the equations for the scalar degrees of freedom in the following form:

1. $\mu\nu$-component of the metric fluctuation:

$$
\begin{aligned}
&\frac{1}{4}\left(\partial_\mu\partial_\nu\tilde{h} - 2\partial_\mu\partial_\nu h_{44}\right) + \frac{1}{4}\gamma_{\mu\nu}\left(-\partial_\sigma\partial^\sigma\tilde{h} + 2\partial_\sigma\partial^\sigma h_{44} + \frac{3}{2}\partial_4\partial_4\tilde{h}\right) - \\
&-\frac{1}{2}\gamma_{\mu\nu}A'\left(4\partial_4\tilde{h} + 3\partial_4 h_{44}\right) + \frac{1}{2}\gamma_{\mu\nu}(A')^2\left(12h_{44} + \tilde{h}\right) - \\
&-\frac{1}{4}\gamma_{\mu\nu}A''\left(\tilde{h} + 6h_{44}\right) + \frac{1}{\sqrt{2M}}\gamma_{\mu\nu}\left[\phi_0'^{+}f' + \left(\phi_0''^{+} - 4A'\phi_0'^{+}\right)f\right] = 0; \qquad (15)
\end{aligned}
$$

2. μ4-component of the metric fluctuation:

$$\frac{3}{4}\partial_4\partial_\mu\tilde{h} - 3A'\partial_\mu h_{44} + \sqrt{\frac{2}{M}}\phi_0'^{+}\partial_\mu f = 0; \tag{16}$$

3. 44-component of the metric fluctuation:

$$\frac{3}{4}\partial_\mu\partial^\mu\tilde{h} + 3A'\partial_4\tilde{h} + \frac{1}{2M^2}Vh_{44} + \sqrt{\frac{2}{M}}\left[\phi_0'^{+}f' - \frac{1}{2}\left(\frac{dV}{d\phi}f + f^{+}\frac{dV}{d\phi^{+}}\right)\right] = 0; \tag{17}$$

4. equation for the field f:

$$\begin{aligned} &M\left(\partial_M\partial^M f + 4A'f' + \frac{d^2V}{d\phi^{+}d\phi}f + f^{+}\frac{d^2V}{(d\phi^{+})^2}\right) - \\ &-\frac{1}{\sqrt{2M}}\left[\frac{1}{2}\phi_0'\left(\partial_4\tilde{h} + \partial_4 h_{44}\right) + \left(\phi_0'' - 4A'\phi_0'\right)h_{44}\right] + \\ &+\left(\frac{d^2\lambda_1}{d\phi^{+}d\phi}f + f^{+}\frac{d^2\lambda_1}{(d\phi^{+})^2}\right)\delta(y) + \left(\frac{d^2\lambda_2}{d\phi^{+}d\phi}f + f^{+}\frac{d^2\lambda_2}{(d\phi^{+})^2}\right)\delta(y-L) + \\ &+\frac{1}{2\sqrt{2M^3}}\left[\frac{d\lambda_1}{d\phi^{+}}\delta(y) + \frac{d\lambda_2}{d\phi^{+}}\delta(y-L)\right]h_{44} = 0. \end{aligned} \tag{18}$$

The remarkable result here is that the fluctuations of the scalar components of the 5D gauge fields, namely A_4 and B_4, vanish due to the equations of motion and the gauge condition choice. Hence there are no extra light scalars, which allows one to correctly reproduce the electroweak sector of the Standard Model in the effective four-dimensional theory.

4 Radion-like interactions of the Higgs boson

Since the bulk Higgs field in the theory is the stabilizing Goldberger-Wise field, the Higgs boson here is the radion at the same time. The radion as a separate particle does not exist. The Higgs boson inherits the interaction with the energy-momentum tensor, which is intrinsic to the radion. The interaction term can be written as [6]

$$S \supset -\frac{1}{\sqrt{8M^3}}\int dx\int_{-L}^{L} dyT^{MN}h_{MN}\sqrt{\gamma} = \frac{1}{\sqrt{8M^3}}\int dx\int_{-L}^{L} dy\left(-\frac{1}{2}T_\mu^\mu - T_{44}\right)e^{-2A}g, \tag{19}$$

where $g = e^{-2A(y)}h_{44}(x,y)$. The field g, in fact, represents the Higgs field f because h_{44} and f are connected by the gauge condition choice and thereby represent the same degree of freedom.

Let us consider the bulk vector field Z_M interacting with the field g. Its Lagrangian is as follows:

$$L_Z = -\frac{1}{4}Z_{MN}Z^{MN} + \frac{1}{2}\phi_0^{+}\phi_0 Z_M Z^M. \tag{20}$$

Here we have ϕ_0 squared instead of m_Z^2 since we consider the situation previous to the electro-weak symmetry breaking. Performing the mode decomposition

$$Z_\mu(x,y) = \sum_n z_\mu^n(x)\,Z_n(y) \tag{21}$$

(where the gauge condition $Z_4(x, y) = 0$ is chosen), one obtains the interaction term as follows:

$$S \supset \frac{1}{\sqrt{8M^3}} \sum_n \int dx \int_{-L}^{L} dy \left(\frac{1}{4} \eta^{\rho\sigma} z^n_{\mu\rho}(x) z^n_{\nu\sigma}(x) - z^n_\mu(x) \sum_m z^m_\nu(x) B_{nm} + \right.$$
$$\left. + m^2_{Zn} z^n_\mu(x) z^n_\nu(x) \right) \eta^{\mu\nu} e^{-2A} Z^2_n(y) g, \tag{22}$$

where $B_{nm} = \int_{-L}^{L} dy e^{-2A} \phi_0^+ \phi_0 Z_n(y) Z_m(y)$, $Z^n_{\mu\nu}(x, y) = z^n_{\mu\nu}(x) Z_n(y)$. Note that the interaction Lagrangian turns out to be non-diagonal, which means that there is KK number non-conservation. Moreover, the couplings of the excited bulk states to the Higgs boson are not characterized by one coupling constant, as in the case of brane localized fields, but depend on the overlap integrals of their wave functions and the Higgs wave function. A similar interaction Lagrangian can be obtained for the W-boson field.

5 Specific example of a non-perturbative background solution

Let us choose the following ansatz:

$$V = \frac{1}{4} \frac{dW}{d\phi} \frac{dW}{d\phi^+} - \frac{1}{24M^2} \left(W(\phi^+\phi)\right)^2, \tag{23}$$

$$\phi'(y) = \text{sign}(y) \frac{1}{2} \frac{dW}{d\phi^+}, \quad \phi'^+(y) = \text{sign}(y) \frac{1}{2} \frac{dW}{d\phi}, \tag{24}$$

$$A'(y) = \text{sign}(y) \frac{1}{24M^2} W(\phi^+\phi). \tag{25}$$

In this case the equations of motion (7)–(10) are valid everywhere, except the branes. In order the equations of motion be valid everywhere, one needs to make a consistent choice of the brane potentials $\lambda_{1,2}(\phi^+\phi)$ and the bulk potential $V(\phi^+\phi)$.

Let us take the function W to be

$$W = 24M^2 k - 2u\phi^+\phi, \tag{26}$$

then the brane potentials $\lambda_{1,2}$ should be chosen as follows:

$$\lambda_1(\phi^+\phi) = MW(\phi^+\phi) + \beta_1 \left(\phi^+\phi - \frac{v_1^2}{2}\right)^2, \tag{27}$$

$$\lambda_2(\phi^+\phi) = -MW(\phi^+\phi) + \beta_2 \left(\phi^+\phi - \frac{v_2^2}{2}\right)^2.$$

Here k, u, $\beta_{1,2}$ and $v_{1,2}$ are the model parameters. Note that the potential λ_2 chosen in this simplest form has the same structure as the standard Higgs potential.

The corresponding solution looks like

$$\phi(y) = \begin{pmatrix} 0 \\ \frac{v}{\sqrt{2}} e^{-u(|y|-L)} \end{pmatrix}, \tag{28}$$

 DOI: 10.1051/epjconf/201715805010

$$A(y) = k(|y| - L) + \frac{v^2}{96M^2}\left(e^{-2u(|y|-L)} - 1\right).$$

Here $v = 246$ GeV is the vacuum value of the Higgs field. In this case all the fermion fields localized on the brane will get the same masses as in the interaction with the usual Higgs field.

The inter-brane distance is defined by the boundary conditions for the scalar field and is expressed in terms of the model parameters by the relation:

$$L = \frac{1}{u}\ln\left(\frac{v_1}{v}\right). \tag{29}$$

The size of the extra dimension becomes stabilized.

6 Estimate of the model parameters

Taking the mode decomposition of the previously defined field g we obtain the equations for g_n with the boundary conditions on the branes:

$$\frac{d}{dy}\left(\frac{e^{2A}}{\left(\phi_2'\right)^2}g_n'\right) - \frac{e^{2A}}{6M^2}g_n = -\mu_n^2 g_n\frac{e^{4A}}{\left(\phi_2'\right)^2}, \tag{30}$$

$$\left(\frac{1}{4M}\frac{d^2\lambda_1}{d\phi_2^2} - \frac{\phi_2''}{\phi_2'}\right)g_n' + \mu_n^2 e^{2A}g_n\Bigg|_{y=+0} = 0,$$

$$\left(\frac{1}{4M}\frac{d^2\lambda_2}{d\phi_2^2} + \frac{\phi_2''}{\phi_2'}\right)g_n' - \mu_n^2 e^{2A}g_n\Bigg|_{y=L-0} = 0.$$

Here ϕ_2 is the lower component of a background solution for the Higgs field. These equations define a mass spectrum of the infinite Kaluza-Klein tower of the field ϕ.

In the case $uL \ll 1$ we can get the following expression for the mass of the lowest excitation of the scalar field, which is identified with the Higgs boson [7]:

$$m_H^2 = \frac{v^2u^2}{3M^2}\frac{\beta_2 v^2 - uM}{\beta_2 v^2 + uk}. \tag{31}$$

The correct value of the Higgs mass can be obtained for the following values of the model parameters: $M = 2$ TeV and $\beta_2 \to \infty$, consequently, $u \simeq 1.76$ TeV, $\phi_1 = 345$ TeV, $k \simeq 186$ TeV, $L = 0.2$ TeV^{-1} $\simeq$ $2\cdot 10^{-18}$ cm. The coupling of the Higgs boson to the energy-momentum tensor of the Standard Model, given by

$$\epsilon_H = -\sqrt{\frac{k}{24M^3}}, \tag{32}$$

turns out to be of the order of 1 TeV^{-1}. It must significantly affect the properties of the Higgs boson in this model.

Further excitations of the field ϕ have masses of the order of hundreds of TeV and cannot be observed at the existing colliders.

Unfortunately, this model turns out to be a toy model: it does not allow one to reproduce the Standard Model Higgs interactions. Another background solution is needed, this is the matter of future research.

 DOI: 10.1051/epjconf/201715805010

7 Conclusion

In the present paper it has been shown that the bulk scalar field stabilizing the extra dimension size in the Randall-Sundrum model can act simultaneously as the Higgs field on the brane, where our world is supposed to be located, providing the spontaneous symmetry breaking there. The equations of motion for the background field configuration and for the field fluctuations against a vacuum solution are derived.

An important point is that the bulk electroweak gauge fields that necessarily appear in the model do not give rise to extra light scalars in the effective four-dimensional theory. The Higgs boson here is the radion at the same time, so it interacts with the energy-momentum tensor. These interactions with the bulk fields are studied. A remarkable result is that the coupling constants of these interactions depend on a particular excitation and there is no KK number conservation in such interactions.

A specific example of a non-perturbative vacuum solution is found. Based on it, the values of the model parameters are estimated, which give the correct value of the Higgs boson mass in the approximation of a small deviation of the metric of the stabilized model from the metric of the unstabilized one. Unfortunately, this background solution is not good due to its inability to correctly reproduce the Standard Model interactions of the Higgs boson, so it should be treated only as a toy model. Another exact vacuum solution is needed, which is the matter of further research.

8 Acknowledgements

The authors are grateful to E. Boos, S. Keizerov and M. Smolyakov for useful discussions. The work was supported by Grant No. 14-12-00363 of the Russian Science Foundation.

References

[1] L. Randall and R. Sundrum, Phys. Rev. Lett. **83**, 3370-3 (1999)
[2] W.D. Goldberger and M.B. Wise, Phys. Rev. Lett. **83**, 4922-5 (1999)
[3] O. DeWolfe *et al.*, Phys. Rev. D **62**, 046008 (2000)
[4] L. Vecchi, J. High Energy Phys. **1111**, 102 (2011)
[5] M. Geller, S. Bar-Shalom and A. Soni, Phys. Rev. D **89** (9), 095015 (2014)
[6] C. Csaki, J. Hubisz and S.J. Lee, Phys. Rev. D **76**, 125015 (2007)
[7] E.E. Boos *et al.*, Mod. Phys. Lett. A **21**, 1431 (2006)

Description of gravity in the model with independent nonsymmetric connection

N. V. Kharuk[1,⋆], *S. A.* Paston[1,⋆⋆], and *A. A.* Sheykin[1,⋆⋆⋆]

[1]*Saint Petersburg State University, Saint Petersburg, Russia*

Abstract. A generalization of General Relativity is studied. The standard Einstein-Hilbert action is considered in the Palatini formalism, where the connection and the metric are independent variables, and the connection is not symmetric. As a result of variation with respect to the metric Einstein equations are obtained. A variation with respect to the connection leads to an arbitrariness in the determination of connection, i.e. the presence of gauge invariance. Then a matter in a form of point particle which interacts with field of connection is introduced. Also the action is complemented by a kinetic term for field of the connection to avoid incompatible equation of motion. Thus after the variation procedures we obtain the Einstein equations, the geodesic equation and the Maxwell's equations for electromagnetism, where some components of the connection play the role of the electromagnetic potential. Thereby the electromagnetic potential is obtained from the geometry of space-time.

1 Introduction

Nowadays the General Relativity is a modern theory of gravity. It is described by curved 4-dimential pseudo-Riemannian space-time. Usually in the framework of this theory the main objects are the symmetric metric $g_{\mu\nu}$ and the connection $\Gamma^{\xi}_{\mu\nu}$, which also symmetric and depends on the metric. The equivalent approach is the so-called Palatini formalism in which the metric and the connection are considered as independent variables, but both are still symmetric.

In this work we consider a generalization of this theory. The connection is considered as nonsymmetric and independent of the metric object. The metric is symmetric as usual. In these assumptions connection $\Gamma^{\alpha}_{\rho\nu}$ is not equal to the Christoffel symbol. It leads to the fact that such theory is more extensive. For example, the contraction $R^{\alpha}{}_{\alpha\mu\nu}$ is not equal to zero and the Ricci curvature tensor $R_{\beta\nu} \equiv R^{\alpha}{}_{\beta\alpha\nu}$ is not symmetric anymore.

Such approach with the Einstein-Hilbert action for the first time was proposed only in 1978 in the paper [1], which was published only in Russian and remained practically unknown to the scientific community. The authors noticed that this theory can be interpreted as unified theory of gravity and electromagnetism, however they did not introduce an action for the matter. Instead the authors have to used some *ad hoc* physical assumptions to obtain correct equations of motions of test particles. In the

⋆e-mail: natakharuk@mail.ru
⋆⋆e-mail: s.paston@spbu.ru
⋆⋆⋆e-mail: a.sheykin@spbu.ru

same year a very similar theory was considered in the paper [2]. However the purpose of the authors was different: they did not try to connect their theory with electromagnetism. Nowadays the studies of Palatini formalism with nonsymmetric connection have been continuing, see. e.g. [3].

In the present paper we generalize the approach proposed in [1] by including a matter in the form of point particles into the action. We obtain a correct form of the action term for the interaction of such a matter with the connection and discover that it is necessary to identify the electromagnetic potential with the trace of the connection rather with the trace of the torsion (as it was done in [1]).

2 Theory without matter

To begin with, let us consider the Einstein-Hilbert action without matter in a standard form:

$$S_1 = -\frac{1}{2\varkappa}\int d^4x\,\sqrt{-g}g^{\mu\nu}R_{\mu\nu}(\Gamma), \tag{1}$$

where $R_{\beta\nu} = R^{\alpha}{}_{\beta\alpha\nu} = \partial_\alpha\Gamma^{\alpha}_{\nu\beta} - \partial_\nu\Gamma^{\alpha}_{\alpha\beta} + \Gamma^{\alpha}_{\alpha\xi}\Gamma^{\xi}_{\nu\beta} - \Gamma^{\alpha}_{\nu\xi}\Gamma^{\xi}_{\alpha\beta}$ is the Ricci tensor which depends only on connection.

To obtain equations of motion it is necessary to vary the action. It is easy task to vary with respect to the metric $g_{\mu\nu}$ because now action does not depend on the derivative of metric. Thereby one can get

$$R^{\nu\mu} + R^{\mu\nu} - Rg^{\mu\nu} = 0. \tag{2}$$

These equations differ from Einstein equations without matter because here $R^{\mu\nu}$ is not a symmetric tensor, so it contains an additional symmetrization.

Variation with respect to the connection gives more interesting result:

$$D_\rho g^{\sigma\nu} = \frac{1}{3}g^{\sigma\nu}S^{\alpha}_{\rho\alpha} + \frac{1}{3}g^{\sigma\xi}S^{\alpha}_{\xi\alpha}\delta^{\nu}_{\rho} + g^{\xi\sigma}S^{\nu}_{\rho\xi}, \tag{3}$$

where $S^{\rho}_{\mu\nu} = \Gamma^{\rho}_{\mu\nu} - \Gamma^{\rho}_{\nu\mu}$ is a torsion. This equation is solved by

$$\Gamma^{\rho}_{\mu\nu} = \bar{\Gamma}^{\rho}_{\mu\nu} + \frac{1}{4}A_\mu\delta^{\rho}_{\nu}, \tag{4}$$

where $\bar{\Gamma}^{\rho}_{\mu\nu} = \frac{1}{2}g^{\rho\sigma}(\partial_\mu g_{\sigma\nu} + \partial_\nu g_{\mu\sigma} - \partial_\sigma g_{\nu\mu})$ is the Christoffel symbols and A_μ is an arbitrary vector. This formula can be rewritten in a form:

$$\Gamma^{\rho}_{\mu\nu} - \frac{1}{4}\Gamma^{\alpha}_{\mu\alpha}\delta^{\rho}_{\nu} = \bar{\Gamma}^{\rho}_{\mu\nu} - \frac{1}{4}\bar{\Gamma}^{\alpha}_{\mu\alpha}\delta^{\rho}_{\nu}, \tag{5}$$

by substitution trace of relation (4). Now it is clear that the trace of the connection has not any restrictions. In this sense, this theory has additional gauge symmetry.

Taking (4) into account the explicit form of the curvature tensor is

$$R^{\mu}{}_{\nu\lambda\rho} = \bar{R}^{\mu}{}_{\nu\lambda\rho} + \frac{1}{4}(\partial_\lambda A_\rho - \partial_\rho A_\lambda)\delta^{\mu}_{\nu}, \tag{6}$$

where $\bar{R}^{\mu}{}_{\nu\lambda\rho}$ is the usual Riemann curvature tensor which depends on the Christoffel symbols $\bar{\Gamma}^{\rho}_{\mu\nu}$ and thus can be expressed through metric. Taking trace of (6) one can find that

$$R^{\mu}{}_{\mu\lambda\rho} = \partial_\lambda A_\rho - \partial_\rho A_\lambda \tag{7}$$

and

$$R^{\mu}{}_{\lambda\mu\rho} \equiv R_{\lambda\rho} = \bar{R}_{\lambda\rho} + \frac{1}{4}(\partial_\lambda A_\rho - \partial_\rho A_\lambda), \tag{8}$$

where $\bar{R}_{\lambda\rho}$ is a Riemannian and therefore symmetric Ricci tensor. The scalar curvature:

$$R \equiv R_{\lambda\rho} g^{\lambda\rho} = \bar{R}_{\lambda\rho} g^{\lambda\rho} \tag{9}$$

does not change. Thereby the equation (2) can be rewritten as:

$$\bar{R}^{\nu\mu} - \frac{1}{2} R g^{\mu\nu} = 0. \tag{10}$$

So it is exactly the vacuum Einstein's equations.

3 Addition of matter

The next step is an addition of a matter in the theory. We will consider it in the form of a set of relativistic point particles. For the sake of simplicity we will write the formulas for a single particle. This particle with a world line $x^\mu(\tau)$ in the gravitational field with the metric $g_{\mu\nu}$ is described by standard action:

$$S_2 = -m \int d\tau \sqrt{\dot{x}^\mu(\tau)\dot{x}^\nu(\tau) g_{\mu\nu}(x(\tau))}, \tag{11}$$

where m is a mass of particle.

In order to get a more general theory the interaction of classical particles with the connection is introduced. One of the most simplest way is

$$S_3 = -q \int d\tau \dot{x}^\mu(\tau) \Gamma^\nu_{\mu\nu}(x(\tau)), \tag{12}$$

where q is just a parameter[1].

Thus the theory consists of three terms $S_1 + S_2 + S_3$. But it turns out that such theory is self-inconsistent. Indeed, after a variation with respect to the connection the following equations of motion arise:

$$\dot{x}^\mu(\tau) = 0 \tag{13}$$

which are incompatible with the normalization of four-velocity. This result is connected with the presence of gauge invariance in S_1 (see after (5)). To avoid this problem one can add one more term to the total action. It is the kinetic term which can be written in different ways, but again the most simple case is chosen:

$$S_4 = -\frac{1}{16\pi} \int d^4x \sqrt{-g} R^{\mu}{}_{\mu\alpha\beta} R^{\mu}{}_{\mu\delta\gamma} g^{\alpha\delta} g^{\beta\gamma}, \tag{14}$$

where the constant $\frac{1}{16\pi}$ is chosen for convenience. It is worth noting that the same term was proposed in [1], but without above motivation since the action of matter was not considered in their work.

As a result the total action consists of four terms:

$$S = S_1 + S_2 + S_3 + S_4 = -\frac{1}{2\varkappa} \int d^4x \sqrt{-g} R - m \int d\tau \sqrt{\dot{x}^\mu(\tau)\dot{x}^\nu(\tau) g_{\mu\nu}(x(\tau))} -$$
$$- q \int d\tau \dot{x}^\mu(\tau) \Gamma^\nu_{\mu\nu}(x(\tau)) - \frac{1}{16\pi} \int d^4x \sqrt{-g} R^{\mu}{}_{\mu\alpha\beta} R^{\mu}{}_{\mu\delta\gamma} g^{\alpha\delta} g^{\beta\gamma}. \tag{15}$$

[1] Another simple combination is the contraction of $\dot{x}^\mu$ with the trace of torsion $S^\nu_{\mu\nu}$ instead of the trace of connection $\Gamma^\nu_{\mu\nu}$. We will discuss it in the section 4.

This is the final view of the action. It depends on three independent variables: the metric $g_{\mu\nu}$, the connection $\Gamma^{\alpha}_{\beta\gamma}$ and the coordinate of the particle $x^{\mu}(\tau)$.

Let us obtain the complete set of field equations by varying with respect to these variables. Firstly, let us consider the variation with respect to the metric. Only terms S_1, S_2 and S_4 depend on $g_{\mu\nu}$. The contribution of S_1 is already found in (10). The variation of S_2 and S_4 can be easily calculated. Finally equations of motion corresponding to the variation of the metric are:

$$\bar{R}^{\mu\nu} - \frac{1}{2} R g^{\mu\nu} = \varkappa (T_1^{\mu\nu} + T_2^{\mu\nu}), \tag{16}$$

where

$$T_1^{\mu\nu} = \rho_m u^{\mu} u^{\nu} \tag{17}$$

is a stress-energy tensor of a relativistic particle,

$$T_2^{\mu\nu} = -\frac{1}{4\pi} (R^{\xi}{}_{\xi}{}^{\mu\alpha} R^{\beta}{}_{\beta}{}^{\nu}{}_{\alpha} - \frac{1}{4} g^{\mu\nu} R^{\xi}{}_{\xi}{}^{\alpha\beta} R^{\varphi}{}_{\varphi\alpha\beta}), \tag{18}$$

$u^{\mu} = \dot{x}^{\mu} \frac{1}{\sqrt{\dot{x}^{\alpha}\dot{x}^{\beta} g_{\alpha\beta}}}$ is a four-velocity and $\rho_m = m \int ds \delta(x - x(s)) \frac{1}{\sqrt{-g(x(s))}}$ is a mass density. The expression for $T_2^{\mu\nu}$ can be rewritten in a more recognizable form using notation $R^{\varphi}{}_{\varphi\alpha\beta} \equiv F_{\alpha\beta}$:

$$T_2^{\mu\nu} = -\frac{1}{4\pi} (F^{\mu\alpha} F^{\nu}{}_{\alpha} - \frac{1}{4} g^{\mu\nu} F^{\alpha\beta} F_{\alpha\beta}). \tag{19}$$

In this form T_2 reproduces the stress-energy tensor of electromagnetic field with electromagnetic tensor $F_{\alpha\beta}$.

Next the connection $\Gamma^{\alpha}_{\mu\nu}$ is varied. The term S_2 does not depend on the connection. The result for S_1 is already found in (3). After calculations for S_3 and S_4 the following equation is obtained:

$$\Gamma^{\rho}_{\mu\nu} = \bar{\Gamma}^{\rho}_{\mu\nu} + \frac{1}{4} A_{\mu} \delta^{\rho}_{\nu}, \tag{20}$$

$$\bar{D}_{\mu} F^{\mu\nu} = 4\pi j^{\nu}, \tag{21}$$

where $\bar{D}_{\mu}$ is a covariant derivative with the Riemannian connection and $j^{\nu} = q \int ds u^{\nu} \delta(x - x(s)) \frac{1}{\sqrt{-g(x(s))}}$ is a four-current of relativistic particles if q is considered as an electric charge. The equation (21) is nothing but the Maxwell's equation. According to (20) and its consequence (7) the role of electromagnetic potential corresponding to $F^{\mu\nu}$ is played by the quantity A_{ρ}.

Finally, the variation with respect to the particle coordinate is calculated. It is necessary to vary only S_2 and S_3. The result is already known since S_2 has a standard view and S_3 can be treated as interaction term for a particle with electromagnetic potential $\Gamma^{\alpha}_{\beta\alpha}$. So the corresponding equation of motion is

$$m u^{\mu} \bar{D}_{\mu} u^{\alpha} = -q u_{\xi} F^{\xi\alpha}. \tag{22}$$

Thus this equation reproduces the equations of motion of relativistic particle in the gravitational field with metric $g_{\mu\nu}$ and the electromagnetic field again defined by the quantity A_{ρ} in (7) as a potential.

As a result we conclude that the system of field equations (16), (20), (21), (22) corresponding to action (15) reproduce Einstein-Maxwell equations.

4 Discussion

Thereby only geometric objects were introduced such as the metric $g_{\mu\nu}$ and the connection $\Gamma^{\alpha}_{\mu\nu}$. The standard Einstein-Hilbert action was complemented by additional terms, which corresponding to a relativistic point-like particle (11), an interaction (12) and a kinetic term (14). The additional degrees of freedom of the connection A_μ were treated as an electromagnetic potential. As a result, the electrodynamics in gravitational field is constructed. Despite the simplicity of the above theory, is had not been discovered at the times of the most intense search of unified field theory (for the detailed historical survey see [4] and the references therein).

As we said above, for the first time a similar approach was proposed in 1978 [1]. However, they treated the trace of the *torsion* as the electromagnetic potential. If we restrict ourselves to consideration of the matter-free action consisting only of S_1 and S_4, the difference between identification of electromagnetic potential with the trace of the torsion and with the trace of connection (as in our approach) turns out to be negligible. However, the addition of matter changes the picture drastically. While the matter can be coupled with the trace of connection without any troubles, coupling with the trace of torsion

$$S'_3 = -q \int d\tau \dot{x}^\mu(\tau) S^{\,\nu}_{\mu\nu}(x(\tau)) \tag{23}$$

leads to the appearance of matter in the expression for the connection. Instead of (20) we have

$$\Gamma^{\rho}_{\mu\nu} = \bar{\Gamma}^{\rho}_{\mu\nu} + \frac{1}{3}A_\mu\delta^{\rho}_{\nu} + \frac{\varkappa}{4}\left(3j^\rho g_{\mu\nu} - j_\mu\delta^{\rho}_{\nu} - j_\nu\delta^{\rho}_{\mu}\right). \tag{24}$$

Note that the authors of [1] did not consider the variational principle for the matter. Instead they used some additional physical assumptions in order to obtain the equations of motion of test particles which are identical to (22). It is worth noting that the paper [1] was published in the obscure Soviet journal which is almost completely unavailable to the scientific community, so the result remained unknown. The theory proposed in [1] was later rediscovered many times, e.g. in [2], where the authors did not consider the interpretation of the A_μ as an electromagnetic potential, and in [5], where it served as the base for the possible extension of GR.

In the present work the role of matter is played by point particles. The generalization on the case of ideal fluid (i. e. continuous medium consisting on classical particles) is quite simple. The corresponding interaction term was proposed in [4] for a variant of description of ideal fluid which was studied in [6]. However, the interpretation of the A_μ as an electromagnetic potential is possible only for the purely classical description of matter. The generalization such theory for the more physical interesting case, where a matter is considered as a field, is a one way of the further development. For this purpose one could consider the approach of frame bundles. Such attempts, for example, was made by Horie [7].

Acknowledgements. The work was supported by SPbGU grant N 11.38.223.2015.

References

[1] Y.N. Obukhov, V.G. Krechet, V.N. Ponomarev, Gravitaciya i teoriya otnositelnosti [Gravitation and theory of relativity, in Russian] **14-15**, 121 (1978)
[2] F.W. Hehl, G.D. Kerlick, General Relativity and Gravitation **9**, 691 (1978)
[3] A.N. Bernal, B. Janssen, A. Jimenez-Cano, J.A. Orejuela, M. Sanchez, P. Sanchez-Moreno, Physics Letters B **768**, 280 (2017), `arXiv:1606.08756`
[4] N.V. Kharuk, A.A. Sheykin, S.A. Paston (2017), `arXiv:1709.02284`

[5] R.W. Tucker, C. Wang, Classical and Quantum Gravity **12**, 2587 (1995), arXiv:gr-qc/9509011
[6] S.A. Paston (2017), arXiv:1708.03944
[7] K. Horie (1996), arXiv:hep-th/9601066

EPJ Web of Conferences **158**, 06001 (2017) DOI: 10.1051/epjconf/201715806001
QFTHEP 2017

Hybrid tracking detector based on semiconductor and gas technologies

George Bashindzhagyan[1], *Natalia* Korotkova[1,*], *Anatoli* Romaniouk[2], *Nikolai* Sinev[3], *Vladimir* Tikhomirov[2;4]

[1]M.V.Lomonosov Moscow State University, Russia
[2]National Research Nuclear University MEPhI, Russia
[3]University of Oregon, Eugene, USA
[4]P. N. Lebedev Physical Institute of the Russian Academy of Sciences, Russia

Abstract. New Silicon-Gas Pixel Detector (SiGPD) is the result of the further development already proposed Gas-Pixel Detector (GPD). The only disadvantage of GPD was the uncertainty of the particle arriving time. The problem can be solved by additional silicon pixels implementation inside existed electronic chip epitaxial layer during regular chip production. The cost of new Si-Gas Pixel Detector remains practically the same. The new detectors have self-triggering properties and can be used for a first level trigger generation in the particle physics experiments and for many other applications.

1 Gas Pixel Detector – the basis of new detector

Gas Pixel Detector (GPD) [1] is a small flat drift chamber with 10 mm thick gas volume and with a large electronic chip with pixel sensors about 100×200 μm2 as anode plane. A mesh is placed in about 50 μm in front of the sensors. Strong electric field between the mesh and anode sensors (~100kV/cm) allows to obtain gas amplification of the order of 10^3 to register every electron crossing the mesh.

GPD has the following advantages:

- Very low material and relatively cheap structure is combined with good particle track reconstruction.
- Time-projection method application and specially developed algorithms allow 3D particle track segment reconstruction with position accuracy of ~30 μm and angle measurement precision better than 0.5°.
- Because particle track reconstruction is realized on hardware level within chip electronics, it can be done very fast in about 1 μs if 25 ns readout cycle is used. This performance allows GPD to be used for a first level trigger generation in accelerator experiments.

* Corresponding author: korona@mail.cern.ch

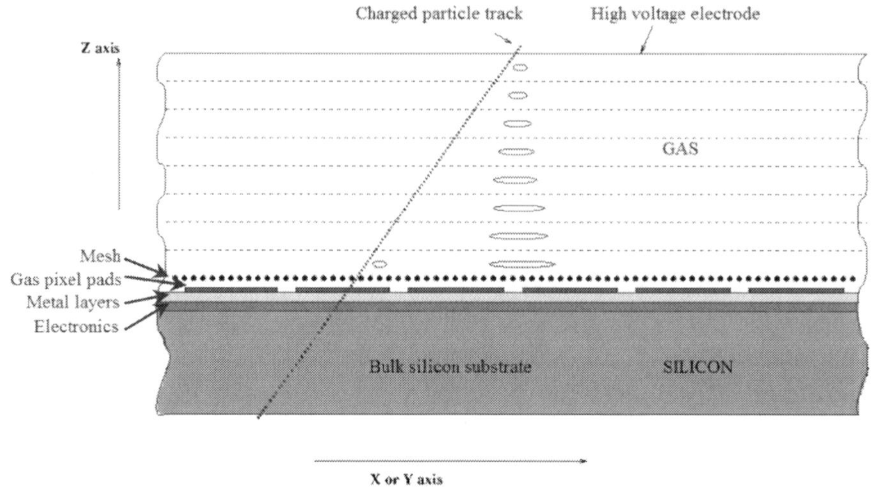

Fig. 1. The schematic view of GPD structure.

2. Time-projection readout

Electrons generated by an ionizing particle in gas drift transversely towards the chip surface, where they are registered by the pixels of the chip every readout cycle.

The process of transverse diffusion spreads electrons during their drift, and up to 50 pixels can be fired during one readout cycle (25 ns). This improves the accuracy of the measurements with relatively large pixels, but creates a problem to transmit and analyze the data online.

With 25 ns readout cycle, 10 mm thick gas volume and maximum drift time of electrons about 200 ns, we can obtain 8 points with X and Y coordinates along the particle track and Z coordinate, which corresponds the number of the cycle when the coordinates were measured (figure 2).

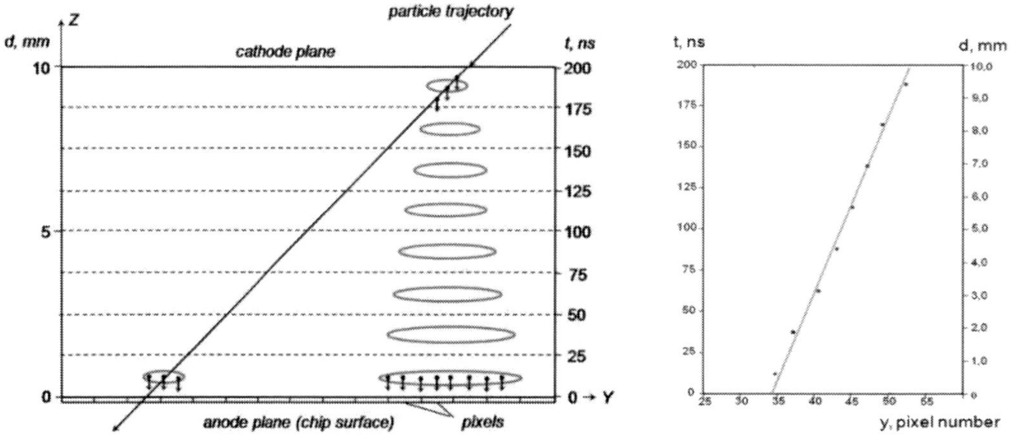

Fig. 2. Electrons collecting on the GPD chip plane (left). The example of the particle trajectory reconstruction (right).

3. Online track reconstruction algorithm

Figure 3 presents a typical cluster structure in XY plane and its projections on both axes for the 3-rd of 8 cycles. To calculate the particle position we use center of mass approach. In each projection the number of the row (n_X^i), is multiplied by the number of fired pixels in this row (R_x^i), and the sum of the products is divided by the total number of fired pixels in the cluster N_P: $X = \sum R_X^i n_X^i / N_P$, $Y = \sum R_Y^j n_Y^j / N_P$.

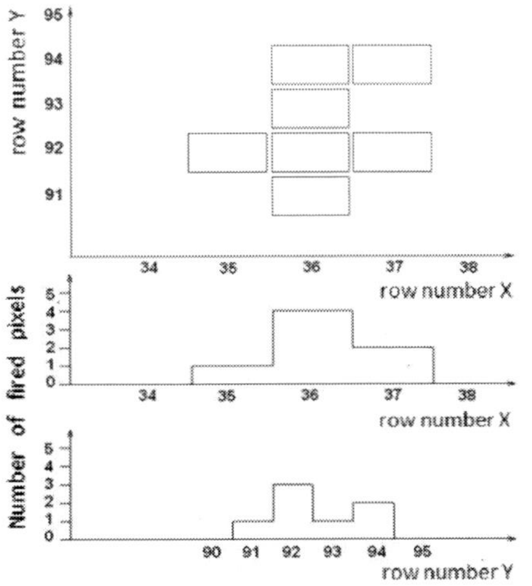

Fig. 3. The example of the cluster structure in XY plane and its projections on X and Y axes.

In experiments operating at high trigger rate the basic data processing must be performed already on chip level. Every pixel has an integrated fast preamplifier, discriminator and flip-flop with two fixed current outputs connected to common X and Y lines going along the particular X_i and Y_j rows correspondingly (figure 4). If a pixel is fired during a cycle, FF turns ON and current flows along the line.

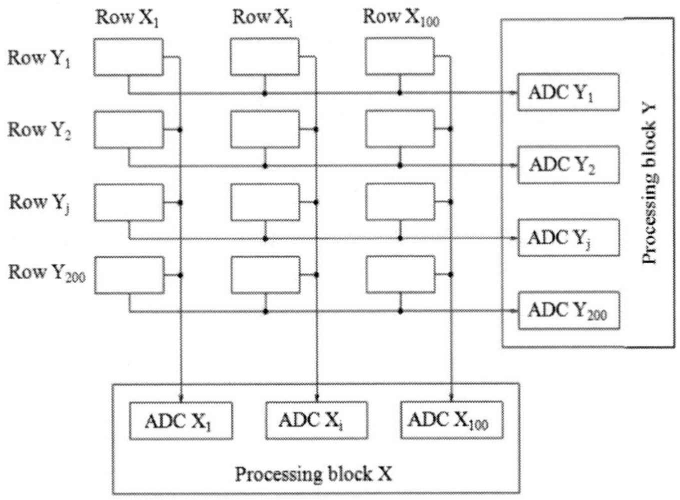

Fig. 4. The block diagram of the readout algorithm.

4. Expected and measured resolution

To estimate the expected resolution of GPD MC simulations for different detector geometries and gas mixtures have been done.

The examples of simulated distributions for 20 GeV pions crossing the detector with 100×200 μm^2 pixels and Xe gas mixture at angles $\theta = 20°$ and $\varphi = 3°$ are shown in figure 5.

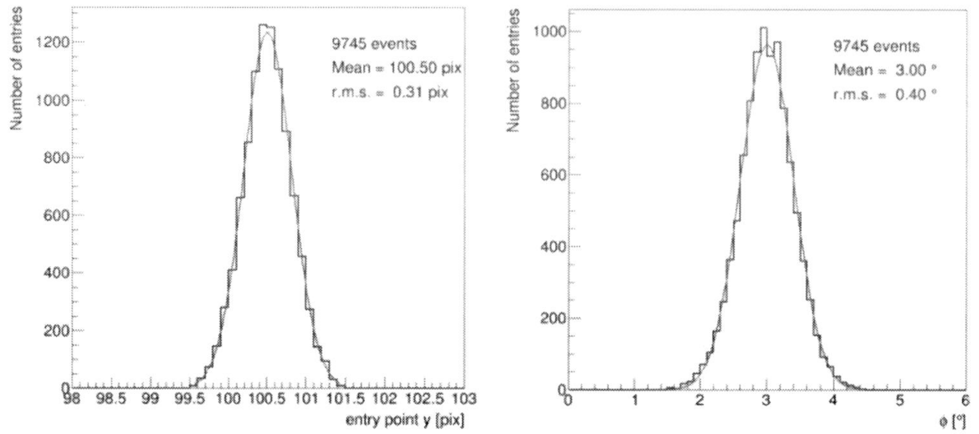

Fig. 5. The distribution of the reconstructed track entry points Y coordinates (left) and incident angles φ (right) for 20 GeV pions crossing the GPD with $\theta = 20°$ and $\varphi = 3.00°$.

Performance of GPD detector with DME/CO2 mixture and 55×55 μm^2 pixels for particles crossing the detector plane the angle 30° is shown in figure 6 [2].

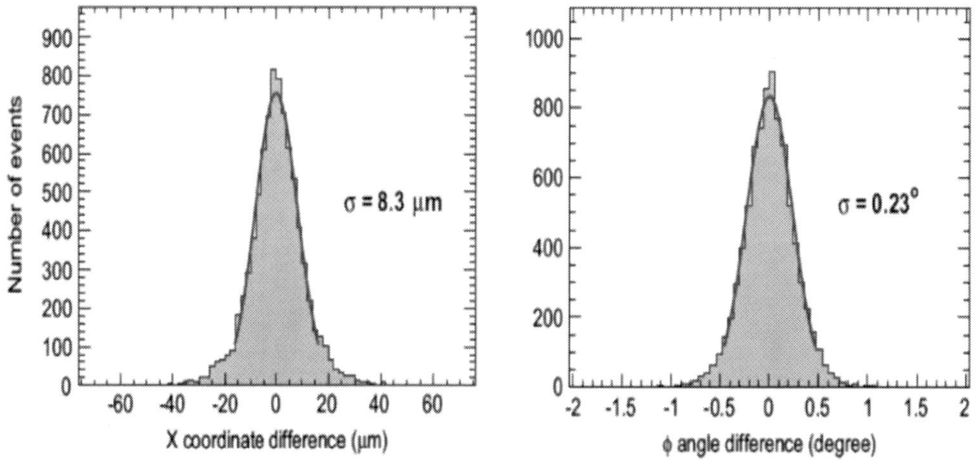

Fig. 6. Position determination accuracy of GPD detector with DME/CO2 mixture and 55×55 μm^2 pixels across the track projection (right plot) and angle measurement accuracy of the track projection on the chip plane (left plot). Solid lines represent Gaussian fits applied to the central parts of distributions.

DOI: 10.1051/epjconf/201715806001

5. Silicon pixels Implementation into GPD electronic chip

GPD assumes permanent data collection and processing with cycles corresponding collider beam crossing cycle. The disadvantage of the GPD is the absence of exact time knowledge when the particle crossed the detector. It becomes more essential if a few particles arrived during 200 ns drift time needed to collect the data.

The simple way to obtain time coordinate is to add a layer of silicon pad or strip detector to existed GPD structure. But it means additional layer of silicon, a lot of interconnections and essential increase of production cost. Therefore we propose to use charged particle detection in epitaxial layer created during regular chip production technology.

Adding silicon sensors below chip electronics within epitaxial layer we can collect electrons generated by the charged particle in the epitaxial layer (~ 7-10 μm) in very short time about 1 ns and obtain exact information about the particle arrival very soon. Despite the charge generated in epitaxial layer is about 700 e, the signal at the amplifier input is sufficient for its registration because of very low capacitance of the silicon pixel (~ 1 pF).

6. Silicon-Gas Pixel Detector

Silicon-Gas Pixel Detector (SiGPD) [3] is a further development of the GPD concept with the determination of the exact time when particle crossed the detector, which can be realized by implementation of silicon pixel layer inside electronic chip without other changes in the existing GPD design and therefore without increasing the production cost of the device.

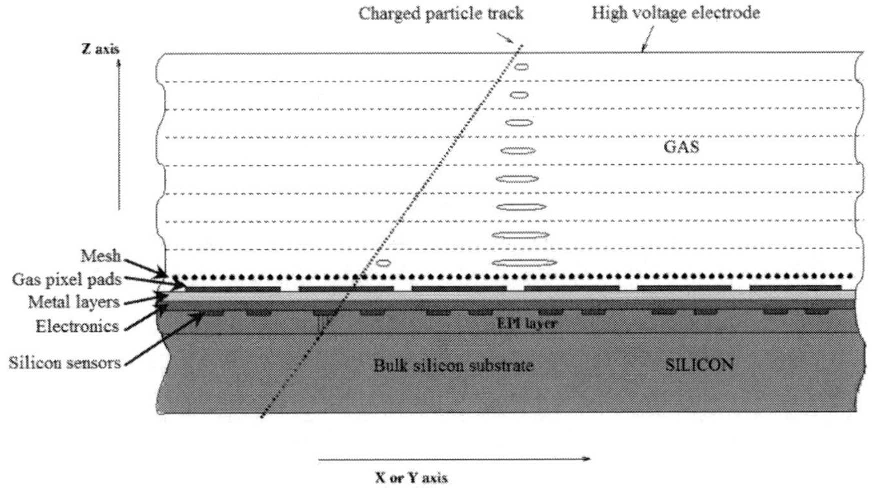

Fig. 7. The schematic view of the SiGPD structure.

In addition to the GPD features SiGPD have the following properties and advantages.
- With implemented silicon pixel layer we exactly know the time when particle crossed the detector and can use SiGPD as a completely self-triggered device.
- SiGPD can be used to generate fast trigger in accelerator experiments.
- SiGPD has an internal generator which starts readout process when particle crossed detector volume. If SiGPD works in accelerator environment its clocks must be synchronized with beam crossings. Suggested electronics can generate output signal for use in the level 1 trigger system. The latency of the output data is about 40 clock periods. With 40 MHz frequency used in LHC, it is 1 μs. If smaller latency is required one can

increase frequency to 80 MHz and decrease latency to about 0.5 μs. If 0.25 μm technology is used for electronics production, 80 MHz looks like reasonable clock speed.
- Silicon pixels in SiGPD does not require high-resistivity silicon and can be implemented into GPD electronic chip within the standard chip production technology without increasing the production cost.
- All the silicon and gas pixel electronics , which measures and analyses signals from X and Y lines, is fully integrated in electronic chip structure and produced within the same chip production process. This means that SiGPD does not require any external electronics for its operation.
- SiGPD have higher radiation hardness compare to regular silicon detector, because of its very thin epitaxial detecting layer.

7. SiGPD readout

As it was described earlier for GPD signals from gas pixels in each row X_i (or Y_j) are added together and transmitted to X (or Y) coordinate processing block, situated at the edges of the electronic chip. We may compare this readout system with two microstrip detectors with perpendicularly going strips. It seriously simplified and speeds up readout procedure. If particle density is high and a few particles can appear at the same clock we can avoid ambiguity adding one more interconnecting system of wires.

Readout from Si sensors operates similar way as for the gas pixels, and even easier because it does not require a number of fired pixels in each row. In other words, we apply "binary" readout for silicon and "analogue" for gas pixels.

With binary silicon pixel readout we can determine XY coordinate of the particle arrival with 7 μm accuracy. If power consumption/dissipation is a critical point, the silicon pixel size can be increased to 50×50 μm. This case XY accuracy is still 15 μm but power dissipation is four times less.

Chip electronics consists of blocks under each gas pixel pad, peripheral X & Y coordinate processing blocks and some communication logic to transfer data to host computer. The layout of SiGPD chip is shown schematically in figure 8.

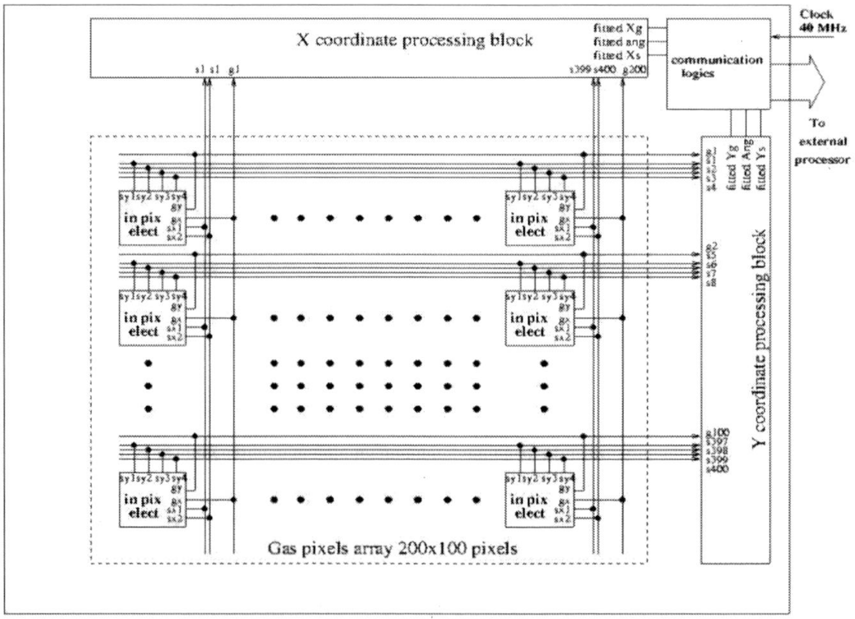

Fig. 8. SiGPD chip layout.

8. Conclusion

A concept of new Silicon-Gas Pixel Detector is developed. This new detector is based on proposed earlier Gas Pixel Detector. But with implemented into existing electronic chip silicon pixels new detector has significantly extended functionality and becomes more convenient and universal device.

Fast online data processing and ability to present full information about charged particle trajectory in ~ 0.5 μs can be very useful for a fast trigger generation.

Very low material (~10 mm gas and ~100 μm Si) and relatively cheap structure are combined with good coordinate (~ 5-10 μm) and angular (~ 0.4°) accuracies of particle track segment reconstruction.

SiGPD works as a completely self-triggered device and can be used in many areas, where particle trajectory reconstruction is required.

This work was partially supported by Russian Foundation for Basic Research (Grant 14-22-0353).

References

1. N. Sinev, G. Bashindzhagyan, N. Korotkova, A. Romaniouk, V. Tikhomirov. Journal of Physics: Conference Series, **675** (2016)
2. A. Boldyrev, F. Hartjes, N. Hessey, M. Fransen, S. Konovalovet et al., Nucl. Instr. and Meth. A **807** (2016)
3. N. Sinev, G. Bashindzhagyan, N. Korotkova, A. Romaniouk, V. Tikhomirov. Journal of Physics: Conference Series, **798** (2017)

EPJ Web of Conferences **158**, 06002 (2017)
QFTHEP 2017 DOI: 10.1051/epjconf/201715806002

Performance of the ATLAS Hadronic Tile Calorimeter in Run-2 and its Upgrade for the High Luminosity LHC

Oleg Solovyanov[1,*] *on behalf of the ATLAS Collaboration*

[1] *State Research Center Institute for High Energy Physics (Protvino), NRC KI*

Abstract. The Tile Calorimeter (TileCal) of the ATLAS experiment at the LHC is the central hadronic calorimeter designed for energy reconstruction of hadrons, jets, tau-particles and missing transverse energy. TileCal is a scintillator-steel sampling calorimeter and it covers the region of pseudo-rapidity up to 1.7, with almost 10000 channels measuring energies ranging from ~30 MeV to ~2 TeV. Each stage of the signal production, from scintillation light to the signal reconstruction, is monitored and calibrated.

The performance of the Tile calorimeter has been studied in-situ employing cosmic ray muons and a large sample of proton-proton collisions, acquired during the operations of the LHC. Prompt isolated muons of high momentum from electroweak bosons decays are employed to study the energy response of the calorimeter at the electromagnetic scale. The calorimeter response to hadronic particles is evaluated with a sample of isolated hadrons. The modelling of the response by the Monte Carlo simulation is discussed. The calorimeter timing calibration and resolutions are studied with a sample of multijets events. Results on the calorimeter operation and performance are presented, including the calibration, stability, absolute energy scale, uniformity and time resolution. TileCal performance satisfies the design requirements and has provided an essential contribution to physics results in ATLAS.

The Large Hadron Collider (LHC) has envisaged a series of upgrades towards a High Luminosity LHC (HL-LHC), delivering five times the LHC nominal instantaneous luminosity. The ATLAS Phase II upgrade, in 2024, will accommodate the detector and data acquisition system for the HL-LHC. In particular, the Tile Calorimeter will undergo a major replacement of its on- and off-detector electronics. All signals will be digitised and then transferred directly to the off-detector electronics, where the signals will be reconstructed, stored, and sent to the first level of trigger at a rate of 40 MHz. This will provide better precision for the calorimeter signals used by the trigger system and will allow the development of more complex trigger algorithms. Changes to the electronics will also contribute to the reliability and redundancy of the system. Three different front-end options are presently being investigated for the upgrade. Results of extensive laboratory tests and with beams of the three options will be presented, as well as the latest results on the development of the power distribution and the off-detector electronics.

*e-mail: Oleg.Solovyanov@ihep.ru

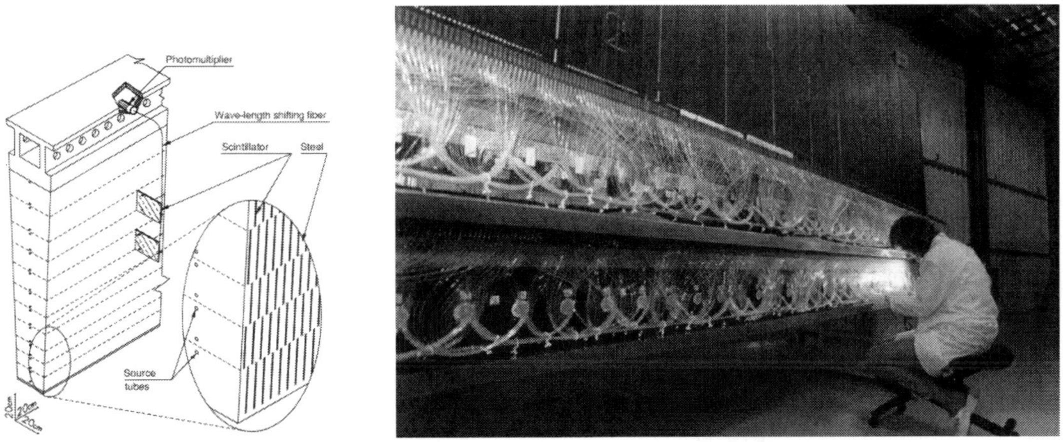

Figure 1. Tile Calorimeter module design and a picture of a long barrel module during instrumentation [2]. Copyright 2017 CERN for the benefit of the ATLAS Collaboration. CC-BY-4.0 license.

1 Introduction

In high energy physics experiments, calorimeters are essential for measurements of particle properties, and the hadronic calorimeters, in particular, are crucial in measuring jets, missing transverse energy, taus and muons, and in providing triggers at all levels.

One of the Large Hadron Collider (LHC) general purpose experiments, ATLAS [1], contains hadronic calorimeter, called Tile Calorimeter (TileCal) [2]. Tile TileCal is a non-compensating sampling calorimeter, with steel as a radiator and scintillating tiles as an active medium. The 3 mm thick tiles are composed of doped polystyrene, and are oriented perpendicular to the beam axis. The light from the tiles is read out via wavelength shifting (WLS) fibres, connected to both short edges of the tiles. The WLS fibres are bundled and read out by photomultiplier tubes, located in a module's girder, as shown in Fig. 1. The calorimeter is divided into 3 cylinders: one Long Barrel (LB) and two Extended Barrels (EB). Each cylinder is further subdivided azimuthally in 64 modules. The calorimeter is 12 m in overall length with a 4.25 m outer radius.

The Long Barrel module covers the pseudo-rapidity region $|\eta| < 1.0$, while the Extended Barrels cover $0.8 < |\eta| < 1.7$. Each module have three longitudinal layers (A,BC,D), with a total thickness in radiation lengths of about 7.4λ. The calorimeter cells are defined by the WLS fibre routing, providing about 5200 cells in the entire calorimeter with 0.1×0.1 $\Delta\eta \times \Delta\phi$ granularity for A and BC cells, and 0.2×0.1 $\Delta\eta \times \Delta\phi$ for the outer D cells. Analogue trigger sums from pseudo-projective towers are provided for the first level trigger. The dynamic range of the PMTs and associated electronics, from 10 MeV to 750 GeV, allows for the measurements of single particles and large jets. The design resolution for jets is $50\%/\sqrt{E} \oplus 3\%$ and the response is linear within 2% for up to 4 TeV jets [2].

The readout is organised as follows. The PMT signals are shaped and amplified with two gains (1:64 ratio); then both gains are digitised in parallel by 40 MHz 10 bits sampling ADCs. The digitised samples are temporarily stored in pipeline memory, until the arrival of first level trigger, when the data of one of the gains are transferred to the back-end electronics via optical links, as shown in Fig. 2.

The signal is reconstructed from 7 samples by applying an optimal filtering (OF) algorithm and the energy, time and quality factors are extracted from signal samples, with the amplitude of the signal proportional to the deposited energy (shaping). To minimise noise, the OF algorithm is using

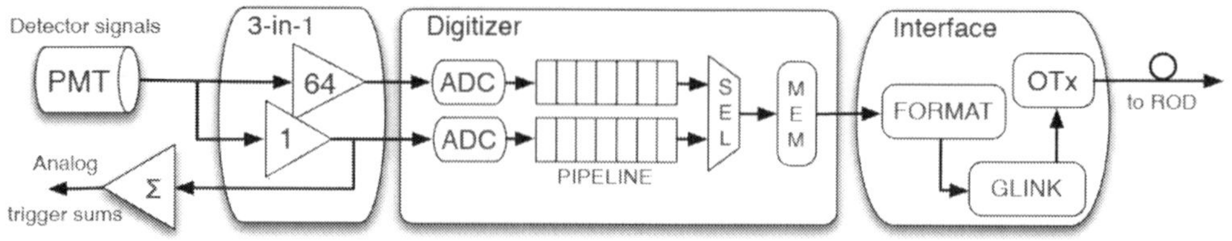

Figure 2. Tile Calorimeter readout scheme [2]. Copyright 2017 CERN for the benefit of the ATLAS Collaboration. CC-BY-4.0 license.

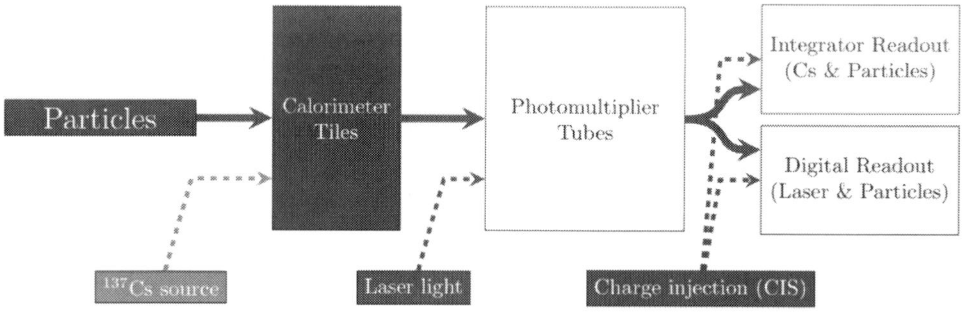

Figure 3. Tile Calorimeter calibration systems chain [3]. Copyright 2017 CERN for the benefit of the ATLAS Collaboration. CC-BY-4.0 license.

weighted sum of samples. With the application of calibration constants, the energy is reconstructed in GeV units.

To provide correct energy and time information for data reconstruction, an elaborate chain of calibration systems [3], shown in Fig. 3, is used. Charge injection system (CIS) is used to calibrate the response of the ADC. Laser calibration system to measure the performance of the PMTs. Cesium-137 radioactive source system (Cs) is used to calibrate the full optical path from scintillating tiles and WLS fibres down to integrated current of the PMT, and a minimum bias monitoring system (MBM) is used to monitor the response of the calorimeter online via integrated currents of PMTs. About 11% of 192 Tile calorimeter modules were calibrated at the test-beams [4] and the electromagnetic (EM) scale (1.05 pC/GeV) was transferred to the final detector with the help of calibration systems mentioned above.

2 Performance

To reduce the number of non-working ("dead") cells, the on-detector electronics is maintained yearly during detector openings. During the extended year end technical stop (EYETS) of 2016–2017, 48 out of 256 electronics “drawers” were opened and most of the problems were fixed, leading to the 0.06% of “dead” cells after the stop.

After the restart of the LHC, the increase of the pile-up conditions lead to an increase of the noise levels. The electronics noise is below 20 MeV for the most of the cells, while the pile-up noise increased up to 160 MeV for the inner layer of the calorimeter.

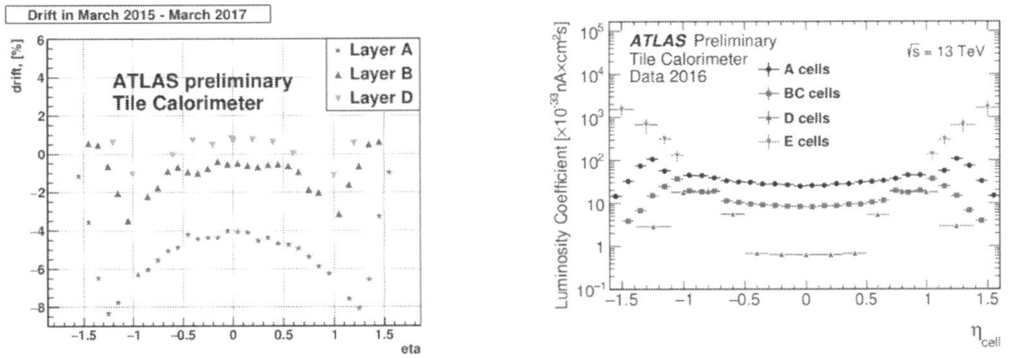

Figure 4. Tile Calorimeter response variation (drift) seen by cesium calibration system vs. cell eta (left) and corresponding cell PMT currents (right) [5]. Copyright 2017 CERN for the benefit of the ATLAS Collaboration. CC-BY-4.0 license.

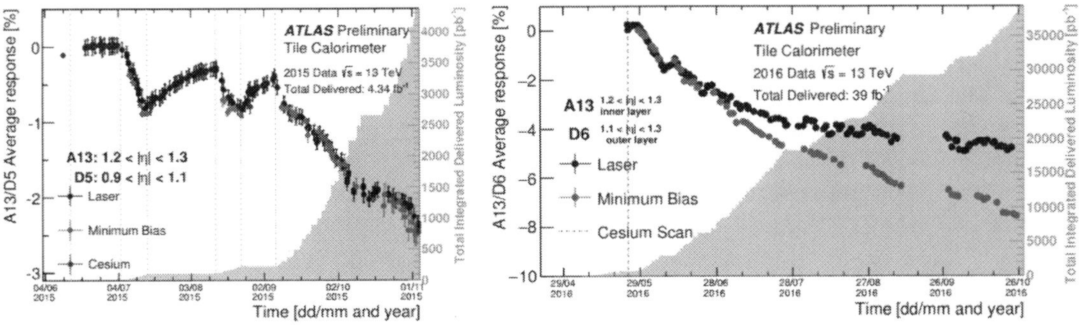

Figure 5. Tile Calorimeter cell response variation in time during 2015 (left) and 2016 (right) pp collisions data taking periods [5]. Copyright 2017 CERN for the benefit of the ATLAS Collaboration. CC-BY-4.0 license.

The cell response is not constant due to the PMT gain variation and scintillator degradation. The response is tracked with the set of calibration systems, described above, as shown in Fig. 4. One of the most exposed cells of the inner calorimeter layer in extended barrel has shown as much as 2% difference between the laser and integrator response ratios. While the was some recovery during the 2015 data taking, there was practically none during the 2016 data taking period, as shown in Fig. 5. While the PMT response drift can be recovered by raising the high voltage, the loss of the response in the scintillator is taken care of by calibration constants.

One of the important Tile Calorimeter characteristics is the ratio of energy to track momentum (E/p) for isolated charged hadrons in minimum bias events that it is used to evaluate calorimeter uniformity and linearity during data taking. The data and simulation do agree, showing linearity and uniformity in detector response, and the dE/dx of minimum ionising muons show data and Monte-Carlo (MC) agreement within 3%.

Muons from cosmic rays, beam halo and collisions are used to study in-situ the electromagnetic energy scale. A 1% response non-uniformity in η is seen in Tile Calorimeter Long Barrel, with a good energy response uniformity in all calorimeter layers with the data and simulation in agreement.

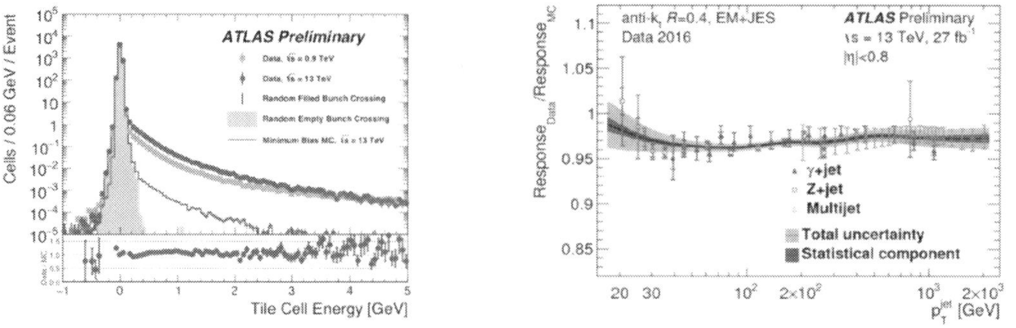

Figure 6. Tile Calorimeter jet performance [5] (left) and ATLAS jet energy data/MC agreement [7] (right).

One of the main uses of the hadronic Tile Calorimeter is the measurement of jet energy and missing transverse energy (MET). The data and MC simulations agree well (see Fig. 6), the jet energy scale is consistent [7] with MC predictions, and the jet energy resolution is below 10% for p_T >100 GeV. The constant term agree with the expected value of 3%.

The Tile Calorimeter contributes to the ATLAS luminosity measurement by providing a calibration transfer from low to high luminosity conditions, and by longer term luminosity monitoring. Its dedicated readout of the anode currents in every channel is fully decoupled from trigger and is intrinsically independent from pile-up. It also allows a cross-check of other luminometers.

3 Upgrade

The upgrade of the LHC machine, the High Luminosity LHC (HL-LHC) will allow almost a factor of 3 increase in integrated luminosity, compared to the simple extension of the running of original accelerator, as it was designed, and almost 10 times more than the data from the first 10 years of LHC operation.

The advanced physics goals and LHC upgrade plans create new challenges for detectors. The upgrade program aims at 5–10 fold luminosity increase, which would mean that more radiation tolerance is required. The ageing electronics, originally planned for 10 years of operation will have to be replaced. Obsolete components will make it impossible to maintain. The high event rates require more efficient trigger algorithms with more primary information from the detectors.

To cope with these issues, a multi-phase upgrade program has been conceived [8]. While the major Tile Calorimeter mechanics and optics will stay, together with associated PMTs. The front-end and back-end electronics, calibration systems, will all undergo major upgrades during Phase-II upgrade stage: complete redesign of the front-end and back-end electronics; a new electronics drawer mechanics; and a fully digital trigger with higher selectivity and finer granularity.

The new readout architecture, shown in Fig. 7, allows the data to be transmitted off-detector at a 40 MHz rate, and to be processed in the off-detector pre-processor boards (PPr). This simplifies the on-detector electronics and provides greater flexibility for the trigger data processing. One of the aims of the upgrade is to improve the redundancy and reliability of the hardware.

The front-end board (FEB) has 3 different options in development [9]. One is based on the well proven original design (3in1) with discrete off-the shelf electronics (see the scheme in Fig. 8), but

DOI: 10.1051/epjconf/201715806002

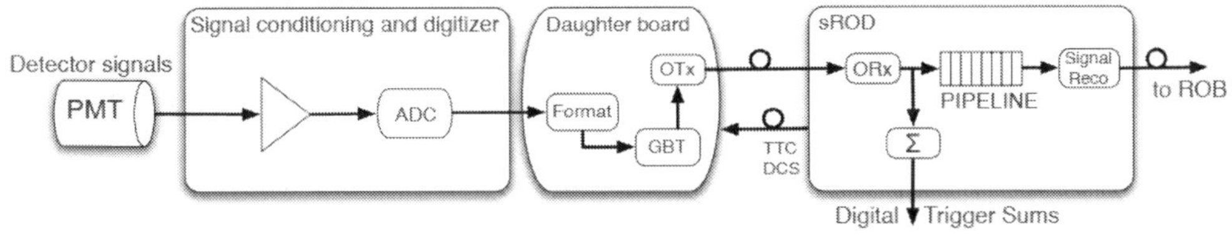

Figure 7. Tile Calorimeter new readout scheme [8]. Copyright 2017 CERN for the benefit of the ATLAS Collaboration. CC-BY-4.0 license.

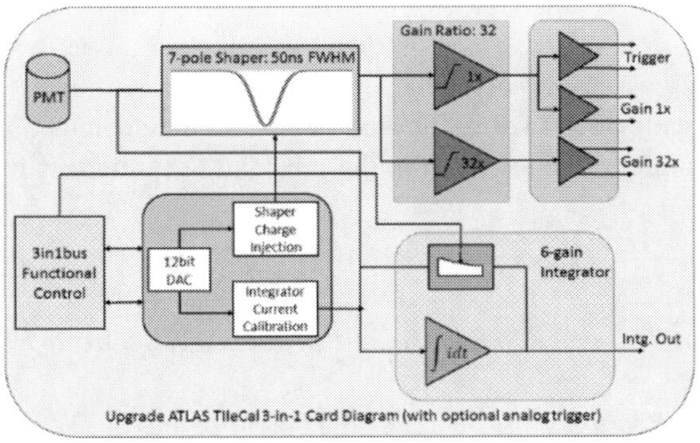

Figure 8. Block diagram of upgraded 3in1 front-end board [9]. Copyright 2017 CERN for the benefit of the ATLAS Collaboration. CC-BY-4.0 license.

with an improved dynamic range and ADC resolution. The other two options (QIE, FATALIC) are based on custom-developed application specific integration circuit (ASIC). Prototypes of all options undergo extensive laboratory and beam tests.

New on-detector electronics requires an upgrade of the low voltage power supply (LVPS). The new design fits the same box, but is simpler, and uses only one voltage level (10 V), while the required voltages are derived in-situ by point of load regulators (POL). The new powering scheme has also an increased redundancy (see Fig. 9).

The high voltage (HV) distribution system, supplying the detector PMTs, has also to be re-designed. Currently, there are two options. One is based on the original design and regulates the high voltage close to the PMTs (internal option). The so-called remote option [11] would simplify the maintenance and does not require radiation tolerance, a suitable multi-wire high voltage cable is being identified.

To study the performance of different front-end and high-voltage options, a large test-beam campaign is being performed. A set of TileCal modules, equipped both with old and new electronics is being studied with electron, hadron and muon beams, and with calibration systems (charge injection, laser, cesium).

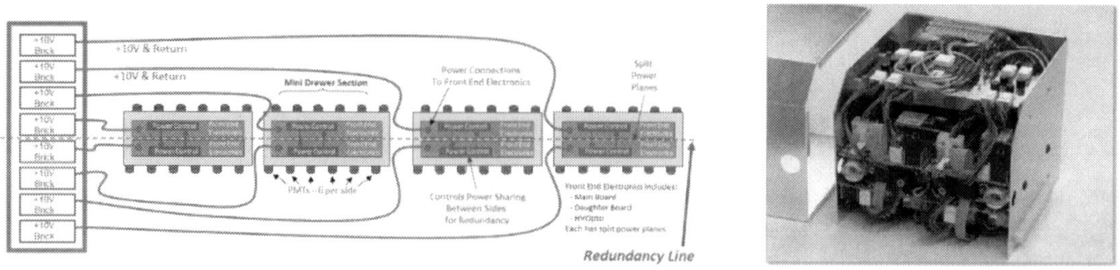

Figure 9. Power distribution scheme of front-end electronics (left) and low voltage power supply box (right) [10].

4 Summary

Being an important part of the ATLAS detector at LHC, the Tile Calorimeter has performed very well in LHC Run-2. The calibration systems have allowed the electromagnetic scale to remain within the required precision. The High Luminosity LHC upgrade poses significant challenges and requires a full redesign and replacement of the front-end and back-end electronics. The prototypes of several components are available and are being tested in the labs and at in the test-beam.

References

[1] ATLAS Collaboration, *The ATLAS Experiment at the CERN Large Hadron Collider*, *JINST* **3** (2008) **S08003**.

[2] ATLAS Collaboration, *Readiness of the ATLAS Tile Calorimeter for LHC collisions*, *Eur. Phys. J. C* **70** *(2010) 1193.*

[3] D. Boumediene *et al.*, *Calibration of the Tile Hadronic Calorimeter of ATLAS at LHC*, J. Phys. Conf. Ser. **587** (2015) no.1, 012009.

[4] P. Adragna *et al.*, *Testbeam studies of production modules of the ATLAS tile calorimeter*, Nucl. Instrum. Meth. A **606** (2009) 362.

[5] ATLAS Tile Calorimeter Subsystem, *Approved Plots*, Accessed 20 September 2017. https://twiki.cern.ch/twiki/bin/view/AtlasPublic/ApprovedPlotsTile.

[6] K. Anderson *et al.*, *Design of the front-end analog electronics for the ATLAS tile calorimeter*, Nucl. Instrum. Meth. A **551** (2005) 469.

[7] M. Aaboud *et al.* [ATLAS Collaboration], *Jet energy scale measurements and their systematic uncertainties in proton-proton collisions at* $\sqrt{s}$ = *13 TeV with the ATLAS detector*, CERN-EP-2017-038, arXiv:1703.09665 [hep-ex].

[8] G. Usai *et al.*, *ATLAS Tile Calorimeter electronics and future upgrades*, JINST **10** (2015) no.04, C04026.

[9] A. Gomes *et al.*, *The new front-end electronics for the ATLAS Tile Calorimeter Phase 2 Upgrade*, JINST **11** (2016) no.02, C02015.

[10] A. Solodkov *et al.*, *Upgrade of the ATLAS hadronic Tile Calorimeter for the High luminosity LHC*, JINST **12** (2017) no.08, C08004.

[11] F. Vazeille *et al.*, *Performance of a remote High Voltage power supply for the Phase II upgrade of the ATLAS Tile Calorimeter*, JINST **11** (2016) no.02, C02050.

EPJ Web of Conferences **158**, 06003 (2017) DOI: 10.1051/epjconf/201715806003
QFTHEP 2017

Beam test results of STS prototype modules for the future accelerator experiments FAIR/CBM and NICA/MPD projects

Petr Kharlamov[1] , *Dmitrii* Dementev[2] , and *Mikhail* Shitenkov[2]

[1]SINP MSU, Department of experimental high energy physics, 119991 Moscow, Russian Federation

[2]LHEP JINR, 141980 Dubna, Moscow region, Russian Federation

Abstract. High-energy heavy-ion collision experiments provide the unique possibility to create and investigate extreme states of strongly-interacted matter and address the fundamental aspects of QCD. The experimental investigation the QCD phase diagram would be a major breakthrough in our understanding of the properties of nuclear matter. The reconstruction of the charged particles created in the nuclear collisions, including the determination of their momenta, is the central detection task in high-energy heavy-ion experiments. It is taken up by the Silicon Tracking System in CBM@FAIR and by Inner Tracker in MPD@NICA currently under development. These experiments requires very fast and radiation hard detectors, a novel data read-out and analysis concept including free streaming front-end electronics. Thermal and beam tests of prototype detector modules for these tracking systems showed the stability of sensors and readout electronics operation.

1 Double-sided strip detectors in high energy physics

In Particle Physics experiments particles like electrons, positrons or protons are accelerated to energies in the GeV to TeV range and brought to collision, either on a fixed target or head on in colliding beams. [1] The particles or their constituents interact. Determining precisely the properties of the final state particles the interaction process can be studied, e.g. cross sections, angular distributions or asymmetries. Often the main interest is not the primary interaction but the properties of new particles created in the interaction, like heavy quarks, W- or Z-bosons, B-mesons etc. They are unstable and decay almost immediately into secondary (or tertiary) particles which are stable and can be detected. Ideally many different quantities need to be determined in order to allow a complete reconstruction of the interaction or a decay: the particle type, its charge (sign), the momentum vector and energy and for short lived particles the production and decay vertex. However, no detector can measure all these quantities together, detector systems made from sub-detectors each specialized in one or several quantities are needed. In colliding beam experiments the detectors have to cover the interaction region completely ("4π detector"). Hence the sub-detectors are arranged in cylindrical layers around the beam pipe, the forward and backward regions are closed with plug-like detectors. In fixed target experiments most of the particles are contained in a narrow cone following the beam direction, hence it is sufficient to have wedge-shaped detectors covering this cone.

One of the main detector components is a tracking detector. In a magnetic field the tracks of charged particles are curved, the curvature is measured by the tracking detector allowing to reconstruct the momentum of the particle. Good momentum resolution requires large detectors, tracking detectors have radii of (1–2) m. Ideally the tracking detectors and the coil of the magnet should give only minimal disturbance to the trajectory of the charged particles and should not cause interactions of neutral particles.

Thus, the subject of this article is tracking detector modules for the future heavy-ion experiments CBM@FAIR, BM@N and MPD@NICA. The modules consist of double-sided silicon strip sensors and self-triggered electronics. So, this paragraph deals with sensors, and the next paragraph covers electronics.

The operation principle of silicon sensors is simple: They are basically diodes with a pn-junction which is depleted applying a reverse voltage from external electrodes. In such a reverse biased diode only a very small thermally generated current is flowing and a large electric field is built up in the depleted zone. Photons and ionizing particles create electron–hole pairs. In the electric field of the depleted diode these charge carriers drift to the electrodes where they induce a signal.

Position information can be obtained by segmenting the electrodes creating many independent diodes. From the known position of the electrode where the signal is induced the position of the ionizing particle can be deduced. In the following sections these aspects will be discussed in more detail.

As electrons and holes in a strip detector are swept by the electric field to opposite sides of the wafer, it is possible to use both types of charge carriers for position measurement by providing charge-collection electrodes on both sides of the wafer (Fig. 1). This double-sided readout brings about the obvious advantage of providing twice the information for the same amount of scattering material.

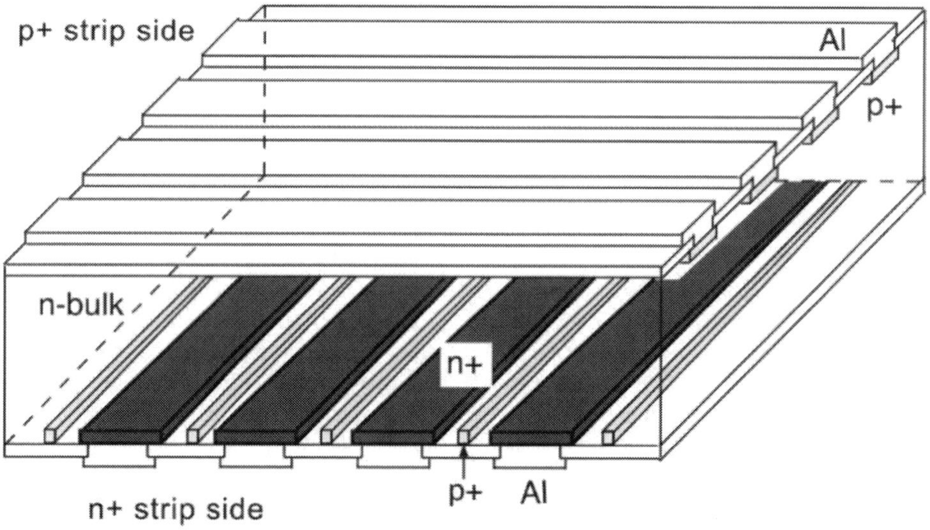

Fig. 1. Sketch of double-sided strip sensor with implanted interleaved p-strips between n^+-charge collection strips

With crossed strips on the two detector faces, projective two-dimensional measurement is obtained from one single detector. For a traversing particle, a spatial point can be reconstructed as both projections are taken from the same initial charge cloud. For absorbed radiation such as x-rays, two-dimensional measurement becomes possible. With analog readout it is furthermore possible (to some degree) to correlate signals from the two sides, making use of Landau fluctuations and the exact equality of positive and negative charge created by each ionizing particle. This can be of interest for resolving ambiguities when several particles simultaneously hit the detector.

A problem in producing the double-sided kind of detector is the insulation of each of the strips from the others simultaneously on both sides of the detector. This problem has several solutions. [2] In our sensors the method of disruption of the electron layer by implantation of p strips between the n^+charge-collection strips is used.

2 Readout electronics

Modern high energy physics call inevitably for carrying out complex and expensive accelerator experiments. Detectors needed for data acquisition are very large and consist of a big amount of sensors. Information as a whole, in turn, is collected from classified information from the separate parts. In such experiments useful data is obtained by successive reduction of the raw data.

As a rule, only a few interactions occurring in high energy particles collisions are of interest for researchers. As a result, modern detectors are created to quickly examine collision data and, for that reason, to quickly make a decision on importance of this collision. Hardware and software systems making such decisions are called "triggers". They divided into "levels"; each subsequent "level" represents more and more limiting set of criteria, which a signal must answer to pass to the next trigger levels. The signal passing through all trigger levels records as *event* corresponding to phenomenon. Readout electronics working with such external systems are called *triggered*.

There is also another type of readout electronics. In such electronics a trigger signal is produced by electronics itself, and so such type of electronics is called *self-triggered*.

Both types have common problems, for example, signal-to-noise ratio. But there are some specific problems. Triggered electronics is used in collider experiments; bunch-crossing time of them is precisely known, so moment of time of possible events is known. And electronics read nothing all other time. Therefore, reliable trigger system is required. In case of fixed-target experiments (e.g. CBM@FAIR) and/or long bunch of accelerated particles (e.g. MPD@NICA) significant fraction of events will be lost inevitably. In that case self-triggered electronics should be used. It is more difficult to obtain high signal-to-noise for self-triggered electronics; also it usually has bigger energy consumption (and thus heat emission). But in general it is more universal solution for experiment in high energy physics.

In our work self-triggered readout electronics n-XYTER (Neutron-X-Y-Time-Energy Read-out) is used now. It integrates 128 channels with low noise preamplifiers and shapers. [3] Each channel has two different shapers with distinct time constants, one optimized for timing resolution, the other one optimized for energy (pulse height) resolution (Fig. 2). A peak detector connected to the slower shaper allows for the application of a spectroscopic amplitude measurement. An internal time stamp generator provides the temporal reference that may be employed to identify time coincidences of signals on different detector channels and thus correlate their spatial point of origin. For testability and calibration purposes, a charge injector with adjustable pulse height was implemented. The bias settings and various other parameters can be controlled via a standard I^2C-interface.

DOI: 10.1051/epjconf/201715806003

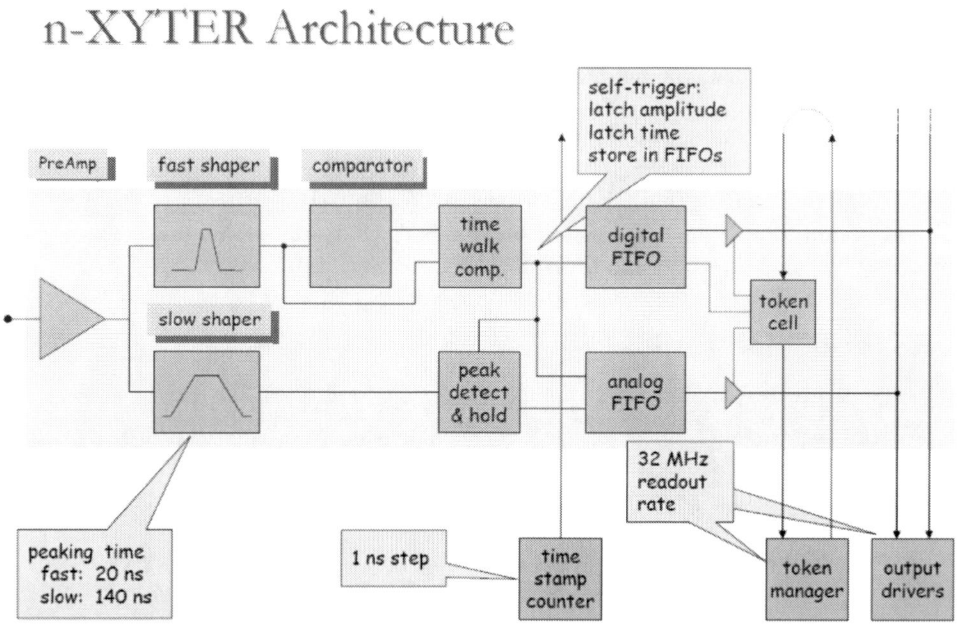

Fig. 2. Flowchart of n-XYTER.

Advantages of n-XYTER: high read rate, works without trigger modules, good energy solution.

Disadvantages of n-XYTER: high heat release (near 6 W) [4], big noise, dependence of DC-signal levels on temperature (in n-XYTER 1.0).

Therefore, n-XYTER can be used in spectrometric measurement and for tracking detectors testing, because it can make high-precision measurements of signal height. And because of self-triggered nature, n-XYTER can also be used in fixed-target experiments.

For the CBM@FAIR a new prototype of the STS read-out is currently under development: STS-XYTER2. [5] As well as n-XYTER, the chip uses a low-noise Charge Sensitive Amplifier (CSA) to aggregate the charge generated in the detector active volume and to convert it into the voltage step. The signal divides into fast and slow paths too. The fast path is optimized for timestamping of events and uses a single-stage shaper with 30 ns shaping time, a discriminator and a 14-bit timestamp latch working with Gray-encoded timestamp counter. The slow path consists of a CR-(RC)^3 shaper working with a 5-bit flash ADC and a digital peak detector. It is optimized for low-noise charge measurement. The shapers' design is very similar to the implementation in the previous prototype chip, STS-XYTER, and n-XYTER.

3 Tracking systems of CBM@FAIR, BM@N and MPD@NICA experiments

High-energy heavy-ion collision experiments provide the unique possibility to create and investigate extreme states of strongly-interacted matter and address the fundamental aspects of QCD such as equation-of-state, structure and properties of nuclear matter and modification of hadrons in nuclear medium. At large baryon chemical potentials QCD phase diagram features structures such as a critical point and transitions between various phases such as hadronic phase, quarkonic phase, quark-gluon plasma and even colour superconductors. The experimental discovery of these landmarks of the QCD phase diagram would be a major breakthrough in our understanding of the properties of nuclear matter.

However, due to luminosity limitations modern experiments suffer from statistics for rare diagnostic probes. In order to overcome this limitation, the new projects are under development at the Facility for Antiproton and Ion Research (FAIR) in Darmstadt and at the Nuclotron-based Ion Collider fAcility in Dubna. And so, the detectors of these experiments are designed to measure rare diagnostic probes such as multy-strange hyperons, charmed particles and vector mesons decaying into lepton pair with unprecedented precision and statistics. Most of these particles will be studied for the first time in the FAIR and NICA energy range. In order to achieve the required precision, the measurements will be performed at reaction rates up to 10 MHz. This requires very fast and radiation hard detectors, a novel data read-out and analysis concept including free streaming front-end electronics, and a high performance computing cluster for online event selection.

The reconstruction of the charged particles created in the nuclear collisions, including the determination of their momenta, is the central detection task in these experiments. In CBM experiment it is taken up by the Silicon Tracking System, an array of eight tracking stations based on silicon micro-strip sensors (Fig. 3) currently under development. [6] In the innermost parts of the first two stations, where particle occupancy is expected to reach 5%, 2 cm high sensors have been chosen: such granularity allows minimizing the number of ghost hits. However, to reduce material budget and to minimize the number of read-out chips, it is possible to use at the outer parts of the stations, where occupancies are less than 5%, up to two 6 cm sensors daisy-chained together, creating effective strip lengths up to about 12 cm. Every sensor has a fixed horizontal width of 6 cm (Fig. 4).

Because of identity of BM@N experiment in essence (it is also used for testing of equipment for CBM) STS is the same as that of CBM but it has less stations. [7] Both of CBM and BM@N are fixed-target experiments and have single-arm spectrometer detectors. MPD@NICA is a collider experiment, and so it has cylinder of many layers detector.

One of the main detector systems of MPD is the Inner Tracker. It consists of four concentric cylinders of silicon strip detectors (Fig. 5). [8] It consists of a silicon barrel and a system of silicon discs. The mechanical support and cooling elements of both the cylinders and the discs are manufactured out of carbon fiber to minimize the material budget of the setup and hence to diminish the impact of the Coulomb multi-scattering on track resolution. The double sided SSDs of the four cylinder-like sensitive layers of the barrel are mounted on the 3-D light Carbon Fiber (CF) space frames similar to those developed for the inner tracker of ALICE (CERN) and STAR (BNL). The stereo angle between the strips on the opposite sides of the SSDs is $\alpha = 7.5^{\circ}$. The optimization of the value of stereo angle is important for minimizing the number of ambiguities in track reconstruction. From typical strip pitch of $h = 58$ μm yields a space resolution of a track close to $\sigma_{\phi} = h/(\sqrt{12}) = 17$ μm in (r,ϕ) direction, and $\sigma_z = h/(\sqrt{12} * \alpha) = 128$ μm in Z the direction along the beam line. To eliminate the effect of insensitive areas at the edges of the sensors, the ladders are positioned with a small overlap.

 DOI: 10.1051/epjconf/201715806003

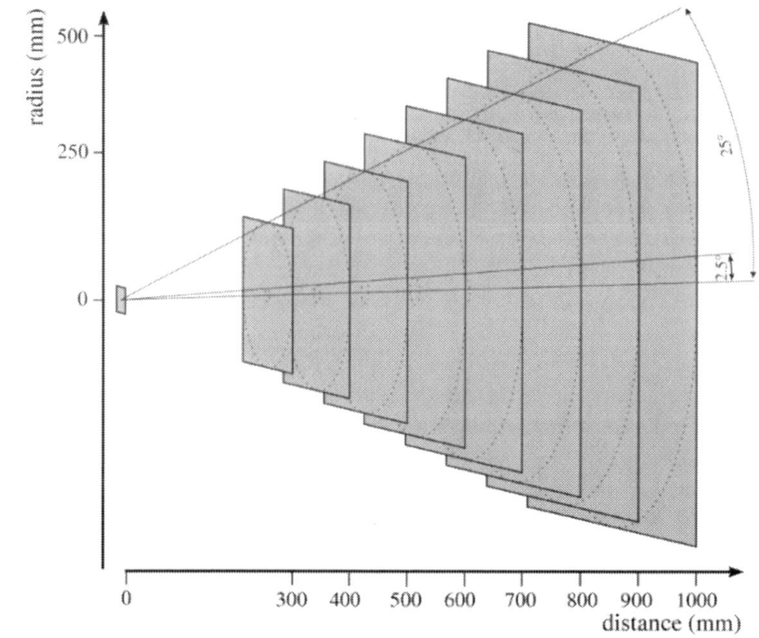

Fig. 3. Concept of STS tracking stations covering the polar angles 2.5° < Θ < 25°.

Fig. 4. Sensors 6×2, 6×4, 6×6 and 6×12 cm^2.

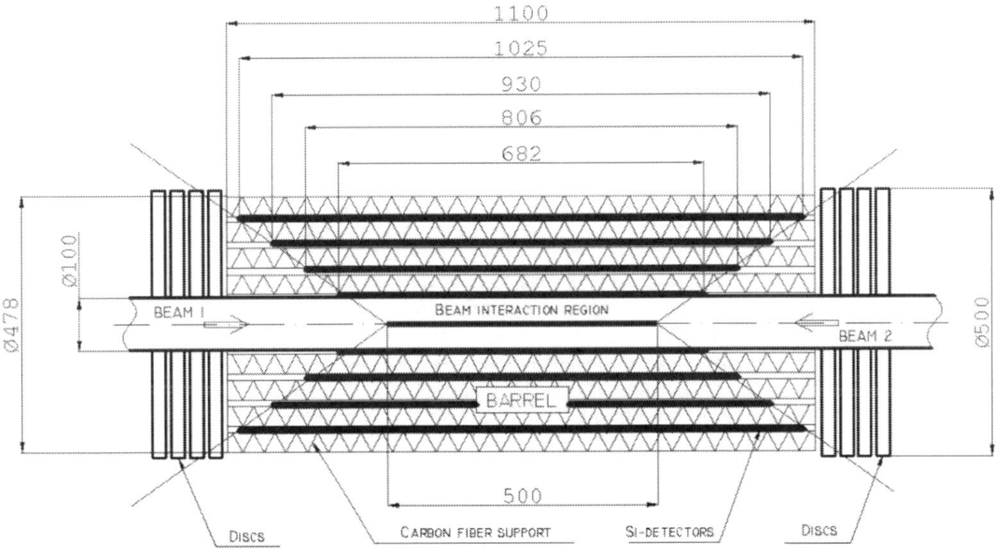

Total number of detectors 806

Fig. 5. Concept of Inner Tracker stations.

4 Tests of tracking modules

Sensors and electronics tests were carried out using experimental setup located in NUCLOTRON NIS-GIBS zone. There were two variants of setup with 6×6 cm^2 sensor and 14.9×14.9 mm^2 baby-sensor (Fig. 6).

Fig. 6. Test station with baby-sensor (in center) and 4 Front-end boards.

 DOI: 10.1051/epjconf/201715806003

Tests were carried out during two accelerating sessions of NUCLOTRON: session №53 had deuterons of kinetic energy 4.5 GeV with intensity I = $2*10^5$ c^{-1} and session №54 had deuterons of kinetic energy 2.94 AGeV and the same intensity.

Signals pickup from two detector modules was performed; the signals characteristics meet expectations. On the left side of spectrum noise of Gaussian form prevails; the form of main signal is the Landau distribution (Fig. 7). The signals were taken for the different angles between sensor and beam in order to determine dependence of average cluster size of sensor on that angle. As expected, cluster-1 event count decreases with increasing of the angle and merges into the noise. For the angle of 19° the signal cannot be distinguished from the noise (Fig. 8).

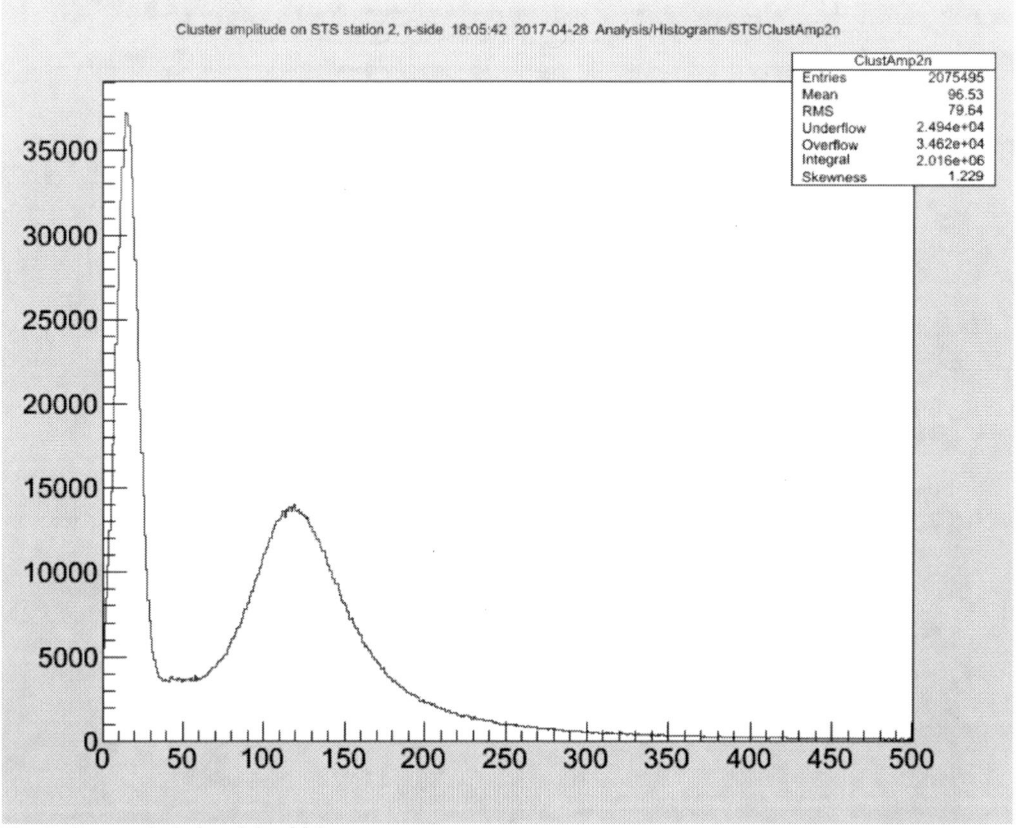

Fig. 7. Sensor 6×6 signal (n-side).

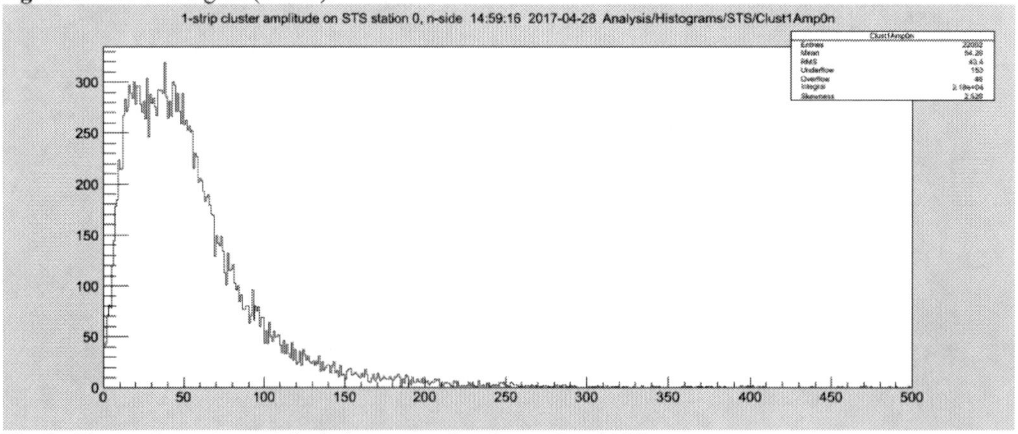

Fig. 8. Cluster-1 event count for baby-sensor (by the angle of 19°).

Average cluster sizes for the angles of 0°, 9°, 19° and 27° were calculated, and dependence on the angle was plotted (Fig. 9). The dependence can be approximated by a quadratic polynomial good enough. Experimental dependence can be compared to that from modeling; experimental cluster sizes are significantly less. This is due to fact that modeling deal with ideal sensors and electronics. In reality on the assumption of big angles energy of ionization divided into about 5 strips, and the signals from the leftmost and rightmost strips swamped by noise and cannot be counted. It leads to real cluster size be considerably decreased.

	Intercept Value	Intercept Standard Er	B1 Value	B1 Standard Er	B2 Value	B2 Standard Er	Statistics Adj. R-Squa
Cluster size	1,3131	0,05672	0,00879	0,01038	0,0021	3,70403E-4	0,99462

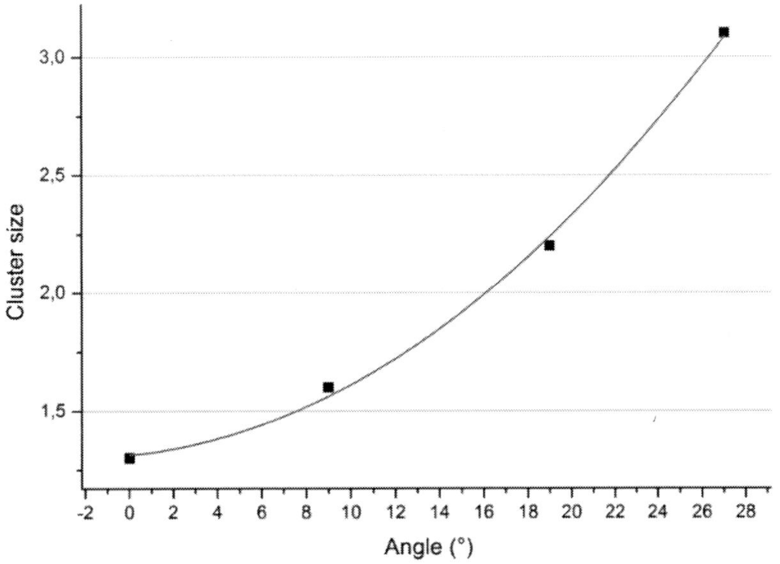

Fig. 9. Cluster size dependence.

Also, it should be noted that temperature tests of detector modules were carried out. [9] After heating, defects were not found. Modules work as expected.

5 Conclusions

All systems of the experimental setup work in normal mode. Used techniques of detector modules manufacturing and assembling are suit well for the future experiments. It will be necessary to test detector modules with new and more appropriate readout electronics (STS-XYTER) and to solve self-triggered electronics problems, such as signal-to-noise ratio and clocks synchronization of different modules and real time of events. Also it is necessary to check algorithms of track reconstruction.

References

1. H.-G. Moser, Progress in Particle and Nuclear, *Physics Silicon detector systems in high energy physics*, p. 186–237, (2009)
2. G. Lutz, *Semiconductor Radiation Detectors*, (2007)
3. A.S. Brogna, S. Buzzetti, C.J. Schmidt et al., *The n-XYTER Reference Manual*, (2009)
4. P. Kharlamov, *Testing of self-triggered readout electronics for double-sided microstrip silicon sensors based on n-XYTER ASIC*, Summer students program at JINR (2015)
5. Kasinski K., Kleczek R. and Szczygiel R., *Front-end readout electronics considerations for Silicon Tracking System and Muon Chamber*, Journal of Instrumentation, V. **11**. (2016)
6. J.M. Heuser, W.F.J. Müller, V. Pugatch et al., *Technical Design Report for the CBM : Silicon Tracking System (STS)*, (2013).
7. T.O.Ablyazimov, P.G.Akishin, V.P.Akishina et al., *Conceptual Design Report BM@N — Baryonic Matter at Nuclotron*, (2013)
8. A.N. Sissakian, A.S. Sorin, V.D. Kekelidze et al., *The MultiPurpose Detector – MPD to study Heavy Ion Collisions at NICA (Conceptual Design Report)*, (2011)
9. Dementyev D., Shitenkov M., and Kharlamov P., *Thermal tests of CBM STS prototype modules on CF-frames*, Compressed Baryonic Matter Experiment at FAIR - Progress Report 2016, (2017)

Betavoltaic device in *por-SiC/Si*

C-Nuclear Energy Converter

Alina Akimchenko[1,⋆], *Victor* Chepurnov[1,⋆⋆], *Mikhail* Dolgopolov[1,2,⋆⋆⋆], *Albina* Gurskaya[1,⋆⋆⋆⋆], *Oleg* Kuznetsov[1,†], *Alikhan* Mashnin[3,‡], *Vitaliy* Radenko[4,§], *Alexander* Radenko[4,¶], and *Oleg* Surnin[5,‖] & *George* Zanin[6,⋆⋆]

[1]*Samara University, Academician Pavlov 1, 443011 Samara, Russia*
[2]*STP «Aviatechnocon», Moscow highway 34, 3b, 443086 Samara, Russia*
[3]*Scientific production Association information technologies LIC, 142620 Moscow region, Russia*
[4]*MEC Quasar Ltd., Mechnikova 3, 443030 Samara, Russia*
[5]*LLC open {code;}, Samara, Russia*
[6]*BetaVoltaics LLC., Samara, Russia*

Abstract. The miniature and low-power devices with long service life in hard operating conditions like the Carbon-14 beta-decay energy converters indeed as eternal resource for integrated MEMS and NEMS are considered. Authors discuss how to create the power supply for MEMS/NEMS devices, based on porous SiC/Si structure, which are tested to be used as the beta-decay energy converters of radioactive C-14 into electrical energy. This is based on the silicon carbide obtaining by self-organizing mono 3C-SiC endotaxy on the Si substrate. The new idea is the C-14 atoms including in molecules in the silicon carbide porous structure by this technology, which will increase the efficiency of the converter due to the greater intensity of electron-hole pairs generation rate in the space charge region. The synthesis of C-14 can be also performed by using the electronically controlled magneto-optic chamber.

1 Introduction

The development of energy-saving technologies, the functioning of the MEMS devices, the reliability of their operation for a long time in offline conditions led to the search of appropriate means of generating energy for them. Especially in the trends of microelectronics on the benchmark 30 micron size active microprocessor board that hosts all the necessary devices.

⋆e-mail: malina-jun@yandex.ru
⋆⋆e-mail: chvi44@yandex.ru
⋆⋆⋆e-mail: mvdolg@yandex.ru
⋆⋆⋆⋆e-mail: a-gurska@yandex.ru
†e-mail: electron97@rambler.ru
‡e-mail: al.mashnin@gmail.com
§e-mail: quasar_ltd@mail.ru
¶e-mail: avradenko@yandex.ru
‖e-mail: surninoleg@gmail.com
⋆⋆e-mail: zager_06@bk.ru

Betavoltaic devices, also known as betavoltaic cells, are generators of electric current, in effect is the form of battery, which uses the energy from radionuclides emitting beta particles (electrons) [1]. A common source used is the hydrogen isotope, tritium, or Nickel-63.

From history it is well known that nuclear battery technology began in 1913, when Henry Moseley first demonstrated the beta cell [2]. And today the global market for active medical implants in 2016 amounted to 24.8 billions US dollars, and the capacity of the potential market of batteries in 2020 is estimated at \$ 40 billions. The annual demand of Russian patients in pacemakers is about 60 thousands [3–5].

In this paper authors will discuss how to create the power source for MEMS/NEMS devices based on por-SiC/Si porous structure, which is tested to be used as the beta-decay energy converter of radioactive Carbon-14 into electrical energy [6–9]. This involves silicon carbide obtaining by self-organizing mono 3C-SiC endotaxy way on the Si substrate [10, 11]. The key interest in the present aspect has the systematic optimization of main parameters that affect the operational efficiency of the betavoltaic current source [12].

Energy converters of beta radiation are based on isotopes transformations reactions with half-lives of more than 100 years. Energy converters are designed to secure operational performance, allowing reinstallation on the objects with the corresponding lifetime. Analogues were brought out to the laboratory samples level for two years. The groundwork of this project is on the part of semiconductor technology: mastered growing por-SiC/Si heterostructures. The groundwork for the development of an energy source not specified. But there is information of methods and businesses with similar activities [12–17]. The use of the product will effectively solve the following problems: the provision of energy for autonomous systems and devices, MEMS devices, sensors in oil capital constructions as example, information systems ⓪ systems, including biosensors of medical direction and equipment.

1.1 Betavoltaics power source for pacemaker

There are a number of ways to use betavoltaics power sources in heart pacemakers [5]. The first way is to replace the lithium-ion battery in the pacemaker to betavoltaic source. The advantage of this method is compact because betavoltaic source is small in size and easy to fit into the pacemaker without changing the design. However, this method has some difficulties: modern betavoltaic sources can not provide sufficient power and stable DC current for simultaneous charging of the capacitor and the power drive circuit of the pacemaker.

The second method is the simultaneous use of a lithium-ion battery and betavoltaic source. Li-ion battery will power the control circuit and to charge the capacitors in the pulse generator, and betavoltaic source will be connected to the lithium-ion battery through the charge controller, and shall provide drip recharging the battery. Drip – charging a small current to compensate for self-discharge or discharging at low external load. Typically, this method of charging is used for lead-acid batteries to extend their lifespan, but can be used for lithium-ion batteries also. Because lithium-ion batteries are sensitive to overcharging, you need to implement drip charging through the charge controller. The pacemaker is a constant drip rechargeable lithium-ion battery will greatly increase its life cycle. The advantage of this method is the lack of necessary changes in the design of the pacemaker - you only need to slightly increase the body to fit two batteries and a charge controller. The downside is the increase in the size of a pacemaker, which may complicate implantation. However, given the size betavoltaic power supplies and charge controller, increasing the hull will be negligible. It is only necessary to make a small charge controller.

DOI: 10.1051/epjconf/201715806004

2 Model of the semiconductor betavoltaic Energy Converter device

Unlike most nuclear power sources, which use nuclear radiation to generate heat, which then is used to generate electricity (thermoelectric and thermionic sources), betavoltaics uses the non-thermal conversion process, converting electron-hole pairs produced by the ionization trail of beta particles traversing the semiconductor [12, 13, 18].

There are two competitive directions of the betavoltaic converters development: the accumulation of charges in the reactions of transformation of elements using supercapacitors; the charge separation in p-n junction of the solar battery analogue [13]; the combination of the two directions. In all cases, the main competitive advantage over other energy sources is the long life cycle of beta-converter. The disadvantage is due to low capacity and low efficiency, which is overcome technological way (the higher degree of enrichment of isotope using the developed nanoporous surfaces, reducing the leakage current of super capacitor and p-n junction, using a combination of several sources of beta-radiation flux; the system microcontroller, which controls the flow of the charge). The ratio 'price/performance' is determined by several factors: semiconductors heterostructures provide a price level conventional semiconductor discrete diodes or transistors, high power, the technology of nuclear fusion, the working fluid used in the operation of medical devices provide a level of price acceptable to the consumer, when there is no alternative. Planning design parameters and performances are following: Voltage range ~ 2 mV, Size of the device $\sim 10 \times 10 \times 0.3 \div 0.5$ mm, Mass of each structural element ≈ 0.17 g, Working current range 0.1 μA, Working temperature range $-50 \div 350^0$C.

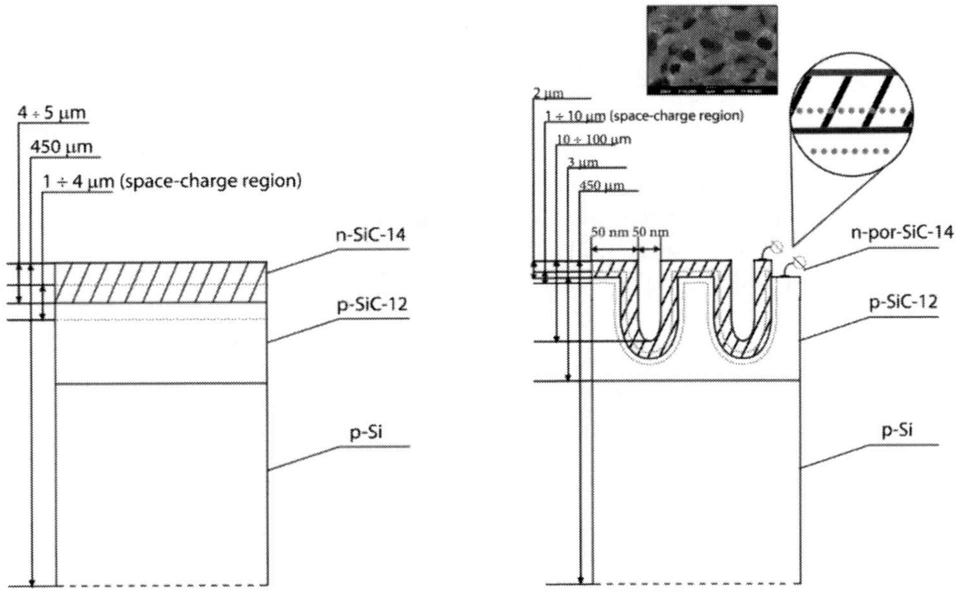

Figure 1. Two model structures variants [19].

The non-porous layer of n-type is determined by under the porous SiC, which is inevitably formed on the bottom of the pores and in fact is non-porous, i. e. p-n junction is the place to be in the right picture in Fig. 1 is still not in the porous layer. and indirectly associated with it and the space-charge region (SCR), between non-porous layers.

3 Carbon-14 choice of as the fuel

The data yield a conclusion on the promise of a new betavoltaic power supply source, namely, the nuclear battery based on Carbon-14, as the active substance would be more useful and less expensive [20]. Additional advantages of such power supply sources can be obtained in a silicon carbide structure based on the technology we developed [19]. Below let us consider the following comparison [20] with a technology that uses Nickel-63.

1. The production of C-14 in the dense ion plasma flow is much cheaper than the production of Ni-63 in Nickel target, due to the low cost of N-14.

2. Ion-plasma implantation C-14 on SiC substrate allows to reduce the crystal process and realize wide production on semiconductor converters on C-14.

3. At the same time, the production of Ni-63 by this technology is also possible, and the cost will be much reduced compared to traditional methods. Because of the greater mass of the ions, the equipment for ion-plasma sputtering with a seal of the flow of Nickel ions will be several times more expensive because of increased working installation area. At this stage, it is more sophisticated to produce C-14.

4. Specific activity of C-14 is different from the Ni-63 about in 10 times per unit volume, due to the huge difference in half-life times.

5. Self-absorption of Ni-63 is larger by approximately three times, which leads to the maximum limit optimal thickness of the layer to 4 microns, and for C-14 this thickness may be up to 60 microns which is better suited. The total quantity of the isotope C-14 may be an order of magnitude greater, therefore, guaranteed more power for the same size of power converters.

6. The specific power of Ni-63 per gram of the substance 5 times (due to more activity) exceeds the power density of C -14. But the maximum and average energies of electrons in the C-14 decay is in 2, and even 3, times more than in the Ni-63 decay.

7. The production cost of 12 *mg* C-14 will take approximately 8 hours, taking into account equipment depreciation, will cost about 170$.

Battery performance characteristics for automotive power applications can be described in terms of two parameters, specific power and specific energy. Specific power is related to the speed of vehicles and acceleration. Specific energy is related to range. Using these two variables, data are presented graphically for a number of commercially available battery systems and battery systems in development [21]. Both axes are logarithmic, which allows comparing performance of very different devices (for example, extremely high and extremely low power). The Ragone plot was first used to compare performance of batteries. It is suitable to compare any energy-storing devices [22].

4 Self-organizing mono 3C-SiC endotaxy technology

Authors present an original technology of manufacture and a model of the isotope energy converter based on the silicon carbide heterostructure, with Carbon-14 as fuel. The radionuclide C-14 is incorporated into the SiC molecule instead C-12 in the por-SiC phase in-por-SiC/SiC/Si heterostructure. Beta-radiation generates nonequilibrium charge carriers, separated by the internal field of p-n junction in the SiC-phase. The concept is based on the solid-phase transformation according to the Si-phase transformation diffusion model into the *SiC*-phase, which allows the buildup of *in*-por-*SiC* heterostructures [10, 11, 23, 24].

DOI: 10.1051/epjconf/201715806004

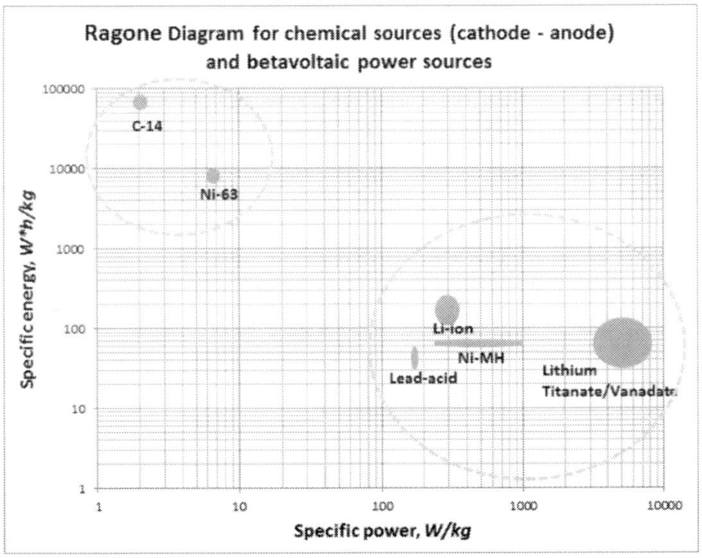

Figure 2. Ragone Diagram.

This is also possible to achieve by ion implantation of Carbon-14 into the silicon carbide phase [25–27]. The latter approach to the direct conversion can increase the effectiveness of beta-voltaic energy converters.

To analyze concentration of Carbon atoms in SiC heterostructure, SIMS method was used. We provide one example (for the left picture case in Fig.1) to illustrate the generation of in SiC/Si heterostructure phase [28].

The method of secondary ion mass spectrometry (SIMS) [29–31] is used to analyze the composition of solid surfaces and thin substrates. The essence of the method is to irradiate the sample surface by a focused beam of ions (primary ions). When hitting the sample, the primary ions undergo multiple collisions, which are emitted from the sample atoms and clusters of atoms in the process of collisions, the latter spontaneously ionized (forming secondary ions). These secondary ions are collected in the secondary beam and analyzed.

The feature of the endotaxy process is the technological method of forming a p-n junction during SiC phase growth by diffusion mechanism. Homotransition is obtained by means of the previously entered extraneous impurities realloying [32]. This is illustrated on the Fig. 3 by the intersection of the concentration distributions of impurities: lines KDB_P and KDB_Ga at point **A** at the depth of 2.5 microns, and/or lines KDB_P and KDB_B at the point **B** of 7.5 μm depth.

The technology of the temperature sequence of phase transitions during solid diffusion allows for the appropriate controlling and stabilization along the way to grow up carbide contacts in the form of nanowhiskers and nanothreads [19].

4.1 Magneto optic synthesis chamber

The synthesis of Carbon-14 can be also performed by using the electronically controlled synthesis of nucleons in the 8-cyclic magneto-optic synthesis chamber [33–35]. Generation proton flux for

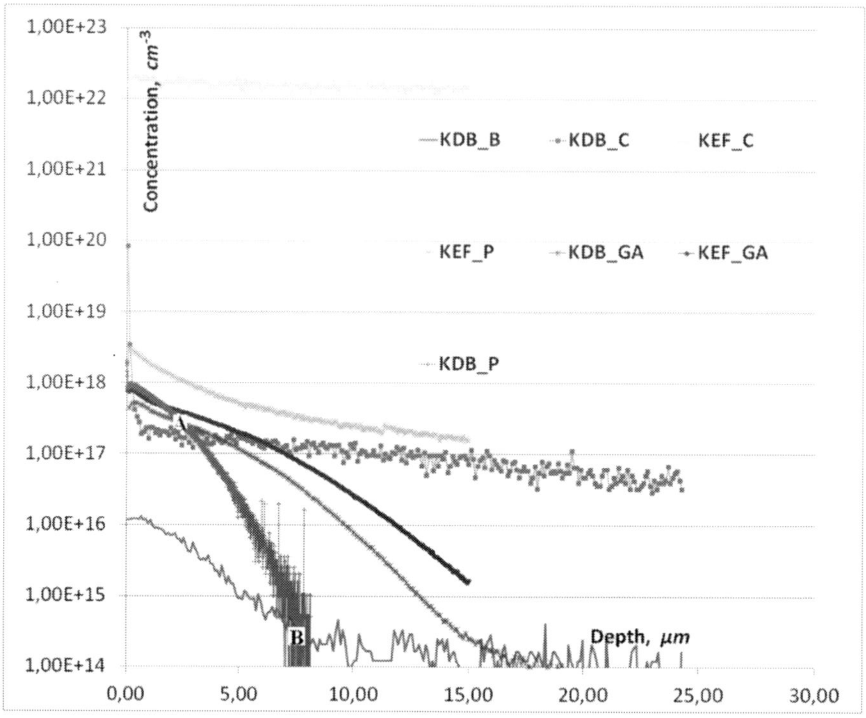

Figure 3. Diagram of concentrations dependencies on depth. The analyzed depth ranges from 5.3×10^{-8} to 24.3×10^{-6} m. The concentration of C-12 atoms (KDB_C curve), depending on the analyzed depth, ranges from 8.4×10^{19} to 3.2×10^{16} [20, 28].

the synthesis of neutrons by lithium ion target as a result of compaction of the primary flux of protons and sampling of parameters specify the density and average energy of the flow. Electronically controllable streams are formed in parameters repetition period T_{sl} and frequency discrete streams ω_{sl}. Generation dense proton flux for the synthesis of neutrons by lithium ion targets:

$$D + Li^6 \rightarrow Be^7 + n^0, \qquad D + Li^7 \rightarrow 2He^4 + n^0. \tag{1}$$

The synthesis reaction at the target dense nitrogen ions:

$$D + Li^7 \rightarrow C^4 + p^+. \tag{2}$$

The Device (see Fig. 4) consists of electronically controlled ion sources with magneto optic flux concentration with the ion energy of up to 50 keV, electronically controlled pulsed accelerator sections that form discrete ion flux with the ion energy from 200 to 600 keV, combined with an 8-cyclic magneto optic synthesis chamber and a controlled magneto-dynamic trap for generating dense neutron fluxes. The synthesis of C-14 by neutrons goes in a 4-cycle magneto optic chamber on the dense ion target. At the final stage, C-14 from the magneto optic storage is deposited on the SiC substrate. Multilayer formation of C-14 on the SiC substrate or on substrates from other materials is also possible.

The total energy of the incident particle and the target is above the Coulomb repulsion energy. Density n_i impinging stream for pulsed operation is equal to or above 10^{22} cm^{-3}. Retention time in

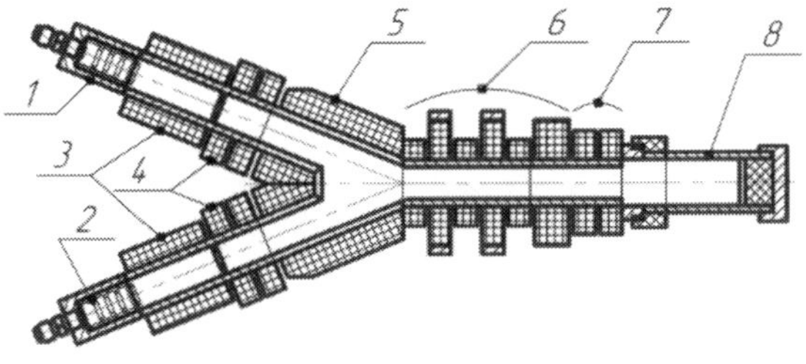

Figure 4. Magneto optic synthesis chamber. *1* – lithium deuteride LiD ionization, *2* – nitrogen N ionization, *3* – injector, *4* – magneto-optical drive-driver, *5* – accelerator-driver, *6* – magneto dynamic eight cyclic camera, *7* – magneto-optical drive-limiter, *8* – the C-14 drainage system. The density of neutron flux n=7.477E+18 cm^{-3}, the energy of neutrons in 1sr reaction (1): 2.72 E+6 MeV, the energy of neutrons is 2nd reaction (1): 12.0 E+6 MeV.

the magneto optical chamber with $\tau \geq 1$ s on one nuclear cycle. There is the ability to generate an 8-cycle synthesis of particles of different grades. Support deuterium-lithium reactions synthesis with a retention time in the system above the magneto optic c a single cycle is of 5 seconds.

4.2 Modeling estimates

As a basis for modeling and calculating of betavoltaic cell, working with the C-14 isotope, the model developed in [13] was chosen. The model is the simple structure of p-n junction in the cell, considered in cross section in Fig. 1 and takes into account such characteristics as: the generation rate, ionization loss energy, the electron-hole pairs formation, and the electron flux density. Betavoltaic cells based on standard silicon solar cells with bilateral coating with beta-radiation sources in the form of 63Ni isotope have been studied experimentally and by numerical simulation [13]. The optimal parameters of the cell, including its thickness, the doping level of the substrate, the depth of the p–n junction on its front side, and the p+ layer on the back side, as well as the activity of the source material, have been calculated in [13]. The limiting theoretical values of the open-circuit voltage (0.26 V), short-circuiting current (2.1 μA), the output power of the cell (0.39 μW), and the efficiency of the conversion of the radioactive energy onto the electric energy (4.8%) have been determined for a beta-source activity of 40 mCi. The results of numerical analysis have been compared with the experimental data in [13].

However, this model [13] does not account for some features of our battery and subsequently would be updated. In the simulation the Fig. 5 shows the comparison of the generation rate for C-14 and Ni-63 depending on the depth.

The Nickel-63 has larger generation rate, but C-14 electrons penetrate deeper. So the total rate of generation can be roughly the same, given the larger surface in the model isotope matrix with C-14 in the form of SiC-14 molecules in por-SiC/SiC/Si .

5 Summary

This development is intended for practical applications in MEMS and NEMS systems and sensors that require long-term autonomous operation. The suitability of C-14 as a source of beta-cells is proposed. Highlights the key parameters affecting the efficiency of beta-cells and conducted a qualitative analysis of the ways to increase the effectiveness of beta-source C-14 in the molecule SiC. Separately it is

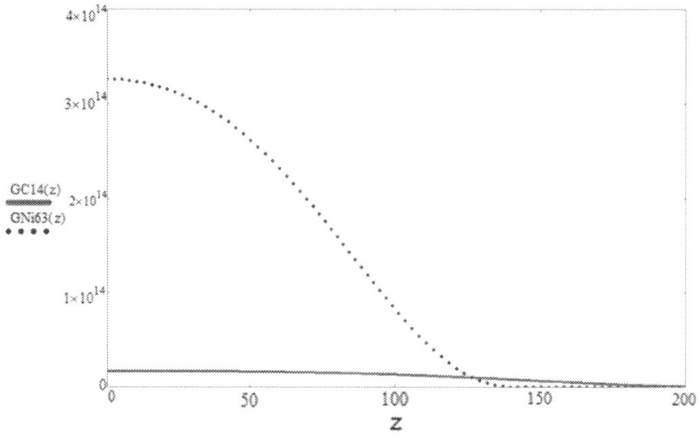

Figure 5. Generation rate comparison for C-14 and Ni-63. Notations for C-14 – solid red curve, Ni-63 – dot blue curve. Parameters: C-14 Specific activity 1.648×10^{11} Becquerel/gram. The energy at the one electron-hole pair of is 3.6 eV. The area of 4 cm^2 (as in [13]).

necessary to solve the problem of isotope enrichment C-14. We need to develop and conduct a series of experiments to verify the theoretical calculations and hypotheses.

The production of Carbon-14 at a cost lower than traditional production. The source of the neutron fluxes for the synthesis of radioactive isotopes for medical purposes. Modification device to vacuum ion-plasma deposition of Carbon-14 on the SiC substrate is developing still.

A method of implantation C-14 in molecule por-SiC is in step preparation heterostructures of p-n-transition. Technology has been developed and a mathematical model of the converter manufacturing: technological (route maps, physical parameters, tools) and design (assembly devices, housing, terminals, electrical insulation, recycling methods) documentation.

Acknowledgements. *M. D. is grateful to Workshop organizers for the attention, creative discussion and the opportunity to present results of research project, also thanks V. Tarala for debates and discussion of experiments. Authors are grateful to collaboration in principle detailed analysis with Bazhanov A.Ya. and Romanov A. (Angstrem) & V. Zinchenko and G. Shestakov (Aviatechnocon).*

References

[1] Rappaport P. The electron-voltaic effect in p-n junctions induced by beta particle bombardment / Phys. Rev. 1953. Vol. 93. P. 246-247.

[2] H. G. J. Moseley, Phil. Mag., 1913, 26, 1025–1034.

[3] YOLE DEVELOPPEMENT Status of the MEMS Industry 2016//i-micronews.com URL: https://www.i-micronews.com/category-listing/product/status-of-the-mems-industry-2016.html

[4] Dolgopolov Mikhail V., Anisimova Valeriya Yu., Akimchenko Alina A., Kuznetsov Oleg V., Zanin George G. Analysis and assessment of the betavoltaics power supply market. P. 452-473. In Economics and management in conditions of nonlinear dynamics / under the editorship of Dr. Econ. Sciences, Professor A. V. Babkin. – SPb. : Publishing house of Polytechnical Institute. University, 2017. – 773 p.

[5] Katz D., Akiyama T. Pacemaker longevity: The World's Longest-Lasting VVI Pacemaker. Annals of Noninvasive Electrocardiology, Volume 12, Issue 3, pages 223-226. 2017.

[6] V.I.Chepurnov, M.V.Dolgopolov, A.V.Gurskaya, S.N.Podgornov, A.N.Scherbakov, International Conference "Electronics-2015" , 13 (2015)

[7] A.V.Gurskaya, M.V.Dolgopolov, V.I.Chepurnov, 22 All-Russian Scientific Conference of physics students and young scientists, 183 (2016)

[8] S.P.Zavodov, V.I.Chepurnov, A.A.Kirpichev, Modern problems of physics and technologies. MEPhI., 189-191 (2016)

[9] The Patent Application 2016129598, 19.07.2016.

[10] The Patent of The RUSSIAN FEDERATION 02370851 C2 Method of self-organizing mono 3c-SiC endotaxi on Si substrate / V.I.Chepurnov, since 15.12.2005.

[11] ROSRID, R & D report, with references to the literature, V.I.Chepurnov, etc. AAAA-B16-216122220098-7.

[12] Tariq R.Alam, Mark A. Pierson. Principles of Betavoltaic Battery Design // J. Energy Power Sources. Vol. 3, No. 1, 2016, pp. 11-41.

[13] Gorbatsevich A.A., Korneev V.I., Danilin A.B., Magomedbekov E.P., Molin A.A. Analysis (Simulation) of Ni-63 beta-voltaic cells based on silicon solar cells. Technical Physics. The Russian Journal of Applied Physics. 2016. T. 61. N 7. P. 1053-1059.

[14] M.A. Polikarpov, E.B. Yakimov Investigation of properties of semiconductor converter based on silicon for beta-voltaic elements. SEMICONDUCTORS. 2015, T. 49, V. 6. P. 763-766.

[15] Abanin I.E. Selection of Active Layers for a Power Supply Device with p-n Junction Excited by β-Radiation. NANO- AND MICROSYSTEMS TECHNOLOGY. N 10 (183). 2015. P. 4-10.

[16] Bulyarskiy S.V., Lakalin A.V., Abanin I.E., Amelichev V.V., Svetuhin V.V. Optimization of the parameters of power sources excited by β-radiation. Semiconductors. 2017. T. 51. N 1. P. 66-72.

[17] Saurov A.N., Bulyarskiy S.V., Risovanniy V.D., Pavlov A.A., Abanin I.E., Kitsyuk E.P., Shamanaev A.A., Lebedev E.A. Nanostructured Current Sources Excited by β-Radiation Based on Carbon Nanotubes. Izvestiya vysshikh uchednykh zavedenii. Elektronika. T 20. N 5. 2015. P. 474-480.

[18] A 25-Year Battery: Long-lived nuclear batteries powered by hydrogen isotopes are in testing for military applications, Katherine Bourzac, Technology Review, MIT, 17 Nov 2009.

[19] A. V. Gurskaya, M. V. Dolgopolov, V. I. Chepurnov. C-14 Beta Converter. Physics of Particles and Nuclei, 2017, Vol. 48, No. 6, pp. 941–944.

[20] V.I.Chepurnov, M.V.Dolgopolov, A.V.Gurskaya, A. A. Akimchenko, O. V. Kuznetsov, A. V. Radenko, V. V. Radenko, A. S. Mashnin. C-BETAVOLTAIC ENERGYCONVERTER IN POR-SIC/SI // MATERIALS SCIENCE. NON-EQUILIBRIUM PHASE TRANSFORMATIONS. 2017. Issue 3. PP. 119-120

[21] Ragone, D., Review of Battery Systems for Electrically Powered Vehicles, SAE Technical Paper 680453, 1968. 8 p.

[22] Christen, Thomas; Carlen, Martin W. (2000). Theory of Ragone plots. Journal of Power Sources. 91 (2). pp. 210–216.

[23] V.I.Chepurnov, M.V.Dolgopolov, A.V.Gurskaya, Sorokov A.S., Oil. Gas. Novatsiya. Volume 12 80-82 (2016).

[24] Pavlikov A.V., Timoshenko V.Y., Latukhina N.V., Chepurnov V.I. Structural and optical properties of silicon-carbide nanowires produced by the high-temperature carbonization of silicon nanostructures // Semiconductors. 2017. T. 51. N 3. P. 402-406.

[25] Vavilov V.S., Guseva M.I., Konorova E.A., Krasnopevtsev V.V., Sergienko V.F., TutovV.V. Semiconducting diamonds produced by ion bombardment // Soviet Phys.-Solid St. 8. 1966. pp.1560-1561.

[26] Vavilov V.S., Guseva M.I., Konorova E.A., Sergienko V.F. Investigation during isochronous multistage annealing of the electrical conductivity of semiconducting n- and p-type diamonds prepared by the ion implantation method // Soviet Phys. Semiconductors 4. 1970. P. 6-11.

[27] Gusev V.M., Titov V.V. Kinetics of thermal annealing of radiation defects in silicon doped by the ion implantation method // Sov. Phys.-Semicond. 3, 1-6. 1969.

[28] Bazhanov A.Ya., Chepurnov V.I., Romanov A. 2017. In preparation.

[29] Thomson J.J. Rays of positive electricity // Phil. Mag. 20: 752-767. 1910.

[30] Herzog R.F.K., Viehboeck F. Ion source for mass spectrography // Phys. Rev. 76 (6): 855-856. 1949.

[31] Honig R.E.. Sputtering of surfaces by positive ion beams of low energy // J. Appl. Phys. 29: 549-555. 1958.

[32] V. I. Chepurnov, Associates of dot defects of various nature in SiCSiC-phase of semiconductor heterostructure of SiC//SiSiC//Si, received by endotaksiya method, Vestnik Samarskogo Gosudarstvennogo Universiteta. Estestvenno-Nauchnaya Seriya, 2014, no. 7(118), 145-162

[33] Physics and Technology of Ion Sources, 2nd, Revised and Extended Edition. Ian G. Brown (Editor). 396 pages. October 2004.

[34] Mesyats Gennady A. Pulsed power. Springer. 2005. 568 p.

[35] A. T. Forrester Large Ion Beams: Fundamentals of Generation and Propagation. Wiley-VCH; 1 edition (April 1988). 325 p.

The evaluation of the systematic uncertainties for the finite MC samples in the presence of negative weights

Petr Mandrik[1,*]

[1] *NRC «Kurchatov Institute» – IHEP, Protvino*

Abstract. The analysis of results from HEP experiments often involves the estimates of the composition of the binned data samples, based on Monte Carlo simulations of various sources. Due to a finite statistic of MC samples they have statistical fluctuation. This work proposes the method of incorporating the systematic uncertainties due to finite statistics of MC samples with negative weights. The possible approximations are discussed and the comparison of different methods are presented.

1 Introduction

Experimental results in high energy physics are often represented as a binned distribution (histogram) of observed events $\boldsymbol{X} = (X_1, X_2, ...)$, where X_i is a number of events in bin i. The usual method to estimate physical parameters such as particle masses or cross sections from this distribution is to perform some of Bayesian or frequentist analyses based on likelihood function. Likelihood function $\mathcal{L}(\boldsymbol{X}|\boldsymbol{m})$ connect the data with a theoretical model and represente how well the observatios are described by the prediction $\boldsymbol{m} = (m_1, m_2, ...)$. This prediction may depend on several different parameters $\boldsymbol{\pi} = (\pi_1, \pi_2, ...)$: nuisance parameters and parameters of interests. In addition if some signals or background processes are known from Monte-Carlo simulations then the likelihood function depends on template distributions $\boldsymbol{t} = (t_1^a, t_2^a, ...t_1^b, t_2^b, ...t_1^c, t_2^c, ...)$, where t_i^k is a number of events for process k in bin i:

$$\mathcal{L}(\boldsymbol{X}|\boldsymbol{m}) = \mathcal{L}(\boldsymbol{X}|\boldsymbol{\pi}, \boldsymbol{t}) = \prod_i P(X_i|\boldsymbol{\pi}, \boldsymbol{t}_i) = \prod_i P(X_i|\pi_1, \pi_2, ..., t_i^a, t_i^b, ...) \tag{1}$$

This is important task to define an adequate likelihood function and take into account all existed statistical and systematic uncertainties present in the analysis.

Template distributions from Monte-Carlo generators are subject to statistical fluctuations due to finite number of events in samples. The influence of these fluctuations can be expected to be significant in regions of low amounts of Monte-Carlo events. For incorporating such uncertainties into likelihood function Barlow and Beeston proposed a method [1] wherein one for every bin i and every process k introduces a new parameter T_i^k corresponding to unknown expected number of events in infinite statistics limit:

$$\prod_i P(X_i|\boldsymbol{\pi}, \boldsymbol{t}_i) \rightarrow \prod_i \left[P(X_i|\boldsymbol{\pi}, \boldsymbol{T}_i) \cdot \prod_k P(t_i^k|T_i^k)\right] \tag{2}$$

*e-mail: Petr.Mandrik@ihep.ru

where for constrain $P(t_i^k|T_i^k)$ Barlow and Beeston assumed a Poisson distribution.

On the other hand several of the modern Monte-Carlo generators [2] produce a weighted events with both negative and positive weights. In this case the transformation (2) is not applicable. In this paper we provide a method of incorporating uncertainties due to the finite statistics of Monte-Carlo samples in the presence of negative weights.

2 Likelihood functions for Monte-Carlo samples with negative weights

In simplified form the algorithm of event production in most important example of Monte-Carlo generator with negative weights MadGraph5_aMC@NLO can be described as follow [3]. A cross section of some process σ_{NLO} is calculated by computing the integrals of two functions $F_H(x)$ and $F_S(x)$:

$$\sigma_{NLO} = \int F_H(x)dx + \int F_S(x)dx \tag{3}$$

By definition, the functions $F_H(x)$ and $F_S(x)$ are finite and $F_H(x)+F_S(x) > 0$, but for some values of x the function $F_S(x)$ is negative. Using the absolute values of the integrands $|F_H(x)|$ and $|F_S(x)|$ two set of events are produced - $\{x\}_H$ and $\{x\}_S$ respectly with weights $w_i^{H,S}$ equal to $+1$ for positive values of functions and weight -1 if function is negative, so:

$$\sigma_{NLO} = \frac{\int |F_H(x)|dx}{N_H} \cdot \sum_i^{N_H} w_i^H + \frac{\int |F_S(x)|dx}{N_S} \cdot \sum_i^{N_S} w_i^S \tag{4}$$

where N_H, N_S are thw number of events in corresponding sets.

In this way in the infinite statistics limit the prediction of any observable in any intervals $[x_i, x_i + \Delta x]$ of histograms from MadGraph5_aMC@NLO can be only positive, but in the case of finite statistics the prediction could get negative values in some bins. On the other hand the events with negative and positive weights should be treated in the same way during analyses and pass the same cuts to keep the correct cross section value (4). Further we assume that the last condition is satisfied, so for the finite number of generated Monte-Carlo events the probability of obtaining a given one in the case of only positive or negative weights is described by multinomial distribution, which is usually approximated by multiplication of independent Poisson distributions (see for example, [4]).

Let us consider a simple case of single bin and only one generated process with total number of event t. If t^+ is a sum of all positively weighted events from Monte-Carlo samples and t^- is a sum of all negative, then t is a difference of two Poissonian quantities t^+ and t^- and described by Skelleman distribution:

$$P(t) = \mathcal{S}(t|T^+, T^-) = \sum_{s=max(0,t)}^{\infty} \mathcal{P}(s|T^+) \cdot \mathcal{P}(s-t|T^-) = e^{-(T^++T^-)}\left(\frac{T^+}{T^-}\right)^{t/2} \mathcal{I}_t(2\sqrt{T^+T^-}) \tag{5}$$

where $\mathcal{I}_t$ is a modified Bessel functions of the first kind, $\mathcal{P}$ is Poisson distribution, T^+ and T^- are the parameters of Skellenam distribution, corresponding to unknown "true" prediction of negative and positive events from MC generator.

Using the equations (5) in (2) we obtain the following transformation rule for taking into account uncertainties due to the finite statistics of Monte-Carlo samples in the presence of negative weights in likelihood function:

$$\prod_i P(X_i|\boldsymbol{\pi}, \boldsymbol{t}_i) \to \prod_i \left[P(X_i|\boldsymbol{\pi}, \boldsymbol{T}_i) \cdot \prod_k \mathcal{S}(t_i^k|T_i^{k+}, T_i^{k-})\right] \tag{6}$$

where the new parameters are related by the equation $T_i^k = T_i^{k+} - T_i^{k-}$.

The constrain on parameters T_i^k, T_i^{k+}, T_i^{k-} in formula (6) may be improved by an independent treatment of values t^+ and t^- in analyses. In this case we get:

$$P(t) = \mathcal{P}(t^+|T^+) \cdot \mathcal{P}(t^-|T^-) \tag{7}$$

and from (2) with (7):

$$\prod_i P(X_i|\boldsymbol{\pi}, \boldsymbol{t}_i) \to \prod_i \Big[P(X_i|\boldsymbol{\pi}, \boldsymbol{T}_i) \cdot \prod_k \mathcal{P}(t_i^{k+}|T_i^{k+}) \cdot \mathcal{P}(t_i^{k-}|T_i^{k-})\Big] \tag{8}$$

The number of extra parameters in the transformation rule (8) is equal to 2× *number of processes* × *number of bins*. We can decrease the number of parameters by using the method of maximum likelihood function Indeed, if $\mathcal{L}$ is a likelihood function with transformation (7) then for bin i one gets:

$$-\ln \mathcal{L}_i = -\ln P(X_i|\boldsymbol{\pi}, \boldsymbol{T}_i) - \sum_k \Big(- T_i^{k+} + t_i^{k+} \cdot \ln T_i^{k+} - \ln (t_i^{k+}!)\Big) - \\ - \sum_k \Big(- T_i^{k-} + t_i^{k-} \cdot \ln T_i^{k-} - \ln (t_i^{k-}!)\Big) \tag{9}$$

The requirement of an extremum gives the following system of equations:

$$\begin{cases} -\dfrac{\partial \ln \mathcal{L}}{\partial T_i^{k+}} = 1 - \dfrac{\partial \ln P(X_i|\boldsymbol{\pi}, \boldsymbol{T}_i)}{\partial T_i^{k+}} - \dfrac{t_i^{k+}}{T_i^{k+}} = 0 \\ -\dfrac{\partial \ln \mathcal{L}}{\partial T_i^{k-}} = 1 - \dfrac{\partial \ln P(X_i|\boldsymbol{\pi}, \boldsymbol{T}_i)}{\partial T_i^{k-}} - \dfrac{t_i^{k-}}{T_i^{k-}} = 0 \end{cases} \tag{10}$$

This system (10) in some cases may be solved analytically for the parameters T_i^{k-}, T_i^{k+}, or numerically with some fixed values of the remaining parameters.

The another way to decrease the number of parameters related to finite statistics of Monte-Carlo is known as Barlow-Beeston "light" transformation [5]. As the statistical uncertainties for each source in each bin are independent they may be combined and be represented approximately by single effective parameter per bin M_i:

$$\prod_i P(X_i|m_i) \to \prod_i P(X_i|M_i) \cdot P(m_i|M_i) \tag{11}$$

Usually, in this approximation for $P(m_i|M_i)$ usually a Gaussian constrain $\mathcal{G}(m_i|M_i, \sigma_i)$ is used, where the value of σ_i are calculated by propagation of the Monte-Carlo statistical uncertainties in bin i with fixed values of the remaining parameters.

For histograms with negative weights the transformation (11) has the form:

$$\prod_i P(X_i|m_i^+ - m_i^-) \to \prod_i P(X_i|M_i^+ - M_i^-) \cdot P(m_i^+|M_i^+) \cdot P(m_i^-|M_i^-) \tag{12}$$

The number of extra parameters is equal to 2× *number of bins in histogram*.

For the likelihood function with transformation (12) a system of equations similar to (10) can be obtained. Using the Gaussian constrain one gets:

$$\mathcal{L}_i = \mathcal{P}(X_i|M_i^+ - M_i^-) \cdot \mathcal{G}(m_i^+|M_i^+, \sigma_i^+) \cdot \mathcal{G}(m_i^-|M_i^-, \sigma_i^-) \tag{13}$$

$$-\ln \mathcal{L}_i = -\Big[-(M_i^+ - M_i^-) + X_i \cdot \ln (M_i^+ - M_i^-) - \ln X_i!\Big] - \\ -\Big[\frac{(M_i^+ - m_i^+)^2}{2(\sigma_i^+)^2} - \ln \sigma_i^+ \sqrt{2\pi}\Big] - \Big[\frac{(M_i^- - m_i^-)^2}{2(\sigma_i^-)^2} - \ln \sigma_i^- \sqrt{2\pi}\Big] \quad (14)$$

$$\begin{cases} -\dfrac{\partial \ln \mathcal{L}}{\partial M_i^+} = 1 - \dfrac{X_i}{M_i^+ - M_i^-} - \dfrac{M_i^+ - m_i^+}{(\sigma_i^+)^2} = 0 \\ -\dfrac{\partial \ln \mathcal{L}}{\partial M_i^-} = 1 + \dfrac{X_i}{M_i^+ - M_i^-} - \dfrac{M_i^- - m_i^-}{(\sigma_i^-)^2} = 0 \end{cases} \quad (15)$$

3 The performance of methods

In this section few results of study the proposed transformations for taking into account uncertainties due to the finite statistics of Monte-Carlo samples are given. The source code was implemented with statistical package SHTA [6].

From the simple single-bin single-channel model:

$$\mathcal{L}_0 = \mathcal{P}(X|\pi \cdot (C + T^+ - T^-)) \cdot \mathcal{P}(t^+|T^+) \cdot \mathcal{P}(t^-|T^+) \quad (16)$$

a set of events (X, t^+, t^-) may be generated for the fixed values of parameters π, T^+, T^-. Here the constant C is introduced in order to avoid a long tail in posterior distribution of π from $T^+ - T^- \sim 0$.

To estimate the parameter of interest π we use three different likelihood functions. First of all a naive approach without incorporating uncertainties due to the finite statistics of Monte-Carlo samples:

$$\mathcal{L}_n = \mathcal{P}(X|\pi \cdot (C + t^+ - t^-)) \quad (17)$$

a likelihood function with transformation (8):

$$\mathcal{L}_p = \mathcal{P}(X|\pi \cdot (C + T^+ - T^-)) \cdot \mathcal{P}(t^+|T^+) \cdot \mathcal{P}(t^-|T^-) \cdot \mathcal{H}(T^+ - T^-) \quad (18)$$

and similar one but with Gaussian approximation for multiplication of two Poissons:

$$\mathcal{L}_g = \mathcal{P}(X|\pi \cdot (C + T)) \cdot \mathcal{G}(t^+ - t^-, T, \sqrt{t^+ + t^-}) \cdot \mathcal{H}(T) \quad (19)$$

where $\mathcal{H}$ is a Heaviside function.

The generated set of toy data from (16) is used to perform a Bayesian inference (see for example [7]) with non-informative flat priors for all parameters. The posterior probability density functions for parameters were obtained from likelihood functions (17), (19), (18) and the confidence intervals were found. The results for two different set of initial values of π, T^+, T^- are presented in the table 1.

From the table 1 we can see that the difference between solution with multiplication of two Poisson and its Gaussian approximation do not exceed one percent. On the other hand without taking into account the uncertainties due to the finite statistics of Monte-Carlo samples in likelihood function (17) the accuracy of measurements falls significantly. For example, only in two out of three experiments the correct interval will be obtained for $CL = 2\sigma$ and first set of parameters.

Larger number of bins in such test models provide additional constrain on the parameter of interest and without considering the increase of the computational complexity may only improve results.

Table 1. Percent of toy data for which the confidence interval with confidence level CL from posterior distribution includes the true value of parameter π. The uncertainties are evaluated by averaging the results from different sets of toy data.

$\mathcal{L}_0$ parameters	CL	$\mathcal{L}_n$	$\mathcal{L}_g$	$\mathcal{L}_p$
$\pi = 3, T^+ = 12, T^- = 4$	1σ	34.71 ± 0.79	67.79 ± 0.75	67.98 ± 0.72
	2σ	62.82 ± 0.70	95.25 ± 0.27	96.14 ± 0.22
$\pi = 3, T^+ = 9, T^- = 7$	1σ	26.41 ± 0.57	77.52 ± 0.63	78.51 ± 0.63
	2σ	49.73 ± 0.71	98.18 ± 0.15	98.21 ± 0.15

4 Conclusion

In this work a method of incorporating the systematic uncertainties due to finite statistics of MC samples with negative weights is presented. The influence of this statistical uncertainty can be expected to be high in regions of low amounts of Monte Carlo events and they must be included into the fit. The proposed transformation (8) and its simplified version (12) can be used to construct the correct likelihood function. While using the Gaussian approximation of multiplication of two Poisson distribution in (8) or (12) leads to the known expressions used in different statistical packages [8][9] in different forms, the choice of specific form of likelihood function depends on the analysis and in some cases the more accurate proposed methods can improve the results.

Acknowledgments

I wish to thank L. Dudko, S. Slabospitskii and G. Vorotnikov for useful discussions.

References

[1] R.J. Barlow, C. Beeston, Comput. Phys. Commun. **77**, 219 (1993)
[2] A. Buckley, J. Butterworth, S. Gieseke, D. Grellscheid, S. Höche, H. Hoeth, F. Krauss, L. Lönnblad, E. Nurse, P. Richardson et al., Physics Reports **504**, 145 (2011)
[3] S. Frixione, B.R. Webber, Journal of High Energy Physics **2002**, 029 (2002)
[4] C. Walck, *Hand-book on statistical distributions for experimentalists* (1996), `http://www.fysik.su.se/~walck/suf9601.pdf`
[5] J.S. Conway, *Incorporating Nuisance Parameters in Likelihoods for Multisource Spectra*, in *Proceedings, PHYSTAT 2011 Workshop on Statistical Issues Related to Discovery Claims in Search Experiments and Unfolding, CERN,Geneva, Switzerland 17-20 January 2011* (2011), pp. 115–120, `1103.0354`, `http://inspirehep.net/record/891252/files/arXiv:1103.0354.pdf`
[6] P. Mandrik, *SHTA - package for experimental statistical data analyses in high energy physics*, `https://github.com/pmandrik/shta/tree/negative_weights` (2017)
[7] G. D'Agostini, CERN-99-03, CERN-YELLOW-99-03 (1999)
[8] T. Müller, J. Ott, J. Wagner-Kuhr, CMS Internal Note CMS-IN 017 (2010)
[9] K. Cranmer, G. Lewis, L. Moneta, A. Shibata, W. Verkerke (ROOT Collaboration), Tech. Rep. CERN-OPEN-2012-016, New York U., New York (2012), `http://cds.cern.ch/record/1456844`

Testing of the VENUS 4.12, DPMJET 2.55, QGSJET II-03 and SIBYLL 2.3 hadronic interaction models via help of the atmospheric vertical muon spectra

L. G. Dedenko[1,2,⋆], *A. V.* Lukyashin[3,4,⋆⋆], *T. M.* Roganova[2,⋆⋆⋆], and *G. F.* Fedorova[2,⋆⋆⋆⋆]

[1] *Faculty of Physics M.V. Lomonosov Moscow State University, Leninskie Gory, 119991 Moscow, Russia*
[2] *Skobeltsyn Institute of Nuclear Physics, Lomonosov Moscow State University, 119234 Moscow, Russia*
[3] *Institute for Theoretical and Experimental Physics named by A.I. Alikhanov of National Research Center "Kurchatov Institute", 117218 Moscow, Russia*
[4] *National Research Nuclear University (MEPhI) Moscow Engineering Physics Institute, 115409 Moscow, Russia*

Abstract.
Uncertainties of the model energy spectra of the most energetic secondary $\pi^{\pm}$ -mesons (and $K^{\pm}$-mesons) are discussed. Computer simulations of the partial energy spectra of the atmospheric vertical muons induced by primary cosmic particles with various fixed energies in interval $10^2 - 10^7$ GeV in terms of the VENUS 4.12, DPMJET 2.55, QGSJET II-03 and SIBYLL 2.3 models had been carried out with help of CORSIKA package. These partial spectra should be convoluted with the contemporary spectra of the primary cosmic particles. Results of simulations are compared with the contemporary atmospheric vertical muon flux data. Comparison shows that all models underestimate production of secondary $\pi^{\pm}$-mesons (and $K^{\pm}$-mesons) by factor of ~ 2 at the most high energies. This underestimation induces more rapid development of extensive air showers in the atmosphere and results in uncertainties in estimates of energy and composition of the primary cosmic particles.

1 Introduction

The extensive air showers (EAS) are the only a tool to understand the origin and composition of cosmic rays, their possible sources and a transport of cosmic particles in various magnetic fields on their way to the Earth at very high energies. All features of the energy spectrum, arrival directions and composition of the primary cosmic particles should be determined through an analysis of the EAS data. These data as some signals in the surface and underground detectors are usually interpreted in terms of various models of hadronic interactions [1-9]. It happened, that such interpretation leads to some inconsistency. As an example, energy of showers calculated in terms of the QGSJET II-03 [3] model with help of the surface detectors signals at Telescope Array [10] happened to be 1.27 times

⋆e-mail: ddn@dec1.sinp.msu.ru
⋆⋆e-mail: lukyashin.anton@physics.msu.ru
⋆⋆⋆e-mail: rogatm@yandex.ru
⋆⋆⋆⋆e-mail: fdr@dec1.sinp.msu.ru

lager than this energy estimated with help of the fluorescence light. To be sure that results of such interpretation are as accurate as possible these models should be thoroughly tested. Usually these models are tested with the help of the accelerator data at small values (~0) of the pseudorapidity η where most of secondary particles (mainly mesons) are produced [11-13]. However, calculations have shown that the maximal energy flow carried by secondary particles occurs at much larger values (~8-10) of the pseudorapidity η [14]. Let us also note that the longitudinal development of EAS depends strongly on the rate of the projectile particle energy fragmentation. The atmospheric muon flux also depends strongly on this production of the most high energy mesons. So, it is of the primary importance to verify a production of the most energetic mesons simulated in terms of various models. This verification may be carried out by comparing model predictions of this muon fluxes with data. We select the classical experiments L3+Cosmic [15], MACRO [16] and LVD [17] and elaborate the smooth approximation of these muon data in the energy interval of 10^2 – 10^5 GeV. The model predictions of muon flux have been estimate as follows. Showers induced by the primary protons and helium nuclei with different fixed energies have been simulated with help of the CORSIKA package [18] and the muon partial energy spectrum in each individual shower have been calculated. Results of these simulations for every type of the primary particles multiplied by intensities of these particles should be integrated on energy of particles. Thus, we need also some expressions for the energy spectra of various primary particles. Inspired by modern results and new precision cosmic rays data base [19], we suggested new approximations of cosmic ray energy spectra for primary protons and helium nuclei. Indeed, there are results of many measurements (e.g.AMS-02 [20], PAMELA [21], ATIC-2 [22], CREAM [23], ARGO-YBJ [24], ARGO-YBJ & FWCTA [25], KASCADE [26], KASCADE-Grande [27], Tunka [28], IceCube [29], Telescope Array [30]) of the fluxes of the primary cosmic nuclei. Besides there are some calculations of spectra of the primary proton and helium nuclei in SNR [31]. We will use these approximations for energy spectra of the primary protons and helium nuclei to estimate convolutions with the partial muon spectra. Thus, with the help of any interaction models [1-9], the package CORSIKA and approximations of data on fluxes of the primary cosmic nuclei [20-30] one can predict the energy spectra of atmospheric vertical high energy muons at sea level. These predictions can be compared with data observed by collaborations L3+Cosmic, MACRO and LVD at energies above 100 GeV. Finally, some conclusion can be drawn about validity of various models.

In fact, some low energy models with the package FLUKA [32] have been tested in such a way. We are sorry that some our results of models testing in [33-35] are not correct. We do apologize for our mistake in input data for the atmosphere.

In this paper models QGSJET II-03, DPMJET 2.55 [8], VENUS 4.12 [9] and SIBYLL 2.3 [5] have been tested. A comparison of muon data observed in [15-17] with results of simulations allows to draw a conclusion about the most energetic mesons production described by these models.

2 Method

To estimate the energy spectra $D(E_\mu)$ of atmospheric vertical muons in the energy range of 10^2 – 10^5 GeV we need to know the energy spectra dI_p/dE and dI_{He}/dE of the primary protons and helium nuclei within the energy interval $10^2 - 10^7$ GeV and the partial energy spectra $S_p(E_\mu, E)$ and $S_{He}(E_\mu, E)$ of the vertical muons in EAS induced by the primary protons and helium nuclei with the various fixed energies E. Simulations of these partial muon spectra have been carried out in terms of the VENUS 4.12, DPMJET 2.55, QGSJET II-03 and SIBYLL 2.3 hadronic interaction models in the same energy range of 10^2 – 10^7 GeV. The smooth approximation of the atmospheric muon data observed by the collaborations L3+Cosmic, MACRO and LVD had been used for comparison with results of simulations.

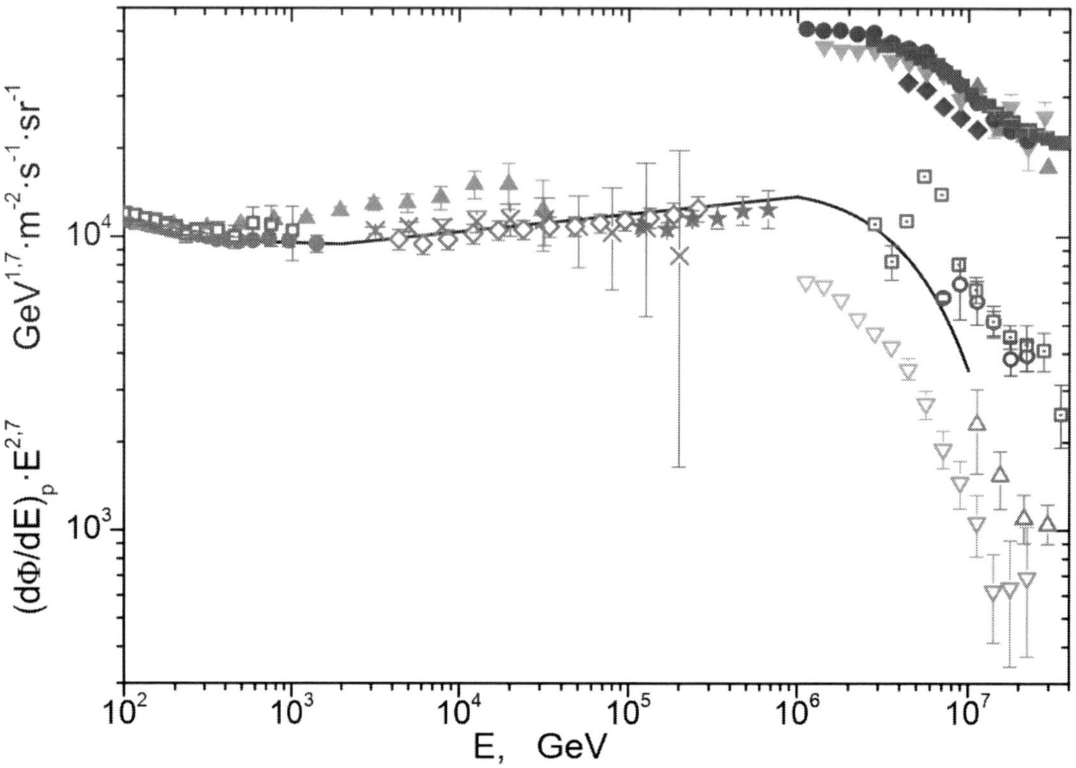

Figure 1. Energy spectra of the primary protons. Solid line - suggested approximation. Data: • - AMS-02; □ - PAMELA; ▲ - ATIC-2; × - CREAM; ◇ - ARGO-YBJ (QGSJETII-03); ★ - ARGO-YBJ & WFCTA (QGSJETII-03); ○ - Tunka (QGSJET01 & QGSJETII-04); ⊡ - IceCube (SIBYLL 2.1); ▽ - KASCADE (QGSJET II); △ - KASCADE-Grande (QGSJET II); ♦ - Telascope Array (QGSJETII-03; All particles); • - Tunka (All particles); ■ - IceCube (All particles); ▼ - KASCADE (All particles); ▲ - KASCADE-Grande (All particles). All particles spectra depends on energy per particle.

Functions $S_p(E_\mu, E)$ and $S_{He}(E_\mu, E)$ are the partial differential energy spectra of muons in showers induced by primary protons and helium nuclei with fixed values of energy E. These spectra were calculated for 24 different fixed values of energy E of the primary protons. The energy distributions of muons induced by primary helium nuclei $S_{He}(E_\mu, E)$ have been also calculated and compared with simulations based on the hypothesis of superposition [36]. As direct results coincides with simulations in terms of the hypothesis of superposition, we will use this hypothesis.

The energy spectra of the primary particles are important ingredients of simulations. As the energy per nucleon is of importance only the energy spectra of the primary protons and helium nuclei should be taken into account. We had used approximations (1) for (dI_p/dE) and (dI_{He}/dE) suggested in [37]. Figures 1 and 2 demonstrate how these approximations fit data [20-30].

To take into account a possible change of primary spectrum above the "knee" at energies above $E_1 \simeq 10^6$ GeV for the primary protons and helium nuclei we had used additional exponential multiplier. The values of parameters for these new approximations are listed in Table 1.

The approximation parameters: $R_1, R_2, \alpha, s, \beta$ and γ are dimensionless. Parameters K_1 and K_2 has dimensional as $[1/(GeV \cdot m^2 \cdot s \cdot sr)]$. The notation E is kinetic energy per nucleon in GeV.

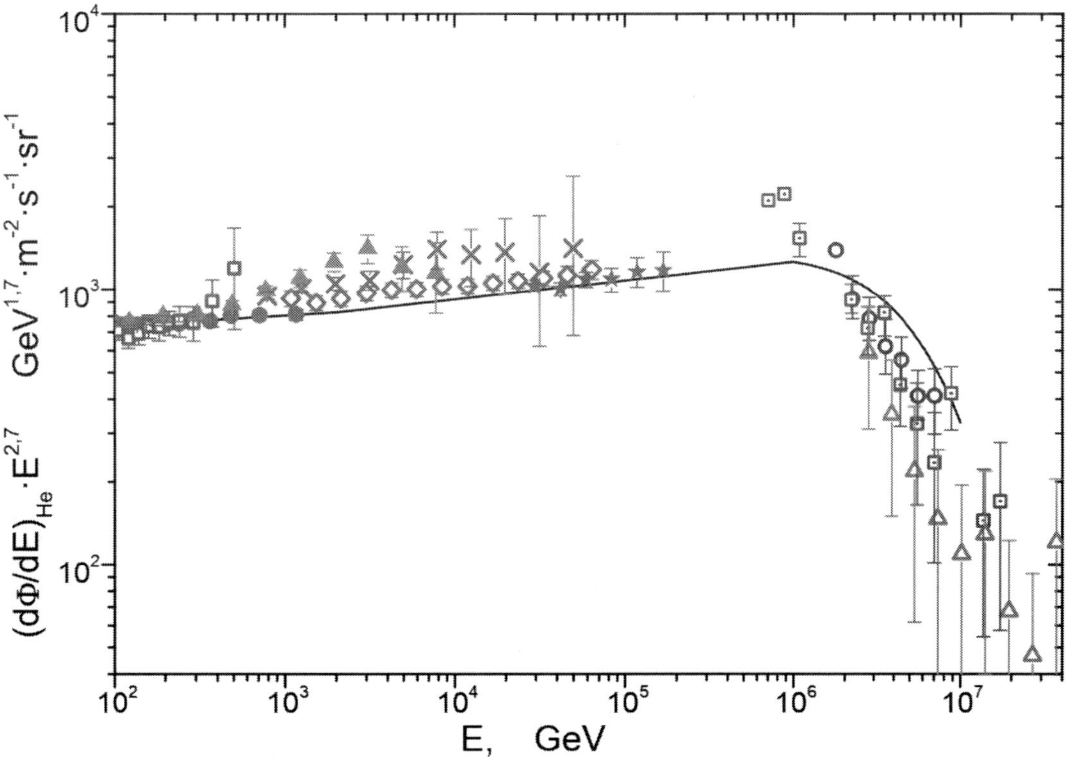

Figure 2. Energy spectra of nucleons contained in the primary helium nuclei (E - energy per nucleon). Solid line - suggested approximation. Data: • - AMS-02; □ - PAMELA; ▲ - ATIC-2; × - CREAM; ◇ - ARGO-YBJ (QGSJETII-03); ★ - ARGO-YBJ & WFCTA (QGSJETII-03); ○ - Tunka (QGSJET01 & QGSJETII-04); ⊡ - IceCube (SIBYLL 2.1); △ - KASCADE-Grande (QGSJET II).

Table 1. Parameters of approximation for the primary nucleon spectra.

Nuclei	K_1	K_2	R_1	R_2	α	s	β	γ
H	0.4544	5833	45	336	-2.849	-2.64	5.5417	0.024
He	0.1896	493	22.5	122.5	-2.78	-2.623	4.4074	0.027

$$\frac{dI}{dE} = \begin{cases} K_1 \cdot (E/R_1)^{\alpha} \cdot \left[1 + (E/R_2)^{\beta}\right]^{\gamma} & E \in 10^2 \div 1.8 \cdot 10^3 \quad GeV \\ K_2 \cdot E^{s} & E \in 1.8 \cdot 10^3 \div 10^6 \quad GeV \\ K_2 \cdot E^{s} \cdot \exp\left[-(E - 10^6)/6 \cdot 10^6\right] & E \in 10^6 \div 10^7 \quad GeV \end{cases} \tag{1}$$

The package CORSIKA 7.4 had been used to simulate the second important ingredients - the partial energy spectra $S_p(E_\mu, E)$ and $S_{He}(E_\mu, E)$ of vertical muons in showers induced by the primary protons and helium nuclei with the various fixed energies E in terms of the VENUS 4.12, DPM-JET 2.55, QGSJET II-03 and SIBYLL 2.3 models in the energy range of $10^2 - 10^5$ GeV with statistics

 DOI: 10.1051/epjconf/201715806006

Table 2. Proton energy E and threshold energy E_{th} in GeV and statistics N.

E	E_{th}	N	E	E_{th}	N	E	E_{th}	N
113,3	100	10^6	$1,778 \cdot 10^4$	100	10^6	$1 \cdot 10^5$	10^3	10^6
177,8	100	10^6	$3,162 \cdot 10^4$	100	10^6	$1,778 \cdot 10^5$	10^3	10^6
237,1	100	10^6	$5,623 \cdot 10^4$	100	10^6	$3,162 \cdot 10^5$	10^3	10^6
316,2	100	10^6	$1 \cdot 10^5$	100	10^5	$5,623 \cdot 10^5$	10^3	10^6
421,7	100	10^6	$1,778 \cdot 10^5$	100	10^5	$1 \cdot 10^6$	10^4	10^6
562,3	100	10^6	$3,162 \cdot 10^5$	100	10^5	$1,778 \cdot 10^6$	10^4	10^6
749,8	100	10^6	$5,623 \cdot 10^5$	100	10^5	$3,162 \cdot 10^6$	10^4	10^6
$1 \cdot 10^3$	100	10^6	$1 \cdot 10^6$	100	10^4	$5,623 \cdot 10^6$	10^4	10^6
$1,778 \cdot 10^3$	100	10^6	$1,778 \cdot 10^6$	100	10^4			
$3,162 \cdot 10^3$	100	10^6	$3,162 \cdot 10^6$	100	10^4			
$5,623 \cdot 10^3$	100	10^6	$5,623 \cdot 10^6$	100	10^4			
$1 \cdot 10^4$	100	10^6	$1 \cdot 10^7$	100	10^3			

10^6 events for the most energetic muons. The Table 2 shows fixed energies E of the primary protons, threshold energy E_{th} and statistics of simulated showers N.

The results of these calculations in the energy range of $10^2 - 10^7$ GeV were interpolated for 100 values of energies E with equal intervals in decimal logarithmic scale. The energy interval $10^2 - 10^5$ GeV of muons was divided into 60 equal bins also in decimal logarithmic scale. So, the width of the bin was equal to $h = \lg(E_{\mu,(i+1)}/E_{\mu,i}) = 0,05$. Let us note that average muon energies for the 1-st, 21-st and 41-st bins we will use later are equal to $1.059 \cdot 10^2$, $1.059 \cdot 10^3$ and $1.059 \cdot 10^4$ GeV accordingly. In fact simulations for helium nuclei have been carried out only for energies 10^4 and 10^6 GeV to test the hypothesis of superposition. As results of simulations for the primary nuclei showed a good agreement with this hypothesis we had used this hypothesis to estimate the flux of the nucleons from the primary helium nuclei.

The energy spectra $D_p(E_\mu)$ and $D_{He}(E_\mu)$ of muons for the primary protons and helium nuclei are calculated as integrals of products of functions $S_p(E_\mu, E)$ and $S_{He}(E_\mu, E)$ with corresponding intensities dI_p/dE and dI_{He}/dE of the primary protons, on energy E of primary nucleons.

$$D_p(E_\mu) = \int \left(\frac{dI_p}{dE}\right) \cdot S_p(E_\mu, E) \cdot dE \tag{2}$$

$$D_{He}(E_\mu) = \int \left(\frac{dI_{He}}{dE}\right) \cdot S_{He}(E_\mu, E) \cdot dE \tag{3}$$

Resulting energy spectrum of atmospheric muons is the sum of these energy spectra of muons produced by primary protons and helium nuclei.

$$D(E_\mu) = D_p(E_\mu) + D_{He}(E_\mu) \tag{4}$$

3 Results of simulations

The partial energy spectra $S_p(E_\mu, E)$ of the atmospheric vertical muons simulated for various fixed energies E of the primary protons in terms of the VENUS 4.12, DPMJET 2.55, QGSJET II-03 and SIBYLL 2.3 models are shown in figure 3. It is seen that statistics of $\sim 10^6$ at the higher energy end of the spectra is not enough.

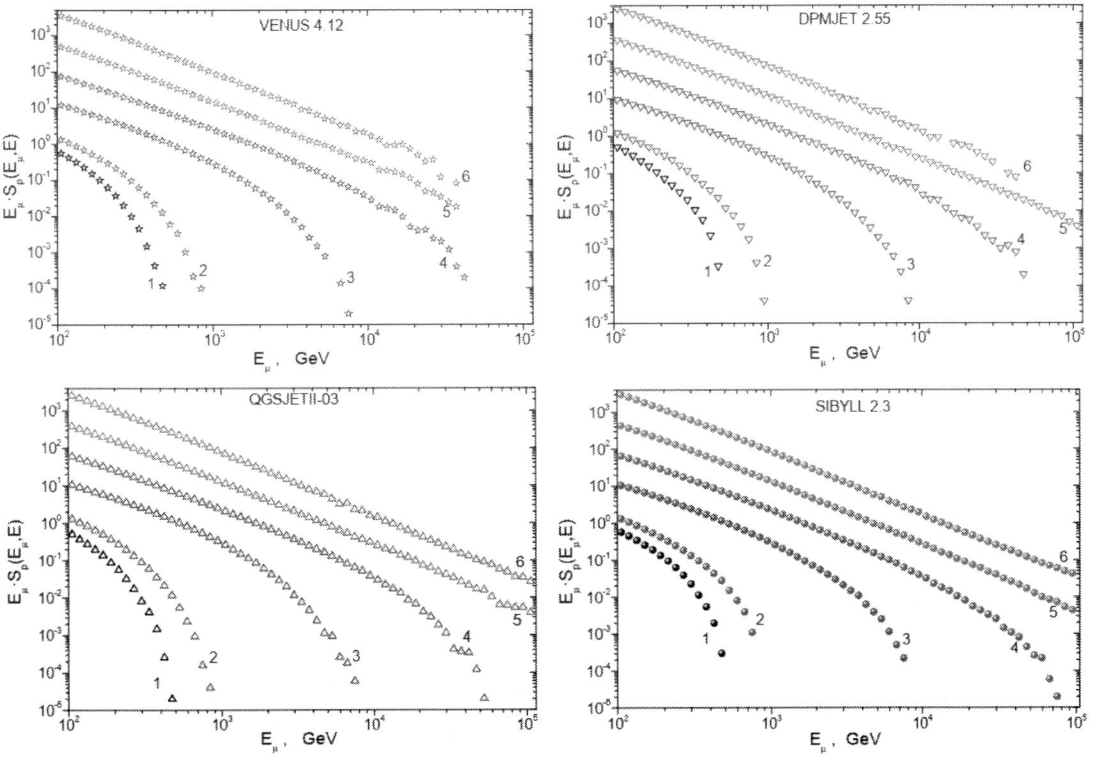

Figure 3. The partial energy spectra of muons in showers induced by primary protons with various fixed energies E: 1 - $5 \cdot 10^2$ GeV; 2 - 10^3 GeV; 3 - 10^4 GeV; 4 - 10^5 GeV; 5 - 10^6 GeV; 6 - 10^7 GeV.

Table 3. Average number of muons with energies above the threshold E_{th} in showers induced by primary protons with energies E.

			$E = 10^5$ GeV		$E = 10^6$ GeV	
Model	Paper	E_{th}	100 GeV	1000 GeV	100 GeV	1000 GeV
VENUS 4.12	[38]		23,5	0,679	153,5	3,932
VENUS 4.12	This work		24,5	0,652	156,2	3,839
QGSJET01	[38]		21,0	0,605	132,3	3,612
QGSJET01	This work		20,9	0,593	132,8	3,490
QGSJET II-03	This work		20,3	0,592	122,8	3,301
DPMJET 2.55	This work		18,7	0,538	116,7	3,150
SIBYLL 2.3	This work		21,4	0,573	136,2	3,461

Table 3 displays the total number of muons with energies above 10^2 and 10^3 GeV in showers induced by the primary protons with energies 10^5 and 10^6 GeV estimated in terms of the VENUS 4.12, DPMJET 2.55, QGSJET II-03 and SIBYLL 2.3 models in our simulations and in [38]. The very reasonable agreement is seen.

The next figure 4 demonstrates a comparison of the partial muon energy spectra $S_p(E_\mu, E)$ calculated in terms of the SIBYLL 2.3 model (left panel) with results in terms of QGSJET II-03, DP-

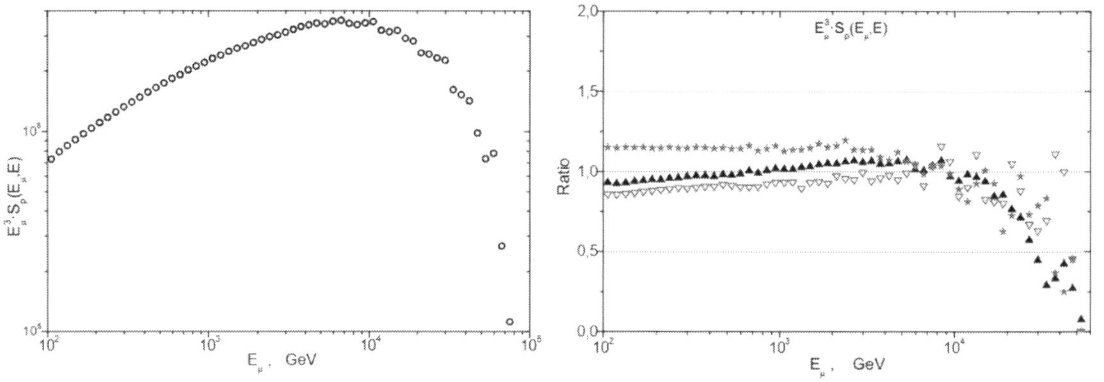

Figure 4. The partial energy spectra of muons in showers induced by primary protons with fixed energy E = 10^5 GeV. Direct result ∘ - SIBYLL 2.3 (left panel) and ratio to SIBYLL 2.3 model: △ - QGSJET II-03, ▽ - DPMJET 2.55, ★ - VENUS 4.12 (right panel).

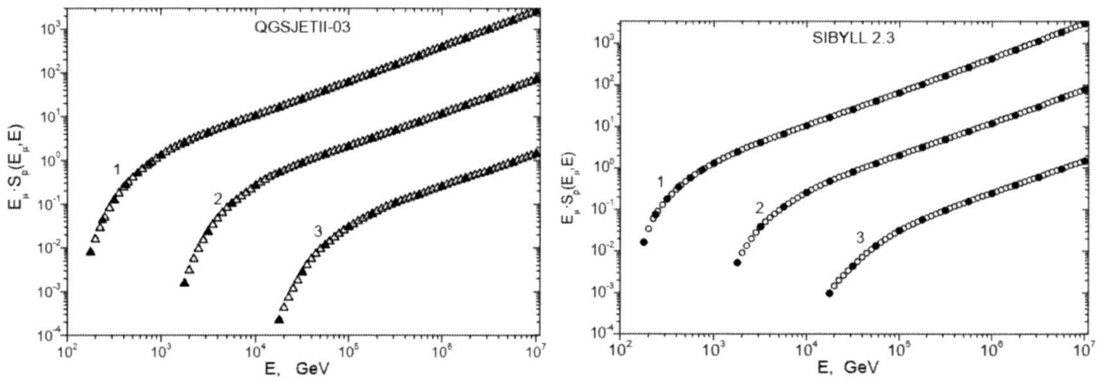

Figure 5. Dependence $E_\mu \cdot S_p(E_\mu, E)$ on E for three muon bins with energies E_μ: 1 - $1.059 \cdot 10^2$ GeV; 2 - $1.059 \cdot 10^3$ GeV; $1.059 \cdot 10^4$ GeV. ▲, • - direct calculation, △, ∘ - interpolated results. Left panel - QGSJET II-03, right panel - SIBYLL 2.3.

MJET 2.55 and VENUS 4.12 models (right panel) for the fixed energy $E = 10^5$ GeV of the primary protons. These results are shown as ratios to the SIBYLL 2.3 data. The results for the VENUS 4.12 model are 15 – 20 % above the QGSJET II-03 model values.

It is of importance to estimate energy intervals of the primary protons which contribute into various bins of the muon energy spectrum. But first some dependence of muon numbers inside the partial bins on energy E should be illustrated.

These dependence on energy E for the 1-st, 21-st and 41-st bins of muon energy spectra $S_p(E_\mu, E)$ are displayed in figure 5 for the QGSJET II-03 and SIBYLL 2.3 model. The good agreement is also seen.

Figure 6 demonstrates distributions of the primary proton energy E for the bins of muon energy spectra for the SIBYLL 2.3 model. Let us remind that average energies of muons for these bins are equal to $1.059 \cdot 10^2$, $1.059 \cdot 10^3$ and $1.059 \cdot 10^4$ GeV accordingly. It is possible to note that nearly two

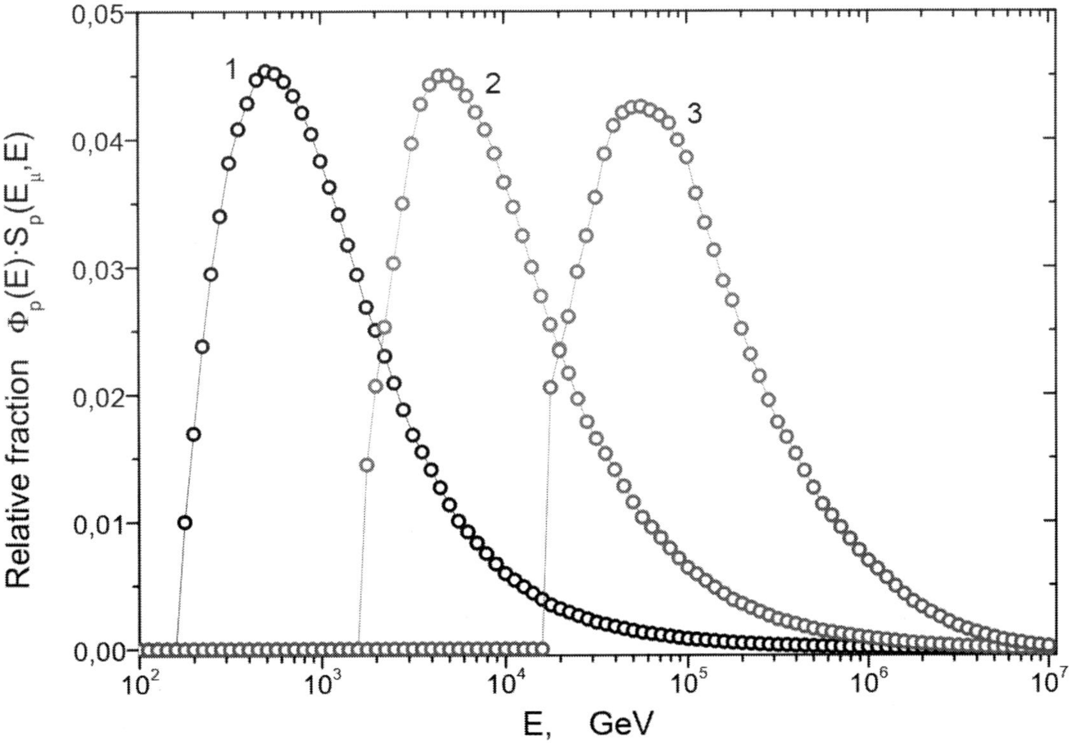

Figure 6. The primary proton energy distributions for three various bins with energies E_μ: 1 - $1.059 \cdot 10^2$ GeV, 2 - $1.059 \cdot 10^3$ GeV and 3 - $1.059 \cdot 10^4$ GeV.

orders of energy E are of importance for any fixed energy E_μ of muons. The maximal contributions occur at energies $\sim 5 \cdot 10^2$ GeV (to the 1-st bin), $\sim 5 \cdot 10^3$ GeV (to the 21-st bin) and $\sim 5 \cdot 10^4$ GeV (to the 41-st bin).

The final results of the muon energy spectra $D(E_\mu)$ calculated in terms of the VENUS 4.12, DPMJET 2.55, QGSJET II-03 and SIBYLL 2.3 models and data [15-17] are shown in figure 7 (left panel) and ratios of MC simulation to data - in the same figure (right panel).

It should be noted that at energies above $\sim$ 100 GeV both the simulated spectra and data are steepened. It is because the decay constant B for the charged mesons is equal to $B \sim 100$ GeV and a probability of decay for charged mesons is decreasing at higher energies.

It is seen that calculated spectra are $\sim$ 2 times below data in case of the QGSJET II-03 model and $\sim$ 1.7 times below data for the SIBYLL 2.3 model. The result of the rest of models are in between of these limits. The main conclusion is quite clear. All considered models demonstrate the valuable deficit of muons.

4 Conclusion

Muons which contribute much to the muon energy spectra are produced in decays of the most energetic $\pi^\pm$-mesons and $K^\pm$-mesons generated in first interactions of the primary particles with nuclei in the atmosphere. As calculated vertical muon energy spectra in case of the VENUS 4.12, DPM-

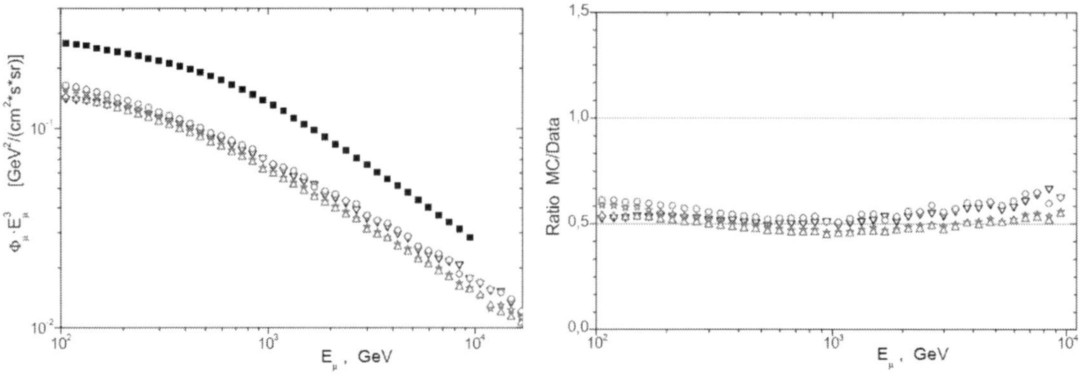

Figure 7. The energy spectra of near vertical muons - left panel. The ratio MC/DATA - right panel. ○ - SIBYLL 2.3, ★ - VENUS 4.12, △ - QGSJET II-03, ▽ - DPMJET 2.55 and (■) - experimental data (L3+Cosmic, MACRO, LVD).

JET 2.55, QGSJET II-03 and SIBYLL 2.3 models are $\sim 1.7 \div 2$ times below data we can conclude that production of the most energetic $\pi^{\pm}$-mesons and $K^{\pm}$-mesons in these models is considerably suppressed. This suppression may induce smaller values of signals in the surface scintillation detectors and will result in larger values of the calculated energy estimates. So, the coefficient 1.27 used by the TA collaboration [10] to decrease the energy estimates of showers calculated on the base of signals in the surface detectors may be understood as a result of this suppression. The increased intensity of the primary particle flux observed at the Yakutsk array at super high energies [39] may be also a result of smaller values of calculated signals in surface scintillation detectors.

Acknowledgement

Authors thanks N. N. Kalmykov who kindly showed alternative calculations and A. A. Lagutin for the important assistance in the verification of the results for the QGSJET01 model.

References

[1] Kalmikov N. N., Ostapchenko S. S., Phys. Atom. Nucl. **Vol. 56**, 346 (1993).
[2] Ostapchenko S. S., Nucl. Phys. B (Proc. Suppl.) **Vol. 151**, 143 (2006).
[3] Ostapchenko S. S., Phys. Rev. D. **Vol. 74**, 014026 ; hep-ph/0505259 (2006).
[4] Ostapchenko S. S., Phys. Rev. D. **Vol. 83**, 014018 (2011).
[5] Ahn E.-J., Engel R., Gaisser T. K., et al., Phys. Rev. D. **Vol. 80**, 094003 (2009).
[6] Werner K., Nucl. Phys. B (Proc. Suppl.) **Vol. 175**, 81 (2008).
[7] Werner K. Liu F. M. and Pierog T., Phys. Rev. C. **Vol. 74**, 044902; hep-ph/0506232(2006).
[8] Ranft J., Phys. Rev. D. **Vol. 51**, 64 (1995); hep-ph/9911213 and hep-ph/9911232 (1999).
[9] Werner K., Phys. Rep. **Vol. 232**, 87 (1993).
[10] Abu-Zayyad T., Aida R., Allen M., et.al., (Telescope Array Collab.) Astrophys. J. Lett., **Vol. 768:L1**, 5 (2013).
[11] Pierog T., EPJ Web of Conf., **Vol. 99**, 09002 (2015).
[12] Ostaphcenko S. S., Progr. of Theor. Phys. Suppl., **Vol. 193**, 204 (2012).

[13] D'Enterria D., Engel R., Pierog T., et. al., Astropart. Phys., **Vol. 35**, 98 (2011).
[14] Engel R., Rebel H., Acta Phys. Polonica B., **Vol. 35**, 321 (2004).
[15] Achard P. Adriani O., Aguilar-Benitez M., et al.,(L3 Collab.), Phys.Lett. B. **598**, 15-32; hep-ex/0408114v1K (2004).
[16] Ambrosio M., Antolini R., Auriemma G., et al., (MACRO Collab.), Phys. Rev. D. **Vol. 52**, 3793 (1995).
[17] Aglietta M., Alpat B., Alieva E. D., et al., (LVD Collab.), Phys.Rev. D. **58**, 092005; hep-ex/9806001v1 (1998).
[18] Heck D., Knapp J., Capdevielle J.-N., et al., Report FZKA 6019 Forschungszentrum Karlsruhe, (1998).
[19] Maurin D., Melot F. and Taillet R., Astronomy and Astrophysics, 569, A32 (2014).
[20] Aguilar M., Aisa D., Alpat B., et al, (AMS Collab.) Phys. Rev. Lett. **Vol. 114**, 171103 (2015). Aguilar M., Aisa D., Alpat B., et al, (AMS Collab.) Phys. Rev. Lett. **Vol. 115**, 211101 (2015).
[21] Adriani O., Barbarino G. C., Bazilevskaya G. A., et. al., (PAMELA Collab.) Advances in Space Research **Vol. 51**, 219-226 (2013).
[22] Panov A. D., Adams J. H., Ahn H. S. et al. (ATIC Collab.), Bull. Russ. Acad. Sci. Phys. **Vol. 71**, 494 (2007); Bull. Russ. Acad. Sci. Phys. **Vol. 73**, 564 (2009).
[23] Ahn H. S. Allison P., Bagliesi M. G., et al (CREAM Collab.), , Astrophys.J., 707, 593 (2009).
[24] Bartoli B., Bernardini P., Bi X., et al (ARGO-YBJ Collab.) Phys. Rev. D **91**, 11, 112017; arXiv:1503.07136 (2015). Di Sciascio G. et. al. (ARGO-YBJ Collab.), J. of Physics: Conf. Series **632**, 012089 (2015).
[25] Bartoli B., Bernardini P., Bi X., et al (ARGO-YBJ Collab.), Chinese Physics C **38**, 4, 045001 (2014).
[26] Antoni T., Apel W. D., Badea A. F., et al., (KASCADE Collab.) Astroparticle Physics **24**, p. 125 (2005).
[27] Apel W. D., Arteaga-Velazquez J. C., Bekk K., et al., (KASCADE-Grande Collab.) Astroparticle Physics **47**, p. 5466 (2013).
[28] Prosin V. V., Berezhnev S. F., Budnev N. M., et al., (Tunka Collab.) Nuclear Instruments and Methods in Physics Research A **756**, p. 94101 (2014).
[29] Rawlins K., Feusels T., et al., (IceCube Collab.) *Proceedings of the 34th International Cosmic Ray Conference*, (The Hague, The Netherlands 2015) PoS(ICRC2015) 334.
[30] Ivanov D., et al., (Telescope Array Collab.) *Proceedings of the 34th International Cosmic Ray Conference*, (The Hague, The Netherlands 2015) PoS(ICRC2015) 349.
[31] Berezhko E.G. Nucl. Phys. B (Proc. Suppl.) **Vol. 256 - 257**, 2335 (2014); Berezhko E.G., Knurenko S.P., Ksenofontov L.T., Astropart. Phys. **Vol. 36** 3136 (2012).
[32] Battistoni G., Ferrari A., Muraro S., et al., Proc. Suppl. Nucl. Phys. B. **Vol. 168**, 286 (2007).
[33] Dedenko L. G., Roganova T. M., Fedorova G. F., JETP Lett. **Vol. 100**, 223 (2014).
[34] Dedenko L. G., Roganova T. M., Fedorova G. F., Phys. Atom. Nucl. **Vol. 78**, N10, 894-902 (2015).
[35] Dedenko L. G., Lukyashin A. V., Fedorova G. F., et al., EPJ Web of Conferences **Vol. 99**, 10003 (2015).
[36] Dedenko L.G., Zatsepin G.T., *Proceedings of the 6-th ICRC, Moscow*, **Vol. II**, *Extensive air showers and cascades process*, 201-208 (1960).
[37] Dedenko L.G., Lukyashin A.V., Roranova T.M. and Fedorova G.F., *Proceedings of International Symposium on Cosmic Rays and Astroparticles 2017*, http://iscra2017.mephi.ru/content/public/files/presentations/June20_Presentation_13_LukyashinAV.pdf

[38] Lagutin A. A., Tyumentsev A. G., Yushkov A. V., J. Phys. G., **Vol. 30**, 573-596 hep-ph/0402070Journal (2004).
[39] Glushkov A. V., Egorova V. P., Ivanov A. A., et al., *Proceedings of the 28-th ICRC, Tsukuba,* **Vol. 1**, 393 (2003).

 DOI: 10.1051/epjconf/201715806007

Monitoring activities of satellite data processing services in real-time with SDDS Live Monitor

Minh Duc Nguyen[1] *

[1] *Skobeltsyn Institute of Nuclear Physics, Lomonosov Moscow State University, Moscow, Russia*

Abstract. This work describes Live Monitor, the monitoring subsystem of SDDS – an automated system for space experiment data processing, storage, and distribution created at SINP MSU. Live Monitor allows operators and developers of satellite data centers to identify errors occurred in data processing quickly and to prevent further consequences caused by the errors. All activities of the whole data processing cycle are illustrated via a web interface in real-time. Notification messages are delivered to responsible people via emails and Telegram messenger service. The flexible monitoring mechanism implemented in Live Monitor allows us to dynamically change and control events being shown on the web interface on our demands. Physicists, whose space weather analysis models are functioning upon satellite data provided by SDDS, can use the developed RESTful API to monitor their own events and deliver customized notification messages by their needs.

1 Introduction

The ultimate goal of research in space weather is creating complex operational magnetosphere-ionosphere models which allow predicting the radiation risk for space satellites in different orbits, and also estimating the occurrence risk of technological disasters due to magnetic storms and charged particle precipitations. Currently, solutions to the problem involve collecting data from all available satellites and ground stations, each of which measures a limited number of parameters of space weather, and generating forecasts based on the result of its analysis. Such task is challenging because data formats, storage methods, datasets and values of measured parameters differ between satellites. To achieve this goal an automated system called SDDS has been created at SINP MSU. SDDS automatically connects to different data storage servers of various satellites, downloads real-time data (both decoded txt-files and binary telemetry) whenever available, decodes binary telemetry, processes decoded data and stores it in a unified database.

Satellite data collected by SDDS is used as the primary source of input data for space weather operational models created at SINP MSU. Since the correctness and preciseness of space weather models depend much on the input data, it is critical to be sure that data is correctly processed. It is also important to detect errors in any step of the data processing cycle as soon as possible so that they could be fixed quickly and their consequences could be prevented in the future. Solution to the problem is monitoring all activities of all components involved in data processing and sending alerts

* e-mail: nguyendmitri@gmail.com

to responsible people (satellite developers, system administrators, workers on duty, etc.) so they could take measures in time. This paper is organized as follows: in the second section, we consider several existing solutions to the problem compared to our approach; in section 3, we give a more detailed view of the overall architecture of the Live Monitor subsystem. A brief description of the backend library is considered in section 4. Also in this section, we explain how events are classified and how alerts are sent via different mechanisms. Section 5 is dedicated to event representation on the web interface. In section 6, we give a brief description of using Live Monitor's API to create a customized monitoring service. In conclusion, we give a short resume of our completed work and describe our vision of the future perspective.

2 Related works

Since SDDS is working on Linux, the first solution came to mind is using an existing open source monitoring solution such as Zabbix [1], Nagios [2], or MMonit [3] to accomplish the task. But the detailed functional analysis showed that these systems were mainly designed for monitoring IT infrastructure (servers, routers, switches, etc.), system and network services. Zabbix and Nagios support application monitoring but this function is commercial. To monitor a custom service (or application) both Zabbix and Nagios assume that one must write a wrapper which runs a number of tests to check the service and produces standardized output data interpretable by the interface of the solution. The server component of Zabbix or Nagios then call this wrapper directly or via a client agent on a regular basis to check the service. Such solutions are not suitable for SDDS because of several reasons. Processing data of different satellites involves various components, the components can change dynamically, and working states of each component also differ. Writing a wrapper for each satellite would lead to a big amount of source code to be maintained. Additional checks on a regular basis for too many services would affect the overall performance of the operating system. Thus, a lightweight event-driven monitoring mechanism would be better.

A better solution is requiring all components of SDDS to inform about their current states in such a way that states could be treated uniformly. Each event of each state can be logged as a record to a journal file. When the log record is produced, we can use a message broadcasting server to either send the record to a web interface to show it to an operator or deliver the record via email or a messenger service directly to him. We have developed a library in three major programming languages Python, PHP, and JavaScript solely for this purpose.

While this approach can be applied perfectly to under development programs and internal components of SDDS being maintained by us, the same is not true for external programs that are used by SDDS to extract scientific data from the raw binary data received from satellites. It is a real challenge because most of these programs were written using different tools, both open source and commercial, and programming languages. The explanation of this fact is that data formats of different satellites differ and physicists use the tools they know best to achieve their goals. Changing these programs is not reasonable, mainly because either many of them were written such a long time ago that no one knows exactly what was implemented inside, or they are such complicated, so changes can lead to unexpected behaviors, resulting in erroneous data. To overcome this difficulty, we have decided to run these programs inside a wrapper which uses the developed library. Our final solution became the Live Monitor subsystem that is used to monitor all activities of SDDS in real-time. We have also developed a RESTful API so that physicists, whose space weather models are working based on the data supplied by SDDS, can use it to create their own monitoring scheme and deliver customized alerts to their customer.

 DOI: 10.1051/epjconf/201715806007

3 Live Monitor's architecture

Live monitor subsystem consists of the following components: a backend logging library, RabbitMQ message broker, a RESTful API backend, a frontend UI library. An illustration of the architecture is shown below in figure 1.

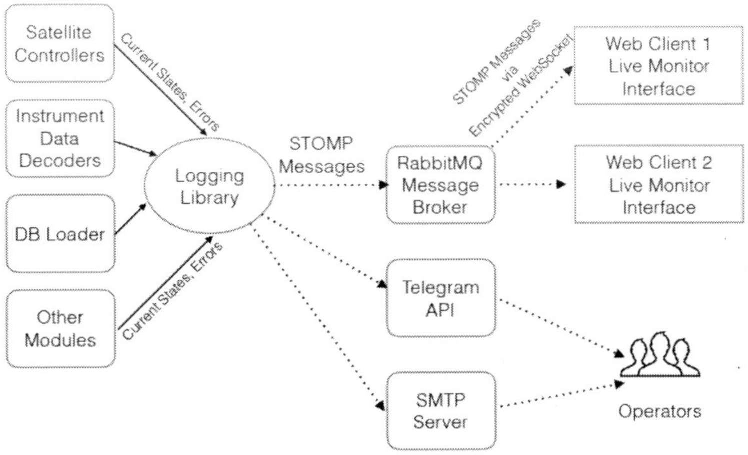

Figure 1. Live Monitor's architecture

The backend logging library is a customization of the popular Python Logging module. Every component of the SDDS system uses the library to inform about its states while running. External programs are executed within a wrapper of the SDDS global controller. When an external program crashes or returns an error code, the wrapper informs about it using the library. Besides the standard behavior which is writing short text messages in different log levels to a log file, the library sends these text messages to the RabbitMQ message broker [4] via a TCP socket. Messages are formatted using the STOMP protocol [5]. RabbitMQ in turn broadcast received text messages to all active web clients. When an error occurs during a certain state of data processing, the library informs operators about the error by sending a corresponding error description directly to the operators via the Telegram messenger service and/or email. Notification features can be turned on and off by editing proper configuration files.

Each data processing cycle is divided into stages. For example, a processing cycle of data from Meteor-M2 satellites consists of the following stages: connecting to data sources, downloading new raw data files, extracting scientific binary data files from raw ones, processing binary data files, adding processed data to the database, moving both raw and binary files to the local storage. Processing stages of different satellites differ. Stages to be displayed on the web interfaces of all components are customizable. One just needs to define a list of all states in the component configuration file; the library automatically does everything else. Since a satellite has a number of instruments on board, the whole processing stage consists of sub-stages each of which belongs to each instrument. An example is shown below in figure 2.

 DOI: 10.1051/epjconf/201715806007

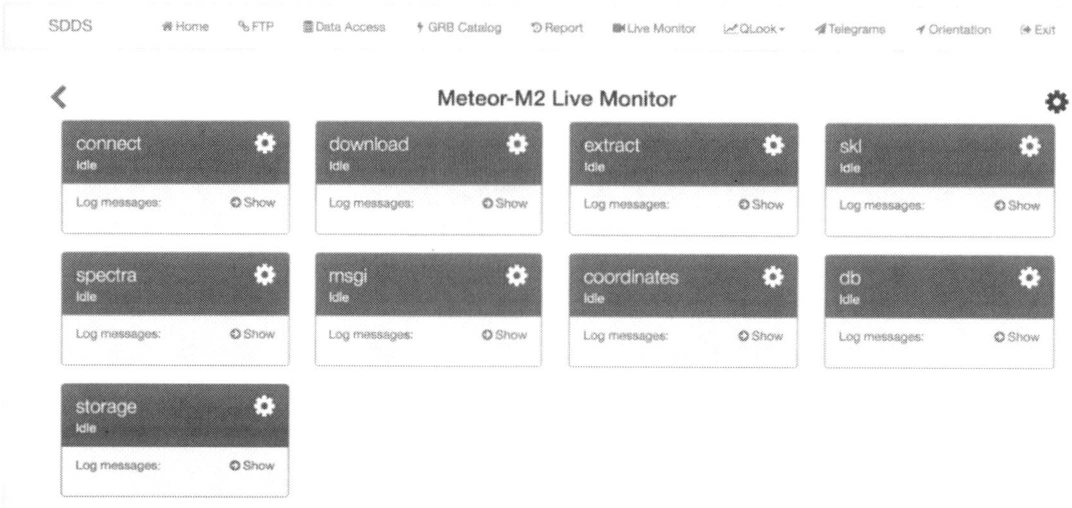

Figure 2. Live Monitor's architecture

4 Live Monitor's backend library

The primary goal of Live Monitor's backend library is logging all events from all components of SDDS as text messages to log files. If an SDDS's component needs to be monitored in real-time, one can change the proper property in the component configuration file, and all logging messages will be passed to RabbitMQ message broker and then delivered to all web clients. To monitor external programs used for decoding scientific data from raw one, SDDS execute them in a wrapper. The wrapper logs all start and stop execution points and also error codes and output of these programs using the logging library.

The logging library supports four levels of logging messages: debug, info, warning, and error. Error messages are logged when an error occurs during the data processing cycle which could lead to incorrectly processed data or cause a component failure. Warning messages are logged for minor errors that do not affect the data correctness and normal functioning. Info messages are just normal text descriptions of events during the data processing cycle. Debug messages include diagnostic information that is helpful in failure investigation. If the Live Monitor property is enabled in the configuration file of a satellite, error messages will be passed directly to operators on duty using the Telegram messenger service.

Another responsibility of Live Monitor's backend is to control what should be shown on the web interface during a data processing cycle of a satellite. When a satellite is created, the satellite operator defines in the configuration what stages the data processing cycle consists of and what instruments are working on the satellite. A RESTful API has been developed to deliver this kind of information to the frontend library flexibly and to allow remote work with Live Monitor subsystem.

RabbitMQ was used as the message broker server to deliver short messages to the frontend web client, primarily, because of its excellent documentation and community support. Also, RabbitMQ provides libraries for all major programming languages, which is a big advantage. In our case, the type of message distribution is "publish-subscribe" with no guaranteed delivery. In this scenario, RabbitMQ demonstrates an impressive performance and stability according to the performance benchmark in the study [6]. Also, RabbitMQ requires less effort in configuration settings and implementation, and it's more reasonable compared with Apache ActiveMQ Apollo [7] or ZeroMQ [8].

5 Live Monitor's frontend library

The main goal of the frontend library is to control how a data processing cycle of a satellite should be shown on the web interface. When a user opens the Live Monitor's page of a satellite, the frontend library sends a request to the backend according to the RESTful API to retrieve necessary information of what should be shown. The answer from the backend is a JSON object that consists of a number of stages and a number of instruments decoders involved in the data decoding stage. After that, the frontend library establishes a WebSocket connection with the RabbitMQ message broker. When a text message of a stage arrives, the frontend library parses its content and changes the visual appearance of the stage. Below is an example of the common "connecting-to-data-sources" stage of each satellite.

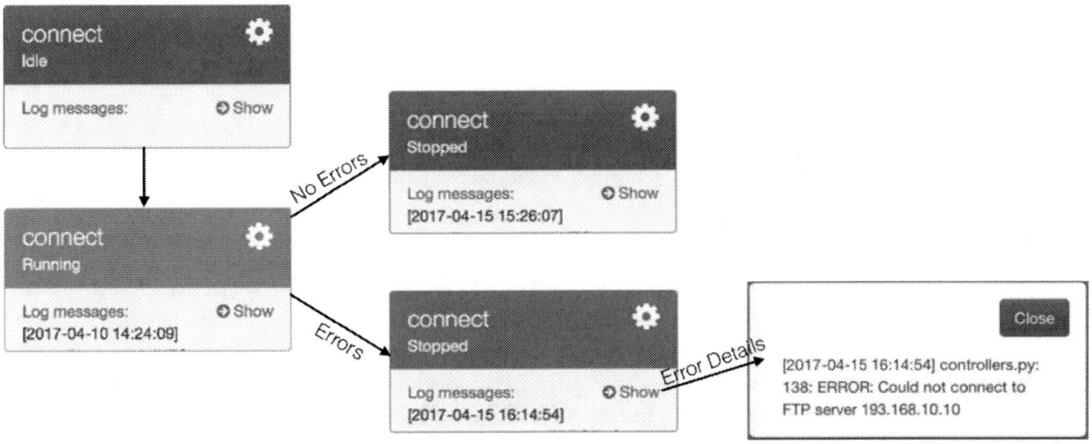

Figure 3. Different states of the "connecting-to-data-source" stage of a satellite

We use WebSocket protocol [9] to deliver messages from RabbitMQ broker to subscribed web clients because of its performance. The big advantage of using WebSocket instead of AJAX long polling requests [10] is that we can push notifications to clients when an event occurs through a bi-directional socket. It significantly reduces the workload of the server. In the case of using AJAX long polling requests, each web client would have to poll the broker regularly for new events. This approach leads to significant performance overheads, especially when we have a large number of clients, due to the unnecessary check data packets sending back and forth between the broker and the clients.

We use Simple (or Streaming) Text Oriented Messaging Protocol (STOMP) as the communication language between the broker and the clients. STOMP provides an interoperable wire format so that STOMP clients can communicate with any STOMP message broker to provide easy and widespread messaging interoperability among many languages, platforms, and brokers. It suits our needs best. STOMP is a very simple and easy to be implemented protocol, coming from the HTTP school of design. It is very easy to write a client to get yourself connected. For example, one can use Telnet to log into any STOMP broker and interact with it. We have managed to write three STOMP clients (in Python, PHP, JavaScript) in just a couple of hours and to integrate them to the RabbitMQ broker.

6 Customized monitoring service based on Live Monitor

At the moment of writing, Live Monitor is used mostly to monitor data processing of satellites. However, it is also possible for one to create his own monitoring service based on Live Monitor's API.

First of all, one must be a registered user of our SDDS system to be authorized to use Live Monitor's API. To create a customized monitoring service, a user must send a POST request with a proper JSON object body which consists of a service name, all its stages, and component names involving in each stage. When such request is received, Live Monitor's backend checks the request body for correctness. If the request is correct, the backend will create a RabbitMQ channel dedicated only to delivering messages of the requested service and a configuration file. As the answer, the user receives a JSON object which consists of a URL to the page of the requested service, a pair of login and password to be used in user's program to deliver messages to the RabbitMQ message broker. To control the behaviour of the service, one must send proper requests to the Live Monitor's backend. Currently, the following operations are supported by Live Monitor:

- create/delete a customized monitoring service;
- switch a monitoring service on/off;
- switch Telegram message delivery for a service on/off.

7 Conclusion

SDDS is the primary satellite data processing system being used at SINP MSU. The Live Monitor subsystem of SDDS system has been deployed from the first day of operation. During 12 months of operation Live Monitor helped us react to any occurred event in data processing in time. Since all stages of data processing are shown in details, Live Monitor allows us to identify and localize the scope of a problem immediately and hence prevent or fix them quickly. In future, we plan to support more operations to control the behavior of monitoring services, such as manipulating a stage name or a component name of a stage, controlling the monitoring behavior of each stage and component of a service independently, and so on.

Acknowledgments

We would like to thank Dr. Vladimir Kalegaev for helpful discussions on satellite data processing and clear problem statements. Also, we would like to thank Dr. Alexander Kryukov for his support in all aspects. This project is supported by RSF grant #16-17-00098.

References

[1] Zabbix SIA, *Zabbix Documentation* (https://www.zabbix.com/documentation/3.0/manual, Version 3.0)

[2] Ethan Galstad, Nagios Core Development Team, and Community Contributors *Nagios Core Documentation* (https://assets.nagios.com/downloads/nagioscore/docs/nagioscore/4/en/index.html, October 4th, 2016)

[3] Tildeslash Ltd, *M/Monit User Manual* (https://mmonit.com/documentation/mmonit_manual.pdf, Version 3.7.1)

[4] Pivotal Software, Inc. *RabbitMQ Server Documentation* (https://www.rabbitmq.com/admin-guide.html, Version 3.6.9)

[5] *STOMP Protocol Specification* (https://stomp.github.io/stomp-specification-1.2.html, Version 1.2)

[6] Hiram Chirino, *Stomp Benchmark* (https://github.com/chirino/stomp-benchmark)

[7] The Apache Software Foundation, *ActiveMQ Apollo User Manual* (https://activemq.apache.org/apollo/documentation/user-manual.html, Version 1.7.1)

[8] iMatix Corporation, *ZeroMQ - The Guide* (http://zguide.zeromq.org/page:all, Version 2.2)

[9] Internet Engineering Task Force, *The WebSocket Protocol* (https://tools.ietf.org/html/rfc6455, December 2011)

[10] J. J. Garrett, *Ajax: A New Approach to Web Applications*, (https://web.archive.org/web/20080702075113/http://www.adaptivepath.com/ideas/essays/archives/000385.php, February 18th, 2005)

EPJ Web of Conferences **158**, 06008 (2017) DOI: 10.1051/epjconf/201715806008
QFTHEP 2017

Deep Learning Neural Networks and Bayesian Neural Networks in Data Analysis

Andrey Chernoded[1], Lev Dudko[1], Igor Myagkov[1], and Petr Volkov[1]

[1]Skobeltsyn Institute of Nuclear Physics M.V. Lomonosov Moscow State University, Moscow 119991, Russian Federation

Abstract. Most of the modern analyses in high energy physics use signal-versus-background classification techniques of machine learning methods and neural networks in particular. Deep learning neural network is the most promising modern technique to separate signal and background and now days can be widely and successfully implemented as a part of physical analysis. In this article we compare Deep learning and Bayesian neural networks application as a classifiers in an instance of top quark analysis.

1 Introduction

Neural networks (NN) are especially useful in the top quark analyses owing to large background contributions to the same detector signature of the signal events. The efficient technique to distinguish signal events is critical to achieve reliable physics results. Different machine learning techniques are used to separate and suppress background processes such as multijet QCD events (QCD), W boson plus jets production (Wjets) and other Standard model (SM) processes. In the searches for the possible deviations from SM different machine learning techniques are used to distinguish Beyond of SM (BSM) effects from the huge SM background. For example in the searches for the anomalous *Wtb* interactions and flavour changing neutral currents (FCNC) [1] Bayesian neural networks (BNN) technique [2, 3] was applied. This technique is one of the most advanced approach compared to conventional Artificial neural networks. BNN allowed authors to successfully conduct a research on single top quark production properties, the analysis scheme is represented in Fig. 1. In the upcoming analysis the Deep neural networks can be used as a successors of BNNs.

In the article [4] it was shown that DNNs are capable to significantly improve precision of the analysis. In this article the suppression of QCD and SM events will be demonstrated using various techniques and packages, such as Markov Chain Sampling and Flexible Bayesian Modeling [5], Tensorflow [6] and Keras [7]. Significant improvement of the sensitivity has been demonstrated in comparison with usual NN methods.

2 Neural network introduction

A brief description of neural networks is presented in this section. Neural network is a set of neurons and their connections. Neurons have an input weights and activation functions. Set

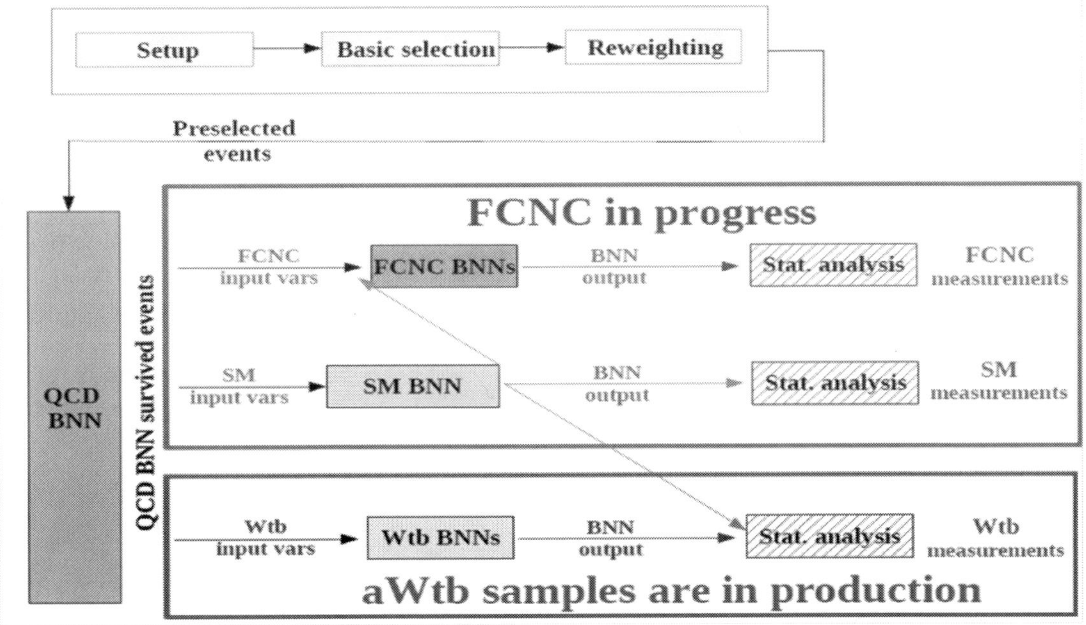

Figure 1. Analysis scheme, Bayesian neural network uses on four analysiss stages.

of neurons and their connections are a neural network architecture which is a very important part of conception. Usually neural networks are schematically described with the help of perceptron. Neural network can fit some function by means of training with simulated data. During the training algorithm minimizes a loss function with the help of changing of weights of neurons. For the productive work of the network, it is necessary to choose efficient architecture and loss function. There are not strict and universal recipe how to find optimal architecture and parameters to train NN, but with some experience one can achieve very good results.

Loss function is the basic element of learning algorithms. In mathematical optimization, statistics, econometrics, decision theory, machine learning and computational neuroscience, a loss function or cost function is a function that maps an event or values of one or more variables onto a real number intuitively representing some "cost" associated with the event. An optimization task tries to minimize a loss function.

Machine learning uses concepts such as algorithm optimizers, training and testing datasets, and criteria to estimate efficiency. They will be described in more details below.

3 Bayesian neural network

Bayesian neural network is an ensemble of neural networks. This approach is more stable and avoids the problem of overfitting one specific network. Programming package for BNN training is Flexible Bayesian Modeling and Markov Chain Sampling package [5]. This software supports Bayesian regression and classification models based on neural networks and Gaussian processes, Bayesian density estimation and clustering using mixture models and Dirichlet diffusion trees. It also supports a variety of Markov chain sampling methods which may be applied to distributions specified by simple formulas, including simple Bayesian models defined by formulas for the prior and likelihood [2]. In addition "bnn-hep" interfacing package is useful for creating training sets and submitting training results. It allows to preprocess input data and combine it into complex variables required by the analysis Fig. 2. The hybrid Monte Carlo algorithm [3], which is also known as Hamiltonian Monte Carlo, is used during network training. It uses Markov Chain Monte Carlo method for obtaining a sequence of random samples from a probability distribution for which the direct sampling is difficult. This sequence can be used to approximate the distribution (i.e., to generate a histogram), or to compute an integral (such as an expected value).

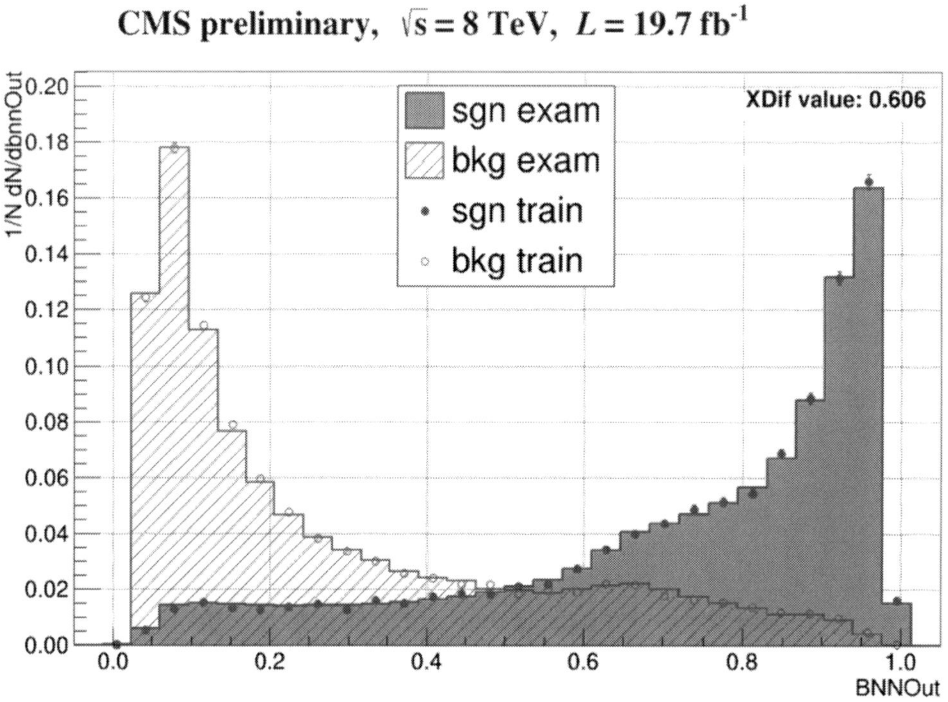

Figure 2. Example of applying Bayesian neural network

4 Deep Learning neural networks

Deep learning neural networks are distinguished from the conventional Artificial neural networks by their depth; that is, the number of node layers through which data is passed in a multistep process of pattern recognition is usually more than 2. Traditional machine learning relies on shallow nets, composed of one input and one output layer, and at most one hidden layer in between. More than three layers (including input and output) qualifies as "deep" learning. In deep-learning networks, each layer of nodes trains on a distinct set of features based on the previous layer's output.

TensorFlow [6] is an open source software package for numerical computation using data flow graphs. Nodes in the graph represent mathematical operations, while the graph edges represent the multidimensional data arrays (tensors) communicated between them. The flexible architecture allows to deploy computation to one or more CPUs or GPUs in a desktop, server, or mobile device with a single API. TensorFlow was originally developed by researchers and engineers working on the Google Brain Team within Google's Machine Intelligence research organization for the purposes of conducting machine learning and deep neural networks research, but the system is general enough to be applicable in a wide variety of other domains as well.

Keras [7] is a high-level neural networks API, written in Python and capable of running on top of TensorFlow, CNTK, or Theano. It was developed with a focus on enabling fast experimentation. Keras allows to create networks with different architecture, loss function, activation function, algorithm optimizers and other features.

Activation functions such as Sigmoid function is a mathematical function having a characteristic "S"-shaped curve or sigmoid curve. Often sigmoid function refers to the special case of the logistic function shown in the first figure and defined by the formula

$$S(x) = \frac{1}{1+e^{-x}} = \frac{e^x}{e^x+1}.$$

In the context of artificial neural networks, the rectifier is an activation function defined as:

$$f(x) = x^+ = \max(0, x),$$

where x is the input to a neuron.

In mathematics, the softmax function, or normalized exponential function, is a generalization of the logistic function that 'squashes' a K-dimensional vector **z** of arbitrary real values to a K-dimensional vector $\sigma(\mathbf{z})$ of real values in the range [0, 1] that add up to 1. The function is given by $\sigma(\mathbf{z})_j = \frac{e^{z_j}}{\sum_{k=1}^{K} e^{z_k}}$ for j = 1, ..., K.

Algorithm optimizers such as Stochastic Gradient Descent (SGD) is a simple and efficient approach to discriminative learning of linear classifiers under convex loss functions such as (linear) Support Vector Machines and Logistic Regression. The advantages of Stochastic Gradient Descent is efficiency. The disadvantages of Stochastic Gradient Descent include: SGD requires a number of hyperparameters such as the regularization parameter and the number of iterations. SGD is sensitive to feature scaling. Adam is an optimization algorithm that can be used instead of the classical SGD procedure to update network weights iteratively based on training data. Keras allows for a number of regulization methods. Regularization is a key component in preventing overfitting. Also, some techniques of regularization can be used to reduce model capacity while maintaining accuracy, for example, to drive some of the parameters to zero. This might be desirable for reducing model size or driving down cost of evaluation in mobile environment where processor power is constrained.

Dropout is a technique where randomly selected neurons are ignored during the training. They are "dropped-out" randomly. This means that their contribution to the activation of downstream neurons is temporally removed on the forward pass and any weight updates are not applied to the neuron on the backward pass [8].

5 Neural network efficiency estimate criteria

The Receiver Operating Characteristic (ROC) curve and the Mean-Square Error (MSE) criteria were taken to estimate the performance of the network.

DOI: 10.1051/epjconf/201715806008

A ROC curve is a way to compare diagnostic tests. It is a plot of the true positive rate against the false positive rate. A ROC curve demonstrates a relationship between sensitivity and specificity. For example, a decrease in sensitivity results in an increase in specificity. Test accuracy: if the graph is closer to the left and to the top borders, the the test is more accurate. Otherwise, if the graph is closer to the diagonal, the test is less accurate. A perfect test would go straight from zero up the top-left corner and then straight across the horizontal. The likelihood ratio: given by the derivative at any particular cut point. Area under ROC curve is usually used as a most simple estimation criteria.

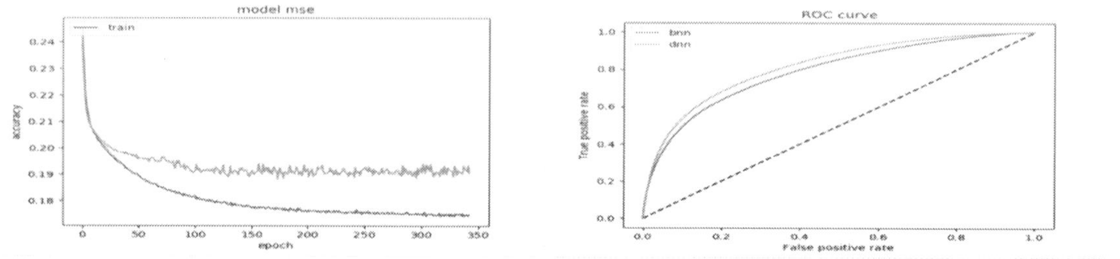

Figure 3. Estimate criteria: MSE and ROC curve

If $\hat{Y}$ is a vector of n predictions, and Y is the vector of observed values corresponding to the inputs of the function which generates the predictions, then the MSE of the predictor can be estimated as

$$\mathrm{MSE} = \frac{1}{n}\sum_{i=1}^{n}(\hat{Y}_i - Y_i)^2$$

6 QCD Network

One of the main backgrounds for the single top is the process of QCD-multijet production. It can be suppressed using classifier which is capable of distinguishing QCD-multijet from other Standard model processes. In QCD-multijet production process muons originate either from leptonic decay of heavy hadrons or from charged hadrons. As a result, these muon candidates are usually surrounded by hadronic activity. This feature is exploited to define a QCD control region in which exactly one muon with passes inverted criterion of isolation from hadronic activity. In order to better reproduce kinematics of the signal region, the jets falling inside a cone of size 0.5 around the selected muon are not considered in the analysis. The surviving jets are subject to the same selection as in the signal region[1]. The neural network is then trained using events from the QCD control region. The training set is composed of 120'000 events with QCD-multijet defined as the background and various Standard Model processes associated with single top quark production as the signal. Bayesian network is formed from an ensemble of 20 networks. Deep network consists of 5 hidden layers with 50 neurons and uses regularization and dropout.

Figures 4-6 show different behavior of separation power and ROC curves in three scenarios. In the first scenario, which results are shown on Fig.4, neural networks were given high-level variables, which had been evaluated prior to training step. Such variables should contain most important information required for particular classification task, their set is a subject to choice of analyst and are selected with respect to underlying physical process. In the second

 DOI: 10.1051/epjconf/201715806008

scenario, shown on Fig.5 only the most basic variables are provided as a neural network input. Thus a network is encouraged to find most suitable features from the most essential vectors of data. In the third scenario, shown on Fig.6, the networks are given a complete list of variables which incorporates both of the previous.

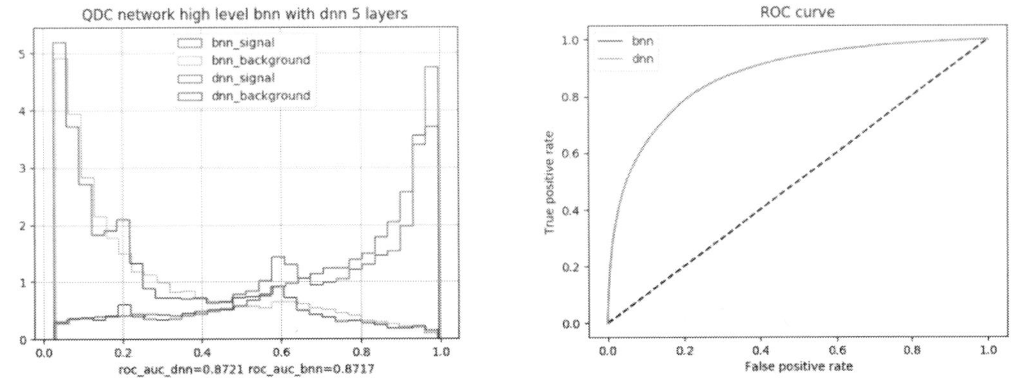

Figure 4. Comparison of QCD networks trained on high level variables.

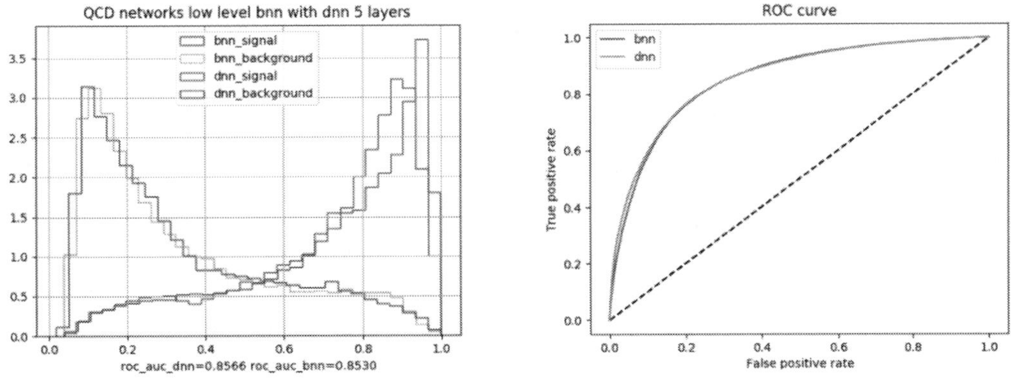

Figure 5. Comparison of QCD networks trained on low level variables.

7 SM network

Distinguishing between single top production process and its background on per-event level is next crucial step of a single top analysis. This classification task is accomplished using either kind of classifier; Bayesian and Deep neural networks will be considered in this section.

After preforming basic initial selection and suppressing QCD-multijet background, as described in previous section, events which pass such filtering are then passed as input to Standard model neural network. The purpose of this classifier is to assign a value of likeliness

DOI: 10.1051/epjconf/201715806008

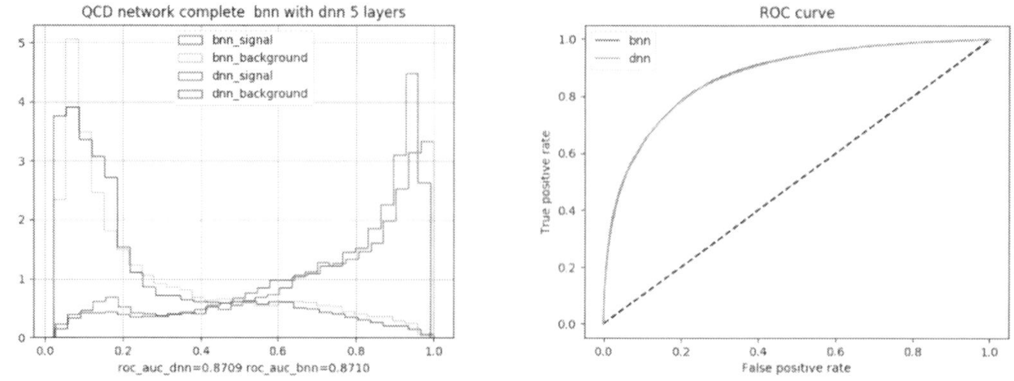

Figure 6. Comparison of QCD networks trained on complete set of variables.

of the event of being produced in t-channel single top production process. Amongst main undesired backgrounds to this process are tW-channel single top process, paired top quark production process, W+jets, diboson and Drell-Yan[1].

As well as a QCD-multijet neural network, Standard Model network takes a set of basic and complex kinematic variables as its input. Training set is composed of 190'000 events for train and test sets. In first approach using Bayesian neural network, the ensemble is composed of 50 networks. Deep network is formed using 5 hidden layers with 100 neurons in each with applied dropout regularization technique.

Figures 7-9 show different behavior of separation power and ROC curves in three scenarios, same as in section 6.

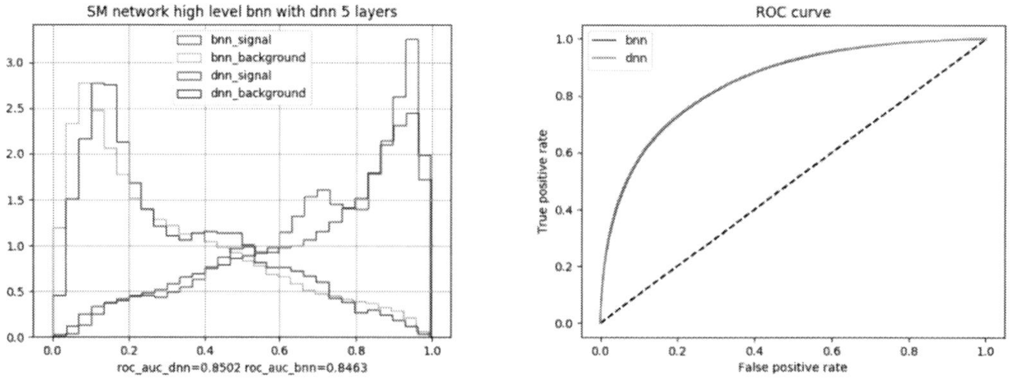

Figure 7. Comparison of SM networks trained on high level variables.

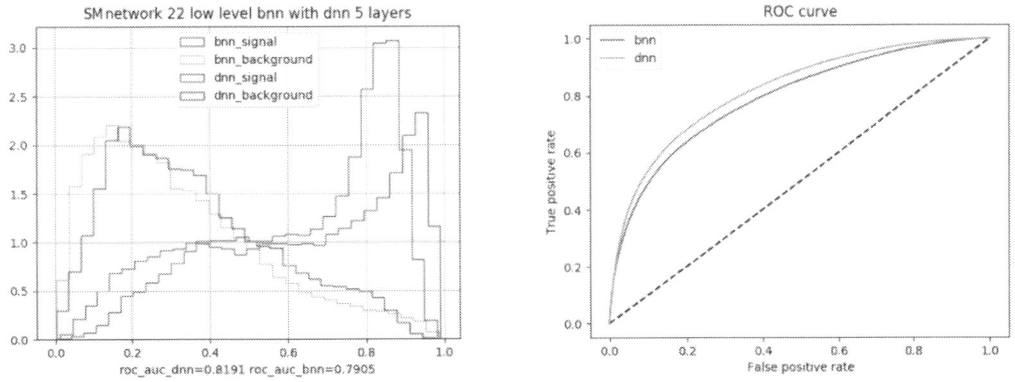

Figure 8. Comparison of SM networks trained on low level variables.

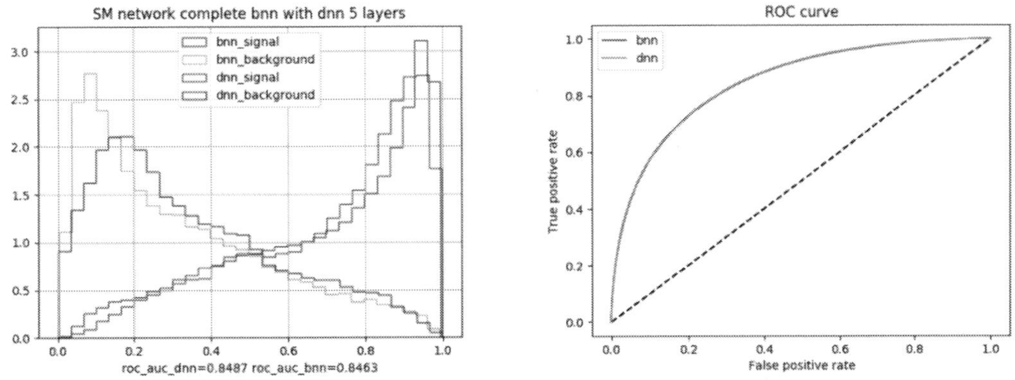

Figure 9. Comparison of SM networks trained on complete set of variables.

8 Conclusion

Neural networks is a powerful technique which allows creating sophisticated classifiers and is widely used in modern physics analysis such as single top quark study considered in this article. In the article authors provide the comprehensive list of packages, regularization and optimization techniques for creating and learning various models of neural networks. Two different techniques for neural network classifiers were implemented, such as Bayesian and deep neural networks. Their separation power and training aspects were considered. The Bayesian neural networks is an improved technique over conventional artificial neural networks, while Deep neural networks allow creating sophisticated multilayer architectures. In combination with modern learning methods and tools DNN is promising technique to significantly improve classifiers sensitivity and overall precision.

References

[1] Search for anomalous Wtb couplings and flavour-changing neutral currents in t-channel single top quark production in pp collisions at sqrt(s)= 7 and 8 TeV. CMS Collaboration. Oct 11, 2016 - 40 pages JHEP 1702 (2017) 028

[2] Neal, R. M. (1994) Bayesian Learning for Neural Networks, Ph.D. Thesis, Dept. of Computer Science, University of Toronto, 195 pages

[3] Neal, R. M. (1992) "Bayesian training of backpropagation networks by the hybrid Monte Carlo method", Technical Report CRG-TR-92-1, Dept. of Computer Science, University of Toronto, 21 pages

[4] Searching for Exotic Particles in High-Energy Physics with Deep Learning - Baldi, Pierre et al. Nature Commun. 5 (2014) 4308 arXiv:1402.4735 [hep-ph]

[5] http://www.cs.toronto.edu/~radford/fbm.software.html

[6] https://www.tensorflow.org/

[7] https://keras.io/

[8] Dropout: A Simple Way to Prevent Neural Networks from Overfitting - Nitish Srivastava, Geoffrey Hinton, Alex Krizhevsky,Ilya Sutskever, Ruslan Salakhutdinov. Journal of Machine Learning Research 15 (2014) 1929-1958

 DOI: 10.1051/epjconf/201715807001

Two-Point Correlators of Fermionic Currents in External Magnetic Field

Alexandra Dobrynina[1,*], *Ilya* Karabanov[1,**], *Alexander* Parkhomenko[1,***], and *Lubov* Vassilevskaya[2]

[1] *P. G. Demidov Yaroslavl State University, Yaroslavl, Russia*
[2] *Fulda University of Applied Sciences, Fulda, Germany*

Abstract. We study the two-point correlation functions under an influence of the constant homogeneous magnetic field. In addition to the correlators of scalar, pseudoscalar, vector and axial-vector fermionic currents, we calculate the non-diagonal one including the tensor and pseudoscalar currents. The tensor current is a fermionic part of the Pauli Lagrangian relevant for the electromagnetic interaction of fermions through the anomalous magnetic moment. Its contribution to the photon polarization operator is briefly discussed.

1 Introduction

The Standard Model — the gauge theory based on the $SU(3)_C \otimes SU(2)_L \otimes U(1)_Y$ group — is intensively testing for its consistency and no significant experimental deviations from theoretical predictions have been found yet [1]. More important, many experiments are quite sensitive to vacuum effects which are usually calculated as radiative corrections to observables within the perturbation theory. In particular, to get a good agreement with experimental data at the Large Hadron Collider (LHC), production cross-sections should be known in the Next-to-Leading Order (NLO) and some of them even in the Next-to-Next-to-Leading Order (NNLO). Moreover, vacuum effects can also generate rare processes which are forbidden at the tree level in the Standard Model. Good examples of such processes are the radiative $H \to \gamma\gamma$ and gluonic $H \to gg$ Higgs decays, already observed at the LHC [1].

In a difference to LHC, electron-positron colliders allow to get detail information about the photon propagation from the energy scan in the electron-positron annihilation process $e^- e^+ \to$ hadrons. In this case, one needs to know the photon polarization function (see, for example, [2, 3]):

$$3Q^2\, \Pi(Q^2) = i \int d^4x\, e^{i(qx)} \langle 0| T \left\{ j^\mu(x) j_\mu(0) \right\} |0\rangle, \tag{1}$$

where $j_\mu(x) = \sum_f \bar{\psi}_f(x) \gamma_\mu \psi_f(x)$ is the flavor-singlet quark vector current and $Q^2 = -q^2$. The Feynman diagram describing the $\gamma \to \gamma$ transition via virtual charged fermions in the one loop approximation is presented in Fig. 1. To get the vacuum average in (1), one needs to put $y^\mu = 0$. The func-

*e-mail: dobrynina@uniyar.ac.ru
**e-mail: karabanovilya@rambler.ru
***e-mail: parkh@uniyar.ac.ru

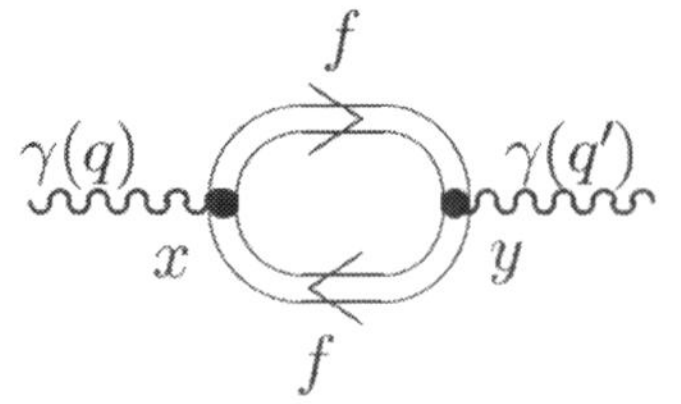

Figure 1. Feynman diagram describing the $\gamma \to \gamma$ transition via virtual fermions. Double lines indicate that effects of an external electromagnetic field are taken into account exactly in the fermion propagators.

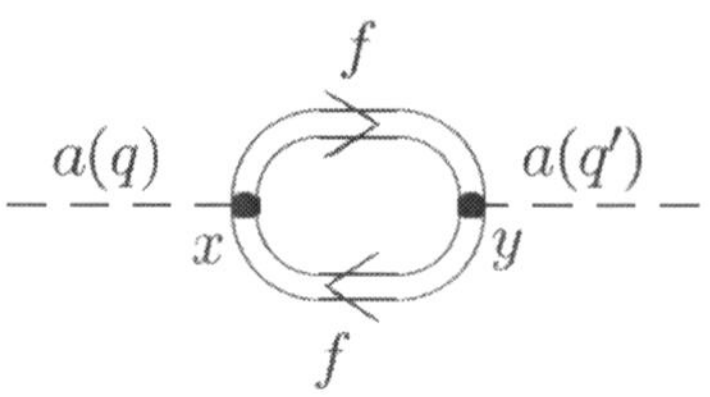

Figure 2. Feynman diagram describing the $a \to a$ transition via virtual fermions. Double lines indicate that effects of an external electromagnetic field are taken into account exactly in the fermion propagators.

tion $\Pi(Q^2)$ can be related with a similar two-point correlation function (the photon polarization function) $\Pi^{\rm em}(Q^2)$ [2, 3] originated by the quark electromagnetic current $j_\mu^{\rm em}(x) = \sum_f q_f \bar{\psi}_f(x)\gamma_\mu\psi_f(x)$, where q_f is the electric charge of the quark. Many physical observables are connected with $\Pi^{\rm em}(Q^2)$ (see, for example, [4]), in particular, the cross-section ratio is proportional to the imaginary part of $\Pi^{\rm em}(Q^2)$ as follows:

$$R(s) \equiv \frac{\sigma(e^-e^+ \to \text{hadrons})}{\sigma(e^-e^+ \to \mu^-\mu^+)} = 12\pi \, \mathrm{Im}\, \Pi^{\rm em}(-s - i\epsilon). \tag{2}$$

In addition to the vector-current correlator, massless correlators of scalar and tensor quark currents at higher orders in α_s were also calculated recently [5].

For more complicated physical systems like astrophysical objects and the early Universe or in heavy-ion collisions one needs to take into account effects of an external medium. Note that such a medium can significantly modify a wave-function and dispersion properties of a charged fermion and, as a consequence, change substantially the quantum field of the photon. The Feynman diagram describing the $\gamma \to \gamma$ transition via virtual charged fermions is presented in Fig. 1, where double lines indicate that fermion propagators are modified by an external electromagnetic field and plasma. Detail discussions of the photon polarization operator under external conditions can be found in the books [6, 7] and references therein. In what follows, we neglect the plasma effects on the fermion propagator and restrict ourselves by accounting effects of the background electromagnetic field only.

The matrix element of the $a \to a$ transition in an external electromagnetic field background also exemplifies the two-point correlation function of two axial-vector currents. The corresponding Feynman graph is shown in Fig. 2. This diagram determines the electromagnetic correction to the axion mass squared m_a^2 [8–10]. The effective neutrino-photon interaction, $\bar{\nu}\nu\gamma$-vertex, in an external electromagnetic field [7] is the other example of the two-point correlations which, in addition to the diagonal vector correlator responsible for the photon polarization operator and diagonal axial-vector correlator, includes the non-diagonal correlator of vector and axial-vector electron currents.

The general case of the two-point correlator modified by the constant homogeneous magnetic field was studied in Ref. [11]. The Lagrangian density of a local fermion interaction with a generalized current $J_A(x)$ which can be, for example, a photon field, neutrino current, derivative of the axion field and etc. (with the corresponding constants), has the following form [7, 11]:

$$\mathcal{L}_{\rm int}(x) = \left[\bar{f}(x)\Gamma^A f(x)\right] J_A(x), \tag{3}$$

where $f(x)$ is the quantum fermion field and Γ_A is any of the γ-matrices from the standard set $\{1, \gamma_5, \gamma_\mu, \gamma_\mu\gamma_5\}$ [12]. The two-point correlation function based on the fermionic currents of the form in (3) can be presented as follows [7, 11]:

$$\Pi_{AB} = \int d^4X\, \mathrm{e}^{-i(qX)}\, \mathrm{Sp}\,\{S_{\mathrm{F}}(-X)\,\Gamma_A\, S_{\mathrm{F}}(X)\,\Gamma_B\}\,, \tag{4}$$

where $S_{\mathrm{F}}(X)$ is the Lorentz-invariant part of an exact fermion propagator calculated in an external field background [13]. We assume that the external field is equal to the constant homogeneous magnetic field. Among the existing representations of $S_{\mathrm{F}}(X)$ for this field, we accept the so-called Fock-Schwinger one [13, 14] in which the fermion propagator has the explicitly Lorentz covariant form. Correlations among the scalar, pseudoscalar, vector and axial-vector fermionic currents in the Fock-Schwinger formalism were already studied [7, 11] but correlators of the tensor fermionic current $J_{\mu\nu}(x) \equiv \left[\bar{f}(x)\sigma_{\mu\nu}f(x)\right]$ with the other currents $J_A(x)$ introduced above, under an influence of the magnetic field was not considered in this approach, on the best of our knowledges.

In this paper, we present the propagator of a charged fermion in the constant homogeneous magnetic field in the Fock-Schwinger representation, show some selected results for the two-point correlation functions and conclude with a discussion of further applications of the formalism considered.

2 Propagator in Constant Homogeneous Magnetic Field

The equation for the fermion propagator in the electromagnetic field is as follows [13]:

$$\left[i\hat{\partial} - e\,Q_f\,\hat{A}(x) - m_f\right]G_{\mathrm{F}}(x,y) = \delta^{(4)}(x-y), \tag{5}$$

where Q_f and m_f are the relative charge and mass of the fermion, $\hat{\partial} = \partial_\mu\,\gamma^\mu$, and $\hat{A} = A_\mu\,\gamma^\mu$ with $A^\mu(x)$ being the external-field four-potential. The general representation of the charged fermion propagator in the Fock-Schwinger representation is known well [13, 14]:

$$G_{\mathrm{F}}(x,y) = \mathrm{e}^{i\Omega(x,y)}\, S_{\mathrm{F}}(x-y), \tag{6}$$

where the Lorentz non-invariant phase $\Omega(x,y)$ can be written in the form [13, 14]:

$$\Omega(x,y) = -eQ_f \int_y^x d\xi^\mu \left[A_\mu(\xi) + \frac{1}{2}F_{\mu\nu}(\xi - y)^\nu\right]. \tag{7}$$

Here $F_{\mu\nu}$ is the strength tensor of the external electromagnetic field. In the two-point correlation function, the phase factors of the two propagators cancel each other:

$$\Omega(x,y) + \Omega(y,x) = 0, \tag{8}$$

and the Lorentz-invariant part of the fermion propagator is required only [see Eq. (4)].

Let us consider a pure constant homogeneous magnetic field, $\mathbf{B} = (0,0,B)$. The corresponding four-potential can be written in the exactly Lorentz-covariant form, $A_\mu(x) = -F_{\mu\nu}x^\nu/2$. Minkowski space filled with the constant homogeneous magnetic field is divided into two subspaces: the Euclidean one with the metric tensor $\Lambda_{\mu\nu} = (\varphi\varphi)_{\mu\nu}$ which is nothing else but the plane orthogonal to the field direction, and pseudo-Euclidean one with the metric tensor $\tilde{\Lambda}_{\mu\nu} = (\tilde{\varphi}\tilde{\varphi})_{\mu\nu}$. The metric tensor of Minkowski space can be written as a difference of the subspace metrics, $g_{\mu\nu} = \tilde{\Lambda}_{\mu\nu} - \Lambda_{\mu\nu}$. In each subspace there are also the Levi-Civita symbols $\varphi_{\mu\nu}$ and $\tilde{\varphi}_{\mu\nu}$ which are the dimensionless tensor of the external magnetic field and its dual:

$$\varphi_{\mu\nu} = \frac{F_{\mu\nu}}{B}\,, \qquad \tilde{\varphi}_{\mu\nu} = \frac{1}{2}\,\varepsilon_{\mu\nu\rho\sigma}\varphi^{\rho\sigma}, \tag{9}$$

where $\varepsilon_{\mu\nu\rho\sigma}$ is the antisymmetric Levi-Civita symbol with the definition $\varepsilon^{0123} = 1$ [12]. An arbitrary four-vector $a^\mu = (a_0, a_1, a_2, a_3)$ can be decomposed into the two orthogonal components:

$$a_\mu = \tilde{\Lambda}_{\mu\nu} a^\nu - \Lambda_{\mu\nu} a^\nu = a_{\parallel\mu} - a_{\perp\mu}. \tag{10}$$

For the scalar product of two four-vectors one has

$$(ab) = (ab)_\parallel - (ab)_\perp, \tag{11}$$

where the terms on r. h. s. are non-trivial in one subspace only, i. e.:

$$(ab)_\parallel = (a\tilde{\Lambda}b) = a^\mu \tilde{\Lambda}_{\mu\nu} b^\nu, \quad (ab)_\perp = (a\Lambda b) = a^\mu \Lambda_{\mu\nu} b^\nu. \tag{12}$$

After these notations are introduced, we can write the Lorentz-invariant part of the fermion propagator as follows [7]:

$$\begin{aligned} S_{\mathrm{F}}(X) &= -\frac{i\beta}{2(4\pi)^2} \int_0^\infty \frac{ds}{s^2} \Big\{ (X\widetilde{\Lambda}\gamma)\cot(\beta s) - i(X\widetilde{\varphi}\gamma)\gamma_5 \\ &- \frac{\beta s}{\sin^2(\beta s)} (X\Lambda\gamma) + m_f s\,[2\cot(\beta s) + (\gamma\varphi\gamma)] \Big\} \\ &\times \exp\left(-i\left[m_f^2 s + \frac{1}{4s}(X\widetilde{\Lambda}X) - \frac{\beta\cot(\beta s)}{4}(X\Lambda X)\right]\right), \end{aligned} \tag{13}$$

where $\beta = eB|Q_f|$. Note that Eq. (4) is written in terms of this part of the charged fermion propagator.

3 Natural Orthogonal Basis in Magnetic Field Background

Correlators having rank different from zero, are convenient to decompose in some set of four independent vectors. In the magnetic field background, such an orthogonal basis naturally exists [7]:

$$\begin{aligned} & b_\mu^{(1)} = (q\varphi)_\mu, \qquad b_\mu^{(2)} = (q\tilde{\varphi})_\mu \\ & b_\mu^{(3)} = q^2 (\Lambda q)_\mu - (q\Lambda q)\, q_\mu, \quad b_\mu^{(4)} = q_\mu, \end{aligned} \tag{14}$$

and an arbitrary four-vector a_μ can be presented as follows:

$$a_\mu = \sum_{i=1}^{4} a_i \frac{b_\mu^{(i)}}{(b^{(i)} b^{(i)})}, \qquad a_i = a^\mu b_\mu^{(i)} = (ab^{(i)}). \tag{15}$$

The norms of the vectors (14) can be calculated easily:

$$\begin{aligned} & (b^{(1)} b^{(1)}) = -(q\Lambda q), \qquad (b^{(2)} b^{(2)}) = -(q\tilde{\Lambda} q), \\ & (b^{(3)} b^{(3)}) = -q^2 (q\tilde{\Lambda} q)(q\Lambda q), \quad (b^{(4)} b^{(4)}) = q^2. \end{aligned} \tag{16}$$

An arbitrary second-rank tensor $T_{\mu\nu}$ can be decomposed similarly:

$$T_{\mu\nu} = \sum_{i,j=1}^{4} T_{ij} \frac{b_\mu^{(i)} b_\nu^{(j)}}{(b^{(i)} b^{(i)})(b^{(j)} b^{(j)})}, \qquad T_{ij} = b_\mu^{(i)} T^{\mu\nu} b_\nu^{(j)} = (b^{(i)} T b^{(j)}). \tag{17}$$

This procedure can be extended to higher rank tensors.

As all the technical details are already discussed, let us present several examples of the two-point correlators.

4 Examples of Correlators

Here, we present several examples of two-point correlation functions.

4.1 Photon Polarization Operator

The photon polarization operator is the typical example of the two-point correlation function.

The Lagrangian density of the fermion-photon interaction in QED is well-known [12, 15]:

$$\mathcal{L}_{\text{QED}}(x) = \sum_f eQ_f \left[\bar{f}(x)\gamma_\mu f(x)\right] A^\mu(x), \tag{18}$$

where $f(x)$ and $A^\mu(x)$ are quantized fermion and photon fields, respectively, and Q_F is the electric charge of the fermion in units of the elementary charge, $e > 0$. Comparison of (18) with the generalized Lagrangian (3) shows that the generalized current is $J_\mu(x) = eQ_f A_\mu(x)$.

In the momentum space, the matrix element of the $\gamma \to \gamma$ transition induced by the Lagrangian (18) can be written in the form [12, 15]:

$$iM_{\gamma\to\gamma} = \varepsilon^*_\mu(q)\,\Pi^{\mu\nu}(q)\,\varepsilon_\nu(q), \tag{19}$$

where $\varepsilon_\mu(q)$ is the polarization vector of the photon with the momentum q^μ. The Lorentz tensor $\Pi^{\mu\nu}(q)$ is the photon polarization operator, very well studied within QED both in vacuum [12, 15] and under an influence of an active media [6, 7].

4.2 Axion Self-Energy

The other example of the two-point correlation function is the axion self-energy. In the magnetic field, it was calculated in Refs. [8–10].

The Lagrangian density of the fermion-axion interaction is as follows [6]:

$$\mathcal{L}_{af}(x) = \sum_f \frac{g_{af}}{2m_f}\left[\bar{f}(x)\gamma^\mu\gamma_5 f(x)\right]\partial_\mu\, a(x), \tag{20}$$

where $f(x)$ and $a(x)$ are quantum fermion and axion fields, respectively, $\partial_\mu\, a(x) = \partial a(x)/\partial x^\mu$, $g_{af} = C_f m_f/f_a$ is the dimensionless Yukawa constant, C_f is the dimensionless factor dependent on the axion model [6], and $\gamma_5 = -i\gamma^0\gamma^1\gamma^2\gamma^3$ [12]. This Lagrangian is also of the type (3) with the generalized current $J_\mu(x) = \left(g_{af}/2m_f\right)\partial_\mu\, a(x)$.

The $a \to a$ transition amplitude, $M_{a\to a}$, calculated in the momentum space determines the shift of the axion mass squared due to the vacuum effects [10]:

$$M_{a\to a} = -\delta m_a^2. \tag{21}$$

The corresponding diagram is shown in Fig. 2 where double solid lines indicate that effects of the external magnetic field on the fermion propagator are taken into account exactly. So, one can see that $M_{a\to a}$ is the two-point correlation function determined by the time-ordered product of two axial-vector fermionic currents. Our calculations confirm the results presented in Ref. [10]:

$$M_{a\to a}(q^2, q_\perp^2, \beta) = \sum_f \frac{g_{af}^2\beta}{8\pi^2}\int\limits_0^\infty \frac{dt}{\sin(\beta t)}\int\limits_0^1 du\left[q_\parallel^2\cos(\beta t) - q_\perp^2\cos(\beta t u)\right]\times \tag{22}$$

$$\times\exp\left\{-i\left[m_f^2 t - \frac{q_\parallel^2}{4}\,t\,(1-u^2) + q_\perp^2\,\frac{\cos(\beta t u) - \cos(\beta t)}{2\beta\sin(\beta t)}\right]\right\},$$

where $q^2 = q_\parallel^2 - q_\perp^2$. Note that two proper-time variables s_1 and s_2 entering each of the propagators have been replaced by the other two: $t = s_1 + s_2$ and $u = (s_1 - s_2)/t$.

The invariant matrix element (22) contains both pure vacuum and field-induced parts and the field-induced contribution $\Delta M(q^2, q_\perp^2, \beta)$ to $M_{a\to a}$ is as follows:

$$\Delta M(q^2, q_\perp^2, \beta) = M_{a\to a}(q^2, q_\perp^2, \beta) - M_{a\to a}(q^2, 0, 0). \tag{23}$$

Note that this quantity is free from the ultraviolet divergences [10].

4.3 Correlator of Pseudoscalar and Tensor Currents

The non-diagonal correlator of pseudoscalar and tensor currents is the second-rank tensor. Taking into account its antisymmetry under permutations, from six independent coefficients in the basis decomposition (17), three only are non-trivial:

$$\Pi_{12}^{(PT)}(q^2, q_\perp^2, \beta) = \frac{\beta}{8\pi^2} q_\parallel^2 q_\perp^2 \int_0^\infty dt \int_0^1 du \, \frac{\sin(\beta t u)}{\sin(\beta t)} \left[\cot(\beta t) - u \cot(\beta t u)\right]$$
$$\times \exp\left\{-i\left[m_f^2 t - \frac{q_\parallel^2}{4} t(1-u^2) + q_\perp^2 \frac{\cos(\beta t u) - \cos(\beta t)}{2\beta \sin(\beta t)}\right]\right\},$$

$$\Pi_{23}^{(PT)}(q^2, q_\perp^2, \beta) = -\frac{\beta}{8\pi^2} q_\parallel^2 \int_0^\infty \frac{dt}{t \sin^2(\beta t)} \int_0^1 du \left\{4\sin^2(\beta t)\left[m_f^2 t + i\right]\right.$$
$$\left. + q_\parallel^2 t\left[2\cos(\beta t)\cos(\beta t u) + 2u \sin(\beta t)\sin(\beta t u) + \left(1 - u^2\right)\sin^2(\beta t) - 2\right]\right\}$$
$$\times \exp\left\{-i\left[m_f^2 t - \frac{q_\parallel^2}{4} t(1-u^2) + q_\perp^2 \frac{\cos(\beta t u) - \cos(\beta t)}{2\beta \sin(\beta t)}\right]\right\},$$

$$\Pi_{24}^{(PT)}(q^2, q_\perp^2, \beta) = \frac{\beta}{32\pi^2} q_\parallel^2 q_\perp^2 \int_0^\infty \frac{dt}{t \sin^2(\beta t)} \int_0^1 du \left\{4\sin^2(\beta t)\left[m_f^2 t + i\right]\right.$$
$$\left. + q_\parallel^2 t\left(1 - u^2\right)\sin^2(\beta t) - 2q_\perp^2 t\left[\cos(\beta t)\cos(\beta t u) + u \sin(\beta t)\sin(\beta t u) - 1\right]\right\}$$
$$\times \exp\left\{-i\left[m_f^2 t - \frac{q_\parallel^2}{4} t(1-u^2) + q_\perp^2 \frac{\cos(\beta t u) - \cos(\beta t)}{2\beta \sin(\beta t)}\right]\right\}.$$

The coefficients $\Pi_{21}^{(PT)}(q^2, q_\perp^2, \beta)$, $\Pi_{32}^{(PT)}(q^2, q_\perp^2, \beta)$, and $\Pi_{42}^{(PT)}(q^2, q_\perp^2, \beta)$ differ by the sign from $\Pi_{12}^{(PT)}(q^2, q_\perp^2, \beta)$, $\Pi_{23}^{(PT)}(q^2, q_\perp^2, \beta)$, and $\Pi_{24}^{(PT)}(q^2, q_\perp^2, \beta)$ because of the tensor anti-symmetry.

Correlators of other generalized currents with the tensor one will be presented elsewhere [16].

5 Applications of Correlators

As discussed in Subsec. 4.1, the standard polarization operator is related with the correlator of two vector currents. Models beyond the Standard Model can effectively modify the QED Lagrangian with an extra term called the Pauli Lagrangian density [17–19]:

$$\mathcal{L}_{\rm AMM}(x) = \frac{\mu_f}{4}\left[\bar{f}(x)\sigma_{\mu\nu} f(x)\right] F^{\mu\nu}(x), \tag{24}$$

where μ_f is the anomalous magnetic moment (AMM) of the fermion. Note that this term is gauge- and Lorentz-invariant but non-renormalizable within the modified QED. The existence of (24) in the effective QED Lagrangian should also change the photon polarization operator. A contribution linear in the fermion AMM is related with the correlator of the vector and tensor fermionic currents. Its influence on the photon properties requires a detail discussion and will be presented elsewhere [16].

The strong-field limit can be also important as the resulting expressions are simplified drastically and allow detail understanding of results without cumbersome numerical analysis. It is worth to note the strong-magnetic-field formalism suggested by Loskutov and Skobelev [20, 21] which allow to perform calculations similar to the conventional ones developed for vacuum. Unfortunately, some loop diagrams are uncertain and a clear procedure is required to fix them. Let us mentioned in connection with this the analysis of the axion self-energy in the magnetic field [8–10]. The correct result was obtained only after the strong-field limit was found explicitly [10].

The technique we presented here can be extended for calculations of three-point correlation functions. The well-known example of such a correlator is the three-photon vertex in the external constant homogeneous magnetic field calculated by Adler [22]. Two of us have already an experience of calculating the three-point correlation functions connected with the effective $a\gamma\gamma$-vertex of the axion interacting with two photons in the crossed [23, 24] and magnetic [25] field configurations of the external background. Note that the same vertex calculated in the crossed field later by Skobelev [26] differs from ours but a reason still remains unclear. One should also mentioned the recent analysis of the magnetic-field influence on the radiative Higgs decay [27]. Some other three-point vertices, like the one describing an interaction of neutrinos with two photons, are also of special importance for astrophysics [6, 7].

6 Conclusions

Two-point correlators in presence of a constant homogeneous external magnetic field are considered. This analysis extended the previous one [11] by including the fermionic tensor current into consideration. With new correlators, modifications of the photon polarization operator induced by the Pauli Lagrangian density can be studied. The computer technique developed for two-point correlators is planned to be applied for three-point correlation functions.

Acknowledgements

The paper is partially supported by the *Russian Foundation for Basic Research* (Project No. 15-02-06033-a) and by the *YSU Initiative Scientific Research Activity* (Project No. AAAA-A16-116070610023-3). A. D. acknowledges financial support by the *Russian Foundation for Basic Research* (Project No. 16-32-00066 мол-а) and by the *Dynasty Foundation*.

References

[1] C. Patrignani et al. (Particle Data Group), Chin. Phys. **C40**, 100001 (2016)
[2] P.A. Baikov, K.G. Chetyrkin, J.H. Kuhn, J. Rittinger, JHEP **07**, 017 (2012), 1206.1284
[3] P.A. Baikov, K.G. Chetyrkin, J.H. Kuhn, J. Rittinger, Phys. Lett. **B714**, 62 (2012), 1206.1288
[4] K. Chetyrkin, J. Kuehn, PoS **LL2016**, 047 (2016), 1607.08374
[5] K.G. Chetyrkin, A. Maier, Nucl. Phys. **B844**, 266 (2011), 1010.1145
[6] G.G. Raffelt, *Stars as Laboratories for Fundamental Physics* (University of Chicago Press, Chicago & London, 1996)

[7] A.V. Kuznetsov, N.V. Mikheev, *Electroweak Processes in External Active Media*, Vol. 252 of *Springer Tracts in Modern Physics* (Springer-Verlag, New York, 2013)
[8] V.V. Skobelev, Phys. Atom. Nucl. **61**, 2123 (1998), [Yad. Fiz. 61, 2236 (1998)]
[9] A.V. Borisov, P.E. Sizin, J. Exp. Theor. Phys. **88**, 1 (1999), [Zh. Eksp. Teor. Fiz. 115, 3 (1999)], hep-ph/9904356
[10] L.A. Vasilevskaya, N.V. Mikheev, A.Ya. Parkhomenko, Phys. Atom. Nucl. **64**, 294 (2001), [Yad. Fiz. 64, 342 (2001)]
[11] M.Yu. Borovkov, A.V. Kuznetsov, N.V. Mikheev, Phys. Atom. Nucl. **62**, 1601 (1999), [Yad. Fiz. 62, 1714 (1999)]
[12] M.E. Peskin, D.V. Schroeder, *An Introduction to Quantum Field Theory* (Addison-Wesley, Reading, Massachusetts, 1995)
[13] C. Itzykson, J.B. Zuber, *Quantum Field Theory* (McGraw-Hill, New York, 1980)
[14] J.S. Schwinger, Phys. Rev. **82**, 664 (1951)
[15] V.B. Berestetsky, E.M. Lifshitz, L.P. Pitaevsky, *Course of Theoretical Physics. Vol. 4. Quantum Electrodynamics* (Pergamon, Oxford, 1982)
[16] L.A. Vasilevskaya, A.A. Dobrynina, I.V. Karabanov, A.Ya. Parkhomenko, in preparation (2017)
[17] P.M. Lavrov, J. Phys. **A18**, 3455 (1985)
[18] H.K. Lee, Y. Yoon, JHEP **03**, 086 (2007), hep-th/0611134
[19] A. Barducci, R. Giachetti, G. Pettini, Eur. Phys. J. **C70**, 445 (2010), 1006.3871
[20] V.V. Skobelev, Russ. Phys. J. **18**, 1481 (1975)
[21] Y.M. Loskutov, V.V. Skobelev, Phys. Lett. **A56**, 151 (1976)
[22] S.L. Adler, Annals Phys. **67**, 599 (1971)
[23] L.A. Vasilevskaya, N.V. Mikheev, A.Ya. Parkhomenko, Phys. Atom. Nucl. **60**, 2041 (1997), [Yad. Fiz. 60, 2224 (1997)]
[24] N.V. Mikheev, A.Ya. Parkhomenko, L.A. Vassilevskaya, Phys. Rev. **D60**, 035001 (1999), hep-ph/9903415
[25] N.V. Mikheev, A.Ya. Parkhomenko, L.A. Vasilevskaya, Phys. Atom. Nucl. **63**, 1046 (2000), [Yad. Fiz. 63, 1122 (2000)]
[26] V.V. Skobelev, J. Exp. Theor. Phys. **88**, 209 (1999), [Zh. Eksp. Teor. Fiz. 115, 385 (1999)]
[27] N.K. Nielsen, Phys. Rev. **D90**, 016010 (2014), 1312.6280

EPJ Web of Conferences **158**, 07002 (2017) DOI: 10.1051/epjconf/201715807002
QFTHEP 2017

The correction of hadronic nucleus polarizability to hyperfine structure of light muonic atoms

A.V. Eskin[1,*], *A.P.* Martynenko[1,**], and *E.N.* Elekina[2]

[1] *Samara U., 443086, Samara, Moskovskoe shosse 34*
[2] *Samara State U. of Architecture and Civil Engineering, 443001, Samara, Molodogvardeiskaya Str. 194*

Abstract. The calculation of hadronic polarizability contribution of the nucleus to hyperfine structure of muonic hydrogen and helium is carried out within the unitary isobar model and experimental data on the polarized structure functions of deep inelastic lepton-proton and lepton-deuteron scattering. The calculation of virtual absorption cross sections of transversely and longitudinally polarized photons by nucleons in the resonance region is performed in the framework of the program MAID.

1 Introduction

Precise investigation of the Lamb shift and hyperfine structure of light muonic atoms is a fundamental problem for testing the Standard model and establishing the exact values of its parameters, as well as searching for effects of new physics. At present, the relevance of these studies is primarily related to experiments conducted by the collaboration CREMA (Charge Radius Experiments with Muonic Atoms) [1–4] with muonic hydrogen and deuterium by methods of laser spectroscopy. So, as a result of measuring the transition frequency $2P^{F=2}_{3/2} - 2S^{F=1}_{1/2}$ a more accurate value of the proton charge radius was found to be $r_E = 0.84087(39)$ fm, which is different from the value recommended by CODATA for 7σ [5]. The CODATA value is based on the spectroscopy of the electronic hydrogen atom and on electron-nucleon scattering. The measurement of the transition frequency $2P^{F=1}_{3/2} - 2S^{F=0}_{1/2}$ for the singlet $2S$ of the state (μp) allowed to obtain the hyperfine splitting of the $2S$ energy level in muonic hydrogen, and also the values of the Zemach's radius $r_Z = 1.082(37)$ fm and magnetic radius $r_M = 0.87(6)$ fm. The first measurement of three transition frequencies between energy levels $2P$ and $2S$ for muonic deuterium $(2S^{F=3/2}_{1/2} - 2P^{F=5/2}_{3/2})$, $(2S^{F=1/2}_{1/2} - 2P^{F=3/2}_{3/2})$, $(2S^{F=1/2}_{1/2} - 2P^{F=1/2}_{3/2})$ allowed to obtain in 2.7 times the more accurate value of the charge radius of the deuteron, which is also less than the value recommended by CODATA [5], by 7.5σ [4]. As a result, a situation emerges when there is an inexplicable discrepancy between the values of such fundamental parameters, like the charge radius of a proton and deuteron, obtained from electronic and muonic atoms. In the process of searching for possible solutions of the proton charge radius "puzzle" various hypotheses were formulated, including the idea of the nonuniversality of the interaction of electrons and muons with nucleons. Preliminary experimental data for muonic helium ions show that there is no large discrepancy in obtained charge radii in comparison with CODATA.

*e-mail: eskin33@mail.ru
**e-mail: a.p.martynenko@samsu.ru

In the experiments of the CREMA collaboration one very important task is solved: to obtain an order of magnitude more accurate values of the charge radii of the simplest nuclei (proton, deuteron, helion, alpha particle ....) that enter into one form or another into theoretical expressions for intervals of fine or hyperfine structure of the spectrum. In this case, high sensitivity of characteristics of the bound muon to distribution of charge density and magnetic moment of the nucleus is used. Successful realization of this program is possible only in combination with precise theoretical calculations of various energy intervals, measured experimentally. In this way, the problem of a more accurate theoretical construction of the particle interaction operator in quantum electrodynamics, the calculation of new corrections in the energy spectrum of muonic atoms acquires a special urgency [6–8]. The aim of this work consists in the calculation of the deuteron, helion and triton polarizability correction to the hyperfine splitting (HFS). We perform a calculation of hadronic polarizability contribution using the isobar model describing the processes of photo- and electroproduction of π, η mesons, nucleon resonances on the nucleon in the resonance region, and experimental data on the nucleon and deuteron polarized structure functions obtained in non-resonance region.

2 General formalism

The leading order polarizability contribution to HFS is determined by two-photon exchange diagrams, shown in Fig. 1. The corresponding amplitudes of virtual Compton scattering on the nucleus can be represented as a convolution of antisymmetric parts of the lepton and hadron tensors which have the following form [9, 10]:

$$L^A_{\mu\nu} = \frac{1}{4}Tr\Big\{(1+\gamma^0)\gamma_5\hat{s}_1\Big[\gamma_1^\mu\frac{\hat{p}_1+\hat{k}+m_1}{(p_1+k)^2-m_1^2}\gamma_1^\nu+\gamma_1^\nu\frac{\hat{p}_1-\hat{k}+m_1}{(p_1-k)^2-m_1^2}\gamma_1^\mu\Big]\Big\}, \quad (1)$$

$$W^A_{\mu\nu} = i\epsilon_{\mu\nu\alpha\beta}k^\alpha\Big\{s_2^\beta\frac{H_1(\nu,Q^2)}{(p_2\cdot k)}+\frac{[(p_2k)s_2^\beta-(s_2k)p_2^\beta]}{(p_2\cdot k)^2}H_2(\nu,Q^2)\Big\}, \quad (2)$$

where m_1, m_2 are the lepton and nucleus masses, the nucleus four-momentum $p_2 = (m_2, 0)$, $\epsilon_{\mu\nu\alpha\beta}$ is the totally antisymmetric tensor in four dimensions. s_1, s_2 are spin four vectors of the lepton and nucleus. H_1, H_2 are the structure functions of polarized scattering. The invariant quantity $p_2\cdot k$ is related to the energy transfer ν in the nucleus rest frame: $p_2\cdot k = m_2\nu$. The invariant mass of the electroproduced hadronic system W is then $W^2 = m_2^2+2m_2\nu-Q^2 = m_2^2+Q^2(1/x-1)$. After computing the convolution of two tensors (1) and (2), we obtain:

$$L^A_{\mu\nu}W^A_{\mu\nu} = \frac{4}{3}(\mathbf{s}_1\mathbf{s}_2)\frac{m_2k^2}{k^4-4m_1^2k_0^2}\left[(k_0^2+2k^2)\frac{H_1}{(p_2\cdot k)}+3k_0^2k^2\frac{H_2}{(p_2\cdot k)^2}\right]. \quad (3)$$

According to the optical theorem the imaginary part of the forward Compton amplitude is related to the cross section of inelastic scattering of off-shell photons from protons: $ImH_1(\nu,Q^2) = g_1(\nu,Q^2)/\nu$, $ImH_2(\nu,Q^2) = m_2g_2(\nu,Q^2)/\nu^2$. As a result, neglecting the lepton mass the nucleus polarizability contribution to HFS can be presented in the form [11–14]:

$$\Delta E^{hfs}_{pol} = \frac{Z\alpha m_1}{2\pi m_2\mu_N}E_F(\Delta_1+\Delta_2) = (\delta_1^p+\delta_2^p)E_F = \delta_{pol}E_F, \quad (4)$$

$$\Delta_1 = \int_0^\infty\frac{dQ^2}{Q^2}\Big\{\frac{9}{4}F_2^2(Q^2)-4m_2\int_{\nu_{th}}^\infty\frac{d\nu}{\nu^2}\beta_1(\theta)g_1(\nu,Q^2)\Big\}, \quad (5)$$

DOI: 10.1051/epjconf/201715807002

$$\Delta_2 = -12m_2 \int_0^\infty \frac{dQ^2}{Q^2} \int_{\nu_{th}}^\infty \frac{d\nu}{\nu^2} \beta_2(\theta) g_2(\nu, Q^2), \tag{6}$$

where ν_{th} determines the pion-nucleus threshold:

$$\nu_{th} = m_\pi + \frac{m_\pi^2 + Q^2}{2m_2}, \tag{7}$$

and the functions $\beta_{1,2}$ have the form:

$$\beta_1(\theta) = 3\theta - 2\theta^2 - 2(2-\theta)\sqrt{\theta(\theta+1)}, \tag{8}$$

$$\beta_2(\theta) = 1 + 2\theta - 2\sqrt{\theta(\theta+1)},\ \theta = \nu^2/Q^2. \tag{9}$$

$F_2(Q^2)$ is the Pauli form factor of the nucleus. The dependence on the mass of the lepton in (5)-(6) can also be taken into account, which leads to a certain modification of the functions $\beta_i(\nu, Q^2)$ [10, 14]. This is important for increasing the accuracy of calculations in the case of muonic atoms.

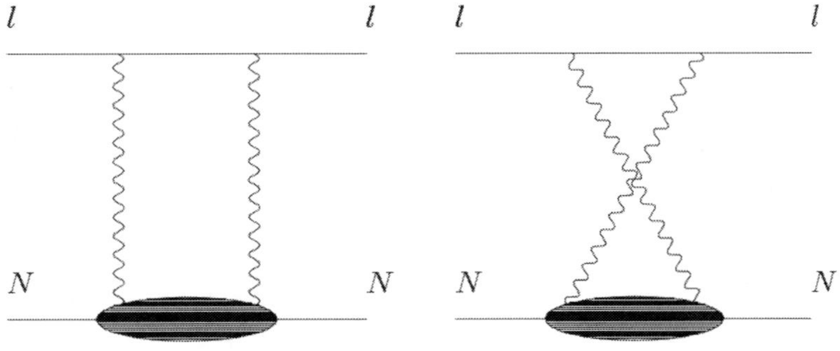

Figure 1. Two-photon Feynman amplitudes determining the correction of the nucleus polarizability to the hyperfine splitting of muonic atom.

Having information on the polarization structure functions of the nuclei, one can integrate into (4) and obtain this correction. The nucleus spin-dependent structure functions $g_1(\nu, Q^2)$, $g_2(\nu, Q^2)$ can be measured in the inelastic scattering of polarized leptons on polarized nuclei. Accurate measurements of proton and deuteron polarized structure functions were made in SLAC, CERN and DESY [15–20]. These experimental data can be used for the calculation the contribution in the nonresonance region where the invariant mass W must be greater than the mass of any resonance N^* in the reaction $\gamma^* + N \to N^*$. The threshold between the resonance region and the deep-inelastic region is not well defined, but it is usually taken to be at about $W^2 = 4\,GeV^2$. On the other hand in the resonance region we need theoretical model describing polarized structure functions $g_{1,2}(\nu, Q^2)$ since experimental data in this region are clearly insufficient. First of all, we can express functions $g_{1,2}(\nu, Q^2)$ in terms of virtual photon absorption cross sections as follows:

$$g_1(\nu, Q^2) = \frac{m_2 \cdot K}{8\pi^2\alpha(1 + Q^2/\nu^2)} \left[\sigma^T_{1/2}(\nu, Q^2) - \sigma^T_{3/2}(\nu, Q^2) + \frac{2\sqrt{Q^2}}{\nu}\sigma^{TL}_{1/2}(\nu, Q^2)\right], \tag{10}$$

$$g_2(\nu, Q^2) = \frac{m_2 \cdot K}{8\pi^2\alpha(1 + Q^2/\nu^2)} \left[-\sigma^T_{1/2}(\nu, Q^2) + \sigma^T_{3/2}(\nu, Q^2) + \frac{2\nu}{\sqrt{Q^2}}\sigma^{TL}_{1/2}(\nu, Q^2)\right], \tag{11}$$

where K is the kinematical flux factor for virtual photons. The virtual photon absorption cross sections have superscripts referring to the initial and final photon polarization being longitudinal L or transverse T. The superscript TL is for the case when the photon polarization direction changes during the interaction. The subscripts refer to the total spin of the photon-nucleus system.

Let us briefly describe the basic formulas that underlie the numerical results. To construct the polarized structure functions (10)-(11) in resonance region we use the Breit-Wigner parameterization for the photoabsorption cross sections [21–28]. In the considered region of the variables k^2, W the most important contribution is given by five resonances: $P_{33}(1232)$, $S_{11}(1535)$, $D_{13}(1520)$, $P_{11}(1440)$, $F_{15}(1680)$. Accounting for the resonance decays to the $N\pi-$ and $N\eta-$ states we can express the absorption cross sections $\sigma^T_{1/2}$ and $\sigma^T_{3/2}$ as follows:

$$\sigma^T_{1/2,3/2} = \left(\frac{k_R}{k}\right)^2 \frac{W^2\Gamma_\gamma\Gamma_{R\to N\pi}}{(W^2 - M_R^2)^2 + W^2\Gamma^2_{tot}} \frac{4m_N}{M_R\Gamma_R}|A_{1/2,3/2}|^2, \tag{12}$$

where $A_{1/2,3/2}$ are transverse electromagnetic helicity amplitudes,

$$\Gamma_\gamma = \Gamma_R\left(\frac{k}{k_R}\right)^{j_1}\left(\frac{k_R^2 + X^2}{k^2 + X^2}\right)^{j_2}, \quad X = 0.3\,GeV. \tag{13}$$

The resonance parameters Γ_R, M_R, j_1, j_2, Γ_{tot} are taken from [23, 24, 29]. In accordance with [23, 25] the parameterization of one-pion decay width is

$$\Gamma_{R\to N\pi}(q) = \Gamma_R\frac{M_R}{M}\left(\frac{q}{q_R}\right)^3\left(\frac{q_R^2 + C^2}{q^2 + C^2}\right)^2, \quad C = 0.3\,GeV \tag{14}$$

for the $P_{33}(1232)$ and

$$\Gamma_{R\to N\pi}(q) = \Gamma_R\left(\frac{q}{q_R}\right)^{2l+1}\left(\frac{q_R^2 + \delta^2}{q^2 + \delta^2}\right)^{l+1}, \tag{15}$$

for resonances $D_{13}(1520)$, $P_{11}(1440)$, $F_{15}(1680)$. l is the pion angular momentum and $\delta^2 = (M_R - m_N - m_\pi)^2 + \Gamma_R^2/4$. Here q (k) and q_R (k_R) denote the c.m.s. pion (photon) momenta of resonances with the mass M and M_R respectively. In the case of $S_{11}(1535)$ we take into account πN and ηN decay modes [25]:

$$\Gamma_{R\to\pi,\eta} = \frac{q_{\pi,\eta}}{q}b_{\pi,\eta}\Gamma_R\frac{q_{\pi\eta}^2 + C_{\pi,\eta}^2}{q^2 + C_{\pi,\eta}^2}, \tag{16}$$

where $b_{\pi,\eta}$ is the π (η) branching ratio.

The cross section $\sigma^{TL}_{1/2}$ is determined by an expression similar to (12), containing the product $(S^*_{1/2}\cdot A_{1/2} + A^*_{1/2}S_{1/2})$ [15]. The calculation of helicity amplitudes $A_{1/2}$, $A_{3/2}$ and longitudinal amplitude $S_{1/2}$, as functions of Q^2, was done on the basis of constituent quark model (CQM) in [30–35]. The program of numerical calculation of cross sections (12) was successfully realized by the authors of [36, 37] within the unitary isobar model framework known as the MAID package (http://www.kph-uni-mainz.de/MAID). In the unitary isobar model [36, 37] accounting for the Born terms, the vector meson, nucleon resonance contributions and interference terms we calculate the cross sections $\sigma^T_{1/2,3/2}$, $\sigma^{TL}_{1/2}$ by means of numerical program MAID in the resonance region as the functions of two variables W and Q^2. The obtained nucleon polarized structure functions $g_{1,2}(W, Q^2)$ are then used for a construction of nucleus structure functions and calculation the polarizability contribution. In Figs. 2,3 we show the obtained structure functions $g_{1,2}(W, Q^2)$.

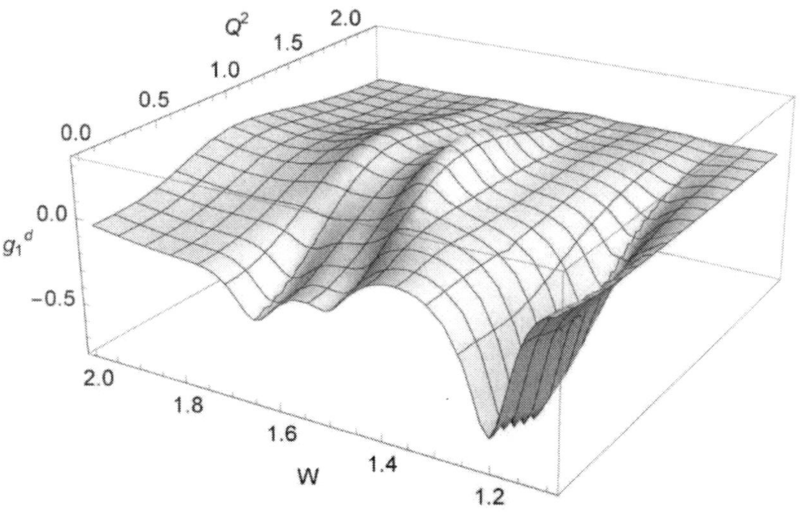

Figure 2. The deuteron polarized structure function $g_1^d(W, Q^2)$ as function of Q^2 ($0 \div 2.0$. GeV^2 and W ($1.1 \div 2.0$) GeV.

Our calculation of polarizability contribution in deep inelastic region is based on experimental data from [15–20]. As was shown in previous paper [10] nucleon polarized structure functions can be expressed through polarized quark and gluon distributions which obey to the evolution equations [38, 39]. Solving Q^2 evolution equations we can construct nucleon functions $g_1(\nu, Q^2)$, $g_2(\nu, Q^2)$ which agree well with experimental data and the following parameterization [15–20, 40]:

$$g_1^{p,d}(x, Q^2) = a_1 x^{a_2}(1 + a_3 x + a_4 x^2)[1 + a_5 f(Q^2)]F_1^{p,d}(x, Q^2), \tag{17}$$

where the superscript index p,d corresponds to the proton or deuteron. Numerical integration is performed with $f(Q^2) = -\ln Q^2$ corresponding to the perturbative QCD behaviour. The calculation of the second part of the correction δ_{pol} in (4) in nonresonance region is carried out by means of Wandzura-Wilchek relation as in [10].

3 Numerical results

Most part of numerical calculation is devoted to muonic deuterium. To the hadronic contribution we include contributions that are determined by the nuclear reactions of the production of π-, η, and K-mesons on nucleons, the production of nucleon resonances. In the approximation, which is then used for the calculation the deuteron appears as a loosely coupled system of the proton and neutron, so the deuteron polarized structure function can be presented as a sum of the proton and neutron structure functions:

$$g_i^d(W, Q^2) = g_i^p(W, Q^2) + g_i^n(W, Q^2). \tag{18}$$

The MAID program allows us to calculate separately the proton and neutron structure functions $g_i^{p,n}(W, Q^2)$.

For the integration in (4) an important role is played by the Gerasimov-Drell-Hern (GDH) sum rule [41–45] which connects an energy-weighted integral of the difference of the helicity dependent

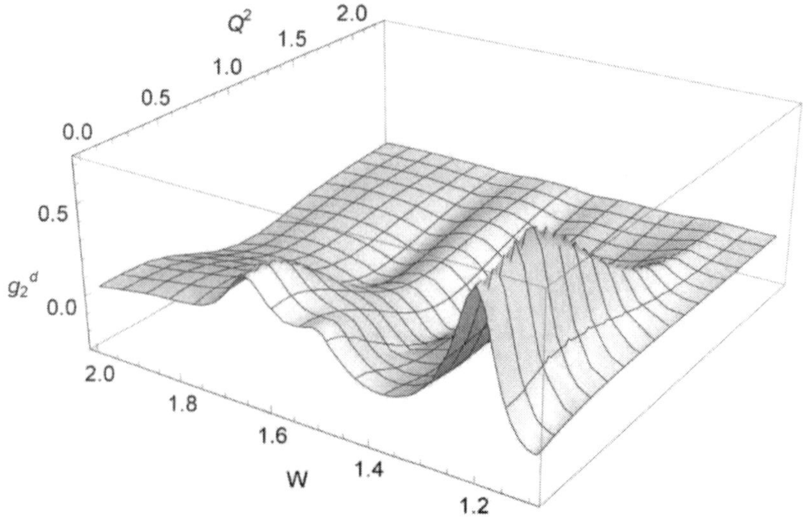

Figure 3. The deuteron polarized structure function $g_2^d(W, Q^2)$ as function of Q^2 ($0 \div 2.0$. GeV^2 and W ($1.1 \div 2.0$) GeV.

real-photon absorption cross sections with the anomalous contribution $\kappa = \frac{\mu_N m_2}{e s_2} - Z$ to the magnetic moment μ of the nucleus:

$$I_{GDH} = \int_{\nu_{th}}^{\infty} \frac{\sigma_p - \sigma_a}{\nu} d\nu = 4\pi^2 \kappa^2 \frac{\alpha s_2}{m_2^2}, \tag{19}$$

where ν is the photon energy, σ_p and σ_a are the total photoabsorption cross sections for parallel and antiparallel orientation of photon and nucleus spins, respectively. The lower limit of the integral, ν_{th}, corresponds to pion production and photodisintegration threshold for a nucleonic and nuclear target, respectively. Strictly speaking, in order for the integral in (4) to converge over the variable Q, the sum rule (19) must exactly be satisfied. The GDH sum rule is satisfied for any nucleus, including a proton and a neutron. Since the deuteron structure function satisfies (18), it is necessary to achieve the sum rules for the proton and neutron. Since the neutron and the proton have large anomalous magnetic moments $\kappa_p = 1.79$, $\kappa_n = -1.91$), we obtain large values for the integral $I^{p,n}_{GDH}$ (19). In turn, in the case of a deuteron, a small value of AMM ($\kappa_d = -0.143$) leads to a small value $I^d_{GDH} = -0.65\ \mu b$ (19), which is two orders of magnitude smaller than the corresponding values for nucleons. When the deuteron is represented in the form of a state of two almost free nucleons, we find that the quantity in right part of (19) is equal to 437 μb. There are also other channels ($\gamma d \to pn$) which can not be treated in the quasi-free approximation but which contribute to (19). An analysis carried out in [46, 47] showed that this value decreases significantly when taking into account the photodisintegration channel, which is not considered in this paper. Therefore, representing the deuteron in the form of the sum of a proton and a neutron, we add in (4) two terms with the Pauli form factors of the proton and the neutron to ensure the fulfilment of the sum rule and the subsequent integration into (4). Calculations in the MAID show that the sum rule for a proton is satisfied with a sufficiently high accuracy, whereas for a neutron the difference between the left and right parts in (19) reaches 30% (see Table 1).

To avoid this difficulty and obtain however an estimate of hadronic contribution to the neutron and deuteron polarizability we introduce a cutoff of the momentum integral over Q in (4) at some value $\kappa \approx 0.01$ GeV, supposing that the GDH sum rule for the neutron holds exactly and the region of small

Table 1. The contributions to the GDH sum rule for the proton and neutron.

Contribution to GDH integral	$I^p_{GDH}, \mu b$	$I^n_{GDH}, \mu b$
Contribution of $N\pi$ states	165.9	133.2
Contribution of $N\eta$ states	-8.9	-5.7
Contribution of K-mesons	-1.8	-3.0
Contribution of $N\pi\pi$ states	47.8	50.5
Total contribution	203.0	175.0
The GDH value	204.8	232.5

Q $(0 \div 0.01)$ GeV does not give essential contribution to general value of correction (4). Similar cutoff procedure at small values Q was used in [48] with $\kappa \approx 0.045$ GeV, in which different corrections to deuterium hyperfine structure were considered from the two-photon exchange amplitudes. As a result total value of the polarizability correction for muonic deuterium including the resonance and nonresonance regions is equal to 0.13 meV. In the case of He-3 total polarizability contribution 0.06 meV is determined by unpaired neutron because two protons have opposite spins and do not contribute to hyperfine splitting. The similar situation occurs for the triton in which two neutrons form closed shell. In muonic tritium the polarizability correction is equal to 0.05 meV. Total error of our calculation is estimated in 30 % which is determined mainly by the uncertainty from two-pion contribution and the cutoff procedure used above. The obtained values of polarizability corrections should be used for obtaining total values of hyperfine splitting in light muonic atoms [6, 49–51].

References

[1] R. Pohl, A. Antognini, F. Nez et al., Nature **466**, 213 (2010).
[2] A. Antognini et al., Science **339**, 417 (2013).
[3] A. Antognini et al., Ann. Phys. (NY) **331**, 127 (2013).
[4] R. Pohl, F. Nez, L. M. P. Fernandes et al., Science **353**, 669 (2016).
[5] P. J. Mohr, D. B. Newell, and B. N. Taylor (CODATA Recommended Values of the Fundamental Physical Constants: 2014) Rev. Mod. Phys. **88**, 035009 (2016).
[6] A. P. Martynenko, F. A. Martynenko, R. N. Faustov, J. Exp. Theor. Phys. **124**, 895 (2017).
[7] A. V. Eskin, A. P. Martynenko, E. N. Elekina, EPJ Web Conf. **132**, 03055 (2017).
[8] R. N. Faustov, A. P. Martynenko, G. A. Martynenko, V. V. Sorokin, Phys. Lett. B **733**, 354 (2014).

[9] G. M. Zinov'ev, B. V. Struminsky, R. N. Faustov, V. L. Chernyak, Sov. J. Nucl. Phys. **11**, 715 (1970).
[10] E. V. Cherednikova, R. N. Faustov, A. P. Martynenko, Nucl. Phys. A **703**, 365 (2002).
[11] R. N. Faustov and A. P. Martynenko, Eur. Phys. J. C **24**, 281 (2002).
[12] R. N. Faustov and A. P. Martynenko, Phys. Atom. Nucl. **65**, 265 (2002).
[13] C. E. Carlson, Can. J. Phys. **85**, 429 (2007).
[14] C. E. Carlson, V. Nazaryan, K. Griffioen, Phys. Rev. A **83**, 042509 (2011).
[15] K. Abe, T. Akagi, P. L. Anthony et al., Phys. Rev. D **58**, 112003 (1998).
[16] K. Abe et al., Phys. Rev. Lett. **78**, 815 (1997).
[17] P. L. Anthony et al., Phys. Lett. B **458**, 529 (1999).
[18] G. S. Mitchell, Preprint SLAC-PUB-**8104**, 1999.
[19] D. Adams et al., Phys. Rev. D **56**, 5330 (1997).
[20] D. Adeva et al., Phys. Rev. D **60**, 072004 (1999).
[21] R. L. Walker, Phys. Rev. **182**, 1729 (1969).
[22] R. A. Arndt, R. L. Workman, Z. Li et al., Phys. Rev. C **42**, 1864 (1990).
[23] S. Teis, W. Cassing, M. Effenberger et al., Z. Phys. A **356**, 421 (1997).
[24] M. Effenberger, A. Hombach, S. Teis et al., Nucl. Phys. A **613**, 353 (1997).
[25] B. Krusche, J. Ahrens, G. Anton et al., Phys. Rev. Lett. **74**, 3736 (1995).
[26] N. Bianchi, V. Muccifora, E. Sanctis et al., Phys. Rev. C **54**, 1688 (1996).
[27] D. Drechsel, O. Hanstein, S. S. Kamalov et al., Nucl. Phys. A **645**, 145 (1999).
[28] Y.-B. Dong, Eur. Phys. Jour. A **1**, 347 (1998).
[29] C. Patrignani et al. (Particle Data Group), Chin. Phys. C **40**, 100001 (2016).
[30] Z. Li and Y.-B. Dong, Phys. Rev. D **54**, 4301 (1996).
[31] R. Koniuk and N. Isgur, Phys. Rev. D **21**, 1888 (1980).
[32] F. E. Close, Z. Li, Phys. Rev. D **42**, 2207 (1990).
[33] S. Capstick, Phys. Rev. D **46**, 2864 (1992).
[34] Zhenping Li, V. Burkert, Zhujun Li, Phys. Rev. D **46**, 70 (1992).
[35] M. Warns, W. Pfeil, H. Rollnik, Phys. Rev. D **42**, 2215 (1990).
[36] W.-T. Chiang, S. N. Yang, L. Tiator, D. Drechsel, Nucl. Phys. A **700**, 429 (2002).
[37] D. Drechsel, S. S. Kamalov, G. Krein, L. Tiator, Phys. Rev. D **59**, 094021 (1999).
[38] G. Altarelli, R. D. Ball, S. Forte et al. Nucl. Phys. B **496**, 337 (1997).
[39] M. Hirai, S. Kumano, M. Miyama, Comp. Phys. Comm. **108**, 38 (1998).
[40] R. D. Erbacher, SLAC-Report-546 (1999).
[41] S. B. Gerasimov, Sov. J. Nucl. Phys. **2**, 430 (1966).
[42] S. D. Drell and A. C. Hern, Phys. Rev. Lett. **16**, 908 (1966).
[43] D. Drechsel and G. Krein, Phys. Rev. D **58**, 116009 (1998).
[44] S. Costanza, J. of Phys.: Conf. Ser. **349**, 012011 (2012).
[45] A. Airapetian et al. [The HERMES Collaboration], Preprint DESY 02-172, (2002).
[46] H. Arenhövel, G. Kreß, R. Schmidt, P. Wilhelm, Phys. Lett. B **407**, 1 (1997).
[47] H. Arenhövel, Phys. Lett. B **595**, 223 (2004).
[48] I.B. Khriplovich and A.I. Milstein, J. Theor. Exp. Phys. **125**, 205 (2004).
[49] R. N. Faustov and A. P. Martynenko, J. Theor. Exp. Phys. **98**, 39 (2004).
[50] A. P. Martynenko, J. Exp. Theor. Phys. **106**, 691 (2008).
[51] F. A. Martynenko, R. N. Faustov, and A. P. Martynenko, EPJ WoC **132**, 03032 (2017).

U(1) gauged Q-balls and their properties

Emin Nugaev[1,⋆], *Alexander* Panin[1,⋆⋆], and *Mikhail* Smolyakov[2,1,⋆⋆⋆]

[1] *Institute for Nuclear Research of the Russian Academy of Sciences, 60th October Anniversary prospect 7a, Moscow 117312, Russia*
[2] *Skobeltsyn Institute of Nuclear Physics, Lomonosov Moscow State University, Moscow 119991, Russia*

Abstract. In the present manuscript, we briefly discuss the differences between ordinary Q-balls and U(1) gauged Q-balls. We focus on the differences in the existence of Q-ball solutions for appropriate values of the model parameters, as well as on the differences in classical stability of such Q-balls.

1 Ordinary Q-balls

Q-balls are localized solutions in the theory of a complex scalar field, described by the action

$$S = \int d^4x \left(\partial^\mu\phi^*\partial_\mu\phi - V(\phi^*\phi)\right), \tag{1}$$

such that

$$\phi(t,\vec{x}) = e^{i\omega t} f(r), \qquad f(r)|_{r\to\infty} \to 0, \qquad \partial_r f(r)|_{r=0} = 0, \qquad r = \sqrt{\vec{x}^2}. \tag{2}$$

Without loss of generality one can take $f(r)$ to be real and such that $f(r) > 0$. Initially, such solutions were proposed in [1] and became popular after the well-known paper [2].

The Q-ball charge is defined as

$$Q = -i\int d^3x\left(\phi^*\dot{\phi} - \dot{\phi}^*\phi\right) = 8\pi\omega\int\limits_0^\infty f^2 r^2 dr, \tag{3}$$

whereas the Q-ball energy is defined as

$$E = \int d^3x\left(\dot{\phi}^*\dot{\phi} + \partial_i\phi^*\partial_i\phi + V(\phi^*\phi)\right). \tag{4}$$

It is easy to show that the relation

$$\frac{dE}{dQ} = \omega \tag{5}$$

⋆e-mail: emin@ms2.inr.ac.ru
⋆⋆e-mail: panin@ms2.inr.ac.ru
⋆⋆⋆e-mail: smolyakov@theory.sinp.msu.ru

DOI: 10.1051/epjconf/201715807003

holds for Q-balls. Since there exists the symmetry

$$\omega \to -\omega \quad \Rightarrow \quad Q \to -Q, \quad E \to E, \tag{6}$$

without loss of generality one can examine only the case $\omega \geq 0$.

As a simple example of the Q-ball, we can consider a model with the scalar field potential

$$V(\phi^*\phi) = M^2\phi^*\phi\,\theta\left(1 - \frac{\phi^*\phi}{v^2}\right) + M^2v^2\theta\left(\frac{\phi^*\phi}{v^2} - 1\right), \tag{7}$$

where θ is the Heaviside step function (see Figure 1). For Q-balls, this piecewise scalar field potential was introduced in a more general form in [1]. The corresponding solution for the scalar field takes the

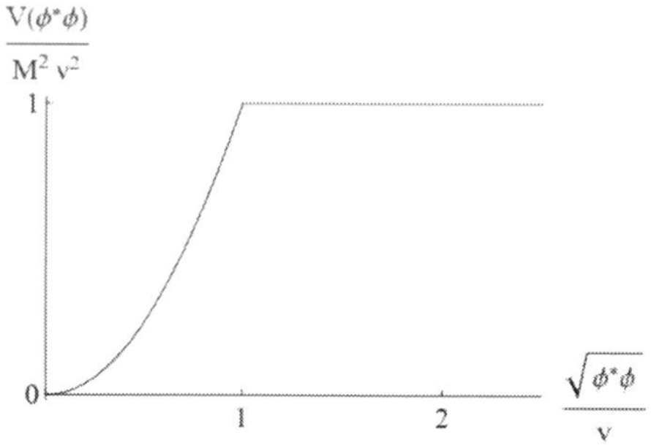

Figure 1. The scalar field potential (7).

form [1]

$$f = vR\frac{e^{-\sqrt{M^2-\omega^2}(r-R)}}{r}, \quad \text{for} \quad f < v \quad (r > R), \tag{8}$$

$$f = \frac{vR}{\sin(\omega R)}\frac{\sin(\omega r)}{r}, \quad \text{for} \quad f > v \quad (r < R). \tag{9}$$

with

$$R = \frac{1}{\omega}\left(\arctan\left(-\frac{\omega}{\sqrt{M^2 - \omega^2}}\right) + \pi\right), \quad 0 < \omega < M. \tag{10}$$

The solution for $\frac{\omega}{M} = 0.8$ is presented in Figure 2. The $Q(\omega)$ and $E(Q)$ dependencies for these Q-balls are presented in Figure 3 (see [3] for a more detailed discussion of Q-balls in the model with the scalar field potential (7)).

2 U(1) gauged Q-balls

Now we turn to U(1) gauged Q-balls, which were initially proposed in [4], see also [5]. To describe such Q-balls, let us consider the action

$$S = \int d^4x\left((\partial^\mu\phi^* - ieA^\mu\phi^*)(\partial_\mu\phi + ieA_\mu\phi) - V(\phi^*\phi) - \frac{1}{4}F_{\mu\nu}F^{\mu\nu}\right). \tag{11}$$

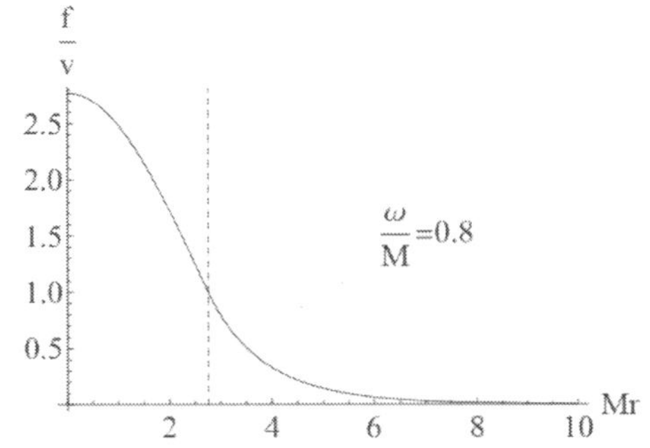

Figure 2. The scalar field profile for $\frac{\omega}{M} = 0.8$. The vertical line indicates the point $r = R$ where $f(r) = v$.

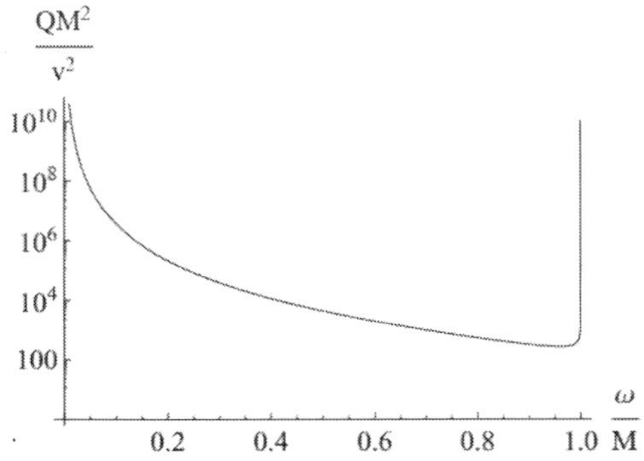

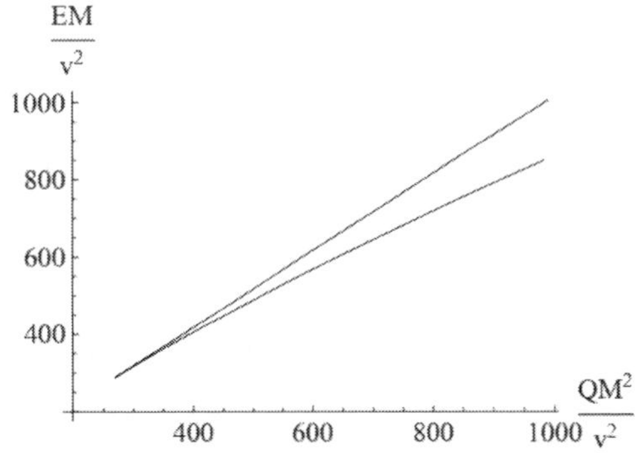

Figure 3. The $Q(\omega)$ and $E(Q)$ dependencies for the Q-ball defined by (8)–(10) (see [3] for details).

The U(1) gauged Q-ball is a localized solution to the equations of motion for the fields, following from action (11), of the form

$$\phi(t,\vec{x}) = e^{i\omega t} f(r), \qquad f(r)|_{r\to\infty} \to 0, \qquad \partial_r f(r)|_{r=0} = 0, \tag{12}$$

$$A_0(t,\vec{x}) = A_0(r), \qquad A_0(r)|_{r\to\infty} \to 0, \qquad \partial_r A_0(r)|_{r=0} = 0, \tag{13}$$

$$A_i(t,\vec{x}) \equiv 0. \tag{14}$$

Again, without loss of generality we take $f(r) > 0$. The charge of U(1) gauged Q-ball is defined as

$$Q = 8\pi \int (\omega + eA_0) f^2 r^2 dr, \tag{15}$$

whereas its energy is defined as

$$E = 4\pi \int_0^\infty \left((\omega + eA_0)^2 f^2 + \partial_r f \partial_r f + V(f) + \frac{1}{2} \partial_r A_0 \partial_r A_0 \right) r^2 dr. \tag{16}$$

It is possible to show [6] that the relation

$$\frac{dE}{dQ} = \omega \tag{17}$$

holds for U(1) gauged Q-balls too.

U(1) gauged Q-balls possess the following properties:

- $\omega \to -\omega \quad \Rightarrow \quad A_0 \to -A_0$.
- The sign of $\omega + eA_0$ coincides with the sign of ω [5].
- If $A_0 \equiv 0$, then $\omega = 0$ [4].

Thus, there exists the symmetry

$$\omega \to -\omega \quad \Rightarrow \quad Q \to -Q, \quad E \to E, \tag{18}$$

and without loss of generality one can examine only the case $\omega \geq 0$.

After this brief introduction, we are ready to turn to examining the differences between ordinary Q-balls and U(1) gauged Q-balls.

3 Behavior of Q-balls at $r \to \infty$

Let us consider the scalar field potential $V(\phi^*\phi)$ such that

$$V(\phi^*\phi) \to M^2\phi^*\phi \qquad \text{for} \qquad \phi^*\phi \to 0. \tag{19}$$

For $f(r) \to 0$, the equation of motion for ordinary Q-balls reduces to

$$(\omega^2 - M^2)f + \frac{1}{r}\frac{d^2}{dr^2}(rf) \approx 0, \tag{20}$$

leading to $f(r) \sim \frac{e^{-\sqrt{M^2-\omega^2}r}}{r}$. It is clear that there are no Q-balls with $\omega = M$, because $Q \to \infty$ (as well as $E \to \infty$) for $\omega \to M$.

In the case of U(1) gauged Q-balls the situation is completely different. Indeed, suppose that there exists a U(1) gauged Q-ball with the charge Q. Then, for $f(r) \to 0$ the equation of motion for the scalar field reduces to

$$(\omega^2 - M^2)f - \frac{2\,\omega e^2 Q}{4\pi r}f + \frac{1}{r}\frac{d^2}{dr^2}(rf) \approx 0. \tag{21}$$

Now we have two cases:

- For $\omega < M$

$$f(r) \sim \frac{\mathrm{e}^{-\sqrt{M^2-\omega^2}\,r}}{r^{1+\frac{\omega e^2 Q}{4\pi\sqrt{M^2-\omega^2}}}} \quad \text{for} \quad r \to \infty. \tag{22}$$

- For $\omega = M$

$$f(r) = C\frac{K_1\left(\sqrt{\frac{2Me^2Q}{\pi}}\,r\right)}{\sqrt{r}}, \tag{23}$$

where C is a constant and $K_1(b,z)$ is the modified Bessel function of the second kind, leading to

$$f(r) \sim \frac{\mathrm{e}^{-\sqrt{\frac{2Me^2Q}{\pi}}\,r}}{r^{\frac{3}{4}}} \quad \text{for} \quad r \to \infty. \tag{24}$$

We see that in the gauged case there may exist a solution possessing the finite charge and energy even for $\omega = M$. This conclusion is supported by the results of numerical simulations (see [7] for details) for the scalar field potential (7) (Figures 4 and 5) and for the scalar field potential $V(f) = M^2f^2 - \lambda f^4$ (Figures 6 and 7).

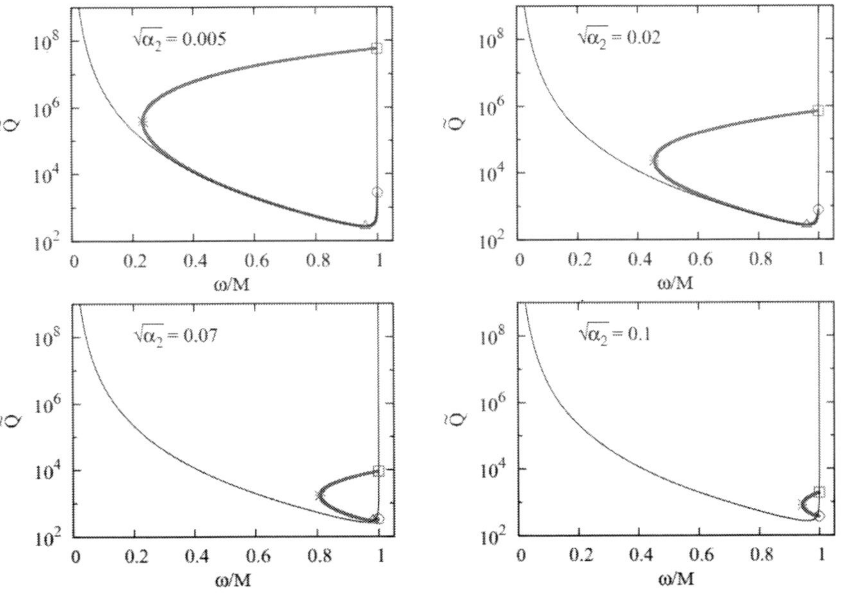

Figure 4. $\tilde{Q} = \frac{M^2}{v^2} Q$ for different values of the parameter $\alpha_2 = \frac{e^2 v^2}{M^2}$ (thick lines). The thin lines stand for the nongauged case. The figure is taken from [7].

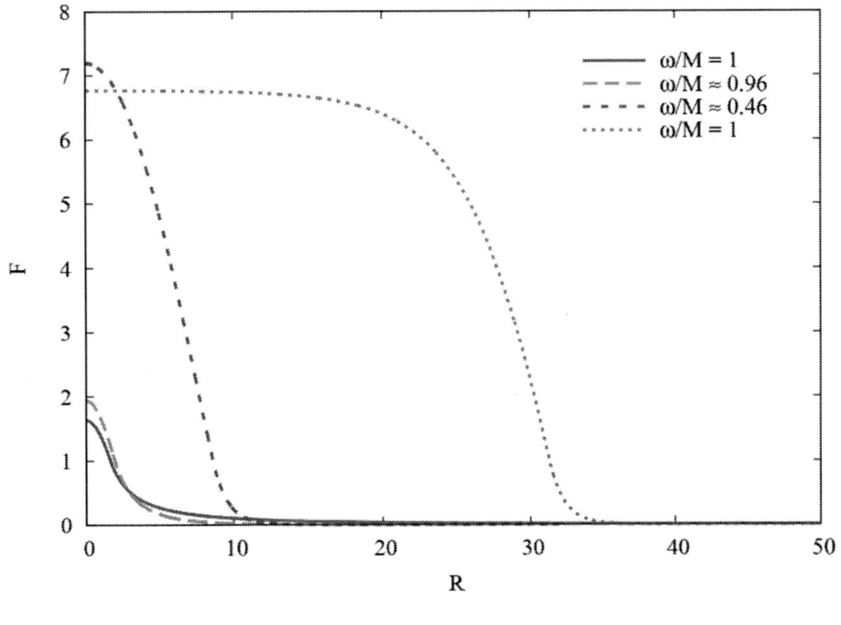

Figure 5. Profiles of the scalar field for different values of $\frac{\omega}{M}$. Here $\sqrt{\alpha_2} = 0.02$, $R = Mr$, $F(R) = \frac{1}{v} f(r)$. The figure is taken from [7].

DOI: 10.1051/epjconf/201715807003

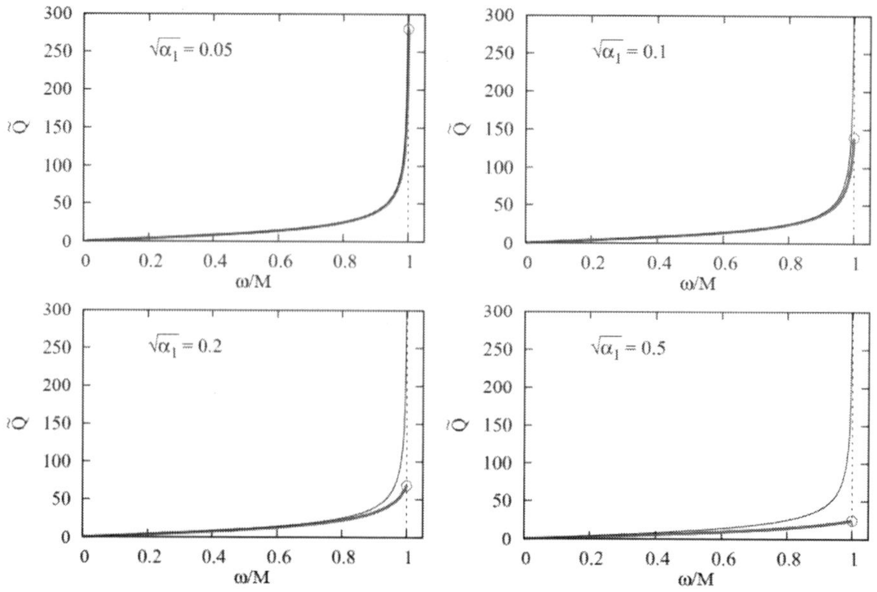

Figure 6. $\tilde{Q} = \lambda Q$ for different values of the parameter $\alpha_1 = \frac{e^2}{\lambda}$ (thick lines). The thin lines stand for the nongauged case. The figure is taken from [7].

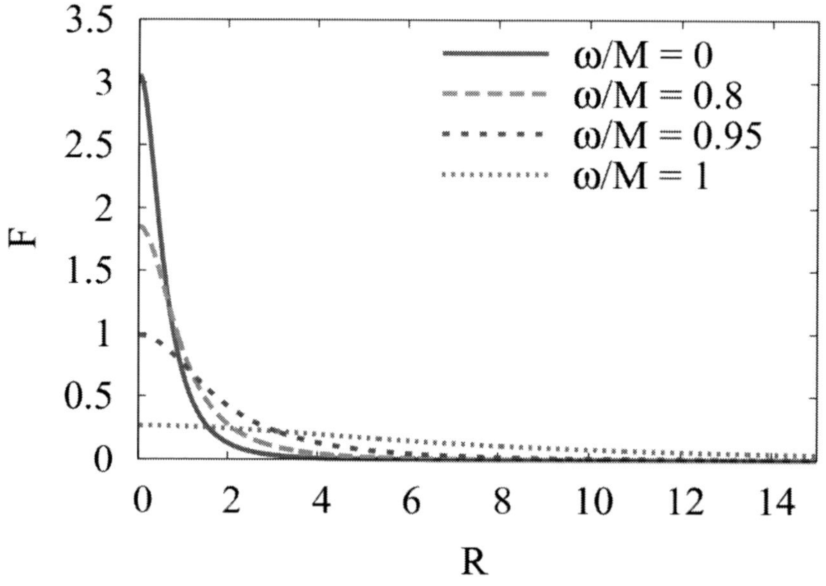

Figure 7. Profiles of the scalar field for different values of $\frac{\omega}{M}$. Here $\sqrt{\alpha_1} = 0.05$, $R = Mr$, $F(R) = \frac{\sqrt{\lambda}}{M} f(r)$. The figure is taken from [7].

4 Classical stability of U(1) gauged Q-balls: theoretical considerations

Now we turn to examining another difference between ordinary and U(1) gauged Q-balls — the classical stability, i.e., stability of the Q-ball solution with respect to small perturbations of the fields.

It is known that ordinary Q-balls are stable with respect to small perturbations if the following conditions hold:

1.
$$\frac{dQ}{d\omega} < 0 \tag{25}$$

2. The operator
$$\hat{L} = -\Delta + \left.\frac{dV}{d(\phi^*\phi)}\right|_{\phi^*\phi=f^2(r)} + 2\left.\frac{d^2V}{d(\phi^*\phi)^2}\right|_{\phi^*\phi=f^2(r)} f^2(r) - \omega^2, \tag{26}$$
where $\Delta = \partial_i\partial_i$, has *only one* negative eigenvalue.

The proof of the classical stability criterion based on the use of the energy functional of the system can be found in [8], the proof based on the use of the linearized equations of motion along the lines of the Vakhitov-Kolokolov method proposed in [9] can be found in [10]. It should be noted that we are not aware of any explicit example of ordinary Q-ball, for which the operator L has more than one negative eigenvalues and which is classically unstable for $\frac{dQ}{d\omega} < 0$.

In order to examine the gauged case, we consider the perturbations above the U(1) gauged Q-ball solution in the form [10]

$$\phi(t,\vec{x}) = e^{i\omega t}f(r) + e^{i\omega t}e^{\gamma t}\left(u(\vec{x}) + iv(\vec{x})\right), \tag{27}$$
$$A_0(t,\vec{x}) = A_0(r) + e^{\gamma t}a_0(\vec{x}), \tag{28}$$
$$A_i(t,\vec{x}) = e^{\gamma t}a_i(\vec{x}), \tag{29}$$

where γ is real, which passes through the linearized equations of motion. It is possible to show (see the detailed derivation in [10]) that U(1) gauged Q-balls are classically stable if the following conditions hold:

1.
$$\frac{dQ}{d\omega} < 0 \tag{30}$$

2. The corresponding operator $\hat{L}$ has *only one* negative eigenvalue.

Here the operator $\hat{L}$ takes the form

$$\hat{L} = \begin{pmatrix} -\Delta + W(r) & -2e(\omega + eA_0)f & 0 \\ -2e(\omega + eA_0)f & \frac{\Delta}{2} - e^2f^2 & 0 \\ 0 & 0 & \left(\frac{\Delta}{2} - e^2f^2 - \frac{\gamma^2}{2}\right)I_{3\times3} \end{pmatrix}, \tag{31}$$

where $I_{3\times3}$ is the 3×3 unit matrix and

$$W(r) = \left.\frac{dV}{d(\phi^*\phi)}\right|_{\phi^*\phi=f^2(r)} + 2\left.\frac{d^2V}{d(\phi^*\phi)^2}\right|_{\phi^*\phi=f^2(r)} f^2(r) - (\omega + eA_0)^2. \tag{32}$$

One can see that there always exist negative eigenvalues of the operator $\hat{L}$, corresponding to the perturbations a_i (because of the elements $\left(\frac{\Delta}{2} - e^2f^2 - \frac{\gamma^2}{2}\right)I_{3\times3}$). Thus, the corresponding negative part of the spectrum is defined only by the spectrum of the operator $\left(\frac{\Delta}{2} - e^2 f^2 - \frac{\gamma^2}{2}\right)$. Moreover, due to the fact that $f(r)$ rapidly tends to zero at $r \to \infty$, one expects that the operator $\left(\frac{\Delta}{2} - e^2 f^2 - \frac{\gamma^2}{2}\right)$ has a negative continuous spectrum starting at $-\frac{\gamma^2}{2}$. Thus, fulfillment of the condition $\frac{dQ}{d\omega} < 0$ does not provide any information about possible stability regions of the U(1) gauged Q-ball, at least within the framework of the generalized Vakhitov-Kolokolov approach used in [10]. At the moment it is not clear what is the origin of the problem described above. On the one hand, it is possible that the existence of the negative spectrum is just a technical artifact of the approach used in [10]. On the other hand, it is possible, due to the repulsive nature of the electromagnetic field, that nonspherically symmetric perturbations destroy any $U(1)$ gauged Q-ball. It is necessary to note that the question remains open as to whether it is possible to have only one negative eigenvalue of the operator $\hat{L}$ even in the case of spherically symmetric perturbations, in which the operator $\hat{L}$ takes a simpler form (see the detailed discussion in [10]).

5 Classical stability of U(1) gauged Q-balls: numerical simulations

In order to examine the classical stability regions of U(1) gauged Q-balls, the numerical simulations in the spherically symmetric case were performed (see the details in [10]). Here we present the results for two different scalar field potentials. The first one is the potential (7). We see from Figure 8

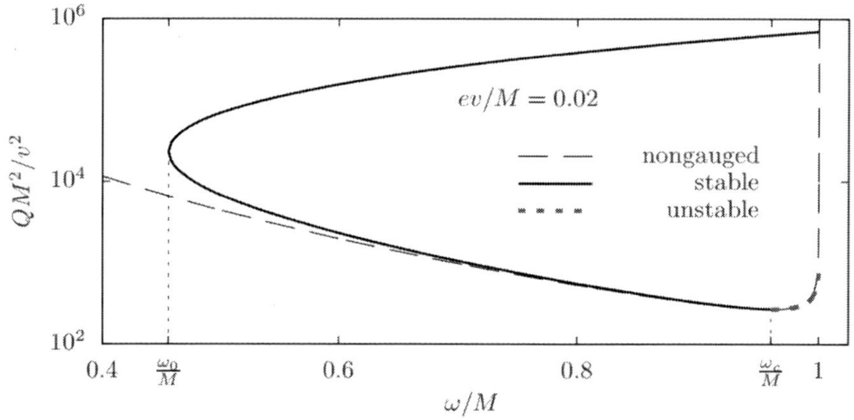

Figure 8. The stability and instability regions of Q-balls in the model with the scalar field potential (7). The long-dashed line stands for the nongauged Q-balls. The solid line stands for the classically stable gauged Q-balls. The short-dashed line stands for the classically unstable gauged Q-balls. The figure is taken from [10].

that the inequality $\frac{dQ}{d\omega} > 0$ holds for classically unstable solutions, which formally is in agreement with the classical stability criterion. Meanwhile, the inequality $\frac{dQ}{d\omega} > 0$ also holds for gauged Q-balls on the upper branch of the $Q(\omega)$ dependence, which appear to be stable at least with respect to spherically symmetric perturbations. Although formally this fact does not contradict the classical stability criterion (in the case of ordinary Q-balls the classical stability criterion does not guarantee that there exist unstable modes in the case $\frac{dQ}{d\omega} > 0$, it just states that there are no classical instabilities if $\frac{dQ}{d\omega} < 0$), this situation differs from the case of ordinary Q-balls, for which usually there exist spherically symmetric unstable modes for Q-balls with $\frac{dQ}{d\omega} > 0$.

The second scalar field potential has the form

$$V(\phi^*\phi) = -\mu^2\phi^*\phi\ln(\beta^2\phi^*\phi). \tag{33}$$

In the nongauged case this scalar field potential was proposed in [11]. An interesting feature of this model is that the linearized equations of motion for perturbations can be solved exactly in the nongauged case [12]. The $\frac{dQ}{d\omega} < 0$ stability criterion is valid for ordinary Q-balls in this model — Q-balls with $\frac{dQ}{d\omega} < 0$ are classically stable (and Q-balls with $\frac{dQ}{d\omega} > 0$ are classically unstable).

The results of the numerical simulation in the gauged case can be found in Figure 9. One can easily

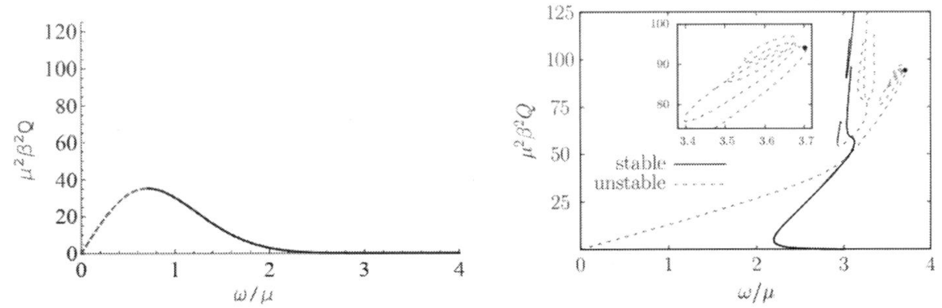

Figure 9. Ordinary (left plot) and U(1) gauged Q-balls for $e/\beta\mu = 1.1$ (right plot) for the logarithmic scalar field potential. The solid lines stand for the classically stable Q-balls. The short-dashed lines stand for the classically unstable Q-balls. The right plot is taken from [10].

see from Figure 9 that there are classically unstable U(1) gauged Q-balls with $\frac{dQ}{d\omega} < 0$ (for example, the Q-ball marked by the dot in Figure 9). This differs considerably from the case of ordinary Q-balls. An example of time evolution of the perturbed U(1) gauged Q-ball is shown in Figure 10, where the

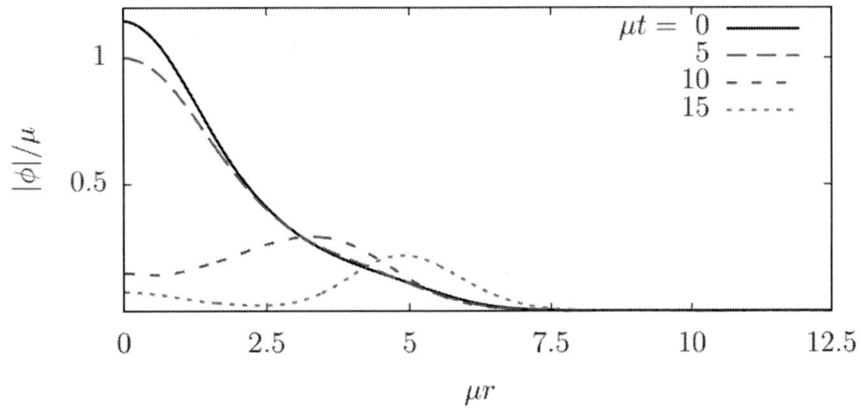

Figure 10. The scalar field profile of the classically unstable gauged Q-ball at different moments of time. The initial solution (at $\mu t = 0$) is marked by the dot in the upper right corner of Figure 9. The figure is taken from [10].

scalar field profile is presented at different moments of time.

6 Conclusion

In the present manuscript, we briefly discussed the differences between ordinary and U(1) gauged Q-balls. It is shown that U(1) gauged Q-balls may exist for the values of the parameters ($\omega = M$), for which ordinary Q-balls do not exist. It is also shown that the classical stability criterion for ordinary Q-balls cannot be applied to U(1) gauged Q-balls. An explicit example of classically unstable U(1) gauged Q-ball with $\frac{dQ}{d\omega} < 0$ is presented. These results demonstrate that addition of the gauge field changes drastically the properties of Q-balls.

The work was supported by the Grant 16-12-10494 of the Russian Science Foundation.

References

[1] G. Rosen, J. Math. Phys. **9**, 996 (1968)
[2] S. R. Coleman, Nucl. Phys. B **262**, 263 (1985) [Erratum-ibid. B **269**, 744 (1986)]
[3] I. E. Gulamov, E. Y. Nugaev and M. N. Smolyakov, Phys. Rev. D **87**, 085043 (2013)
[4] G. Rosen, J. Math. Phys. **9**, 999 (1968)
[5] K.-M. Lee, J. A. Stein-Schabes, R. Watkins and L. M. Widrow, Phys. Rev. D **39**, 1665 (1989)
[6] I. E. Gulamov, E. Y. Nugaev and M. N. Smolyakov, Phys. Rev. D **89**, 085006 (2014)
[7] I. E. Gulamov, E. Y. Nugaev, A. G. Panin and M. N. Smolyakov, Phys. Rev. D **92**, 045011 (2015)
[8] T. D. Lee and Y. Pang, Phys. Rept. **221**, 251 (1992)
[9] N. G. Vakhitov and A. A. Kolokolov, Radiophys. Quantum Electron. **16**, 783 (1973)
[10] A. G. Panin and M. N. Smolyakov, Phys. Rev. D **95**, 065006 (2017)
[11] G. Rosen, Phys. Rev. **183**, 1186 (1969)
[12] G.C. Marques and I.Ventura, Phys. Rev. D **14**, 1056 (1976)

EPJ Web of Conferences **158**, 07004 (2017) DOI: 10.1051/epjconf/201715807004
QFTHEP 2017

About renormalization of the Yang – Mills theory and the approach to calculation of the heat kernel

Aleksandr Ivanov[1,2,⋆]

[1] *Department of Mathematics and Mathematical Physics, Physics Faculty of St. Petersburg State University*
[2] *St. Petersburg Department of V.A. Steklov Institute of Mathematics of the Russian Academy of Sciences 27 Fontanka, St. Petersburg, Russia*

Abstract. The quantum theory of Yang – Mills in four-dimensional space – time plays an important role in modern theoretical physics. Currently, this model contains many open problems, therefore, it is of great interest to mathematicians. This work consists of several parts, however, it only offers a new approach and, therefore, it is methodological. First of all, the diagram technique and the mathematical basis will be recalled in the first part. Then the process of renormalization will be explained. It is based on momentum cut-off regularization and described in [1] and [2]. However, this type of the regularization has several problems, as a result, only the first correction is calculated. After common constructions and observations, the first correction will be described in detail. Namely, the heat kernel will be considered since it plays a main role in this formalism. In particular, the method for calculating of coefficients of arbitrary order will be proposed.

1 Introduction to Yang – Mills theory

Suppose G is a compact group of charges, $\mathfrak{G}$ is the Lie algebra. Assume that t^a are generators of the Lie algebra and f^{abc} are structural constants, such that the equalities

$$[t^a, t^b] = f^{abc} t^c$$

are fulfilled. Further, the Killing form $\mathrm{tr}[\cdot,\cdot]$ may be introduced on the Lie algebra with the normalization conditions

$$\mathrm{tr}[t^a, t^b] = \mathrm{C(N)}\delta^{ab},$$

where N is the number of generators. It is convenient to operate with space $\mathbb{R}^4$, where x^μ , $\mu = 1, \ldots, 4$ are notation of coordinates.
The Yang – Mills field is a 1-form and has view in local coordinates

$$A(x) = A^a_\mu(x) t^a dx^\mu. \qquad (1)$$

In turn, the 2-form

$$F = F^a_{\mu\nu} t^a dx^\mu \wedge dx^\nu \qquad (2)$$

⋆e-mail: regul1@mail.ru

plays a role of the curvature, where

$$F^a_{\mu\nu} = \partial_\mu A^a_\nu - \partial_\nu A^a_\mu + f^{abc} A^b_\mu A^c_\nu.$$

Based on the constructions (1) and (2), the classical action for the Yang – Mills theory has the form

$$S(A) = \frac{1}{\alpha} \int \mathrm{tr}[\, F \wedge F^*],$$

where α is the dimensionless coupling constant. Two differential operators can be introduced:

$$M_1 = \nabla_\sigma \nabla^\sigma \delta_{\mu\nu} + 2[F_{\mu\nu}, \cdot\,], \tag{3}$$

$$M_0 = \nabla_\mu \nabla^\mu, \tag{4}$$

where

$$D_\mu = \partial_\mu + [B_\mu, \cdot\,]. \tag{5}$$

Following the formalism of the background field after shifting

$$A_\mu(x) = B_\mu(x) + \frac{1}{2}\sqrt{\alpha} a_\mu(x),$$

the effective action can be presented in the form

$$W(B) = \frac{1}{\alpha} W_{-1}(B) + \sum_{k=0}^{\infty} \alpha^k W_k(B), \tag{6}$$

where the first and the second corrections are

$$W_{-1}(B) = \int \mathrm{tr}[F_{\mu\nu} F^{\mu\nu}] d^4x,$$

$$W_0(B) = -\frac{1}{2} \ln\det M_1 + \ln\det M_0, \tag{7}$$

and corrections W_n for $n > 0$ are related to the strongly connected vacuum corrections with $n+1$ loops.

2 Approach to renormalization

In the paper [2] a scenario was proposed. If momentum cut-off regularization Λ will be introduced then the expansion terms for the effective action (6) will have the form

$$W_0 = W_{0,0} + L W_{0,1}$$
$$W_i = \sum_{k=0}^{i} L^k W_{i,k}, \ i \geqslant 1$$

where $L = \log\frac{\Lambda}{\mu}$. In this case the problem of the renormalization can be formulated in the form of three conditions:

$$W_{reg}(\alpha(\Lambda), \Lambda, \mu) = \frac{1}{\alpha(\Lambda)} W_{-1} + L W_{0,1} + \sum_{n=0}^{\infty} \sum_{k=0}^{n} W_{n,k} \alpha^n(\Lambda) L^k,$$

$$\Lambda\frac{d\alpha(\Lambda)}{d\Lambda} = \beta(\alpha(\Lambda)),$$
$$\frac{dW_{reg}(\alpha(\Lambda),\Lambda,\mu)}{d\Lambda} = 0,$$

where

$$\beta(x) = \sum_{i=0}^{\infty} \beta_{i+1}x^{i+2}.$$

The first equality is connected with the structure of the effective action after regularization. The second condition is the Gell-Mann-Low equation for the coupling constant and the third one is condition of renormalizability. It is interesting that this problem is mathematically correct and after some calculations the answer has the form for the new coupling constant

$$\alpha_r(\mu) = e^{-L\beta(\alpha)\partial_\alpha}\alpha|_{\alpha=\alpha(\Lambda)},$$

and for the renormalized action

$$W_{ren}(\alpha_r(\mu),\mu) = e^{-L\beta(\alpha)\partial_\alpha}\left(\sum_{n=-1}^{\infty} W_{n,0}\alpha^n\right)\Bigg|_{\alpha=\alpha(\Lambda)}.$$

The decision also contains an equality for the coefficients $W_{m,n}$. Namely, the coefficients $W_{m,n}$, $m \geqslant 1$, $n = 1\ldots m$ can be expressed through the coefficients W_{-1} and $W_{i,0}$ and the coefficients of the β-function.

3 Heat kernel

The scenario which is described in chapter 2 has some problems. For example, only the first correction is calculated because momentum cut-off regularization needs improvement. But now it is possible to discuss the logarithm of the determinant for (3) and (4).

Assume that A_0 is the Laplace operator and A_1 is the covariant Laplace operator, where the covariant derivative is equal to (5). Using the method of proper time [3] it is affordable to write

$$\det A_1/A_0 = -\int_0^{\infty}\frac{d\tau}{\tau}\mathrm{tr}\left(e^{-A_1\tau} - e^{-A_0\tau}\right),$$

where $e^{-A_1\tau}$ and $e^{-A_0\tau}$ are operators with kernels $K_1(x,y;\tau)$ and $K_0(x,y;\tau)$ respectively, which satisfy the problem

$$\begin{cases}\left(\frac{\partial}{\partial\tau} + A_j\right)K_j(x,y;\tau) = 0\\ K_j(x,y;0) = \delta(x-y)\end{cases} \tag{8}$$

for $j = 0, 1$. Usually, the ansatz for $K_1(x,y;\tau)$ has the form

$$K_1(x,y;\tau) = K_0(x,y;\tau)\sum_{n=0}^{\infty}\tau^n a_n(x,y), \tag{9}$$

and after substitution (9) in (8), the relations for coefficients $a_n(x,y)$ have the forms

$$(x-y)^\mu D_\mu a_0(x,y) = 0,$$
$$\Big((n+1) + (x-y)^\mu D_\mu\Big) a_{n+1}(x,y) = A_1 a_n(x,y).$$

Usually it is very difficult to calculate all coefficients and find the formula for the heat kernel (9). However, in some special cases it is possible. In some models it is possible to consider the case of covariant constant field (for example in [4]):

$$D_\mu F_{\nu\rho}(x) = 0\,,\ \forall \mu,\nu,\rho, x\in\mathbb{R}^4.$$

But in the case of the Fock-Schwinger calibration

$$x_\mu B_\mu(x) = 0\,,\ \forall x\in\mathbb{R}^4$$

the answer for the gauge field has the form

$$B_\mu(x) = \frac{1}{2} x_\rho F_{\rho\mu}(x)\,,\ \forall x\in\mathbb{R}^4.$$

After the calculations it is not difficult to obtain the answer for the heat kernel

$$K_1(x,y;\tau) = \frac{1}{(4\pi\tau)^{d/2}} \exp\left\{\frac{1}{2} x_\mu F_{\mu\nu} y_\nu - \frac{1}{4\tau}(x-y)_\mu F_{\mu\sigma}\tau(\coth F\tau)_{\sigma\nu}(\tau)(x-y)_\nu + L(\tau)\right\},$$

where

$$L(\tau) = -\frac{1}{2}\ln(F^{-1}_{\mu\nu}\tau^{-1}(\sinh F\tau)_{\mu\nu}).$$

4 General case of heat kernel

4.1 Path-ordered exponential

The main purpose of this chapter is to refuse the constraints for the background fields and to learn how to obtain the coefficients of the heat kernel. In this case it is necessary to remember the definition for the path-ordered exponential

$$\Phi(x,y) := 1 + \sum_{n=1}^{\infty}(-1)^n \int_0^1\int_0^1\ldots\int_0^1 ds_1\ldots ds_n \frac{dz_1^{\nu_1}}{ds_1}\ldots\frac{dz_n^{\nu_n}}{ds_n} B_{\nu_1}(z_1)\ldots B_{\nu_n}(z_n), \tag{10}$$

where parametrization is by line. It is essential object because it satisfies the differential equation with the boundary condition

$$(x-y)^\mu D_\mu \Phi(x,y) = 0\,,\ \Phi(x,x) = 1\,.$$

But how to use this object? For this purpose, let's look at the formula

$$D_\mu\Phi(x,y) = \int_0^1 ds s\frac{dz_\nu}{ds}\Phi(x,z)F_{\nu\mu}(z)\Phi(z,y)\,, \tag{11}$$

which was proved in the article [5].

 DOI: 10.1051/epjconf/201715807004

4.2 Main idea

Suppose that our goal is to calculate the coefficient $a_n(x, y)$ with the arguments $y = x$. To do this, we must apply the composition of the integration operator and the covariant Laplace operator $-D_\mu D_\mu$ to the path-ordered exponential $\Phi(x, y)$ and then take the point $y = x$.

It's very interesting to remember the formula (11) because after taking $y = x$:

- $\Phi(x, y) \xrightarrow{y \longrightarrow x} 1$,
- the formula contains only the tension $F(x)$ and its derivatives ,
- the formula contains the numerical coefficients in the form $\int_0^1 ds s^p$.

The first aim is to construct the diagram technique, one element of which is the path-ordered exponential

$$\Phi(x, y) = x \;\text{———}\; y$$

and when $y \longrightarrow x$ then a line goes to a dot and one element is related to the tension and its derivatives.

The second aim is to control the numerical coefficients and the degree of parameterization parameters. In this case, it is possible to calculate the coefficients and to do mathematical estimates of the convergence of the series.

5 Acknowledgments

This research is supported by a grant from the Russian Science Foundation (Project No. 14-11-00598).

References

[1] L. D. Faddeev, Int. J. Mod. Phys. A **31**, 1630001 (2016)
[2] S. E. Derkachev, A. V. Ivanov, L. D. Faddeev, Theoret. and Math. Phys. **192**:2, 1134 (2017)
[3] V. Fock, Phys. Z. Sowjetunion **12**, 404 (1937)
[4] J. P. Bornsen, A. E. M. van de Ven, Nucl. Phys. B **657**, 257 (2003)
[5] Graham M. Shore, Annals of physics **137**, 262 (1981)

EPJ Web of Conferences **158**, 07005 (2017) DOI: 10.1051/epjconf/201715807005
QFTHEP 2017

Symanzik approach in modeling of bound states of Dirac particle in singular background

Yu. M. Pismak[1] and *D. Yu.* Shukhobodskaia[2]

[1] *Saint-Petersburg State University, Ulyanovskaya street 1, Petrodvorets, 198504 St-Petersburg, Russia*
[2] *The University of Sheffield, Hicks Building, Hounsfield Road, Sheffield, S3 7RH, UK*

Abstract. In the model of interaction of spinor field with homogeneous isotropic material plane constructed in framework of Symanzik approach, the bound states are studied. For localized near plane Dirac particle the expression for current, charge and density are presented. For bound state with massless dispersion law the current, charge and density are calculated for simplified model with 2 parameter exactly.The model can find application to a wide class of phenomena arising by the interaction of fields of quantum electrodynamics with two-dimensional materials.

Investigation of macroscopic quantum phenomena is an important direction in experimental and theoretical physics which has attracted recently the attention of many researchers. In this area, the main problem of modelling employing the principles of quantum field theory is that the space-time homogeneity and isotropy, which is typical for usual quantum field theory models of elementary particles [1], must be broken. The symmetry properties of the vacuum may change essentially as a result of interactions of quantum fields with macroscopic objects. Therefore quantum macro effects may appear in dynamics of material bodies which cannot be explained both in classical physics and in framework of non relativistic quantum mechanics.

Theoretically, this problem was first considered in 1948 by Casimir [2]. He showed that the fluctuations of the quantum vacuum generate attraction between ideally conducting plates of a planar uncharged capacitor. This phenomenon, called the Casimir effect (CE), is observed experimentally [3–6], and the empirical results obtained for materials with high conductivity, with a high degree of accuracy, are in agreement with theoretical ones.

At typical for CE distances of 10 to 1000 nm both classical and quantum properties of the system turn out to be essential. Their combination forms a special nanophysics. Investigation of it is not solely of theoretical interest, understanding its particular feature is also important for developing new technological devices, in view of the increasing trend to their miniaturization.

Although theoretical investigations of the CE are the subject of numerous works [6, 7], they are often based on simplified models ignoring usually specificity of quantum electrodynamics. Such models are not suitable for comprehensively describing a wide range of nanophysical phenomena occurring in a system as a result of the interaction of its quantum degrees of freedom with material bodies of a given shape (classic defect).

In our work we used the the Symanzik's approach [8], modelling the interaction of quantized fields with a spatial inhomogeneity with the aid of an additional action functional (defect action), concentrated in the spatial domain where this inhomogeneity -a macroscopic object -is located. It is important also that the standard requirements for quantum field models (locality, renormalisability, symmetry properties) are assumed to be fulfilled.

In this work we use Symanzik's approach for modification of the standard QED in (3+1)-dimensional space-time for description of interaction of a two-dimensional material object with a Dirac field. Such a problem for a spinor field was considered in [15]. In accordance with the basic principles of QED, the defect action functional is the sum of the photon Chern-Simons action and a purely Dirac field defect action [17]. It is possible in this model framework, to investigate both scattering of spinor particles on a material plane and also the properties bound states localised in its vicinity. For simple model with 4 parameter in the Dirac defect action it was studied in [17]. The most general model with 8 parameters, describing interaction of Dirac field with homogeneous isotropic material plane was considered in [18]. For this model the dispersion relation for bound state was obtained, and it was shown that by a special choosing parameters the Dirac particle being massive in volume is massless in the bound state moving along the material plane with arbitrary Fermi-velocity less then velocity of light. In this work we study in more details the properties of such bound state and present results of estimation of theirs characteristics which was calculated exactly for a simplified model with 2 independent parameters.

1 Formulation of model

Within the Symanzik approach, the general form of the action functional modelling the interaction of the quantum field with material object (defect) can be written as:

$$S(\varphi) = S_V(\varphi) + S_{def}(\varphi).$$

Here, S_V is the action of the basic quantum field model, and S_{def} is the defect action:

$$S_V(\varphi) = \int L(\varphi(x))d^D x, \; S_{def}(\varphi) = \int_\Gamma L_{def}(\varphi(x))d^{D'} x,$$

where Γ is a subspace of dimension $D' < D$ in a D-dimensional space [8]. In the quantum electrodynamic (QED) the action $S(\varphi)$ is the functional of photon vector field $A_\mu(x)$, Dirac spinor fields $\bar{\psi}(x), \psi(x)$, and

$$S_V(\varphi) = S(\bar{\psi}, \psi, A) = -\frac{1}{4}F_{\mu\nu}F^{\mu\nu} + \bar{\psi}(i\hat{\partial} - m + ie\hat{A})\psi,$$

where $F_{\mu\nu} = \partial_\mu A_\nu - \partial_\nu A_\mu$. The gauge invariance, locality, and renormalisability as the basic principles of QED impose strong constraints on the possible form of the defect action $S_{def}(\varphi)$. Taking them into account by modelling the interaction of the QED fields with a two-dimensional surface without charges and currents, one obtains the following result: $S_{def}(\varphi) = S_{def}(A) + S_{def}(\bar{\psi}, \psi)$. If the surface is defined by equation $\Phi(x) = 0$, $x = (x^0, x^1, x^2, x^3)$, then $S_{def}(A)$ is the Chern-Simons action

$$S_{def}(A) = \frac{a}{2}\int \varepsilon^{\lambda\mu\nu\rho}\partial_\lambda\Phi(x)A_\mu(x)F_{\nu\rho}(x)\delta(\Phi(x))\, d^4x,$$

where $\varepsilon^{\lambda\mu\nu\rho}$ is the totally antisymmetric tensor ($\varepsilon^{0123} = 1$), and the parameter a is a dimensionless coupling constant. This expression is the most general form of a gauge-invariant functional of A_μ concentrated on a defect surface being invariant in respect to reparameterisation of one and not having any negative-dimension parameters. The the most general form of the defect action $S_{def}(\bar{\psi}, \psi)$ is written as

$$S_{def}(\bar{\psi}, \psi) = \sum_{j=1}^{16} \int \alpha_j \bar{\psi}(x)\Gamma_j\psi(x)\delta(\Phi(x))\, d^4x,$$

where Γ_j are the 16 basis Dirac matrices and α_j are dimensionless coupling constants.

We consider the plane $x^3 = 0$ as the material defect. In this case

$$S(\overline{\psi}, \psi) = \int \overline{\psi}(x)(i\hat{\partial} - m + \Omega(x^3))\psi(x)dx. \tag{1}$$

The matrix $\Omega(x_3) = Q\delta(x^3)$ describes the coupling of Dirac fields with the plane. Because $\Omega(x^3)$ and $\delta(x^3)$ have the dimension of mass, the matrix Q is dimensionless. We suppose that the defect plane is isotropic and homogeneous. It means that $S(\overline{\psi}, \psi)$ must be be invariant under translation in x^0-, x^1-, x^2-directions and rotation around x^3-axe. The most general form of the matrix Q fulfilling this condition is

$$Q = r_1\mathbf{1} + ir_2\gamma_5 + r_3\gamma_3 + r_4\gamma_5\gamma_3 + r_5\gamma_0 + r_6\gamma_5\gamma_0 + ir_7\gamma_0\gamma_3 + ir_8\gamma_1\gamma_2, \tag{2}$$

where $\mathbf{1}$ is the unit 4×4 matrix, $i = \sqrt{-1}$, γ^j, $j = 0, 1, 2, 3$ are the Dirac matrices and $\gamma^5 = i\gamma^0\gamma^1\gamma^2\gamma^3$. For the γ matrices we use the following representation

$$\gamma^0 = \begin{pmatrix} 1 & 0 & 0 & 0 \\ 0 & 1 & 0 & 0 \\ 0 & 0 & -1 & 0 \\ 0 & 0 & 0 & -1 \end{pmatrix}, \quad \gamma^1 = \begin{pmatrix} 0 & 0 & 0 & 1 \\ 0 & 0 & 1 & 0 \\ 0 & -1 & 0 & 0 \\ -1 & 0 & 0 & 0 \end{pmatrix},$$

$$\gamma^2 = \begin{pmatrix} 0 & 0 & 0 & -i \\ 0 & 0 & i & 0 \\ 0 & i & 0 & 0 \\ -i & 0 & 0 & 0 \end{pmatrix}, \quad \gamma^3 = \begin{pmatrix} 0 & 0 & 1 & 0 \\ 0 & 0 & 0 & -1 \\ -1 & 0 & 0 & 0 \\ 0 & 1 & 0 & 0 \end{pmatrix}.$$

Taking the functional derivatives of the action (1) with respect to $\overline{\psi}(x)$ and $\psi(x)$ we obtain the Euler-Lagrange equations:

$$(i\hat{\partial} - m + \Omega(x^3))\psi(x) = 0, \; (\partial_\mu\overline{\psi}(x))\gamma^\mu + \overline{\psi}(x)(m - \Omega(x^3)) = 0. \tag{3}$$

It can be easily verified that $\overline{\psi}(x) = \psi^*(x)\gamma_0$, if $\gamma_0\Omega^+(x^3) = \Omega(x^3)\gamma_0$. This condition is fulfilled if parameters r_j, $1 \le j \le 8$ are real. In what follows we assume that this is the case.

Let us introduce the following notations

$$\overline{x} = (x^0, x^1, x^2), \; \overline{p} = (p^0, p^1, p^2), \; \overline{px} = p^0x^0 - p^1x^1 - p^2x^2,$$

for parts of 4-component vectors, and represent $\psi = \psi(\bar{x}, x^3)$ as $\psi(x) = \psi_s(x) + \psi_a(x)$, where

$$\psi_s(x) = \frac{1}{2}(\psi(\bar{x}, x^3) + \psi(\bar{x}, -x^3)), \; \psi_a(x) = \frac{1}{2}(\psi(\bar{x}, x^3) - \psi(\bar{x}, -x^3)).$$

Analysis of the possible singularities of spinor $\psi(x)$ at the plane $x^3 = 0$, made in [17], yields: for $x^3 \neq 0$, the field $\psi(x)$ satisfies the Dirac equation

$$(i\hat{\partial} - m)\psi(x) = 0, \tag{4}$$

and the equation

$$i\gamma^3\psi_a(\bar{x}) + \frac{Q}{2}\psi_s(\bar{x}) = 0 \tag{5}$$

is fulfilled at $x^3 = 0$, within $\psi_a(\bar{x}) = \lim_{x^3\to0}\psi_a(x)$ and $\psi_s(\bar{x}) = \lim_{x^3\to0}\psi_s(x)$.

It is shown in [17] that the general solution of (4) by $x^3 \neq 0$ can be represented as

$$\psi(x) = \frac{1}{(2\pi)^3}\int e^{i\overline{px}}\psi_+(\bar{p}, x^3)d\bar{p}, \ \text{for } x^3 > 0, \ \psi(x) = \frac{1}{(2\pi)^3}\int e^{i\overline{px}}\psi_-(\bar{p}, x^3)d\bar{p}, \ \text{for } x^3 < 0$$

where

$$\psi_\pm(\bar{p}, x^3) = U(\bar{p}, x^3)\chi_\pm(\bar{p}), \ U(\bar{p}, x^3) = e^{i\gamma^3(\hat{\bar{p}}+m)x^3} = e^{i\kappa(\bar{p})x^3}P^+(\bar{p}) + e^{-i\kappa(\bar{p})x^3}P^-(\bar{p}).$$

Here, $\chi_\pm(\bar{p})$ are arbitrary spinors depending only on $\bar{p}$, and we used the notations

$$\kappa(\bar{p}) = \sqrt{\bar{p}^2 - m^2}, \ \hat{\bar{p}} = \overline{\gamma p} = \gamma^0 p^0 - \gamma^1 p^1 - \gamma^2 p^2, \ P^\pm(\bar{p}) = \frac{1}{2}\left(\mathbf{1} \pm \frac{\gamma^3(\hat{\bar{p}} + m)}{\kappa(\bar{p})}\right).$$

It follows from (5) that $\chi_+(\bar{p})$, $\chi_-(\bar{p})$ satisfy the equation

$$i\gamma^3(\chi_+ - \chi_-) + \frac{Q}{2}(\chi_+ + \chi_-) = 0, \tag{6}$$

which implies that

$$\chi_- = S\chi_+, \tag{7}$$

where

$$S = \left(1 - \frac{i\gamma^3 Q}{2}\right)^{-1}\left(1 + \frac{i\gamma^3 Q}{2}\right). \tag{8}$$

The matrices $P^\pm(\bar{p})$ are projectors:

$$P^\pm(\bar{p})P^\pm(\bar{p}) = P^\pm(\bar{p}), \ P^+(\bar{p}) + P^-(\bar{p}) = 1, \ P^+(\bar{p})P^-(\bar{p}) = P^-(\bar{p})P^+(\bar{p}) = 0.$$

Each of them has two-dimensional eigensubspaces L^+ and L^-, which are represented as linear combinations of two mutually orthogonal eigenvectors $v_1^\pm$ and $v_2^\pm$:

$$L^\pm(a_1^\pm, a_2^\pm) = a_1^\pm v_1^\pm + a_2^\pm v_2^\pm,$$

$$P^\pm(\bar{p})L^\pm(a_1^\pm, a_2^\pm) = L^\pm(a_1^\pm, a_2^\pm), \ P^\pm(\bar{p})L^\mp(a_1^\mp, a_2^\mp) = 0.$$

Here, $a_i^\pm$, $i = 1, 2$, are complex parameters, and $v_i^\pm$, $i = 1, 2$, are spinors, which we choose as

$$v_1^+ = n_1^+(-h_1, -h_1h_2^-, 1, h_2^-), \ v_2^+ = n_2^+(h_1, h_1h_2^+, 1, h_2^+),$$

$$v_1^- = n_1^-(-h_1, h_1h_2^+, 1, -h_2^+), \ v_2^- = n_2^-(h_1, -h_1h_2^-, 1, -h_2^-)$$

$$h_1 = \frac{\omega}{m + p_0}, \ h_2^\pm = \frac{\kappa(\bar{p}) \pm \omega}{p_1 - ip_2}, \ \omega = \sqrt{p_0^2 - m^2},$$

with normalisation constants $n_1^\pm, n_2^\pm$ defined by conditions $\bar{v}_1^\pm v_1^\pm = \bar{v}_2^\pm v_2^\pm = 1$. These spinors are eigenstates of the helicity operator

$$\sigma(\vec{p}) = \frac{i}{2|\vec{p}|}(\vec{p}\,\vec{s}), \ \vec{s} = (\gamma_2\gamma_3, -\gamma_1\gamma_3, \gamma_1\gamma_2),$$

and satisfy the relations

$$\sigma(\vec{p})|_{p_3=\mp\kappa(\bar{p})}v_1^\pm = -\frac{1}{2}v_1^\pm, \ \sigma(\vec{p})|_{p_3=\mp\kappa(\bar{p})}v_2^\pm = \frac{1}{2}v_2^\pm$$

We represent the spinors $\chi_\pm(\overline{p})$ as

$$\chi_\pm(\overline{p}) = b_\pm^1(\overline{p})v_1^-(\overline{p}) + b_\pm^2(\overline{p})v_2^-(\overline{p}) + c_\pm^1(\overline{p})v_1^+(\overline{p}) + c_\pm^2(\overline{p})v_2^+(\overline{p}). \tag{9}$$

If $\kappa > 0$ the fields $\bar{\psi}, \psi$ describe the scattering processes of Dirac particles on the defect plane. For the model with $r_5 = r_6 = r_7 = r_8 = 0$ they are studied in [17]. In this paper we consider the properties of bound states localized in vicinity of $x_3 = 0$ without this restriction.

2 Bound states

The state is localised nearly of the plane $x_3 = 0$ if $\bar{p}^2 - m^2 < 0$ and $\kappa(\bar{p}) = i|\kappa(\bar{p})|$ is imaginary. In this case the spinors $\chi_\pm(\bar{p})$ (9) must fulfil the conditions $P^-(\bar{p})\chi_+ = 0$, $P^+(\bar{p})\chi_- = 0$. Other case the asymptotic of the fields $\bar{\psi}(x), \psi(x)$ cold be infinite by $x_3 \to \pm\infty$. Thus, we can present χ_+, χ_- as follows

$$\chi_+(\bar{p}) = c_+^1(\bar{p})\, v_1^+(\bar{p}) + c_+^2(\bar{p})\, v_2^+(\bar{p}), \quad \chi_-(\bar{p}) = c_-^1(\bar{p})\, v_1^-(\bar{p}) + c_-^2(\bar{p})\, v_2^-(\bar{p}). \tag{10}$$

Substituting in (8) the spinors $\chi_+(\bar{p}), \chi_-(\bar{p})$ presented in (10), we obtain the equation

$$c_-^1(\bar{p})\, v_1^-(\bar{p}) + c_-^2(\bar{p})\, v_2^-(\bar{p}) = S\,(c_+^1(\bar{p})\, v_1^+(\bar{p}) + c_+^2(\bar{p})\, v_2^+(\bar{p})). \tag{11}$$

For any c_+^1, c_+^2, we can present the spinor $d = S\chi_+(\bar{p})$ as $d = d^1 v_1^+ + d^2 v_2^+ + d^3 v_1^- + d^4 v_2^-$ and express the coefficients d^1, d^2, d^3, d^4 in terms of c_+^1, c_+^2 in the following way

$$d^i = \sum_{j=1}^{2} M_j^i(\bar{p}) c_+^j, \quad 1 \le j \le 4. \tag{12}$$

Thus, the coefficients c_+^1, c_+^2 satisfying the equation (12) must fulfil the homogeneous equations $d^1 = 0$, $d^2 = 0$, which have nontrivial solution if

$$M_1^1(\bar{p}) M_2^2(\bar{p}) - M_2^1(\bar{p}) M_1^2(\bar{p}) = 0. \tag{13}$$

It means that the determinants of the 2×2 matrix M with elements $\{M\}_{ij} = M_j^i$, $i, j = 1, 2$ must be equal to zero. In this case, one of the coefficients c_+^1, c_+^2 remains to be arbitrary, and the other one it is expressed in its term. If c_+^1 is considered as an arbitrary amplitude, parameterising the solution of equations (12), then

$$c_+^2 = L(\bar{p})\, c_+^1, \quad L(\bar{p}) = -\frac{M_1^2(\bar{p})}{M_2^2(\bar{p})} = -\frac{M_1^1(\bar{p})}{M_2^1(\bar{p})}. \tag{14}$$

The amplitudes c_-^1, c_-^1 expressed in terms of c_+^1 are obtained by the substitution of c_+^1 and $c_+^2 = L(\bar{p})\, c_+^1$ in the right hand sides of equations (12) for d^3, d^4 :

$$c_-^1 = (M_1^3(\bar{p}) + M_2^3 L(\bar{p}))\, c_+^1, \quad c_-^2 = (M_1^4(\bar{p}) + M_2^4 L(\bar{p}))\, c_+^1. \tag{15}$$

The solvability condition describes the correlation between components of the moment $\bar{p} = (p^0, p^1, p^2) = (p^0, \vec{p})$ of the bound state. The relation of such a kinds are called the dispersion laws. For the considered model the equation (13) reads

$$(R_1\lambda - 2(m r_{18}^- - p^0 r_{45}^+))(R_2\lambda - 2(m r_{18}^+ + p^0 r_{45}^-)) - R_3 \vec{p}^2 = 0. \tag{16}$$

where we used the notations $r_{jk}^\pm = (r_j \pm r_k)/2$, $\vec{p}^2 = (p^1)^2 + (p^2)^2$ $\lambda = \sqrt{m^2 + \vec{p}^2 - p^{0\,2}} = -i\kappa$, and

$$R_1 = 1 + {r_{18}^-}^2 + {r_{27}^-}^2 + {r_{36}^+}^2 - {r_{45}^+}^2, \; R_2 = 1 + {r_{18}^+}^2 + {r_{27}^+}^2 + {r_{36}^-}^2 - {r_{45}^-}^2, \; R_3 = r_6^2 + r_7^2 + r_8^2 - r_4^2.$$

For bound states, $\lambda = |\kappa| > 0$ and can be considered as an index of its localisation near the plane $x_3 = 0$. Simple essential characteristics of properties of bound states could be also the scalar density $D(\bar{p}, x^3) = \exp\{-|x^3|\lambda\}\, d(\bar{p})$ and the 4-component current $J(\bar{p}, x^3) = \exp\{-|x^3|\lambda\}\, J(\bar{p})$, $J(\bar{p}) = (j^0(\bar{p}), j^1(\bar{p}), j^2(\bar{p}), j^3(\bar{p}))$ with

$$d_\pm(\bar{p}) = \bar{\chi}_\pm(\bar{p})\chi_\pm(\bar{p}), \quad j_\pm^\alpha(\bar{p}) = \bar{\chi}_\pm(\bar{p})\gamma^\alpha\chi_\pm(\bar{p}), \quad \alpha = 0, 1, 2, 3.$$

We denoted $d_+(\bar{p})$, $J_+(\bar{p})$ the function $d(\bar{p})$ and the vector $J(\bar{p})$ in subspace $x^3 > 0$, and $d_-(\bar{p})$, $J_-(\bar{p})$ for $x^3 < 0$. They are expressed in terms of amplitudes $c^1_\pm, c^2_\pm$ in the following form:

for $p_0 > m$

$$d_\pm = -\frac{4}{p^0+m}\left\{mC^+_\pm + p^0\lambda\left(\frac{c^1_\pm c^{2*}_\pm}{\lambda \mp i\omega} + \frac{c^2_\pm c^{1*}_\pm}{\lambda \pm i\omega}\right)\right\}, \tag{17}$$

$$j^0_\pm = \frac{4}{p^0+m}\left\{p^0C^+_\pm + m\lambda\left(\frac{c^1_\pm c^{2*}_\pm}{\lambda \mp i\omega} + \frac{c^2_\pm c^{1*}_\pm}{\lambda \pm i\omega}\right)\right\}, \tag{18}$$

$$j^1_\pm = \frac{4\omega\,(C^+_\pm\,\omega\, p^1 \pm C^-_\pm\,\lambda\, p^2)}{(p_0+m)(\omega^2+\lambda^2)},\ j^2_\pm = \frac{4\omega\,(C^+_\pm\,\omega\, p^2 \mp C^-_\pm\,\lambda\, p^1)}{(p_0+m)(\omega^2+\lambda^2)},\ j^3_\pm = 0, \tag{19}$$

$$C^+_\pm = |c^1_\pm|^2 + |c^2_\pm|^2,\ C^-_\pm = |c^1_\pm|^2 - |c^2_\pm|^2; \tag{20}$$

for $p_0 < m$

$$d_\pm = -\frac{4}{p^0+m}\left\{m\tilde{C}^+_\pm + p^0\lambda\left(\frac{c^1_\pm c^{1*}_\pm}{\lambda \mp i\omega} + \frac{c^2_\pm c^{2*}_\pm}{\lambda \pm i\omega}\right)\right\}, \tag{21}$$

$$j^0_\pm = \frac{4}{p^0+m}\left\{p^0\tilde{C}^+_\pm + m\lambda\left(\frac{c^1_\pm c^{1*}_\pm}{\lambda \mp i\omega} + \frac{c^2_\pm c^{2*}_\pm}{\lambda \pm i\omega}\right)\right\}, \tag{22}$$

$$j^1_\pm = \frac{4\omega\,(\tilde{C}^+_\pm\,\omega\, p^1 \pm \tilde{C}^-_\pm\,\lambda\, p^2)}{(p_0+m)(\omega^2+\lambda^2)},\ j^2_\pm = \frac{4\omega\,(\tilde{C}^+_\pm\,\omega\, p^2 \mp \tilde{C}^-_\pm\,\lambda\, p^1)}{(p_0+m)(\omega^2+\lambda^2)},\ j^3_\pm = 0, \tag{23}$$

$$\tilde{C}^+_\pm = c^1_\pm c^{2*}_\pm + c^2_\pm c^{1*}_\pm,\ \tilde{C}^-_\pm = c^1_\pm c^{2*}_\pm - c^2_\pm c^{1*}_\pm. \tag{24}$$

We see that the currents flow along the plane $x_3 = 0$ and are parallel to the moment $\vec{p} = (p_1, p_2)$ if $C^-_\pm = \tilde{C}^-_\pm = 0$. The currents above and below of the plane $x_3 = 0$ are parallel if $C^-_-/C^+_- = C^-_+/C^+_+$ by $p0 > m$, and $\tilde{C}^-_-/\tilde{C}^+_- = \tilde{C}^-_+/\tilde{C}^+_+$ by $p0 < m$. From (17-24) it follows that

$$m\,d_\pm + p^0 j^0_\pm = p^1 j^1_\pm + p^2 j^2_\pm = 4(p^0-m)C^+_\pm,\ p^2 j^1_\pm - p^1 j^2_\pm = 4\lambda\sqrt{\frac{p^0-m}{p^0+m}}C^-_\pm \text{ for } p^0 > m, \tag{25}$$

$$m\,d_\pm + p^0 j^0_\pm = p^1 j^1_\pm + p^2 j^2_\pm = 4(p^0-m)\tilde{C}^+_\pm,\ p^2 j^1_\pm - p^1 j^2_\pm = 4\lambda\sqrt{\frac{p^0-m}{p^0+m}}\tilde{C}^-_\pm \text{ for } p^0 < m, \tag{26}$$

and

$$m\,d_\pm + p^0 j^0_\pm - p^1 j^1_\pm - p^2 j^2_\pm = 0. \tag{27}$$

If $R_1 = R_2 = r^+_{18} = r^-_{18} = 0$ by choosing the special parameters in the model , then the dispersion relation (16) has the form

$$p_0^2 - v_F^2\vec{p}^{\,2} = 0 \tag{28}$$

where $\vec{p}$ denotes the two-component vector $\vec{p} = (p^1, p^2)$, $\vec{p}^{\,2} = (p^1)^2 + (p^2)^2$ and v_F is a constant, which is expressed in the term of parameters r_k, $1 \le k \le 8$ of the model. The dispersion relation (28) describe moving of massless particle with Fermi-velocity v_F. The motion of such particles explains numerous effects in graphene.

The constants $R_1, R_2, r^+_{18}, r^-_{18}$ are equal to zero by $r_1 = r_8 = r_3 = 0$ and $r^+_{45} = \pm\sqrt{1 + {r^-_{27}}^2 + r_6^2/4}$, $r^-_{45} = \pm\sqrt{1 + {r^+_{27}}^2 + r_6^2/4}$. For such choosing of parameters, the Fermi-velocity

v_F can obtains two different values: $v_F = v_F^+$ or $v_F = v_F^-$,

$$v_F^{\pm} = \sqrt{\frac{1}{2}\left(1 \pm \frac{4 + 4r_{27}^+ r_{27}^- - r_6^2}{\sqrt{4 + 4{r_{27}^-}^2 + r_6^2}\sqrt{4 + 4{r_{27}^+}^2 + r_6^2}}\right)}.$$

The Fermi-velocity v_F^+ (v_F^-) corresponds to the positive (negative)value of $r_{45}^+ r_{45}^-$. By changing the values of parameters r_2, r_6, r_7 one can obtain each value of the Fermi-velocity in the interval (0,1). The current flowing along the plane $x_3 = 0$ is parallel to the moment $\vec{p}$ only by $r_6 = 0$.

3 Simple example

To understand better the properties of the bound state with massless dispersion law (28), let us consider a simplified model choosing the parameters r_j in the following way: $r_1 = r_2 = r_3 = r_5 = r_8 = 0$, $r_4 = \sqrt{4 + r_6^2 + r_7^2}$. In this case the dispersion relation has the form (28) with Fermi velocity

$$v_F = \frac{2}{\sqrt{4 + r_6^2 + r_7^2}}.$$

The matrix $i/2\gamma^3 Q$ and S are of the form

$$i/2\gamma^3 Q = \frac{1}{2}\begin{pmatrix} ir_6 - r_7 & 0 & i\frac{2i}{v_F} & 0 \\ 0 & -ir_6 - r_7 & 0 & \frac{2i}{v_F} \\ \frac{2i}{v_F} & 0 & ir_6 - r_7 & 0 \\ 0 & \frac{2i}{v_F} & 0 & -ir_6 - r_7 \end{pmatrix},$$

$$S = \begin{pmatrix} \frac{-ir_7}{r_6+2i} & 0 & -\frac{2}{(r_6+2i)v_F} & 0 \\ 0 & \frac{ir_7}{r_6-2i} & 0 & \frac{2}{(r_6-2i)v_F} \\ -\frac{2}{(r_6+2i)v_F} & 0 & \frac{ir_7}{r_6+2i} & 0 \\ 0 & \frac{2}{(r_6-2i)v_F} & 0 & \frac{-ir_7}{r_6-2i} \end{pmatrix}.$$

Solving the equation (11), and employing equations (17- 24) one can obtain the following result

$$c_+^2 = \frac{c_+^1 p^0 (2\lambda + r_6\omega)}{(i\omega - \lambda)(2m + ir_7\omega\, v_F)}, \quad c_-^1 = \frac{c_+^1 p^0}{(i\omega - \lambda)\, v_F}, \quad c_-^2 = \frac{c_+^1 (2\lambda + r_6\omega)\, v_F}{(2m + ir_7\omega\, v_F)}$$

and $d_\pm(\bar{p}) = j_\pm^3(\bar{p}) = 0$. For $p^0 > m$

$$j_\pm^0 = \frac{16|c_+^1|^2 \lambda(2\lambda + \omega r_6)(p^0 - \omega)v_F^2}{(4m^2 + \omega^2 r_7^2 v_F^2)p^0},$$

$$j_\pm^1 = \frac{8|c_+^1|^2 \lambda(p^0 - m)(2p^1 \mp p^2 r_6)}{(2\lambda - \omega r_6)(\lambda^2 + \omega^2)}, \quad j_\pm^2 = \frac{8|c_+^1|^2 \lambda(p^0 - m)(2p^2 \pm p^1 r_6)}{(2\lambda - \omega r_6)(\lambda^2 + \omega^2)},$$

For $p^0 < m$

$$j_\pm^0 = \frac{16|c_+^1|^2 \lambda(m - p^0)}{(\lambda - i\omega)(2m + i\omega r_7 v_F)},$$

$$j_\pm^1 = -\frac{8|c_+^1|^2 p^0 \lambda(p^0 - m)(2p^1 \mp p^2 r_6)}{(\lambda - i\omega)(2m + i\omega r_7 v_F)(\lambda^2 + \omega^2)}, \quad j_\pm^2 = -\frac{8|c_+^1|^2 p^0 \lambda(p^0 - m)(2p^2 \pm p^1 r_6)}{(\lambda - i\omega)(2m + i\omega r_7 v_F)(\lambda^2 + \omega^2)}$$

We see that in the considered simplified model the current $\vec{j}(\bar{p})$ can be presented as the sum $\vec{j}_{\pm}(\bar{p}) = \vec{p}f_{\pm}(p^0/m) + \vec{q}g_{\pm}(p^0/m)$, where $\vec{q}$ is orthogonal to $\vec{p}$ and $f_{+}(p^0/m) = f_{-}(p^0/m)$, $g_{+}(p^0/m) = -g_{-}(p^0/m)$. The charge $j^0_{\pm}$ is also a function of p^0/m, and $j^0_{+}(p^0/m) = j^0_{-}(p^0/m)$. Each of the functions $f_{\pm}(p^0/m), g_{\pm}(p^0/m), j^0_{\pm}(p^0/m)$ is continue in the point $p^0 = m$, but not analytical.

4 Conclusions

We considered a model of interaction of spinor field with material plane. It was constructed in framework of Symanzik approach, by the modification the usual spinor action functional which describes the interaction of Dirac field with the homogeneous isotropic material plane. It is realised by taking into account the basic principles of quantum electrodynamic. The most general additional action functional contains 8 real dimensionless parameters, which characterize the property of material of plane. This model enables to investigate both bound states of particles localised near the plane and the scattering processes on it. Analysis of dispersion relation for the bound states shows that by a choosing of the special model parameter it describes the motion of massless particle with Fermi-velocity $0 < v_F < 1$. We studied a simplified model with 2 parameters and calculated some essential characteristic of bound states behaviour. The scalar density appeared to be equal to zero. The current J^3 in the orthogonal to the plane $x^3 = 0$ direction is absent. The current flows parallel to the plane $x_3 = 0$ and has two components. One of them is the current flowing parallel to the moment, in the same direction on both side of the plane $x^3 = 0$. The other one flows orthogonal to the moment direction, and its directions on different sides of the plane $x^3 = 0$ are opposite ones. It disappears only by choosing the special model parameters. The distributions of charge in the space is symmetric in respect to reflection $x^3 \to -x^3$. The charge and intensities of currents are functions the relation p^0/m. They are continuous in the point $p^0 = m$, but not analytical.

The presented results demonstrate the possibility of used approach for theoretical investigation the effects of interaction of Dirac fields with macroscopic objects . One can hope that its development and application in this research region especially for different phenomena in 2-dimensional materials attracted the attention of many researchers [16] appears to be fruitful.

Acknowledgments. Yu.M.P. is grateful to Franz Wegner for fruitful discussions which help to better understand the nature of the problem undo study and University of Heidelberg for kind hospitality. His work was supported in part by RFBR grant 16-02-00943-a.

References

[1] N. N. Bogoliubov and D. V. Shirkov, *Quantum Fields*, Nauka, Moscow (1993);
C. Itzykson and J.-B. Zuber *Quantum Fields Theory*, McGraw-Hill, New York (1980).

[2] H. B. G Casimir, Proc. K. Ned. Akad. Wet B. **51**, 793-795 (1948).

[3] U. Mohideen and A. Roy, Phys. Rev. Lett. **81**, 4549-4552 (1998); arXiv:physics/9805038v2 (1998);
A. Roy, C.-Y. Lin and U. Mohideen, Phys. Rev. D **60**, 111101(R) (1999); arXiv:quant-ph/9906062v3 (1999).

[4] B. W. Harris, F. Chen and U. Mohideen, Phys. Rev. A **62**, 052109 (2000); arXiv:quant-ph/0005088v1 (2000).

[5] G. Bressi, G. Carugno, R. Onofrio and G. Ruoso, Phys. Rev. Lett. **88**, 041804 (2002); arXiv:quant-ph/ 0203002v1 (2002).

[6] M. Bordag, G. L. Klimchitskaya, U. Mohideen, and V. M. Mostepanenko *Advances in the Casimir Effect*, (Intl. Ser. Monogr. Phys., Vol. 145), Oxford Univ. Press, Oxford (2009); G. L.

 DOI: 10.1051/epjconf/201715807005

Klimchitskaya, U. Mohideen and V. M. Mostepanenko, Rev. Mod. Phys. **81**, 1827-1885 (2009); arXiv:0902.4022v1 [cond-mat.other] (2009).

[7] K. A. Milton, J. Phys. A: Math. Gen. **37**, R209-R277 (2004).

[8] K. Symanzik, Nucl.Phys. B **190**, 1-44 (1981). .

[9] V. N. Markov and Yu. M. Pis'mak, J. Phys. A: Math. Gen. **39**, 6525-6532 (2006); arXiv: hep-th/0505218v3 (2005).

[10] V. N. Marachevsky and Yu. M. Pis'mak, Phys. Rev. D **81**, 065005 (2010); arXiv:0907.1985v2 [hep-th] (2009).

[11] D.Yu. Pis'mak and Yu. M. Pis'mak, Theor. Math. Phys. **175**,816-826 (2013); Phys. Part. Nuclei **44**, 450-461 (2013); Theor. Math. Phys. **166**, 1423-1431 (2011).

[12] D.Yu. Pis'mak, Yu. M. Pis'mak and F. J. Wegner, Phys. Rev. E **92**, 013204 (2015); arXiv:1406.1598v1 [hep-th] (2014).

[13] K. I. Kondo, T. Ebihara, T. Iizuka and E. Tanaka, Nucl.Phys. B **434**, 85-108 (1995); arXiv:hep-ph/9404361v3 (1994); T. Appelquist, M. J. Bowick, D. Karabali and L. C. R. Wijewardhana, Phys. Rev. D **33**, 3774-3776 (1986).

[14] A. N. Grigorenko, M. Polini and K. S. Novoselov, Nature Photonics **6**, 749-758 (2012); arXiv:1301.4241v1 [cond-mat.mes-hall] (2013); A. H. Castro Neto, F. Guinea, N. M. R. Torres, K. S. Novoselov, and A. K. Geim, Rev. Mod. Phys. **81**, 109-162 (2009); I. V. Fialkovsky and D. V. Vassilevich, J. Phys. A: Math. Gen. **42**, 442001 (2009); Wang-Kong Tse and A. H. MacDonald, Phys. Rev. B **84**, 205327 (2011); arXiv:1108.3858v1 [cond-mat.mes-hall] (2011); Liang Chen and Shaolong Wan, Phys. Rev. B **84**, 075149 (2011); J. González, JHEP **1307**, 175 (2013); arXiv:1211.3905v1 [cond-mat.mes-hall] (2012); V. N. Kotov, B. Uchoa, V. M. Pereira, F. Guinea and A. H. Castro Neto, Rev. Mod. Phys. **84**, 1067-1125 (2012); arXiv:1012.3484v2 [condmat. str-el] (2010); I. V. Fialkovsky, V. N. Marachevsky and D. V. Vassilevich, Phys. Rev. B **84**, 035446 (2011); arXiv:1102.1757v2 [hep-th] (2011); G. L. Klimchitskaya, V. M. Mostepanenko and Bo E. Sernelius, Phys. Rev. B **89**, 125407 (2014); arXiv:1401.6886v2 [cond-mat.other] (2014); G. L. Klimchitskaya and V. M. Mostepanenko, Phys. Rev. A **89**, 052512 (2014); arXiv:1404.7057v2 [quant-ph] (2014); J. E. Moore, Nature **464**, 194-198 (2010).

[15] I. V. Fialkovsky, V. N. Markov and Yu. M. Pis'mak, J. Phys. A: Math. Gen. **39**, 6357-6363 (2006); Int. J. Mod. Phys. A **21**, 2601-2616 (2006); arXiv:hep-th/0311236v2 (2003).

[16] G. Bracco and B. Holst (Eds.), *Surface Science Techniques*, (Springer Ser. Surf. Sci., Vol. 51), Springer, Berlin (2013); M. L. Brongersma and P. G. Kik (Eds.), *Surface Plasmon Nanophotonics*, (Springer Ser. Opt. Sci., Vol. 131), Springer, Berlin (2007); S. A. Maier *Plasmonics: Fundamental and Applications*, Springer, Berlin (2007).

[17] D. Yu. Pismak, Yu. M. Pismak, Theor. Math. Phys. **184**, 3, 1329-1341 (2015).

[18] D. Yu. Pismak, D. Yu. Shukhobodskaia1, 126 EPJ Web of Conferences 05012 (2016) ICNFP 2015, 125 EPJ Web of Conferences 05022 (2016) QUARKS-2016

AUTHOR INDEX

AUTHOR INDEX